D. Smitley

Handbook of Sampling Methods for Arthropods in Agriculture

Edited By

Larry P. Pedigo
Department of Entomology
Iowa State University
Ames, Iowa

and

G. David Buntin
Department of Entomology
Georgia Experiment Station
Griffin, Georgia

CRC Press
Boca Raton   Ann Arbor   London   Tokyo

Library of Congress Cataloging-in-Publication Data

Handbook of sampling methods for arthropods in agriculture/edited by
  Larry P. Pedigo and G. David Buntin.
    p. cm.
  Includes bibliographical references and index.
  ISBN 0-8493-2923-X
  1. Arthropod pests—Research—Statistical methods—Handbooks,
manuals, etc. 2. Arthropod populations—Research—Statistical
methods—Handbooks, manuals, etc. 3. Arthropod populations-
Measurement—Handbooks, manuals, etc. 4. Sampling (Statistics)-
Handbooks, manuals, etc. I. Pedigo, Larry P. II. Buntin, G.
David.
SB933.14.H36 1993
632'.652—dc20

93-24175
CIP

This book represents information obtained from authentic and highly regarded sources. Reprinted material is quoted with permission, and sources are indicated. A wide variety of references are listed. Reasonable efforts have been made to publish reliable data and information, but the author and the publisher cannot assume responsibility for the validity of all materials or for the consequences of their use.

Neither this book nor any part may be reproduced or transmitted in any form or by any means, electronic or mechanical, including photocopying, microfilming, and recording, or by any information storage or retrieval system, without prior permission in writing from the publisher.

All rights reserved. Authorization to photocopy items for internal or personal use, or the personal or internal use of specific clients, may be granted by CRC Press Inc., provided that $.50 per page photocopied is paid directly to Copyright Clearance Center, 27 Congress Street, Salem, MA 01970 USA. The fee code for users of the Transactional Reporting Service is ISBN 0-8493-2923-X/9 $0.00 + $.50. The fee is subject to change without notice. For organizations that have been granted a photocopy license by the CCC, a separate system of payment has been arranged.

CRC Press, Inc.'s consent does not extend to copying for general distribution, for promotion, for creating new works, or for resale. Specific permission must be obtained in writing from CRC Press for such copying.

Direct all inquiries to CRC Press, Inc., 2000 Corporate Blvd., N.W., Boca Raton, Florida 33431.

© 1994 by CRC Press, Inc.

No claim to original U.S. Government works

International Standard Book Number 0-8493-2923-X

Library of Congress Card Number 93-24175

Printed in the United States of America 1 2 3 4 5 6 7 8 9 0

Printed on acid-free paper

PREFACE

The development of sampling methods and measures to assess the status of pest populations has been critical to the development and advancement of pest control technology. Actually, these activities set modern integrated pest management (IPM) apart from the "identify-and-spray" strategies of the past, and they produce the rudimentary data that governs all other activities in ecology and pest management programs.

In dealing with arthropods, sampling allows insights on habitat invasion, long-range migration, local movement, feeding, and reproduction. Depending on the kind of pest (native, newly introduced, potential invader), sampling gives information on species presence and population density, dispersion, and dynamics.

To develop acceptable arthropod sampling programs, researchers face an array of decisions. Some of the most crucial to these involve where to sample, when to sample, how many samples to take, and the sampling technique for gathering information. Initially, these questions may seem absurdly simple; however, they are not. These are very difficult questions, if accurate information is to be obtained. Indeed, arbitrary and inaccurate decisions on these sampling parameters, can seriously impede the progress of arthropod ecology and IPM development.

Given the central importance of sampling methodology to arthropod ecology and IPM, suprisingly few publications are dedicated exclusively to the subject. Probably the most comprehensive publication on arthropod sampling has been *Ecological Methods* by T. R. E. Southwood. This superb book has served as the ultimate, authoritative reference for over two decades. Unfortunately, the latest edition, published in 1978, lacks many recent advances in sampling procedures. Another dedicated book on arthropod sampling is *Sampling Methods in Soybean Entomology*, edited by M. Kogan and D. C. Herzog. This book also is an excellent reference, giving detailed information on sampling design and techniques. However, this book also is dated (published in 1980) and is specific to pests of a single crop. Other frequently cited references to sampling include *Introduction to Experimental Ecology* (1967) by T. Lewis and L. R. Taylor and *Sampling Techniques* (1977) by W. G. Cochran. The Lewis and Taylor book is comprised mainly of sampling exercises, and the Cochran book explains sampling principles, referencing the social sciences. Arthropod pest sampling has been treated most recently in the second edition of the *Handbook of Pest Management in Agriculture* (1991) in a chapter by A. M. Shelton and J. T. Trumble and another chapter by J. P. Nyrop and M. Binns. Both chapters are useful, with the former concentrating on practical aspects of program development and the latter emphasizing sampling theory. However, no chapters of the book address sampling techniques (sampling tools), per se, and, as in other general references, detail is kept to a minimum. Other treatments on arthropod sampling are scattered through the literature, forcing researchers and students to consult many different references, with the risk of omitting significant advancements pertinent to their particular sampling problem. Consequently, our purpose was to develop a single, practical work dedicated to sampling of arthropod pests and natural enemies.

With the involvement of many talented and experienced entomologists, we have synthesized a comprehensive book that deals with the principles and practicalities of developing accurate sampling programs for arthropod pests and natural enemies. The book emphasizes the latest developments in the major aspects of sampling, with the goal of becoming a useful reference to the topic for researchers, students, and agriculturalists. An important focus of this work is on developing sampling plans that provide accurate population estimates at a reasonable cost. The book discusses sampling theory, as appropriate, but emphasis is placed on practicality and application to IPM.

The basic plan of the book includes an introduction that presents an overview of sampling, and the remainder of the book is divided into sections that include sampling principles (Chapters 2 through 4), sampling techniques and initial program development (Chapters 5 through 7), improving sampling program efficiency (Chapters 8 through 12), sampling programs for selected commodities (Chapters 13 through 21), and implementing sampling programs (Chapters 22 and 23).

In developing the topics, we have attempted to maintain a broad scope. The authors deal with both technique descriptions and programs that apply to some of the most important IPM systems. Specific examples are given for the most economically important agroecosystems, and we believe these can serve as models for sampling development in other agroecosystems. With this, we hope the book will be useful to a wide audience.

Larry P. Pedigo
Editor

G. David Buntin
Editor

THE EDITORS

Larry P. Pedigo, Ph.D., is Professor of Entomology in the College of Agriculture of Iowa State University in Ames.

Dr. Pedigo received his B.S. degree (With Distinction) in biology from Fort Hays Kansas State University, Hays, Kansas in 1963. He obtained his M.S. and Ph.D. degrees in 1965 and 1967, respectively, from the Department of Entomology, Purdue University, in West Lafayette, Indiana. He joined the faculty at Iowa State upon graduation.

Dr. Pedigo is a member of the Entomological Society of America, the Kansas Entomological Society, the South Carolina Entomological Society, and the honorary societies Phi Eta Sigma, Phi Kappa Phi, Gamma Sigma Delta, and Sigma Xi.

He has been the recipient of the Researchers Recognition Award of the American Soybean Association, the C. V. Riley Award for Outstanding Contributions to Entomology, the J. E. Bussart Memorial Award for Outstanding Achievements in Crop Protection Entomology, the Faculty Citation for Outstanding Service to Iowa State University, The Margaret Ellen White Graduate Faculty Award for Excellence as an Educator of Graduate Students, and the Burlington Northern Foundation Award for Career Achievement in Graduate Teaching.

Dr. Pedigo conducts research with insect population dynamics and management and teaches both undergraduate and graduate courses in entomology. He is the author or coauthor of over 150 research publications and has written or edited three previous books on entomology and pest management.

G. David Buntin, Ph.D. is Associate Professor of Entomology, Department of Entomology, University of Georgia, Georgia Experiment Station, Griffin, Georgia.

Dr. Buntin graduated from the University of Delaware with a B.S. degree in Entomology and Applied Ecology in 1977. He received a M.S. degree in Entomology in 1980 and a Ph.D. in Entomology in 1984 from Iowa State University. Dr. Buntin was assistant professor of Entomology from 1984 to 1990 and is currently associate professor of Entomology at the University of Georgia.

Dr. Buntin is a member of the Entomological Society of America, Georgia Entomological Society, Central States Entomological Society, American Society of Agronomy, Crop Science Society of America and the South Carolina Entomological Society. Dr. Buntin currently is associate editor of the Journal of Entomological Science.

Dr. Buntin has authored more than 53 research papers and has published 70 additional scientific publications. He also has presented over 60 papers at professional meetings. His current major research interests include insect-induced plant stress, pest management systems, and sampling methodology for insect pests of field and forage crops.

CONTRIBUTORS

Edward John Bechinski
Plant, Soil & Entomology Sciences
University of Idaho
Moscow, Idaho

Elizabeth H. Beers
Department of Entomology
Washington State University
Wenatchee, Washington

Richard Berberet
Department of Entomology
Oklahoma State University
Stillwater, Oklahoma

Michael R. Binns
Agriculture Canada
Central Experimental Farm
Ottawa, Ontario

G. David Buntin
Department of Entomology
University of Georgia
Griffin, Georgia

Dennis D. Calvin
Department of Entomology
Penn State University
University Park, Pennsylvania

Gerrit W. Cuperus
Department of Entomology
Oklahoma State, University
Stillwater, Oklahoma

Paula M. Davis
Department of Entomology
Cornell University
Ithaca, New York

Norman C. Elliott
Plant Science Research Laboratory
USDA, Agriculture Research Service
Stillwater, Oklahoma

J. Daniel Hare
Department of Entomology
University of California
Riverside, California

Gary L. Hein
Panhandle Research & Extension Center
University of Nebraska
Scottsbluff, Nebraska

Leon G. Higley
Department of Entomology
University of Nebraska
Lincoln, Nebraska

Larry A. Hull
Department of Entomology
Penn State University
Biglerville, Pennsylvania

Scott H. Hutchins
Insect and Disease Management Group
DowElanco
Indianapolis, Indiana

William D. Hutchison
Department/Entomology
University of Minnesota
St. Paul, Minnesota

Vincent P. Jones
Department of Entomology
University of Hawaii
Honolulu, Hawaii

Tom H. Klubertanz
Department of Biology
Peru State College
Peru, Nebraska

Douglas A. Landis
Entomology & Pesticide Research Center
Michigan State University
E. Lansing, Michigan

David E. Legg
Department/Plant, Soil & Insect Science
University of Wyoming
Laramie, Wyoming

Timothy J. Lysyk
Department/Livestock Sciences
Agriculture Canada Research Station
Lethbridge, AB Canada

Roger D. Moon
Department of Entomology
University of Minnesota
St. Paul, Minnesota

Jan P. Nyrop
Department of Entomology
NYSAES/Cornell University
Geneva, New York

Larry P. Pedigo
Department of Entomology
Iowa State University
Ames, Iowa

Robert K. D. Peterson
Department of Entomology
University of Nebraska
Lincoln, Nebraska

B. Merle Shepard
Coastal Research & Education Center
Clemson University
Charleston, South Carolina

Jon James Tollefson
Department of Entomology
Iowa State University
Ames, Iowa

John T. Trumble
Department of Entomology
University of California
Riverside, California

Wopke van der Werf
Department of Entomology
Cornell University
Geneva, New York

L. T. Wilson
Department of Entomology
Texas A & M University
College Station, Texas

Michael R. Zeiss
Department of Entomology
Iowa State University
Ames, Iowa

ACKNOWLEDGMENTS

It pleases us to express our deep gratitude to the many authors involved in the writing of this volume. Without their hard work and diligence in providing high quality copy, our task would have been much more difficult and much less enjoyable. Thanks are extended also to CRC Press, in general, and to Monique Power, Managing Editor, in particular, for advice, patience, and pleasant responses to our questions as the manuscript was developed.

DEDICATION

To R. F. Morris, theoretician and practitioner, whose research and innovative analyses laid the foundation for modern approaches to arthropod sampling and population study.

TABLE OF CONTENTS

Chapter 1. Introduction to Sampling Arthropod Populations 1
Larry P. Pedigo

SECTION I. SAMPLING PRINCIPLES

Chapter 2. Arthropod Sampling in Agricultural Landscapes:
Ecological Considerations 15
Douglas A. Landis

Chapter 3. Statistics for Describing Populations 33
Paula M. Davis

Chapter 4. Bias and Variability in Statistical Estimates 55
David E. Legg and Roger D. Moon

SECTION II. SAMPLING TECHNIQUES AND INITIAL PROGRAM DEVELOPMENT

Chapter 5. Techniques for Sampling Arthropods in Integrated
Pest Management 73
Scott H. Hutchins

Chapter 6. Developing a Primary Sampling Program 99
G. David Buntin

SECTION III. IMPROVING SAMPLING PROGRAM EFFICIENCY

Chapter 7. Initiating Sampling Programs 119
Leon G. Higley and Robert K. D. Peterson

Chapter 8. Sequential Sampling for Classifying Pest Status 137
Michael R. Binns

Chapter 9. Sequential Estimation and Classification Procedures
for Binomial Counts 175
Vincent P. Jones

Chapter 10. Sequential Sampling to Determine Population
Density 207
William D. Hutchison

Chapter 11. Sampling to Predict or Monitor Biological Control 245
Jan P. Nyrop and Wopke van der Werf

Chapter 12. Time-Sequential Sampling for Taking Tactical
Action 337
Larry P. Pedigo

SECTION IV. SAMPLING PROGRAMS

Chapter 13. Sampling Methods for Insect Management in Alfalfa 357
Richard C. Berberet and William D. Hutchison

Chapter 14. Sampling Pest and Beneficial Arthropods of Apple 383
Elizabeth H. Beers, Larry A. Hull, and Vincent P. Jones

Chapter 15. Sampling Arthropod Pests in Citrus 417
J. Daniel Hare

Chapter 16. Sampling Arthropod Pests in Field Corn 433
Jon J. Tollefson and Dennis D. Calvin

Chapter 17. Estimating Abundance Impact, and Interactions
Among Arthropods in Cotton Agroecosystems 475
L. T. Wilson

Chapter 18. Sampling Arthropods in Livestock Management
Systems 515
Timothy J. Lysyk and Roger D. Moon

Chapter 19. Sampling Programs for Soybean Arthropods 539
Michael R. Zeiss and Thomas H. Klubertanz

Chapter 20. Sampling Arthropod Pests in Vegetables 603
John T. Trumble

Chapter 21. Sampling Arthropod Pests of Wheat and Rice 627
Norman C. Elliott, Gary L. Hein, B. Merle Shepard

SECTION V. IMPLEMENTATION OF SAMPLING PROGRAMS

Chapter 22. Training Specialists in Sampling Procedures 669
Gerrit W. Cuperus and Richard C. Berberet

Chapter 23. Designing and Delivering In-The-Field Scouting
Programs 683
Edward J. Bechinski

Chapter 1

INTRODUCTION TO SAMPLING ARTHROPOD POPULATIONS

Larry P. Pedigo

TABLE OF CONTENTS

I. Background Information 2

II. Nature of Populations 3

III. Measures of Population Density 4
    A. Types of Population Estimates 4
    B. Absolute and Related Estimates 4
        1. Absolute Estimates Proper 4
        2. Population Intensity Estimates 5
        3. Basic Population Estimates 5
    C. Relative Estimates 5
    D. Converting Relative Estimates to Absolute Estimates 6
    E. Population Indices 6

IV. Sampling Concepts 8
    A. Sampling Units and Samples 8
    B. Sampling Universe 9
    C. Sampling Techniques and Sampling Programs 9

Acknowledgments 10

References 10

I. BACKGROUND INFORMATION

Sampling populations to determine kinds and to estimate numbers of living species are the most fundamental research activities in ecology. In fact, it is difficult to envisage any searching study of natural populations or communities that does not involve some sort of sampling activity. This is true because most ecological questions focus on distribution and abundance of organisms as influenced by biological and physical environments.

Sampling and numbers estimation are particularly central to population ecology. Here, sampling provides a foundation for the research program,[1] furnishing data on density, dispersion, age structure, reproduction, and migration. A synthesis of these data ultimately yields an understanding of population dynamics. Without a well-developed sampling framework, ecological studies are unsound and destined to fail.

Just as sampling is an indispensable part of population ecology, it serves as the primary basis of integrated pest management (IPM)[2] (Figure 1). This stems from IPM being an ecologically based approach that relies on current information about the status of pests and crops for decisions on appropriate courses of action. Typically, the aims of IPM are most effectively accomplished by employing preventive tactics such as biological control, plant resistance, and cultural practices to maintain pest populations below the economic injury level.[3] However, if these tactics fail, therapeutics (curatives) are applied, which most often involve the use of quick-acting suppressants such as conventional or microbial pesticides. The use of therapeutics, according to IPM principles, calls for a protocol of pest population sampling, estimation of density,

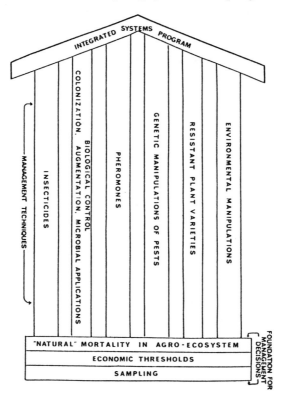

FIGURE 1. Schematic diagram showing the key role of sampling in the overall scheme of IPM. (Reprinted with permission by Tall Timbers Research, Inc. from *Proceedings: Tall Timbers Conference on Ecological Animal Control by Habitat Management*, 1971, Tall Timbers Research Station, Tallahassee, FL.)

consultation of appropriate decision rules such as economic thresholds, and judicious use of a curative.[4] In the therapy protocol, an effective and practical sampling program is a mandatory prerequisite for all other activities.

Although many advances in sampling methodology have been made recently,[5] adequate sampling programs for IPM are lacking for many important pests in cropping systems.[3] Indeed, the absence of inexpensive, user-friendly sampling protocols may well be the greatest impediment to the implementation of IPM practice in agriculture today. Consequently, this book was written to provide researchers and practitioners with a comprehensive reference for direct use in arthropod IPM programs, as well as to encourage the development of new programs and improvement of existing ones. Information is provided on sampling principles, techniques, existing programs, and implementation. In addition to the information herein, the reader is encouraged to consult related speciality publications on the topic, especially those of Morris,[6] Strickland,[7] Southwood,[8] Kogan and Herzog,[9] Shelton and Trumble,[10] Nyrop and Binns,[3] Kuno,[5] and Binns and Nyrop.[11]

The purpose of this introductory chapter is to give the reader an overview of sampling arthropod pests in agricultural systems and provide an appropriate orientation to subsequent chapters. To this end, population concepts, types of estimates, and sampling categories are discussed and primary terms defined.

II. NATURE OF POPULATIONS

The IPM paradigm is based on the study of pest populations. Although the paradigm has its roots in biological and integrated control, primarily it is the product of ecological principles. Geier and Clark[12] outlined these principles and suggested the term "protective population management" or "pest management" for the concept. Their concept differed from earlier pest-control approaches in scope, synthesis of tactics, and the inclusion of basic population theory in the design.[13] Although pest management or IPM (as it has come to be known) has acquired an expanded perspective since its inception, the ideas of species population and population dynamics remain the prime considerations in developing new programs.

Sampling in IPM is done on a pest population or populations usually to determine specific characteristics, e.g., density and dispersion. Although the concept of population is treated differently in various disciplines,[14] biologists frequently follow the definition of Pearl,[15] viz., "a group of living individuals set in a frame that is limited and defined in respect to both time and space." Most entomologists further restrict the meaning to individuals of the same species. In some instances the meaning of population is extended to include the whole species over its entire geographical range (the "natural population" of Andrewartha and Birch[16]). For both scientific and practical purposes, however, local populations become the units of study. In this book, the meaning of an arthropod population will follow that of Pearl.

Arthropod populations usually are sampled to determine their attributes or properties. Some of the major properties estimated through population sampling include the following:[14]

1. Density—an expression of a species' abundance in an area
2. Dispersion—the spatial distribution of individuals of a species
3. Natality—birth rate
4. Mortality—death rate
5. Age structure—relative proportions of individuals in different age classes
6. Growth form—shape of population growth curves

It is from repeated measures of these characteristics and determination of population movements (immigration and emigration) that inferences about population dynamics are made.

III. MEASURES OF POPULATION DENSITY

Of the foregoing properties, density is probably the most basic one measured, and many other properties are derived from the density measure. Depending on the objectives of sampling, density is considered in one of two ways, absolute and relative.[17] *Absolute density* refers to a total count or estimate of all the arthropods in a given area. Absolute densities are expressed in terms of a unit that does not vary, usually land surface area, e.g., numbers per square meter.[6] The absolute measure is indispensable in studies of arthropod population dynamics, particularly life tables. This is because many arthropods occupy different habitats during their life cycle. For example, insect egg and larval stages may inhabit foliage, with the pupal stage inhabiting the soil surface, and the adult stage inhabiting air and foliage. A common reference for all of these habitats is the land surface, and measures based on surface area allow comparisons of densities for calculating stage-specific survival.

Conversely, *relative density* refers to a count of arthropods in a sampling unit that has no direct relationship to land surface area. Instead of land area being the constant, the sampling technique is the constant. Therefore, relative densities can be compared only amongst themselves, using the same sampling procedure. Usually, different arthropod stages cannot be sampled with the same procedure, which precludes relative density measures from being used directly in life table studies. However, these measures can be used in assessing density changes of a stage over time or space. Therefore, they are useful in IPM for detecting trends and classifying pest populations for decision-making purposes.[13]

A. TYPES OF POPULATION ESTIMATES

Counting all the individuals in a population is the most direct and accurate method of determining population density.[14,17] This type of count, called a *census*, is not used frequently with arthropods because of their large numbers, small individual size, and secretive habits.[3] Instead, representative units of the population are selected and numbers in these units counted to derive an *estimate* of the total population.[18] The approaches to estimating arthropod population density vary, depending on objectives of a study. Of the classifications of estimates proposed, the classification of Morris[1] is one of the most meaningful and therefore one of the most widely accepted. Southwood[8] grouped Morris' estimate types into three convenient categories, viz., absolute and related estimates, relative estimates, and population indices.

B. ABSOLUTE AND RELATED ESTIMATES

Absolute estimates present information on numbers of arthropods usually per unit of land surface area in the habitat (e.g., number per square meter) or provide data that can be easily converted to such. In IPM, these estimates are most often obtained by taking total counts from a known proportion of the habitat.

1. Absolute Estimates Proper

Estimates in this category are necessary for a detailed understanding of arthropod population dynamics. Absolute estimates quantify actual numbers in the arthropod population according to surface-area units. Often, these are some of the most difficult estimates to make and frequently are the most costly.[13] Absolute estimates are of utmost importance and used widely in research. Because of high cost, however, they are not used as widely in IPM practice. Some examples of procedures to obtain absolute estimates include using (1) fumigants to kill all arthropods within an enclosure,[19] (2) extraction funnels where efficiency of extraction is known,[20] (3) removal of vegetation and laboratory processing to count all arthropods,[21] and (4) suction traps where the volume of air passing through the trap is known.[22]

2. Population Intensity Estimates

Morris[1] describes population intensity estimates as expressions of arthropod numbers per unit of the arthropod's available food supply. Expressions such as number per leaf, number per fruit, or number per plant are common population intensity measures. Population intensity estimates are used frequently in pest management because they relate closely to crop injury. They have been particularly useful in research to establish economic injury levels (e.g., Mailloux et al.[23]), and consequently, they are found in recommendations for IPM decision making.[13]

Both Morris[1] and Southwood[8] caution that intensity estimates lack stability because of changes in habitat. For example, an arthropod defoliator with a high intensity early in the growing season may give the unrealistic impression that absolute density is high simply because of small plant size. As plants grow, the population may also increase but seem less dense because of the greater amount of foliage and, consequently, lower intensity estimates. Because of this, care must be taken in selecting the appropriate intensity measure and determining the relationship of the measure to land surface area, e.g., when counting number of arthropods per plant, a count of the number of plants per square meter also is advised.

3. Basic Population Estimates

Basic population estimates are related to absolute estimates and are used often in IPM. The basic population is expressed as number per standard unit of habitable space. In forestry, Morris[1] used numbers of spruce budworms per 10 ft^2 of branch surface area. With row crops the basic population is often recorded in number of arthropods per foot or meter of row. The standard unit may be established formally, as through a group of researchers, or informally through traditional practice. The basic population estimate is another useful way of expressing economic injury levels for decision making. As with population intensity estimates, numbers of standard units per land surface area should be determined along with arthropod counts when the sampling objective is to understand population dynamics.

C. RELATIVE ESTIMATES

Relative estimates differ greatly from absolute and related estimates because they have no direct relationship to land surface area. Rather, they are referenced to the type of sampling technique used. The techniques invariably used are either counts per unit effort or counts per trap.[8]

Commonly, arthropod counts per unit effort may involve direct observations in the habitat for a set period of time or collections from a set number of swings of a sweep net. Indeed, relative estimates based on sweep net sampling units are probably the most common types of sampling methods in entomology, particularly for decision making in IPM.[13] Very frequently, numbers of insects per sweep with a 38-cm (15-in.) diameter sweep net are used to report the current status of pests or to recommend population densities where management tactics are appropriate.

Relative estimates do not directly assess the number of arthropods per hectare, per plant, or per meter of row. Rather, they are used for comparison of population density in time or space. For example, potato leafhopper, *Empoasca fabae* (Harris), numbers per sweep-net swing through alfalfa could be compared between 1991 and 1992 in Smith County or between Smith and Baker Counties in 1992, but number of leafhoppers per hectare of alfalfa would not be known from the data.

Accurate relative estimates are necessary for valid time/space comparisons. This requires faithful repetition of sampling procedures in similar settings through time at the same location or at one time in different locations. Although this criterion is rarely satisfied completely, conscientious attempts at uniformity can make the estimates useful for comparing gross changes in population abundance and determining the need for management activities.

The great advantage of relative estimates pertains to their cost. Compared to absolute and related estimates, relative estimates are usually inexpensive,[1] making their use most practical for IPM decision making. Indeed, many economic thresholds in management recommendations are expressed in terms of relative estimates. Lower cost also makes increases in sampling-unit number possible when population densities are low, a particularly difficult time to obtain precise estimates.

D. CONVERTING RELATIVE ESTIMATES TO ABSOLUTE ESTIMATES

As Southwood[8] has pointed out, there is no hard and fast distinction among methods to achieve absolute and relative estimates. This is because most absolute procedures seldom yield 100% of all arthropods in an area and relative estimates can be transformed to estimate the absolute population. These transformations can be accomplished in one of two ways, statistically or experimentally.

A statistical transformation can be achieved by making paired relative and absolute estimates from a series of populations, followed by analyses that regress the relative estimates on the absolute. This procedure has been used frequently with sweep net, vacuum net, and shake cloth methods.[19,21,24-26]

The statistical approach is exemplified in a study of two grasshopper species, *Melanoplus femurrubrum* (DeGeer) and *M. differentialis* (Thomas), in soybean to transform sweep-net estimates into estimates of the absolute population.[27] Assessments of absolute density were made at night by using a plant removal method, and these were compared to both day and night sweep-net samples using regression analysis (Figure 2). The resulting regression models were evaluated on the degree of fit to the data (R^2). The models make it possible to predict absolute population densities from either day or night sweep samples.

Although the statistical approach focuses on studies of natural populations, the experimental approach utilizes artificially established populations with known density. This approach is exemplified in a study of grasshoppers in burned and unburned tallgrass prairie.[28] Here, a night trap (Figure 3) was used to capture marked grasshoppers placed inside. Marked individuals constituted the experimental population, and it was determined that about 75% of the marked individuals were captured. Such a procedure allows calibration of the method to give estimates of absolute density. Because density of the experimental population is known exactly, statistical error is not involved in determining the absolute density. Therefore, experimental populations represent one of the few methods available to substantiate efficiency of methods designed to give absolute estimates.

E. POPULATION INDICES

Population indices are related to relative population estimates in that the absolute density of the arthropod population is not known directly from these. The term *population index* has had different uses in ecological studies. In many references,

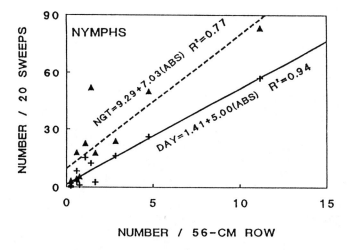

FIGURE 2. Sweep-net regression models for grasshopper nymphs and adults using mean catch of *M. femurrubrum* and *M. differentialis*. ABS = cage-bag technique; DAY = day sweeps; NGT = night sweeps. (From Browde et al.[27] With permission.)

population index means any representation that gives a relative estimate of the population.[6,29] However, a distinction can be made between counting arthropods themselves and quantifying their products or effects as indicators of population abundance. Defining a population index in terms of products or effects[1,8] is a useful distinction for IPM, and subsequently, the term will be used in this way.

Population indices have been based on arthropod products such as webs, nests, exuviae, and frass droppings.[8] Frass droppings of forest insect pests particularly have received attention as population indices, but the method is not acceptable for all forest defoliators.[6]

FIGURE 3. Night trap for obtaining absolute estimates of grasshopper populations in tallgrass prairie. Trap constructed from a 68-l plastic trash can (covering 0.152 m^2). Cut-away shows a crawling ramp leading to transparent-plastic receptacle with soapy water. (From Evans et al.[28] With permission.)

Probably the most practical population indices in agricultural IPM are based on rankings or other quantifications of insect injury. In some instances, regression analyses of feeding injury and absolute density have been done[30] and the resulting model used to predict density from injury. This approach is specially useful when absolute estimates are difficult to obtain compared to injury estimates.

Although it may be possible to measure population size from insect injury, quite frequently, density is less the question than is the injury/host-yield relationship. Therefore, in many instances decisions on management activities are made directly from the injury estimate,[9,31,32] and no estimation of density is attempted.

IV. SAMPLING CONCEPTS

The statistical principles and constraints of sampling are detailed in Chapters 3 and 4. However, a brief overview of some elementary terms and concepts will allow an orientation to many succeeding topics and serve to standardize some of the terminology in the remainder of the book.

A. SAMPLING UNITS AND SAMPLES

Just as the type of estimate needed depends on the objective of the sampling study, the nature of a population sample depends on the sampling units selected. A *sampling unit* is a proportion of the habitable space from which arthropod counts are taken. Therefore, the population can be envisioned as being composed of a finite number of distinct sampling units. Among other criteria,[1] the size of the habitable proportion is determined by the individual worker, however, the units themselves must be distinct and not overlap.[18] Together, all sampling units contain the population.

The sampling-unit concept is most easily explained with absolute estimates, where total counts are taken from a unit-area of land surface.[33] For example, direct counts of all the caterpillars in 1 m^2 of alfalfa could be considered a sampling unit. If the caterpillar population occupies 100 m^2, the habitable space is composed of 100 sampling units.

It is somewhat more difficult to visualize sampling units with relative estimates. For instance, a sampling unit based on sweeping looper caterpillars in cotton with a sweep net may consist of taking 20 sweeps down the row. Yet, the sweep net makes contact with only a portion (e.g., the upper third) of each plant that is swept. However, the whole canopy, or the part inhabited by loopers, can be divided into a finite number of nonoverlapping 20-sweep units. Even though not all loopers are captured when a unit of habitable space is swept, a relative estimate can be made. However, as discussed earlier, the estimate is useful only in relation to the sampling procedure.

Another example of sampling units related to relative estimates involves the use of insect light traps. Here, the light trap has a certain effective range of attracting and capturing cutworm adults, and the total area occupied by flying adults encompasses a finite number of spaces for trap placement such that one trap does not influence the catch of another. Consequently, the sampling unit becomes the trap area-of-capture and duration-of-operation. The resulting estimate obtained is used to compare numbers in time and space with other traps of the same configuration and operating specifications.

Because it is usually impractical to count all arthropods in all the sampling units, a group of such units is delineated, which subsequently is used to characterize the whole population. This group of sampling units is referred to as a *sample*, and it is from the sample that an estimate is made. Both sampling-unit size and number taken for a sample are dictated by the sampling design. Information on design and procedures to establish sampling-unit size and number is presented in Chapter 6.

B. SAMPLING UNIVERSE

A frequently used term in the sampling literature is the "sampling universe". In statistics the term often is used to refer to the whole population from which samples are taken, i.e., universe and population are synonymous. However, in arthropod sampling, the *sampling universe* has come to represent the habitat in which the population occurs.[1] Consequently, sampling units and samples are taken within the sampling universe. Therefore, if we determine that essentially all of a population of bean leaf beetle, *Cerotoma trifurcata* (Forster), eggs are found in the soil within 7.6 cm on either side of a soybean row and no deeper than 3.8 cm,[34] then the sampling universe can be designated as a band 15.2 cm wide × 3.8 cm deep, over the row, and soil sampling units can be drawn from this location. Precise determination of sampling universe dimensions ensures that no samples are taken in uninhabited areas. In this manner, variability among egg counts and, subsequently, sampling costs are reduced. The nature of sampling universes in agricultural landscapes is discussed from an ecological perspective in Chapter 2.

C. SAMPLING TECHNIQUES AND SAMPLING PROGRAMS

To achieve the objectives of sampling, it is necessary to count arthropods in sampling units from a sampling universe and generate an estimate of population density. To achieve this, both a sampling technique and a sampling program are required. Although these terms are often used interchangeably in the literature, it is useful to distinguish between them.

A *sampling technique* is the method used to collect information from a single sampling unit.[14] Therefore, the focus of a sampling technique is on equipment and/or the way an arthropod count is accomplished. Examples of sampling techniques are taking direct counts of all face flies, *Musca autumnalis* De Geer, on the face of a Hereford, and taking 20 sweeps, in a pendulum fashion, in alfalfa to obtain a sampling unit of adult ladybird beetles. Details on equipment or so-called tools[10] used as part of the sampling techniques are presented in Chapter 5 as well as in other publications.[8,13]

In contrast to the sampling technique, a *sampling program* is the procedure for employing the sampling technique to obtain a sample and make an estimate.[13] Sampling programs direct how a sample is to be taken, including (1) sampling-unit size, (2) sampling-unit number, (3) spatial pattern of obtaining sampling units, and (4) timing of samples.[8]

In general, sampling programs comprise two types, extensive programs and intensive programs.[6,8] *Extensive programs* are conducted over broad areas for the purpose of determining such information as species distributions or the status of injurious arthropod stages. Usually, in extensive programs only a single arthropod stage is sampled, and only one or a few samples are taken per season. Quite often, only moderate levels of precision are required for extensive programs, with the primary emphasis on low cost. Conversely, *intensive programs* usually are conducted as part of research in population ecology and dynamics. Here, sampling is done frequently, often more than once per week, in a small area (a few fields). Usually, all or most stages in the life cycle of the arthropod are sampled, and a high degree of precision is sought.

Much of this book is focused on sampling programs. The development of new or primary programs is discussed in Chapter 6, followed in Chapter 7 by a treatment of how sampling activities are timed. Subsequently, Chapters 8 through 12 discuss methods of improving efficiency of sampling programs for decision making. Information on established programs in IPM are presented, by commodity, in Chapters 13 through 21, inclusive. Finally, Chapters 22 and 23 focus on methods of educating individuals in the skill of sampling and delivery of sampling programs to users, respectively.

ACKNOWLEDGMENTS

Many thanks are extended to D. Buntin of the University of Georgia, and my graduate students J. Browde, T. DeGooyer, T. Klubertanz, and M. Zeiss for their comments and suggestions on this chapter.

REFERENCES

1. Morris, R. F., The development of sampling techniques for forest insect defoliators with particular reference to the spruce budworm, *Can. J. Zool.*, 33, 225, 1955.
2. Gonzalez, D., Sampling as a basis for pest management strategies, in *Proc. Tall Timbers Conf. Ecological Animal Control Habitat Mgt.*, No. 2, Komarek, E. V., Ed., Tall Timbers Res. Stn., Tallahassee, FL, 1971, 83.
3. Nyrop, J. P. and Binns, M., Quantitative methods for designing and analyzing sampling programs for use in pest management, in *Handbook of Pest Management in Agriculture*, 2nd ed., Pimentel, D., Ed., CRC Press, Boca Raton, FL, 1991, 67.
4. Pedigo, L. P., Integrating preventive and therapeutic tactics in soybean insect management, in *Pest Management of Soybean*, Copping, L. G., Green, M. B., and Rees, R. T., Eds., Elsevier, London, 1992, 10.
5. Kuno, E., Sampling and analysis of insect populations, *Annu. Rev. Entomol.*, 36, 285, 1991.
6. Morris, R. F., Sampling insect populations, *Annu. Rev. Entomol.*, 5, 243, 1960.
7. Strickland, A. H., Sampling crop pests and their hosts, *Annu. Rev. Entomol.*, 6, 201, 1961.
8. Southwood, T. R. E., *Ecological Methods*, Chapman & Hall, London, 1978.
9. Kogan, M. and Herzog, D. C., *Sampling Methods in Soybean Entomology*, Springer-Verlag, New York, 1980.
10. Shelton, A. M. and Trumble, J. T., Monitoring insect populations, in *Handbook of Pest Management in Agriculture*, 2nd ed., Pimentel, D., Ed., CRC Press, Boca Raton, FL, 1991, 45.
11. Binns, M. R. and Nyrop, J. P., Sampling insect populations for the purpose of IPM decision making, *Annu. Rev. Entomol.*, 37, 427, 1992.
12. Geier, P. W. and Clark, L. R., An ecological approach to pest control, *Proc. 8th Tech. Meet. Int. Union for Conservation of Nature and Natural Resources, Warsaw, 1960*, 1961, 10.
13. Pedigo, L. P., *Entomology and Pest Management*, Macmillan, New York, 1989.
14. Allee, W. C., Emerson, A. E., Park, O., Park, T., and Schmidt, K. P., *Principles of Animal Ecology*, W. B. Saunders, London, 1949.
15. Pearl, R., On biological principles affecting populations: human and other, *Am. Nat.*, 71, 50, 1937.
16. Andrewartha, H. G. and Birch, L. C., *The Ecological Web*, University of Chicago, Press, Michigan, 1984.
17. Andrewartha, H. G., *Introduction to the Study of Animal Populations*, University of Chicago Press, Michigan, 1961.
18. Elliot, J. M., *Some Methods for the Statistical Analysis of Samples of Benthic Invertebrates*, Sci. Publ. 25, 2nd ed., Freshwater Biology Association, 1977.
19. Pedigo, L. P., Lentz, G. L., Stone, J. D., and Cox, D. F., Green cloverworm populations in Iowa soybean with special reference to sampling procedure, *J. Econ. Entomol.*, 65, 414, 1972.
20. Macfadyen, A., Soil arthropod sampling, *Adv. Ecol. Res.*, 1, 1, 1962.
21. Hammond, R. B. and Pedigo, L. P., Sequential sampling plans for the green cloverworm in Iowa soybeans, *J. Econ. Entomol.*, 69, 181, 1976.

22. Lewis, T. and Taylor, L. R., Diurnal periodicity of flight by insects, *Trans. R. Entomol. Soc. Lond.*, 116, 393, 1965.
23. Mailloux, G., Binns, M. R., and Bostanian, N. J., Density yield relationships and economic injury level model for Colorado potato beetle larvae on potatoes, *Res. Popul. Ecol.*, 33, 101, 1991.
24. Marston, N. L., Morgan, C. E., Thomas, G. D., and Ignoffo, C. M., Evaluation of four techniques for sampling soybean insects, *J. Kan. Entomol. Soc.*, 49, 389, 1976.
25. Hillhouse, T. L. and Pitre, H. N., Comparison of sampling techniques to obtain measurements of insect populations on soybeans, *J. Econ. Entomol.*, 67, 411, 1974.
26. Ruesink, W. G. and Haynes, D. L., Sweep net sampling for the cereal leaf beetle, *Oulema melanopus*, *Environ. Entomol.*, 2, 161, 1972.
27. Browde, J. A., Pedigo, L. P., DeGooyer, T. A., Higley, L. G., Wintersteen, W. K., and Zeiss, M. R., Comparison of sampling techniques for grasshoppers (Orthoptera: Acrididae) in soybean, *J. Econ. Entomol.*, 85, 2270 1992.
28. Evans, E. W., Rogers, R. A., and Opfermann, D. J., Sampling grasshoppers (Orthoptera: Acrididae) on burned and unburned tallgrass prairie: night trapping vs. sweeping, *Environ. Entomol.*, 12, 1449, 1983.
29. Odum, E. P., *Fundamentals of Ecology*, 3rd ed., Saunders, Philadelphia, 1971.
30. Ewan, H. G., The use of host size and density factor in appraising the damage potential of a plantation insect, *Proc. Int. Congr. Entomol., 10th Meeting, Montreal, Quebec, 1956,* 4, 363, 1958.
31. Ruesink, W. G. and Kogan, M., The quantitative basis of pest management: sampling and measuring, in *Introduction to Insect Pest Management*, 2nd ed., Metcalf, R. L. and Luckmann, W. H., Eds., John Wiley & Sons, New York, 1982, 315.
32. Bellinger, R. G. and Dively, G. P., Development of sequential sampling for insect defoliation on soybeans, *J. N.Y. Entomol. Soc.*, 86, 278, 1978.
33. Snedecor, G. W. and Cochran, W. G., *Statistical Methods*, 7th ed., Iowa State University Press, Ames, 1980.
34. Waldbauer, G. P. and Kogan, M., Position of bean leaf beetle eggs in soil near soybeans determined by a refined sampling procedure, *Environ. Entomol.*, 4, 375, 1975.

Section I: Sampling Principles

Chapter 2

ARTHROPOD SAMPLING IN AGRICULTURAL LANDSCAPES: ECOLOGICAL CONSIDERATIONS

Douglas A. Landis

TABLE OF CONTENTS

I. Introduction to Landscapes and Agricultural Ecosystems 16
  A. Landscape Defined 16
  B. Agroecosystem Defined 18
  C. Pattern vs. Process 18

II. Characteristics of Agricultural Landscapes Influencing Arthropods 18

III. Characteristics of Agroecosystems Influencing Arthropods 20

IV. Landscape Characteristics and Arthropod Sampling 21
  A. Patches 21
    1. Patch Size 23
    2. Patch Composition 23
    3. Plant Architecture 23
    4. Patch Shape 23
    5. Patch Topography 23
  B. Patch Associations 24
    1. Spatial Associations 25
    2. Temporal Associations 25
    3. Linking of Temporal and Spatial Associations 26
  C. Corridors 26
    1. Movement in Corridors 26
    2. Corridors as Barriers to Movement 27
    3. Interactions of Corridors with Associated Patches 27

V. Summary 28

Acknowledgments 28

References 28

I. INTRODUCTION TO LANDSCAPES AND AGRICULTURAL ECOSYSTEMS

The abundance and spatial distribution of insects in agricultural landscapes is seldom static, with populations typically flowing through multiple habitats over the course of a season. When sampling or examining hypotheses regarding the processes that affect arthropod numbers or distributions, entomologists frequently concentrate on a specific management unit (e.g., crop field, orchard block). While this approach alone may be sufficient to understand the dynamics of some species, it may overlook important aspects of other insects' interactions with their surroundings and in the worst situation lead to misleading or erroneous results.

The corn earworm, *Helicoverpa zea* in the southeastern U.S. is a good illustration of the importance of understanding a pest's interactions with multiple habitats to effectively sample and manage it.[1] Overwintering female *H. zea* emerge from the soil and begin to lay eggs of the first generation which are deposited on seedling corn and to a lesser extent tobacco and wild hosts.[2,3] Second generation eggs are laid almost exclusively within corn where the larvae develop in the ears. Due to the limited number of ears per plant and their confined size, larval cannibalism is high and a maximum of 50,000 individuals per hectare (20,355/acre) are produced.[4] Adult *H. zea* of this generation leave the now-senescing corn and lay eggs of the third generation on a variety of hosts including cotton, soybean, peanut, tobacco and sorghum. It is in this generation that the larvae cause the most extensive damage. Sampling and managing *H. zea* in any one of these crops depends on an understanding of the interactions of the insect with the mosaic of past and potential hosts in the agricultural landscape.[5] As an example, Stinner et al.[4] described how hot, dry conditions occasionally result in the second generation being produced in soybeans vs. corn. Less restricted feeding sites reduce cannibalism so that larval densities may commonly exceed 250,000/ha (101,175/acre). These conditions result in a much larger third generation that must be managed on a variety of hosts.

Recognition of the dynamic temporal and spatial patterns of insect abundance and distribution in agricultural systems is not new, however, examples typically have been examined as unique cases without the benefit of an underlying theoretical framework. Recently, many ecologists have begun to view the dynamic patterns of many ecological phenomena in the context of landscape ecology.[6-11] Within this framework, landscape structure is viewed as the basic template upon which all ecological interactions take place.[12] This chapter will attempt to address sources of variability in arthropod abundance, distribution, and diversity in agricultural systems from an ecological perspective. The goal is to provide a framework for viewing agricultural landscapes and agroecosystems in order to better understand some of the sources of variability in arthropod population size, distribution, and community structure and to use this knowledge to design effective sampling programs.

A. LANDSCAPE DEFINED

Most persons have an intuitive sense of the word "landscape" as it refers to the structure of our natural surroundings. Increasingly, the term is used to define many types of environments including the "urban landscape" or even the "political landscape". However, landscape has a particular ecological meaning and lexicon which is useful in understanding the interactions that take place within agricultural systems. Forman and Godron defined a landscape as, "a heterogeneous land area composed of a series of interacting ecosystems that is repeated in similar form throughout."[9] Figure 1 shows a portion of a Midwest agricultural landscape. By examining it closely

FIGURE 1. Portion of the agricultural landscape of Ingham County, Michigan, showing various landscape elements including: (a) crop fields, (b) orchards, (c) woodlots, (d) fencerows, (e) hedgerows, (f) wetlands, (g) farm yards, and (h) roads. (Used with permission, Michigan Department of Natural Resources.)

we can pick out many of the characteristics which define the ecological meaning of landscape. Quickly apparent is the heterogeneous nature of this land area. It is made up of a large number of individual units of land that vary in size, shape, and composition. These landscape elements are defined as the smallest, relatively homogeneous ecological units that make up the landscape. In general, individual elements have dimensions on the order of tens to thousands of meters. While at the field level we are able to discern objects as small as individual plants or animals, at the landscape level we will concern ourselves with the entire crop field as a single element. Other elements in the landscape of Figure 1 are the woodlots, fencerows, hedgerows, wetlands, farm yards, orchards, and roads.

Each of these elements can be considered a separate ecosystem. However, it is obvious that these ecosystems have a considerable amount of interaction. Plants, animals, nutrients, and materials flow between these ecosystems with ease. In an agricultural landscape many of these flows or rates of flows are influenced by humans, either directly or indirectly. For example, when a crop is harvested, insects may move into neighboring landscape elements such as fields, hedgerows, or woodlots. While some movement would naturally take place, the rate is greatly accelerated by agricultural activity.

Another feature of landscapes is that the same types of ecosystems (elements) are repeated throughout the landscape. In Figure 1, a point placed anywhere on the landscape would probably be surrounded by corn, soybean, small grain, and alfalfa fields as well as woodlots, fencerows, roads, and human habitations. Any number of other random points in this landscape would also be surrounded by a similar cluster of elements. This grouping of ecosystems is diagnostic for this particular landscape and serves to distinguish it from another landscape.

B. AGROECOSYSTEM DEFINED

The term "ecosystem" is used to describe the sum total of the living organisms (biotic) in a given area, their interactions with other organisms, and with the physical (abiotic) components of the environment. Similarly, the term "agroecosystem" describes the biotic and abiotic features and interactions that take place within agricultural fields or systems. A small subset of these interactions and processes would include nutrient cycling, predator-prey interactions, competition, commensalism, and succession.[13] Agroecosystems exhibit all of the same interactions that occur in natural ecosystems but differ in that they are to some degree designed and managed by humans to yield a desired output. They are typically converted native ecosystems and show characteristics that are intermediate between natural ecosystems and those that are highly human managed such as urban/industrial areas. An agroecosystem can be defined at a number of scales. A single field can be considered a complete agroecosystem.[14,15] Alternatively, one may focus on similar farms within a small geographic area or one may consider an agroecosystem with regional or global interactions.[13,16]

C. PATTERN VS. PROCESS

The concepts of agricultural landscapes and agroecosystems are complimentary in understanding the abundance, distribution, and diversity of arthropods in agriculture and in developing efficient sampling efforts. While both concepts can be used to describe the same physical land area, each provides a slightly different perspective. The concept of landscape deals primarily with spatial patterns in the physical structure of the environment.[17] These spatial patterns influence the biology of arthropods both directly and indirectly. The concept of agroecosystems deals primarily with the processes at work within the agricultural environment. These processes interact with and contribute to shaping larger scale landscape spatial patterns. The agroecosystem concept has been dealt with extensively and will receive comparatively less attention here.[13,16,18,19] The influence of landscape pattern on arthropods is an area that has received less attention and will be the primary focus of the remainder of this chapter.

II. CHARACTERISTICS OF AGRICULTURAL LANDSCAPES INFLUENCING ARTHROPODS

Not all elements within a landscape are similar in physical structure. Differences in structure and origin of landscape elements can alter the function of the landscape. Patches are common elements in most landscapes and are normally composed of an assemblage of plant and animal species (e.g., the woodlot), but may also be essentially lifeless (e.g., a large parking lot). The most abundant type of patch in agricultural landscapes is that which arises from the introduction of species (plants, animals, humans). This type may be represented by either planted patches, as in the case of agricultural production fields, or habitation patches (e.g., farm yards) in which the primary introduced species is human. Corridors are distinguished from patches primarily in that they are narrow strips of land that differ from the matrix on either side.[9] Several types of corridors can be seen in Figure 2. Line corridors are narrow corridors that are dominated by edge species. Fencerow, hedgerow, and road corridors are examples of this type. Strip corridors are wider and contain some interior species (e.g., utility right of way, interstate highway). A unique type of corridor is the stream or river corridor. These corridors are composed of the water and the surrounding strips of vegetation on either side (Figure 2).

One of the principal distinguishing characteristics of agricultural landscapes is the nature and frequency of their disturbance regimes. Perhaps the most significant and long lasting of these disturbances is the initial act of land clearing or cultivation. Even

FIGURE 2. Corridor elements in an agricultural landscape. Line corridors include the (a) hedgerow networks and (b) secondary roads. Strip corridors are illustrated by the (c) interstate highway and (d) utility right of way. River (e) and (f) stream corridors are also present in this landscape. (Used with permission, Michigan Department of Natural Resources.)

without further disturbance, most native ecosystems would require tens to hundreds or perhaps thousands of years to resume a natural appearance following initial disturbance. Periodic tillage and planting of agricultural fields is a more frequent form of disturbance, occurring as infrequently as once every 8 to 10 years or as frequently as several times per year. The outcome of this activity is to functionally revert the tilled area to an earlier stage of succession. Even more frequent disturbance results from agronomic activities such as cultivation, pesticide applications, fertilization, and harvest. Annual crops may commonly receive two to ten such disturbances yearly, depending on the crop, region, and production system. Natural disturbances such as floods or fire also occur in and shape agricultural landscapes, however, their periodicity in most areas is very rare in comparison to human-induced disturbance.

The spatial pattern of disturbance is equally as important as its frequency. The pattern of this initial disturbance produces legacies which can persist for extremely long periods of time. In many parts of the U.S. early agricultural systems frequently included an animal component. The requirement for confinement of livestock resulted in small fields, each bordered by a fence or hedgerow.[20] Natural succession in these noncultivated areas often resulted in the growth of trees which further delineated the field border. In some areas, field size is little changed from earlier times due to the difficulty of removing these barriers. However, increasingly they are being removed to make room for irrigation and/or larger farm equipment.[21,22]

A general characteristic of agricultural landscapes is their geometric regularity, in contrast to natural landscapes. The patterns of linearity have arisen for a variety of reasons, including land ownership patterns and mechanization. Rectangular fields are more efficient to farm as equipment becomes increasingly larger. By management of drainage, irrigation, and fertility patterns, producers are able to farm over topographic or soil geomorphologic features that previously inhibited production. Geometrization is also apparent in the increased abundance of line corridors in agricultural landscapes, including roads, hedgerows, drainage ditches, grass waterways, etc.[9] Stream corridors, while typically wide and irregular in natural landscapes, are increasingly narrowed and linearized in agricultural landscapes due to the pressure to increase tillable land. The most striking example is the dredging and straightening of streams and rivers to improve drainage.

Agricultural landscapes thus evolve from undisturbed natural landscapes over time. Initially, the matrix of the landscape was the native ecosystems dominated by disturbance patches of natural origins (fire, flood). As human influence increases, introduced patches (cultivated ecosystems) become more prevalent. As this occurs, the total number of patches in the landscape (patch density) tends to increase, and the variability of patch sizes tends to decrease.[9] The total area of native ecosystems declines, but the number of remnant patches (i.e., native ecosystem types) increases as the habitat becomes more fragmented. The structure of the agricultural landscape that emerges has a distinct influence on the types, abundance, and distribution of the organisms that live there. Overall, species diversity declines in agricultural landscapes, and increasingly so does genetic diversity, as large patches of genetically similar crops appear. These changes have implications for sampling and managing insects within agricultural landscapes.

III. CHARACTERISTICS OF AGROECOSYSTEMS INFLUENCING ARTHROPODS

Odum described some of the major characteristics of agroecosystems (Table 1).[23] Both natural ecosystems and agroecosystems derive their primary energy from the

TABLE 1
Characteristics of Agroecosystems in Contrast to Natural Ecosystems[23]

1. The auxiliary energy sources that enhance productivity are processed fuels (along with human and animal labor) rather than natural energies.
2. Diversity is greatly reduced by human management in order to maximize yield of specific food and other products.
3. The dominant plants and animals are under artificial rather than natural selection.
4. Control is external and goal oriented rather than internal via subsystem feedback.

sun. However, in agroecosystems there is a significant auxiliary input of energy in the form of processed fuels and animal and human labor not present in natural ecosystems. Species diversity is greatly reduced in agroecosystems through human management aimed at increasing the productivity of certain specific products (crops, animals). The result is fewer and shorter food chains and reduced complexity of the overall food web within these systems.[13] The dominant plants and animals are under artificial rather than natural selection, resulting in greatly reduced genetic diversity especially when few crop types or large monocultures of genetically identical individuals (e.g., single-cross hybrid corn) are grown. This homogeneity has implications for the type and numbers of arthropods that can inhibit the system.

Finally, regulation of ecosystem processes is largely external and directed toward specific production goals, rather than the internal feedback controls present in natural ecosystems. This is evident in the need for external pest control actions vs. the natural controls on populations in native ecosystems. Clearly, the farmer is an essential ecological variable in the agroecosystem, influencing all of the characteristics discussed here and ultimately, determining the agroecosystem's composition, functioning, and stability.[13,16]

Cox summarized many of the interactions that occur in mechanized agroecosystems vs. a natural ecosystem (Figure 3).[24] Agroecosystems contain fewer species and more empty niche space than natural systems. Additionally, within any niche, the within-species diversity in terms of age, size, genetic structure, etc., is low. In natural ecosystems, much of the output (energy or materials) is recycled and used to maintain ecosystem organization functions such as nutrient cycling or biotic regulation of species. In agroecosystems, most of the output is exported (harvested) and removed from the system, leaving little to reinvest in ecosystem organization. Therefore, humans must substitute labor and material inputs to maintain the stability and productivity of the system.

The rate of disturbance in agroecosystems is high, resulting in a functional reversion of these systems to an earlier state of succession.[25] The early successional state of these habitats favors their colonization by "r-selected" species.[26] An r-selected species maximizes its fitness by reproducing rapidly in uncrowded environments, vs. k-selected species which maximize fitness through competitive mechanisms in a population that is close to its carrying capacity.[27] Many of the characteristics that we associate with arthropod pests (high reproductive capacity, good dispersal capabilities, rapid colonization of new habitats, etc.) are characteristic of r-selected species.[25]

IV. LANDSCAPE CHARACTERISTICS AND ARTHROPOD SAMPLING

A. PATCHES

In agricultural landscapes, the management unit, for example, a particular crop field or orchard block, is considered a landscape element. Typically these elements

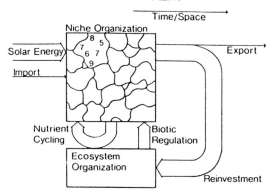

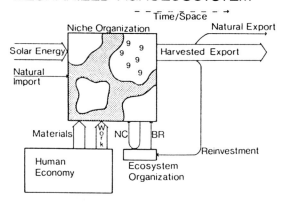

FIGURE 3. A comparison of natural ecosystems with mechanized agroecosystems. Natural ecosystems contain a more diverse biotic community (number of cells in niche space box) with more varied individual characteristics such as age, health, and genetic makeup (indicated by numbers within each species' cell) and which more fully exploit the available niche space. In agroecosystems the large export (food, fiber) allows a comparatively small portion of the systems' production to be reinvested into ecosystem organization. The agroecosystem is thus dependent on inputs of labor and materials from the human economy to maintain functioning. (From Cox 1984, reprinted by permission of John Wiley & Sons.)

are introduced patches, as discussed earlier; however, in some instances patch shape could characterize them as corridors (e.g., strip cropping). It is typically at the level of the whole field that we want to make decisions regarding the arthropod species composition or population size, particularly if pest management decisions are the ultimate goal of sampling. This section will explore characteristics of patches in terms of size, shape, topography, and the plant forms that make up the patch, to determine how these characteristics may influence the arthropod community.

The influence of patch characteristics on arthropod populations has been extensively explored in the ecological literature, particularly in relation to predictions of the resource concentration hypothesis.[28] This hypothesis predicts that specialized herbivores are more likely to find and remain in areas where host plants are concentrated, i.e., where they grow in pure, large, or dense stands. This hypothesis has obvious implications for agriculture, where large monocultures of crop plants are common.

1. Patch Size

Herbivorous insects show a variable response to patch size.[29,30] Increasing patch size may increase,[31,32] decrease,[33,34] or not alter herbivore abundance.[35,36] Bach demonstrated that patch size effects can be significantly altered by the diversity of the surrounding vegetation.[37] Patch size can also interact with natural enemies to alter the distribution and abundance of herbivorous insects. In a field of mixed crucifers, van Emden found that predators of the cabbage aphid, *Brevicoryne brassicae*, were more abundant at field margins, causing a significant reduction in aphid density.[38] In the 0.71-ha (1.75-acre) field, all areas were close enough to an edge that predators reduced densities over the entire field. He hypothesized that predators would cause less of a reduction in the center of larger fields.

2. Patch Composition

Increasing diversity of plants within a patch has been shown to frequently lower the abundance of herbivores. Risch et al.[39] found that in 150 studies they reviewed, 53% of the species showed a decrease in abundance in response to increasing within-field plant diversity, 18% increased in abundance, 9% showed no difference, and 20% a variable effect. Baliddawa reviewed many of the same studies looking for the factors affecting herbivore populations.[40] In crop-weed diversified systems, natural enemies were most often cited as the reason for herbivore decline while in crop-crop diversified systems (intercropping), natural enemy activity was cited in less than one third of the cases reviewed.

3. Plant Architecture

The architecture of the plants within a patch can also influence the diversity of the arthropod fauna.[41,42] Lawton reviewed the subject and found that for herbivorous insects,[43] species richness is generally greatest in trees and declines with decreasing complexity of plant architecture. Thus, woody shrubs contain more herbivores than perennial herbs, which have more than weeds and other annuals, which are greater than monocots. This pattern can generally be seen in agricultural systems when one compares the extent of the pest complex in orchard vs. annual crops vs. grasslands.

4. Patch Shape

Patch shape has rarely been addressed in studies with arthropods. Stanton discusses some of the possible implications of patch shape in agriculture primarily as it relates to herbivore host finding.[30] Given that many pests use plant odors in seeking host habitats, the shape of the patch from which the odor is emanating can greatly influence the area from which pests are attracted (Figure 4a to c). Long, thin patches oriented perpendicular to the prevailing wind would spread host odors over a wider area, although perhaps in lower concentration, than a square patch of the same area. Forman and Godron discuss a similar general result for flying insects and birds regardless of mechanism of attraction.[9] They point out that both patch shape and orientation in regard to direction of the flight path will influence an organism's chance of encountering a host patch (Figure 4d to f).

5. Patch Topography

Topography is also a factor that can influence the distribution of organisms within a patch. In agriculture, this could occur for two reasons. First, topography may mediate the composition of the plant community which in turn affects the insect, or alternatively, topographic features may directly influence the insect. Alfalfa fields are good examples of the first case. In low areas of a field, increased soil moisture, surface

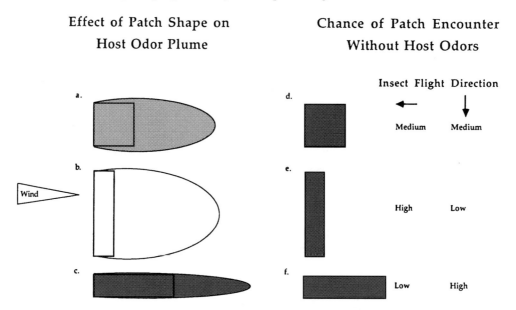

FIGURE 4. Influence of patch shape on chance of arthropod patch encounter (all patches, a to f are of equal area). For species that use host odors to locate patches, patch shape (a to c) can affect the shape and perhaps concentration of the host odor plume altering the area from which herbivores will be attracted (after Stanton 1983). From arthropods that search without use of olfaction, patch shape and orientation in relation to direction of flight can alter the relative chance of patch encounter (d to e).

ice sheets, and frost heaving interact with certain alfalfa diseases and result in a decreased stand of alfalfa.[44] As space is opened, weeds invade and shift the plant community to a polyculture. As a result, low areas may have a much different insect community than upland parts of the field.

Topographic features may also directly affect insect behavior. Aggregations of male insects at particular landscape features where they more frequently encounter potential mates, is a common characteristic of many insect mating systems. Thornhill and Alcock list six orders of insects for which this behavior has been observed.[45] Frequently the landscape feature is associated with increased height such as hilltops, tops of trees, bushes, or rock faces, but the feature can also be low areas such as gullies.[46] Nontopographic features such as sunny areas, bare patches, or dark areas in light soils also serve as landmarks for aggregation.[45]

The influence of landscape-scale crop diversity on arthropod populations has been poorly studied but is likely of significance.[47] Andow reviewed 17 studies where changes in regional cropping patterns (increase or decrease in degree of monoculture) were discussed.[48] Of the 26 insect populations studied, 16 increased with increasing degree of monoculture and 10 decreased. Notable is that all of those insects that increased were monophagous on the crop that was increasing in monoculture while nine of the ten species that decreased were oligophagous or polyphagous. Host range may thus play a significant role in the response of herbivores to changing landscape structure.

B. PATCH ASSOCIATIONS

The association of one or more patches is often critical for a particular insect's life system. For example, many insects require one type of habitat for overwintering and one or more additional habitats for feeding, mating, resting, and oviposition. The association of these habitat patches in both time and space is a critical feature of the functioning of an agricultural landscape. It is critical that these aspects are considered when designing arthropod sampling systems.

1. Spatial Associations

Patches in close spatial association frequently constitute a source/sink relationship for insect populations. Crops of the same species which are planted or harvested at different times form an obvious example. Alfalfa field harbor a truly impressive insect fauna.[49] When a particular field is harvested, there is potential for mobile species to disperse from the cut field into other nearby alfalfa. Potato leafhopper (*Empoasca fabae*) adults are quite mobile and an increase in numbers of leafhoppers is frequently associated with the cutting of nearby alfalfa.[50] These immigrants accumulate in the exterior margins of the field and tend to be predominantly female.[51] There is a gradual shift from a female-biased sex ratio early in the spring to approximately a 1:1 ratio in the fall. This change in sex ratio has been used to predict if a population is in the initial phase of colonization or is already established.[52] Field to field movement of potato leafhoppers thus has obvious implications for pest management.

The potato leafhopper also demonstrates movement from one crop to another. Poston and Pedigo found that potato leafhoppers moved from recently harvested alfalfa and accumulated in the margins of nearby soybean fields.[53] Flanders and Radcliffe examined the early season invasion of potato leafhopper into potato and snap beans.[54] They concluded that the most likely source of immigrants was from individuals produced in local, undisturbed habitats, possibly soybeans.

Movement of insects between cultivated and uncultivated habitats is also common and can have important implications for sampling. The potato leafhopper has a wide host range and can exist in many types of uncultivated habitats in association with crop fields.[55] Wise documented the emigration of potato leafhoppers from a recently cut alfalfa field into wooded and fencerow vegetation.[56] Re-colonization of the new alfalfa growth was not uniform, but occurred first in proximity to these transitional habitats.

Aphids are another well-known example of movement from uncultivated to crop hosts. Eastop states that while only about 10% of all aphids are heteroecious,[57] these host-alternating species comprise 42% (63 out of 151) of the species regarded as pests. Many of these species must shift habitats in order to find their alternate host. The bird cherry-oat aphid, *Rhopalosiphum padi*, spends the winter in the egg stage on bird cherry, *Prunus padus*, in noncrop habitats.[58] After feeding on the buds and developing leaves of bird cherry in the spring, it migrates to crop fields where it colonizes cereals and other various grasses which are its summer hosts. In cereal grains it is considered an important vector of the barley yellow dwarf virus.[59] In September, aphids return to the primary host where breeding occurs and overwintering eggs are deposited.[58]

2. Temporal Associations

Insects are also affected by the temporal association of patches. This is particularly true in agricultural landscapes where the patch turnover rate is extremely high (e.g., for annual crops one to three times per year).[9] Over the course of several years, an insect population may or may not be able to exist in a particular field due to the temporal association of the crops grown there. Crop rotation can be used as an example of both outcomes. The western corn rootworm, *Diabrotica virgifera virgifera*, is the major pest of corn throughout the Midwest.[60] This insect, however, can be essentially eliminated as a corn pest in a given field by the practice of annual crop rotation.[61] Female western corn rootworms lay eggs in the soil of corn fields in late

July to September. After overwintering, the eggs hatch the following June and the larvae seek corn roots on which to feed. If a field is in an annual crop rotation (e.g., corn-soybean-corn) where a nonhost crop other than corn is present, when the eggs hatch the larvae will die. Only in fields where corn is grown for 2 or more years in a row is the rootworm a potential pest of corn roots. The northern corn rootworm (*D. barberi*) has adapted to the practice of annual crop rotation by developing a prolonged egg diapause. In certain populations, eggs of the northern corn rootworm may pass from 2 to 5 years in the soil before hatching. This adaptation allows them to spread the risk of hatching in the "wrong" year, when the host plant is not available. Successful sampling programs aimed at managing both of these insects, which typically occur together as a complex, must incorporate a knowledge of the temporal crop associations.[62,63]

3. Linking of Temporal and Spatial Associations

Some of the most interesting challenges in sampling comes when an insect utilizes several crop habitats within a landscape at different times of the season. Both crop pests and their natural enemies exhibit dispersal within agricultural landscapes.[64,65] The two-spotted spider mite *Tetraynachus urticae*, is a pest of several crops in the southeastern U.S. Brandenburg and Kennedy studied the spatial and temporal associations of the spider mite in corn and peanuts.[66] In the spring, mites were present along the margins of crop fields on wild hosts such as red clover and blackberry. Studies showed that while equal density of mites existed in the margins of both corn and peanuts, only corn became infested during the early season. The primary mode of mite dispersal into corn fields was by crawling. Mite populations first developed on corn plants near the edges of fields, gradually increasing throughout the field during June and July. A subsequent decline in host quality of the corn crop prompted dispersal behavior where mites became positively phototactic and congregated in large masses near the top of corn plants.[67,68] From this elevated position, mites dispersed on the wind. Peanut fields in close association with heavily infested corn fields frequently showed a sudden massive infestation of mites at the time of mite dispersal from corn. Mark-recapture studies indicated that the invasions were originating in the nearby corn.

C. CORRIDORS

Landscape elements that act as corridors can also serve to influence the abundance and distribution of arthropods within agricultural landscapes. Corridors can act as a conduit for the movement of species along the feature, or they may act as a barrier to the movement of a species across the corridor. In addition, since they are in essence, highly elongated patches with a large surface to volume ratio, they may be expected to have a high degree of interaction with closely associated elements.[9] In regards to arthropod distribution in agroecosystems, we can find examples of all of these effects.

1. Movement in Corridors

There are many types of corridors in agricultural landscapes. An incomplete list would include highways, roads, lanes, ditches, streams, rivers, fencerows, hedgerows, shelterbelts, powerlines, irrigation travel lanes, and canals, etc. As mentioned earlier, even the crop field itself can be considered a corridor, as in the case of contour or strip farming systems. Hedgerows have received significant attention, particularly in European countries, regarding their effects on arthropod distribution and abundance in agricultural landscapes.[69,70] Burel studied the impact of a hedgerow network on the distribution of forest carabid beetles in France.[71] She found that some forest-dwelling carabids were never found more than 50 m (55 yd) from a forest, and termed these forest-core species. Others were found both in the forest and along hedgerows up to 550 to 700 m (602 to 766 yd) from the forest edge (forest-peninsula species). Finally, hedgerows served as corridors, allowing some forest carabids (corridor-forest species) to penetrate as much as 20 km (12 mi) into nonforested agricultural landscapes.

The screwworm, *Cochliomyia hominivorax*, is an example of an insect that uses landscape corridors for short- and long-distance migration. Because of its impact on livestock, massive efforts aimed at eradication of the screwworm utilized the sterile male technique.[72,73] In 1962, a program was initiated to eradicate the screwworm from the southwestern U.S. Historically, the fly would invade much of the Southwest yearly from its overwintering area in Mexico and south Texas. Estimates of natural dispersal northward of 48 to 80 km (30 to 50 mi)/week previously had been measured.[74] It was known that during periods of hot, dry weather, the screwworm concentrated along watercourses to obtain water and shelter.[75] Hightower and Alley found that the flies used these landscape features, as well as other corridors such as dry gullies and fencelines, for local dispersal of up to several miles.[76] Long-distance dispersal along watercourses was examined following the initial year of eradication efforts in which infestations of flies were found far north of the eradication zone. Hightower et al.[77] found that marked flies released along the Colorado River dispersed upstream as much as 290 km (180 mi) in less than 2 weeks. This finding led to a critical modification of the program whereby targeted releases of sterile males were made along major watercourses, both within and beyond the eradication zone.[72]

2. Corridors as Barriers to Movement

Corridor elements can also serve to impede the movement of arthropods around the landscape. Crawling insects are unlikely to readily cross streams or ditches with flowing water, and certain corridors may simply contain inhospitable conditions as with asphalt roads that can reach high temperatures. Flying insects may be channeled by corridor elements, with breaks in the structure providing limited areas for through passage. Townes recommended this type of feature as a very productive location for trapping flying insects.[78] It is also likely that corridors may be locations of high predation for insects, as insectivorous birds and mammals frequently use corridors as stations for hunting and nesting.[20,22]

3. Interactions of Corridors with Associated Patches

Due to their extensive edge to volume ratio, organisms in corridor elements tend to have extensive interactions with nearby elements.[79] Hedgerows, ditches, and windbreaks are corridor elements that typically border fields and may increase or decrease pest density in the crop.[80] These changes result from both biological and physical effects.[81] These uncultivated habitats can serve as a reservoir of pest species and thus, as a source of initial infestation for adjoining crop fields and orchards.[38,82,83] The modified wind patterns and microclimate that hedgerows and windbreaks cause can alter the distribution and severity of insect attack.[84-89] Alternatively, association with wooded edge habitats may reduce the infestation of certain pests. Cromartie showed that *Pieris rapae* failed to colonize host plants that were placed next to wooded edges, although plants in nearby meadows were attacked.[90] The cause was attributed to a reduced likelihood of discovery of plants occurring in partial or complete shade.

Increased abundance of natural enemies and more effective biological control is often found where uncultivated habitats, frequently corridors, occur in association with crops.[91] These habitats may be important as overwintering sites for predators,[92-95] or they may provide increased resources such as pollen and nectar from flowering plants,[38,80,96] or a combination of factors.[97] Manipulation of corridor elements has been attempted to increase the effectiveness of biological control of cereal aphids. Sotherton found higher populations of aphid predators (primarily Carabidae and Staphalinidae) in the raised grassy banks of field borders.[98] Creation of artificial grassy banks within cereal fields resulted in successful overwintering of these predators at numbers up to $1600/m^2$ ($1333/yd^2$).[99,100] Re-colonization of the field in the spring from these sites was more rapid than where predators had to disperse from more distant field edges.[101]

V. SUMMARY

Sampling arthropods in agricultural landscapes is a challenging task. Systems of sampling for either pest management decision making or for research purposes must both balance the need for accuracy vs. requirements for efficiency. Understanding the underlying factors that drive changes in arthropod distribution and abundance in these habitats can help researchers and pest managers alike in developing sampling plans that meet both of these competing needs. An understanding of how agricultural landscape structure influences patterns of arthropod abundance is critical to the development and practice of efficient and reliable sampling.

ACKNOWLEDGMENTS

I wish to thank Lawrence Dyer, Katherine Gross, Mike Haas, Joy Landis, and Paul Marino who each provided useful comments on earlier versions of this manuscript. I am also grateful to my colleague Stuart Gage for many conversations through which some of the ideas expressed in this manuscript have developed. Funding from the National Science Foundation, BSR 89-06618, Rackham Foundation, and USDA Low Input Sustainable Agriculture Program has fostered our research examining the impact of agricultural landscape structure on insect natural enemies. I also wish to thank Alice Kenady for assistance in the final preparation of the manuscript.

REFERENCES

1. Stinner, R. E., Rabb, R. L., and Bradley, J. R., Jr., Natural factors operating in the population dynamics of *Heliothis zea* in North Carolina, *Proc. Int. Congr. Entomol.*, 15, 622, 1977.
2. Isley, D., *Relation of Hosts to Abundance of Cotton Bollworm*, Bull. No. 320, Arkansas Agric. Exp. Stn., 1935.
3. Neunzig, H. H., *The Biology of the Tobacco Budworm and the Corn Earworm in North Carolina*, Tech. Bull. No. 196, North Carolina Agric. Exp. Stn., Raleigh, 1969.
4. Stinner, R. E., Regniere, J., and Wilson, K., Differential effects of agroecosystem structure on dynamics of three soybean herbivores, *Environ. Entomol.*, 11, 538, 1982.
5. Stinner, R. E., Bradley, J. R., Jr., and Van Duyn, J. W., Sampling Heliothis spp. on soybean, in *Sampling Methods in Soybean Entomology*, Kogan, M. and Ruesink, W., Eds., Springer-Verlag, New York, 1980, chap. 21.
6. Forman, R. T. T., Interactions among landscape elements: a core of landscape ecology, in *Perspectives in Landscape Ecology*, Tjallingii, S. P. and de Veer, A. A., Eds., Pudoc, Wageningen, The Netherlands, 1981, 35.
7. Forman, R. T. T., An ecology of the landscape, *BioScience*, 33, 535, 1983.
8. Forman, R. T. T. and Godron, M., Patches and structural components for a landscape ecology, *BioScience*, 31, 733, 1981.

9. Forman, R. T. T. and Godron, M., *Landscape Ecology*, John Wiley & Sons, New York, 1986.
10. Naveh, Z. and Lieberman, A. S., *Landscape Ecology: Theory and Application*, Springer-Verlag, New York, 1984.
11. Risser, P. G., Karr, J. R., and Forman, R. T. T., Landscape ecology: directions and approaches, Illinois Natural History Survey Special Publ. No. 2, Illinois Natural History Survey, 1984.
12. Urban, D. L., O'Neill, R. V., and Shugart, H. H., Jr., Landscape ecology, a hierarchical perspective can help scientists understand spatial patterns, *BioScience*, 37, 119, 1987.
13. Tivy, J., *Agricultural Ecology*, Longman Scientific and Technical, Essex, 1990.
14. Crossley, D. A., Jr., House, G. J., Snider, R. M., Snider, R. J., and Stinner, B. R., The positive interactions in agroecosystems, in *Agricultural Ecosystems*, Lowrance, R., Stinner, B. R., and House, G. J., Eds., John Wiley & Sons, New York, 1984, 73.
15. Hecht, S. B., The evolution of agroecological thought, in *Agroecology: The Scientific Basis of Alternative Agriculture*, Altieri, M., Ed., Westview Press, Boulder, 1987, chap. 1.
16. Altieri, M., *Agroecology: The Scientific Basis of Alternative Agriculture*, Westview Press, Boulder, 1987.
17. Krummel, J. R. and Dyer, M. I., Consumers in agroecosystems: a landscape perspective, in *Agricultural Ecosystems*, Lowrance, R., Stinner, B. R., and House, G. J., Eds., John Wiley & Sons, New York, 1984, 55.
18. Lowrance, R., Stinner, B. R., and House, G. J., *Agricultural Ecosystems*, John Wiley & Sons, New York, 1984.
19. Carroll, C. R., Vandermeer, J. H., and Rosset, P. M., *Agroecology*, McGraw-Hill, New York, 1990.
20. Dambach, C. A., *A Study of the Ecology and Economic Value of Crop Field Borders*, The Ohio State University Press, Columbus, 1948.
21. Baltensperger, B. H., Hedgerow distribution and removal in nonforested regions of the Midwest, *J. Soil Water Conserv.*, January-February, 60, 1987.
22. Shalaway, S. D., Fencerow management for nesting birds in Michigan, *Wildl. Soc. Bull.*, 13, 302, 1985.
23. Odum, E. P., Properties of agroecosystems, in *Agricultural Ecosystems*, Lowrance, R., Stinner, B. R., and House, G. J., Eds., John Wiley & Sons, New York, 1984, chap. 3.
24. Cox, G. W., The linkage of inputs to outputs in agroecosystems, in *Agricultural Ecosystems*, Lowrance, R., Stinner, B. R., and House, G. J., Eds., John Wiley & Sons, New York, 1984, 187.
25. Price, P. W. and Waldbauer, G. P., Ecological aspects of pest management, in *Introduction to Insect Pest Management*, Metcalf, R. L. and Luckmann, W. H., Eds., John Wiley & Sons, New York, 1975, chap. 2.
26. Odum, E. P., The strategy of ecosystem development, *Science*, 164, 262, 1969.
27. MacArthur, R. H. and Wilson, E. O., *The Theory of Island Biogeography*, Princeton University Press, Princeton, NJ, 1967.
28. Root, R. B., Organization of a plant-arthropod association in simple and diverse habitats: the fauna of collards (*Brassica oleracea*), *Ecol. Monogr.*, 43, 95, 1973.
29. Kareiva, P., Influence of vegetation texture on herbivore populations: resource concentration and herbivore movement, in *Variable Plants and Herbivores in Natural and Managed Systems*, Denno, R. F. and M. S. McClure (eds.) Academic Press, New York, 1983, chap. 8.
30. Stanton, M. L., Spatial patterns in the plant community and their effects upon insect search, in *Herbivorous Insects*, Ahmad, S., Ed., Academic Press, New York, 1983, chap. 4.
31. Thompson, J. N., Within-patch structure and dynamics in *Pastinaca sativa* and resource availability to a specialized herbivore, *Ecology*, 59, 443, 1978.
32. Kareiva, P., Finding and losing host plants by *Phyllotreta*: patch size and surrounding habitat, *Ecology*, 66, 1809, 1985.
33. Cromartie, W. J., Jr., The effect of stand size and vegetational background on the colonization of cruciferous plants by herbivorous insects, *J. Appl. Ecol.*, 12, 517, 1975.
34. Maguire, L. A., Influence of collard patch size on population densities of Lepidopteran pests (Lepidoptera: Pieridae, Pluttellidae), *Environ. Entomol.*, 12, 1415, 1983.
35. Raupp, M. J. and Denno, R. F., The influence of patch size on a guild of sap-feeding insects that inhabit the salt marsh grass *Spartina patens*, *Environ. Entomol.*, 8, 412, 1979.
36. MacGarvin, M., Species-area relationships of insects on host plants: herbivores on rosebay willowherb, *J. Anim. Ecol.*, 51, 207, 1982.
37. Bach, C. E., A comparison of the responses of two tropical specialist herbivores to host plant patch size, *Oecologia*, 68, 580, 1986.
38. van Emden, H. F., The effect of uncultivated land on the distribution of cabbage aphid (*Brevicoryne brassicae*) on an adjacent crop, *J. Appl. Biol.*, 2, 171, 1965.
39. Risch, S. J., Andow, D., and Altieri, M., Agroecosystem diversity and pest control: data, tentative conclusions, and new research directions, *Environ. Entomol.*, 12, 625, 1983.

40. Ballidawa, C. W., Plant species diversity and crop pest control, *Insect Sci. Appl.*, 6, 479, 1985.
41. Murdoch, W. W., Evans, F. C., and Peterson, C. H., Diversity and pattern in plants and insects, *Ecology*, 53, 819, 1972.
42. Strong, D. R., Jr. and Levin, D. A., Species richness of plant parasites and growth form of their hosts, *Am. Nat.*, 114, 1, 1979.
43. Lawton, J. J., Plant architecture and the diversity of phytophagous insects, *Annu. Rev. Entomol.*, 28, 23, 1983.
44. McKenzie, J. S., Paquin, R., and Duke, S. H., Cold and heat tolerance, in *Alfalfa and Alfalfa Improvement*, Agron. Monog., No. 29, Hanson, A. A., Barnes, D. K., and Hill, R. R., Jr., Eds., American Society of Agronomy, Inc. Crop Science Society of America, Soil Science Society of America, Inc. Publishers, Madison, WI. 1988.
45. Thornhill, R. and Alcock, J., *The Evolution of Insect Mating Systems*, Harvard University Press, Cambridge, MA, 1983.
46. Scott, J. A., Adult behaviour and population biology of two skippers (Hesperiidae) mating in contrasting topographic sites, *J. Res. Lepidop.*, 12, 181, 1973.
47. Powers, A. G. and Kareiva, P., Herbivorous insects in agroecosystems, in *Agroecology*, Carroll, C. R., Vandermeer, J. H., and Rosset, P. M., Eds., McGraw-Hill, New York, 1990, chap. 10.
48. Andow, D., The extent of monoculture and its effects on insect pest populations with particular reference to wheat and cotton, *Agric. Ecosyst. Environ.*, 9, 25, 1983.
49. Pimentel, D. and Wheeler, A. G., Jr., Species and diversity of arthropods in the alfalfa community, *Environ. Entomol.*, 2, 659, 1973.
50. Pienkowski, R. L. and Medler, J. T., Potato leafhopper trapping studies to determine local flight activity, *J. Econ. Entomol.*, 59, 837, 1966.
51. Flinn, P. W., Hower, A. A., and Taylor, R. A., Immigration, sex ratio, and local movement of the potato leafhopper (Homoptera: Cicadellidae) in a Pennsylvania alfalfa field, *J. Econ. Entomol.*, 83, 1858, 1990.
52. Drake, D. C. and Chapman, R. K., Part 1. Evidence for long distance migration of the six-spotted leafhopper into Wisconsin, in *Migration of the Six-Spotted Leafhopper Macrosteles fascifrons Into Wisconsin*, Res. Bull. No. 261, Wisconsin Agric. Exp. Stn., Madison, WI, 1965, 5.
53. Poston, F. L. and Pedigo, L. P., Migration of plant bugs and the potato leafhopper in a soybean-alfalfa complex, *Environ. Entomol.*, 4, 8, 1975.
54. Flanders, K. L. and Radcliffe, E. B., Origins of potato leafhoppers (Homoptera: Cicadellidae) invading potato and snap bean in Minnesota, *Environ. Entomol.*, 18, 1015, 1989.
55. Lamp, W. O., Morris, M. J., and Armbrust, E. J., *Empoasca* (Homoptera: Cicadellidae) abundance and species composition in habitats proximate to alfalfa, *Environ. Entomol.*, 18, 423, 1989.
56. Wise, J. C., Local Dispersal of the Potato Leafhopper (Homoptera: Cicadellidae) in Response to Alfalfa Cuttings, M.S. thesis, Michigan State University, East Lansing, 1990.
57. Eastop, V. F., The wild hosts of aphid pests, in *Pests, Pathogens, and Vegetation*, Thresh, J. M., Ed., Pitman, London, 1981, 285.
58. Dixon, A. F. G., The life-cycle and host preferences of the bird cherry-oat aphid, *Rhopalosiphum padi* L., and their bearing on the theories of host alternation in aphids, *Ann. Appl. Biol.*, 68, 135, 1971.
59. Wiese, M. V., *Compendium of Wheat Diseases*, 2nd ed., American Phytopathological Society Press, St. Paul, MN, 1987.
60. Metcalf, R. L., Forward, in *Methods for the Study of Pest Diabrotica*, Krysan, J. L. and Miller, T. A., Eds., Springer-Verlag, New York, 1986, vii–xv.
61. Levine, E. and Oloui-Sadeghi, H., Management of Diabroticite rootworms in corn, *Annu. Rev. Entomol.*, 36, 229, 1991.
62. Landis, D. A., Levin, E., Haas, M. J., and Meints, V., Detection of prolonged diapause of northern corn rootworm in Michigan (Coleoptera: Chrysomelidae), *Great Lakes Entomol.*, 24, 215, 1992.
63. Levine, E., Oloui-Sadeghi, H., and Ficher, J. R., Discovery of a multiyear diapause in Illinois and South Dakota northern corn rootworm (Coleoptera: Chrysomelidae) eggs and incidence of the prolonged diapause trait in Illinois, *J. Econ. Entomol.*, 85, 262, 1992.
64. Bunce, R. G. H., *Species Dispersal in Agricultural Habitats*, CRC Press, Boca Raton, FL, 1990.
65. Maredia, K. M., Gage, S. H., Landis, D. A., and Scriber, J. M., Habitat use patterns by the seven-spotted lady beetle (Coleoptera: Coccinellidae) in a diverse agricultural landscape, *Biol. Control*, 2, 159, 1992.
66. Brandenburg, R. L. and Kennedy, G. G., Intercrop relationships and spider mite dispersal in corn/peanut agroecosystem, *Entomol. Exp. Appl.*, 32, 269, 1982.
67. Suski, Z. W. and Naegele, J. A., Light response in a two-spotted spider mite. II. Behavior of the "sedentary" and "dispersal" phases. Symposium paper, *Recent Advances in Acarology*, Cornell University Press, Ithaca, NY, 1966.
68. McEnroe, W. D., Spreading and inbreeding in the spider mite, *J. Hered.*, 60, 343, 1969.

69. Mader, H. J., Effects of increased spatial heterogeneity on the biocenosis in rural landscapes, *Ecol. Bull.*, 39, 169, 1988.
70. Park, J. R., *Environmental Management in Agriculture*, Belhaven Press, New York, 1988.
71. Burel, F., Landscape structure effects on carabid beetles spatial patterns in western France, *Landscape Ecol.*, 2, 215, 1989.
72. Baumhover, A. H., Eradication of the screwworm fly, *JAMA*, 196, 240, 1966.
73. Bushland, R. C., Screwworm eradication program, *Science*, 184, 1010, 1974.
74. Barrett, W. L., Jr., Natural dispersion of *Cochliomyia americana*, *J. Econ. Entomol.*, 30, 873, 1937.
75. Bushland, R. C., Screwworm research and eradication, *Entomol. Soc. Am. Bull.*, 23, 1975.
76. Hightower, B. G. and Alley, D. A., Local distribution of released laboratory-reared screwworm flies in relation to water sources, *J. Econ. Entomol.*, 56, 798, 1963.
77. Hightower, B. G., Adams, A. L., and Alley, D. A., Dispersal of released irradiated laboratory-reared screwworm flies, *J. Econ. Entomol.*, 58, 373, 1964.
78. Townes, H., A light-weight malaise trap, *Entomol. News*, 83, 239, 1972.
79. Baudry, J., Hedgerows and hedgerow networks as wildlife habitats in agricultural landscapes, in *Environmental Management in Agriculture*, Belhaven Press, New York, 1988, chap. 11.
80. van Emden, H. F., The role of uncultivated land in the biology of crop pests and beneficial insects, *Sci. Hort.*, 17, 121, 1965.
81. van Emden, H. F., Wild plants in the ecology of insect pests, in *Pests, Pathogens and Vegetation*, Thresh, J. M., Ed., Pitman Advanced Publishing Program, Boston, 1981, 251.
82. Wainhouse, D. and Coaker, T. H., The distribution of carrot fly (*Psila rosae*) in relation to the flora of field boundaries, in *Pests, Pathogens and Vegetation*, Thresh, J. M., Ed., Pitman Advanced Publishing Program, Boston, 1981, 263.
83. Solomon, M. G., Windbreaks as a source of orchard pests and predators, in *Pests, Pathogens and Vegetation*, Thresh, J. M., Ed., Pitman Advanced Publishing Program, Boston, 1981, 273.
84. Brandle, J. R., Hintz, D. L., and Sturrock, J. W., *Windbreak Technology*, Elsevier, New York, 1988.
85. Lewis, T., The effects of an artificial windbreak on the aerial distribution of flying insects, *Ann. Appl. Biol.*, 55, 503, 1965.
86. Lewis, T., The effect of an artificial windbreak on the distribution of aphids in a lettuce crop, *Ann. Appl. Biol.*, 55, 513, 1965.
87. Lewis, T., The effects of shelter on the distribution of insect pests, *Sci. Hort.*, 17, 74, 1965.
88. Lewis, T., Windbreaks, shelter and insect distribution, *Span* 11 (reprint), 4 pages, 1968.
89. Pasek, J. E., Influence of wind and windbreaks on local dispersal of insects, in *Windbreak Technology*, Brandle, J. R., Hintz, D. L., and Sturrock, J. W., Eds., Elsevier, New York, 1988, chap. 30.
90. Cromartie, W. J., Jr., Influence of habitat on colonization of collard plants by *Pieris rapae*, *Environ. Entomol.*, 4, 783, 1975.
91. Wratten, S. D., The role of field boundaries as reservoirs of beneficial insects, in *Environmental Management in Agriculture*, Belhaven Press, New York, 1988, chap. 15.
92. Desender, K., Ecological and faunal studies on Coleoptera in agricultural land. II. Hibernation of Carabidae in agroecosystems, *Pedobiologia*, 23, 295, 1982.
93. Klinger, Von K., Effects of margin-strips along a winter wheat field on predatory arthropods and the infestation by cereal aphids, *J. Appl. Entomol.*, 104, 47, 1987.
94. Sotherton, N. W., The distribution and abundance of predatory arthropods overwintering on farmland, *Ann. Appl. Biol.*, 105, 423, 1984.
95. Coombes, D. S. and Sotherton, N. M., The dispersal and distribution of polyphagous predatory Coleoptera in cereals, *Ann. Appl. Biol.*, 108, 461, 1986.
96. Leius, K., Influence of wild flowers on parasitism of tent caterpillar and codling moth, *Can. Entomol.*, 88, 444, 1967.
97. Landis, D. A. and Haas, M. J., Influence of landscape structure on abundance and within-field distribution of European corn borer (Lepidoptera: Pyralidae) larval parasitoids in Michigan, *Environ. Entomol.*, 21, 409, 1992.
98. Sotherton, N. W., The distribution and abundance of predatory Coleoptera overwintering in field boundaries, *Ann. Appl. Biol.*, 106, 17, 1985.
99. Thomas, M. B., Wratten, S. D., and Sotherton, N. W., The creation of 'island' habitats in farmland to manipulate populations of beneficial arthropods: predator densities and species composition, *J. Appl. Ecol.*, 29, 524, 1992.
100. Wratten, S. D. and Powell, W., Cereal aphids and their natural enemies, in *The Ecology of Temperate Cereal Fields*, Firbank, L. G., Carter, N., Darbyshire, J. F., and Potts, G. R., Eds., Blackwell Scientific, Boston, 1990, chap. 12.
101. Thomas, M. B., Wratten, S. D., and Sotherton, N. W., The creation of 'island' habitats in farmland to manipulate populations of beneficial arthropods: predator densities and emigration, *J. Appl. Ecol.*, 28, 906, 1991.

Chapter 3

STATISTICS FOR DESCRIBING POPULATIONS

Paula M. Davis

TABLE OF CONTENTS

I. Introduction 34

II. Population Density 34
   A. Sample Mean 34
   B. Binomial Sampling 34

III. Dispersion 34
   A. Use of Mathematical Distributions to Describe Dispersion 36
      1. Poisson Distribution 36
      2. Negative Binomial Distribution 38
      3. Other Mathematical Distributions 40
   B. Indices for Classifying Dispersion Patterns 42
      1. Variance-to-Mean Ratio 43
      2. k as an Index 43
      3. Fitting a Common k 43
      4. Lloyd's Mean Crowding, $\overset{*}{x}$ 45
      5. Morisita's I_δ Index 45
      6. Green's Index 46
      7. Equivalence of Indices 47
   C. Regression Techniques for Evaluating Dispersion 47
      1. Iwao's Patchiness or Mean Crowding Regression 47
      2. Taylor's Power Law 48
   D. Use of Distance Measurements 49
   E. Geostatistics 51

IV. Conclusions 52

Acknowledgment 52

References 53

ISBN 0-8493-2923-X/94/$0.00 + .50
© 1994 by CRC Press, Inc.

I. INTRODUCTION

In developing sampling programs for either research or pest management purposes, we are faced with several key questions. How can I estimate an insect population with a given degree of precision? How many samples are required to make a decision? How can I design an efficient sampling program? Before any of these questions can be answered, one must determine two of the characteristic features of any population, its density and dispersion.

II. POPULATION DENSITY

A. SAMPLE MEAN

Density and dispersion are largely statistical in nature. As discussed in Chapter 1, population density may be defined in terms of the number of individuals per unit area, such as insects per square meter, or in terms of individuals per sampling unit, such as per leaf or per sweep. Because of the time and cost involved, censusing every insect in a defined population usually is impractical for either research or pest management programs. Population density most often is estimated by collecting a series of sample units and recording the number of individuals per sample unit.

Define N as the number of samples collected and x_i as the number of individuals counted in the ith sample. An appropriate estimate of the population density is the sample mean ($\bar{x}$):

$$\bar{x} = \frac{1}{N}(x_1 + x_2 + \cdots + x_N) = \frac{1}{N}\sum_{i=1}^{N} x_i \qquad (1)$$

B. BINOMIAL SAMPLING

In some cases, determining population density by counting the number of individuals per sample unit may be extremely tedious and too time consuming to be used in a pest management program. Presence/absence sampling, also termed binomial sampling, may be a viable alternative.[1] The relationship between the sample mean and the proportion of samples containing one or more individuals is first established. As shown in Figure 1, this relationship is an asymptotic one. Sampling programs based on binomial sampling are usually faster and less costly on a per sample basis and may be the only feasible field sampling method. A major disadvantage of binomial sampling is that more samples are required to estimate a population density for a given precision level. In addition, the asymptotic relationship between population density and proportion infested may limit the usefulness of the sampling program to low density populations. Mathematical formulas and a more detailed discussion of binomial sampling are presented in Chapter 9.

III. DISPERSION

In addition to information related to population density, the inherent variability in a set of samples can provide important information on the spatial distribution or *dispersion* of a population. Traditionally, spatial patterns have been classified into three categories: uniform, random, and aggregated (Figure 2). The observed dispersion pattern for a particular species is largely determined by behavior. A uniform or regular dispersion pattern indicates some degree of repulsion between individuals

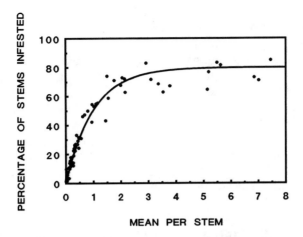

FIGURE 1. Asymptotic relationship between the mean number of larvae per stem and the proportion of alfalfa stems within a 0.09 m² quadrat infested with alfalfa weevils.

which tends to equalize the number of individuals per sample.[2] The low variability in sample counts from a regularly dispersed population results in a sample variance (s^2) which is smaller than the sample mean. Here, the sample variance is calculated as:

$$s^2 = \frac{1}{N-1}\left[\sum_{i=1}^{N} x_i^2 - \left(\sum_{i=1}^{N} x_i\right)^2 \bigg/ N\right] \quad (2)$$

where N is the number of sample units and x_i is the number of individuals in the ith sample unit. Cannabalism, use of oviposition markers, and food requirements may cause uniform dispersion patterns.

The mean equals the variance in a randomly dispersed population. In a random population, there is an equal probability of an organism occupying any point in space, and the presence of one individual does not influence the distribution of another.[3] For most populations, however, some degree of aggregation or clumping occurs. In a typical series of samples from an aggregated population, many samples contain few or no individuals of a particular species while some samples may contain a high number of individuals. In this case, the sample variance exceeds the mean. Two factors influencing aggregation are oviposition (eggs laid in masses vs. singly) and food preferences (feeding on top or terminal growth of plants only).

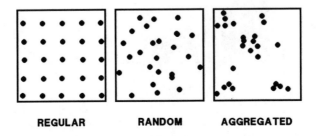

FIGURE 2. Examples of regular, random, and aggregated dispersion patterns.

There are several difficulties encountered when studying dispersion. The strong dependence on behavior generally restricts research to natural populations rather than populations established artificially or under experimental conditions. Because dispersion is spatially dependent, the sampling method and size of the sample unit can affect the observed spatial pattern. In addition, movement and rapid changes in numbers over time influence dispersion. Finally, the study of dispersion is largely theoretical in nature and has given rise to numerous methods for evaluating and interpreting spatial patterns.

Although evaluating dispersion may be challenging, knowledge of spatial patterns provides information vital to understanding basic behavior of a species. From a research standpoint, information on dispersion is used to transform data prior to analysis, determine optimal sampling pattern and sample size, and construct sequential sampling programs.

A. USE OF MATHEMATICAL DISTRIBUTIONS TO DESCRIBE DISPERSION

Information on the number of individuals occurring within a series of sample units may be summarized in a frequency distribution or histogram (Figure 3). Graphically, a frequency distribution plots the number of individuals present in a single sample on the x-axis vs. the corresponding observed frequency on the y-axis. For data sets containing more than 30 sample units, one method for evaluating the observed spatial pattern of a species is to fit discrete mathematical distributions to the frequency data. The two most commonly tested mathematical distributions are the Poisson distribution[4] and the negative binomial distribution,[5] which are useful for describing random and aggregated distributions, respectively.

1. Poisson Distribution

Random distributions are best described by the Poisson distribution in which the variance equals the mean. The concept of total randomness in nature has been debated in the literature. Although Taylor and others have made claims that true randomness is rare in nature,[6] several insects have spatial patterns well described by the Poisson distribution, such as green cloverworm (*Plathypena scabra* (F.)) eggs in soybeans.[7] Very often, low density populations will seem to fit a Poisson distribution.

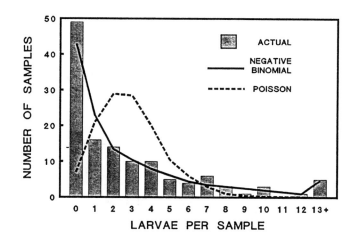

FIGURE 3. Frequency distribution of the number of *Papaipema nebris* larvae in randomly sampled quadrats measuring 900 cm^2 and collected from grass terraces. Observed counts were compared to expected frequencies predicted by the Poisson and negative binomial distributions.

The probability (P) of finding a certain number (x) of individuals in a sample from a population with a given mean (μ) and Poisson distribution is given by Equation 3:

$$P(x) = \frac{e^{-\mu}\mu^x}{x!} \qquad x = 0, 1, 2, \ldots \qquad (3)$$

The procedures for fitting any discrete distribution to frequency data include four basic steps. First, the parameters of the distribution need to be estimated from the available data set. For the Poisson distribution, the parameter, μ, is estimated by the sample mean, $\bar{x}$. After the parameters of the distribution are estimated, the second step is to calculate the expected probabilities, $P(x)$, associated with $x = 0, 1, 2, 3, \ldots$ individuals per sample. This is done by substituting parameter estimates into the probability function and calculating the probability series for each value of x. The Poisson series is generated by substituting the sample mean into the probability function:

$$P(0) = e^{-\bar{x}}$$

$$P(1) = e^{-\bar{x}}\bar{x}/1! = P(0)\bar{x}/1$$

$$P(2) = e^{-\bar{x}}\bar{x}^2/2! = P(1)\bar{x}/2$$

$$\vdots$$

$$P(r) = e^{-\bar{x}}\bar{x}^r/r! = P(r-1)\bar{x}/r \qquad (4)$$

Expected frequencies, E_x, are calculated by multiplying each corresponding probability by the total number of samples (N):

$$E_0 = N \cdot P(0)$$

$$E_1 = N \cdot P(1)$$

$$E_2 = N \cdot P(2)$$

$$\vdots$$

$$E_r = N \cdot P(r) \qquad (5)$$

Because discrete distributions form an infinite series, the question arises as to how many terms in the series should be included. Consequently, the third step is to determine the number of classes to use to evaluate the fit of the distribution. As a general rule, no expected frequency should be less than 1 or much less than 5.[8,9] Typically minimum expected frequencies of 1 or 3 are used. Classes which contain expected frequencies below the selected cutoff value are pooled. Because the sum of all expected frequencies equals the total number of insects in the sample (T), the expected frequency in the final class (E_r) is readily calculated by subtraction:

$$E_r = T - \sum_{x=0}^{r-1} E_x \qquad (6)$$

The final step is to evaluate how well the selected discrete distribution fits the data by using a χ^2 goodness-of-fit test. The χ^2 statistic, χ^2, is calculated as:

$$\chi^2 = \sum_{x=0}^{r} \frac{(O_x - E_x)^2}{E_x} \quad (7)$$

where O_x and E_x are the observed and expected number of samples containing x individuals, respectively, and r is the maximum frequency class determined after selected classes are pooled.

The null hypothesis being tested is that the observed data set follows the selected discrete distribution. The calculated χ^2 statistic is compared with tabulated probabilities for a χ^2 distribution. The appropriate degrees of freedom (df) for the test are

$$\text{df} = n - p - 1 \quad (8)$$

where n and p are the number of frequency classes and number of estimated parameters, respectively. Because one parameter was estimated for the Poisson distribution, the appropriate degrees of freedom for the test are $n - 2$. If the calculated χ^2 statistic exceeds the tabulated χ^2 at a selected probability level (say 5%), the null hypothesis is rejected. Sample calculations for fitting a Poisson distribution are given in Table 1.

2. Negative Binomial Distribution

Commonly in ecological studies, the variance is larger than the mean. In this case, the negative binomial distribution, with parameters μ and k may adequately describe the observed frequencies. The probability distribution is obtained by using a Taylor's series expansion of $(q - p)^{-k}$ where $p = \mu/k$ and $q = 1 + p$:

$$(q - p)^{-k} = q^{-k} \left[1 + k\frac{p}{q} + \frac{k(k + 1)}{2!}\left(\frac{p}{q}\right)^2 + \right.$$

$$\left. \cdots + \frac{k(k + 1) \cdots (k + r - 1)}{r!}\left(\frac{p}{q}\right)^r + \cdots \right] \quad (9)$$

TABLE 1
Sample Calculations for Fitting a Poisson Distribution

Counts	Observed frequency	$P(x)$	Expected frequency	$(O - E)^2 / E$
0	106	0.5143	102.86	0.096
1	65	0.3420	68.40	0.169
2	22	0.1137	22.74	0.024
3	5 ⎫	0.0252	5.04 ⎫	
4	1 ⎬ 7	0.0042	0.84 ⎬ 6.00	0.167
5	1 ⎭	0.0006	0.12 ⎭	
Totals	200	1.0000	200.0	$\chi^2 = 0.456$

Note: classes containing expected frequencies <1 were pooled with the preceding class before calculating the χ^2 statistic. Sample mean = 0.665; sample variance = 0.753. df = $4 - 2 = 2$; probability of larger χ^2 = 0.80.

The first term in the series, q^{-k}, corresponds to the probability that the sample contains zero individuals. The second term in the series, $kpq^{-(1+k)}$ corresponds to the probability that the sample contains a single individual. Consequently, the probabilities for the series can be expressed by Equation 10:

$$P(x) = \frac{(k+x-1)!}{x!(k-1)!} p^x q^{-(x+k)} \qquad x = 0, 1, 2, \ldots \qquad (10)$$

For the negative binominal distribution, the mean and variance are

$$\mu = kp \qquad (11)$$

$$\sigma^2 = kp + kp^2 \qquad (12)$$

In fitting frequency data to the negative binomial distribution, two parameters need to be estimated. The population mean, μ, is estimated by the sample mean, $\bar{x}$. Several methods exist for estimating k, described as a parameter related to the degree of clumping. One of the simplest methods, the method of moments, involves equating the sample moments to their theoretical moments and solving the set of equations for the unknown parameters. For example, in the two-parameter negative binomial model, the first and second moments correspond to $\bar{x}$ and sample variance (s^2), respectively, and are set equal to their expected values. The two equations are solved giving an estimate of k:[10]

$$\hat{k} = \frac{\bar{x}^2}{s^2 - \bar{x}} \qquad (13)$$

The efficiency of the method of moments estimator of k varies with density. Here, efficiency is defined in terms of the relative size of the mean squared error and gives an indication of an estimator's variability in comparison to other types of estimators.[11] The most efficient estimator is the least variable. The method of moments has an efficiency of 90% or greater if one of the following conditions holds:[5,10]

(a) For small $\bar{x}$, $k/\bar{x} > 6$

(b) For large $\bar{x}$, $k > 13$

(c) For intermediate $\bar{x}$, $(k + \bar{x})(k + 2)/\bar{x} \geq 15$ (14)

A second relatively simple method for calculating k is based upon the proportion of zeros obtained in the sample.[5,10] Let N be the total number of samples and N_0 is the number of samples with zero individuals. Then k can be estimated by using the following iterative equation:

$$\log(N/N_0) = \hat{k} \log(1 + \bar{x}/\hat{k}) \qquad (15)$$

First, the left-hand side of Equation 15 is calculated. The value of the right-hand side initially is determined by inserting an estimate of k, such as obtained using the method of moments. Subsequent estimates of k are inserted into the equation until an estimate of k with the desired precision is achieved. The proportion of zeros method has an efficiency of 90% or more when at least one third of the samples are empty if the mean is greater than or equal to 10 or if

$$(\bar{x} + 0.17)(N_0/N - 0.32) > 0.20 \quad \text{for} \quad \bar{x} < 10 \tag{16}$$

A more complicated, but widely used method for estimating parameters, is based on the method of maximum likelihood.[11,12] This method generally is preferred because parameter estimates are considered the most efficient. Bliss and Fisher[5] describe a method for the efficient calculation of k. Let A_x be the sum of all frequencies of sample units containing more than x individuals. Then:

$$N \ln(1 + \bar{x}/\hat{k}) = \sum_{x=0}^{\infty} \left(A_x/(\hat{k} + x)\right) \tag{17}$$

The method of moments often is used to get an initial estimate of k. Estimates of k are increased or decreased until the two sides of the equation balance. One limitation of the maximum likelihood method is that for highly aggregated data sets containing some samples in excess of 30 individuals, parameter estimation may be computationally difficult if not impossible because of the nature of the calculations.[5,13]

Once a suitable estimate of k is determined, expected probabilities for each class are computed by using Equation 10. To eliminate the need to calculate factorials, Equation 10 can be rearranged to provide the following set of equations for calculating the expected probabilities predicted by the negative binomial distribution:

$$P(0) = \left[1 + (\bar{x}/\hat{k})\right]^{-\hat{k}}$$

$$P(1) = P(0)\left[\bar{x}/(\bar{x} + \hat{k})\right](\hat{k}/1)$$

$$P(2) = P(1)\left[\bar{x}/(\bar{x} + \hat{k})\right]\left[(\hat{k} + 1)/2\right]$$

$$\vdots$$

$$p(r) = P(r - 1)\left[\bar{x}/(\bar{x} + \hat{k})\right]\left[(\hat{k} + r - 1)/r\right] \tag{18}$$

Expected frequencies are determined by multiplying the above probabilities by N, the total number of samples in the set. Because very small expected values can inflate the χ^2 statistic used to test the fit of the negative binomial distribution, the expected number series should be carried out until an expected frequency below the selected cutoff value is calculated. The residual is pooled with the last class. In testing the fit of the negative binomial distribution, use a χ^2 with $n - 3$ degrees of freedom (n is the effective number of classes). Sample calculations for fitting a negative binomial distribution to counts for alfalfa weevil larvae (*Hypera postica* ((Gyllenhal))) are given in Tables 2 and 3.

3. Other Mathematical Distributions

In addition to the Poisson and negative binomial distributions, numerous other discrete distributions have been used to evaluate spatial patterns. The most common one- and two-parameter distributions include the positive binomial, Thomas double-Poisson, Neyman Type A, Poisson-binomial, Poisson with zeros, and logarithmic with zeros.[4,14-18] Probability functions for these distributions are given in Table 4. Gates[13] developed a computer program, DISCRETE, which fits the previously mentioned distributions by using maximum likelihood methods for estimating parameters.

TABLE 2
Sample Calculations for Fitting a Negative Binomial Distribution to Data for Number of Alfalfa Weevil Larvae per Stem

Number per stem	Observed frequency	A_x [a]	Expected frequency	$(O - E)^2 / E$
0	127	161	126.85	0.00
1	63	99	65.84	0.22
2	42	57	37.86	0.45
3	25	32	22.47	0.29
4	14	18	13.55	0.02
5	10	8	8.24	0.37
6	1	7	5.05	3.24
7	1	6	3.10	1.43
8	3	3	1.91	0.62
9	0	3	1.18	
10	0	3	0.73	
11	0	3	0.46	
12	1 } 3	2	0.28 } 3.12	0.00
13	0	2	0.18	
14	1	1	0.11	
15	1	0	0.07	
Totals	288	—	288.00	$\chi^2 = 6.64$

Note: Expected frequencies used the maximum likelihood method for estimating k. Classes with expected frequencies below 1 were pooled with the preceding class. Sample mean: 1.406; sample variance: 4.186. Degrees of freedom: $10 - 3 = 7$; probability of larger $\chi^2 = 0.47$.

[a] Sum of all frequencies of sample units containing more than x individuals.

TABLE 3
Calculation of $\hat{k}$ for Alfalfa Weevil Data in Table 2

1. Method of Moments

$$\hat{k} = (1.406)^2 / (4.186 - 1.406) = 0.711$$

2. Frequency of Zeros Method

$$\log(288/127) = 0.3556$$

| | Iteration | | | |
	1	2	3	4
$\hat{k}$	0.711	0.800	0.850	0.8201
$\hat{k} \log(1 + 1.406/\hat{k})$	0.3369	0.3524	0.3603	0.3556
Difference	−0.0187	−0.0032	0.0047	0.0000

3. Maximum Likelihood Method (values for A_x found in Table 2)

| | Iteration | | | | |
	1	2	3	4	5
$\hat{k}$	0.711	0.800	0.850	0.824	0.8228
$288 \ln(1 + 1.406/\hat{k})$	314.230	292.125	281.120	286.729	286.994
$\Sigma A_x/(\hat{k} + x)$	322.555	293.504	279.643	286.659	286.993
Difference	−8.325	−1.379	1.477	0.070	0.001

TABLE 4
One and Two Parameter Discrete Distributions That Have Been Used to Evaluate Insect Dispersion

Distribution	Parameter(s)	Probability function	Conditions
1. Binomial[4]	p	$P(x) = \dfrac{n!}{(n-x)!\,x!} p^x q^{n-x}$	for $x = 0, 1, \ldots, n$
2. Thomas Double Poisson[14]	m, λ	$P(x) = e^{-m}$	$x = 0$
		$P(x) = \sum_{r=1}^{x} \dfrac{m^r e^{-m}}{r!} \dfrac{(r\lambda)^{x-r} e^{-r\lambda}}{(x-r)!}$	for $x = 1, 2, 3, \ldots$
3. Neyman Type A[15]	μ, ν	$P(x) = \dfrac{e^{-\mu}\nu^x}{x!} \sum_{r=0}^{\infty} \dfrac{r^x(\mu e^{-\nu})^r}{r!}$	for $x \geq 0$ and $\mu, \nu > 0$
4. Poisson-binomial[16]	α, p	$P(x) = \exp(-\alpha(1-q^n))$	for $x = 0$
		$P(x) = F_x/x!$	for $x > 0$
		where:	
		$F_x = \alpha p n \sum_{i=0}^{x-1} \dfrac{(x-1)!}{i!(x-1-i)!}$	
		$(n-1)^{[x-i-1]} p^{x-i-1} q^{n-x+i} F_i$	for $n = 2, 3, 4, \ldots$; $x > 0$; and $[j]$ indicates factorial moments: $(n-1)^{[0]} = 1$ $(n-1)^{[1]} = n - 1$ $(n-1)^{[2]} = (n-1)(n-2)$, etc.
5. Poisson with zeros[17]	λ, Θ	$P(x) = 1 - \Theta$	for $x = 0$
		$P(x) = \dfrac{\Theta e^{-\lambda}\lambda^x}{(1 - e^{-\lambda})x!}$	for $x = 1, 2, 3, \ldots$
6. Logarithmic with zeros[18]	λ, Θ	$P(x) = 1 - \lambda$	for $x = 0$
		$P(x) = \dfrac{\lambda \Theta^x}{-x \ln(1 - \Theta)}$	for $x = 1, 2, \ldots$

Source: (modified from Gates[13]).

Continuous mathematical distributions also have been used to evaluate dispersion. The Adès family of distributions was advocated by Perry and Taylor[19] as an alternative to the negative binomial distribution to evaluate insect dispersion. Three parameters (τ, r, λ_i) describe each member of an Adès family. If X_i is a random variable from a gamma distribution with coefficient of variation, r, and other parameter λ_i, then Y_i has an Adès distribution if the following relationship between Y_i and X_i holds:

$$Y_i = (\ln X_i)^{[2/(2-\tau)]} \qquad X_i > 1, \ \tau < 2, \ i = 1, \ldots, n$$
$$Y_i = 0 \qquad X_i \leq 1 \qquad (19)$$

Probability functions and methods for fitting data sets to the Adès distribution are described elsewhere.[19]

B. INDICES FOR CLASSIFYING DISPERSION PATTERNS

In many instances, fitting mathematical distributions may not be practical or necessary. For instance, sample sizes may be relatively small (less than 30) or one may be interested in distinguishing between random, aggregated, or regular spatial patterns. Using information based on the sample mean, sample variance, and size of the sampling unit, several indices have been developed for classifying dispersion patterns.

1. Variance-to-Mean Ratio

Recall that the mean and variance would be equal in a randomly distributed population. Dispersion of a population can be classified by calculating the *variance-to-mean ratio*

$$s^2/\bar{x} > 1 \quad \text{Aggregated}$$
$$= 1 \quad \text{Random}$$
$$< 1 \quad \text{Regular} \qquad (20)$$

Departure from a random distribution can be tested by calculating the index of dispersion, I_D, where n is the number of samples:[3]

$$I_D = (N - 1)s^2/\bar{x} \qquad (21)$$

I_D is approximately distributed as a χ^2 with $n - 1$ degrees of freedom. Values of I_D which fall outside a confidence interval bounded by χ^2 with $n - 1$ degrees of freedom and selected probability levels of 0.95 and 0.05, for instance, would indicate a significant departure from a random distribution.

The variance-to-mean ratio is the simplest, most fundamental index and is useful for assessing agreement of the data set to the Poisson series.[20] However, Pielou[21] has pointed out that the reliability and usefulness of this index is limited by several inherent flaws such as (1) measured pattern is highly dependent on quadrat size, (2) different patterns can give exactly the same index, and (3) patterns inherently the same can be judged to be different.

2. k as an Index

The negative binomial k was suggested by Waters[22] as a measure of spatial pattern:

$$\frac{1}{k} = \frac{\sigma^2 - \mu}{\mu^2} = 0 \quad \text{Random}$$
$$> 0 \quad \text{Aggregated}$$
$$< 0 \quad \text{Regular} \qquad (22)$$

Any of the methods described in the previous section on fitting a negative binomial distribution may be used for estimating k. In general, use of k as an index is more feasible than use of a frequency distribution. However, the use of k as an index has been sharply criticized because k is not always consistent and changes with the mean, as well as lacks any biological interpretation.[23]

3. Fitting a Common k

For transforming data or developing a sequential sampling program, it is desirable to determine a single value which would be an acceptable estimate for every population of a particular species. Bliss and Owen[24] developed a regression method for calculating a common k (k_c):

1. Calculate $x' = \bar{x}^2 - s^2/N$ and $y' = s^2 - \bar{x}$. N is the number of individual counts on which $\bar{x}$ is based.
2. Plot y' vs. x'. Regression line intercepts 0 and has slope $1/k$ (Figure 4A).
3. An approximate estimate of common k is

$$1/\hat{k}_c = \Sigma y'/\Sigma x'.$$

4. A graphical test for k_c consists of plotting $1/k$ ($= y'/x'$) vs. the mean $\bar{x}$ for each sub-area or group of samples. If there is neither trend nor clustering, the fit of a common k is justified (Figure 4B).

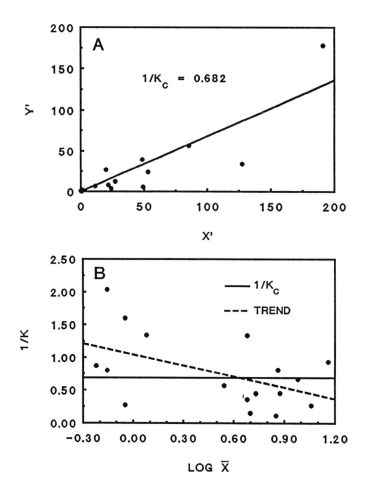

FIGURE 4. Fitting a common k for western corn rootworm (*Diabrotica virgifera virgifera*) larvae. Means and variances are based on 10 soil samples per field and sampling date. (A) Method of moments regression method for estimating k_c were $x' = \bar{x} - (s^2/N)$ and $y' = s^2 - \bar{x}$. (B) Graphical test to determine if the fit of a common k is justified. Slope of regression line (-0.56 ± 0.25) differs from 0, indicating that a common k may not be justified.

4. Lloyd's Mean Crowding, $\overset{*}{x}$

A third index, termed mean crowding ($\overset{*}{x}$) was proposed by Lloyd[25] to indicate the possible effect of mutual interference or competition among individuals. Theoretically, mean crowding is the mean number of other individuals per individual in the same quadrat:

$$\overset{*}{x} = \bar{x} + \frac{s^2}{\bar{x}} - 1 \tag{23}$$

As an index, mean crowding is highly dependent upon both the degree of clumping and population density. To remove the effect of changes in density, Lloyd introduced the *Index of Patchiness*, expressed as the ratio of mean crowding to the mean. As with the variance-to-mean ratio, the index of patchiness is dependent upon quadrat size

$$\begin{aligned} \overset{*}{x}/\bar{x} &< 1 \quad \text{Regular} \\ &= 1 \quad \text{Random} \\ &> 1 \quad \text{Aggregated} \end{aligned} \tag{24}$$

5. Morisita's I_δ Index

Reasoning that the diversity of numbers of individuals per quadrat could be used as a measure of spatial pattern, Morisita[26] developed the index I_δ. Suppose n quadrats are sampled and x_i represents the number of individuals in the ith quadrat. Define δ as the probability that individuals of a randomly drawn pair will come from the same quadrat. If the x_i came from a random or Poisson distribution, the expected value of δ is $1/n$. Then the index I_δ can be defined as the ratio of δ to its expected value assuming a random distribution. Values for I_δ may range from 0 to n. Letting N be the total number of individuals sampled in n quadrats, I_δ may be calculated as follows and used to classify populations:

$$\begin{aligned} I_\delta = n\frac{\Sigma x_i(x_i - 1)}{N(N - 1)} &> 1 \quad \text{Aggregated} \\ &= 1 \quad \text{Random} \\ &< 1 \quad \text{Regular} \end{aligned} \tag{25}$$

To determine if the sampled population significantly differs from random, the following large sample test of significance can be used:[27]

$$Z = (I_\delta - 1)/(2/n\bar{x}^2)^{1/2} \tag{26}$$

Compare values of Z with tabulated values for a normal distribution and reject the hypothesis that the sampled population is dispersed randomly if $|Z| > z(\alpha/2)$.

I_δ is not dependent upon density if the following conditions hold: (1) the population is divided into subareas, (2) individuals are randomly distributed in each subarea, and (3) plots are small relative to the size of the subarea. Because I_δ is affected by the size of the quadrat, we could sample larger and larger quadrats and calculate a separate I_δ value for each size. In a random distribution, I_δ will not change. In a nonrandom distribution, a rapid drop indicates a clump size that corresponds approximately to the quadrat size at which the rapid drop occurred (Figure 5). However, one

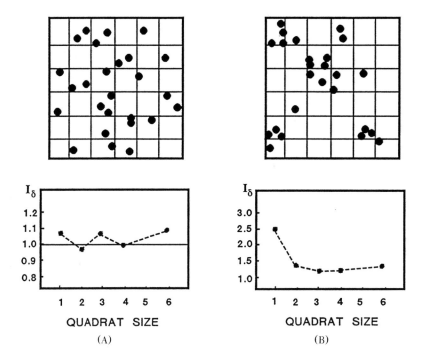

FIGURE 5. Evaluating the effect of quadrat size on Morisita's I_δ index. (A) Values of I_δ change little with quadrat size for random distributions. (B) A rapid decrease in I_δ indicates an aggregated distribution.

may encounter difficulty in using I_δ to determine clump size because time or money may restrict the number of various sized quadrats that can be sampled. In addition, data interpretation can be very difficult if only a small number of quadrats are sampled. Finally, natural variability in clumps and discreteness of clumps can cause complications.[21]

6. Green's Index

A modification of the variance-to-mean ratio was developed by Green.[28] Green's index (C_x) has the advantage of being independent of sample size and is calculated as follows:

$$C_x = \frac{(s^2/\bar{x}) - 1}{n - 1} = 0 \quad \text{Random}$$

$$> 0 \quad \text{Aggregated}$$

$$< 0 \quad \text{Regular} \quad (27)$$

Because the range of C_x is bounded by $-1/(\Sigma x - 1)$, which indicates maximum uniformity, and a value of one, which indicates maximum clumping (all individuals in a single sample), C_x is a useful index for measuring the degree of aggregation. The upper value for a test of randomness at the significance level of α is

$$C_{x,(1-\alpha)} = \frac{\left(\chi^2_{(1-\alpha)}/(n-1)\right) - 1}{n\bar{x} - 1} \quad (28)$$

where χ^2 has $n - 1$ degrees of freedom.[28] Calculated values of C_x are compared to $C_{x,(1-\alpha)}$ to determine if the data set differed from random.

7. Equivalence of Indices

Although differing in origin, the negative binomial parameter k, the patchiness index $\overset{*}{x}/\bar{x}$, and Morisita's I_δ have been shown to be essentially equivalent as measures of aggregation[29]

$$\frac{\overset{*}{x}}{\bar{x}} = 1 + \frac{s^2 - \bar{x}}{\bar{x}^2} \approx 1 + \frac{1}{k} \approx E(I_\delta) \tag{29}$$

C. REGRESSION TECHNIQUES FOR EVALUATING DISPERSION

Two regression techniques have been widely used to evaluate dispersion, develop sampling protocols, and to normalize data for statistical analysis: Iwao's mean crowding regression and Taylor's power law.[2,23,30,31]

1. Iwao's Patchiness or Mean Crowding Regression

Linear regression has been used to examine the relationship between Lloyd's mean crowding ($\overset{*}{x}$) and mean density ($\bar{x}$)[2]

$$\overset{*}{x} = \alpha + \beta\bar{x} \tag{30}$$

By rearranging and combining Equations 23 and 30, the sample variance can be expressed as a quadratic function of the sample mean

$$s^2 = (\alpha + 1)\bar{x} + (\beta - 1)\bar{x}^2 \tag{31}$$

The intercept, α, is termed the index of basic contagion and has been interpreted as the average number of other individuals living together in the same quadrat with a given individual. Thus, $\alpha + 1$ is a measure of clump size. When the basic component is a single individual, the regression passes through the origin, i.e., α equals zero. If the basic component is a colony rather than an individual, α is greater than zero. However, if α is less than zero, the regression indicates a tendency for repulsion between individuals.

The slope β is defined as the density contagiousness coefficient and is a measure of the spatial distribution of the clumps. Clump dispersion is classified as random when β equals one. Values of β greater than one or less than one indicate either aggregated or regular dispersion patterns, respectively.

Student t-tests can be used to determine whether the colony is composed of single individuals and if colonies are dispersed randomly:[4]

$$\text{Test } \alpha = 0: \quad t = \alpha/s_\alpha \tag{32}$$

where s_α is the standard error of the intercept for the mean crowding regression.

$$\text{Test } \beta = 1: \quad t = (\beta - 1)/s_\beta \tag{33}$$

where s_β is the standard error of the slope for the mean crowding regression. Calculated values are compared with tabulated t-values with $n - 2$ degrees of freedom.

The three major types of aggregation patterns are listed below along with their respective causes.

Possible Causes

1. $\alpha > 0$, $\beta = 1$ Poisson distribution of clumps with constant clump size.
2. $\alpha = 0$, $\beta > 1$ Series of distributions with constant tendency for aggregation—k_c.
3. $\alpha > 0$, $\beta > 1$ Organism's response to characteristic modes of reproduction and dispersal.
               Response to heterogeneous conditions in habitat.

In Figure 6, the mean crowding regression was used to evaluate dispersion of alfalfa weevil larvae. In this example, the intercept, α, was not significantly different from 0 ($\alpha = 1.4 \pm 4.0$; $t = 0.35$; df = 14; $p > 0.5$) indicating that colonies were composed of single individuals. The slope of the regression, β, indicated a significant degree of clumping of the colonies ($\beta = 1.133 \pm 0.016$; $t = 8.31$; df = 14; $p < 0.001$).

2. Taylor's Power Law

For many insect and animal species, Taylor[23,32] found that a power law function could be used to model the relationship between the mean and variance (Figure 7):

$$s^2 = a\bar{x}^b \tag{34}$$

We can solve for the coefficients with linear regression if a log transformation (base 10 or natural log) is used:

$$\log(s^2) = \log(a) + b \log(\bar{x}) \tag{35}$$

The coefficient, a, has been described as a scaling factor related to sample size. Taylor[32] proposed that b is a constant dependent upon species' behavior or the environment. The slope b can be used to classify dispersion patterns as random ($b = 1$), aggregated ($b > 1$), or uniform ($b < 1$). As with the mean crowding coeffi-

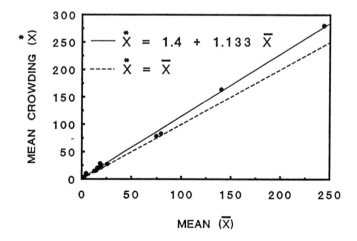

FIGURE 6. Use of Iwao's mean crowding regression to assess dispersion of alfalfa weevil larvae. Means and variances were based on a sample size of eight quadrats (0.09 m^2) per field and sample date. The intercept (α) was not significantly different from 0 and the slope (β) was significantly different from 1, indicating an aggregated distribution composed of single individual colonies.

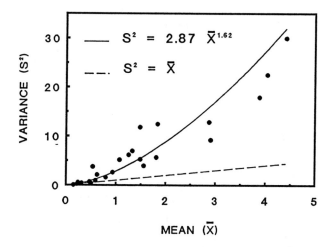

FIGURE 7. Use of Taylor's power law, $s^2 = a\bar{x}^b$, to evaluate between-plant dispersion of greenhouse whitefly, *Trialeurodes vaporariorum* infesting poinsettias. Each data set consisted of the average number of whiteflies per leaf per 40 plants. The slope, b, significantly differed from 1, indicating an aggregated distribution.

cient β, t-tests may be used to detect departures from random if b significantly differs from 1.

Considerable debate has ensued over the superiority of Iwao's mean crowding regression and Taylor's power law.[2,6,23,33-35] In a recent review article, Kuno[36] summarized the merits and drawbacks of each technique. In terms of descriptive ability, both functions tend to fit most data sets very well. However, in general, correlation coefficients tend to be slightly higher for the power law. Both functions are simple to fit using linear regression to estimate two parameters. For the mean crowding regression, linear regression occasionally may provide biased estimates if the range in means is very large. The main advantage of the mean crowding regression lies in parameter interpretation. This equation was originally derived with close reference to theoretical distribution models, thus allowing the ecological implications of the parameter to be interpreted. On the other hand, Taylor's power law is an empirical model which lacks definite theoretical background. Although this model fits most data sets extremely well, earning the name "universal" law, the constancy of the power law coefficients has been challenged. For instance, power law coefficients have been shown to vary for different insect instars and quadrat size.[37-39]

D. USE OF DISTANCE MEASUREMENTS

A different class of indices is based upon distance measurements between individuals instead of the number of individuals in a sample. Nearest-neighbor approaches have been used most often in ecology to examine spatial patterns in plants. However, this method has potential for assessing spatial patterns for stationary insects, insect damage, or location of colonies such as ant hills, tent caterpillar webs in trees, etc.

The simplest use of distance to nearest neighbor assumes that populations are arranged in two-dimensional space.[40] A series of distances from each individual to its nearest neighbor in a given population or randomly selected sample of individuals is taken. Define $\bar{r}_A$ as the mean of the series of N distances to nearest neighbor:

$$\bar{r}_A = \sum_{i=1}^{N} r_i / N \tag{36}$$

If A is the total area expressed in the same terms as r (such as square meters and meters), the expected mean distance to nearest neighbor ($\bar{r}_E$) in an infinitely large random distribution of density N/A is given by Equation 37:

$$\bar{r}_E = \frac{1}{2(N/A)^{1/2}} \qquad (37)$$

Then the ratio (R) of the actual mean distance to the expected mean distance assuming a random distribution can be calculated and used to evaluate dispersion.

$$\begin{aligned} R = \bar{r}_A/\bar{r}_E &= 1 \quad \text{Random} \\ &< 1 \quad \text{Aggregated} \\ &> 1 \quad \text{Uniform} \end{aligned} \qquad (38)$$

Values of R range from 0 (completely aggregated) to 2.1491 (perfectly uniform). Departures from random can be detected with the following significance test.

$$c = \frac{\bar{r}_A - \bar{r}_E}{\sigma_{\bar{r}_E}} \qquad (39)$$

$\sigma_{\bar{r}_E}$ is the standard error of the mean distance to nearest neighbor in a randomly distributed population of density N/A and has a value of $0.26136/(NA^{-1/2})$. Values of c are compared with tabulated values from a standard normal distribution.

The nearest-neighbor approach has several advantages and disadvantages. One of the key advantages is that the effect of quadrat size is eliminated. Because the theoretical range in values for R is explicitly defined, the degree of aggregation or uniformity can be determined. One of the biggest limitations of using distance measurements is one must know the exact location of each individual. A second limitation is that the nearest-neighbor approach may not distinguish between certain types of patterns. For example, a population consisting of scattered pairs of individuals will have the same R value as a population whose individuals are clustered in one spot (Figure 8).

To improve the ability to distinguish between certain pattern types, Thompson[41] derived an extension of the nearest-neighbor approach, the distance to the nth neighbor. Define r_n as the distance to the nth nearest neighbor and λ as the mean number of individuals in a circle of unit radius. r_n can be transformed as follows:

$$x_n = 2\lambda r_n^2 \qquad (40)$$

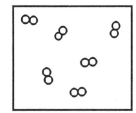

FIGURE 8. The nearest neighbor approach may not be able to distinguish between scattered pairs of organisms (right) and a tightly clumped group of individuals (left).

Then $N\bar{x}_n$ is distributed as a χ^2 random variable with $2Nn$ degrees of freedom. A value of $N\bar{x}_n$ greater than $\chi^2_{0.95}$ indicates significant aggregation while a value less than $\chi^2_{0.05}$ indicates that the population is more uniform than expected.

E. GEOSTATISTICS

A final method for evaluating spatial structure incorporates information on distances between samples and sample counts. Geostatistics relies on spatial variation to determine the degree of association and dependence of spatially related data. Geostatistical theory was first developed for geology and mining to estimate ore and mineral quantities at specific locations.[42,43] Recently, Schotzko and O'Keeffe evaluated dispersion of *Lygus hesperus* Knight in lentils[44] with geostatistics and evaluated the effect of sample unit size and sample placement on geostatistical analysis.[45]

Evaluating insect spatial patterns with geostatistics requires intensive sampling of the specified area. First, a grid is superimposed over the sampling universe (e.g., field) subsequently subdividing the sampling universe into a series of plots or sample units. Samples are then collected from all of the plots or from a subset of plots. As a general rule of thumb, a minimum of 30 sample units should be collected for effective spatial analysis.[44] Information recorded includes sample-unit coordinates and insect counts.

The most commonly used method for geostatistical analysis involves the construction of the semivariogram. The semivariogram is a graph of the spatial dependence of an organism and plots the sample variance of sample pair differences against the distance between sampling points. These sample variances may be defined in terms of Gamma as a function of the separation distance between points (h):

$$\text{Gamma}(h) = \tfrac{1}{2}n(h) \sum_{i=1}^{n(h)} [Z(x_i) - Z(x_{i+h})]^2 \qquad (41)$$

where $Z(x_i)$ is the measured sample value (count) at point x_i, $Z(x_{i+h})$ is the sample value at point x_{i+h}, and $n(h)$ is the total number of sample pairs for any separation distance.[44,45] Figure 9 illustrates key features of a semivariogram. The y intercept, or localized discontinuity, is the measure of microdistributional effect and measurement error.[46] Thus, the localized discontinuity estimates the proportion of the total varia-

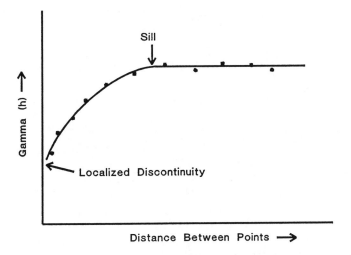

FIGURE 9. Example of a semivariogram used in geostatistical analysis.

tion that is below the sampling scale. A second feature of the semivariogram is the sill, which is the point at which the semivariance or sample variance no longer increases and becomes random.[46] The distance to the sill defines the range of spatial dependence.

Numerous mathematical functions have been used to model a semivariogram. Semivariograms of random and uniform spatial patterns exhibit a linear relationship between sample variances and distance between sample units. In this situation, the slope of the line is approximately zero, thus making the localized discontinuity and sill equal in value. The variability about the regression line is greater for random distributions than for uniform distributions. Aggregated spatial patterns may exhibit a wide range of spatial dependence. Two common functions for modeling clumped patterns are the spherical and power models.[44] A spherical model would be appropriate for semivariograms similar to Figure 9 where there is a gradual reduction in dependence with increased distance until the sill or point of spatial dependence is attained. A power model would better fit semivariograms that appear more j-shaped, indicating a strong spatial continuity over short distances with a rapid reduction in dependence at longer distances.[45] More detailed procedures for conducting geostatistical analyses and computer programs for calculating semivariograms are found elsewhere.[42-46]

Schotzko and O'Keeffe[45] point out several limitations of using geostatistics for assessing insect dispersion. One major limitation is the large number of sample units that must be collected in order to accurately estimate sample pair variances. In addition, the orientation, scale, and spatial arrangement of sample units influence the results of the analysis. In particular, the spatial structure of a phenomenon will be missed if the scale of the study is too large or too small.

IV. CONCLUSIONS

Numerous indices, regression models, and distributions have been used to evaluate dispersion. However, none of them are without some degree of criticism. If the "perfect coefficient" could be found, it would possess these criteria:

1. Should provide real and continuous values for the complete range of spatial distributions.[28]
2. Should be uninfluenced by number of sample units, the size of the unit sample, and number of individuals.[28]
3. Should be easy to calculate.[28]
4. Values for the index appropriate to some theoretical expectation should be central in position.[47]
5. Tests of significance should be available.[47]
6. Its descriptive function should be clearly separated, both in application and interpretation, from any supposed theoretical justification.[23]

For now, which approach is best depends upon both the objectives of the sampling study as well as on the type and quantity of data collected. In most cases, it is best to evaluate dispersion with several different techniques.[48]

ACKNOWLEDGMENT

The author thanks John Sanderson, Cornell University, for providing data for greenhouse whitefly.

REFERENCES

1. Kuno, E., Evaluation of statistical precision and design of efficient sampling for the population estimation based on frequency of occurrence, *Res. Popul. Ecol.*, 28, 305, 1986.
2. Iwao, S. and Kuno, E., An approach to the analysis of aggregation pattern in biological populations, in *Statistical Ecology*, Vol. 1, Patil, G. P., Pielou, E. C., and Waters, W. E., Eds., Penn State University Press, Philadelphia, 1971, 461.
3. Southwood, T. R. E., *Ecological Methods with Particular Reference to the Study of Insect Populations*, Chapman & Hall, London, 1978, chap. 2.
4. Steel, R. G. D. and Torrie, J. H., *Principles and Procedures of Statistics*, McGraw-Hill, New York, 1960, chap. 9 and 20.
5. Bliss, C. I. and Fisher, R. A., Fitting the negative binomial distribution to biological data and note on the efficient fitting of the negative binomial, *Biometrics*, 9, 176, 1953.
6. Taylor, L. R., Woiwod, I. P., and Perry, J. N., The density-dependence of spatial behavior and the rarity of randomness, *J. Anim. Ecol.*, 47, 383, 1978.
7. Buntin, G. D. and Pedigo, L. P., Dispersion and sequential sampling of green cloverworm eggs in soybeans, *Environ. Entomol.*, 10, 980, 1981.
8. Snedecor, G. W. and Cochran, W. G., *Statistical Methods*, 7th ed., Iowa State University Press, Ames, 1980, chap. 5.
9. Poole, R. W., *An Introduction to Quantitative Ecology*, McGraw-Hill, New York, 1974.
10. Anscombe, F. J., The statistical analysis of insect counts based on the negative binomial distribution, *Biometrics*, 5, 165, 1949.
11. Lindgren, B. W., *Statistical Theory*, 3rd ed., Macmillan, New York, 1976. chap. 5.
12. Gurland, J., Some applications of the negative binomial and other contagious distributions, *Am. J. Publ. Health*, 49, 1388, 1959.
13. Gates, C. E., DISCRETE, a computer program for fitting discrete frequency distributions, in *Estimation and Analysis of Insect Populations*, McDonald, L., Manly, B., Lockwood, J., and Logan, J., Eds., Springer-Verlag, Berlin, 1989, 459.
14. Thomas, M., A generalization of Poisson's binomial limit for use in ecology, *Biometrica*, 36, 18, 1949.
15. Douglas, J. B., Fitting the Neyman type A (two parameter) contagious distribution, *Biometrics*, 11, 149, 1955.
16. McGuire, J. U., Brindley, T. A., and Bancroft, T. A., The distribution of European corn borer larvae, *Pyrausta nubilalis* (Hbn.), in field corn, *Biometrics*, 13, 65, 1957.
17. Cohen, A. C., An extension of a truncated Poisson distribution, *Biometrics*, 16, 446, 1960.
18. Chakravarti, I. M., Laha, R. G., and Roy, J., *Handbook of Methods of Applied Statistics*, Vol. 1, Part 2, John Wiley & Sons, New York, 1967, chap. 3.
19. Perry, J. N. and Taylor, L. R., Adès: new ecological families of species-specific frequency distributions that describe repeated spatial samples with an intrinsic power-law variance-mean property, *J. Anim. Ecol.*, 54, 931, 1985.
20. Ludwig, J. A. and Reynolds, J. F., *Statistical Ecology: A Primer on Methods and Computing*, John Wiley & Sons, New York, 1988, chap. 3.
21. Pielou, E. C., *Mathematical Ecology*, John Wiley & Sons, New York, 1977, chap. 6.
22. Waters, W. E., A quantitative measure of aggregation in insects, *J. Econ. Entomol.*, 52, 1180, 1959.
23. Taylor, L. R., Assessing and interpreting the spatial distributions of insect populations, *Annu. Rev. Entomol.*, 29, 321, 1984.
24. Bliss, C. I. and Owen, A. R. G., Negative binomial distributions with a common k, *Biometrika*, 45, 37, 1958.
25. Lloyd, M., Mean crowding, *J. Anim. Ecol.*, 36, 1, 1967.
26. Morisita, M., I_δ-index, a measure of dispersion of individuals, *Res. Popul. Ecol.*, 4, 1, 1962.
27. Hutcheson, K. and Lyons, N. I., A significance test for Morisita's index of dispersion and the moments when the population is negative binomial and Poisson, in *Estimation and Analysis of Insect Populations*, McDonald, L., Manly, B., Lockwood, J., and Logan, J., Eds., Springer-Verlag, Berlin, 1989, 335.
28. Green, R. H., Measurement of nonrandomness in spatial distributions, *Res. Popul. Ecol.*, 8, 1, 1966.
29. Patil, G. P. and Stiteler, W. M., Concepts of aggregation and their quantification: a critical review with some new results and applications, *Res. Popul. Ecol.*, 15, 238, 1974.
30. Iwao, S. and Kuno, E., Use of the regression of mean crowding on mean density for estimating sample size and the transformation of data for the analysis of variance, *Res. Popul. Ecol.*, 10, 210, 1968.
31. Healy, M. J. R. and Taylor, L. R., Tables for power-law transformations, *Biometrika*, 49, 557, 1962.
32. Taylor, L. R., Aggregation, variance and the mean, *Nature*, 189, 732, 1961.

33. Itô, Y. and Kitching, R. L., The importance of nonlinearity: a comment on the views of Taylor, *Res. Popul. Ecol.*, 28, 39, 1986.
34. Iwao, S., A new regression method for analyzing the aggregation pattern of animal populations, *Res. Popul. Ecol.*, 10, 1, 1968.
35. Taylor, L. R., Aggregation as a species characteristic, *Stat. Ecol.*, 1, 357, 1971.
36. Kuno, E., Sampling and analysis of insect populations, *Annu. Rev. Entomol.*, 36, 285, 1991.
37. Banerjee, B., Variance to mean ratio and the spatial distribution of animals, *Experimentia*, 32, 993, 1976.
38. Davis, P. M. and Pedigo, L. P., Analysis of spatial patterns and sequential count plans for stalk borer (Lepidoptera: Noctuidae), *Environ. Entomol.*, 18, 504, 1989.
39. Sawyer, A. J., Inconstancy of Taylor's b: simulated sampling with different quadrat sizes and spatial distributions, *Res. Popul. Ecol.*, 31, 11, 1989.
40. Clark, P. J. and Evans, F. C., Distance to nearest neighbor as a measure of spatial relationships in populations, *Ecology*, 35, 445, 1954.
41. Thompson, H. R., Distribution of distance to nth neighbour in a population of randomly distributed individuals, *Ecology*, 37, 391, 1956.
42. Vieira, S. R., Hatfield, J. L., Nielsen, D. R., and Biggar, J. W., Geostatistical theory and application to variability of some agronomical properties, *Hilgardia*, 51, 1, 1983.
43. Davis, J. C., *Statistics and Data Analysis in Geology*, 2nd ed., John Wiley & Sons, New York, 1986.
44. Schotzko, D. J. and O'Keeffe, L. E., Geostatistical description of the spatial distribution of *Lygus hesperus* (Heteroptera: Miridae) in lentils, *J. Econ. Entomol.*, 82, 1277, 1989.
45. Schotzko, D. J. and O'Keeffe, L. E., Effect of sample placement on the geostatistical analysis of the spatial distribution of *Lygus hesperus* (Heteroptera: Miridae) in lentils, *J. Econ. Entomol.*, 83, 1888, 1990.
46. Cressie, N., Spatial prediction and ordinary kriging, *Math. Geol.*, 17, 563, 1985.
47. Lefkovitch, L. P., An index of spatial distribution, *Res. Popul. Ecol.*, 8, 89, 1966.
48. Myers, J. H., Selecting a measure of dispersion, *Environ. Entomol.*, 7, 619, 1978.

Chapter 4

BIAS AND VARIABILITY IN STATISTICAL ESTIMATES

David E. Legg and Roger D. Moon

TABLE OF CONTENTS

I.	Introduction	56
	A. Definitions and Examples	56
	B. Importance of Variance and Bias to Pest Management	59
II.	Variance and Bias of Sample Estimators	60
	A. Sources of Inconsistent Error: Variance	60
	1. Biological Error	60
	2. Sampling Error	60
	a. Random Sampling	61
	b. Nonrandom Sampling	62
	B. Sources of Consistent Error: Bias	62
III.	Precision	63
	A. Definitions and Some Measures of Precision	63
	B. Importance of Precision and Its Relation to Sample Size	65
	1. Increasing Precision: Collecting More Samples	65
	2. Increasing Precision: Modifying the Sampling Plan	67
References		69

I. INTRODUCTION

Determining the level of arthropod pest infestation and/or degree to which arthropod natural enemies are present on a commodity at a given point in time is essential to effective pest management. Without doubt, the best way to determine the exact level of pest infestation is to conduct a census if labor, time, and monetary sources were not limiting; i.e., to count each and every arthropod pest and natural enemy found on all sample units. Assuming that no units are overlooked, and no errors were incurred when counting, the census will provide an exact and true assessment of the infestation level. In most agricultural settings, however, a census would be out of the question as counting arthropod pests or natural enemies inhabiting all plants in a field, animals in a herd, or grains in a storage facility would require far more labor, time, and monetary support than is normally available. In lieu of a census, sampling should be conducted. Sampling offers many advantages over alternative methods such as casual observation or guessing in that, if conducted properly, it will provide a good *estimate* of the actual level of infestation.

There are many sampling plans to choose from and selecting that which is best suited for a specific purpose can be difficult. In general, this difficulty can be minimized if the goal of the sampling effort is first clearly stated. Then, only those plans that accommodate that goal should be considered. For example, one goal of research is to obtain the best possible estimate of the average number of arthropods per sample unit or proportion of infested sample units. This goal places a premium on collecting and examining many samples, which is necessary when researchers develop sampling plans for pest management programs. For pest management practitioners, however, a common goal is to assess the level of arthropod infestation as rapidly as possible. This minimizes the number of samples that must be taken.

When a sampling plan is selected, it is important to carefully research that plan to ensure that it performs according to expectations and to identify its weaknesses. Weaknesses of a plan can be manifested in many ways, but they generally fall into two categories: those with relatively high variation and those that produce biased results.

This chapter explores the fundamentals of variance and bias and identifies sources for each. It ends with a short discussion on how to minimize the variance of the average level of arthropod infestation.

A. DEFINITIONS AND EXAMPLES

Variation can be thought of as inconsistent, nondirectional error while bias is consistent, unidirectional error. Variance may be demonstrated by supposing that we have sampled a winter wheat field to determine the presence and abundance of the Russian wheat aphid, *Diuraphis noxia* (Mordvilko), on 20 randomly selected, small grain tillers. As each tiller is chosen, it is carefully examined for the presence of Russian wheat aphids, and, if any are found, they are counted. Table 1 shows the results of 15 such sampling efforts. Note that for each effort, the actual number of Russian wheat aphids often differs from one sample unit to the next. Assuming that bias was not involved, these differences are due to variance. The variance of a set of examples is estimated according to a specific formula for the sampling plan that is used. One of the simplest formulas for estimating the variance is as follows:

$$s^2 = \sum_{i=1}^{n} \left(X_i - \bar{X} \right)^2 / (n - 1) \tag{1}$$

TABLE 1
Computer-Simulated Russian Wheat Aphid Counts on Small Grain Tillers for Each of 15 Sampling Efforts

							sampling effort							
1	2	3	4	5	6	7	8	9	10	11	12	13	14	15
1	2	6	0	2	0	5	0	0	0	0	0	0	0	0
5	0	1	0	1	1	0	0	1	1	3	0	0	0	0
0	5	0	0	2	1	0	0	0	0	0	1	0	0	0
0	0	1	1	3	0	0	1	1	0	0	1	0	1	0
0	0	3	0	1	0	0	1	1	3	1	0	9	0	0
4	0	0	0	3	0	2	2	3	2	0	2	0	0	3
0	4	4	0	0	1	5	0	0	1	1	0	0	2	0
0	0	1	0	0	2	2	1	2	1	13	0	2	2	1
0	1	0	2	0	0	1	0	0	0	2	0	0	0	0
0	0	0	1	0	2	3	3	1	0	0	0	1	2	1
1	2	0	1	1	0	0	3	1	0	0	0	1	9	2
0	0	1	0	0	0	0	1	0	0	0	1	3	5	0
0	2	1	0	0	0	3	0	2	1	1	0	0	2	0
2	1	2	0	0	0	3	0	0	0	5	0	0	1	3
0	1	0	1	1	0	0	2	0	0	0	0	1	0	1
0	1	0	0	1	1	2	2	2	0	0	0	0	0	1
0	1	0	0	1	0	0	1	1	1	0	1	0	0	0
1	1	2	0	2	0	1	1	1	1	0	0	1	1	0
0	0	0	10	0	0	4	0	3	0	1	0	0	1	0
0	0	0	1	0	0	0	1	2	0	2	0	0	1	0
Mean 0.7	1.05	1.1	0.85	0.9	0.4	1.55	0.95	1.05	0.55	1.45	0.3	0.9	1.35	0.6

where X_i is the ith sample (in this case, i ranges from 1 to 20), $\overline{X}$ is the sample mean (for the first sampling effort, $\overline{X} = 0.7$), and n is the total number of samples collected (in this example, $n = 20$).[1] Therefore, to estimate the variance for the first sampling effort, we perform the following calculations: $[(1 - 0.07)^2 + (5 - 0.7)^2 + \cdots + (0 - 0.7)^2]/(19) = 2.0105$. Note also that the means from one sampling effort to another are not identical, indicating that they too are influenced by variance.

As with sample variances, variances for the mean are calculated using formulas that are specific for each sampling plan. One of the simplest formulas to calculate the variance of an estimated mean is as follows:

$$s_{\overline{x}}^2 = s^2/n \qquad (2)$$

To continue with our example, variance of the mean from the first sampling effort is $2.0105/20 = 0.1005$.[1]

If we score the samples for Russian wheat aphids as being "aphids present" or "aphids absent", the data would look like those shown in Table 2. Here, a score of "1" indicates that one or more Russian wheat aphids were present, and a score of "0" indicates that no Russian wheat aphids were present. Note that the characters "aphids present" or "aphids absent" are attributes used to make the data conform to the binomial distribution. Other attributes, such as none or one aphid present could have been assigned a score of "0," with two or more aphids present being assigned a score of "1." With the data shown in Table 2, we estimated the proportion of infested samples by cumulating the number of samples on which one or more Russian wheat aphids were found and then dividing that sum by the number of samples that were

TABLE 2
Computer-Simulated Russian Wheat Aphid Presence / Absence Counts on Small Grain Tillers for Each of 15 Sampling Efforts

						Sampling Effort								
1	2	3	4	5	6	7	8	9	10	11	12	13	14	15
1	1	1	0	1	0	1	0	0	0	0	0	0	0	0
1	0	1	0	1	1	0	0	1	1	1	0	0	0	0
0	1	0	0	1	1	0	0	0	0	0	1	0	0	0
0	0	1	1	1	0	0	1	1	0	0	1	0	1	0
0	0	1	0	1	0	0	1	1	1	1	0	1	0	0
1	0	0	0	1	0	1	1	1	1	0	1	0	0	1
0	1	1	0	0	1	1	0	0	1	1	0	0	1	0
0	0	1	0	0	1	1	1	1	1	1	0	1	1	1
0	1	0	1	0	0	1	0	0	0	1	0	0	0	0
0	0	0	1	0	1	1	1	1	0	0	0	1	1	1
1	1	0	1	1	0	0	1	1	0	0	0	1	1	1
0	0	1	0	0	0	0	1	0	0	0	1	1	1	0
0	1	1	0	0	0	1	0	1	1	1	0	0	1	0
1	1	1	0	0	0	1	0	0	0	1	0	0	1	1
0	1	0	1	1	0	0	1	0	0	0	0	1	0	1
0	1	0	0	1	1	1	1	1	0	0	0	0	0	1
0	1	0	0	1	0	0	1	1	1	0	1	0	0	0
1	1	1	0	1	0	1	1	1	1	0	0	1	1	0
0	0	0	1	0	0	1	0	1	0	1	0	0	1	0
0	0	0	1	0	0	0	1	1	0	1	0	0	1	0
Proportion 0.3	0.55	0.5	0.35	0.55	0.3	0.55	0.6	0.65	0.4	0.45	0.25	0.35	0.55	0.35

taken. For the first sampling effort, the sum of infested tillers was 6, and the proportion of infested samples was 6/20 = 0.3. When this proportion is multiplied by 100, we obtain the percentage of infested samples. Note that the proportions from one sampling effort to another are not identical, indicating that variance affects the estimates of proportions as well. A common formula for calculating the variance of a proportion is as follows:

$$s_p^2 = pq/n \qquad (3)$$

where p is the estimated proportion (0.3 for the first sampling effort), q is $1 - p$ (0.7 for the first sampling effort), and n is as before (20 in this example).[1] Therefore, the variance of the first estimated proportion is $(0.3)(0.7)/20 = 0.0105$.

Bias may be best demonstrated through a series of illustrations. Suppose that a marksperson was shooting a firearm at a target. If the sights were aligned properly, several rounds of fire may yield the results shown in Figure 1A, with the majority of hits being scattered all around and very near the center of the target. If, however, the sights were improperly aligned, then the resulting hits would miss the center of the target by a certain distance in one principal direction, as shown in Figure 1B. In this example, all hits are to the right of center. Note that in each example, there is a certain amount of scatter in the hits (variance) and, if each could be mathematically scored for their distance from center, then an average center for the hits could be estimated ($\bar{X}$). If we repeated this for several targets, we could compute an average of these averages, m (note that m is the sum of the $\bar{X}$ divided by the number of trials, or, in this case, targets). Note that we know the value for the true center of the target,

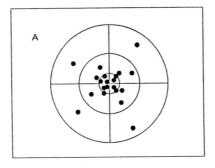

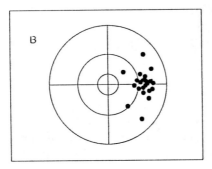

FIGURE 1. Illustrations of unbiased (A) and biased (B) sampling efforts. Note that all of the hits for Figure 1B are to the right of center (true mean).

hence, we know the true mean, u. We may therefore measure bias as follows:[2]

$$B = u - m \qquad (4)$$

Of course, in agriculture and pest management, bias can be manifested in just one of two directions: it can be greater or less than the true level of arthropod infestation.

B. IMPORTANCE OF VARIANCE AND BIAS TO PEST MANAGEMENT

Why are variance and bias important to the pest manager? Because variance can effectively mask the level of pest infestation and bias will lead to an under- or overestimation of infestation levels.

A sampling plan associated with relatively low variance will yield arthropod counts that will not differ greatly from one sample to the next. As variance increases, however, arthropod counts will begin to differ by increasing amounts; the greater the variance, the greater the potential for differences in these counts. Therefore, as variance increases, so does the uncertainty about estimated levels of arthropod infestation. As the level of uncertainty increases, the sampler must either accept a dubious estimate of the infestation or increase the number of samples to some point where the estimate becomes more certain. Stated in other terms, variance incurs a cost in time, money, and human power; the greater the variance, the greater the demands on these resources.

Bias is a potentially serious problem for both the researcher and pest manager. For the researcher, the problem of bias is clear in that estimated parameters for various sampling models that are less or greater than they should be are obviously wrong. For the pest management practitioner, sampling plans that are based on biased models will also be incorrect, but their full impact on management decisions is not clear. This is because arthropod pest infestations range from uninfested to heavily infested relative to a decision-making level, and, unless many of those infestations are very near that level, then sampling plans that are just slightly biased will have little impact on the ultimate pest management decision. In addition, the uncertainty that is always associated with commodity values, which are used to calculate decision-making levels, tend to mitigate minor imperfections of most sampling plans. Finally, Cochran has stated that, as a working rule, "the effect of bias on the accuracy of an estimate is negligible if the bias is less than one tenth of the standard deviation of the estimate."[2]

II. VARIANCE AND BIAS OF SAMPLE ESTIMATORS

A. SOURCES OF INCONSISTENT ERROR: VARIANCE

Where does variance come from? Generally, it originates from biological sources (biological error) and from sample selection (sampling error).

1. Biological Error

Biological error may be thought of as random error that we know little to nothing about but can presume its association with such factors as microclimate, field, and crop characteristics.[3] Many potential sources of biological error are possible, and these include the fitness and physical condition of the sample unit, the microenvironment, and range in genetic composition of the arthropod being sampled. Physical condition of the sample unit, as measured by such factors as stage of growth, degree of leaf senescence, and nutritional value of the host may play an important role when an arthropod selects a host for feeding. Moreover, due to a high degree of genetic uniformity in most crop cultivars, there may be greater differences between the physical condition of younger and older leaves for a given host plant than between any two host plants of a given field. However, situations do arise where fields are planted, with several cultivars having a range of attributes, such as maturation dates. In these situations, marked pest presence and/or abundance may occur on some plants but not on others.

Another potential source for biological error is the microenvironment, which includes temperature, light intensity, and humidity and moisture. Some arthropods, particularly insects, may at times prefer to inhabit or rest on plants or plant parts that are in a given temperature range or they may alternate between shaded and sunlit parts of the plant to maintain an approximate body temperature. If such temperature-driven or thermoregulatory behavior is an important aspect of the arthropod being sampled, and, if the pool of potential sample units at the time of sampling occupy a range of temperatures, light intensities, and/or aspects to the sun, the arthropod counts may vary accordingly. Much the same argument can be made for arthropod preferences for humidities or moisture levels that could enhance or hinder their development or survival.

The basis for arthropod biological error may be due to genetic predispositions toward or against certain behaviors and microenvironments. Such predispositions potentially differ from one individual of a population to another, and, the greater the range of genetic differences, the greater the potential for biological error.

2. Sampling Error

Unlike biological error, sampling error is understood and arises solely from the selection of some sample units over others. Here, the selection process is important. This occurs in one of two ways: random and nonrandom sampling.

a. Random Sampling

Random sampling occurs when the selection of one sample unit in no way influences the choice of another sample unit. Theoretically, this is a relatively simple concept, although it has been debated whether a given selection of sample units can be proven as originating from a random process.[4]

How is a random set of sample units selected? In agriculture, two methods are common: sample listing or random coordinate selection.[5]

Sample listing is practical and widely used in agricultural experiments, where the number of sample units is not too great to be counted and labeled with a number. Random number tables or random selection routines can be used to select a subset of numbers without replacement, thereby selecting the associated sample unit. However, enumerating all sample units in many agricultural settings, such as plants in a field, leaves in a tree, grains in a bin, or animals in a large herd, is not possible because the number of sample units is too great to be reliably counted and have a number assigned to each. Under these conditions, random coordinate selection can be conducted.

With random coordinate selection, it is important to develop a basis for the coordinate system. For crops or orchards that are planted to rows, the number of rows across one end of the field or orchard can be counted and labeled with a number. If a similar pattern of rows exists on the adjacent side of the field, then these too can be counted and labeled. Then, a random row from one side of the field and a random row from the adjacent side can be selected. The plant or tree at or nearest the point where these two rows meet (coordinate) is the selected sample unit. If the crop is planted to rows on just one side of the field or is not planted to rows at all, as is the case for alfalfa, one or both sides of the field can be measured by determining the number of steps or paces the sampler must take to walk that or both sides of the field. Then, random coordinate selection can be conducted based on a random selection of rows and/or paces. As before, the point at which the selected number of paces and/or rows meet is where the sample should be taken. A random selection of samples taken from the entire field, orchard, grain bin, or animal herd without restriction is referred to as unrestricted random sampling.

Once a set of random coordinates is selected throughout a field or orchard, given that the sampling effort does not involve sequential sampling, the sampler can spatially arrange the order of visits to those coordinates so they may be visited in an efficient manner.[5] For sampling irregularly shaped fields or orchards, it is possible to modify published algorithms to account for these shapes, but this must be done on a field-by-field basis. If that is not desirable, then one could apply a "bounce-back" rule, where the sampler turns back into the field or orchard upon encountering a border to complete the required number of paces or rows for that coordinate.[5] For sampling grain bins, the coordinate selection method should account for width, breadth, and depth of storage facilities that have various shapes and sizes. For situations in which the sample units are readily apparent (e.g., alfalfa stems, small grain tillers, or fruits), it is important to look away from the sample units prior to their selection, as some research has shown that the sampler may be tempted to select those that show excessive damage or damage symptoms.[6]

Variance associated with the random selection process depends entirely on the "chance" of the draw as the choice of any one sample unit is just as likely to occur as any other.[1] Through this process, a group of, say, 100 randomly selected sample units may contain 10 that are heavily infested with an arthropod. If another group of 100 samples is randomly selected, however, there may be none that is heavily infested.

Obviously, selecting sample units via a random process requires some work and preparation; this may lead one to wonder why they should bother with random sampling when it is easier to do otherwise. Simply put, estimates of infestation levels that are based on random sampling, provided that counting and identification errors do not occur, guarantees that the results are not biased. In addition, the standard errors (i.e., square root of sample variances) of those estimates can be calculated using published and well-known formulas.[1] These two claims cannot be made with estimates that are based on nonrandom sample selection.

b. Nonrandom Sampling

Nonrandom sampling occurs when the selection of one sample unit influences the selection of another sample unit. Nonrandom sampling in agriculture is common, especially in pest management programs, and most, if not all, can be thought of as sampling along predetermined routes or patterns in fields, orchards, or forests.[5]

There are many ways to conduct a nonrandom sampling effort. Some involve walking a given number of steps or rows into a field before taking the first sample, then walking through that field along a predefined pattern and stopping at systematic intervals to select a sample unit.[7] Patterns that are commonly used are straight line-, zig-zag-, or "U"-shaped. Some sampling plans may involve stopping at random intervals along predetermined sampling patterns, while still others make use of several straight line patterns for a given area.[8] Some sampling plans involve dividing the field into quadrats of a given size and collecting one sample from the exact center of each (centric systematic area sampling).[9]

As with random sampling, variance is associated with arthropod enumerations and presence/absence counts that are made with nonrandom sampling. Unlike random sampling, however, variance associated with these enumerations or counts does not depend on the chance of the draw. Rather, it depends on the specifics of the sampling plan that is used and the spatial arrangement of the arthropods over the area being sampled.

B. SOURCES OF CONSISTENT ERROR: BIAS

Bias in arthropod enumerations or presence/absence counts can occur in many ways, but it may be associated with one of three general sources: nonrandom sampling, sampling technique, and interpersonal error. Nonrandom sampling allows for the potential of bias to occur in two ways: (1) the selected sample units which lie on the chosen sampling pattern may be entirely within a heavily infested area of a field when the remainder is just slightly or completely uninfested (or vice versa), and (2) the systematic collection of samples along the selected sampling pattern may unexpectedly coincide with the distribution pattern of the arthropod being sampled.[7,9] Regarding the first point, knowledge of the biology and behavior of the arthropod being sampled may elucidate the potential for unusually "spotty" infestations so that the sampler can take precautions to guard against this source of bias. Regarding the second, it has been argued that arthropod spatial distributions are almost never arranged so their densities form geometric shapes or their variances are coincidentally periodic with the straight line-, zig-zag-, "U"-shaped, or centric systematic area sampling patterns that are commonly used in pest management.[9] Rather, a more imminent source of bias may be the superimposition of human activities on the area that is being sampled. Examples include nonuniform manuring or fertilizer applications, partial flooding, incomplete herbicide applications, widely varying planting dates, or planting several distinct cultivars in the same field at the same time.[9] To uncover such activities, the sampler should check the agronomic history of the field prior to initiating any nonrandom sampling effort.

Bias that arises from nonrandom sampling may be difficult to detect because the usual method of its estimation (Equation 4) may not be available as it is difficult to know the true mean density or proportion of infested sample units in most agricultural situations. Alternative methods have been suggested, but these may require tremendous numbers of samples to detect any bias that exists.[7,10]

Bias may also occur when the sampler collects, identifies, and counts the arthropod of interest.[2] Collecting arthropods can be done with many different tools, but those most commonly used in pest management involve the sweep net, drop cloth, light trap, pheromone trap, suction trap, or some form of colored and/or specially shaped sticky trap.[10,11] Each of these collecting tools have their virtues, but their effectiveness may depend on such factors as wind velocity, larval stage, and moisture.[12-15]

Pest identification is, of course, an important aspect of pest management, yet evidence exists that some producers may not be minimally proficient in this endeavor.[16] In pest management programs where pest identification skills are suspect, it may be possible to teach pest managers how to identify arthropod pests under field conditions.[6]

Counting errors occur when the sampler is not being thorough or careful when examining or collecting the sample unit.[17] Counting errors can also occur when the sampler attempts to count very fast-moving arthropods, particularly at moderate to high infestation levels.[18] To circumvent the first source of error, an acceptable level of sampler diligence may be maintained or enhanced through techniques that alleviate fatigue. Accounting for the second source of bias may be possible through the application of carefully researched correction factors.[18]

Related to the counting error problem is the bias associated with extracting arthropods from plants that are shaken into a pail, beaten over a cloth, or placed into short-term storage. Sometimes, sampling or arthropods in pest management requires the use of simple extraction techniques such as shaking or beating the sample unit into a container or over a drop cloth. When using such techniques, care should be taken to ensure that all arthropods inhabiting the selected sample units are dislodged. If some remain, it may be possible to compensate for them by using a correction factor.[19] If a sampler collects a number of sample units that require extensive examination and/or extraction techniques, the samples may be placed in sealed containers, properly labeled, and put aside for short-term storage. Care should be taken, however, to ensure that the storage process does not reduce the eventual arthropod enumerations or presence/absence counts or both.

Interpersonal errors sometimes occur when sampling for arthropod pests in agriculture. These may reflect a lack of sampler experience, a tendency to select heavily damaged or infested sample units, or a fundamental difference between samplers in their temperaments, sampling abilities, and interest in sampling.[19,20] Careful screening of candidates for pest management scouting positions, coupled with adequate training, should help minimize this form of bias.

III. PRECISION

A. DEFINITIONS AND SOME MEASURES OF PRECISION

Precision, a characteristic of the *estimated* average level of pest infestation, indicates how well the mean is estimated. Specifically, precision refers to the distance that each estimated mean ($\bar{X}$) deviates from the average of a group of means (m).[2] This may be represented as

$$\text{Precision} = \bar{X} - m \tag{5}$$

As was the case for bias, precision may be best demonstrated through example. Suppose that a marksperson was shooting a firearm at a target and was fairly erratic in their aim. If the sights were aligned properly, the *average* hits from each of several targets may be as shown in Figure 2A. Here, we use the symbol "+" instead of a solid circle to differentiate between mathematical average of the individual hits per target and the individual hits. Note the excessive amount of scatter in the average hits. This represents a relatively low level of precision. If, however, the marksperson was more proficient in their aim, the average hits from several targets may appear as shown in Figure 2B. Note the tight grouping of averages near the true center of the target. This represents a relatively high level of precision. Can sampling efforts of relatively low or high precision be biased? Yes, as is shown in Figures 2C and 2D.

Table 3 shows the results of 20 separate computer-simulated sampling assessments of an arthropod pest infestation that was distributed randomly across a "field" of 2000 sample units. Each assessment was based on 100 randomly selected samples. Column 3 represents the deviations of each estimated mean from the average of those means.

It is apparent that measures of precision involve estimates of the variance of the mean (i.e., Equation 2). These measures may be of scale or they may be scale free. Measures of scale involve estimating the variance of the mean, $s_{\bar{x}}^2$ for enumerative scale or s_p^2 for presence/absence counts, and taking the square root of these estimates. These are referred to as the standard error of the mean ($s_{\bar{x}}, s_p$). They are measures of scale used to calculate confidence intervals of sample means or proportions.[1]

Scale-free measures of precision involve calculating the coefficient of variation of the mean:

$$\mathrm{CV}_{\bar{x}} = \sqrt{s^2/n}\,/\bar{X} \qquad (6)$$

and the coefficient of variation of the proportion of infested sample units:[21]

$$\mathrm{CV}_p = \sqrt{pq/n}\,/p \qquad (7)$$

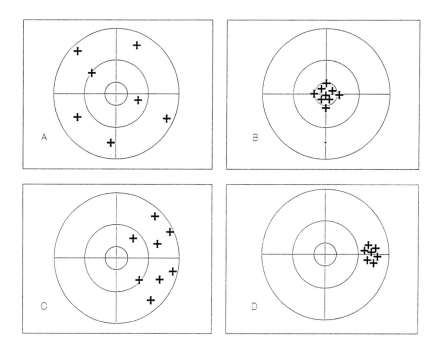

FIGURE 2. Illustrations of relatively low precision and no bias (A), relatively high precision and no bias (B), relatively low precision with bias (C), and relatively high precision with bias (D).

TABLE 3
Estimated Mean Densities (Mean), Deviations from the Grand Mean (m), Variances, and Percent Relative Variations (RV) for Each of 20 Computer-Simulated Arthropod Sampling Trials[a]

Trial	Mean	Mean − m	Variance	RV (%)
1	3.93	−0.046	3.6617	4.8691
2	4.05	0.074	4.4116	5.1861
3	3.96	−0.016	3.8368	4.9464
4	3.79	−0.186	3.9858	5.2676
5	4.13	0.154	3.1849	4.3212
6	4.15	0.174	3.7045	4.6379
7	3.76	−0.216	4.1034	5.3875
8	3.92	−0.056	4.2764	5.2754
9	3.85	−0.126	3.5631	4.9029
10	4.14	0.164	3.2731	4.3700
11	3.85	−0.126	2.7753	4.3270
12	4.00	0.024	3.8788	4.9237
13	4.39	0.414	3.9171	4.5083
14	3.90	−0.076	3.9293	5.0827
15	3.87	−0.106	3.7708	5.0177
16	4.08	0.104	2.8016	4.1025
17	4.18	0.204	4.7349	5.2057
18	3.75	−0.226	4.4722	5.6394
19	4.01	0.034	3.9090	4.9305
20	3.81	−0.166	5.6302	6.2278
Sum	79.520	3.33786E-06	77.8206	99.1294
Average	3.976[b]	1.66893E-07	3.89103	4.9565

[a] Estimated means and variances were based on a sample size of 100. The true mean (u) was 3.9975 and the true variance was 3.977496.
[b] Grand mean (m).

These terms often appear in the literature as relative variation (RV) when they are multiplied by 100.[22] Both $CV_{\bar{x}}$ and CV_p can be used to develop sampling models that are based on precision.[21] Southwood suggests that when sampling for the purposes of pest management, a RV of 25% or less is sufficiently precise; for research purposes, however, a RV of 10% or less may be desired.[10] Column 5 in Table 3 represents the corresponding RVs for each of the 20 sampling assessments on a randomly distributed arthropod population.

B. IMPORTANCE OF PRECISION AND ITS RELATION TO SAMPLE SIZE

Why is precision important to the researcher and pest manager? Because a high level of precision assures the sampler that the estimated pest density or proportion of infested samples is relatively close to m, the average of the sample averages. Conversely, a low level of precision suggests that the estimate is not close to m. How may a sampler increase the precision of the estimated pest density or proportion of infested samples? By increasing the number of samples or modifying the sampling plan.

1. Increasing Precision: Collecting More Samples

The relationship between $s_{\bar{x}}$, and n, and RV and n for arthropod infestations that are both randomly distributed and aggregated may be seen in Figures 3A and 3B. Note that for both kinds of distributions, $s_{\bar{x}}$ and RV decreased with increasing n. Had

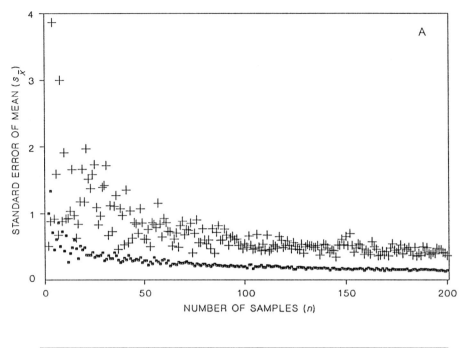

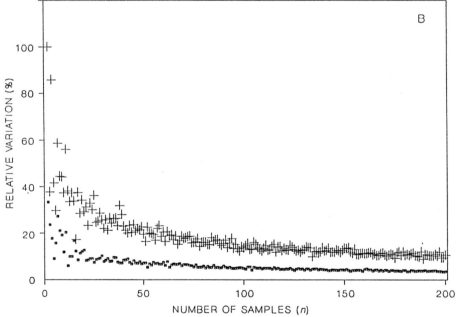

FIGURE 3. Response of the standard error of the mean (A) and relative variation of those means (B) to sample size (n) for a simulated aggregated arthropod population (+) and a randomly distributed population (■).

we carried out these computer-simulated sampling efforts for n of 200 or more, we would have applied the finite population correction for density estimates

$$s_{\bar{x}} = \sqrt{s^2/n} * \sqrt{1 - n/N} \tag{8}$$

and for proportion estimates

$$s_p = \sqrt{pq/n} * \sqrt{1 - n/N} \qquad (9)$$

where N represents the total number of sample units in the field.[1] As a general rule, the finite population correction is applied only when n/N is greater than or equal to 0.1.[1] It is not difficult to see that, as n approaches N, the ratio n/N approaches 1, $1 - n/N$ approaches 0, and $\sqrt{1 - n/N}$ becomes numerically smaller. This makes $s_{\bar{x}}$ and RV numerically smaller. For the special case of a census, $\sqrt{1 - n/N} = 0$ and, therefore, $s_{\bar{x}}$ and RV = 0. These same results can be seen in presence/absence counts for an arthropod population that is near complete infestation ($p = 0.98$) and one that is 2/3 infested ($p = 0.67$) (Figures 4A and 4B). If the sampling plan is not responsible for introducing a great deal of variation into the arthropod counts, then the sampler should use published formulas to determine the number of samples that should be taken to obtain a given level of precision.[21,23]

2. Increasing Precision: Modifying the Sampling Plan

If the sampling plan is suspected of introducing excessive variation into the arthropod counts, then the sampler may consider using a different sample unit, change the way that samples are selected, or use a biased sampling plan that is known to have a high level of precision. If the sample unit is not defined on an overriding ecological principle,[24] then the sampler may experiment with different sample units to determine that which minimizes the variance in arthropod counts. Such experimentation could involve a change from sampling trifoliates of a legume crop to individual leaflets, or from sampling 1 m² quadrats to 0.1 m² quadrats.

The way that sample units are selected from a given field, orchard, or grain bin may contribute to excessive variation in arthropod pest density counts. If the sampler recognizes a pattern of excessively high or low counts being associated with certain areas of the field, orchard, or grain bin, then those areas may be isolated from each other and sampled separately. This is referred to as stratified sampling, where the strata are the identified areas of high or low infestations and the sample units are randomly selected from each stratum.[1] Such stratification may produce a higher level of precision than unrestricted random sampling for the field that is being sampled.[2]

If the sampler finds that a selected sampling plan for a given arthropod yields a low level of precision, it may be possible to use another sampling plan that is known to produce biased results but provides an increased level of precision. When using such a plan, the sampler adjusts for the amount of bias after the samples are collected and the arthropods are counted.

Finally, the sampler may determine that the arthropod being sampled provides mean density estimates that are inherently associated with low precision. This appears to be the case for the Russian wheat aphid, which tends to be extremely aggregated in wheat fields.[25] In this case, the sampler may wish to transform the arthropod pest enumerations into presence/absence counts which, when presented as proportions, may be associated with a higher level of precision. This phenomenon may be observed by contrasting the RV values for the aggregated arthropod population in enumerative scale (Figure 3B) with those for that same population in presence/absence scale (Figure 4B). For the enumerative scale, the RVs of Figure 3B become relatively steady at or just less than 20%, whereas those of the presence/absence scale (Figure 4B) did so at or just less than 10%.

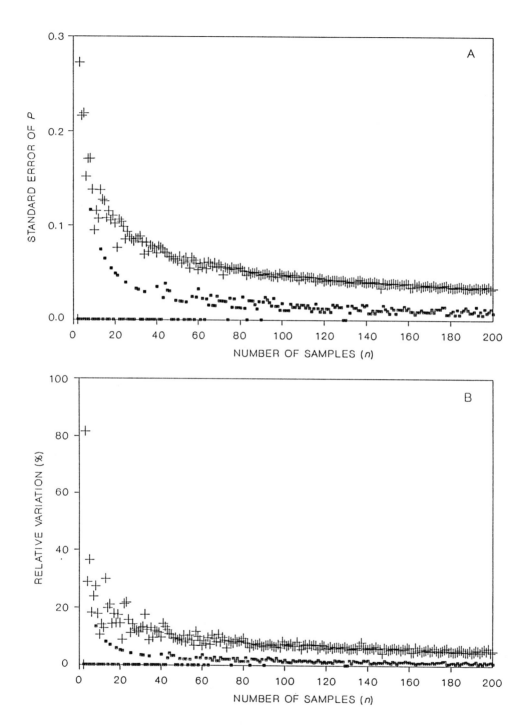

FIGURE 4. Response of the standard error of a proportion (A) and relative variation of those proportions (B) to sample size (n) for a simulated aggregated arthropod population (+) and a randomly distributed population (■).

REFERENCES

1. Snedecor, G. W. and Cochran, W. G., *Statistical Methods*, 6th ed., Iowa State University Press, Ames, 1967, chap. 2, 8, 17.
2. Cochran, W. G., *Sampling Techniques*, 2nd ed., John Wiley & Sons, New York, 1963, chap. 1, 5.
3. Schaalje, G. B., Butts, R. A., and Lysyk, T. J., Simulation studies of binomial sampling: a new variance estimator and density predictor, with special reference to the Russian wheat aphid (Homoptera: Aphididae), *J. Econ. Entomol.*, 84, 140, 1991.
4. Kac, M., What is random?, *Am. Sci.*, 71, 405, 1983.
5. Legg, D. E. and Yeargan, K. V., Method for random sampling insect populations, *J. Econ. Entomol.*, 78, 1003, 1985.
6. Barney, R. J., Legg, D. E., and Christensen, C. M., Intersampler variability and scouting for larvae of the alfalfa weevil (Coleoptera: Curculionidae), *Trans. Kentucky Acad. Sci.*, 49, 2, 1988.
7. Legg, D. E., Shufran, K. A., and Yeargan, K. V., Evaluation of two sampling methods for their influence on the population statistics of alfalfa weevil (Coleoptera: Curculionidae) larva infestations in alfalfa, *J. Econ. Entomol.*, 78, 1468, 1985.
8. Osborne, J. G., Sampling errors of systematic and random surveys of cover-type areas, *J. Am. Stat. Assoc.*, 37, 256, 1942.
9. Milne, A., The centric systematic area-sample treated as a random sample, *Biometrics*, 15, 270, 1959.
10. Southwood, T. R. E., *Ecological Methods*, 2nd ed., Chapman & Hall, New York, chap. 2, 4, 5, 6, 7, 8.
11. Kogan, M. and Pitre, H. N., Jr., General sampling methods for above-ground populations of soybean arthropods, in *Sampling Methods in Soybean Entomology*, Kogan, M. and Herzog, D. C., Eds., Springer-Verlag, New York, 1980, chap. 2.
12. DeLong, D. M., Some problems encountered in the estimation of insect populations by the sweeping method, *Ann. Entomol. Soc. Am.*, 25, 13, 1932.
13. Romney, V. E., The effect of physical factors upon catch of the beet leafhopper (*Eutettix tenellus* (Bak.)) by a cylinder and two sweep net methods, *Ecology*, 26, 135, 1945.
14. Saugstad, E. S., Bram, R. A., and Nyquist, W. E., Factors influencing sweep-net sampling of alfalfa, *J. Econ. Entomol.*, 60, 421, 1967.
15. Cothran, W. R. and Summers, C. G., Sampling for the Egyptian alfalfa weevil: a comment on the sweep-net method, *J. Econ. Entomol.*, 65, 689, 1972.
16. Chiang, H. C., Ecological considerations in developing recommendations for chemical control of pests: European corn borer as a model, *FAO Plant Prot. Bull.*, 21, 30, 1973.
17. Higgins, R. A., Rice, M. E., Blodgett, S. L., and Gibb, T. J., Alfalfa stem-removal methods and their efficiency in predicting actual numbers of alfalfa weevil larvae (Coleoptera: Curculionidae), *J. Econ. Entomol.*, 84, 650, 1991.
18. Fleischer, S. J. and Allen, W. A., Field counting efficiency of sweep-net samples of adult potato leafhoppers (Homoptera: Cicadellidae) in alfalfa, *J. Econ. Entomol.*, 75, 837, 1982.
19. Barney, R. J. and Legg, D. E., Conversion factors for use with shake-bucket technique for sampling alfalfa weevil (Coleoptera: Curculionidae) larvae, *J. Econ. Entomol.*, 82, 825, 1989.
20. Shufran, K. A. and Raney, H. G., Influence of inter-observer variation on insect scouting observations and management decisions, *J. Econ. Entomol.*, 82, 180, 1989.
21. Karandinos, M. G., Optimum sample size and comments on some published formulae, *Bull. Entomol. Soc. Am.*, 22, 417, 1976.
22. Pedigo, L. P., Lentz, G. L., Stone, J. D., and Cox, D. F., Green cloverworm populations in Iowa soybean with special reference to sampling procedure, *J. Econ. Entomol.*, 65, 414, 1972.
23. Ruesink, W. G., Introduction to sampling theory, in *Sampling Methods in Soybean Entomology*, Kogan, M. and Herzog, D. C., Eds., Springer-Verlag, New York, 1980, chap. 3.
24. Christensen, J. B., Guitierrez, A. P., Cothran, W. R., and Summers, C. G., The within field spatial pattern of the Egyptian alfalfa weevil, *Hypera brunneipennis* (Coleoptera: Curculionidae): an application of parameter estimates in simulation, *Can. Entomol.*, 109, 1599, 1977.
25. Legg, D. E., Hein, G. L., Nowierski, R. M., Feng, M. G., Peairs, F. B., Karner, M., and Cuperus, G. W., Binomial regression models for spring and summer infestations of the Russian wheat aphid (Homoptera: Aphididae) in the southern and western plains states and Rocky Mountain region of the United States, *J. Econ. Entomol.*, 85, 1779, 1992.

Section II: Sampling Techniques and Initial Program Development

Chapter 5

TECHNIQUES FOR SAMPLING ARTHROPODS IN INTEGRATED PEST MANAGEMENT

Scott H. Hutchins

TABLE OF CONTENTS

I. Introduction 74

II. Techniques for Surface and Above-Ground Arthropods 74
    A. Direct (*In Situ*) Techniques 75
    B. Knockdown Techniques 78
    C. Netting Techniques 81
    D. Trapping Techniques 82

III. Techniques for Soil Arthropods 86
    A. Dry Extraction Techniques 89
    B. Wet Extraction Techniques 90

IV. Techniques for Indirect Assessment of Arthropod Populations 91
    A. Arthropod-Induced Injury 91
    B. Arthropod Products 93
    C. Arthropod Acoustic Signatures 93

V. Conclusions and Future Directions 95

References 96

I. INTRODUCTION

Integrated pest management (IPM) was founded on the premise that pests should be managed as populations based on the collective potential injuriousness to their host.[1] Inherent to this premise and the successful implementation of IPM is the recognition that researchers, specialists, and practitioners must assess (1) population size and structure, (2) potential injuriousness per individual comprising the population, (3) host physiological response to injury, and (4) socioeconomic (e.g., monetary) consequence of the pest-host relationship. Indeed, each of these four assessments is critical to the successful development of an IPM strategy.[2]

The injuriousness per individual pest, host physiological response to injury, and socioeconomic consequences all are addressed with the use of *economic injury levels* (EILs). This decision tool, first proposed in concept by Stern et al.[1] and later refined for use by Stone and Pedigo[3] and Pedigo et al.,[2] defines the key economic and biologic variables necessary to objectively critique benefits and costs related to a pest control decision. Although the EIL actually represents a *level of injury* leading to economic damage, a critical *population of arthropods* frequently is cited as the measurable unit for sampling. A primary focus for IPM, therefore, is to discover, develop, and utilize sampling techniques and procedures that provide accurate and consistent estimates of arthropod populations as a proxy for estimating host injuriousness.[4]

As part of an overall IPM strategy, sampling techniques should be compatible with the objective of the decision maker. For example, if the management situation calls for a binary decision on whether or not to apply a prophylactic suppression tactic, then the sampling technique must provide discrete predictive data within the appropriate time horizon to accommodate the decision. If, however, the situation calls for continuous monitoring to assess population growth and development, then the sampling technique(s) must account for population dynamics and host susceptibility in order to make effective curative control decisions.[5] The design and use of arthropod sampling techniques, therefore, helps to *enable* IPM by providing a means by which to make informed decisions. In addition, because the cost of sampling is integral to the cost (and hence, feasibility) of IPM, every effort should be made to develop a high level of efficiency with the design and use of the technique(s). Otherwise, sampling and subsequent curative control decision making will prove to be a less desirable management alternative, resulting in either overuse or underuse of suppression tactics.[6]

Inasmuch as the challenge of establishing an effective IPM strategy largely is a challenge of developing an effective and efficient sampling technique, the objective of this chapter is to review several basic techniques currently employed for the surveillance and sampling of arthropods. Primary categories include: techniques for surface and above-ground arthropods, techniques for soil arthropods, and techniques for indirect assessment of arthropods. Although reference to, or review of, all adaptations to these primary sampling techniques is not pragmatic, representative examples for categories are provided for illustrative purposes.

II. TECHNIQUES FOR SURFACE AND ABOVE-GROUND ARTHROPODS

Life stages of species that attack or inhabit their host(s) in a manner that makes them readily visible and/or externally accessible are candidates for sampling techniques designed for surface and above-ground arthropods. Specifically, arthropods are sampled along with a determined area or volume of habitat. The result is a measurement of arthropod *density* (i.e., number of individuals per land area) or arthropod *intensity* (i.e., number of arthropods per host area or volume). The distinction between density and intensity is an important one. Samples relating to arthropod intensity generally will provide more useful data in situations where host response or injury assessment is the primary consideration. The EIL identifies injury per arthropod as a critical variable which relates the pest in terms of the host's physiology (e.g., leaf consumption (cm^2) per arthropod). Where population dynamics of the species through time is of primary concern (e.g., to construct ecological life tables), measurements of arthropod density may provide the most useful information.[7]

A. DIRECT (*IN SITU*) TECHNIQUES

Sampling techniques that involve in-place or *in situ* direct counting of arthropods per unit of land or habitat are common in agricultural and horticultural IPM situations. The technique generally requires careful observation of conspicuous arthropods as a means to estimate arthropod density or intensity in a predetermined area. When extrapolated, the population size of a larger management area can be estimated.

In situations where the number of arthropods is relatively low and their habitat is easily observed, all individuals may be directly counted as part of the estimate. Arthropod intensity may be determined by dividing the number of individuals in a sample by the appropriate host measurement (e.g., number of seedlings in the sample) to estimate individuals per unit of habitat. Mathematical conversion to arthropod density will be based on the amount of habitat per row meter or square meter, with subsequent conversion to larger areas (e.g., hectare) possible (see Chapter 1). Because dispersion characteristics vary among arthropod species, several samples typically are collected to estimate the central tendency of the population.

One example of *in situ* sampling involves estimating grasshopper density in range and forage crops. To make direct estimates of a larger population, a metal or wood frame of specific size (e.g., 1 m^2) typically is used to delineate a habitat size. After the frame has been in place for a period of time adequate to allow normal behavior, the sample area is carefully approached, and all individuals vacating the area are counted as they fly or hop out.[8] For this particular adaptation of direct sampling to be effective, the arthropod species must be somewhat mobile and large enough to be conspicuous in movement. If the arthropods are less mobile and (or) inconspicuous to the eye, then it may be necessary to establish a barrier to movement outside the sample area until counting can be completed.[9]

In many instances, a specific host area or biological organ is the focus for the direct sample (see Figure 1). In order to determine the intensity of face flies on cattle, for example, direct observations of adult flies resting on the frontal region of the animal are made.[10] Arthropods inhabiting plants frequently are sampled from specific plant parts. Alfalfa weevil larvae (*Hypera postica* [Gyllenhal]), for example, are collected from a random selection of alfalfa stems and counted to estimate intensity (see Figure 2).[11] Adult corn rootworm beetles also are counted directly on a plant organ. Because feeding and potential damage is focused on the corn silks, adults are sampled in the ear zone on field and sweet corn.[12] With cotton, *in situ* counts of corn earworm and tobacco budworm neonates are specifically made within the terminal region of the plants. Scale insects may be counted directly on the leaves, branches, or fruit based on the species-specific feeding site.[13] In situations where the habitat or arthropods are not readily accessible, but are visible (e.g., a forest canopy), a magnifying instrument may be used to sample specified areas of the habitat (see Figure 3).[14]

FIGURE 1. Photograph of corn earworm observed *in situ* feeding within the terminal region of a cotton plant. Several terminals will be sampled to determine percent of plants with live larvae.

FIGURE 2. Photograph of alfalfa weevil larva (left) and adult (right) observed *in situ* feeding on an alfalfa stem. Several stems will be sampled to determine feeding intensity.

FIGURE 3. Photograph of corn earworm (second generation) observed feeding *in situ* on corn ears. Several plants will be sampled to determine percent of damaged ears and number of live larvae per plant.

When habitats are such that stratification of the population may occur, caution should be taken to ensure that representative samples are obtained. This is important for both arthropod and host reasons. Arthropods will tend to develop preferred feeding locations where they obtain the greatest degree of nourishment or protection. Because leaves and stems in the lower strata of agronomic crops tend to become lignified relatively early, they become less palatable to arthropods.[15] The result may be movement of the individuals toward less lignified foliage more distally located on the stem. In this case, although neither the intensity per total plant nor the density per area has changed, a sampling technique focused only on the upper strata might indicate otherwise. When sampling several injurious species within an overall arthropod population, the species-specific distribution within the canopy also may affect the sampling effectiveness. Prior knowledge of the bionomics and feeding behavior of the various arthropod species along with knowledge of the host response to injury at different strata are important considerations when designing a sampling technique.[16] It may become necessary to develop specialized direct sampling techniques to assess population intensity at several strata, and, essentially, treat each as a separate crop for IPM purposes.

The various specific techniques for using direct or *in situ* counts to assess arthropod populations are too extensive to review here. As one of the simplest and most accurate techniques, direct sampling is a primary consideration when first assessing sampling options for surface and above-ground species. Despite its general simplicity in technique, care should be taken to ensure that direct observations (sample units) are not only accurate when extrapolated (i.e., adequate sample-unit number and size based on knowledge about dispersion), but also meaningful (i.e., associated with a pest management or ecological objective with knowledge of host response per unit of injury).

Successful use of direct observation techniques may include modification of the habitat to provide contrast. *Selective staining* of the plant material, for example, has been useful in providing contrast to arthropod eggs inserted in plant stems or petioles[7] (e.g., potato leafhopper [*Empoasca fabae* (Harris)] eggs in alfalfa stems). Staining techniques provide options for arthropod stages embedded (and hence, invisible) in plant material, albeit in a manner that typically is more laborious than surface observation.

B. KNOCKDOWN TECHNIQUES

Sampling techniques that dislodge arthropods from their habitat or environment before counting are commonly used to assess population size when *in situ* techniques are either not feasible or not desirable.[10] Several techniques for dislodgment have been developed for use in an array of IPM sampling situations. One of the oldest and still commonly used methods for dislodgment is physically *jarring* arthropods from their habitat in order to make counting easier and more accurate. Specifically, a collection device (e.g., cloth, tray, polyethylene sheet) is positioned so as to collect arthropods dislodged from the impact of a blunt object to the host.[7] The utility of this technique generally is limited to collection of relatively immobile species or stages of arthropods in order to allow time for collection or counting before escape. Lepidopterous or coleopterous larvae, adult leaf beetles, and adult weevils may be candidates for use of jarring techniques. Lepidopterous larvae on soybeans, for example, can be jarred using a ground cloth technique (see Figure 4).[17] Here, a light colored cloth of predetermined length (e.g., 1 m) is carefully placed between two rows of plants to collect arthropods dislodged from plants vigorously shaken on both sides

FIGURE 4. Photograph of "drop cloth" knockdown sampling technique. Plant rows on either side of the cloth are bent and shaken to dislodge the arthropods for counting. Several areas will be sampled for an average number of larvae per row meter.

of the cloth. Another example of where jarring is helpful involves sampling thrips on cotton seedlings.[12] A small white sheet of paper beside seedlings that are shaken will dislodge adults and nymphs for counting (although the size of the individuals is too small to allow for accurate species identification) (see Figure 5). In each instance, the resultant counts can be segregated by species and reported as a measure of intensity (e.g., larvae per row meter or per plant) or converted to density based on plant row spacing (e.g., larvae per square meter).

Where samples of very small arthropod species are required, *brushing* may be an appropriate knockdown technique. Here, an apparatus of two revolving spiral brushes will carefully dislodge the arthropods onto a collection plate or receptacle for counting.[18] Frequently an adhesive is used to secure the specimens in place for future counting. Examples include sampling of phytophagous mites on various orchard or row crops and collection of ectoparasites (e.g., lice) from livestock. Small arthropod species also may be dislodged by *washing* from the host surface.[19] In some instances, washing is the most effective technique to remove eggs of moths and other species from bark or otherwise inaccessible organs. Regardless of the specific removal technique, a counting grid often is used to make an estimate for the population of very small arthropods extracted by brushing or washing (see Figure 6).[20] The technique involves placing a petri dish of settled arthropods (in solution) over a grid that allows visibility of only a portion of the known surface area. By counting the number of individuals in the visible area a total sample size can be calculated and related to the sample unit.

Arthropods that are somewhat mobile once dislodged often are candidates for sampling techniques that include a *sucking* apparatus (see Figure 7). Several types of machines have been designed based on the premise of creating a level of suction

FIGURE 5. Photograph of thrips being dislodged from cotton seedlings onto a white index card and counted. Several seedlings will be sampled to determine the average number of thrips per plant.

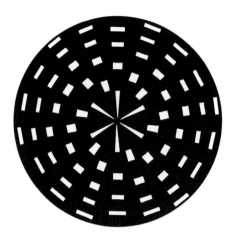

FIGURE 6. Drawing of a counting grid used to estimate the mite population from a sample without counting each individual. Only the mites seen on the white areas will be counted (represents one sixth of the total area).

intense enough to overcome the hopping and flying capacity of the target arthropods, but not so strong as to destroy the fragile bodies of small species.[21] Effective sucking devices also frequently are used to obtain absolute samples of arthropods for both IPM decision making as well as for research. In addition, these devices represent a quick and effective method for collecting large volumes of arthropods for reinfestation[22] or, on a larger scale, as a removal technique. A consideration of all suction techniques is the indiscriminate manner arthropods are collected, which necessitates sorting for target species prior to counting. This nonselectivity makes suction-based sampling an attractive option for arthropod community sampling (e.g., to assess predator/prey ratios).

The use of *heat*, generally through intense light, has been successful in dislodging arthropods from plant material. Inasmuch as arthropods are sensitive to desiccation, they tend to move toward more humid habitats. When fresh plant material is placed in a container and subjected to directed heat (presumably from above), the arthropods migrate away from encroaching desiccation until eventually they fall into a collection device and can be counted. These devices have been successful with many stored-grain applications and increasingly are used with separating arthropod species inhabiting herbage. More detail on these extraction techniques will be provided with the discussion of sampling techniques for soil arthropods.

Knockdown sampling techniques utilizing the anesthetizing or lethal effects of chemicals are commonly employed for absolute sampling of mobile arthropods or entire communities.[7] Indeed, the use of *chemical knockdown* tactics is widely accepted as a method to manipulate arthropod species. A tree covered by a screen and treated with a quick knockdown insecticide (e.g., pyrethrum) will allow for the dead arthropods to be collected after dislodging.[23] Similarly, a plant or group of plants contained in a gas impermeable plastic and subsequently flooded with CO_2 will provide an absolute sample for the contained community of arthropods (if not embedded within the plant) (see Figure 8).[24] In some instances the use of a systemic insecticide will kill phytophagous arthropods for eventual dislodging and counting. The use of chemical knockdown may be selectively employed for specific plant organs so long as the desired sampling area is contained to prevent escape of the target species (or escape of the gas if used). *Fumigation* of animal hosts is a viable sampling

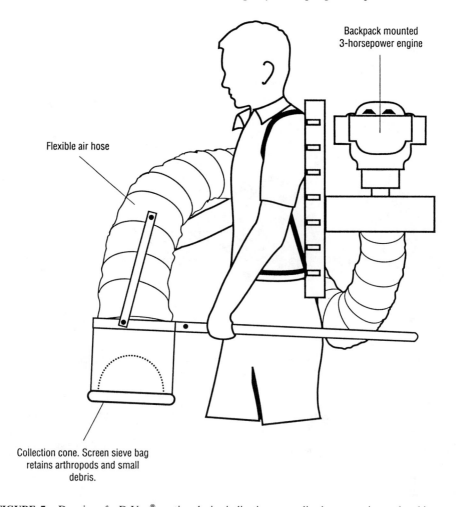

FIGURE 7. Drawing of a D-Vac® suction device indicating generalized construction and architecture.

tactic to dislodge ectoparasites for counting.[25] For internal parasites (bot maggots), certain ingested chemicals will allow for dislodging and eventual voiding. If a known number of hosts are treated, the overall infestation level can be estimated to determine whether or not the entire host population should be treated.

C. NETTING TECHNIQUES

Techniques focused on netting of arthropods are employed in many sampling situations because of their ease of use and relative low cost. In many ways, nets are utilized as collection instruments for techniques that actually utilize a pattern of systematic jarring. In agronomic crops such as alfalfa and soybeans, repeated pendulum movements of the net through the foliage canopy essentially dislodges arthropod species for collection.[17] Once any foliage is removed from the net, the remaining arthropods can be counted and recorded (see Figure 9). Although the sweep net collection technique does have many advantages, it also can be somewhat variable in efficiency from one individual to the next. In addition, care should be taken to calibrate this sampling technique with absolute population collection effectiveness in order to make accurate IPM decisions. For example, because the EIL typically is

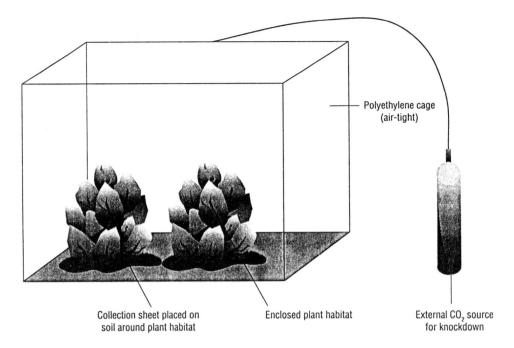

FIGURE 8. Drawing of polyethylene cage over a specified area of habitat with associated infusion of CO_2 for knockdown of arthropods.

based on arthropod intensity or density, some conversion or accounting of arthropods per "sweep" must be made.[17]

In addition to the sweep net, a *vacuum net* attached to a suction device also represents a netting technique. Because the function of the suction is critical to dislodge the arthropods or retard their escape, this method was discussed previously with knockdown techniques.

Netting methods that quickly cover a known area or unit of habitat are efficient but have the tradeoff of complexity and time for set up and collection. An example of the *covering method* includes the preparation of a net either at the base of a canopy or suspended above the canopy prior to sampling. After the arthropod community has stabilized from the positioning of the net, an operator quickly raises (or lowers) the net for capture.[26] Chemical or other dislodging techniques can be employed to separate the arthropods from the sampled foliage.

Aerial nets may be utilized for surveying arthropods at various distances above the ground. These techniques may include attaching two nets to fixed-wing aircraft, helicopters, weather balloons, or other devices which may be particularly useful in monitoring the progress of migratory species. For estimation of absolute populations, aerial netting techniques should include a determination of air volume.[7]

D. TRAPPING TECHNIQUES

Several variations of traps have been developed as a means to capture and hold arthropods when left unattended. In order for trapping to be considered as a viable sampling technique, therefore, the target species must be mobile.[10]

In cases where there is active movement of arthropods *toward* the trap, some form of directional stimulus or taxis is necessary. Predominant examples of attractants include visual, bait, and pheromone stimulants (see Figure 10). *Visual traps* include

FIGURE 9. Photograph of netting technique using a sweep net and pendulum sweeps. Populations will be determined as number of arthropods per sweep.

attraction with short wavelength light (e.g., blacklights) known to generally attract flying arthropods, especially moths. "New Jersey" traps combine the attractive stimulus of CO_2 with light to collect adult mosquitos.[10] Other types of visual traps depend upon the physical shape for attraction. The Manitoba trap, for example, attracts horse flies by suspending a black or red sphere under a collection cone. A visual stimulus frequently is part of a trap design, even if the primary taxis is not visual in nature (see Figure 11). Missouri cutworm traps utilize wheat seedlings and wheat bran to attract moths for oviposition, but a vertical screen has been shown to enhance the attraction by moths.

Baits frequently are used in association with a trapping device and provide a chemotaxis stimulus. An example introduced earlier is the olfaction associated with food substances attractive to the target species. The use of bone meal and molasses has been shown to attract ovipositing seedcorn maggot flies. Although sampling for IPM purposes may not be the intended use, planting susceptible varieties of crops will attract certain arthropod species for collection and redistribution, or even for control as a trap crop.

84 *Handbook of Sampling Methods for Arthropods in Agriculture*

FIGURE 10. Photograph of blacklight trap used primarily for collecting moths and other flying arthropods.

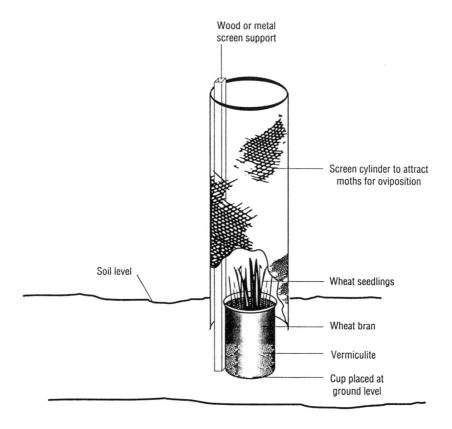

FIGURE 11. Drawing of Missouri cutworm trap using both visual (vertical screen) and bait (wheat seedlings) stimuli to attract moth oviposition for sampling.

Traps that utilize the instinct of reproduction or mating as an attractant typically incorporate actual or synthetic chemical *sex pheromones* (see Figure 12). These substances generally are economical to use as part of an attractant tactic but only attract either the male or female of the species, depending upon the specific pheromone. One advantage is the extreme specificity for target species, which makes counting relatively easy. Pheromone traps have been particularly useful as a tool for detection and survey of the introduction of exotic species[5,10] such as the Mediterranean fruit fly and Asian gypsy moth.

Traps that do not include a stimulus for active movement rely on collection through random interception. Malaise traps, for example, collect arthropods after flying species strike a sheet and instinctively move vertically up to a collection device. Other types of *interception traps* include sticky traps that rely upon contact to retain arthropods. Many times, however, sticky traps are not truly passive as they frequently are colored to provide a visual stimulus (either directional or omnidirectional). Other types of passive interception traps include window traps that rely on collision with a transparent surface, resulting in flying arthropods falling in a solution (e.g., soapy water) for counting.[7,12]

FIGURE 12. Photograph of boll weevil pheromone trap used to monitor adult populations on an area-wide basis.

Emergence traps are effective at capturing adults from a medium (e.g., soil) where pupae are located (see Figure 13). These traps consist of placing a mesh screen above a known area of soil and waiting for emergence of mobile adults. Counts can be related to previous larval populations and provide an indication of the subsequent larval population if applicable. Examples of where emergence traps have been used include seedcorn maggot (*Delia platura* [Meigen]) adult flies and corn rootworm (*Diabrotica* spp.) adult beetles.

Suction traps work by "pulling" flying arthropods that pass over the suction area into a collection device using a mechanical fan (see Figure 14). Although these types of traps do not attract or visually stimulate the arthropods, they do actively disrupt the atmosphere and therefore could be considered as a special knockdown technique. Regardless, suction traps are important for sampling small migratory arthropods such as certain species of aphids and leafhoppers.

Traps that capture ground-dwelling arthropods are referred to as *pitfall traps* (see Figure 15). Although several designs exist, the basic approach utilizes funnels buried such that the rim is flush with the soil surface. As ground arthropods (e.g., carbid beetles) forage, they "slip" through the funnel into a solution (e.g., ethylene glycol) for capture. Pitfall traps are nonselective so several mature and immature stages of arthropods will be collected.

III. TECHNIQUES FOR SOIL ARTHROPODS

Sampling of arthropods within the soil profile generally is more difficult than sampling surface or above-ground arthropods due to the density of the soil environment.[7] Indeed, the efficiency for removing arthropods varies with the target species (or community of species), the soil texture (e.g., clay vs. sand), and the water/organic matter content of the sample. In spite of the potential difficulties, design and use of

FIGURE 13. Photograph of pecan weevil emergence traps used to assess adult populations.

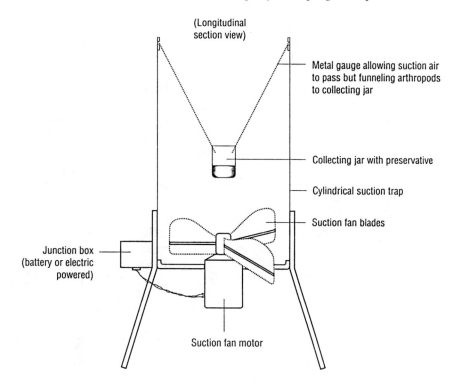

FIGURE 14. Drawing of aerial suction trap used to assess populations of migrating arthropods within a specific volume of air.

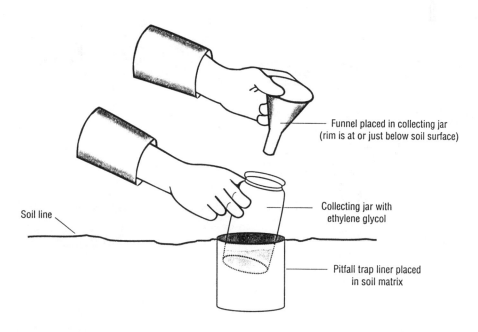

FIGURE 15. Drawing of pitfall trap used to assess populations of arthropods on the soil surface.

soil-sampling techniques is critical to making accurate IPM decisions for pests affecting the root systems of plants. The importance of assessing soil populations is strengthened further by the fact that over 80% of all species utilize the soil environment for at least one of their developmental stages.[10]

Sampling of soil arthropods generally includes two phases: *soil habitat collection* and *arthropod extraction*. Collection of soil samples typically is accomplished with a core cutter (several versions), shovel, trowel, or similar device. For situations where arthropods may be injured or killed by compaction as the soil is pushed through and removed from the extraction device, a corer, modified to disassemble, may be used. These *split corers* minimize compression and also allow for separation of the soil layers (e.g., humus, topsoil) for sampling within the soil profile (see Figure 16).[27] In situations where the soil habitat is not conducive to manually operated corers or where the volume of sampled soil is large, trenching or soil removal with mechanical devices might be beneficial. From a pragmatic IPM sampling standpoint, however, use of mechanized equipment for soil extraction is not common.

Arthropod extraction procedures generally are categorized as either *dry extraction* or *wet extraction* techniques. To manage the many sources of soil-sampling variability, however, situation-specific adaptations of these basic techniques frequently are developed. Although adaptations are necessary to account for the fact that no single

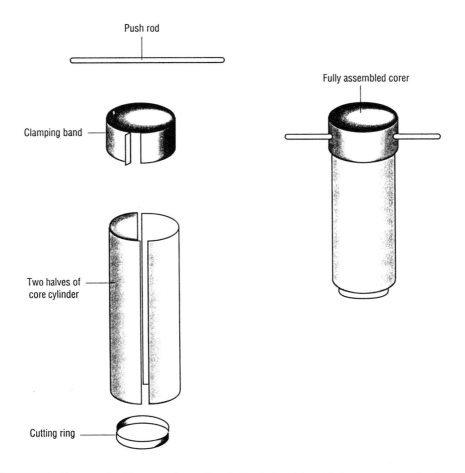

FIGURE 16. Drawing of split corer soil sampling device designed to minimize compaction of arthropods contained within the sample.

technique will provide absolute population estimates under all edaphic conditions, a complete listing or accounting of the adaptations is not possible here.

A. DRY EXTRACTION TECHNIQUES

One of the most basic forms of arthropod extraction from the soil is through *dry sieving*. This approach has one or more sieves positioned for sequential separation of particle sizes until the target arthropods can be collected together and counted. Practically, however, dry sieving has many drawbacks for efficient IPM sampling. Specifically, much time typically is needed to hand sort the sieved material and "break up" soil aggregates (especially with clay soil types). Dry sieving does provide viable techniques, however, for collecting and counting immobile arthropod stages (e.g., mosquito larvae) under some edaphic conditions.[28]

The majority of *dry* arthropod extraction techniques rely upon a stimulus to produce active and directed movement of the sampled species to a collection point (see Figure 17). The most basic and common techniques include *Berlese* (and Berlese-like) directed repulsion and collection methods.[29,30] Although many variations of Berlese extraction devices exist, the simplest type has a funnel holding the soil sample under a light or heat lamp, which repulses (with encroaching soil desiccation) the arthropods lower into the funnel and, eventually, into a collection device.

The advantage of Berlese techniques for IPM can be found in the fact that because arthropods find their own way to the collection point, soil samples generally can be left unattended. One disadvantage is the fact that migratory movement is necessary within the soil sample, precluding the possibility of eggs and other immobile stages from being sampled. Another disadvantage is the possibility of variability in efficiency depending on the climate, moisture content, and well being of the arthropods. Also, soil samples may not be stored via freezing, which requires enough Berlese devices be available to handle current sample requirements.

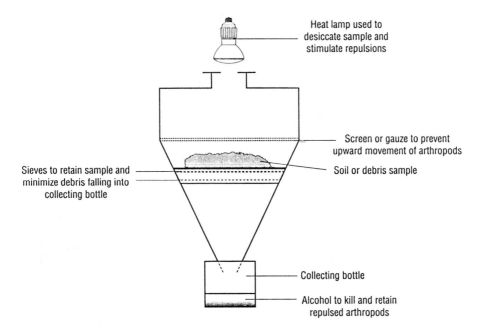

FIGURE 17. Drawing of *Berlese*-type funnel used to repulse soil arthropods from a soil sample and into a collecting bottle. Arthropods migrate away from the encroaching desiccation created by the heat lamp.

B. WET EXTRACTION TECHNIQUES

Equally basic to the dry sieving techniques presented previously is the *wet sieving* technique. Despite the need to have access to pressurized water, wet sieving generally has been found to be an efficient method to mechanically collect arthropods from soil. The action of the water usually makes soil particles move through the sieve(s) easier and more efficiently. As with dry sieving, many sizes of sieves may be used in sequence based on the size of the target species. Wet sieving is a viable sampling technique when the target arthropods and the soil particles are of distinctly different sizes. Frequently, however, wet sieving is used as one extraction phase in conjunction with one or more other wet extraction techniques.

Flotation is a viable wet extraction method to mechanically separate target arthropods from their soil substratum based on differential specific gravity. Biotic substances tend to be of lower specific gravity than mineral substances, so, in the absence of extraneous biotic debris, the arthropod life stages tend to float for easy collection and counting (see Figure 18).[7] Corn rootworm larvae will readily float to a top of a separatory device when the soil sample is mixed with water and allowed to stand.[10] Frequently, a salt solution is used to accentuate the differences in specific gravity among biotic and abiotic substances. Mosquito eggs, for example, can be more easily separated from soil through flotation in a magnesium sulfate solution.[31] Collection of corn rootworm eggs from soil samples combines wet sieving and flotation methods.[10] Here, field-collected soil samples are initially put in solution and wet sieved for collection of eggs and similar size soil particles. After the sieved material is placed in a separatory funnel with magnesium sulfate solution, the rootworm eggs float to the

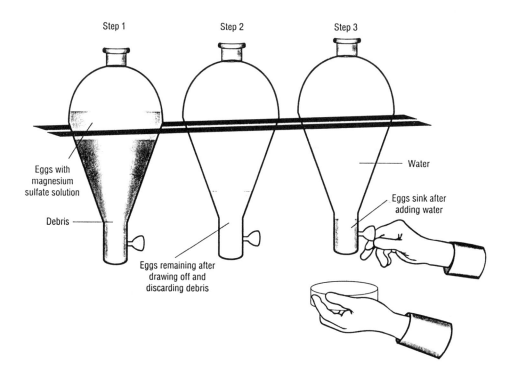

FIGURE 18. Drawing of corn rootworm egg flotation device used to determine egg density in a soil sample. Stages include (1) flotation of eggs in magnesium sulfate solution, (2) removal of soil particles through funnel stopcock, and (3) inversion of eggs with water in funnel column and subsequent drawing off of eggs into petri dish through stopcock.

top of the funnel, allowing soil particles to be drawn off and discarded. Adding freshwater to the funnel inverts the specific gravity relationship of the eggs such that they can be drawn off and counted.

Other mechanical techniques for wet extraction include *centrifugation*, *sedimentation*, and *elutrification*. These techniques build on the basic premises of sieving and flotation to collect very small arthropods, or, more frequently, other soil organisms. From an IPM perspective, these techniques are valuable for collection and identification of various species of plant-parasitic nematodes, but they generally find limited applicability to arthropod sampling.

Sampling techniques that rely on active movement of arthropods from their soil habitat also have been developed. The *Baermann funnel*, for example, submerges a soil sample wrapped in muslin within the top of a separatory funnel. Nematodes and small arthropod larvae may evacuate the muslin and fall to the bottom of the funnel for collection.[32] Wet extraction techniques that require migration generally are not viable for sampling arthropods because of their need for atmospheric oxygen.

Many adaptations of both dry and wet soil extraction techniques have been developed to meet specific sampling needs. When developing a specific application, consideration of the substratum type (e.g., clay vs. sand), target arthropod species and life stage (e.g., mobility, size), and cost (e.g., value of sample information vs. cost of obtaining it) all are critical considerations. At present, because of the time and cost involved with sampling soil arthropods, techniques have been designed primarily for research purposes, with only limited wide-scale use in IPM programs.

IV. TECHNIQUES FOR INDIRECT ASSESSMENT OF ARTHROPOD POPULATIONS

When techniques for direct sampling of arthropods either are not available or not feasible, indirect methods of assessing arthropod populations may be utilized. Techniques for indirect assessment can be considered in two primary categories, measuring arthropod-induced injury and measuring arthropod products, and a third category that recently has begun development, acoustic imaging.

A. ARTHROPOD-INDUCED INJURY

Often the emphasis of IPM is too centered on pests rather than the injury and resultant damage that they cause.[4] Because IPM should focus on maintaining host injury below levels that result in economic damage, an assessment of arthropod-induced injury may be an effective means to determine the injuriousness (vs. density) of a pest population. If the purpose of the sample is to determine the need for an arthropod suppression tactic, a direct accounting of the existing population also should be made in order to assure the pest is still contributing to host damage. This is particularly important in circumstances where a pest population is prone to natural control and may "crash" in population size. With this proviso, however, assessing injury directly has many applications in practical IPM.

Assessment of corn root pruning is an example of where indirect assessment of corn rootworm larvae has been helpful (the impracticality of sampling larvae in the soil was addressed previously) (see Figure 19). Unfortunately, the assessments must be made after control decisions have been made and therefore do not lead to curative IPM decision making (see Figure 20). Another example of indirect assessment is the determination of the number of damaged terminals (%) and damaged squares (buds; %) as an indication of corn earworm-induced injury to cotton. In this case, the indirect assessment provides information on current feeding status while a direct

FIGURE 19. Photograph of root injury with associated root rating index (Iowa 16-scale). The index of root injury serves as an indirect measurement of arthropod presence.

FIGURE 20. Photograph of boll weevil-damaged square showing frass "plug" where the adult weevil oviposited. The number of damaged squares indicate the severity of infestation.

FIGURE 21. Photograph of alfalfa leaf showing inverted chlorotic "V" typical of potato leafhopper-induced injury.

assessment of "number of live larvae" provides data on current population intensity/density (see Figure 21). The combined plant and arthropod assessments likely will lead to a more accurate control decision. Sampling techniques for soybean foliar pests also rely upon both arthropod density/intensity as well as evidence of host injury in the form of "percent defoliation". Control recommendations use both inputs to base recommendations (see Figure 22).

In each of the above examples a population or damage index has been created to reflect the size of a pest population. These indices prove very effective as a means to make quick assessments of arthropod damage potential, but, as mentioned previously, should not be used solely as the basis for making control decisions.

B. ARTHROPOD PRODUCTS

When arthropods are concealed or otherwise difficult to sample, often a determination of their presence can be made by observing by-products of their feeding or inhabitation. Identifying the number of "spittles" in a given area of alfalfa, for example, provides an indication of population size without actually counting the individual spittle bugs.[33] Identifying wood shavings or frass of wood-boring insects is an effective way to detect their presence within a tree or structure (see Figure 23).

Many times arthropod products will provide a technique for detection, but will not necessarily confirm that the population currently exists. As with assessments of arthropod-induced injury, therefore, some accounting of actual presence must be made prior to determining whether or not management is necessary.

C. ARTHROPOD ACOUSTIC SIGNATURES

Electronic technology has, in recent years, advanced to the point where remote sensing and sampling of arthropods via acoustic imaging may be possible. Currently,

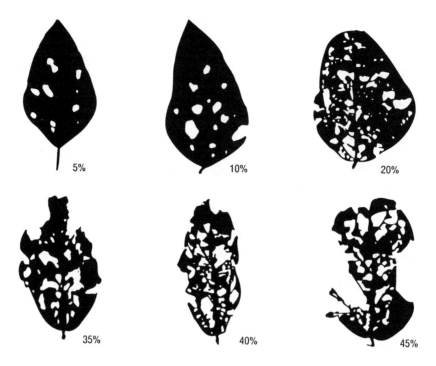

FIGURE 22. Drawing of soybean leaves showing different levels of defoliation as an indirect measure of lepidopteran larval feeding.

FIGURE 23. Photo of dry wood termite frass as evidence of active infestation using an arthropod product.

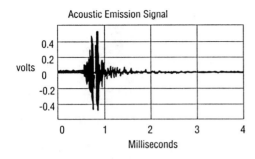 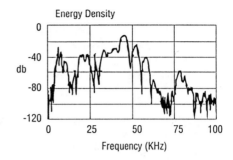

FIGURE 24. Drawing of acoustic frequency for mandibular movements of subterranean termites actively feeding.

acoustic sensing devices that positively identify the frequency and wavelength of mandibular feeding are being developed for stored grain pests as well as pests of structures (e.g., termites). The precision and practicality of such devices currently is under investigation, but evidence suggests that a new capability for remote species-specific sampling will be possible.

Once the electronic algorithms are sorted out for individual species such that acoustic signatures are identified for critical biological activities of arthropods (e.g., chewing, flying, crawling), the possibilities for use are widespread (see Figure 24). Consider the possibility of remotely sampling mosquito populations based on wing beat signatures to determine abatement options, remotely sampling European corn borer larval populations in corn based on mandibular chewing signatures in corn stalks, remotely sampling arthropod communities to develop life tables without disturbing the experimental area. Although all these applications are many years away, the further progress of electronic engineering certainly will accelerate development.

V. CONCLUSIONS AND FUTURE DIRECTIONS

Development of effective and efficient arthropod sampling techniques is paramount to the success of an IPM program. Although many techniques have been discussed here, each has or could have dozens of modifications consistent with the sampling objective. The goal in designing a specific application for area-wide use should be to build in as much accuracy and precision of sampling balanced with the cost and labor involved to make the technique a realistic alternative.

The future of sampling technology development will continue to evolve in directions consistent with the needs of specific IPM programs. Two trends, however, will affect development in significant ways. First, as referenced previously, electronic technology will alter the paradigm of sampling and move us toward remote sensing as a pragmatic and more efficient alternative to current manual methods. Second, sampling will begin to account for and make provisions to measure the influence of pest complexes rather than just single species. Guilds of arthropods, established based on their physiological mode of injury to the host, will be managed collectively. This change in management philosophy, therefore, will necessitate the development of sampling techniques that collect information on arthropod populations which affect a common host physiological function.

Development of sampling techniques based upon a clear understanding of insect bionomics, host interaction, and management goals will continue to be critical aspects for developing an IPM program.

REFERENCES

1. Stern, V. M., Bosch, R., and Hagen, K. S., The integrated control concept, *Hilgardia*, 29, 81, 1959.
2. Pedigo, L. P., Hutchins, S. H., and Higley, L. G., Economic-injury levels in theory and practice, *Ann. Rev. Entomol.*, 31, 341, 1986.
3. Stone, J. D. and Pedigo, L. P., Development and economic-injury level of the green cloverworm on soybean in Iowa, *J. Econ. Entomol.*, 65, 197, 1972.
4. Hutchins, S. H., Higley, L. G., and Pedigo, L. P., Injury equivalency as a basis for developing multiple-species economic-injury levels, *J. Econ. Entomol.*, 81, 1, 1988.
5. Metcalf, C. L., Flint, W. P., Metcalf, R. L., *Destructive and Useful Insects*, 4th ed., McGraw-Hill, New York, 1962.
6. Hutchins, S. H. and Gehring, P. J., A perspective on the value, regulation, and objective utilization of pest control technology, *Am. Entomol.*, 39: 12–15.
7. Southwood, T. R. E., *Ecological Methods with Special Reference to the Study of Insect Populations*, 2nd ed., Chapman & Hall, New York, 1978.
8. Richards, O. W. and Waloff, N., Studies on the biology and population dynamics of British grasshoppers, *Anti-Locust Bull.*, 17, 1954.
9. Balogh, J. and Loksa, I., Untersuchungen Uber die Zoozonose des Luzernenfeldes. Strukurzonologische Abhandlung, *Acta. Zool. Hung.*, 2, 1956.
10. Pedigo, L. P., *Entomology and Pest Management*, Macmillan, New York, 1989, chap. 6.
11. Berberet, R. C. and McNew, R. W., Reduction in yield and quality of leaf and stem components of alfalfa forage due to damage by larvae of *Hypera postica* (Coleoptera: Curculionidae), *J. Econ. Entomol.*, 79, 212, 1986.
12. Metcalf, R. L. and Luckmann, W. H., *Introduction to Insect Pest Management*, John Wiley & Sons, New York, 1982.
13. Wildholtz, T., Vogel, W., Straub, A., and Gesler, B., Befallskontrolle an apfelbaumen im Fruhjahr, *Schwiez. Z. Obst. Weinb.*, 65, 85, 1955.
14. Moore, N. W., Intra- and interspecific competition among dragonflies (Odonata), *J. Anim. Ecol.*, 33, 49, 1964.
15. Buxton, D. R., Hornstein, J. S., Wedin, W. F., and Marten, G. C., Forage quality in stratified canopies of alfalfa, birdsfoot trefoil, and red clover, *Crop Sci.*, 25, 273, 1985.
16. Hutchins, S. H. and Pitre, H. N., Differential mortality response of Lepidopteran defoliators to insecticides deposited within three strata of wide- and narrow-row soybean, *J. Econ. Entomol.*, 80, 1244, 1987.
17. Kogan, M. and Herzog, D. C., Eds., *Sampling Methods in Soybean Entomology*, Springer-Verlag, New York, 1980.
18. Chant, D. A., A brushing method for collecting mites and small insects from leaves, *Prog. Soil Zool.*, 1, 222, 1962.
19. Taylor, E. A. and Smith, F. F., Three methods for extracting thrips and other insects from rose flowers, *J. Econ. Entomol.*, 48, 767, 1955.
20. Strickland, A. H., An aphid counting grid, *Plant Pathol.*, 3, 73, 1954.
21. Dietrick, E. J., Schlinger, E. I., and Bosch, R., A new method for sampling arthropods using a suction collecting machine and modified Berlese funnel separator, *J. Econ. Entomol.*, 52, 1085, 1959.
22. Hutchins, S. H., Buxton, D. R., and Pedigo, L. P., Forage quality of alfalfa as affected by potato leafhopper feeding, *Crop Sci.*, 29, 1541, 1989.
23. Gibbs, D. G., Pickett, A. D., and Leston, D., Seasonal population changes in cocoa capsids (Hemiptera: Miridae) in Ghana, *Bull. Entomol. Res.*, 58, 279, 1968.
24. Dempster, J. P., A sampler for estimating populations of active insects upon vegetation, *J. Anim. Ecol.*, 30, 425, 1961.
25. Janion, S. M., Quantitative dynamics in fleas (Aphaniptera) infesting mice of Puszcza Kampinoska Forest, *Bull. Acad. Pol. Sci. II*, 8, 213, 1960.
26. Turnbull, A. L. and Nicholls, C. F., A "quick trap" for area sampling or arthropods in grassland communities, *J. Econ. Entomol.*, 59, 1100, 1966.
27. O'Conner, F. B., An ecological study of the Enchytraeid worm population of a coniferous forest soil, *Oikos*, 8, 162, 1957.
28. Stage, H. H., Gjullin, C. M., and Yates, W. W., Mosquitoes of the Northwestern States, *Agric. Handb.*, *U.S.D.A.*, 46, 1952.

29. Macfadyen, A., A comparison of methods for extracting soil arthropods, in *Soil Zoology*, Kevan, D. K. McE., Ed., University of Nottingham School of Agriculture, Nottingham, England, 1955, 315.
30. Macfadyen, A., Soil arthropod sampling, *Adv. Ecol. Res.*, 1, 1, 1962.
31. Iversen, T. M., The ecology of the mosquito population (*Aedes communis*) in a temporary pool in a Danish beech wood, *Arch. Hydrobiol.*, 69, 309, 1971.
32. Peters, B. G., A note on simple methods of recovering nematodes from soil, in *Soil Zoology*, Kevan, D. K. McE., Ed., University of Nottingham School of Agriculture, Nottingham, England, 1955, 373.
33. Parman, V. R. and Wilson, M. C., Alfalfa crop responses to feeding by the meadow spittlebug (Homoptera: Cercopidae), *J. Econ. Entomol.*, 75, 481, 1982.

Chapter 6

DEVELOPING A PRIMARY SAMPLING PROGRAM

G. David Buntin

TABLE OF CONTENTS

I. Introduction 100
   A. Definitions and Terminology 100
   B. Objectives and Types of Sampling Programs 101
   C. The Sampling Universe 102

II. Preliminary Sampling 102
   A. Selection of a Sample Unit 102
   B. Comparison of Sampling Techniques 104
   C. Population Dispersion 106

III. The Primary Sampling Program 107
   A. Pattern and Timing of Sampling 107
   B. Stratification of the Sampling Universe 108
   C. Number of Sample Units 109

IV. Program Evaluation and Improvement 113

References 113

ISBN 0-8493-2923-X/94/$0.00 + .50
© 1994 by CRC Press, Inc.

I. INTRODUCTION

Sampling is the cornerstone of integrated pest management (IPM) because IPM requires current information concerning pest status to make timely decisions about management activities.[1] Sampling information traditionally is used in IPM to make decisions about therapeutic or curative actions such as use of conventional insecticides to prevent pests from causing significant damage. Reliable and practical sampling programs are essential for the effective utilization of control tactics in IPM systems.

Sampling also is an essential activity in entomological and ecological research because of the need for accurate and precise estimates of species abundance. A well-developed sampling protocol is indispensable in population ecology where an understanding of population dynamics is desired. The importance of accurate and reliable sampling procedures was recognized in a number of early studies in insect population ecology.[2,3]

A. DEFINITIONS AND TERMINOLOGY

Pedigo defined a number of specific sampling terms in Chapter 1 of this book. He defines a sampling program as "the procedure for employing the sampling technique to obtain a sample and make an estimate". Sampling programs delineate the sampling universe, sampling technique, sample unit size, number of sample units, pattern of collecting sample units, and timing of samples. A sample unit is the proportion of the habitable space from which arthropods are counted, and the sampling technique is the method by which information is collected from a sample unit.[4] The type of population estimate desired dictates the type of sampling method that can be used.[2] Morris[3] categorized estimates as being absolute and related estimates or relative. Absolute estimates describe arthropod density on a basis of unit surface land area and represent a census of arthropods in each sample unit. Population intensity and basic population estimates describe estimates according to available food supply or as a standard unit of habitable space, respectively, and usually can be easily converted to a unit area measurement. However, relative estimates are referenced to the sampling technique and are not directly related to land surface area.

A number of statistical characteristics of population estimates should be considered while developing a sampling program. These include:

- *Accuracy*—the extent to which an estimated mean deviates from the true population mean.
- *Bias*—the degree of systematic error in estimating the mean.[5]
- *Efficiency*—the level of precision per unit of cost of sampling effort.
- *Fidelity*—the accuracy with which population estimates over time reflect actual changes in population numbers.
- *Precision*—the degree of statistical error in making estimates of population number.

Accuracy is of the utmost importance when attempting to estimate the absolute or true mean density. Indeed, the concept of accuracy in a statistical sense becomes almost meaningless with relative techniques when population estimates are referenced to the technique. Accuracy of relative techniques often refers to the correct identification of the target species and correct enumeration of the target species in each sample unit.[6] In IPM programs, fidelity rather than statistical accuracy probably is a more important issue. Decision rules can be expressed in relative terms as long as the relative scale accurately reflects true population changes. The term "reliability" often is used to describe precision.[7,8] It is desirable to minimize random sampling error to improve sampling precision, but the tradeoff of increased cost and effort must be weighed against the benefit of improved precision. However, biological error that causes systematic bias by underrepresenting or sometimes overrepresenting part of the population can undermine the proficiency of a sampling program, particularly if the bias changes during and between sampling times. Although the definition of sampling efficiency is limited to sampling cost effectiveness, the term also is used in a more general sense to include the accuracy and precision of sampling procedures.[9]

B. OBJECTIVES AND TYPES OF SAMPLING PROGRAMS

To develop a usable and effective sampling program, the objectives of sampling must be clearly defined because they will largely determine the methods used.[2] Two issues that must be addressed are how fine a delineation of population change is desired and what will be the intensity of sampling effort in each habitat.[2] In other words, is an intensive effort in few habitats desired or should sampling effort in each habitat be limited so that many habitats in a wide geographical area can be sampled? Southwood[2] terms these two types of sampling programs as extensive and intensive, respectively. Extensive programs limit sampling activity in each habitat so that many habitats can be sampled over a large geographic area to provide information on species occurrence or prevalence. Intensive sampling programs often are designed for population ecology research where many samples are collected frequently from a few habitats so that mean density can be estimated with a great degree of precision.

Sampling programs can be classified as essentially having three general objectives. These are (1) simply detecting presence of a target species, (2) providing information on the status of a target species (pest) which would include most IPM programs, and (3) providing accurate density estimates with a high level of precision. Categories (1) and (3) are clearly extensive and intensive programs, respectively, as defined by Southwood,[2] with category (2) being more extensive than intensive.

Programs designed to detect a target species are used to prevent the spread of undesirable species into new areas or habitats and to monitor the spread of exotic species once established in new areas. Detection sampling is used in quarantine efforts to prevent exotic species from being imported into new geographic areas.[9] Sampling efforts often are concentrated on habitats where the target species is most likely to occur such as the preferred host plant.[2,4] Detection sampling also is widely used in IPM to prevent introduction of infested materials into new habitats. Examples would include inspection of grain for grain beetles to prevent contamination of uninfested grain or sampling of plants to prevent introduction of infested plants into an uninfested greenhouse. Detection sampling in IPM often relies on visual inspection for detecting infestations and on relative sampling techniques such as pheromone traps or sticky cards that actively attract mobile stages of the target species.[9] Susceptible indicator plants also are particularly useful for the detection of pathogenic diseases.

Sampling to determine species or pest status is widely used in IPM. Here, the objective is not necessarily the estimation of mean density. Instead, we may simply wish to know if the population is above or below one or more critical densities such as an economic threshold (i.e., the density at which control measures should be implemented to prevent economic damage from occurring)[10] so that the population can be classified. This type of sampling is designed to direct real-time decisions concerning management activities. Species status sampling usually uses techniques that provide relative or sometimes population intensity estimates. Depending on a species' biology, sampling may or may not focus on the damaging stage(s) of the target species. Typically, it may be much easier to sample the mobile adult stage that is attracted to traps or easily observed in the field than to directly sample damaging immature stages that feed inside plants or inhabit soil. For example, larvae of corn rootworms (*Diabrotica* spp.) feed on corn roots in soil and are very difficult to sample.[11] Consequently, IPM programs in corn use the results of sampling adults, which can be easily sampled by direct counts on corn ears to make decisions concerning control measures.[11,12] Emphasis on this type of sampling program is placed on minimizing sampling complexity, costs, and effort so that many habitats can be sampled inexpensively in a short period. This causes a sacrifice in sampling precision; consequently, an important consideration in developing a sampling program usually is minimizing the risk of not controlling a damaging pest population. This is usually accomplished by using a conservative decision rule.

The objective of intensive sampling programs is to provide mean density estimates with a high degree of precision so that small differences or changes in population density can be measured. Sampling often is confined to a few, small homogeneous habitats and occurs frequently enough so that a species cannot pass through a developmental stage without being sampled.[2,13] Techniques used typically provide absolute or population intensity estimates that can be easily converted to a unit area basis. Furthermore, several techniques may be utilized so that most or all the life stages of a species are accurately sampled. Unfortunately, relative techniques such as the sweep net often are used in agricultural research when more accurate techniques really are needed to achieve the sampling objectives. Clearly sweep netting would not be an appropriate technique for a life table study where population estimates per unit area are needed. Morris,[3] Harcourt,[14] and Southwood[2] provide detailed discussion of many aspects of developing intensive sampling programs.

C. THE SAMPLING UNIVERSE

Consideration of the sampling universe is important in determining the appropriate sample unit and technique. Although the sampling universe in statistics refers to the whole population from which samples are collected,[7] the sampling universe in field biology and agriculture usually represents the habitat in which the population occurs and from which samples are collected.[3] In pest management systems where sampling and therapeutic measures are done on a field-by-field basis, the plants within a field or some portion of the field become the logical sampling universe. If initial studies reveal that the target species occurs in a portion of the habitat, then the sampling universe can be restricted to the portion of the habitat where the species is most likely to occur. For example, if eggs of a species are laid almost entirely on the buds of a host plant, then the sampling can be restricted to the buds. Restricting the sampling universe usually reduces sampling variability and cost.

II. PRELIMINARY SAMPLING

A. SELECTION OF A SAMPLE UNIT

A sample unit can be defined as the proportion of habitable space from which arthropod counts are taken.[3,4] A sampling technique is the method by which information is collected from a single sample unit. Although the sample unit and sampling technique are distinct attributes of a sampling program, both are intimately related because the sampling technique often dictates the form of the sample unit. For example, a single plant or tiller in a plant stand cannot be easily sampled with a sweep net.

Selection of a sample unit and sampling technique depends on the objective of the sampling effort and the available resources for sampling. A sample unit should be representative of the target arthropod and the sample universe, and sample units must be distinct and not overlap.[3,13] Morris[3] outlined a number of criteria for selecting a sample unit:

1. All units of the universe should have an equal chance of selection (see Section III.A).
2. The unit should be stable and easily delineated in the field (if it is not stable changes should be easily documented such as changes over time in plant number per unit area or leaves per plant).
3. The proportion of the insect population using the sample unit must remain constant at least within each sample period.
4. The sample unit size should provide a reasonable balance between cost and variance (i.e., precision).
5. The sampling unit should be scaled to the arthropod's size, mobility, and relative abundance. (If the unit is too small this increases edge effect bias, or if the target species is uncommon an unacceptably large proportion of the sample unit may be empty, thereby providing limited information about species numbers. Conversely, if the sample unit is too large for the size and abundance of the species only a few units can be processed in a reasonable period of time.)
6. The sample unit should be convertible to units of area. This last criterion almost always is important for population research, but is not necessary for most extensive programs and for the purposes of pest management.

The size of a sample unit can have an important influence on the cost and precision of a sampling program. The optimal size of a sample unit can be calculated using the results of a nested analysis of variance where the variance within and between sample units is measured.[15] In this analysis, the sample unit is subdivided into a number of subunits and the variance between subunits (i.e., within sample unit) is measured. Examples would include single sucks with an insect vacuum machine, single leaves in a sample with a number of leaves, or a subdivided area or time measure such as 10 cm^2 in a 100-cm^2 sample unit or 1 min of a 5-min timed sample unit. Variance within (s_{ws}^2) and between (s_{bs}^2) sample units is estimated with a nested analysis of variance[15] where a reasonable number of sample units are collected and a representative set of fields is sampled. The optimal number of subdivisions of the sample unit (N_s) can be calculated by solving:[2,15]

$$N_s = \sqrt{(s_{bs}^2/s_{ws}^2)(c_{bs}/c_{ws})} \qquad (1)$$

where s_{ws}^2 and s_{bs}^2 are previously defined, c_{ws} is the sampling cost usually measured in human-minutes to examine one subunit, and c_{bs} is the cost of locating the next sampling unit.

Hein et al.[16] used this approach to determine the optimal number of soil cores to collect per sample site using a frame sample method to sample corn rootworm (*Diabrotica* spp.) eggs in soil in cornfields. They found that sampling was most efficient when one or two soil cores were collected per site with a large number of sites being collected per field. Bechinski and Pedigo[17] also used the results of a nested analysis of variance to calculate the optimal sample unit size to sample green cloverworm [*Plathypena scabra* (F.)] pupae in soybean. Pupae occur primarily on the soil surface, consequently, sampling consisted of inspecting a known area of the soil surface for pupae. These authors subdivided a 60 × 60 cm sample unit into six subunits measuring 10 × 60 cm, and found that five subunits or a sampling unit measuring 50 × 60 cm was the optimal compromise between precision and cost.

The optimal sample unit size also can be determined by comparing the relative efficiency of sample units of various sizes. Southwood[2] suggested that the variance (s_u^2) and costs (c_u) of the different sample units such as 1, 3, and 5 plants per unit or 5, 10, and 15 sweeps per unit could be determined from a preliminary sample. The variance and cost are divided by the smallest unit to reduce s_u^2 and c_u to a common basis. The relative cost of each sample unit is calculated as $(c_u)(s_u^2)$ with a higher value indicating greater precision for the same cost.

Pieters[18] compared the mean estimates and precision of different sample unit sizes for collecting arthropods in cotton using a D-Vac® insect vacuum machine and found that precision was similar for all sample unit sizes, but mean estimates increased as sample unit size decreased. He concluded that smaller sample unit sizes should be employed. A similar approach of selecting a sample unit size using relative net precision as a measure of efficiency has been used in a number of recent studies.[6,19-21] Calculation of relative net precision is discussed in the next section on sampling techniques. Zehnder et al.[20] determined the optimal sample unit size for sampling larvae and adults of the Colorado potato beetle [*Leptinotarsa decemlineata* (Say)] in potato by comparing the relative net precision of sample units consisting of one to five stems. They found that the five-stem sample unit was most efficient, but that a three- or four-stem unit would reduce costs with only a small reduction in sampling precision. Studies[6,19] on sampling potato leafhopper [*Empoasca fabae* (Harris)] in alfalfa using a sweep net found that a sample unit of ten sweeps was more efficient than larger sample unit sizes (20 and 25 sweeps per unit). Efficiency varied with mean density because accuracy of counting leafhoppers in larger sample units declined in the field at higher densities,[6] although larger sample units are useful for detecting leafhoppers at low densities.[22] One potential limitation of this approach for selecting sample unit size is that the range of unit sizes often are selected arbitrarily and may or may not represent the most efficient range of sample unit size.

B. COMPARISON OF SAMPLING TECHNIQUES

Selection of a sampling technique is determined by the objectives and resources of the sampling program. An intensive, expensive sampling technique usually is not appropriate for pest decision making, and likewise an inexpensive, yet imprecise relative technique that cannot be easily converted to a unit-area basis probably will not be suitable for intensive sampling programs. The selection of a sampling technique is conceptually distinct from selection of a sample unit, but usually the two decisions are interdependent and are determined concurrently. Unfortunately, choice of sampling technique often is based on arbitrary or subjective criteria.[23] An acceptable sampling technique should maximize precision while minimizing cost,[7] therefore, information on technique variability and cost is essential to select the best technique.

Estimates of absolute arthropod density per unit area are required to assess the accuracy and fidelity of other techniques and to calibrate relative techniques.[2,23] A truly absolute sampling technique should provide an accurate, unbiased measure of arthropod numbers per unit area. Absolute techniques in agricultural crops typically involve some sort of cage and fumigation procedure and removal of vegetation, followed by extraction of arthropods in the laboratory. These procedures which are described in Chapter 5 and elsewhere[2,24] often are difficult to obtain without sampling bias and are always labor intensive and therefore costly.

Other sampling techniques provide either population intensity estimates, basic population estimates, or relative population estimates.[3] Population intensity estimates express arthropod numbers in terms of the food supply such as number per leaf, number per fruit, or number per plant.[3] Basic population estimates express numbers in terms of a standard unit of habitat such as number per foot or meter of row or number per unit weight of plant material. Although these estimates are not absolute, they usually can be easily converted to a unit area of land. Relative sampling techniques provide estimates that are not directly related to land surface area.[3] Instead, the technique itself provides the reference for estimates such that counts are expressed as number per unit of the technique. The best example is the sweep net where counts are expressed as number per sweep. Other examples are pitfall traps, pheromone traps, light traps, and insect counts per unit of time. Relative sampling techniques are commonly used in pest management programs because of their low cost and minimal labor requirements. Specific sampling techniques are described in Chapter 5.

An essential step in the development of a primary sampling program is the comparison and evaluation of potential sampling techniques. Typically, this is done by randomly locating sites in a habitat such as a field and collecting one sample unit by each sampling technique being considered. Care should be taken so that one technique does not interfere or bias another technique. This process is repeated in a number of representative habitats or fields. Sometimes accuracy and precision of a technique may change with species population density[6,19] and habitat conditions.[25] Therefore, comparisons representing a range of target species density and habitat characteristics, such as plant size or maturity, should be conducted under the conditions in which the sampling will occur. The accuracy, precision, and efficiency of techniques can be evaluated using the count mean, variance (s^2), and cost estimates, where cost of each technique usually is measured as time (human-minutes). The procedures used to determine optimal sample unit size also are useful in evaluating sampling techniques. Technique accuracy can be evaluated by comparing means of each technique with an absolute density estimate. If several stages of the target arthropod are sampled, such as all larval stages, the accuracy of sampling each stage should be evaluated separately. For example, an insect vacuum machine may accurately sample small lepidopteran larvae, but not large larvae.[26] Furthermore, the accuracy of given technique may change with changes in the habitat such as occurs in the growth and development of annual field crops.[25] Consequently, technique accuracy should be evaluated during different growth periods. These comparisons with absolute population estimates will reveal habitat and species age biases in a less than absolute technique.

Precision is most commonly measured by relative variation (RV)[8] where variability is expressed as a percentage of the ratio of the standard error of the mean (SE) to the mean (m):

$$RV = (SE/m)(100) \qquad (2)$$

where $SE = s/\sqrt{n}$. Precision also can be expressed as the coefficient of variation (CV), which is the percentage of the ratio of the standard deviation (s) to the mean:

$$CV = (s/m)(100) \qquad (3)$$

Smaller RV and CV indicates greater precision.

Sampling efficiency refers to precision per unit cost and can be measured by relative net precision (RNP)[26] which is calculated as:

$$\text{RNP} = [1/(\text{RV}_m)(c_u)](100) \qquad (4)$$

where RV_m = mean relative variation calculated from a number of data sets and c_u is the cost in human-minutes or -hours to collect and process one sample unit. Greater RNP indicates better sampling efficiency. In this analysis, RNP as a measure of sampling efficiency implies that equal consideration be given to precision and cost. However, depending on the sampling objectives and resources, placing more emphasis on precision compared with cost or vice versa may be appropriate. For pest management programs, selecting a technique with substantially reduced costs may be worth a loss in precision as long as precision is sufficient to make correct decisions.

Comparison of m and RV between techniques can be made using an analysis of variance, with means being separated using least significant difference or some other appropriate multiple range test. Because mean RV is used to calculate RNP, statistical analysis of RNP may not be possible. Comparison of these values provides information for the selection of the most accurate and precise sampling technique for a given cost. One of the original applications of this approach compared techniques for sampling the green cloverworm [*Plathypena scabra* (F.)] in soybean.[26] Many studies too numerous to list have been conducted on comparing sampling techniques for arthropods of agricultural importance. Kogan and Herzog[24] provide an extensive review of techniques for sampling arthropods in soybean.

Caution, however, may be needed in selecting a sampling technique based solely upon sampling precision and efficiency. Simplicity and subjective social factors can be important considerations in the selection of a sampling technique that will be used primarily by growers and other nontechnical personnel (see Chapter 22). Techniques that selectively collect the target species generally are easier and less costly to process than similar nonselective techniques. This can be important in extensive programs because it reduces the complexity and level of training needed to use the technique. Furthermore, a program designed around a technique that the end users are unwilling or unable to use is doomed to failure. Although sweep netting is a cost-effective technique for sampling many insects in annual row crops, many growers are reluctant to use a sweep net because of the perceived unfavorable social image by the general public for insect netting activities. There is no easy solution to this problem.

C. POPULATION DISPERSION

Information about the biology and dispersion of the target arthropod is useful in determining the pattern and number of samples. Populations can be distributed in space as either a uniform (= regular, underdispersion), random, or aggregated (= clumped, contagious, overdispersed) pattern (Chapter 3, Figure 2).[2,27] Uniform dispersion patterns occur when individuals occur in a systematic pattern throughout a habitat, that is, they are spaced approximately equidistant. Trees planted in an orchard is an example of a uniform dispersion pattern. Uniform dispersion patterns almost never occur in arthropod population in the field.[27,28] A random dispersion occurs when there is an equal probability of an organism occupying a point in space or sample unit and that the presence of one individual does not influence the occurrence of another individual.[2] Aggregated dispersion patterns occur when occurrence of one individual increases the probability of occurrence of another individual in the same sample unit. Dispersion patterns of arthropod populations in agricultural settings commonly exhibit some degree of aggregation.[27,28]

A simple statistical measure of population dispersion is provided by the ratio of sample variance (s^2) to sample mean (m).[2,29] Uniform, random, and aggregated dispersion patterns are indicated when $s^2 < m$, $s^2 = m$, and $s^2 > m$, respectively. Basically two approaches have been used to statistically describe dispersion patterns of arthropods. The first is to compare observed frequency distributions with theoretical frequency distributions. The two most commonly used theoretical distributions are the Poisson and negative binomial distributions.[8] The Poisson series assumes that $s^2 = m$, and therefore effectively describes random dispersion patterns. The negative binomial distribution has been widely used to describe arthropod dispersion patterns because it often describes aggregated patterns well.[30-33] The negative binomial is based on two parameters, the mean (m) and an exponent k, which is a measure of the amount of clumping; k becomes smaller as the degree of clumping increases. The negative binomial also can describe a random distribution if k is large ($k > 10$).[8,30] Procedures for determining k are discussed by Bliss[30] and Southwood.[2] A number of other mathematical distributions also have been used to describe dispersion patterns of arthropods.[34]

A limitation of using theoretical distribution to describe population dispersion is that the "best" theoretical distribution can vary with population density and with changes in sample unit size.[29] Furthermore, a major limitation of the negative binomial is that k often varies with mean density. Consequently, more recently a number of aggregation indices have been developed for describing dispersion patterns using models based on the relationship between population variance and mean that are designed to be independent of mean density.[35-39]

The most useful index is Taylor's power law, which is based on a logarithmic relationship between variance and m such that:

$$s^2 = a(m)^b \qquad (5)$$

In practice a and b are estimated by linear regression of $\log_e(s^2)$ and $\log_e(m)$, where b is the index of aggregation. The value of b is independent of mean density and is constant for a species in a particular environment.[39] The a value varies with sample unit size[39] and location.[40] Uniform, random, and aggregated dispersion patterns are indicated when $b < 1$, $b = 1$, and $b > 1$, respectively. A t-test can be used to test whether b is significantly different than 1.0. Another useful index of aggregation is Iwao's regression procedure,[36] which assumes a quadratic relationship between s^2 and m, and regresses Lloyd's[41] mean crowding index (m^*) on m (see Chapter 3). Taylor's power law tends to provide a better fit to most data, particularly at a low range of mean densities.[29,42] The analysis of population dispersion is discussed in detail in Chapter 3.

III. THE PRIMARY SAMPLING PROGRAM

After selecting a sample unit and sampling technique, the next step in developing a sampling program is determining the pattern and timing of sampling and the optimal number of sample units.

A. PATTERN AND TIMING OF SAMPLING

A basic tenant of all sampling is that samples are collected randomly so that every sampling unit has an equal chance of being sampled.[7] Random samples are essential to provide unbiased estimates of the population mean. Morris[3,13] recognized the importance of randomness when he listed his first criterion for selecting a sample unit, as all units should have an equal chance of being selected. The most reliable procedure of insuring that sample units are collected randomly is to divide the habitat into a grid and use a random number generator to select coordinates of sample units to be sampled.[28,43]

To reduce time and effort needed to locate sample units randomly, sampling programs used in pest management often collect sample units in some predetermined systematic pattern such as U-, V-, X-, or Σ-shaped patterns within the field.[4,27] Such patterns tend to under-sample the field margin and may miss localized infestations, if an area in the field is consistently avoided during repeated sampling periods. Some randomness can be introduced by collecting sample units at a randomly selected interval within the predetermined pattern on each sample date.[27] Additionally, changing the shape and direction of the predetermined pattern on successive sampling dates will help reduce sampling bias. Generally, throwing an object while looking in the other direction may help reduce personal bias in selecting a sampling site.[4,27] If random samples are not collected, sample units at least should be representative of the habitat.[44]

Timing of sampling can have a major effect on the accuracy and fidelity of population estimates. Many arthropods exhibit diel patterns of activity, with differences often being greatest between crepuscular and diurnal periods.[45,46] Sampling in the morning may bias counts because cool temperatures reduce arthropod activity. Arthropods also may move up and down in the plant canopy at different times of the day, thereby affecting sampling results. Furthermore, abiotic factors such as temperature, light intensity, wind, etc., can dramatically affect sampling accuracy and fidelity.[24,46,47] Generally, sampling activity should occur under conditions and timing of program development. If sampling is to occur at other times of the day, an initial comparison of samples at different times should reveal potential sampling bias and provide information to calculate a correction factor.[8]

B. STRATIFICATION OF THE SAMPLING UNIVERSE

An unrestricted series of random samples from the entire sample universe may or may not be the best pattern for collecting samples. Stratification of the habitat and randomly collecting sample units within strata is referred to as stratified random sampling.[7] Stratification can improve sampling efficiency by reducing sample variation for a given sampling effort and by insuring that samples are collected from all areas of the habitat.[2,7] The benefit of stratification can be assessed from a preliminary sampling study where the universe is divided into strata, and results are analyzed with a nested analysis of variance (Table 1).[15] In this analysis, the sampling habitat is subdivided into increasingly smaller units, with the sample unit as the smallest subdivision. The habitat initially is divided into a number of equal-sized strata which in turn can be subdivided into a number of smaller strata. Sample units are collected randomly within the smallest stratum. Although any number of strata can be used, usually two or three is the most that is practical. Furthermore, the sample unit can be subdivided in this analysis to determine optimal sample unit size as part of the same study. The nested analysis of variance tests whether each strata accounts for a statistically significant amount of the total variation between sample units within the habitat. If a stratum does not account for a large proportion of the total variation, then adding that stratum to the sampling program does not improve sampling efficiency. If a stratum does account for a significant amount of the variance, then stratifying the collection of sample units can improve sampling efficiency. Table 1 shows the results examining the effect of stratification on sampling green cloverworm eggs in soybean, which were randomly distributed among four-leaflet sample units.[42]

TABLE 1
Nested Analysis of Variance for Stratified Random Sampling of Green Cloverworm Eggs in Soybean[a]

	Analysis	df	Expected SEM	Variance component	% of total variation
1st	Fields	30	$\sigma_L^2 + 26.7\sigma_s^2 + 133.5\sigma_P^2 + 534\sigma_Q^2 + 1068\sigma_F^2$	0.0001315	0.56
	Quadrats/F	31	$\sigma_L^2 + 26.7\sigma_s^2 + 133.5\sigma_P^2 + 534\sigma_Q^2$	0.0000036	0.03
	Plots/Q/F	186	$\sigma_L^2 + 26.7\sigma_s^2 + 133.5\sigma_P^2$	0.0000011	0.01
	Samples/P/Q/F	992	$\sigma_L^2 + 26.7\sigma_s^2$	0.0004351	3.83
	Leaves/S/P/Q/F	31,868	σ_L^2	0.0108610	95.57
	Total	33,107		0.0116423	100.00
2nd	Fields	30	$\sigma_L^2 + 26.7\sigma_s^2 + 1068\sigma_F^2$	0.0001315	1.14
	Samples/F	1,209	$\sigma_L^2 + 26.7\sigma_s^2$	0.0005060	4.40
	Leaves/S/F	31,868	σ_L^2	0.0108610	94.46
	Total	33,107		0.0114985	100.00

[a] Buntin and Pedigo.[42]

Stratifying into quadrats per field and plots per quadrat accounted for a very small proportion of the total variance; consequently, stratification of sample units was not recommended. This approach has been used in developing sampling programs for several cabbage insect pests,[14] the alfalfa blotch leafminer [*Agromyza frontella* (Rondani)] in alfalfa,[48] and corn rootworm eggs[16] and adults[11] in corn. Usually, stratification is not necessary if the population is randomly dispersed or only slightly aggregated. However, stratification usually will explain a significant amount of variation when a population is aggregated.

The benefit of stratified random sampling may be worthwhile in ecological and research studies, but the added cost of locating strata may not be cost effective in pest management sampling programs. However, in agricultural settings one habitat division that should be considered is the division of habitat (field) margin or edge from the center. It is not uncommon for samplers to avoid the field edge (such as the first 5 m) while collecting samples, especially if a systematic sampling pattern is employed.[4] A field margin of 5 m accounts for 19% of the area of a field measuring 1 ha and 6.4% of the area of a field measuring 10 ha. Consequently, avoiding the edge can avoid a substantial portion of habitat in small fields. Stratifying samples between field edge and center is particularly important if the pest species normally moves into the field from the edge or is important in the field margin. Examples included grasshoppers and armyworms in many crops,[4] flea beetles (*Phyllotreta* sp.) in oilseed *Brassica* crops,[49] and common stalkborer [*Papaipema nebris* (Guenée)] in corn.[50]

C. NUMBER OF SAMPLE UNITS

A major component in developing a sampling program is determining the number of sample units to collect. The optimum number of sample units, which is referred to as the optimum sample size, is the smallest number of sample units that would satisfy the objectives of the sampling program and achieve the desired precision of estimates.[51] Reliability or precision of estimates increases as the number of samples increases. However, the collection of too large a sample can be costly and reduces the efficient use of sampling resources.

Karandinos[51] and more recently Ruesink[8] provide excellent reviews of procedures for calculating optimum sample size, and the information presented below is mostly derived from these sources. Calculation of optimum sample size depends on the statistical dispersion of the target population and how precision is defined. Two approaches have been used to calculate optimum sample size and depend on how precision is defined. The first approach defines precision in terms of the RV such that the standard error of the mean should be within a certain value of the mean. If this value is expressed as a proportion of the mean then the optimum sample size (N) can be calculated by the formula:

$$N = s^2/(m^2)(c^2) \qquad (6)$$

where s^2 = variance, m = expected mean, and c = standard error of the mean expressed as a proportion of m. For example, if one wishes to collect enough sample units to estimate m with a SE of $\pm 20\%$ of the mean, then $c = 0.2$.

If the dispersion pattern of the target population can be described by the Poisson or the negative binomial distribution, then optimum sample size can be calculated as:

Poisson: $\qquad N = 1/(c^2 m) \qquad (7)$

Negative binomial: $\quad N = (k + m)/(c^2(k)m) \qquad (8)$

where m and c are as previously defined and k is the aggregation factor of the negative binomial. If the underlying mathematical distribution is not known, but the relationship between population mean and variance can be described by Taylor's Power Law ($s^2 = am^b$),[39] then optimum sample size can be derived by the formula:

$$N = (am^{b-2})/(c^2) \qquad (9)$$

The second approach to calculating optimal sample size is to express precision as a confidence interval such that the estimate of the mean should be within a certain value of the true mean with a given probability.[8,51] With this approach, the confidence interval, which is calculated as $t_{\alpha/2}(s/\sqrt{n})$, can be expressed as a fixed proportion (d) of the mean so that the mean is estimated with a confidence interval $\pm dm$. If the confidence interval is defined as dm, then $dm = t_{\alpha/2}(s/\sqrt{n})$, which when solved for n becomes

$$N = (t_{\alpha/2}/d)^2(s^2/m^2) \qquad (10)$$

where s^2 = sample variance, m = mean density, d = desired fixed proportion of the mean and $t_{\alpha/2} = t$ value for a given probability (α). In most instances, $\alpha = 0.05$ (i.e., 95% probability level) and for a large sample $t_{\alpha/2} \approx 2.0$. For example, if preliminary sampling finds that $s^2 = 5$ and $m = 5$ and we wish to calculate optimal sample size so that the mean can be estimated with a confidence interval $\pm 10\%$ ($d = 0.1$), with 95% probability ($\alpha = 0.05$), then $N = (2/0.1)^2(5/5^2) = 80$ sample units. If the 20% level ($d = 0.2$) is satisfactory then $N = 20$ sample units.

If the underlying distribution can be described by the Poisson or negative binomial or if Taylor's power law adequately describes the variance to mean relationship, then various expressions of s^2 can be substituted in the general formula and solved for n.

The resulting formulas are

Poisson: $N = (t_{\alpha/2}/d)^2(1/m)$ (11)

Negative binomial: $N = (t_{\alpha/2}/d)^2((k+m)/km)$ (12)

Taylor's power law: $N = a(t_{\alpha/2}/d)^2(m^{b-2})$ (13)

In the previous formulas, precision is defined as a percentage of the mean. Optimum sample size also can be determined where the standard error or confidence interval represent a fixed positive value (h) such that population density is estimated as $m \pm h$. The derivation and use of formulas with precision defined as a fixed positive value are presented by Karandinos[51] and Ruesink.[8]

Initially when using optimal sample size formulas, the question arises as to what are appropriate values for c and d. The main practical difference between formulas based on the standard error and on probabilistic statement is the inclusion of the t statistic in the latter approach. If in the second approach, optimal sample size is calculated using the 95% probability level, then $t_{\alpha/2} \approx 2$ and $d = 2c$.[50] If some other probability level is used then $d = t_{\alpha/2}/c$.[29] Southwood[2] suggests a standard error to mean ratio of 5% for most intensive sampling, which would provide a confidence interval of $\pm 10\%$. Therefore, $c = 0.05$ and $d = 0.1$ for the 95% probability level. This level of precision can result in a rather large number of samples for a given mean, particularly for aggregated populations. Consequently, standard error-to-mean ratios of 10, 15, and 25% where $c = 0.1, 0.15$, and 0.25, respectively, and $d = 0.2, 0.3$, and 0.5, respectively, may be more suitable for most sampling programs. Usually, the 25% level is used for pest management sampling programs, but the 15 or 20% levels may be more suitable if greater precision is desired.

The optimal sample size changes with mean density. When precision is defined as a percentage of the mean, sample size declines as mean density increases for a given level of precision (Figure 1). Usually, the optimal sample size becomes quite large as mean density becomes small and may become infinitely large at very low densities. If population abundance normally dictates very large sample sizes, then reconsideration of the desired level of precision or of the sample unit size and technique may be necessary (see Chapter 4). Although rarely used in pest management, if precision is defined as a fixed value, sample size increases as mean density increases for a given level of precision (see Karandinos[51]).

Because of the stability of Taylor's power law, Formulas (9) and (13), using the power law to express variance, have been widely used to develop sampling programs.[28] However, Trumble et al.[40] showed that Taylor's a value can have a substantial impact on the calculated number of sample units. The a value is affected by sampling conditions and sample unit size,[27] and a change of the a value can have a large impact on the calculated number of samples. For example, changing $a = 1.0$ to $a = 2.0$ can double the optimal number of sample units.[28] This underscores the importance of careful selection of sample unit size. Another peculiar attribute of formulas that use Taylor's power law is that sample unit number declines as mean density increases when $b < 2.0$, but when $b > 2.0$ sample unit size increases with mean density (Figure 2). This inversion of the relationship between sample size and mean is contrary to the curves developed by Karandinos.[51] Shelton and Trumble[28] concluded that formulas based on Taylor's power law are useful only when $b < 2$, which brings into question the validity of sample unit size estimates when $b < 2.0$. Additional clarification of this problem is needed. An alternative approach would be to use Formulas (8) and (12), based on the negative binomial distribution, and use information from Taylor's power

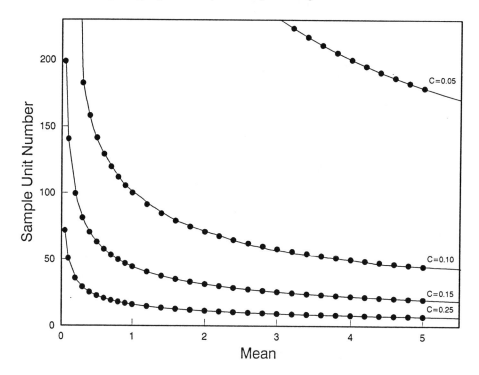

FIGURE 1. Relationship between the optimum number of sample units and mean density for four levels of precision where precision is expressed as relative variation (C). Values were calculated using Equation 8 where Taylor's $a = 1.0$ and $b = 1.5$.

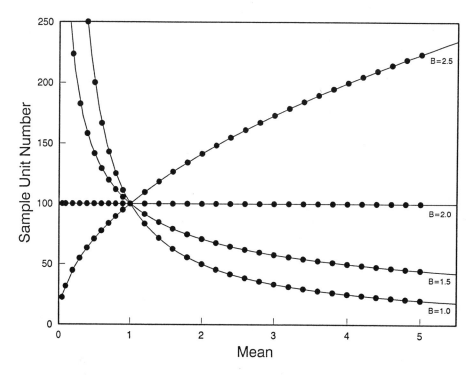

FIGURE 2. Relationship between sample unit size and mean density as determined by Equation 8 for four Taylor's b-values. Taylor's intercept value was $a = 1.0$ and precision was held constant at $c = 0.10$.

law to estimate k by solving the formula proposed by Ruesink:[8]

$$k = m/((a(m)^{b-1}) - 1) \tag{14}$$

There is no solution to this formula if $a = 1.0$ and $b = 1.0$, but $b = 1$ indicates a random distribution and $k \to \infty$.[2,32] Equation 14 provides an estimate of k when $b > 1.0$.

IV. PROGRAM EVALUATION AND IMPROVEMENT

In practice, optimal sample size formulas are used to calculate a fixed sample unit number that will estimate mean density with a desired precision for most expected mean densities. The proficiency of a given sampling program can be compared with a more intensive program with a large number of sample units which presumably would provide a more accurate estimate of the true mean density.[2] Legg and Moon in Chapter 4 discuss procedures for reducing sampling error and bias and improving sampling precision.

One major shortcoming of the optimal sample size formulas is the requirement of an initial estimate of mean density. An initial estimate can be obtained from a preliminary sample of three or four sample units and the optimal number of sample units calculated. This is cumbersome and, if this initial mean estimate is inaccurate, too few or too many sample units will be collected for mean estimates with a desired level of precision.

Sequential sampling procedures avoid this problem by recomputing the mean estimate after each sample so that no more samples are collected then necessary to estimate mean density with a desired level of precision.[29] Sequential count procedures vary sample unit number to provide mean density estimates with fixed precision levels.[52,53] Other sequential sampling procedures which are more suitable for pest management sampling programs classify populations as being above or below some critical value such as an economic injury level.[54,55] Sequential sampling procedures are reviewed in Chapters 8, 9, and 10, and Binns and Nyrop[56] and Nyrop and Binns[29] discuss a number of procedures for improving sampling programs.

Finally, it is useful in IPM to evaluate the value of using sampling information for management decisions by comparing the cost of sampling and associated crop losses and savings in control measures with the cost of controlling or not controlling a pest and associated crop losses without sampling information. It can be argued that using sampling information for pest control decisions is economically viable only if using sampling information reduces both economic and environmental costs of pest control.[57] Sampling may not be cost effective when control measures are inexpensive and preventive or possibly when treatment thresholds are very low and commodity value is high.[58] Forster et al.,[57] Nyrop et al.,[58,59] and Nyrop and Binns[29] discuss the methodology and application of assessing the value of sample information in IPM.

REFERENCES

1. Gonzalez, D., Sampling as a basis for pest management strategies, in *Proc. Tall Timber Conf. Ecol. Animal Control Habitat Mgt.*, No. 2, Komarek, E. V., Ed., Tall Timbers Res. Stn., Tallahassee, FL, 1971, 83–101.
2. Southwood, T. R. E., *Ecological Methods*, 2nd ed., Chapman & Hall, New York, 1978.
3. Morris, R. F., The development of sampling techniques for forest insect defoliators, with particular reference to the spruce budworm, *Can. J. Zool.*, 33, 225–294, 1955.

4. Pedigo, L. P., *Entomology and Pest Management*, Macmillan, New York, 1989.
5. Eisenhart, C., Expression of uncertainties of final results, *Science*, 160, 1201–1204, 1968.
6. Fleischer, S. J. and Allen, W. A., Field counting efficiency of sweep-net samples of adult potato leafhoppers (Homoptera: Cicadelidae) in alfalfa, *J. Econ. Entomol.*, 75, 837–840, 1982.
7. Cochran, W. G., *Sampling Techniques*, 3rd ed., John Wiley & Sons, New York, 1977.
8. Ruesink, W. G., Introduction to sampling theory, in *Sampling Methods in Soybean Entomology*, Kogan, M. and Herzog, D. C., Ed., Springer-Verlag, New York, 1980, 61–78.
9. Ruesink, W. G. and Kogan, M., The quantitative basis of pest management: sampling and measuring, in *Introduction to Insect Pest Management*, 2nd ed., Metcalf, R. L. and Luckman, W. H., Eds., John Wiley & Sons, New York, 1982, chap. 9.
10. Stern, V. M., Economic thresholds, *Annu. Rev. Entomol.*, 18, 259–280, 1973.
11. Steffey, K. L., Tollefson, J. J., and Hinz, P. N., Sampling plan for population estimation of northern and western corn rootworm adults in Iowa cornfields, *Environ. Entomol.*, 11, 287–291, 1982.
12. Foster, R. E., Tollefson, J. J., and Steffey, K. L., Sequential sampling plans for adult corn rootworms (Coleoptera: Chrysomelidae), *J. Econ. Entomol.*, 75, 791–793, 1982.
13. Morris, R. F., Sampling insect populations, *Annu. Rev. Entomol.*, 5, 243–264, 1960.
14. Harcourt, D. G., Design of a sampling plan for studies on the population dynamics of the diamondback moth, *Plutella maculipennis* (Curt.) (Lepidoptera: Plutcllidae), *Can. Entomol.*, 93, 820–831, 1961.
15. Snedecor, G. W. and Cochran, W. G., *Statistical Methods*, Iowa State University Press, Ames, 1967.
16. Hein, G. L., Tollefson, J. J., and Hinz, P. N., Design and cost considerations in the sampling of northern and western corn rootworm (Coleoptera: Chrysomelidae) eggs, *J. Econ. Entomol.*, 78, 1495–1499, 1985.
17. Bechinski, E. H. and Pedigo, L. P., Microspatial distribution of pupal green cloverworm, *Plathypena scabra* (F.) (Lepidoptera: Noctuidae), in Iowa soybean fields, *Environ. Entomol.*, 12, 273, 1983.
18. Pieters, E. P., Comparison of sample-unit sizes for D-Vac sampling of cotton arthropods in Mississippi, *J. Econ. Entomol.*, 71, 107–108, 1977.
19. Sheilds, E. J. and Specker, D. R., Sampling for potato leafhopper (Homoptera: Cicadellidae) on alfalfa in New York: relative efficiency of three sampling methods and development of a sequential sampling plan, *J. Econ. Entomol.*, 82, 1091–1095, 1989.
20. Zehnder, G. W., Kolondny-Hirsch, D. M., and Linduska, J. J., Evaluation of various potato plant sample units for cost-effective sampling on Colorado potato beetle (Coleoptera: Crysomelidae), *J. Econ. Entomol.*, 83, 428–433, 1990.
21. Stewart, J. G. and Sears, M. K., Quarter-plant samples to detect populations of Lepidoptera (Noctuidae, Pieridae, and Plutellidae) on cauliflower, *J. Econ. Entomol.*, 82, 829–832, 1989.
22. Luna, J. M., Fleischer, S. J., and Allen, W. A., Development and validation of sequentila sampling plans for potato leafhopper (Homoptera: Cicadelidae) in alfalfa, *J. Econ. Entomol.*, 12, 1690–1694, 1983.
23. Bechinski, E. H. and Pedigo, L. P., Evaluation of methods for sampling predatory arthropods in soybeans, *Environ. Entomol.*, 11, 735–741, 1982.
24. Kogan, M. and Herzog, D. C., Ed., *Sampling Methods in Soybean Entomology*, Springer-Verlag, New York, 1980.
25. Saugstad, E. S., Bram, R. A. and Nyquist, W. E., Factors influencing sweep-net sampling of alfalfa, *J. Econ. Entomol.*, 60, 421–426, 1967.
26. Pedigo, L. P., Lentz, G. L., Stone, J. D., and Cox, D. F., Green cloverworm populations in Iowa soybean with special reference to sampling procedures, *J. Econ. Entomol.*, 65, 414–421, 1972.
27. Taylor, L. R., Assessing and interpreting the spatial distributions of insect populations, *Annu. Rev. Entomol.*, 29, 321–357, 1984.
28. Shelton, A. M. and Trumble, J. T., Monitoring insect populations, in *Handbook of Pest Management in Agriculture*, Vol. 2, 2nd ed., Pimentel, D., Ed., CRC Press, Boca Raton, FL, 1991, 45–62.
29. Nyrop, J. P. and Binns, M., Quantitative methods for designing and analyzing sampling programs for use in pest management, in *Handbook of Pest Management in Agriculture*, 2nd ed., Pimentel, D., Ed., CRC Press, Boca Raton, FL, 1991, 67–132.
30. Bliss, C. I., *Statistics in Biology*, Vol. 1, McGraw-Hill, New York, 1967.
31. Bliss, C. I. and Fisher, R. A., Fitting the negative binomial distribution to biological data, *Biometrics*, 9, 176–200, 1953.
32. Bliss, C. I. and Owen, A. R. G., Negative binomial distributions with a common k, *Biometika*, 45, 37–58, 1958.
33. Waters, W. E. and Henson, W. R., Some sampling attributes of the negative binomial with special reference to forest insects, *Forest Sci.*, 5, 1959.
34. Myers, J. H., Selecting a measure of dispersion, *Environ. Entomol.*, 7, 619–621, 1978.
35. Green, R. H., Measurement of nonrandomness in spatial distributions, *Res. Popul. Ecol.*, 8, 1–7, 1966.

36. Iwao, S., A new regression method for analyzing the aggregation pattern of animal populations, *Res. Popul. Ecol.*, 10, 1–20, 1968.
37. Morisita, R. F., I_δ-index, a measure of dispersion of individuals, *Res. Popul. Ecol.*, 4, 1–7, 1962.
38. Morisita, R. F., Composition of the I_δ-index, *Res. Popul. Ecol.*, 13, 1–27, 1971.
39. Taylor, L. R., Aggregation, variance, and the mean, *Nature*, 189, 732, 1961.
40. Trumble, J. T., Brewer, M. J., Shelton, A. M., and Nyrop, J. P., Transportability of fixed-precision level sampling plans, *Res. Popul. Ecol.*, 31, 325–342, 1989.
41. Lloyd, M., 'Mean crowding', *J. Anim. Ecol.*, 36, 1–30, 1967.
42. Buntin, G. D. and Pedigo, L. P., Dispersion and sequential sampling of green cloverworm eggs in soybeans, *Environ. Entomol.*, 10, 980–985, 1981.
43. Legg, D. E., and Yeargan, K. V., Method for random sampling insect populations, *J. Econ. Entomol.*, 78, 1003, 1985.
44. Hurlbert, S. H., Pseudoreplication and the design of ecological field experiments, *Ecol. Monogr.*, 54, 187–211, 1984.
45. Mathews, R. W. and Mathews, J. R., *Insect Behavior*, John Wiley & Sons, New York, 1978.
46. Schotzko, D. J. and O'Keefe, L. E., Comparison of sweep net D-Vac, and absolute sampling, and diel variation of sweep net sampling estimates in lentils for pea aphid (Homoptera: Aphididae), nabids (Hemiptera: Nabidae), lady beetles (Coleoptera: Coccinellidae), and lacewings (Neuroptera: Chrysopidae), *J. Econ. Entomol.*, 82, 491–506, 1989.
47. Beall, G., Study of arthropod populations by the method of sweeping, *Ecology*, 16, 216–225, 1935.
48. Harcourt, D. G. and Binns, M. R., A sampling system for estimating egg and larval populations of *Agromyza frontella* (Diptera: Agromyziidae) in alfalfa, *Can. Entomol.*, 112, 375, 1980.
49. Wylie, H. G., Observations on distribution, seasonal history and abundance of flea beetles (Coleoptera: Chrysomelidae) that infest rape crops in Manitoba, *Can. Entomol.*, 111, 1345–1353, 1979.
50. Davis, P. M. and Pedigo, L. P., Analysis of spatial patterns and sequential count plans for stalk borer (Lepidoptera: Noctuidae), *Environ. Entomol.*, 18, 504–509, 1989.
51. Karandinos, M. G., Optimum sample size and comments on some published formulae, *Bull. Entomol. Soc. Am.*, 22, 417–421, 1976.
52. Kuno, E., A new method of sequential sampling to obtain the population estimates with a fixed level of precision, *Res. Popul. Ecol.*, 11, 127–136, 1969.
53. Green, R. H., On fixed precision level sequential sampling, *Res. Popul. Ecol.*, 12, 249–251, 1979.
54. Wald, A., *Sequential Sampling*, John Wiley & Sons, New York, 1948.
55. Waters, W. E., Sequential sampling in forest insect surveys, *Forest Sci.*, 1, 68–79, 1955.
56. Binns, M. R. and Nyrop, J. P., Sampling insect populations for the purpose of IPM decision making, *Annu. Rev. Entomol.*, 37, 427–453, 1992.
57. Foster, R. E., Tollefson, J. J., Nyrop, J. P., and Hein, G. L., Value of sample information in pest decision making, *J. Econ. Entomol.*, 79, 303, 1986.
58. Nyrop, J. P., Shelton, A. M., and Theunissen, J., Value of a control decision rule for leek moth infestations in leek, *Entomol. Exp. Appl.*, 53, 167, 1989.
59. Nyrop, J. P., Foster, R. E., and Onstad, D. W., Value of sample information in pest control decision making, *J. Econ. Entomol.*, 79, 1421, 1986.

Section III : Improving Sampling Program Efficiency

Chapter 7

INITIATING SAMPLING PROGRAMS

Leon G. Higley and Robert K. D. Peterson

TABLE OF CONTENTS

I. Introduction 120

II. Deciding When to Begin Sampling 120
    A. Timing and Monitoring 120
    B. Host, Pest, and Environmental Considerations 121

III. Techniques 121
    A. Initiation Based on Life History and Environment 121
        1. Calendar Date 121
        2. Migration 122
        3. Thermal Development 124
            a. Why Does Temperature Matter? 124
            b. Approaches for Estimating Development 127
            c. Limitations 129
            d. Using Estimates of Thermal Development in Sampling .. 129
    B. Initiation Based on Ongoing Monitoring 130
        1. Hosts 130
        2. Alternate Hosts and Attraction Sites 131
        3. Trapping 131
        4. Surveys 132
        5. Remote Sensing 132
    C. Initiation Based on Pest Impact 133

IV. Practical Considerations 134

References 134

I. INTRODUCTION

Initiating sampling programs involves questions of efficiency and of risk. It is inefficient to begin programs too early, before sufficient numbers of insects are present to allow accurate estimates of population density. On the other hand, waiting to sample until pest populations reach their highest numbers presents the risk that significant injury will occur before sampling and management action can be taken. Indeed, the use of economic injury levels (EILs) is based on the premise that injury is preventable; EILs are not accurate when used where substantial injury has occurred already.[1] Thus, sampling programs must be initiated before significant pest injury can occur but not much before pest populations are established.

Generally, issues involving the timing of sampling programs are less studied than other aspects of sampling. For many situations, timing is a simple question that does not require much effort. For other instances, timing may be a critical consideration. We present some of the general approaches for determining when to begin sampling and discuss some considerations of host, pest, and environment that pertain to this question. Next, we explore some of the techniques associated with initiating sampling programs. In particular, because of its overriding importance to many sampling programs, we discuss in detail measuring thermal development of insects to predict pest occurrence. Finally, we conclude with a discussion of practical considerations in initiating sampling programs.

II. DECIDING WHEN TO BEGIN SAMPLING

A. TIMING AND MONITORING

The two fundamental approaches for determining when to initiate sampling programs are timing and monitoring. Timing refers to activities used to predict the occurrence of pests, independent of direct observations or counts of pests. Monitoring, as we define it, refers to broad, often qualitative, sampling of pests or their effects used to trigger more precise, quantitative sampling programs. Timing includes considerations of seasonality, weather, and especially temperature. Monitoring includes surveys, trapping, and possibly remote sensing. These approaches are not mutually exclusive. For example, the identification of favorable weather systems for insect migration is a timing approach, using pheromone traps to confirm insect migration from such systems is a monitoring approach.

Methods for initiating sampling based on timing are common in regions with strongly variable climates. Probably the most common timing procedure is the simple use of calendar dates to begin sampling. Such an approach often succeeds because of the strong relationships between many insect life histories and seasonal cycles. Similarly, temperature is crucially important in determining insect development times; therefore, using temperature to predict the occurrence of damaging pest species is one of the most important considerations in timing sampling programs for many species.

Because monitoring is a form of sampling, most of the issues and considerations of sampling discussed elsewhere in this book apply. Indeed, some entire sampling programs, such as those used to detect new pest species introductions, represent little more than ongoing monitoring. In the context of initiating sampling programs, monitoring is a low cost, qualitative approach used to trigger more quantitative, and expensive, sampling efforts. Evaluation criteria from monitoring may include pest occurrence, pest impact, estimates of pest density, or the recognition of specific pest forms, such as gregarious phenotypes of grasshopper species or alate aphids. (The occurrence of such forms is an indication of high population levels.) Typically, monitoring is of greatest utility for pests whose occurrence is not predictable from seasonality, weather patterns, or estimates of thermal development. Additionally, monitoring is valuable as a confirmation of predictions made from timing approaches.

Both of these methods for determining when to start a sampling program operate within a universe set by specific attributes and relationships of the pest, its host, and the environment. All pest management activities, including sampling, depend upon the interactions of these three factors.

B. HOST, PEST, AND ENVIRONMENTAL CONSIDERATIONS

Host and pest characteristics have an important bearing on sampling. Host susceptibility to pest attack is the most basic parameter. Usually, sampling is only required when the host is at a stage susceptible to the pest. Typically, seedling and reproductive stages are most severely affected by pest injury, therefore, sampling may be critical at these times.[2] In addition to host phenology, pest life history characteristics, particularly synchrony between injurious life stages of the pest and susceptible stages of the host, are important. Sampling may be initiated when such synchronies arise. For example, phenological delays in plant development may render plants susceptible to pest attack that ordinarily might be avoided.

The environment influences various aspects of host and pest relationships. Water stress frequently alters pest and host phenologies and renders plants more susceptible to pest injury. Indeed, the occurrence of drought may itself be an indicator of the need to monitor pest species such as grasshoppers and spider mites. Another environmental influence is the use of cultural practices which may reduce or increase the likelihood of pest occurrence. For example, crop rotations or certain tillage practices may eliminate the need for sampling some potential pests. In contrast, other practices, such as incorporation of a cover crop or poor weed control may increase pests and trigger sampling efforts.

Interactions between these factors can alter host susceptibility to pest attack, pest population levels, host or pest phenologies, or pest occurrence in a host. If these relationships are sufficiently predictable, the occurrence of a given situation (late season water stress, for instance) may be sufficient to initiate sampling efforts for some species. More commonly, this interplay of host, pest, and environment represents a moving surface upon which timing and monitoring approaches for sampling initiation must contend.

III. TECHNIQUES

A. INITIATION BASED ON LIFE HISTORY AND ENVIRONMENT
1. Calendar Date

Pest life history events and their consistent correlations to calendar dates often provide an indication of when to initiate sampling. The timing of biological events (e.g., oviposition or appearance of first instars) and their fidelity to dates has often been determined through several years of observation. In many instances, the phenological information has been gained without intensive research. Moreover, using calendar dates is one of the most inexpensive techniques for determining when to begin sampling.

In Nebraska, recommendations to begin sampling for second-generation European corn borer, *Ostrinia nubilalis*, are based, in part, on the periodicity of adult flight activity—approximately mid-July each year.[3] For corn rootworm, *Diabrotica* spp., management and sampling for adults are initiated in early July, when adults typically begin to emerge.[3]

In many instances, using calendar dates is the only viable technique available to determine when to begin sampling. Monitoring techniques such as trapping or surveying may be impractical or prohibitively expensive. Although calendar dates may be adequate for predicting seasonality, degree-day accumulations typically are more precise than calendar dates. However, degree-days have not been determined for many pests or for some pest life stages. Therefore, calendar dates often are used because degree-day information is nonexistent or incomplete.

2. Migration

Migration is an important biological phenomenon for many arthropods, including several pest species. Arthropod migration can be defined as an adaptive behavior that results in the displacement of large groups of individuals.[4] In the Noctuidae alone, several important pests migrate very long distances each year and cause significant damage to crop plants in regions where they cannot overwinter. Further, migration within Noctuidae may be more widespread than previously reported.[5]

The characterization of arthropod migration continues to be an active research area. In recent years, the ability to accurately predict both the temporal and spatial aspects of migration has greatly enhanced the capacity to manage these pests. Moreover, arthropod migration events, when predicted or observed, are used to initiate sampling programs.

Insect migrations have been characterized using a number of meteorological techniques.[6] Probably the simplest method is to observe persistent airflows from specific directions. These airflows can indicate that migration events are likely to occur (Figure 1). Migration also can be predicted when appropriate synoptic weather systems occur.[7-9] Synoptic weather systems are atmospheric circulation patterns that influence the weather over an area of 1000 km or greater.[6] Trajectory analysis, the projection of the path of an air parcel, provides a sophisticated technique for predicting the movement of migratory insects that are transported above their typical activity layer.[9,10] In recent years, trajectory models have been developed that are very accurate in projecting migratory movements within specific time intervals.[9-11]

Regardless of the meteorological methodology used, monitoring of individuals usually is needed to verify that migration has occurred. Monitoring of migrants typically is accomplished by trapping or observing individuals on hosts or attraction sites. However, unconventional techniques, such as radar imaging and observation by plane or helicopter, also have been used to identify migrants and to predict infestation locations. Direct observations of insects that migrate over relatively short distances also have been used to initiate sampling. Examples include chinch bug, *Blissus leucopterus leucopterus*, Mormon cricket, *Anabrus simplex*, and true armyworm *Pseudaletia unipuncta*.[4]

In the midwestern U.S., some of the most important insect pests of field crops are migratory species. Species such as the black cutworm, *Agrotis ipsilon*, the green cloverworm, *Plathypena scabra*, and the potato leafhopper, *Empoasca fabae*, cannot overwinter in the upper Midwest. These species recolonize northern areas by migrating from southern overwintering regions each spring. Migrants, or, more commonly, their progeny can significantly injure crops from localized areas to entire regions, depending on the magnitude of the migration. The observation of immigrant green cloverworm moths in Iowa has been used to initiate a time-sequential sampling program developed by Pedigo and van Shaik[12] to determine if subsequent generations will cause economic damage to soybean. Similarly, the observation of potato leafhop-

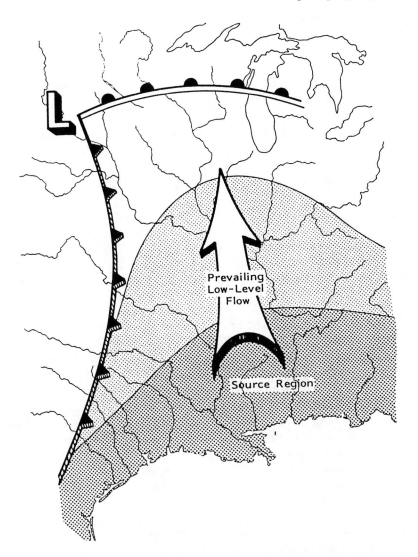

FIGURE 1. A typical weather pattern exhibited during insect migration events in the Spring in the U.S. A northward-moving airflow (arrow) is east of an approaching cold front (line with triangles) and south of a warm front (line with semicircles). Insects are transported to the midwestern and northern U.S. from source regions in the southern U.S. (Reprinted with permission of the Entomological Society of America, from *Bull. Entomol. Soc. Am.*, 34, 9, 1988.)

per immigrants late in the first growth cycle of alfalfa may initiate a sampling program during the second and third growth cycles.[6]

An advanced system of migration and outbreak prediction has been developed for the black cutworm in the midwestern U.S.[6,13] A meteorologically based trajectory model has been used to effectively predict black cutworm outbreak locations. Moreover, the model predicts the magnitude of the outbreaks using a numerical rating system based on trajectory proximity to black cutworm source regions in the southern and southeastern U.S.[6,13] Predictions of black cutworm immigration have made it possible to initiate and intensify sampling at relatively specific locations.

Ongoing research will lead to a better understanding of arthropod migration as an ecological phenomenon and will result in better predictions of migration events. More accurate predictions will then improve decisions to initiate sampling programs for immigrant arthropods.

3. Thermal Development

Because pest management concerns are typically directed at the injurious stage in an insect's life cycle, it is important to predict when this stage will be present. Ideally, sampling will be started at the first presence of the injurious insect. Consequently, it often is necessary to track common developmental events such as emergence or activity after overwintering, oviposition, egg hatch, and larval development. These developmental events may be observed through ongoing monitoring or sometimes predicted based on calendar dates. However, these approaches may be infeasible or inappropriate for many situations.

Insects are poikilothermic organisms; their body temperatures depend upon ambient temperatures. Homeothermic organisms, such as humans and other mammals, also have temperature-dependent development, but by maintaining constant body temperatures the impact of temperature on development is constant. Describing insect development often requires a description of temperature as well as time. This principle of using time and temperature to describe poikilothermic development has been known since the early 1700s,[14] but many fundamental aspects of these relationships are uncertain. Remaining questions include: what is the physical basis for effects of temperature on development, what are appropriate mathematical models for describing temperature development relationships, how does environmental variability, especially fluctuating temperature, influence development, and what are the most accurate methods for estimating development in the field?

Proper timing of sampling programs for many insects depends upon accurate estimates of insect development based on ambient temperatures. Although some questions mentioned above can be neglected in practical applications, others may not. In practical applications, we need only employ that degree of complexity necessary to provide a suitably accurate prediction of development. Estimates of development for timing sampling programs typically need to achieve 85 or 90% accuracy for most insect species. Also, relating temperature and development is of trivial importance if temperatures are constant or developmental events are driven by other factors, such as photoperiod. Thus, establishing temperature-development relationships typically is most important for early season insects in temperate regions, where temperatures are variable and limit development.

There is a large volume of literature on questions of temperature and insect development. We present key issues as they relate to initiating sampling programs and refer to major reviews that can guide the reader to additional information as desired. The points we consider are (1) the physical basis for temperature-development relationships, (2) approaches for estimating development, (3) limitations to existing methods, and (4) using estimates of development to start sampling programs.

a. Why Does Temperature Matter?

This seemingly simple question actually encompasses substantial biological complexity. The basic answer is that growth depends on temperature because biochemical reactions that are essential to growth also depend on temperature. However, the details of how temperature affects physiology and growth is more complicated.

At a macro level, the relationship between temperature and growth rate is well established. As indicated in Figure 2, at low temperatures there is a curvilinear response; at higher temperatures growth rates increase linearly with temperature; at still higher temperatures, the relationship becomes curvilinear again and maximum

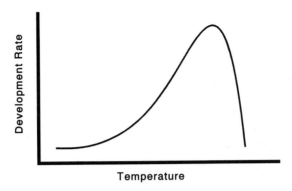

FIGURE 2. The thermal development curve—the generalized relationship between temperature and rate of development.

rates are achieved; and finally, at temperature beyond the optimum, growth rates decrease precipitously. Usually, temperatures at the upper limit of development are near the lethal temperatures for a species.

Why the temperature development curve has this shape is more problematic. Undoubtedly, the extremes of the curve represent situations in which temperature restrains development. At high temperatures, respiration may be increased, potential dessication may be greater, and enzyme activities may be diminished. At low temperatures, membrane permeabilities may be reduced, insufficient heat may be available for ready formation of enzyme-substrate complexes, and, again, enzyme activities may be suboptimal.

One explanation suggested for the shape of the low temperature portion of the curve is that nonlinearity is a reflection of genetic differences between individuals and that this, coupled with low temperature mortality, produces an apparently nonlinear relationship.[15] Lamb[16] addressed this question directly by examining development rates of pea aphid, *Acyrthosiphon pisum*, clones at low temperature. Lamb demonstrated that curvilinear responses between temperature and development rates are not a function of genetic differences. Instead, temperature and development rates have a nonlinear relationship at low temperatures, which seems to be a reflection of the intrinsic physiology of development.

A simple model for understanding temperature and growth begins by recognizing that growth requires the production of new cellular material. (Strictly speaking, growth involves differentiation and cell enlargement, as well as the production of new cells, but these other aspects of growth follow from increases in cellular material.) Production of new materials depends, in turn, on synthetic biochemical reactions. These reactions require heat; they are temperature dependent. Besides heat of activation, rates of diffusion for enzymes and substrates also depend on temperature, as may membrane permeabilities, which also influence substrate and enzyme availability. Temperature also may influence metabolic rates and partitioning of resources between growth and maintenance. In addition to temperature, any factor that influences substrate availability or enzyme availability and function will have a corresponding influence on development rates. This observation helps explain why nutrition, photoperiod, stress conditions, and similar factors all influence development rates.

Some approaches to modeling thermal development are based on the assumption that one or more enzyme-catalyzed reactions are "rate limiting" for growth,[17] which tends to ignore the impact of temperature on other factors such as membrane function. This notion of rate-limiting reactions has provided a loose physical explanation for some models of insect development. More sophisticated explanations have been proposed and are the basis for some influential approaches to modeling thermal development. Sharpe and DeMichele[18] proposed a biophysical model for poikilothermic development based on the premise that reversible inactivation of control enzymes occurs at the upper and lower temperature limits for a species. This model was subsequently reparameterized to improve its ease of fit and other statistical properties.[19]

The Sharpe and DeMichele model has been very influential and widely used to describe development.[20] Adoption of the Sharpe and DeMichele model did not arise out of ease of use, because determining parameters for the model is a somewhat involved procedure (although a computer program has been published for making these estimates[20]). Instead, its appeal probably derives from its theoretical basis in reflecting a biochemical mechanism for describing development. Unfortunately, as Lamb[16] has pointed out, the biophysical basis of the Sharpe and DeMichele model is unfounded. The premise of the Sharpe and DeMichele model is that upper and lower temperature control enzymes regulate development through reversible thermal inactivation. However, Lowry and Ratkowsky[21] showed that this assumption is not true for some bacterial strains. Hilbert and Logan[22] argued that describing development through a rate-controlling enzyme ignores the important influence of temperature on phase changes in cuticular waxes. Also, Lamb et al.[23] pointed out that different life stages of insects have different developmental rate parameters, which would not be the case were a single rate-controlling enzyme operating. There is no direct evidence that specific enzymes regulate developmental rates at temperature extremes. Consequently, although Sharpe and DeMichele's model may provide a good empirical fit to some data, it does not reflect an underlying biochemical reality, as some of its users propose.[20]

A further complication to our understanding of how temperature affects development arises when we consider development under fluctuating temperatures. Typically, development is faster for fluctuating temperatures around a low temperature mean than under constant conditions at the same mean temperature. In contrast, development is slower for fluctuating temperatures around a high temperature mean than under constant conditions at that temperature.[24,25] Why this is so, involves mathematical, and possibly, physiological explanations.

The mathematical aspect is a reflection of the shape of the temperature-development curve (Figure 2) in what is called the rate summation effect.[25] Tanigoshi et al.[24] provide a lucid explanation of the effect; essentially, fluctuations around a low temperature mean will have the effect of producing rapid acceleration in development rates because of the concave shape of the curve. (Figure 3 presents this effect graphically.) In the low temperature portion of the development curve, the magnitude of the decrease in rates with temperature fluctuations below the mean is much less than the magnitude of the increase in rates associated with fluctuations above the mean. Therefore, temperature fluctuations above a given mean will increase development more than comparable fluctuations below a mean decrease development. The reverse is true for the convex, higher temperature portion of the development curve. On the linear portion of the curve, fluctuations above and below a mean temperature cause equally offsetting changes in development.

The other possible explanation for changes in development rate is that temperature fluctuations may cause physiological changes that stimulate or retard development. Although various mechanisms have been suggested, after reviewing the available evidence, Worner[25] concluded that present experimental findings could reflect

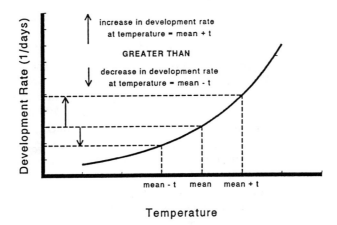

FIGURE 3. The rate summation effect at low temperatures. Temperature fluctuations (t) above and below a given mean produce a development rate greater than that predicted by the mean temperature, because fluctuations above the mean increase development rates proportionally more than fluctuations below the mean reduce development rates.

inadequate developmental models, currently unknown physiological mechanisms, or a combination of both factors.

As this discussion illustrates, we cannot as yet describe in a detailed, physiological sense how temperature influences development. Temperature-development curves have been described for many insect species, albeit based on constant temperature development. Undoubtedly, better physiological understanding of how temperature influences development would offer the prospect of improved models for thermal development. However, the absence of such understanding is not an impediment to developing acceptable development models for use in sampling.

b. Approaches for Estimating Development

Three basic methods have been used for modeling development: physiologically based models; degree-day, or linear, models; and curvilinear models. The one example of a physiologically based model is that of Sharpe and DeMichele,[18] but as we pointed out, its physiological basis is not well supported. Consequently, the Sharpe and DeMichele model probably is better characterized as a curvilinear model. Both the linear and curvilinear approaches are descriptive rather than explanatory. They are mathematical expressions of the temperature-development curve.

Degree-day, or linear summation, models are the most widely used for insect sampling. They have the virtues of being easily developed and easily used. They have the failing of being obviously incorrect. The temperature-development curve clearly is a curve, rather than a straight line. Because the degree-day approach is based on a linear relationship between temperature and development, the lower and upper portions of the development curve are discounted. However, this inaccuracy may not matter if daily temperatures do not often enter the curvilinear portions of the curve. Additionally, the degree of curvilinearity differs among species. Often degree-day models provide sufficient accuracy for practical uses in timing sampling. As we will discuss in more detail, weighing complexity and biological accuracy against simplicity and practical acceptability is an important issue in modeling thermal development, as well as other approaches for timing sampling programs.

Degree-days (also called heat units, thermal units, or growing degree-days) represent a measure of physiological time through a combination of time and temperature.

Degree-days are the number of degrees above the minimum temperature acceptable for growth multiplied by time in days. Ten degrees above the minimum for 5 d represents 50 degree-days (10° × 5 d) just as does 2° above the minimum for 25 d (2° × 25 d). Both instances are taken to represent the same amount of physiological time; an insect would have grown the same amount under either circumstance.

Using degree-days to estimate development requires that a minimum developmental threshold, the lowest temperature at which development will proceed, be determined. Additionally, the total degree-day requirements for specific life history events (such as egg hatch, larval development, pupation, adult emergence, and preoviposition period) must be calculated. For some species, a maximum developmental threshold also may be determined.

At least three important approximations are associated with degree-days. First, degree-days are based on constant relationships between temperature and development, in other words, linearity. Second, degree-days do not address changes in development associated with fluctuating temperature. Third, degree-days use an estimate for the minimum developmental threshold. The first two of these approximates are intrinsic to the technique. However, approaches for improving estimates of the minimum developmental threshold for a species are possible. Commonly, a minimum threshold is established by extending the linear portion of the development curve to where development is zero (the x-intercept method).[26] Another approach is to iteratively test thresholds and identify that threshold producing the least variability in estimates of development times.[5,26] Lamb[16] proposed a new procedure based on relationships between low temperature development and maximum development. He presents evidence indicating that a minimum developmental threshold can be set as that temperature where development is 8% of the maximum rate, which both improves the estimate and avoids conventional difficulties in determining threshold temperatures.

The practical use of degree-days involves making estimates of daily temperature accumulations. These represent the degrees above the minimum threshold and below the daily temperature cycle. In practice, the actual daily temperature cycle rarely is available; instead, it is approximated based on maximum and minimum temperatures. Higley et al.[27] summarize approaches for estimating temperature accumulations and limitations to their use. In brief, daily accumulations may be calculated as a rectangle (average temperature-minimum developmental threshold) or as the area under a sine wave and above the minimum developmental threshold. Although sine wave estimates frequently are held to be more accurate, both approaches involve assumptions about the shape of daily temperature cycles that are not always valid (particularly when the movement of weather fronts produces rapid temperature shifts).[27]

Many nonlinear models have been developed to describe thermal development, and Wagner et al.[20] reviewed most of these approaches. Other than the Sharpe and DeMichele model, no individual model has been widely adopted. The key problems with curvilinear models are the substantial research required for their development and greater complexity in use. Also, the validity of any given model tends to vary depending upon the individual species being modeled. Nevertheless, curvilinear models clearly represent a more accurate representation of thermal development curve than linear approaches.

Models that reflect the influence of fluctuating temperatures on development represent an extension of basic curvilinear models. Hagstrum and Milliken[28] summarize efforts to develop such models and present their own method for describing development under fluctuating temperatures. Worner[25] presents an insightful, comprehensive discussion on the problems of modeling development under fluctuating temperature regimes. Her analysis suggests that existing approaches based on laboratory and field data are largely inadequate. She concludes that more research is needed, particularly with field data under extreme climate conditions for a species, before accurate predictions of development under fluctuating temperatures is likely to be possible.

c. Limitations

The most obvious limitations in techniques for predicting development relate to their biological validity. None of the techniques, including more advanced approaches, provides a definitive, accurate description of development across most environmental conditions. However, models do differ in their accuracy, although absolute accuracy is not a requirement for initiating sampling programs. Beyond the theoretical validity of any given model, a number of additional factors will limit the accuracy of virtually all models. Indeed, variability from extrinsic factors may overwhelm intrinsic differences between models.

Higley et al.[27] discuss many of these limitations. Among these are effects on development of insect nutrition, availability of water, and hormonal regulation of development (modified by mechanisms of diapause and physiological responses to stimuli such as photoperiod).[29] Approximations in laboratory estimates of development, particularly constant temperature estimates, and estimates of developmental thresholds for degree-days are other potential sources of error. Also, development curves and thresholds differ among the life stages of an insect species, but most often development curves are averaged across life stages for ease of use. This is one aspect of thermal development where simulation models are useful to provide stage-specific development estimates.

Probably the most important limitation to all types of thermal development estimates are the temperature data used to drive those estimates. Short of directly measuring insect body temperatures, field estimates of ambient temperature only provide a gross picture of the actual thermal environment an insect experiences. Baker et al.[30] provide a valuable discussion of the many factors producing discrepancies between weather station temperature data and actual temperatures in the field. These factors include observation time, latitude, surface structure, topography, and urbanization. Additionally, many insect species can actively modify their body temperatures through thermoregulation.[31]

Some limitations may preclude the use of predictions of thermal development for a species. More often, these factors add to variability in developmental estimates. Consequently, accuracy in predictions of thermal development is a function not only of the choice of estimation technique, but also of the extrinsic circumstances in which an estimate is made.

d. Using Estimates of Thermal Development in Sampling

Predictions of pest occurrence based on thermal development usually are of greatest value where temperature limits development. In temperate regions, spring development of insect pests is a common situation where thermal estimates of development are of value. Just as such estimates are needed for initiating sampling activities, so must estimates for thermal development be initiated. Typically, degree-day accumulations are started by calendar date, but specific incidents, such as oviposition or a migration event, also may start thermal development estimates.

Choice of a specific estimation technique reflects a consideration of acceptable accuracy weighed against increasing complexity. The great virtue of the degree-day technique over others is its simplicity of calculation and use. Despite its many theoretical failings, the degree-day approach may be the method of choice, if it can provide sufficient accuracy for a given pest situation. However, degree-days do not work with many species and for these curvilinear models are a necessity. Similarly, the need to consider fluctuating temperatures will differ among species and environments.

Often, complexity of a developmental model mistakenly is assumed to provide greater accuracy. For example, the assumption that simulation models provide better estimates of development than mathematical models is based on a misconception. Actually, simulation models do not represent an alternative to degree-day or curvilinear approaches. Instead, such models incorporate degree-day or curvilinear approaches within the simulation (although simulation models do allow for the use of stage-specific developmental thresholds). Even between degree-day and curvilinear approaches to modeling development, increasing complexity of the model is of itself a poor guide to the accuracy of the model. For instance, degree-day models have proved as good or better predictors than curvilinear models in a number of instances.[32-34] Occam's razor (the principle that given two explanations that fit the data the simple should be chosen over the complex) clearly applies to models of insect development. Consequently, the guiding principle in selecting models to describe insect development is to begin with the simplest model or procedure possible and only add complexity as a specific pest situation requires.

B. INITIATION BASED ON ONGOING MONITORING
1. Hosts

Monitoring of arthropods on their hosts is a direct and commonly practiced technique to determine if sampling should begin. The level of resources devoted to monitoring the host often is related to its economic value and the type of feeding injury. For example, apple usually is monitored more intensively than soybean.

Host monitoring can occur throughout the season, or during a short time interval, depending on pest biology and host susceptibility to injury. In the midwestern U.S., alfalfa is monitored for alfalfa weevil larvae, *Hypera postica*, only during the first growth cycle, and possibly during the early regrowth of the second cycle because adults or larvae are not injurious on subsequent cuttings. Sorghum is monitored for greenbug in Nebraska to indicate arrival of migrants and to indicate time of sampling for assessing mid-to-late season populations.[3] Soybean often is monitored throughout the growing season in the southern U.S. because of the ubiquitous presence of several pest species.

Occasionally, only specific hosts or areas of fields need to be monitored. For example, first generation European corn borer moths initially prefer to oviposit on the tallest corn plants in an area; therefore, monitoring for egg masses may be limited to fields where the corn stand is taller than surrounding areas of the field or surrounding fields.[3] Similarly, corn plants that have green silks may be monitored for second generation European corn borer egg masses because females prefer to oviposit on these plants. Grasshoppers initially feed in weedy, grassy field margins, so monitoring this type of field margin is a common technique to determine if sampling should begin.[35]

Several examples have been provided regarding monitoring of insects on plant hosts. However, host monitoring is not limited to plant hosts. Indeed, pests such as the stable fly, *Stomoxys calcitrans*, face fly, *Musca autumnalis*, and horn fly, *Haematobia irritans*, are observed on cattle throughout the summer. In some cases, medical pests, such as various species of ticks, tsetse flies, *Glossina* spp., and mosquitoes have been monitored on human hosts.[36]

2. Alternate Hosts and Attraction Sites

Many arthropod species use alternate hosts or attraction sites for one or more activities, such as resting, mating, oviposition, or feeding. Consequently, the presence of a species on an alternate host or attraction site may initiate sampling on the primary host. Moreover, the technique for monitoring the alternate host may be similar to monitoring the primary host.

In the spring, newly eclosed European corn borer adults are attracted to grassy or weedy areas at field margins. These areas, termed action sites, have been used to provide an indication that oviposition on corn will soon begin; therefore, moths at action sites indicate that sampling of corn should begin for first-generation egg masses, first instars, or first instar feeding injury.[3] Black cutworm moths prefer to oviposit in low, wet portions of fields that are weedy or grassy. Developing larvae cut crop stalks in these areas or move to crop plants after the weeds have been destroyed by cultivation or herbicides. Monitoring of these areas is important for sampling and management programs.

Chinch bugs overwinter in attractive sites, and feed on alternate hosts, such as wheat, oat, rye, and barley before migrating to corn and sorghum, where they are most injurious and economically damaging. The observation of chinch bugs in alternate host areas (e.g., in wheat fields adjacent to corn or sorghum fields) may initiate sampling to estimate potential economic populations.[3]

3. Trapping

Traps are used to provide direct evidence of the presence of pest species in a specific area. Additionally, the number of individuals captured may indicate the likelihood that an economic population currently exists or may exist during the season. Specimens captured in traps may indicate the emergence of adults or immatures from overwintering sites, the emergence of adults from immature stages, or the immigration of individuals from other regions.

Traps used to capture arthropods generally are left unattended and specimens are collected by humans at specific time intervals.[4] Because they do not require continuous human labor to operate, they usually provide a very cost-efficient technique for arthropod monitoring. Traps vary greatly in size, shape, and especially in mode of collection. In general, traps are divided into two main categories: those that incidentally capture individuals, and those that attract and then capture individuals.[36] Some traps, however, both attract and randomly capture individuals, so the delineation between the two types can be obscure. Traps that randomly capture arthropods include Malaise, window, cone, and pitfall. Traps that attract and then capture arthropods include light, pattern and/or color, Manitoba, shelter, carbon dioxide, tethered animal, boxed animal, bait, kairomone, and pheromone. Traps that both randomly capture and attract arthropods include sticky, color, and water. Detailed descriptions of arthropod traps are provided by Southwood[36] and in Chapter 5 of this book.

The different trapping mechanisms vary in their ability to attract specific arthropod species. For example, most conventional light traps attract many night-flying insects from several orders. Indeed, light trap receptacles can become so cluttered with specimens that identification of the species of interest may be difficult. Conversely, some traps use pheromone lures that are specific to only one species and one sex.

The presence of individuals in traps can lead to the decision to initiate sampling. The decision to begin sampling for several noctuid pests is dependent on their presence in light traps. Pheromone traps are used to assist in the decision to begin sampling or to use a management tactic for pests such as European corn borer, codling moth, *Cydia pomonella*, Mediterranean fruit fly, *Ceratitis capitata*, oriental fruit fly, *Dacus dorsalis*, melon fly, *Dacus curcubitae*, and corn earworm, *Helicoverpa zea*.[4]

4. Surveys

Surveys and surveillance programs usually are organized programs with detailed, established protocols for monitoring arthropods.[4] The programs may either be intensive or extensive, but, at the very least, they rely on established protocols. Federal, state, and private organizations conduct surveillance programs to identify arthropods, and sometimes to quantify pest populations. At the federal level in the U.S., the USDA, through APHIS and PPQ, detects agricultural pests introduced into the U.S., determines the spread of introduced pests into new regions, and determines the abundance of established pests.[4] Individual states also conduct pest surveys, using state entomologists and Cooperative Extension Service specialists. Private consulting companies also engage in survey activities that identify important pests.

Surveillance programs often use methods such as monitoring hosts, monitoring alternate hosts, monitoring attraction sites, and trapping (described above). Arthropods surveyed in the U.S. include gypsy moth, *Lymantria dispar*, Japanese beetle, *Popillia japonica*, *Ixodes damini*, and the Asian tiger mosquito, *Aedes albopictus*. After a pest, whether introduced or indigenous, is identified in a surveillance program, decisions typically are made to begin more formalized and thorough sampling, or to initiate a management tactic.

5. Remote Sensing

Remote sensing in entomology is a relatively new field, initiated by advances in radar, satellite, and optical technologies. Remote sensing, in a broad definition, is the "observation of a target by a device some distance from it".[37] The uses for remote sensing in entomology include the direct observation of insects, the detection of insect-induced injury to plants, and the characterization of environmental factors that affect insects. Remote sensing techniques include photography, videography, satellite photography, multispectral scanning, thermal imaging, radar, and acoustic sounding. Radar imaging, photography, and videography for entomological applications have been conducted both from the ground and from aircraft. Riley[38] provides an extensive review of remote sensing techniques used in entomology.

Remote sensing techniques have yet to be used extensively for determining favorable conditions to begin sampling. However, for many years, forest entomologists have used remote sensing methods to rapidly identify stands of trees injured by several insect pests, such as hemlock looper, *Lambdina fiscellaria fiscellaria*, spruce budworm, *Choristoneura fumiferana*, bark beetles, *Dendroctonus* spp., gypsy moth, and forest tent caterpillar, *Malacosoma disstria*.[38] After injured trees are discovered by remote sensing techniques, sampling then can be initiated to determine if management actions are necessary.[39] Additionally, plants that experience significant abiotic stress, such as drought stress, may be detected by optical and thermal imaging. Because host plants typically are more susceptible to insect injury when stressed by other abiotic and biotic factors, early detection of these conditions with remote sensing techniques may indicate where and when sampling for pests should begin.

In fruit orchards, aerial remote sensing has been used to indirectly detect infestations of insects by identifying injured trees or products produced by insects, such as honeydew.[38] For example, infestations by the European red mite, *Panonychus ulmi*, were identified in peach orchards because low-altitude infrared photographs detected differences between normal and mite-induced chlorotic leaves.[40] Defoliation of pecan by the walnut caterpillar, *Datana integerrima*, has been detected by infrared techniques.[41] In citrus groves, infestations of brown soft scale, *Coccus hesperium*, were detected with aerial photography from the blackened foliage caused by a sooty mold fungus that uses honeydew as a growth medium.[42]

In field crop agroecosystems, the effects of insects have been detected successfully by aerial remote sensing using color and infrared film. Infestations of corn leaf aphids, *Rhopalosiphum maidis*, and sweetpotato whiteflies, *Bemisia tabaci*, have been identified by the photographic detection of sooty mold on leaves.[43,44]

Environmental factors that influence insect pest abundance and location can be detected by aerial and satellite remote sensing techniques. This has been especially useful in detecting short-lived environmental characteristics that dramatically influence pests. For example, both satellite and aerial techniques have been used to identify ephemeral vegetation used by the desert locust, *Shistocerca gregaria*, in Africa.[38] Mosquito breeding habitats in Africa and the U.S. have been readily identified using aerial photography.[45,46] Rainstorms that are responsible for the concentration and deposition of African armyworm adults, *Spodoptera exempta*, have been detected using thermal imaging techniques from satellite systems.[47]

Direct observation of insects with remote sensing has been successful primarily because of advances in radar technology.[38] Migrating moths, locusts, aphids, planthoppers, and leafhoppers have been observed using radar systems. Although radar imaging of migrating insects has not been used to determine when and where to begin sampling, remote sensing of frontal systems responsible for the transport of migrating pests has been used in this manner.

Remote sensing offers tremendous potential for pest management programs. Specifically, by remotely sensing insect activity, the effects of insects, and the environmental factors that affect insects, it is possible to determine when and where to begin sampling. An overriding consideration regarding the use of remote sensing to initiate sampling is that evidence of insect activity or favorable environmental factors is necessary before economic injury occurs.

Remote sensing technology is progressing at a rapid rate.[38] Currently, most remote sensing techniques are relatively expensive, precluding substantive use on all but the most devastating insect pests. Moreover, satellite-based systems have not, as yet, proven as beneficial as aerial-based systems for entomological uses.[38] Advances in optical, thermal, and audio technology undoubtedly will improve the precision and cost-effectiveness of remote sensing systems used for entomological purposes.

C. INITIATION BASED ON PEST IMPACT

Using symptoms of pest impact to trigger sampling and other pest management actions commonly occurs, particularly with occasional pest species. Because the occurrence of such pests is unexpected, monitoring programs for injury often are likely to be the only proactive procedures used. The risk in using injury symptoms to initiate sampling is the prospect that excessive injury will occur before a management decision is made. Therefore, successfully initiating sampling based on pest impact is largely a function of injury rates. If the development of injury is slow, initiating sampling based on symptoms of injury may be acceptable but not if the injury rate is rapid.

In Europe, spruce bark beetle, *Ips typographus*, infestations initially are identified based on symptoms of injured trees. The beetle vectors the fungal pathogen, *Ceratocystis polonica*, which may cause substantial tree mortality. Stands of discolored trees that are dropping their needles and/or the presence of dead trees may indicate an outbreak population. Identifications of injured trees lead to intense sampling activities over a larger area to determine the extent of the outbreak.

For insect larvae, injury rates (consumption rates) are closely tied to development rates. Additionally, injury rates for young larvae are much lower than for later stages.

For example, more than 90% of injury by most defoliating caterpillar species occurs in the last two larval stages.[48] Consequently, monitoring injury or numbers of young larvae can be successful because relatively little injury occurs at this stage. However, environmental conditions that rapidly accelerate development (warm temperatures) or extremely high numbers of insects, even young larvae, dramatically increase injury growth rates. A further potential problem is that young larvae and their injury may not be readily apparent in a field. Some insect behaviors may help conceal both insects and their injury. For example, yellow woollybear, *Spilosoma virginica*, larvae are gregarious in early stages, so their presence is more difficult to detect than if they were widely dispersed.[49] Older, more injurious, larvae rapidly disperse, and sampling must be initiated before this dispersion occurs.

Direct measures of pest impact on a host of itself may be valuable in triggering sampling or pest management action. For example, a key impact of insect defoliation of soybean is to reduce the efficiency of canopy light interception.[50] Consequently, monitoring light interception or canopy leaf area development (which is highly correlated with light interception) would provide a mechanism for identifying when sampling for defoliators might be initiated. Monitoring light interception or canopy development has the potential virtues of relating more directly to potential yield losses then insect numbers and of potentially being an inexpensive form of monitoring (because instrumentation, including some remote sensing procedures, can be substituted for labor).[50]

IV. PRACTICAL CONSIDERATIONS

Choices between timing and monitoring approaches, and between individual techniques, depend upon the specific host, pest, and environmental conditions. Certain hosts or pests will necessitate a given procedure. Other host and pest combinations may present more flexibility.

As is true in all sampling, the more accuracy demanded, the greater the cost. Consequently, the guiding principle for initiating sampling programs is to employ only that degree of effort necessary to start sampling before pest injury becomes unacceptable. The key question for initiating sampling programs is how much accuracy is needed? This in turn directly depends upon how susceptible the crop is to pest attack. Susceptibility is a function of the value of the host and the injuriousness of the pest. Less accuracy is needed for lower value hosts and for less injurious pests. Additionally, sampling is conducted amidst considerable background variability: variability in environmental conditions, particularly weather, variability in host phenology, and, most importantly, variability in pest occurrence and numbers. Therefore, whatever approach is chosen, the approach must accommodate this natural variation. Fortunately, timing and monitoring do not need to provide precise estimates of pest densities, that is, the function of the sampling programs they initiate. Consequently, approaches for starting sampling programs tend to be more forgiving and less quantitative than other features of sampling.

REFERENCES

1. Pedigo, L. P., Hutchins, S. H., and Higley, L. G., Economic injury levels in theory and practice, *Annu. Rev. Entomol.*, 31, 341, 1986.
2. Funderburk, J. E. and Higley, L. G., Managing arthropod pests, in *Sustainable Agriculture: The New Conventional Agriculture,* Hatfield, J. L. and Karlen, D. L., Ed., CRC Press, Boca Raton, FL, 1993.
3. Wright, R. J., Danielson, S. D., Witkowski, J. F., Hein, G. L., Campbell, J. B., Jarvi, K. J., Seymour,

R. C., and Kalisch, J. A., Insect Management Guide for Nebraska Corn and Sorghum, Nebraska Cooperative Extension Service EC 92-1509, Lincoln, 1992.

4. Pedigo, L. P., *Entomology and Pest Management*, Macmillan, New York, 1989.
5. Peterson, R. K. D., Higley, L. G., and Bailey, W. C., Phenology of the adult celery looper, *Syngrapha falcifera* (Lepidoptera: Noctuidae), in Iowa: evidence for migration, *Environ. Entomol.*, 17, 679, 1988.
6. Hutchins, S. H., Smelser, R. B., and Pedigo, L. P., Insect migration: atmospheric modeling and industrial application of an ecological phenomenon, *Bull. Entomol. Soc. Am.*, 34, 9, 1988.
7. Pienkowski, R. L. and Medler, J. T., Synoptic weather conditions associated with long range movement of potato leafhopper, *Empoasca fabae*, into Wisconsin, *Ann. Entomol. Soc. Am.*, 57, 588, 1964.
8. Shaw, M. W. and Hurst, G. W., A minor immigration of the diamond-back moth, *Plutella xylostella* (L.) (*maculipennis* Curtis), *Agric. Meteorol.*, 6, 125, 1969.
9. Domino, R. P., Showers, W. B., Taylor, S. E., and Shaw, R. H., Spring weather pattern associated with suspected black cutworm (Lepidoptera: Noctuidae) introduction to Iowa, *Environ. Entomol.*, 12, 1863, 1983.
10. Wolf, R. A., Pedigo, L. P., Shaw, R. H., and Newsom, L. D., Migration/transport of the green cloverworm, *Plathypena scabra* (F.) (Lepidoptera: Noctuidae), into Iowa as determined by synoptic-scale weather patterns, *Environ. Entomol.*, 16, 1169, 1987.
11. Reap, R. M., Techniques Development Laboratory Three Dimensional Trajectory Model, Technical Procedures Bulletin 255, U.S. Department of Commerce, NOAA, National Weather Service, Washington, D.C., 1978.
12. Pedigo, L. P. and van Schaik, J. W., Time-sequential sampling: a new use of the sequential probability ratio test for pest management decisions, *Bull. Entomol. Soc. Am.*, 30, 588, 1984.
13. Smelser, R. B., The Use of Trajectory Analysis to Project the Long-Range Migration of the Black Cutworm Moth into the Midwest, M.S. thesis, Iowa State University, Ames, 1986.
14. Reamur, R. A. F. de, Observation du thermomètre, faites à Paris pendant l'année 1735, comparees avec cells qui ont été faites sous là ligne à Isle de France, à Alger et an quelque-unes de nos isles de l'Amerique, *Mem. Acad. Sci.*, Paris, 545, 1735.
15. Campbell, A. B., Fraser, B. D., Gilbert, N., Gutierrez, A. P., and Mackauer, M., Temperature requirements of some aphids and their parasites, *J. Appl. Ecol.*, 11, 431, 1974.
16. Lamb, R. J., Development rate of *Acyrthosiphon pisum* (Homoptera: Aphididae) at low temperatures: implications for estimating rate parameters for insects, *Environ. Entomol.*, 21, 10, 1992.
17. Barnes, T., *Textbook of General Physiology*, P. Blakiston & Sons, London, 1937.
18. Sharpe, J. H. and DeMichele, D. W., Reaction kinetics of poikilotherm development, *J. Theor. Biol.*, 64, 649, 1977.
19. Schoolfield, R. M., Sharpe, P. J. H., and Magnuson, C. E., Nonlinear regression of biological temperature-dependent rate models based on absolute reaction-rate theory, *J. Theor. Biol.*, 88, 719, 1981.
20. Wagner, T. L., Wu, H., Sharpe, P. J. H., Schoolfield, R. M., and Coulson, R. N., Modeling insect development rates: a literature review and application of a biophysical model, *Ann. Entomol. Soc. Am.*, 77, 208, 1984.
21. Lowry, R. K. and Ratkowsky, D. A., A note on models of poikilotherm development, *J. Theor. Biol.*, 105, 453, 1983.
22. Hilbert, D. W. and Logan, J. A., Empirical model of nymphal development for the migratory grasshopper, *Melanoplus sanguinipes* (Orthoptera: Acrididae), *Environ. Entomol.*, 12, 1, 1983.
23. Lamb, R. J., Gerber, G. H., and Atkinson, G. F., Comparison of development rate curves applied to egg hatching data of *Entomoscelis americana* Brown (Coleoptera: Chrysomelidae), *Environ. Entomol.*, 13, 868, 1984.
24. Tanigoshi, L. K., Hoyt, S. C., Browne, R. W., and Logan, J. A., Influence of temperature on population increase of *Tetranychus mcdanieli* (Acarina: Tetranychidae), *Ann. Entomol. Soc. Am.*, 69, 712, 1976.
25. Worner, S. P., Performance of phenological models under variable temperature regimes: consequences of the Kaufmann or rate summation effect, *Environ. Entomol.*, 21, 689, 1992.
26. Arnold, C. Y., The determination and significance of the base temperature in a linear heat unit system, *Proc. Am. Soc. Hortic. Sci.*, 74, 430, 1959.
27. Higley, L. G., Pedigo, L. P., and Ostlie, K. R., DEGDAY: a program for calculating degree-days, and assumptions behind the degree-day approach, *Environ. Entomol.*, 15, 999, 1986.
28. Hagstrum, D. W. and Milliken, G. A., Modeling differences in insect development times between constant and fluctuating temperatures, *Ann. Entomol. Soc. Am.*, 84, 369, 1991.
29. Tauber, M. J., Tauber, C. A., and Masaki, S., *Seasonal Adaptations of Insects*, Oxford University Press, New York, 1986.
30. Baker, D. G., Kuehnast, E. L., and Zandlo, J. A., Climate of Minnesota. Part XV. Normal temperatures (1951–80) and their application, *Univ. Minnesota Agric. Exp. Stn.*, AD-SB-2777.

31. May, M. L., Insect thermoregulation, *Annu. Rev. Entomol.*, 24, 313, 1979.
32. Hochberg, M. E., Pickering, J., and Getz, W. M., Evaluation of phenology models using field data: case study for the pea aphid, *Acyrthosiphon pisum*, and the blue alfalfa aphid, *Acyrthosiphon kondoi* (Homoptera: Aphididae), *Environ. Entomol.*, 15, 227, 1986.
33. Stinner, R. E., Rock, G. C., and Bacheler, J. E., Tufted apple budmoth (Lepidoptera: Tortricidae): simulation of postdiapause development and prediction of spring adult emergence in North Carolina, *Environ. Entomol.*, 17, 271, 1988.
34. McClain, D. C., Rock, G. C., and Stinner, R. E., San Jose scale (Homoptera: Diaspididae): simulation of seasonal phenology in North Carolina orchards, *Environ. Entomol.*, 19, 916, 1990.
35. Danielson, S. D., Wright, R. J., Witkowski, J. F., Hein, G. L., Campbell, J. B., Jarvi, K. J., Seymour, R. C., and Kalisch, J. A., Insect Management Guide for Nebraska Alfalfa, Soybeans, Wheat, Range, and Pasture, Nebraska Cooperative Extension Service EC 92-1511, Lincoln, 1992.
36. Southwood, T. R. E., *Ecological Methods with Special Reference to the Study of Insect Populations*, 2nd ed., Chapman & Hall, New York, 1978.
37. Barrett, E. C. and Curtis, L. F., *Introduction to Environmental Remote Sensing*, 2nd ed., Chapman & Hall, New York, 1982.
38. Riley, J. R., Remote sensing in entomology, *Annu. Rev. Entomol.*, 34, 247, 1989.
39. Ciesla, W. M., Operational remote sensing for forest insect and disease management—some challenges and opportunities, *Proc. Remote Sensing Nat. Resour., Moscow, Idaho, 1979*, 1980, 39.
40. Payne, J. A., Hart, W. G., Davis, M. R., Jones, L. S., Weaver, D. J. et al., Detection of peach and pecan pests and disease with color infrared aerial photography, *Proc. Bienn. Workshop Color Aerial Photogr. Plant Sci.*, Gainesville, FL, 1971, 216.
41. Harris, M. K., Hart, W. G., Davis, M. R., Ingle, S. J., and Van Cleave, H. W., Aerial photographs show caterpillar infestation, *Pecan Q.*, 10, 12, 1978.
42. Hart, W. G. and Ingle, S. J., Detection of arthropod activity on citrus foliage with aerial infrared color photography, *Proc. Workshop Aerial Color Photogr. Plant Sci.*, Gainesville, FL, 1969, 85.
43. Wallen, V. R., Jackson, H. R., and MacDiarmid, S. W., Remote sensing of corn aphid infestation, 1974 (Hempitera: Aphididae), *Can. Entomol.*, 108, 751, 1976.
44. Nuessly, G. S., Meyerdirk, D. E., Hart, W. G., and Davis, M. R., Evaluation of color-infrared aerial photography as a tool for the identification of sweetpotato whitefly induced fungal and viral infestations of cotton and lettuce, *Proc. Bienn. Workshop Color Aerial Photogr. Plant Sci.*, Weslaco, Tex., Am. Soc. Photogramm. Remote Sensing, 1987, 141.
45. Muirhead-Thomson, R. C., The use of vertical aerial photographs in rural yellow fever mosquito surveys, *Mosq. News*, 33, 241, 1973.
46. Wagner, V. E., Hill-Rowley, R., Norlock, S. A., and Newson, H. D., Remote sensing: a rapid and accurate method of data acquisition for a newly formed mosquito control district, *Mosq. News*, 39, 283, 1979.
47. Garland, A. C., The Use of Meteosat in the Analysis of African Armyworm Outbreaks in East Africa, Ph.D. thesis, University of Bristol, Bristol, U.K.
48. Hutchins, S. H., Higley, L. G., and Pedigo, L. P., Injury equivalency as a basis for developing multiple-species economic injury levels, *J. Econ. Entomol.*, 81, 1, 1988.
49. Peterson, R. K. D., Higley, L. G., Buntin, G. D., and Pedigo, L. P., Flight activity and ovarian dynamics of the yellow woollybear, *Spliosoma virginica* (Lepidoptera: Arctiidae), in Iowa, *J. Kan. Entomol. Soc.*, 66, 97, 1993.
50. Higley, L. G., New understandings of soybean defoliation and their implications for pest management, in *Pest Management of Soybean*, Copping, L. G., Green, M. B., and Rees, T. R., Eds., Elsevier, Amsterdam, 1992.

Chapter 8

SEQUENTIAL SAMPLING FOR CLASSIFYING PEST STATUS

Michael R. Binns

TABLE OF CONTENTS

I.	Introduction	138
II.	The Sequential Probability Ratio Test or SPRT	140
III.	Iwao's Confidence Interval Method	143
IV.	Comparison of SPRT and Iwao	144
V.	Three-Decision SPRT	146
VI.	Sequential Sampling in Practice	148
VII.	Computer Programs	150
VIII.	Conclusion	150
	Acknowledgments	151
	Appendix I	151
	Appendix II	171
	References	173

ISBN 0-8493-2923-X/94/$0.00+.50
© 1994 by CRC Press, Inc.

I. INTRODUCTION

From at least as far back as the invention of the wheel and the axe, human beings have been concerned, consciously or subconsciously, with quality. Before mass production of manufactured goods, people (manufacturers or consumers) could rely on their own judgment before entering into a contract to buy or sell, but with the introduction of mass production (especially in the early 1900s), this was in general no longer possible. Buyers were still concerned with quality, and manufacturers soon realized that even if they set their production lines to attain a prescribed quality level, they could not ensure that random or other effects would not interfere with production and degrade the quality; thus, a serious possibility existed that buyers might refuse to accept the product. Quality control methods were introduced early this century, so that manufacturers could recognize a deterioration of quality in time to readjust the settings before products left the factory. Improvements and extensions of the original ideas to cover the whole manufacturing process continue to prove important today.[1,2]

The simplest kind of quality control is (1) take samples from an ongoing process, (2) infer from them what the process is doing, and (3) make a decision either to leave well enough alone or reset the process. Before implementing this procedure, it is necessary to devise a sampling plan, which should give the required information, be simple and cheap to operate, and not interfere with the process. Early work in quality control sampling was done collaboratively between Bell Telephone Laboratories and the Western Electric Company.[3] It should be noted that the goal of quality control sampling is to make a decision, not estimate a property of the process, and that generally this results in simpler sampling protocols and fewer samples (on average).

In agriculture pest management, the "process" might be the set of fields or orchards which could be sampled by a commercial scouting operation, or the development over time of pest status in a field. When enough information on the types of distribution of the pests and their potential for damage is obtained, the operator can choose a sampling and decision-making protocol, balancing economy of sampling with the expected quality of the decisions to be made. The latter is best summarized by the operating characteristic (OC) function which is equal to the probability of deciding not to intervene (e.g., spray with insecticide), as a function of pest abundance (or a surrogate of pest abundance, such as an indicator of pest damage) (Figure 1a).

The ideal OC curve would be equal to unity up to a critical threshold, m_t, and then be equal to zero, but this would require a complete census of the population. In general, the steeper the slope of the OC curve around m_t, the better the quality control. The value m_t below which spraying is not desirable and above which it is desirable must be obtained from an analysis of the economic damage which the pest could inflict if left alone. These analyses might include other functions related to the OC curve: the average outgoing quality (AOQ) and the average outgoing quality limit (AOQL). The AOQ is illustrated for a simple case in Figure 1b by the expected pest density with and without sampling and/or spraying. If spraying is done without sampling, the expected number of pests surviving depends only on the efficiency of the pesticide. If a decision-making protocol is used, the average number of pests remaining is represented by a curve which starts out following the "no spray" line, but moves gradually to the "always spray" line in a manner dependent on the OC curve. The AOQL is the maximum of the AOQ curve. The average long-term economic effect of the protocol can be calculated from the AOQ if the effect of various pest densities on the crop is known. The potential gains and losses of competing management sampling protocols should be assessed before any one (or none) is chosen.[4,5]

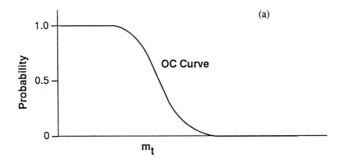

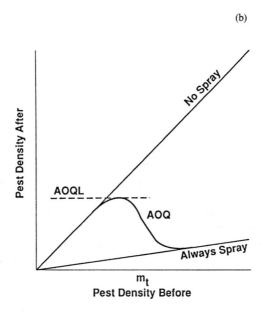

FIGURE 1. (a) A typical operating characteristic (OC) curve, relating pest density to the probability of deciding not to intervene; m_t represents the threshold above which pest density is economically injurious to the crop. (b) The effect on pest density of implementing a (87.5% efficient spray) decision-making scheme with the OC curve shown in (a) is represented by the average outgoing quality (AOQ) curve; the AOQ limit (AOQL) is the maximum of the AOQ curve.

The area to be sampled in any particular instance may or may not be homogeneous with respect to the species being sampled, and costs of field sampling vs. species identification (possibly in a laboratory) can vary considerably. How to obtain optimum sampling procedures covering these problems and in general is discussed in earlier chapters, but sequential sampling is ideally suited to situations where early stopping is not expected to introduce bias, and fewer samples do, in fact, lead to worthwhile *cost* reductions. For decision-making, sequential sampling is often the only method to be seriously considered because it allows sampling to cease as soon as it becomes obvious which decision should be made. It can result in a 50% reduction in sampling effort relative to fixed size plans,[6] and occasionally much more.[7] Sample units (leaves, twigs, stems, etc.) are selected, the cumulative total number of pests found on them is plotted on a graph as samples are accumulated, and sampling stops when this path crosses a predetermined boundary (Figure 2). Two sequential sampling protocols are

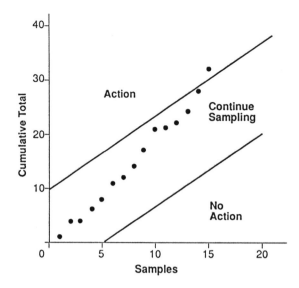

FIGURE 2. Example implementation of a sequential probability ratio test (SPRT), stopping after 15 samples with the "action" recommendation.

commonly used in entomology: the sequential probability ratio test (SPRT)[8] and a confidence interval method.[9] If sequential sampling is used, another important property of the decision-making protocol must be obtained, namely the average sample number (ASN) curve, which shows the expected number of samples required for each value of the pest density.

Sequential sampling has been used in pest management for many years,[6,10] with earliest applications appearing in forestry. Shepard[11] gives a good practical summary of the application of sequential methods in entomology. Although the text by Armitage[12] was written for medical researchers, much of it may be useful for agriculturalists, because it was written for those with little mathematical training. Texts such as that of Siegmund[13] were written for mathematicians. Wald[8] and Wetherill and Glazebrook[14] are mathematical but not as much as Siegmund. Schilling[15] has written an encyclopedic book on acceptance sampling in quality control for those with some mathematical background.

Sequential sampling for one-time decision-making is described in this chapter. Binomial sequential sampling for decision making is discussed in Chapter 9. Sequential sampling for estimation is discussed in Chapter 10. Sequential sampling for biological control decision making is discussed in Chapter 11. Sampling over time is discussed in Chapter 12.

II. THE SEQUENTIAL PROBABILITY RATIO TEST (SPRT)

Sequential sampling for decision making in the form we know it today originated in the Statistical Research Group (SRG), based at Columbia University, New York, during the period 1942 to 1945.[16] In a conversation with the director of SRG (Wallis), a Navy captain stated that part way through a long test comparing firing procedures he could often easily tell that one procedure was inferior... so why continue the test? Similar observations had been made in different contexts for many years, but this time the mathematical ability to solve the problem was available. After mulling over the problem for a few days with some of his colleagues, Wallis finally posed the problem to Wald. For a short while the test, assuming it existed, was called "super colossal", because it was going to be more powerful than so-called "most powerful tests" (statisticians enjoy using extreme language to describe their work). However, the more classical term "sequential test" was eventually coined, and within a few months the basic theory was completed (Wald,[8] Introduction). Wald proved that the intuitive idea of applying the well-established theory of "likelihood ratio tests" to the data as they are gathered has very satisfactory properties, and showed how to estimate the OC and ASN curves.[8]

The likelihood is defined as the probability function with the data value(s) inserted. For example, if the distribution of the data $x_1, x_2, \ldots, x_n$ is known to be Poisson with mean m (see Chapter 3), the likelihood for the first data point, x_1, is

$$L(x_1|m) = e^{-m} m^{x_1}/x_1! \tag{1}$$

and its logarithm is

$$l(x_1|m) = \ln[L(x_1|m)] = -m + x_1 \ln(m) - \ln(x_1!) \tag{2}$$

The log-likelihood for all n values of x is the sum over all the data:

$$l(x_1, x_2, \ldots, x_n|m) = -nm + \Sigma x_i \ln(m) - \Sigma \ln(x_i!)$$

$$= -nm + T_n \ln(m) - \Sigma \ln(x_i!) \tag{3}$$

where Σ mean "the summation of" and $T_n = \Sigma x_i$. The log of the likelihood ratio for testing m_1 against m_0 ($m_1 > m_0$) is defined as

$$\ln\{L(x_1, x_2, \ldots, x_n|m_1) \div L(x_1, x_2, \ldots, x_n|m_0)\}$$

$$= l(x_1, x_2, \ldots, x_n|m_1) - l(x_1, x_2, \ldots, x_n|m_0)$$

$$= -n(m_1 - m_0) + T_n \ln(m_1/m_0) \tag{4}$$

A common notation for the likelihood ratio for the ith data point and any distribution is

$$z_i = \ln\{L(x_i|m_1) \div L(x_i|m_0)\} = l(x_i|m_1) - l(x_i|m_0) \tag{5}$$

Wald showed that to get a test of m_1 against m_0 with probabilities α and β of wrongly deciding for m_1 or m_0, respectively, sampling should continue as long as

$$\ln[\beta/(1-\alpha)] < \Sigma z_i < \ln[(1-\beta)/\alpha] \tag{6}$$

and stop, choosing m_0 if the left-hand inequality is broken, or choosing m_1 if the right-hand inequality is broken. For a wide range of distributions, this complicated double inequality can be transformed into simple inequalities for cumulative sums of the data. For example, if the Poisson distribution is assumed,

$$z_i = -n(m_1 - m_0) + T_n \ln(m_1/m_0) \tag{7}$$

so sampling is continued as long as

$$\ln[\beta/(1-\alpha)] < -n(m_1 - m_0) + T_n \ln(m_1/m_0) < \ln[(1-\beta)/\alpha] \quad (8)$$

or

$$\ln[\beta/(1-\alpha)]/\ln(m_1/m_0) + n(m_1 - m_0)/\ln(m_1/m_0) < T_n$$
$$< \ln[(1-\beta)/\alpha]/\ln(m_1/m_0) + n(m_1 - m_0)/\ln(m_1/m_0) \quad (9)$$

This can be rewritten as

$$h_1 + sn < T_n < h_2 + sn \quad (10)$$

where

$$h_1 = \ln[\beta/(1-\alpha)]/\ln(m_1/m_0) \quad (11)$$

$$h_2 = \ln[(1-\beta)/\alpha]/\ln(m_1/m_0) \quad (12)$$

$$s = (m_1 - m_0)/\ln(m_1/m_0) \quad (13)$$

have geometrical meanings: in a graph of the cumulative sum (T_n) against the sample size (n), s, h_1, and h_2 are the slope and intercepts of boundary stopping lines (Figure 2).

These formulas, and corresponding ones for the binomial, negative binomial, and normal distributions (Chapter 3) are given in Table 1. Cumulative counts can be plotted against the sample size until the path crosses either of two parallel straight lines, at which point sampling stops and a decision is made: if the upper boundary is crossed, a decision is made corresponding to belief that the mean is m_1 (e.g., spray with insecticide); if the lower boundary is crossed, a decision is made corresponding to the belief that the mean is m_0 (e.g., do nothing). Plans such as that shown in Figure 2 are called "open," because there is no upper limit on the sample size.

Because the likelihood can be calculated only if the underlying distribution of individuals on sample units is known, it is important to assess whether any of these distributions actually fits the spatial dispersion of the species to be sampled (Chapter 3). Some distributions (e.g., Neyman Type A), have SPRTs which cannot be repre-

TABLE 1
Formulas for Slopes and Intercepts of SPRT Plans

Distribution	Slope	Intercept factor[a]
Binomial[b]	$\dfrac{\ln(q_0/q_1)}{\ln\{(p_1 q_0)/(p_0 q_1)\}}$	$\dfrac{1}{\ln\{(p_1 q_0)/(p_0 q_1)\}}$
Poisson	$\dfrac{(m_1 - m_0)}{\ln(m_1/m_0)}$	$\dfrac{1}{\ln(m_1/m_0)}$
Negative binomial	$\dfrac{k \ln\{(k+m_1)/(k+m_0)\}}{\ln\{[m_1(k+m_0)]/[m_0(k+m_1)]\}}$	$\dfrac{1}{\ln\{[m_1(k+m_0)]/[m_0(k+m_1)]\}}$
Normal	$(m_1 + m_0)/2$	$\sigma^2/(m_1 - m_0)$

[a] To be multiplied by $\ln[\beta/(1-\alpha)]$ for the lower intercept, and $\ln[(1-\beta)/\alpha]$ for the upper intercept.

[b] $q_1 = 1 - p_1$ and $q_0 = 1 - p_0$.

sented by straight line boundaries as in Figure 2, which may be why they are generally avoided! However, in some instances, either a more complicated distribution must be considered[17,18] or large samples and a normal approximation must be assumed (see Section III).

OC and ASN curves can be calculated from standard formulas.[19] A computer program to do this for the Poisson, binomial, negative binomial, and normal distributions is reproduced in Appendix I. The formulas for OC and ASN curves were obtained by Wald assuming that the sample path always ended *on* the boundary. It should be noted, however, that in practice, sample paths rarely end *on* a boundary, but usually beyond it. Because of this, the formulas for OC and ASN curves are only approximations (the actual OC curve tends to be steeper than the formula suggests, while the actual ASN tends to be higher than the ASN formula). The differences are usually slight, but the ability to use a computer simulation program to check on the difference is important.

Although the theory for SPRTs was developed in terms of the four parameters α, β, m_0, and m_1 and their meaning in terms of hypothesis testing, it is probably more useful for pest management decision-making to regard these simply as parameters of OC and ASN curves. More precisely, the four parameters define the slope and intercepts of the boundary lines. The slope, which, for the distributions commonly used, is just a function of m_0 and m_1 (Table 1), roughly defines the position of the center of the OC and ASN curves [near $(m_0 + m_1)/2$]. The vertical distance between the lines, which, for the same distributions, is the intercept factor of Table 1 multiplied by $\ln[(1 - \alpha)(1 - \beta)/\alpha\beta]$ (Table 1), roughly defines the steepness of the OC curve and the height of the ASN curve. A practical approach, therefore, is to estimate OC and ASN curves for different sets of these four parameters and choose the one that has the most attractive properties, such as steepness of the OC curve around the critical mean density or lower ASN.[4]

In practice, the use of open plans means that very large sample sizes can occur when the field being sampled has mean infestation about midway between the two critical mean values (m_0 and m_1). For such mean densities there is usually a degree of ambivalence on whether intervention is or is not necessary, so excessive time spent on sampling there is wasted. Setting a limit on the maximum sample size helps reduce such waste. If the limit is not too low there may be only small changes to the OC and ASN curves (e.g., Siegmund,[13] Chapter 3). However, if the limit is low (e.g., less than twice the ASN), it is advisable to check the effect on the OC and ASN curves by simulation. More complicated adaptations of the SPRT with converging boundaries, such as those used in clinical trials,[12] are not likely to be necessary in agricultural pest management.

III. IWAO'S CONFIDENCE INTERVAL METHOD

Whatever the distribution of x, if the sample size (n) is large, the distribution of

$$\Sigma x_i/n = T_n/n \qquad (14)$$

is approximately normal, so an approximate confidence interval for the mean (m) can be written as

$$T_n/n - z_\alpha \sigma/\sqrt{n} \leq m \leq T_n/n + z_\alpha \sigma/\sqrt{n} \qquad (15)$$

where z_α is the $100(1 - \alpha)\%$ normal deviate and σ is the population standard deviation. Iwao[9] suggested turning this into a test for a critical threshold m_t, using a standard relationship between m_t and σ [$\sigma = \sigma(m_t)$] to estimate σ: continue sampling as long as

$$m_t - z_\alpha \sigma(m_t)/\sqrt{n} < T_n/n < m_t + z_\alpha \sigma(m_t)/\sqrt{n} \qquad (16)$$

or

$$nm_t - z_\alpha \sigma(m_t)\sqrt{n} < T_n < nm_t + z_\alpha \sigma(m_t)\sqrt{n} \qquad (17)$$

and stop when one of the boundaries is crossed.

There are three problems which must be addressed before using this procedure: (1) the normal distribution assumption, (2) imperfect knowledge of σ, and (3) use of a fixed sample size procedure in a sequential context.

Non-normality of the distribution is generally satisfied if a minimum number of samples is taken; 30 samples should be enough, but even 20 has been suggested.[20] Taking all samples in batches of, say, 20 to satisfy the normality assumption is unlikely to be necessary and would result in extra sampling effort. The most popular formulas for $\sigma = \sigma(m_t)$ are those of Taylor[21] and Iwao.[22] There has been much discussion on the relative merits of these formulas in terms of applicability,[23,24] and transportability[25] (see also Chapter 3). Therefore, some care is required before either is used in practice. The most fundamental consequence of using a fixed sample size testing procedure is that the expected significance rate tends to be greater than the nominal significance probability, α.[12] Although mathematical approximations are available for certain distributions,[13] Nyrop and Simmons[20] came to the conclusion that the simplest way to obtain realistic properties of Iwao's scheme was by simulation, as with SPRT. Nyrop and Binns[4] presented a Fortran program to do this, and an updated version is included in Appendix I.

IV. COMPARISON OF SPRT AND IWAO

The often-quoted main attraction of Iwao's method is that it can be applied even if no distribution has been found which fits the data. A more accurate statement is that it relies on large sample sizes and the central limit theorem[26] so that the *normal* distribution can be assumed. Because an SPRT exists for the normal distribution, a direct comparison of the two methods can be made.

Suppose that the critical density is m_t and $\sigma^2 = am^b$. Then Iwao's boundaries are

$$nm_t - z\sqrt{(nam_t^b)} < T_n < nm_t + z\sqrt{(nam_t^b)} \qquad (18)$$

Letting $m_t = (m_1 + m_0)/2$, the SPRT boundaries are (Table 1)

$$n(m_1 + m_0)/2 + \ln[\beta/(1-\alpha)]am_t^b/(m_1 - m_0) < T_n$$
$$< n(m_1 + m_0)/2 + \ln[(1-\beta)/\alpha]am_t^b/(m_1 - m_0) \qquad (19)$$

Because the distance between Iwao's limits increases with n, they must intersect the SPRT limits. This is exemplified in Figure 3 for $a = 4$, $b = 1.5$, $z = 1.5$, $m_t = 2.5$, $m_0 = 2$, $m_1 = 3$, $\alpha = 0.1$, $\beta = 0.05$. The OC and ASN curves for the two procedures were compared based on the negative binomial distribution (Table 2); k was estimated from Taylor's power law (TPL) variance-mean relationship by the method of moments (i.e., $k = m^2/[am^b - m]$). The SPRT parameters α and β were chosen to

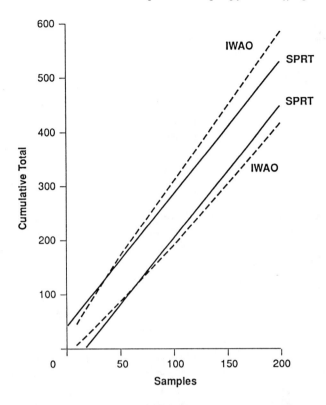

FIGURE 3. Boundaries for comparable SPRT and Iwao procedures; normal SPRT with $m_0 = 2$, $m_1 = 3$, $\alpha = 0.1$, and $\beta = 0.05$ and Iwao variance formula $\sigma^2 = 4m^{1.5}$, $z = 1.5$, and $m_t = 2.5$.

TABLE 2
Comparison between the Sequential Procedure of Iwao and the SPRT by Simulation[a]

| | No limit on sample size | | | | Sample size ≤ 100 | | | |
| | OC | | ASN | | OC | | ASN | |
Mean	Iwao[b]	SPRT[c]	Iwao	SPRT	Iwao[b]	SPRT[c]	Iwao	SPRT
1.4	0.998	0.999	33	43	0.999	0.999	33	42
1.6	0.995	0.997	44	52	0.992	0.995	41	51
1.8	0.984	0.983	63	65	0.970	0.973	52	61
2.0	0.957	0.946	108	84	0.915	0.914	63	70
2.2	0.896	0.833	223	107	0.783	0.788	71	76
2.4	0.745	0.601	1056	123	0.613	0.609	74	78
2.6	0.231	0.340	1094	118	0.415	0.421	72	75
2.8	0.106	0.141	233	97	0.259	0.254	67	68
3.0	0.049	0.053	119	74	0.151	0.130	60	60
3.2	0.025	0.019	76	58	0.072	0.063	52	52
3.4	0.016	0.007	55	48	0.035	0.026	45	45
3.6	0.010	0.002	44	40	0.018	0.012	40	40

[a] Simulation parameters: negative binomial with k estimated from $\sigma^2 = 4m^{1.5}$, 10,000 trials for each m, minimum sample size = 20.
[b] Iwao parameters: $\sigma^2 = 4m^{1.5}$, $z = 1.5$, $m_t = 2.5$.
[c] Normal distribution SPRT parameters: $\alpha = 0.1$, $\beta = 0.05$, $m_0 = 2$, $m_1 = 3$, $\sigma^2 = 4(2.5)^{1.5} = 15.81$, intercepts = -45.7, and 35.6, slope = 2.5.

make the two OC curves nearly the same, especially at $m = 2$ and $m = 3$. When the boundaries were kept open, the ASN for Iwao was generally much greater than that for SPRT from m_0 to m_1, which reflects the optimum property of the SPRT: any sequential test with the same error probabilities (α, β at m_0, m_1) as the SPRT has higher expected sample sizes there.[27] However, between the SPRT critical values ($2.2 \leq m \leq 2.8$), the Iwao OC curve was steeper than the SPRT OC curve. When a maximum sample size of 100 was imposed, there was little difference between the OCs or ASNs. The author has repeated these simulations using the Adès distribution,[28] using parameters $\tau = 1.5$ and $r = 3.3$, which correspond closely to the TPL values $a = 4$ and $b = 1.5$ (in general, τ is essentially equivalent to b). For the closed tests, the results were similar to those shown in Table 2, except that at $m = 2.4$ and 2.6, the OC curves were steeper; for the open tests the results were again mostly similar, but the Iwao ASN was lower at these values *and* the OC was steeper.

These results suggest that there may be no need to use Iwao's procedure but this can be checked by simulation in any application.

V. THREE-DECISION SPRT

In some situations, a simple decision to intervene or not is not useful enough, especially because pest abundance is rarely static. In economic terms there may be a range of pest densities which are regarded as endemic, where there is no immediate danger to the crop but where there is a possibility that a damaging outbreak may occur quickly and unexpectedly. Below that range there is little chance of the pest attaining seriously high densities at least for a reasonable period of time. Although this problem might seem to call for an estimation procedure, a simpler procedure involving only decision-making sampling is available:

- Density is very high: intervention necessary
- Density is very low: it will be a long time before it could reach dangerous levels, so wait some time before resampling
- Density moderate: nothing need be done now, but keep a close watch

Management of phytophagus mites in apples falls into this category.[29] See also Chapter 11 for a discussion of monitoring prey and predator species together.

It is fairly easy to set up two SPRTs, one of which is designed to decide between very high and moderate densities, and the other to decide between moderate and very low densities. If these are superimposed, a three-decision SPRT is created with four boundaries, such as that of Waters[30] shown in Figure 4.

There are, however, some practical difficulties in implementing such a scheme. If a sample path crosses the (dotted) line AB, sampling should not stop as if only the lower SPRT were working, because the path may eventually come out in favor of "high". However, if it then crosses the (dotted) line AC, should sampling stop and a decision for "medium" be made? The consensus is that it should,[14] but if the sampling protocol is set up in such a way that this does not happen, it may not make much difference, except to the ASN. (The same observations apply *mutatis mutandis* for sample paths crossing AC and then AB.) Another problem is what to do if the sample path crosses a (dotted) boundary below C. Crossing the first of these cannot be taken as acceptance of "low", and a subsequent crossing of the second cannot be taken as acceptance of "medium". Various methods have been proposed to deal with this situation. Armitage[31] suggested adding a third SPRT to test "high" vs. "low" directly, thus adding a third pair of boundary lines; Billard and Vagholkar[32] effec-

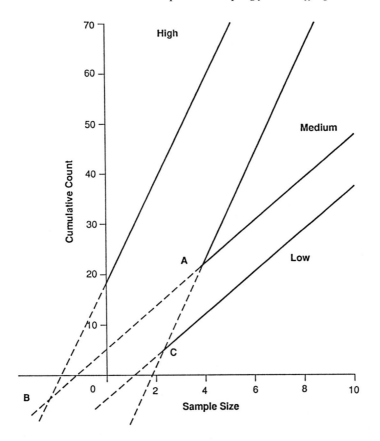

FIGURE 4. Boundaries for a three-decision SPRT, based on the parameters in Waters (1955), indicating the important points of intersection of the two simple SPRT boundaries.

tively suggested that a minimum sample size should be imposed beyond C, thus eliminating the problem altogether. It is possible also, depending on the parameters of the two SPRTs, that B (like C here) could be reached by a sample path: the same types of solutions are still available. The computer program calculates and prints the positions of A, B, and C, giving a warning if B or C is potentially an accessible point, subsequently insisting that a sufficiently high minimum sample size is used; it also stops in favor of "medium" if both AB and AC are crossed.

No satisfactory mathematical formulas exist for calculating the OC and ASN curves for such a scheme; the best way of finding them is by simulation. The computer program for SPRT is designed to do this. Using this program, it is possible to estimate the OC and ASN curves for Waters'[30] three-decision plan based on an underlying negative binomial distribution. Because there are three possible outcomes, one OC curve is not enough to describe the properties of a three-decision SPRT. Waters presented the theoretical OCs for the two simple SPRTs, but then the "medium vs. high" OC curve cannot be interpreted as the probability of deciding for "medium", because all of the decisions for "low" are also included in that OC.

In Table 3, the (simulation) probabilities of reaching any one of the three decisions are all presented, any two of which would be a good summary. The ASN for a three-decision SPRT has two peaks corresponding to the two pairs of parallel lines, and is exemplified in Table 3 also. Simulations were run for each of the two simple SPRTs separately, to investigate how the OC curves differed. For this example, there

TABLE 3
Simulation Results for Three-Decision SPRT Proposed by Waters (1955)[a]

	Three-decision SPRT[b]				Two-decision SPRTs			
	OC[c]				OC[d]		ASN	
Means	Low	Medium	High	ASN	Low	High	Low	High
1	1.000	0.000		4.0	1.000		2.2	2.3
2	0.999	0.001		4.2	0.999		2.9	2.8
3	0.968	0.032		5.4	0.954		4.6	3.1
4	0.669	0.331		7.7	0.660		6.6	3.5
5	0.230	0.770		7.3	0.234		5.7	4.1
6	0.054	0.946	0.000	6.4	0.063	0.000	4.0	5.0
7	0.012	0.987	0.001	6.9	0.019	0.002	2.9	6.3
8	0.004	0.989	0.007	9.0	0.006	0.008	2.3	8.8
9	0.002	0.934	0.064	13.4	0.003	0.067	2.0	13.2
10	0.000	0.674	0.326	18.2	0.001	0.319	1.7	18.3
11		0.277	0.723	17.2	0.001	0.720	1.6	17.3
12		0.079	0.921	12.4	0.000	0.919	1.5	12.2
13		0.020	0.980	9.0		0.974		8.5
14		0.006	0.994	7.1		0.995		6.6
15		0.001	0.999	5.9		0.997		5.2
16		0.001	0.999	5.1		1.000		4.4

[a] Using the original SPRT parameters: m_0, m_1 = 3, 6 for the first SPRT and 9, 12 for the second; $\alpha = 0.1$, $\beta = 0.1$; $k = 7.228$; open plans with 10,000 simulations each.
[b] Minimum sample size = 3 (see Figure 4).
[c] Simulation probabilities of accepting that the density is in the low, medium, or high category, in the three-decision SPRT.
[d] Simulation probabilities of accepting that the density is in the lower category in the first SPRT (low), and in the higher category in the second SPRT (high), when running the two SPRTs independently.

was very little difference (Table 3). Although this is but one example, the effect of running two SPRTs together should make little difference to the probabilities of deciding for either of the extreme densities, and therefore to the probability of making the intermediate decision.[14] Thus the OCs of a three-decision SPRT can be approximated by the OCs of the component SPRTs, provided that the second OC is correctly interpreted (see previous and Table 3). Estimating the ASN in the same way is not possible.[33]

VI. SEQUENTIAL SAMPLING IN PRACTICE

Considerations of randomness, bias, etc., that are important for nonsequential sampling are probably even more important for sequential sampling. If, for example, an X or W pattern in the field is optimum for fixed sample size (Chapter 6), the pattern can still be followed, but care must be taken to avoid the bias which could arise if the points on the sequential path (e.g., Figure 2) always follow the field sequence. The estimate of the economic threshold is also more critical for sequential sampling, because it helps define the entire system: if experience shows that it must be changed, the whole sampling protocol, boundaries and all, needs to be changed.

A good practical discussion of the difficulties involved in setting up a sequential sampling plan was given by Sylvester and Cox.[34] Based on 2 years of data, they showed that the negative binomial distribution could describe the spatial pattern of aphids on sugar beets, but that k was not constant. Two solutions were proposed:

binomial sampling (see Chapter 9) or choosing a compromise value of k. The trouble with using a single k, when in reality it varies, is that the formulas for the OC and ASN curves are no longer valid. Warren and Chen[35] discussed the problem and concluded that "light misspecification of the parameter k will not be harmful and, indeed, underestimation of k will result in improved error rates, albeit at the expense of some additional sampling." However, they considered mean values, m, equal to 10 or more and k equal to 1 or more. Hubbard and Allen[36] obtained approximate formulas for the OC and ASN curves when k is misspecified, and showed that for smaller values of k and m than those considered by Warren and Chen, there could be appreciable effects on the OC and ASN. Simulation can, however, be used to check on the effect in any particular instance.

Sylvester and Cox[34] proposed a compromise k-value of 0.81 as being appropriate for m near 1. With α, β, m_0, and m_1 equal to 0.1, 0.1, 0.9, and 1.1, respectively, they calculated intercepts and slope to be ± 24.46 and 0.997. The effects on the OC and ASN curves of k as low as 0.3 or as high as 1.1 (their estimates of k ranged from 0.005 to 1.6) are shown in Table 4. The decision errors at $m = 0.9$ and $m = 1.1$ are approximately halved or doubled, but, although these discrepancies are large, considering the difficulty of the problem and the state of knowledge then, the compromise reached by the original authors could hardly have been improved, at least as far as the OC is concerned. However, the ASN is large and becomes very large if k is underestimated. This was hinted at by the authors ("...something in excess of 200 plants...actual value of 0.9 aphids per plant"). If an estimate of TPL had been available,[21] and the data fitted the negative binomial distribution for all mean values (k not constant), it would have been possible to estimate k by the method of moments through TPL (see previous) and thus estimate the effects of changes in k on the OC and ASN. Using a program such as that in Appendix I, it might therefore be possible to adjust the four SPRT parameters to get realistic and acceptable OC and ASN curves, but there is no guarantee of this. Alternatively, a binomial sampling scheme with tally number greater than zero might be robust with respect to some variability in k, as described by Binns and Bostanian.[37]

TABLE 4
Simulation Results on the Sensitivity to Varying Values of k of Sylvester and Cox's (1961) Sampling Protocol Based on the Negative Binomial Distribution[a]

	OC			ASN		
$m \setminus k$	0.3	0.81	1.1	0.3	0.81	1.1
0.6	0.999	1.000	1.000	62	63	63
0.7	0.990	1.000	1.000	83	84	84
0.8	0.945	0.994	0.998	112	127	127
0.9	0.795	0.916	0.935	151	217	228
1.0	0.511	0.486	0.480	161	288	335
1.1	0.250	0.094	0.068	140	200	212
1.2	0.111	0.012	0.007	108	124	125
1.3	0.046	0.002	0.001	85	87	86
1.4	0.020	0.000	0.000	69	66	65
1.5	0.012	0.000	0.000	57	54	52

[a] SPRT parameters: $\alpha = \beta = 0.1$, $k = 0.81$, $m_0 = 0.9$, $m_1 = 1.1$; simulation parameters: 10,000 trials for each m, minimum sample size = 20.

It may be lack of space or interest, but recently there have not been many such careful and reasoned reports in the literature. On at least one occasion the present author has read that the Poisson distribution did not fit the data..., but proposals for plans based on the Poisson distribution were presented nonetheless.

VII. COMPUTER PROGRAMS

The computer programs mentioned previously (annotated code in Appendix I) are written in standard Fortran and contain optional code to use the Vax/VMS operating system random number generator, the MS DOS random number generator, or one explicitly written in subroutine form. They are interactive with prompts. An example run of the SPRT program is included in Appendix I; the question-answer format for the Iwao program is similar. Simulation can be done using the binomial, Poisson or negative binomial distributions. For the negative binomial distribution, k can be entered as a constant, or it can be estimated from the variance-mean power law of Taylor.[21] In the latter instance, the effect of uncertainty in the parameters of the law (because they themselves would have been estimated) can be included in the simulation also.

Simulation based on bootstrap-type methods[38] has not been considered here. On the one hand it is a more realistic method than that used here (especially when no standard distribution fits the data), but it is restricted to the characteristics of the populations already sampled so it may not cover a wide enough range; it may also be more difficult to implement.

VIII. CONCLUSION

If technology were such that producers were able to manage their crops well enough for pest density always to be below a critical level, there would be no need for quality control sampling. If growers do not or cannot manage their crops well enough, so that pest densities are always too high, there is again no need to sample—intervention is always necessary (unless there is no crop left!). Only in the intermediate situation can quality control sampling be beneficial. Since sampling and corrective action are added costs, continuous improvement in crop management (biological control, plant resistance, cultural practices) is critical. Deming[2] quotes a letter to him dated March 1980 from a Japanese businessman

> "In Europe and America, people are now more interested in cost of quality and in systems of quality-audit. But in Japan, we are keeping very strong interest to improve quality by use of methods which you started...when we improve quality we also improve productivity, just as you told us in 1950 would happen."

However, while the theory and practice of biological control, plant resistance, and cultural practices continue to develop (which will eventually enable quality to improve), cost-effective sampling for decision making remains important for all producers. Sequential sampling is a cost-effective alternative to fixed size sampling, provided that care is taken to eliminate potential problems:

1. There must be a realistic estimate of the critical threshold.
2. Care must be taken to prevent early stopping from giving biased results.
3. If SPRT is used, the spatial pattern corresponding to the samples to be taken must fit the appropriate distribution.

4. If none of the "simple" distributions (binomial, Poisson, negative binomial) fits the spatial pattern, a normal approximation (SPRT or Iwao) can be used provided that a realistic minimum sample size is used.
5. If a three-decision SPRT is used, the stopping boundaries must be checked for potential anomalies.
6. Robustness to departures from the assumptions (e.g., value of k in the negative binomial, or precision of the variance-mean relationship if one is used) should be checked.
7. Properties of the scheme (OC, ASN, costs, thresholds) should be assessed, theoretically or by simulation, and continually checked in practice.

ACKNOWLEDGMENTS

Initial versions of the computer programs appeared in Nyrop and Binns.[4] Thanks are expressed to Dr. Nyrop (NYAES, Geneva, N.Y.) and Dr. Thompson (RPS, Agriculture Canada) for useful comments, and to the Biographic Unit of Research Program Service for the artwork. Contribution number R-129 from Research Program Service, Agriculture Canada, Ottawa.

APPENDIX I

Computer programs for SPRT and Iwao procedures. Both are presented here in MS DOS and Fortran, but changes necessary to use the Vax/VMS random number generator, or a user-defined generator are indicated. The subroutines attached to the SPRT program are also necessary for the Iwao program.

```
c program SPRT

c VARIABLE DEFINITIONS :
c alpha, beta : nominal SPRT error probabilities
c lna, lnb : log{(1 − beta)/alpha}, log{alpha/(1 − beta)}
c m0, m1 : means for null and alternative hypotheses
c m : the true mean value for which results are calculated
c k : negative binomial exponent
c lowi, highi, slope : lower and higher intercepts and slope of the SPRT
c distr : integer indicator of the distribution being used
c h( ) : array of values to calculate theoretical OC and ASN curves
c r, srand, iseeds, nrands, iseedx, nrandx : used for user random
c       number generator
c iseed1, iseed2 : used for VAX random number generator
c ndec : 1 for 2-decision SPRT, 2 for 3-decision SPRT
c normv : (normal distribution only) 1 if sigma entered by user, 2 if
c       sigma to be calculated from TPL
c tpla, tplb : parameters of TPL ( var  =   tpla*m**tplb )
c sigsq : (normal distribution only) variance used in the SPRT
c oc : theoretical OC curve
c asn : theoretical ASN curve
c ax, ay, bx, by, cx, cy : (3-decision SPRT only) x- and y- coordinates
c       of points of intersection of the 2 SPRTs. A is intersection of the
c       2 interior lines, B of the 2 upper lines, C of the 2 lower lines
c nmin, nmax : (simulation) minimum and maximum sample sizes
c
c xbar : (simulation) mean of samples
```

c slu, sll, slu2, sll2 : (simulation) upper and lower stopping values
c count : (simulation) number of samples
c sum : (simulation) cumulative number of individuals
c isensit : (simulation) +1 to use negative binomial distribution, −1 to
c   use binomial or Poisson as appropriate
c sentype : (simulation) if negative binomial used, 1 if k to be fixed,
c   2 if k estimated from TPL
c t, t2 : (simulation) thresholds to be used if maximum sample size
c   reached. t for 2-decision SPRT or lower arm of 3-decision SPRT,
c   t2 for upper arm of 3-decision SPRT.
c terh,terl : (simulation) numbers of runs ending in upper and lower
c   terminal decisions
c seqh,seql : (simulation) numbers of runs ending in upper and lower
c   sequential decisions
c mci : (simulation) number of Monte Carlo runs for each mean
c icrossl : (simulation, 3-decision SPRT) −1 initially, +1 if upper
c   boundary of lower arm SPRT crossed before A
c icrossh : (simulation, 3-decision SPRT) −1 initially, +1 if lower
c   boundary of upper arm SPRT crossed before A
c minm, maxm, increment : (simulation) minimum, maximum mean values and
c   increment defining the range of values of m
c ks : (simulation) negative binomial exponent
c seqoc : (simulation) sequential OC curve
c teroc : (simulation) terminal OC curve
c avgsmp : (simulation) ASN
c totcnt : (simulation) total number of individuals "found" in a
c   Monte Carlo run

        real lna,lnb,m0,m1,k,m,lowi,highi,slope
        real maxm,minm,increment,ks
        integer distr
        character respond,respond2,respond3,respond4
        dimension h(12)
        dimension alpha(2),beta(2),lowi(2),highi(2),slope(2),m0(2),m1(2)
      1   ,terh(2),terl(2),seql(2),seqh(2)
c for user defined random number generator :
c   double precision r, srand
c   common/rand/ iseeds(3), nrands(5), iseedx(3), nrandx(5)
c for Vax/VMS random number generator :
        common /seeds/iseed1,iseed2
c set up input and output file numbers
c VAX declares files as FOR0xx.DAT where xx is the file number.
c MS DOS Fortran asks the user for a name when it first encounters a
c "write" instruction
        COMMON /INOUT/ IN1,IN2,IOUT1,IOUT2,IOUT3
        DATA IN1,IN2,IOUT1,IOUT2,IOUT3/5,5,6,22,25/
        data h/4.,3.5,3.,2.5,2.,1.5,1.,.5,.4,.3,.2,.1/

        write (iout3,801)
        write (iout1,801)
        read (in1,803) i
    803 format (20a4)
        write (iout3,802)
        write (iout1,802)
    801 format (1x,'---'/
        z1x,' SEQUENTIAL PROBABILITY RATIO TEST'//
        11x,'   PURPOSE'/
        21x,' This program computes stop lines, OC and ASN functions for'/
        31x,' Wald"s SPRT using a binomial, Poisson, normal, or'/
        41x,' negative binomial distribution. '/
        51x,' A single SPRT can be used to make one of two decisions.'/
        61x,' Two SPRTs, combined, can be used to make one of three'/

```
          71x,' decisions. Reliable formulae do not exist for the OC and'/
          81x,' ASN of a 3-decision SPRT; this program can simulate them.'/
          91x,'              SIMULATIONS '/
          a1x,' Theoretical OC and ASN curves can be checked by simulation.'/
          b1x,' When the actual distribution of individuals is negative'/
          c1x,' binomial, the sensitivity to deviations in k of a 2- or'/
          d1x,' 3-decision SPRT, based on the Poisson, negative binomial'/
          e1x,' or normal distribution can be determined.'/
          f1x,' The SPRT is truncated at a user specified maximum sample'/
          g1x,' size.'/
          h1x,'-----------------------------------------------'/
          i1x,' type return ...'/)
      802 format(1x,'-----------------------------------------------'/
          z1x,'            INPUT FOR SETTING UP BOUNDARIES'/
          11x,' Distribution, lower and upper critical means, alpha, beta'/
          21x,'            INPUT FOR SIMULATIONS'/
          h1x,' i) a k-value, possibly different from that used to'/
          i1x,'    construct the stop lines, or parameters of TPL from'/
          j1x,'    which k is computed for each mean;'/
          k1x,' ii) the minimum initial sample size, and the maximum size'/
          l1x,' iii) the minimum and maximum mean to be sampled, and an'/
          mls,'    increment for increasing the mean from the minimum to'/
          n1x,'    the maximum,'/
          o1x,' iv) threshold(s) for determining the decision at maximum'/
          p1x,'    sample size,'/
          q1x,' v) the number of Monte Carlo iterations (a minimum of 500'/
          r1x,'    is recommended)'/
          s1x,'              OUTPUT '/
          t1x,' Sequential OC, a terminal OC at the maximum sample size,'/
          u1x,'    and the ASN.'/
          v1x,'-----------------------------------------------'/)

c Set seeds for pseudo-random number generator
c
c MS DOS Fortran
          call gettim(ihr,imin,isec,i100th)
          iseed = 30*i100th + 20*isec + imin
          call seed(iseed)
c
c VAX
c         write (iout1,*) ' read in two integer SEEDS for random number gene
c      1rator :'
c         read (in1,*) iseed1,iseed2
c For systems without a reliable pseudo-random number generator. Use
c SRAND at the end of the listing
c         nrands(1) = 327
c         nrands(2) = 2184
c         nrands(3) = 230


      500 continue
C ndec = 1 for 2-decision SPRT, ndec = 2 for 3-decision SPRT
          ndec = 1
          WRITE (IOUT1,*) ' Choose a distribution by entering a number 1-4'
          WRITE (IOUT1,*) ' 1) Binomial         2) Poisson'
          WRITE (IOUT1,*) ' 3) Negative binomial    4) Normal'
          READ (IN1,*) distr
             if (distr.ne.1.and.distr.ne.2.and.distr.ne.3.and.distr.ne.4)
         1      goto 500
      501 continue
             if (distr.eq. 3.and.ndec.eq.1) then
                WRITE (IOUT1,*) ' Input value for k'
```

```
            READ (IN1,*) k
          end if
          if (distr.eq. 4. and. ndec.eq.1) then
41        continue
          WRITE (IOUT1,*) ' A variance estimate is needed. Enter 1 or 2 :'
          WRITE (IOUT1,*) ' 1) input a variance, or'
          WRITE (IOUT1,*) ' 2) use Taylor''s Power Law '
          READ (IN1,*) normv
            if (normv.eq.1) then
            WRITE (IOUT1,*) ' Input value for the variance'
            READ (IN1,*) sigsq
            else if (normv.eq.2) then
            WRITE (IOUT1,*) ' input alpha and beta values for V/M relation
     1ship'
            WRITE (IOUT1,*) ' input alpha'
            READ (IN1,*) tpla
            WRITE (IOUT1,*) ' input beta'
            READ (IN1,*) tplb
            sigsq = tpla*(0.5*(m1(ndec) + m0(ndec)))**tplb
            else
            go to 41
            end if
          end if
        WRITE (IOUT1,*) ' Input means for null and alternative hypotheses'
        WRITE (IOUT1,*) ' and alpha and beta.'
          if (distr.eq.1) then
          write (iout1,*) ' Means should be entered as proportions (< 1).'
          end if
        WRITE (IOUT1,*) ' Mean for null hypothesis?'
        READ (IN1,*) m0(ndec)
        WRITE (IOUT1,*) ' Mean for alternative hypothesis?'
        READ (IN1,*) m1(ndec)
        WRITE (IOUT1,*) ' Input alpha (type I error) ?'
        READ (IN1,*) alpha(ndec)
        WRITE (IOUT1,*) ' Input beta (type II error) ?'
        READ (IN1,*) beta(ndec)
          if (distr.eq.1) then
          WRITE (IOUT3,1)
          WRITE (IOUT1,1)
1         format (' Binomial distribution used')
          else if (distr.eq.2) then
          WRITE (IOUT3,2)
          WRITE (IOUT1,2)
2         format (' Poisson distribution used')
          else if (distr.eq.3) then
          WRITE (IOUT3,3)
          WRITE (IOUT1,3)
3         format (' Negative binomial distribution used')
          else if (distr.eq.4) then
          WRITE (IOUT3,31)
          WRITE (IOUT1,31)
31        format (' Normal distribution used')
          end if
        WRITE (IOUT3,4)
        WRITE (IOUT1,4)
4       format (' H0 mean   H1 mean    alpha   beta')
        WRITE (IOUT3,5) m0(ndec),m1(ndec),alpha(ndec),beta(ndec)
        WRITE (IOUT1,5) m0(ndec),m1(ndec),alpha(ndec),beta(ndec)
5       format (f7.3,2x,f7.3,3x,f6.4,2x,f6.4)
          if (distr.eq. 3) then
          WRITE (IOUT3,6) k
```

```
         WRITE (IOUT1,6) k
  6      format (' k  =  ' ,f8.4)
         end if
         if (distr.eq. 4) then
           if (normv.eq.1) then
           WRITE (IOUT3,61) sigsq
           WRITE (IOUT1,61) sigsq
  61       format (' variance  =  ' ,f10.3)
           else if (normv.eq.2) then
           WRITE (IOUT3,62) tpla,tplb,sigsq
           WRITE (IOUT1,62) tpla,tplb,sigsq
  62       format (' TPL parameters, var{(m0 + m1)/2}  =  '/1x,3f10.3)
           end if
         end if
 510  WRITE (IOUT1,511)
 511  format (' These are the parameters that will be used in'/
     1 ' the computations; are they correct ("Y" or "N")?')
         READ (IN1,*) respond
           if (respond.ne.'N'.and.respond.ne.'n') then
             if (respond.ne.'Y'.and.respond.ne.'y') goto 510
           end if
         if (respond .eq. 'N' .or. respond .eq. 'n') goto 500


c computations
c variables used for all distributions
         a  =  (1 − beta(ndec))/alpha(ndec)
         lna  =  log(a)
         b  =  beta(ndec)/(1 − alpha(ndec))
         lnb  =  log(b)


c variables used for binomial and negative binomial
         if (distr.eq.1) then
         q1  =  1 − m1(ndec)
         q0  =  1 − m0(ndec)
         bvar1  =  log((m1(ndec)*q0)/(m0(ndec)*q1))
         end if
         if (distr.eq.3) then
         pnb0  =  m0(ndec)/k
         pnb1  =  m1(ndec)/k
         qnb0  =  1 + pnb0
         qnb1  =  1 + pnb1
         vnb1  =  (pnb1*qnb0)/(pnb0*qnb1)
         vnb2  =  qnb0/qnb1
         end if


c calculate and print stop limit parameters
         if (distr.eq.1) then
         lowi(ndec)  =  lnb/bvar1
         highi(ndec)  =  lna/bvar1
         slope(ndec)  =  log(q0/q1)/bvar1
         else if (distr.eq.2) then
         lowi(ndec)  =  lnb/log(m1(ndec)/m0(ndec))
         highi(ndec)  =  lna/log(m1(ndec)/m0(ndec))
         slope(ndec)  =  (m1(ndec) − m0(ndec))/log(m1(ndec)/m0(ndec))
         else if (distr.eq.3) then
         lowi(ndec)  =  lnb/log(vnb1)
         highi(ndec)  =  lna/log(vnb1)
         slope(ndec)  =  k*(log(qnb1/qnb0)/log(vnb1))
         else if (distr.eq.4) then
         lowi(ndec)  =  lnb*sigsq/(m1(ndec) − m0(ndec))
```

```
              highi(ndec) = lna*sigsq/(m1(ndec) − m0(ndec))
              slope(ndec) = (m1(ndec) + m0(ndec))/2.
              end if


       WRITE (IOUT3,7)
       WRITE (IOUT1,7)
    7  format (' low intercept      high intercept      slope')
       WRITE (IOUT3,8) lowi(ndec), highi(ndec), slope(ndec)
       WRITE (IOUT1,8) lowi(ndec), highi(ndec), slope(ndec)
    8  format (6x,f10.3,6x,f10.3,2x,f10.3)
       WRITE (IOUT3,9)
       WRITE (IOUT1,9)
    9  format ('     Mean      OC      ASN')


c OC and ASN computations
          do 20 j = 1,12
          htemp = h(j)
          oc = (a**htemp − 1)/(a**htemp − b**htemp)
            if (distr.eq.1) then
            m = (1 − (q1/q0)**htemp)/
     1         ((m1(ndec)/m0(ndec))**htemp − (q1/q0)**htemp)
            asn = (lnb*oc + (1 − oc)*lna)/(m*bvar1 + log(q1/q0))
            else if (distr.eq.2) then
            m = ((m1(ndec) − m0(ndec))*htemp)/((m1(ndec)/m0(ndec))**htemp − 1)
            asn = (lnb*oc + lna*(1 − oc))/
     1         (m*log(m1(ndec)/m0(ndec)) + m0(ndec) − m1(ndec))
            else if (distr.eq.3) then
            p = (1 − vnb2**htemp)/(vnb1**htemp − 1)
            m = p*k
            asn = (lnb*oc + lna*(1 − oc))/(k*log(vnb2) + k*p*log(vnb1))
            else if (distr.eq.4) then
            m = 0.5*(m1(ndec) + m0(ndec) − htemp*(m1(ndec) − m0(ndec)))
            asn = (lnb*oc + lna*(1 − oc))*2.*sigsq/
     1         (2.*(m1(ndec) − m0(ndec))*m + m0(ndec)**2 − m1(ndec)**2)
            end if
          WRITE (IOUT3,10) m,oc,asn
          WRITE (IOUT1,10) m,oc,asn
    10    format (1x,f7.3,2x,f7.3,2x,f10.2,f8.3)
    20    continue
       oc = lna/(lna − lnb)
         if (distr.eq.1) then
         m = −log(q1/q0)/bvar1
         asn = −lna*lnb/(log(m1(ndec)/m0(ndec))*log(q0/q1))
         else if (distr.eq.2) then
         m = (m1(ndec) − m0(ndec))/log(m1(ndec)/m0(ndec))
         asn = −(lna*lnb)/(m*(log(m1(ndec)/m0(ndec)))**2)
         else if (distr.eq.3) then
         p = log(qnb1/qnb0)/log(vnb1)
         q = 1 + p
         m = p*k
         asn = −(lna*lnb)/(k*p*q*(log(vnb1))**2)
         else if (distr.eq.4) then
         m = 0.5*(m1(ndec) + m0(ndec))
         asn = −(lna*lnb)*sigsq/(m1(ndec) − m0(ndec))**2
         end if
       WRITE (IOUT3,10) m,oc,asn
       WRITE (IOUT1,10) m,oc,asn
          do 30 j = 12,1,−1
          htemp = −h(j)
```

```
            oc = (a**temp - 1)/(a**htemp - b**htemp)
            if (distr.eq.1) then
            m = (1 - (q1/q0)**htemp)/
   1            ((m1(ndec)/m0(ndec))**htemp - (q1/q0)**htemp)
            asn = (lnb*oc + (1 - oc)*lna)/(m*bvar1 + log(q1/q0))
            else if (distr.eq.2) then
            m = ((m1(ndec) - m0(ndec))*htemp)/((m1(ndec)/m0(ndec))**htemp -
               1)
            asn = (lnb*oc + lna*(1 - oc))/
   1            (m*log(m1(ndec)/m0(ndec)) + m0(ndec) - m1(ndec))
            else if (distr.eq.3) then
            p = (1 - vnb2**htemp)/(vnb1**htemp - 1)
            m = p*k
            asn = (lnb*oc + lna*(1 - oc))/(k*log(vnb2) + k*p*log(vnb1))
            else if (distr.eq.4) then
            m = 0.5*(m1(ndec) + m0(ndec) - htemp*(m1(ndec) - m0(ndec)))
            asn = (lnb*oc + lna*(1 - oc))*2.*sigsq/
   1            (2.*(m1(ndec) - m0(ndec))*m + m0(ndec)**2 - m1(ndec)**2)
            end if

            WRITE (IOUT3,10) m,oc,asn
            WRITE (IOUT1,10) m,oc,asn
   30       continue
            WRITE (IOUT3,*) ' These OC and ASN curves are based on standard formulae.'
            WRITE (IOUT1,*) ' These OC and ASN curves are based on standard formulae.'
            if (ndec.eq.1) then
            write (iout1,541)
  541       format (1x,' The above is for a 2-decision SPRT.'/
   1        ' Do you want to set up a 3-decision SPRT ("y" or "n")?')
            read(in1,*) respond
            if (respond.eq.'y'.or.respond.eq.'Y') then
            ndec = 2
            write (iout1,542)
  542       format (1x,' The intercepts and slope for one of the SPRTs is'/
   1        '    stored. Now input details for the second SPRT.')
            go to 501
            end if
            else if (ndec.eq.2) then
c steeper sloped pair must be in second place
            if (slope(1).gt.slope(2)) then
            temp = slope(1)
            slope(1) = slope(2)
            slope(2) = temp
            temp = highi(1)
            highi(1) = highi(2)
            highi(2) = temp
            temp = lowi(1)
            lowi(1) = lowi(2)
            lowi(2) = temp
            end if
c Calculate points of intersection of the two SPRTs. These are important in
c    the operation of a 3-decision SPRT.
c If either the intersection (B) between the two upper stop lines, or the
c    intersection (C) between the two lower stop lines can theoretically be
c    reached by a sample path, the plan MUST be adjusted: by
c i) having a minimum sample size which would prevent any decision before
c      B or C; or
c ii) respecifying the 2-decision SPRTs
c Calculation of the intersection (A) between the two inner stop lines is
c    important, because before A, the middle decision is made if and only if
c    both inner boundaries have been crossed.
```

```
              bx = (highi(1) − highi(2))/(slope(2) − slope(1))
              by = highi(1) + slope(1)*bx
              cx = (lowi(1) − lowi(2))/(slope(2) − slope(1))
              cy = lowi(1) + slope(1)*cx
              ax = (highi(1) − lowi(2))/(slope(2) − slope(1))
              ay = lowi(2) + ax*slope(2)
              write (iout1,700) bx,by,cx,cy,ax,ay
              write (iout3,700) bx,by,cx,cy,ax,ay
   700        format (1x,' Points of intersection of the SPRTs  :'
        a ,12x,'x',11x,'y'/
        1 5x,' intersection point B: two upper lines   :',2f12.3/
        2 5x,' intersection point C: two lower lines   :',2f12.3/
        3 5x,' intersection point A: intermediate      :',2f12.3)
              nmin0 = −1
                if (bx.gt.0) nmin0 = 1 + bx
                if (cy.gt.0) nmin0 = 1 + cx
                if (cy.gt.0.or.bx.gt.0) then
                write(iout3,701) nmin0
   710        write(iout1,701) nmin0
   701        format(1x,' points B and/or C are the wrong side of the axes'/
        1     ' (see, eg, Wetherill and Glazebrook (1986), chapter 3).'/
        2     ' You will have to either (1) re-input the parameters for'/
        3     ' both SPRTs, or (2) make the minimum sample size at least'/
        4     ' equal to ',18,'    Answer 1 or 2 :')
              read(in1,*) j
                if (j.ne.1.and.j.ne.2) go to 710
                if (j.eq.1) go to 500
              end if
            end if

c ********** beginning of if-then block for simulations *****************
   520      continue
            write(iout1,543)
   543      format(1x,' Do you want to simulate the OC and ASN, or do a'/
        1   ' sensitivity analysis on the negative binomial parameter k'/
        2   ' ("Y" or "N") ?')
            read(in1,*) respond2
              if (respond2.ne.'N'.and.respond2.ne.'n') then
                if (respond2.ne.'Y'.and.respond2.ne.'y') goto 520
              end if
              if (respond2.eq.'Y'.or.respond2.eq.'y') then
            isensit = −1
                if (distr.eq.2) then
                write(iout1,544)
   544        format(1x,' Use Poisson or negative binomial for simulation'/
        1     ' ("P" or "N") ?')
              read(in1,*) respond
                 if (respond.eq.'n'.or.respond.eq.'N') isensit = 1
              end if
              if (distr.eq.3.or.distr.eq.4) isensit = 1
   521      continue
              if (isensit.gt.0) then
  5210        continue
              WRITE (IOUT1,*) ' SPRT sensitivity analysis'
              WRITE (IOUT1,*) ' Nominal value of k  = ',k
              WRITE (IOUT1,*) ' Sensitivity analysis can be done with constant'
              WRITE (IOUT1,*) ' or dynamic k modelled using Taylors Power Law'
              WRITE (IOUT1,*) ' Choose a method (enter 1 or 2)'
              WRITE (IOUT1,*) ' 1) constant k'
              WRITE (IOUT1,*) ' 2) dynamic k'
              READ (IN1,*) sentype
                if (sentype.ne.1.and.sentype.ne.2) goto 5210
```

```
522     if (sentype.eq.1) then
        WRITE (IOUT1,*) ' Input k from negative binomial dist.'
        READ (IN1,*) ks
        else
           if (distr.eq.4.and.normv.eq.2) go to 524
        write (IOUT1,*) ' input alpha and beta for V/M relationship'
        WRITE (IOUT1,*) ' input alpha'
        READ (IN1,*) tpla
        WRITE (IOUT1,*) ' input beta'
        READ (IN1,*) tplb
524     continue
        end if
        end if
554   WRITE (IOUT1,*) ' Input minimum sample size'
      READ (IN1,*) nmin
        if (ndec.eq.2.and.nmin.lt.nmin0) then
        write (iout1,*) ' min no of samples must be  > =  ',nmin0
        go to 554
        end if
      WRITE (IOUT1,*) ' Input maximum sample size'
      READ (IN1,*) nmax
      WRITE (IOUT1,*) ' Input threshold (used at max sample size)'
        if (ndec.eq.1) then
        READ (IN1,*) t
        else
        WRITE (IOUT1,*) ' threshold for lower arm '
        READ (IN1,*) t
        WRITE (IOUT1,*) ' threshold for upper arm'
        READ (IN1,*) t2
        end if
      WRITE (IOUT1,*) ' Specify range of means to be sampled'
      WRITE (IOUT1,*) ' Input minimum for mean'
      READ (IN1,*) minm
      WRITE (IOUT1,*) ' Input maximum for mean'
      READ (IN1,*) maxm
      WRITE (IOUT1,*) ' Input increment for means'
      READ (IN1,*) increment
      WRITE (IOUT1,*) ' Input monte carlo iterations'
      READ (IN1,*) mci
530   continue
        if (isensit.gt.0.and.sentype.eq.1) then
        WRITE (IOUT3,52) ks
        WRITE (IOUT1,52) ks
52      format(/,' Constant value of k  =  ' , f10.4)
        else if (isensit.gt.0.and.sentype.ne.1) then
        WRITE (IOUT3,53)
        WRITE (IOUT1,53)
53      format (/,' Parameters for the mean/variance relationship')
        WRITE (IOUT3,54) tpla,tplb
        WRITE (IOUT1,54) tpla,tplb
54      format (' alpha  =  ' , f5.3, ' beta  =  ' , f5.3)
        end if
      WRITE (IOUT3,553) nmin
      WRITE (IOUT1,553) nmin
553   format (' Minimum number of samples  =  ' , I4)
      WRITE (IOUT3,55) nmax
      WRITE (IOUT1,55) nmax
55    format (' Maximum number of samples  =  ' , I8)
        if (ndec.eq.1) then
        t2 = t
        WRITE (IOUT3,56) t
        WRITE (IOUT1,56) t
```

```
      56        format (' Threshold value  =  ' , f6.2)
                else
                WRITE (IOUT3,5600) t,t2
                WRITE (IOUT1,5600) t,t2
    5600        format (' Threshold values  =  ' , 2f6.2)
                end if
                WRITE (IOUT3,57) mci
                WRITE (IOUT1,57) mci
      57        format (' Monte Carlo iterations  =  ' ,I8)
     523        WRITE (IOUT1,*) ' Are these values correct ("Y" or "N")?'
                READ (IN1,*) respond2
                  if (respond2.ne.'N'.and.respond2.ne.'n') then
                    if (respond2.ne.'Y'.and.respond2.ne.'y') goto 523
                  end if
                if (respond2.eq.'n'.or.respond2.eq.'N') goto 521
                if (ndec.eq.2) then
                write (iout3,580)
                write (iout1,580)
     580        format(12x,'accept lowest',5x,'accept moderate',5x,
        1       'accept highest'/9x,3(4x,'infestation',4x)/3x,'mean',4x,
        2       3('Seq OC   Ter OC',4x),5x,'ASN')
                else
                write (iout3,581)
                write (iout1,581)
     581        format(13x,'accept lower',7x,'accept higher'/
        1       9x,2(4x,'infestation',4x)/3x,'mean',4x,
        2       2('Seq OC   Ter OC',4x),5x,'ASN')
                end if
                m = minm
     540        if (m.gt.maxm) goto 550
c set decision counters to zero
                do 5401 i = 1,2
                seql(i) = 0.
                seqh(i) = 0.
                terl(i) = 0.
                terh(i) = 0.
    5401        continue
                totcnt = 0.
c Monte Carlo loop
                do 60 i2 = 1,mci
c for 3-decision SPRT only :
c icrossl = −1 at start; = 1 after lower boundaries crossed
c icrossh = −1 at start; = 1 after upper boundaries crossed
c        AND nmin samples taken
                icrossl = −1
                icrossh = −1
                count = 0.
                sum = 0.
                iexit = −1
                  if (isensit.gt.0.and.sentype.eq.2) then
                  var = tpla*m**tplb
                    if (var.le.m) var = m + m*m/100
                  ks = (m**2)/(var − m)
                  end if
      70        continue
                if (isensit.gt.0) call ngb(m,ks,x)
                if (distr.eq.1) call posb(m,x)
                if (isensit.lt.0.and.distr.eq.2) call pois(m,x)
                sum = sum + x
                count = count + 1
c compute stop lines and compare samples to them
```

```
          slu  = highi(1) + slope(1)*count
          sll  = lowi(1) + slope(1)*count
          slu2 = highi(2) + slope(2)*count
          sll2 = lowi(2) + slope(2)*count
            if (ndec.eq.1) slu2 = slu
            if (count .ge. nmax) then
            totcnt = totcnt + count
            iexit = 1
            xbar = sum/count
              if (xbar.le.t) then
              terl(1) = terl(1) + 1.
              else if (xbar.gt.t2) then
              terh(2) = terh(2) + 1.
              else if (ndec.eq.2) then
              terl(2) = terl(2) + 1.
              end if
            else if ( (sum.gt.slu2.or.sum.lt.sll) .and.
     1              count.gt.nmin) then
              if (sum.lt.sll) then
              totcnt = totcnt + count
              seql(1) = seql(1) + 1.
              iexit = 1
              else
              totcnt = totcnt + count
              seqh(2) = seqh(2) + 1.
              iexit = 1
              end if
            else if ((ndec.eq.2).and.(sum.gt.slu)
     1              .and.(count.gt.nmin)) then
            icrossl = 1
              if ( ((sum.lt.sll2).and.(count.ge.ax)) .or.
     1              (icrossh.gt.0) ) then
              totcnt = totcnt + count
              seql(2) = seql(2) + 1.
              iexit = 1
              end if
            else if ((ndec.eq.2).and.(sum.lt.sll2)
     1              .and.(count.gt.nmin)) then
            icrossh = 1
              if ( ((sum.gt.slu).and.(count.ge.ax)) .or.
     1              (icrossl.gt.0) ) then
              totcnt = totcnt + count
              seqh(1) = seqh(1) + 1.
              iexit = 1
              end if
            end if
            if (iexit.lt.0) goto 70
   60     continue

c compute OC values
          seqoc1 = seql(1)/float(mci)
          seqoc2 = (seqh(1) + seql(2))/float(mci)
          seqoc3 = seqh(2)/float(mci)
          teroc1 = terl(1)/float(mci)
          teroc2 = (terh(1) + terl(2))/float(mci)
          teroc3 = terh(2)/float(mci)
c compute average sample size
          avgsmp = totcnt/float(mci)
c write results
            if (ndec.eq.2) then
            WRITE(IOUT3,59) m,seqoc1,teroc1,seqoc2,teroc2,seqoc3,teroc3,
```

```
      1          avgsmp
          WRITE(IOUT1,59) m,seqoc1,teroc1,seqoc2,teroc2,seqoc3,teroc3,
      1          avgsmp
  59     format (1x,f6.2,3(5x,f6.4,2x,f6.4),7x,f6.2)
          else
          WRITE(IOUT3,590) m,seqoc1,teroc1,seqoc3,teroc3,avgsmp
          WRITE(IOUT1,590) m,seqoc1,teroc1,seqoc3,teroc3,avgsmp
  590    format (1x,f6.2,2(5x,f6.4,2x,f6.4),7x,2f6.2)
          end if
        m = m + increment
        goto 540
  550  continue
        write (iout1,591)
  591  format(1x,' Simulated OC is separated into paths crossing the SP
      1RT boundaries (Seq)'/1x,' and those ending at the maximum sample s
      2ize (Ter)'/)
          if (sentype.eq.1) then
  532    format(' Repeat calculations for new k ("Y" or "N")?')
          WRITE (IOUT1,532)
          READ (IN1,*) respond3
            if (respond3.ne.'N'.and.respond3.ne.'n') then
              if (respond3.ne.'Y'.and.respond3.ne.'y') goto 550
            end if
          if (respond3.eq. 'y' .or. respond3.eq. 'Y') then
            WRITE (IOUT1,*) ' Input k from negative binomial dist.'
            READ (IN1,*) ks
            goto 530
          end if
  533    format(' Do sensitivity analysis using TPL ("Y" or "N")?')
  531    WRITE (IOUT1,533)
          READ (IN1,*) respond3
              if (respond3.ne.'N'.and.respond3.ne.'n') then
                if (respond3.ne.'Y'.and.respond3.ne.'y') goto 551
              end if
            if (respond3.eq.'y'.or.respond3.eq.'Y') then
              WRITE (IOUT1,*) ' input alpha and beta values for TPL'
              WRITE (IOUT1,*) ' input alpha'
              READ (IN1,*) tpla
              WRITE (IOUT1,*) ' input beta'
              READ (IN1,*) tplb
              sentype = 2
              goto 530
              end if
          else
          goto 520
          end if
c ********** end of if-then block for simulations ****************
          end if


  534  format(' Compute OC and ASN for new model ("Y" or "N") ?')
  560  WRITE (IOUT1,534)
          READ (IN1,*) respond4
            if (respond4.ne.'N'.and.respond4.ne.'n') then
              if (respond4.ne.'Y'.and.respond4.ne.'y') goto 560
            end if
            if (respond4.eq.'Y' .or. respond4.eq.'y') goto 500
          stop
          end


c -------------------------------------------------------------------
```

```fortran
      subroutine ngb(mean,k,x)
c     double precision r, srand
      common /seeds/iseed1,iseed2
c     common/rand/ iseeds(3), nrands(5), iseedx(3), nrandx(5)
      real x,mean,k
c MS DOS Fortran
      call random(r)
c VAX
c     call randu(iseed1,iseed2,r)
c user generator
c     r = srand()
      x = 0.0
      px = 1/((1+mean/k)**k)
      ppx = px
      pxsum = px
      if (r .le. px) goto 4
    5 x = x + 1
      px = ((k+x -1.)/x)*(mean/(mean+k))*ppx
      ppx = px
      pxsum = pxsum + px
      if (r .le. pxsum) goto 4
      if (x .gt. 200.) goto 4
      goto 5
    4 continue
      return
      end


c -----------------------------------------------------------------
      subroutine posb(mean,x)
c     double precision r, srand
      common /seeds/iseed1,iseed2
c     common/rand/ iseeds(3), nrands(5), iseedx(3), nrandx(5)
      real x,mean
c MS DOS Fortran
      call random(r)
c VAX
c     call randu(iseed1,iseed2,r)
c user generator
c     r = srand()
      x = 0.
      if (r.le.mean) x = 1.
      return
      end


c -----------------------------------------------------------------
      subroutine pois(mean,x)
c     double precision r, srand
      common /seeds/iseed1,iseed2
c     common/rand/ iseeds(3), nrands(5), iseedx(3), nrandx(5)
      real x,mean
c MS DOS Fortran
      call random(r)
c VAX
c     call randu(iseed1,iseed2,r)
c user generator
c     r = srand()
      x = 0.0
      px = exp(-mean)
      ppx = px
      pxsum = px
      if (r .le. px) goto 4
```

```
      5 x = x + 1
        px = ppx*mean/x
        ppx = px
        pxsum = pxsum + px
        if (r .le. pxsum) goto 4
        if (x .gt. 200.) goto 4
        goto 5
      4 continue
        return
        end
```

```
c -----------------------------------------------------------------
c for systems without a reliable pseudo-random number generator
      DOUBLE PRECISION FUNCTION SRAND( )
c
c insert code here
c to calculate SRAND
c     SRAND = ...
      RETURN
      END
```

```
c -----------------------------------------------------------------
      function gausinv(p,ifault)
c
c algorithm AS 70 appl. statist. (1974) vol. 23, no. 1
c gausinv finds percentage points of the normal distribution
c
      data zero, one, half, alimit /0.0, 1.0, 0.5, 1.0e−20/
      data      p0,      p1,      p2,      p3
     *   / −.322232431088, −1.0, −.342242088547, −.204231210245e−1 /
      data      p4,      q0,      q1
     *   / −.453642210148e−4, .993484626060e−1, .588581570495 /
      data      q2,      q3,      q4
     *   / .531103462366, .103537752850, .38560700634e−2 /
c
      ifault = 1
      gausinv = zero
      ps = p
      if (ps .gt. half) ps = one − ps
      if (ps .lt. alimit) return
      ifault = 0
      if (ps .eq. half) return
      yi = sqrt(alog(one / (ps * ps)))
      gausinv = yi + ((((yi * p4 + p3) * yi + p2) * yi + p1) * yi + p0)
     *      / ((((yi * q4 + q3) * yi + q2) * yi + q1) * yi + q0)
      if (p .lt. half) gausinv = −gausinv
      return
      end
```

```
c -----------------------------------------------------------------
      subroutine normal(xm,stdev,xnml)
c for user generator of pseudo random numbers :
c     double precision r,srand
c     common/rand/ iseeds(3), nrands(5), iseedx(3), nrandx(5)
c MS DOS Fortran
      call random(r)
c VAX
c     call randu(iseed1,iseed2,r)
c user generator
c     r = srand( )
```

```
        if (r .le. 0.000001) r  =  0.000001
        if (r .ge. 0.999999) r  =  0.999999
        y = gausinv(r,ifault)
        xnml  =  xm  +  y*stdev
        return
        end
```

THIS PROGRAM REQUIRES ALL THE SUBROUTINES FOUND IN THE SPRT PROGRAM

```
c Iwao's sequential procedure


c VARIABLE DEFINITIONS:
c tpla, tplb: parameters for Taylor's power law
c rmse,dp,vartplb,regm: mean square error, data points, variance of
c       tplb and mean of ln(m) used in TPL regression
c mci: number of monte carlo iterations
c nsamp,nmax: minimum and maximum number of samples to be taken
c minm, maxm, increment: min and max mean and increment in simulation
c var, vart, stdlnv, ranvar: variance of mean and threshold based on
c       TPL, standard deviation for predicted variance, random variable
c       that models predicted variance
c seql, terl: number of "below threshold" sequential decisions, number
c       of "above threshold" terminal decisions
c totcnt: total number of samples drawn
c rk: k from negative binomial distribution
c count, sum: number of samples, sum of animals in total samples from
c       one run
c cum, d: variables used to construct stop lines
c slu, sll: upper and lower stop lines
c xbar: mean of sample observations
c seqoc, teroc: sequential and terminal oc
c avgsmp: average number of samples
c counter: number of random variances generated
c varlm: number of random variances less than the mean


        real m,maxm,minm,increment
        integer distr,count,totcnt,varlm
        character TPLvarn,correct
        common /seeds/iseed1,iseed2
c for user random number generator
c       double precision r, srand
c       common/rand/ iseeds(3), nrands(5), iseedx(3), nrandx(5)
c
c set up input and output file numbers:
c VAX declares files as FOR0xx.DAT where xx is the file number.
c MS DOS asks the user for a name when it first encounters a "write"
c instruction
        COMMON /INOUT/ IN1,IN2,IOUT1,IOUT2,IOUT3
        DATA IN1,IN2,IOUT1,IOUT2,IOUT3/5,5,6,22,25/

        write (iout1,801)
        write (iout3,801)
801     format(1x,' ---------------------------------------------------'/
       z1x,' IWAO"S SEQUENTIAL PROCEDURE'//
       11X,'          PURPOSE'/
       21X,' This program simulates the OC and ASN functions for Iwao"s'/
       31x,'     sequential classification procedure.'/
       41x,'          INPUT'/
       51x,' Threshold mean and two normal deviates (z) to define the'/
       61x,'     distance between the boundary lines.'/
```

```
      71x,' Minimum and maximum sample sizes (recommended min > 20).'/
      81x,' The distribution to be used for the simulations: binomial,'/
      91x,'     Poisson, negative binomial. If negative binomial, TPL'/
      a1x,'     can be used if desired. The effect of uncertainty in'/
      b1x,'     the parameters of TPL can be examined.'/
      c1x,' Range of means to be sampled.'/
      d1x,' Number of Monte Carlo simulations for each mean value. For'/
      e1x,'     reliable results, this should be > 500 : variance of an'/
      f1x,'     OC probability is p(1 - p)/nmc where p is the OC value'/
      g1x,'     and nmc is the number of simulations.'/
      h1x,'-----------------------------------------------'/)

c Set seeds for pseudo-random number generator
c
c MS DOS Fortran
       call gettim(ihr,imin,isec,i100th)
       iseed = 30*i100th + 20*isec + imin
       call seed(iseed)
c
c VAX
c      write (iout1,*) ' read in two integer SEEDS for random number gene
c     1rator :'
c      read (in1,*) iseed1,iseed2
c For systems without a reliable pseudo-random number generator. Use
c SRAND at the end of the listing
c      nrands(1) = 327
c      nrands(2) = 2184
c      nrands(3) = 230
c      write (iout1,9999)
c 9999 format(' read in preliminary number of calls to the random number
c     1 generator, to get started with a different seed from last run')
c      read (in1,*) n
c      write (iout3,9997) n
c 9997 format (' preliminary number of srand( ) calls :',i15)
c      do 9998 j = 1,n
c 9998 r = srand( )

  500  continue
       write (iout1,*) ' INPUT PARAMETERS TO DEFINE IWAO''S BOUNDARIES'
       write (iout1,*) '     threshold mean :'
       read (in1,*) t
       write (iout1,*) ' z values for stop lines '
       write (iout1,*) '     z value for upper boundary :'
       read (in1,*) zu
       write (iout1,*) '     z value for lower boundary :'
       read (in1,*) zl
       write (iout1,*) '     minimum sample size :'
       read (in1,*) nsamp
       write (iout1,*) '     maximum sample size :'
       read (in1,*) nmax
       write (iout1,*) ' INPUT PARAMETERS FOR THE SIMULATIONS '
       write (iout1,600)
  600  format (1x,' Distribution to be assumed for the simulations :'/
     1    4x,'1) binomial'/4x,'2) Poisson'/4x,'3) negative binomial'/
     2    8x,'Enter 1, 2 or 3')
       read(in1,*) distr
        if (distr.ne.1.and.distr.ne.2.and.distr.ne.3) go to 500
        if (distr.eq.1) then
         write (iout1,*) ' NOTE : binomial means must be entered as proportions'
```

```
          else if (distr.eq.3) then
705       continue
          write (iout1,*) ' k constant (1), or depends on TPL (2) ?'
          read(in1,*) kcalc
            if (kcalc.eq.1) then
            write (iout1,*) ' constant value of k :'
            read (in1,*) rk
            else if (kcalc.eq.2) then
            write (iout1,*) ' input a and b values for TPL: var = a*m**b'
            write (iout1,*) ' input a'
            read (in1,*) tpla
            write (iout1,*) ' input b'
            read (in1,*) tplb
            else
            go to 705
            end if
          end if
          write (iout1,*) '      specify range of means to be sampled'
          write (iout1,*) '            minimum for mean :'
          read (in1,*) minm
          write (iout1,*) '            maximum for mean :'
          read (in1,*) maxm
          write (iout1,*) '            increment for means :'
          read (in1,*) increment
          write (iout1,*) '      monte carlo iterations :'
          read (in1,*) mci

          if (distr.eq.1) write(iout1,*) ' Binomial distribution'
          if (distr.eq.2) write(iout1,*) ' Poisson distribution'
          if (distr.eq.3) then
          write(iout1,*) ' Negative binomial distribution'
            if (kcalc.eq.1) then
            write(iout1,*) ' constant k  = ',rk
            else
            write (iout1,*) ' TPL parameters.: a  = ',tpla,' b  = ',tplb
            end if
          end if
          write (iout1,*) ' Monte Carlo iterations  = ' , mci
          write (iout1,*) ' Threshold  = ' , t
          write (iout1,*) ' z values for stopping boundary  = ',ZL,ZU
          write (iout1,*) ' minimum sample size  = ' , nsamp
          write (iout1,*) ' maximum sample size  = ' , nmax
          write (iout1,*) ' minimum mean  = ' , minm
          write (iout1,*) ' maximum mean  = ' , maxm
          write (iout1,*) ' mean increment  = ' , increment

510       write (iout1,*) ' These are the parameters to be used in the'
          write (iout1,*) '    simulations; are they correct '
          write (iout1,*) '    ("Y" or "N")?'
          read (in1,*) correct
          if (correct.ne.'N'.and.correct.ne.'n') then
            if (correct.ne.'Y'.and.correct.ne.'y') goto 510
          end if
          if (correct .eq. 'N' .or. correct .eq. 'n') goto 500

c placement of 520
   520    continue
              if (distr.eq.3.and.kcalc.eq.2) then
              write (iout1,*) ' do you wish to include variation in TPL '
```

```
              write (iout1,*) '   ("Y" or "N")'
              read (in1,*) TPLvarn
                if (TPLvarn.ne.'N'.and.TPLvarn.ne.'n') then
                  if (TPLvarn.ne.'Y'.and.TPLvarn.ne.'y') goto 520
                end if
                if (TPLvarn .eq. 'N' .or. TPLvarn .eq. 'n') goto 530
530           write (iout1,*) ' details of the TPL regression computations :'
              write (iout1,*) ' input mse '
              read (in1,*) rmse
              write (iout1,*) ' input number of data points '
              read (in1,*) dp
              write (iout1,*) ' input variance of b '
              read (in1,*) vartplb
              write (iout1,*) ' input mean of ln(mean) '
              read (in1,*) regm
              write (iout1,*) ' mse from TPL regression  = ' , rmse
              write (iout1,*) ' data points in TPL regression = ' , dp
              write (iout1,*) ' variance of b from TPL regression = '
     1                , vartplb
              write (iout1,*) ' mean of ln(mean) in the TPL regression = '
     1                , regm


540           write (iout1,*) ' Are these details of the TPL regression '
              write (iout1,*) '   correct ("Y" or "N")?'
              read (in1,*) correct
                if (correct.ne.'N'.and.correct.ne.'n') then
                  if (correct.ne.'Y'.and.correct.ne.'y') goto 540
                end if
                if (correct .eq. 'N' .or. correct .eq. 'n') goto 520
              end if
530   continue


560   continue
         if (distr.eq.1) write (iout3,701)
701      format(//1x,' Binomial distribution')
         if (distr.eq.2) write (iout3,702)
702      format(//1x,' Poisson distribution')
         if (distr.eq.3) then
         write (iout3,703)
703      format(//1x,' Negative binomial distribution')
            if (kcalc.eq.1) write (iout3,704) rk
704         format(/1x,' constant k  = ',f10.4)
            if (kcalc.eq.2) then
            write (iout3,1)
     1      format (/,' Parameters for the TPL relationship')
            write (iout3,2) tpla,tplb
     2      format (' a  =  ' , f5.3, ' b  = ' , f5.3)
               if (TPLvarn .eq. 'y' .or. TPLvarn .eq. 'Y') then
               write (iout3,7) rmse
     7         format (/, ' MSE from TPL regression  = ' , f7.3)
               write (iout3,8) dp
     8         format (/, ' Data points in TPL regression = ' , f7.3)
               write (iout3,9) vartplb
     9         format (/, ' Variance of b in TPL regression  = ' , f7.3)
               write (iout3,10) regm
     10        format (/, ' Mean of ln(mean)  = ' , f7.3)
               end if
            end if
         end if
      write (iout3,3) nsamp
```

```
    3  format(/,' Minimum number of samples  =  ' , I4)
       write (iout3,4) nmax
    4  format (/, ' Maximum number of samples  =  ' , I15)
       write (iout3,5) zl,zu
    5  format (/, ' Values of z for stopping boundary  =  ' , 2f16.3)
       write (iout3,6) mci
    6  format (/, ' Monte Carlo iterations  =  ' , I10)
       write (iout3,11) t
   11  format (/, ' Threshold  =  ' , f7.3)
       write (iout3,12)
       write (iout1,12)
   12  format (/,'    Mean      Seq. OC      Ter. OC     Avg. samples')
       m = minm
  100  if (m.gt.maxm) goto 1000
       if (distr.eq.1) vart = t*(1. − t)
       if (distr.eq.2) vart = t
       if (distr.eq.3.and.kcalc.eq.1) vart = t + t*t/rk
       if (distr.eq.3.and.kcalc.eq.2) then
          var  = tpla*m**tplb
          vart = tpla*t**tplb
       end if


c set counters to zero
       seql   = 0.
       seqh   = 0.
       terl   = 0.
       terh   = 0.
       totcnt = 0
       varlm  = 0
c Monte Carlo loop
       do 20 i2 = 1,mci



          if (distr.eq.3.and.kcalc.eq.2) then
             if (TPLvarn .eq.'y' .or. TPLvarn .eq. 'Y') then
c                calculate standard deviation of a "var" predicted by TPL
c                for each run
                 stdlnv = sqrt(rmse + rmse/dp + vartplb*(log(m) − regm)**2)
                 v1 = log(var)
c                "var" is adjusted for variation in TPL to "ranvar"
                 call normal(v1,stdlnv,xnml)
                 ranvar = exp(xnml)
c     write(iout3,711) rmse,dp,vartplb,regm,m,var,ranvar,xnml
  711        format(1x,' rmse,dp,vartplb,regm / m,var,ranvar,xnml' /
     1            5x,4f15.5/15x,4f15.5)
             else
                 ranvar = var
             end if
             rk = ranvar − m
             if (rk.le.0.001) then
c                if using TPL results in "less than random variation" ...
                 rk = 0.001
                 varlm = varlm + 1
             end if
             rk = (m**2)/rk
          end if


          count = 0
          sum   = 0.
```

c compute stop lines and compare samples to them

```
     50    continue
              if (distr.eq.1) call posb(m,x)
              if (distr.eq.2) call pois(m,x)
              if (distr.eq.3) call ngb(m,rk,x)
           sum   =  sum + x
           count =  count + 1
              if (count.le.nsamp) go to 50
           cum  =  t*count
           du   =  zu*sqrt(count*vart)
           dl   =  zl*sqrt(count*vart)
           slu  =  cum + du
           sll  =  cum – dl
c          write (iout3,710) m,rk,x,sum,count,vart,cum,slu,sll
    710    format (1x,' m,rk,x / sum,count,vart,cum,slu,sll ',
       1   f6.1,f8.3,f6.0/11x,f12.2,i8,4f12.2)
              if (sum.gt.slu.or.sum.lt.sll) then
                 if (sum.lt.sll) then
                 totcnt  =  totcnt + count
                 seql    =  seql + 1.
                 else
                 totcnt  =  totcnt + count
                 end if
              else if (count .ge. nmax) then
                 xbar  =  sum/count
                 if (xbar.lt.t) then
                 terl   =  terl + 1.
                 totcnt =  totcnt + count
                 else
                 totcnt =  totcnt + count
                 end if
              else
              go to 50
              end if
     20    continue


c compute oc
           seqoc  =  seql/float(mci)
           teroc  =  terl/float(mci)


c compute average sample size
           avgsmp  =  float(totcnt)/float(mci)


c write results
           write (iout3,15) m,seqoc,teroc,avgsmp
           write (iout1,15) m,seqoc,teroc,avgsmp
     15    format (1x,f6.2,5x,f5.3,9x,f5.3,6x,f10.2)
              if (varlm .gt. 0) then
              write (iout1,25) varlm,mci
              write (iout3,25) varlm,mci
     25       format (1x,' WARNING: estimating k by moments failed in ',i8/
       1      10x,'of the total ',i8,' simulation runs. Poisson was used.')
              end if
           m = m + increment
           goto 100
   1000    continue
           write (iout1,803)
           write (iout3,803)
```

```
803 format(1x,' Simulated OC is split into paths which crossed the seq
    1uential boundaries'/' (Seq), and those which reached the maximum
    2 sample size (Ter)'/)
    write (iout1,*) ' Choose an action (1-4)'
    write (iout1,*) ' 1) Repeat calculations for new means'
      if (distr.eq.3.and.kcalc.eq.1) then
      write (iout1,*) ' 2) Repeat calculations with new k'
      else if (distr.eq.3.and.kcalc.eq.2) then
      write (iout1,*) ' 2) Repeat calculations with/without'
      write (iout1,*) '    variation in TPL'
      end if
    write (iout1,*) ' 3) Repeat calculations for new model'
    write (iout1,*) ' 4) Quit'
    read (in1,*) ianswer
      if (ianswer .eq. 1) then
      write (iout1,*) ' input minimum for mean'
      read (in1,*) minm
      write (iout1,*) ' input maximum for mean'
      read (in1,*) maxm
      write (iout1,*) ' input increment for means'
      read (in1,*) increment
      goto 560
      else if (ianswer .eq. 2) then
      write (iout1,712)
712   format (/)
        if (distr.ne.3) go to 500
        if (kcalc.eq.1) then
        write (iout1,*) ' constant value of k :'
        read (in1,*) rk
        else if (kcalc.eq.2) then
        write (iout1,*) ' input a and b values for TPL: var = a*m**b'
        write (iout1,*) ' input a'
        read (in1,*) tpla
        write (iout1,*) ' input b'
        read (in1,*) tplb
        end if
      goto 520
      else if (ianswer .eq. 3) then
      goto 500
      end if
    stop
    end
```

APPENDIX II

Example input for SPRT computer program (MS DOS Fortran version). Responses by user are preceded by " > ".

```
Choose a distribution
1) Binomial
2) Poisson
3) Negative binomial
4) Normal
enter a number 1-4
> 3
Input means for null and alternative hypotheses
and alpha and beta.
```

Mean for null hypothesis?
> 2
Mean for alternative hypothesis?
> 4
Input alpha (type I error) ?
> .1
Input beta (type II error) ?
> .1
Input value for k
> 1
Negative binomial distribution used

H0 mean	H1 mean	alpha	beta
2.000	4.000	0.1000	0.1000

k = 1.0000
These are the parameters that will be used in
the computations; are they correct ('Y' or 'N')?
> 'y'

low intercept	high intercept	slope
−12.051	12.051	2.802

Mean	OC	ASN
0.811	1.000	6.05

...

| 12.972 | 0.000 | 1.18 |

These OC and ASN curves are based on standard formulae.
The above is for a 2-decision SPRT.
Do you want to set up a 3-decision SPRT ('y' or 'n')?
> 'n'
Do you want to simulate the OC and ASN, or do a
sensitivity analysis on the negative binomial parameter k
('Y' or 'N') ?
> 'y'
SPRT sensitivity analysis
Nominal value of k = 1.000000
Sensitivity analysis can be done with constant
or dynamic k modelled using Taylors Power Law
Choose a method (enter 1 or 2)
1) constant k
2) dynamic k
> 1
Input k from negative binomial dist.
> 1
Input minimum sample size
> 0
Input maximum sample size
> 99999999
Input threshold (used at max sample size)
> 3
Specify range of means to be sampled
Input minimum for mean
> 1
Input maximum for mean
> 8
Input increment for means
> 1
Input monte carlo iterations
> 1000
Constant value of k = 1.0000
Minimum number of samples = 0
Maximum number of samples = 99999999
Threshold value = 3.00

Monte Carlo iterations = 1000
Are these values correct ('Y' or 'N')?
> 'y'

	accept lower infestation		accept higher infestation		
mean	Seq OC	Ter OC	Seq OC	Ter OC	ASN
1.00	1.0000	0.0000	0.0000	0.0000	7.40
...					
8.00	0.0020	0.0000	0.9980	0.0000	3.98

Note : OC probabilities are split depending on whether the cumulative path crossed a sequential boundary (Seq) or reached the maximum sample size (Ter).

Repeat calculations for new k ('Y' or 'N')?
> 'n'
Do sensitivity analysis using TPL ('Y' or 'N')?
> 'n'
Compute OC and ASN for new model ('Y' or 'N') ?
> 'n'
FORTRAN STOP

REFERENCES

1. Deming, W. E., *Out of the Crisis*, Massachusetts Institute of Technology, Center for Advanced Engineering Study, Cambridge, MA, 1986.
2. Deming, W. E., *Quality, Productivity, and Competitive Position*, Massachusetts Institute of Technology, Center for Advanced Engineering Study, Cambridge, MA, 1982.
3. Dodge, H. F. and Romig, H. G., *Sampling Inspection Tables, Single and Double Sampling*, John Wiley & Sons, New York, 1944.
4. Nyrop, J. P. and Binns, M. R., Quantitative methods for designing and analyzing sampling programs for use in pest management, in *Handbook of Pest Management in Agriculture, Volume II*, Pimentel, D., Ed., CRC Press, Boca Raton, FL, 1991.
5. Binns, M. R. and Nyrop, J. P., Sampling insect populations for the purpose of IPM decision making, *Annu. Rev. Entomol.*, 37, 427, 1992.
6. Fowler, G. W. and Lynch, A. M., Bibliography of sequential sampling plans in insect pest management based on Wald's sequential probability ratio test, *Great Lakes Entomol.*, 20, 165, 1987.
7. Sterling, W. L., Sequential sampling of cotton insect populations, in *Proc. Beltwide Cotton Prod. Res. Conf.*, National Cotton Council of America, Memphis, TN, 1975.
8. Wald, A., *Sequential Analysis*, John Wiley & Sons, New York, 1947.
9. Iwao, S., A new method of sequential sampling to classify populations relative to a critical density, *Res. Popul. Ecol.*, 16, 281, 1975.
10. Pieters, E. F., Bibliography of sequential sampling plans for insects, *Bull. Entomol. Soc. Am.*, 24, 372, 1978.
11. Shepard, M., Sequential sampling plans for soybean arthropods, in *Sampling Methods in Soybean Entomology*, Kogan, M. and Herzog, D. C., Eds., Springer-Verlag, New York, 1980.
12. Armitage, P., *Sequential Medical Trials*, 2nd ed., John Wiley & Sons, New York, 1975.
13. Siegmund, D., *Sequential Analysis: Tests and Confidence Intervals,* Springer-Verlag, New York, 1985.
14. Wetherill, G. B. and Glazebrook, K. D., *Sequential Methods in Statistics*, 3rd ed., Chapman & Hall, London, 1986.
15. Schilling, E. G., *Acceptance Sampling in Quality Control*, Marcel Dekker, New York, 1982.
16. Wallis, W. A., The Statistical Research Group, 1942–1945, *J. Am. Statist. Assoc.*, 75, 320, 1980.
17. Kemp, A. W., Families of discrete distributions satisfying Taylor's power law, *Biometrics*, 43, 693, 1987.
18. Perry, J. N. and Taylor, L. R., Families of distributions for repeated sampling of animal counts, *Biometrics*, 44, 881, 1988.

19. Fowler, G. W. and Lynch, A. M., Sampling plans in insect management based on Wald's sequential probability ratio test, *Environ. Entomol.*, 16, 345, 1987.
20. Nyrop, J. P. and Simmons, G. A., Errors incurred when using Iwao's sequential decision rule in insect sampling, *Environ. Entomol.*, 13, 1459, 1984.
21. Taylor, L. R., Aggregation, variance, and the mean, *Nature*, 189, 732, 1961.
22. Iwao, S., A new regression method for analyzing the aggregation pattern of animal populations, *Res. Popul. Ecol.*, 10, 1, 1968.
23. Taylor, L. R., Assessing and interpreting the spatial distributions of insect populations, *Annu. Rev. Entomol.*, 29, 321, 1984.
24. Kuno, E., Sampling and analysis of insect populations, *Annu. Rev. Entomol.*, 36, 285, 1991.
25. Trumble, J. T., Brewer, M. J., Shelton, A. M., and Nyrop, J. P., Transportability of fixed precision level sampling plans, *Res. Popul. Ecol.*, 31, 325, 1989.
26. Snedecor, G. W. and Cochran, W. G., *Statistical Methods*, 8th ed., Iowa State University Press, Ames, 1989.
27. Wald, A. and Wolfowitz, J., Optimum character of the sequential probability ratio test, *Ann. Math. Statist.*, 19, 326, 1948.
28. Perry, J. N. and Taylor, L. R., Adès: new ecological families of species-specific frequency distributions that describe repeated spatial samples with an intrinsic power-law variance-mean property, *J. Anim. Ecol.*, 54, 931, 1985.
29. Agnello, A. M., Kovach, J., Nyrop, J., Reissig, H., and Wilcox, W., *Simplified Integrated Management Program. A Guide for Apple Sampling Procedures in New York*, Cornell Cooperative Extension, Ithaca, NY, 1991.
30. Waters, W. E., Sequential sampling in forest insect surveys, *For. Sci.*, 1, 68, 1955.
31. Armitage, P., Sequential analysis with more than two alternative hypotheses, and its relation to discriminant function analysis, *J. R. Statist. Soc.* B12, 137, 1950.
32. Billard, L. and Vagholkar, M. K., A sequential procedure for testing a null hypothesis against a two-sided alternative hypothesis, *J. R. Statist. Soc.* B31, 285, 1969.
33. Sobel, M. and Wald, A., A sequential decision procedure for choosing one of three hypotheses concerning the unknown mean of the normal distribution, *Ann. Math. Statist.*, 20, 502, 1949.
34. Sylvester, E. S. and Cox, E. L., Sequential plans for sampling aphids on sugar beets in Kern County, California, *J. Econ. Entomol.*, 54, 1080, 1961.
35. Warren, W. G. and Chen, P. W., The impact of misspecification of the negative binomial shape parameter in sequential sampling plans, *Can. J. For. Res.*, 16, 608, 1986.
36. Hubbard, D. J. and Allen, O. B., Robustness of the SPRT for a negative binomial to misspecification of the dispersion parameter, *Biometrics*, 47, 419, 1991.
37. Binns, M. R. and Bostanian, N. J., Robust binomial decision rules for integrated pest management based on the negative binomial distribution, *Am. Entomol.*, 1, 50, 1990.
38. Hutchison, W. D., Hogg, D. E., Poswal, M. A., Berberet, R. C., and Cuperus, G. W., Implications of the stochastic nature of Kuno's and Green's fixed-precision stop lines: sampling plans for the pea aphid (Homoptera: Aphididae) in alfalfa as an example, *J. Econ. Entomol.*, 81, 749, 1988.

Chapter 9

SEQUENTIAL ESTIMATION AND CLASSIFICATION PROCEDURES FOR BINOMIAL COUNTS

Vincent P. Jones

TABLE OF CONTENTS

I.	Introduction	176
II.	Advantages and Disadvantages of Binomial Sampling Using p_T-to-m Relationships	176
III.	Binomial Estimation Procedures	177
	A. Binomial Distribution	177
	B. Sequential Estimation of a Proportion	178
	C. Estimation of Mean Using Kono-Sugino's Empirical Equation	180
	D. Estimation of a Mean Using the Negative Binomial Probability Model	186
	1. Negative Binomial Distribution	186
	2. Modifications of the Negative Binomial Probability Model	187
	3. Tally Thresholds Greater than 0	187
	4. Estimation of a Mean	188
IV.	Binomial Classification Systems	189
	A. Classification of a Proportion	189
	B. General Classification Using p_T-to-m Relationships	194
	C. Classification Methods Based on Kono-Sugino Equations	195
	D. Methods Based on Negative Binomial Probability Model	198
	E. Variable Intensity Binomial Sampling	200
V.	Future Studies in Binomial Sampling	203
	Acknowledgments	204
	References	204

I. INTRODUCTION

The impetus for the development of binomial sampling programs for agricultural systems has arisen from the need to quickly estimate or classify an arthropod's population density. To meet this need, binomial sampling plans can be easily developed for sequential estimation of a proportion or for sequential decision making using a critical proportion infested. These types of sampling plans are primarily useful for direct pests, where even one animal feeding can cause rejection of the commodity, or for ecological studies where the investigator is interested in the proportion of the population possessing a particular attribute. The primary difference between these types of binomial sampling programs (referred to as "classical" binomial sampling plans in this chapter) and those presented in the other chapters of this book which use complete enumeration methods, is that estimates of the mean and variance are based on categorization of the sampling units as infested/not infested rather than on complete counts.

In many agricultural systems, sampling plans for direct pests are less common than those for indirect pests. This is because even a small amount of direct damage can markedly decrease the value of many commodities and the cost of lowering the pest's population density prophylactically (through pesticides, resistant cultivars, biological control, etc.) is usually low compared to the increase in value of the undamaged commodity. For indirect pests, where the damage (measureable loss of either quality or quantity of yield) per unit injury (effect of pest activities on plant physiology) is generally very low compared to that caused by direct pests, a management approach makes economic and ecological sense. Because damage per unit injury is difficult to assess directly, an estimate of the population density (typically the mean, m) is frequently used as an index of the damage.[1] Therefore, to fit into current integrated pest management (IPM) systems, binomial sampling programs for indirect pests require some method of relating a proportion of the sampling units infested with $> T$ individuals (T is referred to as the tally threshold and p_T refers to the proportion of sampling units infested with $> T$ individuals) to the mean population density. The focus of this chapter is to provide information on binomial sampling for these needs. However, because the methodologies using p_T-to-m relationships are based at least partially on classical binomial theory, I develop the classical binomial methodologies first and then present the modifications required for using p_T-to-m relationships based on an empirical equation first published by Kono and Sugino[2] (K-S equation) and on a negative binomial probability model.[3] Examples of both sequential estimation and classification are presented. Background information on the *Panonychus* and *Tetranychus* data sets used for examples can be found in Jones.[4]

II. ADVANTAGES AND DISADVANTAGES OF BINOMIAL SAMPLING USING p_T-to-m RELATIONSHIPS

At a given sample size, estimates of a mean using binomial sampling have considerably more variation associated with them than an estimate of the mean based on complete enumeration methods.[5] This is because the binomial sample retains only presence/absence information from each sampling unit, and adds to the normal binomial variance (see Equation 3 below) a component associated with the fit of the p_T-to-m model. In some situations, the precision obtained using binomial sampling may never reach that of complete enumeration methods (see the section below entitled "Estimation of a Mean Using Kono-Sugino's Empirical Formula") and thus binomial sampling may be inappropriate. However, in agricultural systems, the increased variation and the requisite increase in sample size needed to predict m can be offset by the ease with which the units are classed as being infested or not infested compared to complete enumeration methods. This advantage of binomial sampling becomes more apparent as the degree of population clumping increases, the density of the population increases (within reason), and the organisms become smaller in relation to the dimensions of the sampling unit. For example, Wilson et al.[6] showed complete counts of spider mites on a single cotton leaf required more than 2 h, whereas classification of the leaf as infested or not could be accomplished in 1 min or less.

The use of different tally thresholds can extend and improve binomial sampling plans. However, by increasing the tally threshold from 0 to 8, the effort required to categorize the sampling unit as infested or not infested increases and some of the advantages of binomial sampling are lost. For example, if the tally threshold is set to 8, then at least nine individuals must be found in the sampling unit to classify the sampling unit as infested (recall the definition of tally threshold is $> T$ individuals). This would lead one to expect that the time required would be approximately eight-fold greater than a threshold of 8. However, the increase in time required to process a sample may not be this excessive,[5] and the advantages (as discussed below) may greatly outweigh the costs.

An advantage of binomial sampling rarely mentioned is that, compared to complete enumeration methods, binomial sampling methods are resistant to the effects of one to a few unusual observations in a sample. This is important because the goal of a sampling plan in agricultural systems is to estimate the population density and use the resulting value as an estimate of the "typical" damage that a "typical" plant experiences.[1] If a complete enumeration method is used to estimate the mean, a single observation can radically change the mean and thus affect the decision reached. For example, if ten leaves are sampled for spider mites, and the ten values are 0, 0, 0, 4, 5, 5, 7, 3, 6, and 10, then $m = 4$ mites per leaf. However, if instead of four mites on the fourth leaf there were 25 (not an unusual occurrence with spider mites), the mean would be 6.1. In this example, a single value has shifted the mean $> 50\%$ of the original estimate. In the same situation, the estimate of the proportion infested cannot shift by a factor of greater than $1/n$ (where n is the sample size) and may not shift at all.

III. BINOMIAL ESTIMATION PROCEDURES

A. BINOMIAL DISTRIBUTION

The binomial distribution arises when there are only two possible outcomes (e.g., organisms present or absent) and the probability of selecting an infested sampling unit with $> T$ individuals present [P(present)] is constant.[7] If P(present) varies because the sampling process removes a significant fraction of the sampling universe, then either a finite population correction or the hypergeometric distribution would be appropriate. However, in most situations, the binomial approximation is adequate and the added precision using the hypergeometric distribution is minimal.[7]

The probability generating function for the binomial distribution is

$$\Pr(A) = \frac{n!}{A!(n-A)!} P_T^A Q_T^{n-A} \qquad (1)$$

where the formula provides the probability of drawing a sample with A units that have $> T$ individuals (i.e., the tally level). In this equation, P_T is the proportion

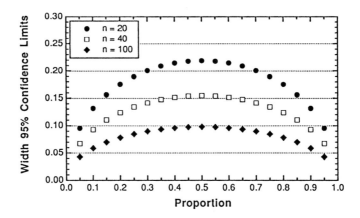

FIGURE 1. Effect of P_T on width of confidence intervals at a given fixed sample size.

infested with $> T$ individuals, $Q_T = (1 - P_T)$, and n is the sample size. In this chapter, sample estimates will be denoted by lower case values and parameters as capitalized letters unless specifically stated (i.e., p_T is a sample estimate and P_T is the true population value).

An unbiased estimate of the mean of a binomial distribution is calculated as

$$p_T = a/n \qquad (2)$$

where a is the number of sampling units which have $> T$ individuals present and n is the sample size. The sample estimate of the variance is

$$\text{Var}(p_T) = \frac{p_T * q_T}{n} \qquad (3)$$

Confidence limits about a given p_T are calculated using the normal (Gaussian) approximation as

$$CL = p_T \pm z_{\alpha/2} \sqrt{\text{Var}(p_T)} \qquad (4)$$

where $z_{\alpha/2}$ is the standard normal deviate. This formula requires at least 10 to 25 sampling units be examined to be a reasonable approximation depending on the author consulted.[8,9] At a fixed n, the confidence intervals are widest at $p_T = 0.5$ and decline about 0.5 in a symmetrical manner (Figure 1).

B. SEQUENTIAL ESTIMATION OF A PROPORTION

Sequential estimation occurs when the number of samples required to estimate the proportion is not fixed in advance and the decision as to whether sampling should continue or not is made after each sampling unit is inspected. The stopping criteria (i.e., the stop line) is usually whether a defined coefficient of variation ($cv = s_{p_T}/p_T$), a pre-set standard deviation ($d_0 = s_{p_T}$), or to a confidence interval equal to a fixed positive number (h) is achieved or not. To obtain a defined coefficient of variation, a rearrangement of the formula given by Karandinos[10] for optimal sample size gives

$$p_T = \frac{1}{1 + cv^2 n} \qquad (5)$$

[this formula and all those listed in this section assume that the normal (Gaussian) approximation to the binomial distribution holds]. A more convenient method for sampling is to rewrite Equation 5 in terms of the total number of sampling units which are infested by substituting a/n (Equation 2) for p_T and solving for a

$$a_n = \frac{1}{(1/n) + cv^2} \qquad (6)$$

where a_n is the total number of sampling units infested with $> T$ organisms in a sample of size n.[11] To use this formula, a cv level is chosen in advance and a_n is calculated for a series of different n values. Sampling is stopped if the number of leaves infested with $> T$ individuals is greater than the a_n calculated (Figure 2A). The mean is calculated using Equation 2, and the standard deviation is obtained by multiplying the resulting mean (p_T) by cv to obtain s_{p_T} (e.g., $(s_{p_T}/p_T) * p_T = s_{p_T}$). The 95% confidence intervals are obtained by multiplying s_{p_T} by $z_{\alpha/2}$. If it is desired to achieve a given d_0 (d_0 = the desired s_{p_T}), rather than cv, by use of the relationship $cv * p_T = d_0$, Kuno[11] provides the equation for the stop line as

$$a_n = \frac{n}{2}\left(1 \pm \sqrt{1 - 4d_0^2 n}\right) \qquad (7)$$

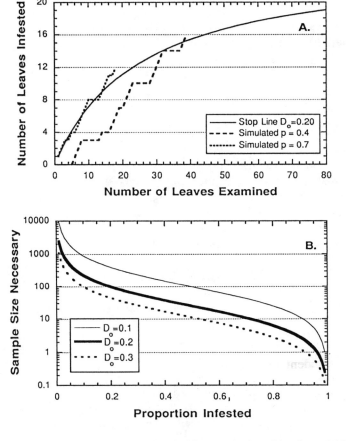

FIGURE 2. Sequential estimation of a proportion at a fixed level of precision ($cv = 0.25$). (A) Sampling chart showing use with two simulated populations. (B) The effect of changing precision levels on the ASN.

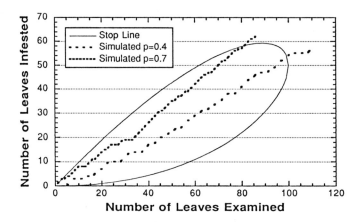

FIGURE 3. Sequential estimation of a proportion to achieve a fixed σ ($d_0 = 0.05$), showing use with two simulated populations.

This equation has two solutions, $a'_n = (n/2)(1 + \sqrt{1 - 4d_0^2 n})$ and $a''_n = (n/2)(1 - \sqrt{1 - 4d_0^2 n})$, which, when graphed, intersect to form a closed curve at $n = 1/4d_0^2$ (note this n is also the maximum sample size). The use of the sampling rule is slightly different and requires that sampling is continued until $a'_n < a_{obs}$ or $a_{obs} > a''_n$ (Figure 3).

If a fixed confidence interval (h) is desired, the last equation can be rearranged using the relationship $h = z_{\alpha/2} d_0$ as

$$a_n = \frac{n}{2}\left(1 \pm \sqrt{1 - 4\left(\frac{h}{z_{\alpha/2}}\right)^2 n}\right) \quad (8)$$

where $z_{\alpha/2}$ is the standard normal deviate.[10] This equation also has two roots and has a closed curve similar to Figure 3.

Rearrangements of these formulas can also be used to calculate the average sample number (ASN) to achieve the desired outcome at a given p_T. Table 1 has these formulas, and example ASN curves using several different cv values are presented (Figure 2B). As a rule, when the precision increases (i.e., cv becomes numerically smaller), the sample size required increases.

C. ESTIMATION OF A MEAN USING KONO-SUGINO'S EMPIRICAL EQUATION

The equation of Kono and Sugino[2] (K-S equation) and Gerrard and Chiang[12] is perhaps the best studied empirical equation to relate p_T to the mean density. They use

$$m = \exp^{a'}[-\ln(1 - p_T)]^{b'} \quad (9)$$

where a' and b' are parameters estimated from the data, and ln signifies the natural logarithm. The most common formulation takes the logarithm of both sides which allows simple linear regression to be used for parameter estimation

$$\ln(m) = a' + b' \ln(-\ln[1 - p_T]) \quad (10)$$

TABLE 1
Formulas and Example Calculations for Sequential Estimation of a Proportion

Sampling attribute	Formula	Example	Average sample number	Example
Fixed cv	$a_n = \dfrac{1}{cv^2 + \dfrac{1}{n}}$	$cv = 0.25, n = 20$ $a_n = \dfrac{1}{(0.25)^2 + \dfrac{1}{20}}$ $= 8.89$	$n = \dfrac{q_T}{p_T cv^2}$	$cv = 0.25, p_T = 0.30$ $n = \dfrac{(1 - 0.30)}{0.30(0.25)^2}$ $= 37.33$
Fixed $\sigma(d_0)$	$a_n = \dfrac{n}{2}\left(1 \pm \sqrt{1 - 4d_0^2 n}\right)$	$d_0 = 0.03, n = 45$ $a_n = \dfrac{45}{2}\left(1 \pm \sqrt{1 - 4(0.03)^2 45}\right)$ $d'_n = 43.1$ $d''_n = 1.90$	$n = \dfrac{p_T q_T}{d_0^2}$	$d_0 = 0.05, p_T = 0.35$ $n = \dfrac{0.35(1 - 0.35)}{(0.05)^2}$ $= 91.0$
Fixed width confidence interval (h)	$a_n = \dfrac{n}{2}\left(1 \pm \sqrt{1 - 4\left(\dfrac{h}{z_{\alpha/2}}\right)^2 n}\right)$	$n = 100, h = 0.08, a_{\alpha/2} = 1.96$ $a_n = \dfrac{100}{2}\left(1 \pm \sqrt{1 - 4\left(\dfrac{0.08}{1.96}\right)^2 100}\right)$ $d'_n = 78.88 \quad d''_n = 21.12$	$n = \left(\dfrac{z_{\alpha/2}}{h}\right)^2 p_T q_T$	$z_{\alpha/2} = 1.96, h = 0.1,$ $p_T = 0.32$ $n = \left(\dfrac{1.96}{0.1}\right)^2 0.32(1 - 0.32)$ $= 83.59$

The fit of this equation is often quite good (Figure 4A) and its formulation in terms of p_T allows tally levels higher than 0 to be estimated without modification of the other formulas presented in this section (Figure 4B). However, examination of the residuals and lack of fit statistics should be performed, because the equation is not strictly linear.[13] For example, in the *Tetranychus* data set, curvature was detected when a tally threshold greater than 4 was used. Where curvature bias is significant, Kuno's bias correction (bias = $mb'(1 - p_T)(b' - \ln p_T - 1)/[2np_T(\ln p_T)^2]$) should be squared and added to mse.

Exact variance estimates for m in the K-S equation do not exist; approximations have been the subject of at least six different papers. Because there are several components of these variance estimates, I use the notation of Schaalje et al.[14] which clearly differentiates each component. Kuno[15] calculated the variance as approximately

$$c1 = p_T b'^2 / n(1 - p_T) \ln[1 - p_T]^2 \qquad (11)$$

where n is the number of sampling units selected and b' is the slope of Equation 10. Unfortunately, Equation 11 is only an estimate of the sampling variance for the estimate of $\ln(-\ln[1 - p_T])$ from the binomial distribution and does not account for any error in estimation of a' and b' in Equation 10. Binns and Bostanian[13] point out that the total variance needs to be the sum of $c1$ and the variance of predicting $\ln m$

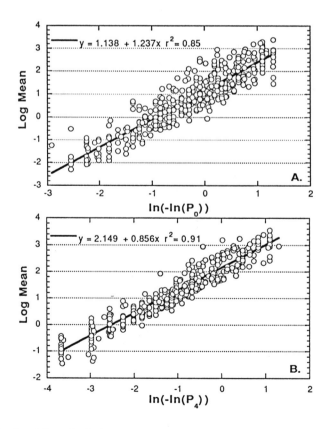

FIGURE 4. The fit of Kono-Sugino's equation for the *Tetranychus* data set using two different tally thresholds. (A) P_0 regression. (B) P_4 regression.

from the estimation of a' and b' using standard regression formulas for prediction of the confidence intervals for an individual case when the population is sampled at p_T:

$$\text{Var}(\ln m)_P = \text{mse} + \frac{\text{mse}}{N} + \{\ln[-\ln(1-p_T)-\bar{p}_T]\}^2 s_b^2 \quad (12)$$

They partition this into two components

$$c2 = \frac{\text{mse}}{N} + \{\ln[-\ln(1-p_T)] - \bar{p}\}^2 s_b^2 \quad (13)$$

and

$$c4 = \text{mse} \quad (14)$$

where mse is the mean square error from the regression of Equation 10, N is the number of data points in the regression used to estimate a' and b', $\bar{p}$ is the mean of the independent variable (i.e., $\ln[-\ln(1-p_T)]$) in the data set used to estimate a' and b', and $s_{b'}^2$ is the sample estimate of the variance of b'. They state that the term mse is generally the dominant term in Equation 12. Nyrop and Binns[5] suggest that mse represents both random sampling error and biological variability ($s_{\text{biological}}^2$) around Equation 10 and ideally, mse should be replaced by a smaller term representing purely biological variability because m is not measured without error. Binns and Bostanian[13] estimate the total variance as

$$\text{Var}(\ln m)_{\text{Total}} = c1 + c2 + c4 \quad (15)$$

Using simulation, Schaalje et al.[14] showed that Equation 15 may overestimate the biological variance because m is not measured without error, and suggested a method based on large sample size theory which uses Taylor's power law[16] (i.e., $s^2 = am^b$, where a and b are derived from empirical data) to estimate sampling variation independently from mse to obtain $s_{\text{biological}}^2$. The formula for the variance component is

$$c3 = \exp\{\ln a + (b-2)[a' + b' \ln(-\ln(1-P_T))]\}/n \quad (16)$$

where the values a and b are the estimates of the coefficients from Taylor's power law (TPL), a' and b' are from fitting Equation 10, and n is the number of sampling units collected. The estimate of sampling variation from TPL is then subtracted from mse. Accordingly, their equation is

$$\text{Var}(\ln m)' = c1 + c2 + (c4 - c3) \quad (17)$$

To estimate the variance of m (i.e., $\text{Var}(m)$, rather than $\text{Var}(\ln m)$), the relationship $\text{Var}(m) = m^2 \text{Var}(\ln m)$ is used as a conversion factor.[14]

Binns and Nyrop[17] suggest that a similar approach to the estimation of $s_{\text{biological}}^2$ could be used. Because the sample error for a mean value m_i based on TPL and a sample size n_i is am_i^b/n_i, the total variance (including biological variance, $s_{\text{biological}}^2$) of a given data point m_i, is approximately $(am_i^b - 2)/n_i + s_{\text{biological}}^2$. They suggest that a weighted regression analysis can be used to provide a maximum-likelihood estimate of $s_{\text{biological}}^2$. This estimate can then be used instead of the approximation suggested by Binns and Bostanian[13] of using mse ($c4$) from the regression as an estimate of biological variance.

Equations 15 and 17 may be more accurate than required for pest management purposes. Nyrop and Binns[5] in an earlier publication suggest that mse is a reasonable approximation for the variance estimate given in Equation 15. Examination of each variance component for the *Tetranychus* data set shows the $c1$ and $c3$ variances generally increase as $p_T \to 0$, $c2$ is very low throughout the range of p_T, and mse is constant (Table 2). If we consider Equation 15 to be the total variance, use of mse by itself as suggested by Nyrop and Binns[5] is an underestimate of the total variance, especially when n and p_T are low (Figure 5A, B). In all situations, $c1 + c4$ is a much better estimate of the total variance. When n is large, the sampling error is minimized and mse is a good estimate of biological variation (Figure 5C, D). This approximation is easier to calculate and provides nearly the same estimates as either Equations 15 or 17. Other more complex approximations of the variance associated with a prediction using the K-S equation have been reported by Schaalje and Butts[18] and Hepworth and Macfarlane.[19] However, the increase in accuracy is generally small provided n is large (in these studies large ≈ 75 to 100 sampling units).

TABLE 2
Importance of Variance Components for *Tetranychus* Data Set. Total Variance Refers to That of Binns ($c1 + c2 + $ mse)

p_T	Mean	$c1$	$c2$	mse	$c3$	Binns	Schaalje	mse + $c1$	mse + $c1$ (% total)	Schaalje (% total)	mse (% total)
						P0 Regression					
0.05	12.13	1.021	0.005	0.256	0.379	1.283	0.904	1.278	99.58	70.49	19.98
0.20	5.62	0.256	0.002	0.256	0.150	0.514	0.364	0.513	99.69	70.84	49.85
0.35	3.31	0.148	0.001	0.256	0.099	0.405	0.306	0.404	99.80	75.55	63.25
0.50	1.98	0.106	0.001	0.256	0.073	0.363	0.290	0.363	99.83	79.78	70.58
0.65	1.10	0.086	0.001	0.256	0.057	0.343	0.287	0.342	99.79	83.52	74.72
0.80	0.49	0.079	0.001	0.256	0.043	0.336	0.293	0.335	99.69	87.15	76.25
0.95	0.08	0.108	0.002	0.256	0.029	0.366	0.337	0.364	99.47	92.03	69.97
						P2 Regression					
0.05	15.63	0.707	0.002	0.173	0.216	0.881	0.665	0.880	99.80	75.45	19.58
0.20	8.24	0.177	0.001	0.173	0.100	0.351	0.250	0.350	99.84	71.44	49.22
0.35	5.31	0.103	0.000	0.173	0.071	0.276	0.205	0.275	99.85	74.25	62.63
0.50	3.46	0.074	0.000	0.173	0.055	0.247	0.191	0.246	99.81	77.58	69.98
0.65	2.12	0.060	0.001	0.173	0.044	0.233	0.188	0.232	99.74	80.89	74.15
0.80	1.08	0.055	0.001	0.173	0.036	0.228	0.192	0.227	99.62	84.41	75.68
0.95	0.24	0.075	0.001	0.173	0.026	0.249	0.223	0.247	99.44	89.68	69.36
						P4 Regression					
0.05	21.94	0.488	0.001	0.106	0.127	0.596	0.469	0.595	99.86	78.67	17.86
0.20	12.89	0.123	0.000	0.106	0.067	0.229	0.162	0.229	99.87	70.79	46.39
0.35	8.94	0.071	0.000	0.106	0.050	0.177	0.127	0.177	99.83	71.67	59.92
0.50	6.27	0.051	0.000	0.106	0.041	0.158	0.117	0.157	99.75	74.06	67.50
0.65	4.17	0.041	0.001	0.106	0.034	0.148	0.114	0.147	99.63	76.96	71.84
0.80	2.38	0.038	0.001	0.106	0.028	0.145	0.116	0.144	99.47	80.45	73.43
0.95	0.67	0.052	0.001	0.106	0.022	0.159	0.138	0.158	99.24	86.43	66.78

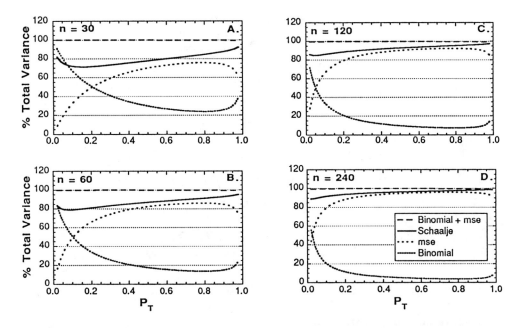

FIGURE 5. Contribution of variance components to the total variance (Equation 15) using Kono-Sugino's equation for the *Tetranychus* data set. (A) $n = 30$. (B) $n = 60$. (C) $n = 120$. (D) $n = 240$.

Using either Equation 15, 17, or the combination of $c1 + c4$ for the variance estimate allows the confidence limits of a prediction to be estimated. The confidence limits on the logarithmic scale are computed as

$$\ln m \pm z_{\alpha/2} \sqrt{\text{Var}(\ln m)} \qquad (18)$$

and for minimum bias can be converted back to the arithmetical scale using the equation of Schaalje[14]

$$m' = \exp(\ln m)[1 - 0.5(\text{Var}(\ln m))] \qquad (19)$$

If Equation 15 is used for estimation of the variance, increasing the number of sampling units used to estimate p_T has little effect on the width of the final confidence limits. This is because the only component with a term for the number of sampling units is $c1$. Examination of Figure 5 and Table 2 shows that this is most important when n is low and/or p_T is low. Therefore, the number of sampling units only needs to be high enough to minimize the sampling error ($c1$) and to provide an accurate estimate of p_T. Using the techniques previously described for sequential estimation of a proportion makes little sense in this situation. Instead, a fixed sample size can be calculated where further increases in n produce little change in the width of the confidence intervals on an arithmetical scale. This is shown with the *Tetranychus* data using P_2 regression (i.e., the tally threshold = 2) where the sample size used to estimate $\hat{p}$ is incremented from $n = 10$ to 600 (Figure 6). For this data set, taking more than ≈ 70 leaves is inefficient because the decrease in confidence intervals between 70 and 600 leaves is minimal considering the extra work (CI for 70 leaves = 2.13 – 10.17, CI for 600 leaves = 2.26 – 9.61). The lower limit on the confidence limit changes more slowly than the upper confidence limit.

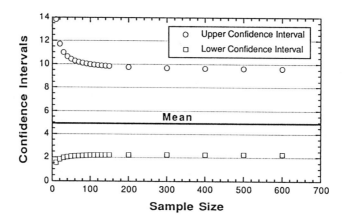

FIGURE 6. Effect of sample size used to predict a mean ($m = 5$) for Kono-Sugino's equation on confidence intervals of an estimate.

D. ESTIMATION OF A MEAN USING THE NEGATIVE BINOMIAL PROBABILITY MODEL

1. Negative Binomial Distribution

Counts of arthropods typically have a clumped distribution where the ratio of variance to mean (s^2/m) is > 1.[20] Because of this, many authors have tested the fit of the negative binomial to their data sets, with varying degrees of success. When the negative binomial is a satisfactory model, binomial sampling can be developed using methods detailed below. However, the negative binomial is only one of many distributions which have a (s^2/m) ratio > 1, and a statistically acceptable fit does not necessarily mean that the counts are actually distributed as a negative binomial.[21] This may become important in calculating Monte Carlo estimates of sampling statistics, because the underlying assumption in these cases is that the population is actually distributed according to a negative binomial (because the simulations use a negative binomial random number generator) and thus the results of these calculations may not reflect the distribution of the organism in question. Methods using bootstrap simulations from actual field data sets should be used whenever possible to check Monte Carlo simulations based on any specific assumed distribution (Hutchison, Chapter 10, this volume).

Using the notation of Bliss and Fisher,[3] the negative binomial can be completely defined by the mean (μ) and an exponent K. The sample mean (m) is estimated in the normal fashion, and estimation of K (i.e., k) is defined by the method of moments as

$$k = \frac{m^2}{(s^2 - m)} \qquad (20)$$

where s^2 is the sample estimate of the population variance. The value of k can also be estimated iteratively from the proportion of empty sampling units or by using the preferred maximum likelihood estimation as discussed by Bliss and Fisher.[3] The basic formula for the estimate of the proportion of sampling units with no animals present [$p(0)$] at a given m is

$$p(0) = \left(1 + \frac{m}{k}\right)^{-k} \qquad (21)$$

The proportion of units infested [$p(I)$] is

$$p(I) = 1 - p(0) \tag{22}$$

and is used by many authors rather than $p(0)$ to correlate to the mean population level.

2. Modifications of the Negative Binomial Probability Model

Wilson and Room[22] use a negative binomial distribution where the variance is constrained to follow TPL[16] (i.e., $s^2 = am^b$, where a and b are derived from empirical data) to develop a binomial sampling plan. This results in the value of k being defined as

$$k = \frac{m}{(am^{(b-1)} - 1)} \tag{23}$$

and by substitution into Equation 20 results in

$$p(0) = \exp\left[-m * \ln_e(am^{(b-1)}) * (am^{(b-1)} - 1)^{-1}\right] \tag{24}$$

Unfortunately, the fit of Equation 23 is often variable and there is often considerable variability in estimates of k which is independent of density.[23,24] In addition, the estimated variance of Equation 24 has not been derived. Nyrop and Binns[23] provide an approximate method for evaluating the effect of the variance of k by assuming the predicted variances are normally distributed with $m = 0$ and $s^2 = $ mse.

3. Tally Thresholds Greater Than 0

Tally thresholds greater than zero are possible by using the expected frequencies from the negative binomial for other frequency classes (e.g., the probability that you expect to find exactly one, two, or x individuals in a sampling unit at a given m and k). The expansion for terms higher than zero can be calculated recursively as

$$p(x) = \frac{(k + x - 1)mk^{-1}/(1 + mk^{-1})}{x} * p(x - 1) \tag{25}$$

where x is the frequency class being predicted.[3] The probability of then finding $> T$ individuals in a sampling unit is then

$$p_T = 1 - \sum_{x=0}^{T} p(x) \tag{26}$$

which is just one minus the sum of the individual probabilities between 0 and T. For example, if our tally threshold is two, then the expansion would be

$$p_2 = 1 - [p(0) + p(1) + p(2)] \tag{27}$$

In the following example, $k = 0.78$ and $m = 6$. The proportion of sampling units with no animals present is

$$p(0) = \left(1 + \frac{6}{0.78}\right)^{-0.78}$$

$$= 0.185$$

and the proportion with only one animal present is

$$p(1) = \frac{(0.78 + 1 - 1) * 6 * 0.78^{-1}/(1 + 6 * 0.78^{-1})}{1} * 0.185$$

$$= 0.128$$

and the proportion with only two animals present is

$$p(2) = \frac{(0.78 + 2 - 1) * 0.78^{-1}/(1 + 6 * 0.78^{-1})}{2} * 0.128$$

$$= 0.202$$

and $p_2 = 1 - (0.202 + 0.128 + 0.185) = 0.485$.

Because Wilson and Room's[22] method is just a constrained negative binomial, the expansion uses the same logic and formulas, with the exception that k is calculated using Equation 23. Note that the basic equation for $P(0)$ is always calculated using Equation 21.

4. Estimation of a Mean

There is frequently a good relationship between the proportion of sampling units infested and the mean population level. When the negative binomial is used to predict this relationship (using k and $p_{observed}$ in Equations 21 or 25), the major problems come from the variability in the estimate of k resulting from changes in mean population density and the variation independent of density.[5,25,26] If the k used in Equations 21 or 25 is an overestimate of K (the true population parameter), the mean is underestimated; if k is an underestimate, then the mean is overestimated[26,27] (Figure 7). Fortunately, k rarely varies excessively over a reasonable range of mean population densities.[26,27] Binns and Bostanian[26] point out that the robustness of the estimation procedure with respect to variation in k can be evaluated by plotting the p_T-to-m relationship using k_{best} and two extreme k values (k_{min}, k_{max}). Nyrop and Binns[23] suggest that the three k values be calculated as the 10th, 50th, and 90th percentiles of the distribution of k values observed under field conditions.

The problems with a variable k can be further reduced by proper selection of the tally threshold.[26,27] This can be seen in Figure 7, where several different tally thresholds are used to estimate the mean with the *Panonychus* data set (k_{best} = 0.72, k_{max} = 1.91, k_{min} = 0.23). For this data set, the P_0 estimate of 6 mites per leaf is highly susceptible to variation in k, but the P_8 and P_{10} estimates are much more stable. Binns and Bostanian[27] further show that rarely does the choice of tally threshold have to exceed m to adequately reduce the bias caused by variation in k. Clearly, the choice of the tally threshold is highly dependent on the sample mean and at least a preliminary estimate of the range of means of interest is required in design of the sampling program. This same relationship between m and T is seen in the classification section.

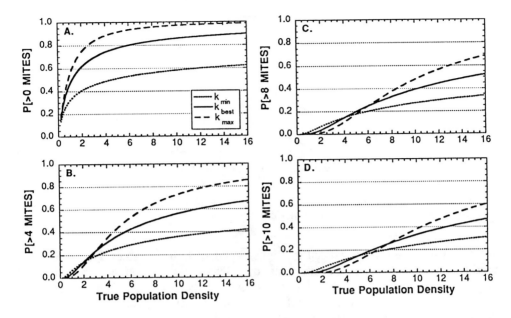

FIGURE 7. Sensitivity of density estimates to variation in k at different tally thresholds for the *Panonychus* data set. (A) P_0. (B) P_4. (C) P_8. (D) P_{10}.

IV. BINOMIAL CLASSIFICATION SYSTEMS

A. CLASSIFICATION OF A PROPORTION

Binns (Chapter 8), provides a complete discussion of classification procedures and the discussion below is for specific points regarding binomial classification. Classification procedures are aimed at determining if the proportion of sampling units is either above or below a critical proportion by a predetermined amount. Statistically, this is described by two hypotheses such as

$$H_1: \theta_1 \leq 0.20 \text{ proportion of leaves infested} > T \text{ individuals}$$

and

$$H_2: \theta_2 \geq 0.30 \text{ proportion of leaves infested} > T \text{ individuals}$$

at α (Type 1) and β (Type 2) error rates. When graphed, these hypotheses are frequently shown as parallel lines or slightly diverging lines depending on the method used for developing the sequential sample. When the population is in the region below the decision line for H_1 or above the line for H_2, then sampling is terminated. When the population density is between the lines, then sampling is continued until a decision is reached (Figure 8). The chief advantage of classification methods is that they minimize sample sizes required to reach the decision compared with estimation procedures and they are easy to explain and implement in the field.[17] The biggest disadvantage is that the estimates of the proportion of sampling units infested are very inaccurate when the population density is greatly different than the thresholds and may not be acceptable for other purposes (e.g., the estimates would be poor for analyzing population density changes through time).

Classification procedures based on Wald's sequential probability ratio test (SPRT)[28] and other methods are best evaluated using the operating characteristic (OC) curves

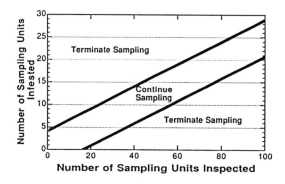

FIGURE 8. Classification of a proportion using a binomial SPRT showing termination and continuation boundaries. $\theta_1 = 0.20$, $\theta_2 = 0.30$, $\alpha = 0.10$, $\beta = 0.10$.

and the ASN curves. As discussed by Binns (Chapter 8, this volume), Wald's SPRT is used because the approximate equations for ASN and OC are simple, and among all sequential tests with equal OC values for θ_1 and θ_2, the ASN value is minimized. The OC curve is a plot of the probability of "no intervention" vs. the true population density, and thus varies from a value of 1 to 0.[29] If the θ_1 and θ_2 thresholds are very close together (note they cannot be identical), then the ideal OC curve would be a step function described by

$$P(\text{no intervention})|\theta = 1, \quad \theta < \theta_1$$

$$P(\text{no intervention})|\theta = 0, \quad \theta > \theta_2$$

where the equations are read as "the probability of no intervention given the true mean is θ equals one when θ is less than θ_1." However, such an ideal curve cannot be obtained unless every sampling unit in the statistical universe under consideration is examined.[29] The typical OC curve has its value ca. 1 when the true density (θ) $\ll \theta_1$, ca. 0.5 when it is near θ, and ca. 0 when $\theta \gg \theta_2$ (Figure 9). The steepness of the OC curve between θ_1 and θ_2 is a function of the precision desired; as the α and β or $\theta_2 - \theta$ are made smaller, the sample size required to make a decision increases and the OC curve becomes steeper.

The ASN curve gives the average number of samples required to reach a decision as the true mean population density changes in relation to the threshold. The ASN is achieved only over the long run and for any given sampling session the sample number obtained may depart significantly from the ASN. The average sample size required is smallest when $\theta_1 \ll \theta$ or $\theta \gg \theta_2$ and is greatest between θ_1 and θ_2 (Figure 9). The ASN at a given θ increases as α and β are made numerically smaller.

Fowler and Lynch[30] provide an excellent discussion and summary of the calculation of OC and ASN functions using SPRT when the distribution is known to be approximately binomial, negative binomial, normal, or Poisson. They point out that OC and ASN functions are only approximations primarily because of overshooting when a sampling termination decision is made. Overshoot occurs because the stopping rules generated by SPRT are expressed as real numbers, whereas in binomial count sampling the sampling units are integers (i.e., you can't have 1.5 leaves counted). The phenomena of overshoot becomes more important as the difference between θ_1 and θ_2 becomes larger or if α and β are much larger than 0.05. The ASN values are also affected by overshoot and the result is that more samples are

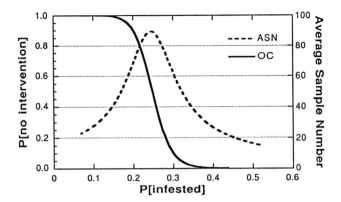

FIGURE 9. OC and ASN curves for a binomial SPRT with $\theta_1 = 0.20$, $\theta_2 = 0.30$, $\alpha = 0.10$, $\beta = 0.10$.

taken than required to accurately reach a decision. Fowler and Lynch[30] and Nyrop and Binns[5] both suggest using Monte Carlo simulations to estimate the OC and ASN curves.

The nominal values of α and β also need to be discussed. Fowler and Lynch[30] and Fowler[31] working with SPRT plans show that these values are only approximations, again because of overshooting the decision boundaries. In practice, the commonly used strategies of requiring a minimum number of samples to be taken before applying the stopping rules and/or terminating sampling when a maximum number of samples are taken, both affect the α and β values obtained. They state that the most common problem with SPRT is that the nominal OC curve underestimates α near θ_1 and overestimates β near θ_2. Nyrop et al.[5,32,33] have suggested that the α and β values should thus be used as parameters which can be adjusted to provide acceptable ASN and OC functions. This idea makes the cost/reliability trade-off an integral part of the sampling program.

A binomial SPRT can be developed using the equations in Tables 3 and 4. The required parameters are θ_1 and θ_2, α and β. For the following example, $\theta_1 = 0.2$, $\theta_2 = 0.3$, $\alpha = 0.10$, and $\beta = 0.10$. The common slope for the decision lines is calculated as

$$\text{slope} = \frac{\ln\left[\dfrac{1 - 0.3}{1 - 0.2}\right]}{\ln\left[\dfrac{0.3(1 - 0.2)}{0.2(1 - 0.3)}\right]}$$

$$= 0.25$$

and the intercept for the upper and lower decision lines are

$$\text{intercept}_1 = \frac{\ln((1 - 0.1)/0.1)}{\ln\left[\dfrac{0.3(1 - 0.2)}{0.2(1 - 0.3)}\right]} \qquad \text{intercept}_2 = \frac{\ln(0.1/(1 - 0.1))}{\ln\left[\dfrac{0.3(1 - 0.2)}{0.2(1 - 0.3)}\right]}$$

$$= 4.08 \qquad\qquad = -4.08$$

If α and β are equal, then the intercepts will be the same magnitude but opposite in sign. If α and β are different the intercepts will be different.

TABLE 3
Formulas for Binomial SPRT Decision Lines

$$\text{slope} = \frac{\ln\left[\dfrac{1-\theta_2}{1-\theta_1}\right]}{\ln\left[\dfrac{\theta_2(1-\theta_1)}{\theta_1(1-\theta_2)}\right]}$$

$$\text{intercept}_1 = \frac{\ln\left(\dfrac{1-\alpha}{\beta}\right)}{\ln\left[\dfrac{\theta_2(1-\theta_1)}{\theta_1(1-\theta_2)}\right]} \qquad \text{intercept}_2 = \frac{\ln\left(\dfrac{1-\beta}{\alpha}\right)}{\ln\left[\dfrac{\theta_2(1-\theta_1)}{\theta_1(1-\theta_2)}\right]}$$

The OC and ASN curves can be constructed as follows. The variable h is a dummy variable (unrelated to h in Equation 8) which can range from 4 to -4 with intervals of 0.2 to provide a smooth curve. The OC at θ corresponding to $h = 1.5$ is defined as

$$\text{OC}(\theta) = \frac{\left[\dfrac{1-0.1}{0.1}\right]^{1.5} - 1}{\left[\dfrac{1-0.1}{0.1}\right]^{1.5} - \left[\dfrac{0.1}{(1-0.1)}\right]^{1.5}}$$

$$= 0.964$$

The θ value corresponding to $h = 1.5$ is defined as

$$\theta(1.5) = \frac{1 - \left[\dfrac{1-0.3}{1-0.2}\right]^{1.5}}{\left[\dfrac{0.3}{0.2}\right]^{1.5} - \left[\dfrac{1-0.3}{1-0.2}\right]^{1.5}}$$

$$= 0.178$$

and the ASN at this θ is

$$\text{ASN}(\theta) = \frac{\ln\left(\dfrac{1-0.1}{0.1}\right)\text{OC}(\theta) + \left[\ln\left(\dfrac{1-0.1}{0.1}\right)\right]\{1 - \text{OC}(\theta)\}}{\theta \ln\left[\dfrac{0.3(1-0.2)}{0.2(1-0.3)}\right] + \ln\left[\dfrac{1-0.3}{1-0.2}\right]}$$

$$= 54.43$$

The complete OC and ASN curves are shown in Figure 9. These calculations are easily implemented using virtually any spreadsheet program.

TABLE 4
Formulas for OC, ASN, and θ for Binomial SPRT

	Estimate of θ (i.e., $\hat{p}$)	OC at θ	ASN at θ
$h = 0$	$\hat{p} = \dfrac{\ln\left[\dfrac{1-\theta_1}{1-\theta_2}\right]}{\ln\left[\dfrac{\theta_2(1-\theta_1)}{\theta_1(1-\theta_2)}\right]}$	$\mathrm{OC}(\theta) = \dfrac{\ln\left(\dfrac{1-\alpha}{\beta}\right)}{\ln\left(\dfrac{1-\alpha}{\beta}\right) - \ln\left(\dfrac{1-\beta}{\alpha}\right)}$	$\mathrm{ASN}(\theta) = \dfrac{-\ln\left[\dfrac{1-\alpha}{\beta}\right]\left[\dfrac{1-\beta}{\alpha}\right]}{\ln\left[\dfrac{p_1}{p_2}\right]\ln\left[\dfrac{1-p_0}{1-p_1}\right]}$
$h \neq 0$	$\hat{p} = \dfrac{1 - \left[\dfrac{1-\theta_2}{1-\theta_1}\right]^{h(p)}}{\left[\dfrac{\theta_2}{\theta_1}\right]^{h(p)} - \left[\dfrac{1-\theta_2}{1-\theta_1}\right]^{h(p)}}$	$\mathrm{OC}(\theta) = \dfrac{\left(\dfrac{1-\beta}{\alpha}\right)^{h(p)} - 1}{\left(\dfrac{1-\beta}{\alpha}\right)^{h(p)} - \left(\dfrac{\beta}{1-\alpha}\right)^{h(p)}}$	$\mathrm{ASN}(\theta) = \dfrac{\ln\left(\dfrac{1-\beta}{\alpha}\right)\mathrm{OC}(p) + \ln\left(\dfrac{1-\alpha}{\beta}\right)[1 - \mathrm{OC}(p)]}{p\ln\left[\dfrac{\theta_2(1-\theta_1)}{\theta_1(1-\theta_2)}\right] + \ln\left[\dfrac{1-\theta_2}{1-\theta_1}\right]}$

B. GENERAL CLASSIFICATION USING p_T-to-m RELATIONSHIPS

If there were no variance in the p_T-to-m relationship, sequential classification systems could rely exclusively on the approximate equations of SPRT. To develop a sampling program, a critical action threshold density (M_a) could be used to calculate a critical P_T value using a p_T-to-m relationship. A binomial SPRT could then be constructed about the critical P_T value as shown previously. However, because there is always variation associated with the empirical p_T-to-m relationship, modifications to this process must be made and the robustness of the procedures must be evaluated.[5,13,26]

Another change from the classical binomial SPRT is that while θ_1 and θ_2 are both calculated on proportion scale, they are eventually converted to a mean number of animals per sampling unit basis.[5] This conversion of θ_1 and θ_2 to m_1 and m_2 must consider the nonlinear nature of the p_T-to-m relationship (Figure 10A). At low levels of θ, the slope of the p_T-to-m relationship is approximately constant and if θ_1 and θ_2 are chosen as $\theta \pm x$, when the proportion is converted to m_1 and m_2 via the p_T-to-m relationship, the values of m_1 and m_2 are approximately equidistant about the desired threshold (M_a) (Figure 10B). However, because the derivative of the p_T-to-m relationship is constantly increasing as θ increases, when θ is large, $\theta \pm x$ results in highly asymmetrical boundaries around m_a. This is easily illustrated using the *Panonychus* data set, where $k = 0.72$, and a simple statistic for determining the degree of asymmetry can be calculated using $\Delta m = (m_2 - M_a)/(M_a - m_1)$. The Δm ratio is highly dependent on the proportion of sampling units infested (θ) and relatively insensitive to T (Figure 10B). However, if two different T values are used, the actual magnitude of $m_2 - m_a$ will be larger at the higher T value. An advantage

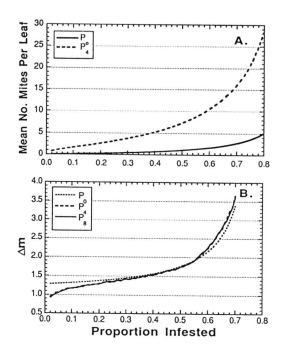

FIGURE 10. Nonlinear relationship between the proportion infested and the mean number of mites per leaf for the *Panonychus* data set using two different tally thresholds. (A) P_0 and P_4 relationships. (B) The effect of higher tally thresholds on asymmetry when using $\theta \pm x$ when converting back to an arithmetical scale.

of using a higher tally threshold is therefore to move the proportion infested to a lower value so that the asymmetry is not as severe. However, even at lower θ levels, a correction for asymmetry must be performed when assigning θ_1 and θ_2.

C. CLASSIFICATION METHODS BASED ON KONO-SUGINO EQUATIONS

The empirical K-S equation can be used as a basis for binomial classification systems as well as for estimation systems. As in estimation, the variance of the p_T-to-m model is of overriding importance in the calculation of the OC and ASN curves.[13,23] A rearrangement of the K-S model allows us to predict p_T for a given m instead of the converse:

$$\ln(-\ln p_T) = c' + d' \ln(m) \qquad (28)$$

and the variance estimate of $\ln(-\ln P_T)$ from standard regression formulas:

$$\text{mse} + \text{mse}/N + (\ln(m) - \overline{m})^2 s_{d'}^2 \qquad (29)$$

where mse is the residual error about the regression line, N is the number of data points in the regression, $\overline{m}$ is the mean value for the independent variable (i.e., the mean of $\ln(m)$), and $s_{d'}^2$ is the variance of the slope from Equation 28. Notice that the variance estimate is different than that needed in estimation by eliminating Equation 11 ($c1$). As before, Binns and Bostanian[13] split this into two variance components, which are

$$\text{mse}/N + (\ln(m) - \overline{m})^2 s_{d'}^2 \qquad (30)$$

and

$$\text{mse}$$

The variance estimate from Equation 29 tends to spread out the OC curves compared to the usual binomial SPRT and its effect needs to be assessed when evaluating the robustness of the procedures.[13]

For the *Panonychus* data set, the contribution of Equation 30 to the total variation is minimal (Table 5). This appears to be a general result and means that the variance estimate from Equation 29 is approximately constant and is equal to mse.[13] The result is that the confidence limits about the best fit p_T-to-m line are approximately parallel lines rather than the standard curving lines.[13] If the p_T-to-m line is representative for the species and crop system, the effect of random variation will be to shift the p_T-to-m relationship observed in a given field to a position parallel to the original. Nyrop and Binns[23] have shown that the variation can reasonably be assumed to be normally distributed with $m = 0$ and $s^2 = \text{mse}$. Binns and Bostanian[13] refer to this vertical displacement as Δh (Figure 11). Note this is not the same h used as a dummy variable in calculation of the SPRT or in the desired width of a confidence interval.

Because the binomial sampling plan relates a critical P_T to the critical mean, M_a, the effect of Δh is to shift M_a to a new value.[13] If Δh is displaced above the best fit line, P_T will correspond to m'_a that is less than M_a and if displaced below, it will correspond to $m''_a > m_a$ (Figure 11). The OC and ASN values obtained using the best fit line and the associated p_T values must be related to the new m'_a values. This

TABLE 5
Importance of Variance Components for *Panonychus* Data Set at Three Different Tally Thresholds When Used for the Purposes of Classification

P_T	Mean	c2	mse	Total	mse / Total
		P0 Regression			
0.95	0.05	0.002	0.106	0.108	0.983
0.80	0.32	0.001	0.106	0.107	0.995
0.65	0.78	0.000	0.106	0.106	0.997
0.50	1.48	0.000	0.106	0.106	0.996
0.35	2.58	0.001	0.106	0.107	0.995
0.20	4.57	0.001	0.106	0.107	0.993
0.05	10.50	0.001	0.106	0.107	0.988
		P2 Regression			
0.95	0.45	0.002	0.173	0.175	0.990
0.80	1.76	0.001	0.173	0.174	0.996
0.65	3.25	0.001	0.173	0.174	0.994
0.50	5.07	0.001	0.173	0.174	0.992
0.35	7.48	0.002	0.173	0.175	0.989
0.20	11.15	0.003	0.173	0.176	0.985
0.05	19.92	0.004	0.173	0.177	0.977
		P4 Regression			
0.95	0.93	0.002	0.160	0.162	0.990
0.80	3.12	0.001	0.160	0.161	0.994
0.65	5.36	0.001	0.160	0.161	0.991
0.50	7.93	0.002	0.160	0.162	0.987
0.35	11.16	0.003	0.160	0.163	0.983
0.20	15.87	0.004	0.160	0.164	0.977
0.05	26.47	0.006	0.160	0.166	0.967

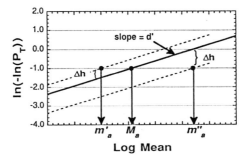

FIGURE 11. Effect of variance in Kono-Sugino's equation where data points lie off the best fit line. If Δh lies above the line, it is equivalent to lying on a parallel line (dotted line) and shifts the estimate of the mean to m'_a which is less than the desired m_a. If Δh is shifted below, the mean is shifted to a higher m''_a.

change results in a spreading out of the OC and ASN functions compared to the optimal value (Figure 12A, B). Binns and Bostanian[13] give the change from M_a as

$$m'_a = M_a / \exp(h/d') \tag{31}$$

where d' is from Equation 29.

Nyrop and Binns[23] show that these relationships can be used to compute expected OC and ASN functions that take into account the variation in the p_T-to-m relationship and allow the robustness of the sampling plan to be determined. They use a variable h which is normally distributed with $m = 0$ and $s^2 = $ mse and construct 11 families of OC and ASN functions. These families are then averaged using a weighted mean to arrive at expected OC and ASN functions:

$$E(OC_j) = \sum_{k=1}^{11} OC_{kj} \, weight_k \quad \text{and} \quad E(ASN_j) = \sum_{k=1}^{11} ASN_{kj} \, weight_k$$

where

$$weight_k = 2 \Pr\{Z \le z_k\} \quad \text{and} \quad \Pr\{Z \le z_k\} = \int_{z_{k-1}+z_k/2}^{z_{k+1}+z_k/2} \phi(x) \, dx \quad \text{and} \quad \phi(x)$$

is the standard cumulative normal density function. These methods are explained in detail by Nyrop and Binns[23] and they provide Fortran programs to compute the expected OC and ASN functions.

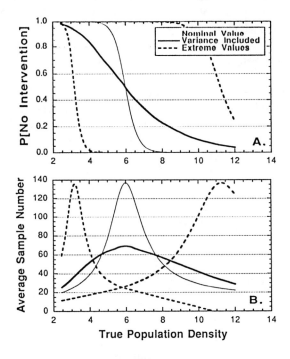

FIGURE 12. Effect of variance associated with the Kuno-Sugino equation on OC and ASN curves. (A) OC. Nominal value is variance according to binomial SPRT equations. Variance added is the expected OC curve using methods described in text. (B) ASN curve.

TABLE 6
Effect of Different Tally Thresholds on Regression Parameters in the Equation $\ln(-\ln(1 - p_T)) = c' + d'\ln(\text{mean})$ for the *Panonychus* and *Tetranychus* Data Sets

Tally threshold	c'	d'	$s_{d'}$	mse	N	$\bar{m}$
Panonychus						
0	−0.657	0.746	0.0126	0.106	288	−0.112
2	−2.104	1.070	0.0231	0.173	229	0.494
4	−2.888	1.214	0.0282	0.160	178	0.866
6	−3.267	1.198	0.0327	0.158	146	1.106
8	−3.663	1.253	0.0398	0.157	121	1.354
Tetranychus						
0	−0.827	0.686	0.0143	0.142	410	0.765
2	−1.493	0.874	0.0143	0.146	418	0.822
4	−2.399	1.069	0.0167	0.133	379	1.139
6	−3.012	1.174	0.0193	0.148	359	1.265
8	−3.460	1.225	0.0222	0.169	335	1.391

As with the estimation procedures mentioned previously, increasing the threshold, T, can have a profound effect on the robustness of the sampling program. The reason for this is that the displacement from M_a is affected by the slope (d') of the p_T-to-m relationship (Equation 31). The slope, d', of the p_T-to-m relationship generally increases as T increases,[13] therefore, increasing the tally threshold should result in less spreading of the OC and ASN functions compared to the optimal value. Binns and Bostanian[13,27] state that increasing the value of T will always decrease the displacement provided mse does not change much. This can only be determined by actually calculating the different p_T-to-m regressions. They show a case of how this can change with their data set of mixed European red mite and twospotted spider mite populations. In both the *Tetranychus* and *Panonychus* data sets this was also found to occur in the range of T investigated (Table 6).

D. METHODS BASED ON NEGATIVE BINOMIAL PROBABILITY MODEL

As mentioned previously, for organisms which fit the negative binomial distribution, the exponent k is the major source of variation. Use of a k value which underestimates the true K value results in ASN and OC curves which are flatter and shifted to the right; the converse is true if the true K is overestimated[26] (Figure 13, 14A). Binns and Bostanian[26] show that if k is fairly constant over the range of m

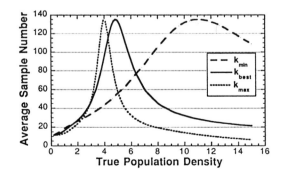

FIGURE 13. Effect of variance in k on ASN using three different k values for *Tetranychus* data set.

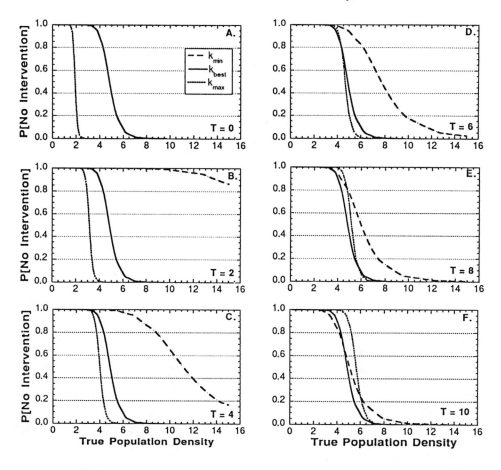

FIGURE 14. Sensitivity of the OC to variation in k at different tally thresholds for the *Tetranychus* data set. $m'_a = 4.5$, $m_a = 6.0$, $m''_a = 7.5$. (A) $T = 0$. (B) $T = 2$. (C) $T = 4$. (D) $T = 6$. (E) $T = 8$. (F) $T = 10$.

values of interest, estimation of k_{best}, k_{max}, and k_{min} from a number of different field samples allows the robustness of the sampling plan to be determined by examination of the OC and ASN curves computed for the different k values. Nyrop and Binns[23] suggest that the k values be calculated using the 10th and 90th percentile values found in these data sets as k_{min} and k_{max} and the median as k_{best}. They choose these levels as a way to prevent the OC and ASN curves from being unrealistic approximations of what is obtained in the field. Fortran programs which calculate the OC and ASN in this manner are provided by Nyrop and Binns.[23]

Another important finding presented by Binns and Bostanian[26,27] is that the effect of variation in k can be reduced by proper selection of T just as it is in estimation procedures. They show that an OC curve developed using k_{min} moves downward and to the right toward the OC curve using k_{best} as T increases and the OC curve using k_{max} moves upward and to the left as T increases (Figure 14A to F). They further show that rarely does the value of T have to be much larger than m for optimal efficiency. Although the increase in T increases the time required to process a sample, Nyrop and Binns[5] show that the effect may not be as large as expected. If the increase in sampling time becomes problematic, then a value of T lower than optimum can be selected which minimizes sampling time, but which also improves the OC and ASN of the sampling plan. For example, using the *Tetranychus* data set

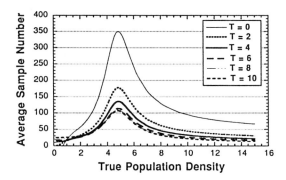

FIGURE 15. Effect of different tally thresholds on ASN for *Tetranychus* data set $m'_a = 4.5$, $m_a = 6.0$, $m''_a = 7.5$.

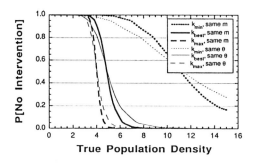

FIGURE 16. Effect of using a constant $\theta \pm x'$ vs. $m_a \pm x'$ on OC.

mentioned above, and $k_{min} = 0.20$, $k_{best} = 0.58$, and $k_{max} = 1.75$, $T = 10$ appears to be the optimal tally threshold at a mean of six mites per leaf, but $T = 8$ may provide an effective sampling program when considering time limitations (Figure 14A to F). Notice that as T increases, the ASN required to reach a decision decreases (Figure 15).

The change in the OC and ASN curves with increasing T should also consider the effect of the nonlinear relationship between P_T and m. When calculating the OC curve, it is tempting to use the same value of x in $\theta \pm x$, for the calculations. However, if this is done, the hypothesis tested on the m scale are different and OC curves are affected (Figure 16).

E. VARIABLE INTENSITY BINOMIAL SAMPLING

Binomial sampling programs designed to give area-wide population estimates or decisions must take into account the population biology of the insect in question. In reading this and other publications, the reader may get the impression that a single sample can be used to predict an area-wide population density or make a decision on an area-wide basis. However, this is frequently untrue because arthropod populations rarely have a homogenous distribution across large areas. In particular, there is often an "edge effect", where the population density is much higher at the edge than at the center of the field.[34-36] Another possible cause for spatial heterogeneity may occur when different cultivars (or even species) are planted in close proximity and some are better hosts and thus harbor higher population densities. Poor spray coverage may likewise affect local population densities. Using the assumptions of a simple random sampling across environmental heterogeneity results in a distortion of the basis of all the binomial sampling programs [i.e., P(present) being constant] so far discussed. This leads to a bias of the population density estimates or, in the case of classification studies, to an increase in the probability of incorrect classification. As mentioned previously, binomial sampling is somewhat resistant to heterogeneity, but not completely. The solution most frequently recommended is to use a nested variance function.[7] However, this may not always give an adequate representation of the total variance if spatial heterogeneity is extreme. This can be the case when the dispersal capabilities of a species are particularly low compared to the size of the area sampled. This can be seen in the *Tetranychus* and *Panonychus* data sets where estimates of $\hat{p}_{tree}$ on individual trees are pooled to get an area-wide estimate of $\hat{p}_{orchard}$. The underlying assumption is that the $\hat{p}_{tree}$ estimates are distributed as a binomial distribution with $\bar{P} = \hat{p}_{orchard}$. When the validity of this assumption is tested using χ^2 analysis, the hypothesis is rejected in 20 of the 22 data sets for the *Tetranychus* data set and 13 of 14 in the *Panonychus* data set. The only solution to this problem is to restrict the sampling universe in ways that minimize spatial heterogeneity and sample each area separately. Ideally, these sampling universes should correspond to the growers' management unit (for decision making), but the dispersal of the species is independent of grower's constraints. Clearly, to reach a decision, a single sample of x leaves is insufficient and many samples of x leaves may be required. A second caution, when using nested variance functions, is that when a minimum number of sampling units per area are taken (for example, the often recommended single leaf from each tree sampled[7]), the result is insensitivity of the sampling program at low population densities. This is because there is a low probability of finding pests on any one sampling unit when the distribution is clumped and the mean is low. This insensitivity can lead the sampler to conclude that the populations "just suddenly appear" between sample dates and provides an unrealistic impression of population growth rates.

The problems with spatial heterogeneity have given rise to variable intensity sampling (VIS).[37,38] In VIS, the field to be sampled is divided into an equal number of segments using a transect or some other convenient method. The number of sampling units taken in each segment is a function of the mean estimated from all previous observations, once a segment is completely sampled. The first segment is sampled at maximum intensity and at least one sampling unit is always chosen in the remaining segments.[37,38]

VIS is designed to reduce bias by selecting sampling units from throughout the sampling universe.[17] A second design criterion is that VIS sampling protocol minimizes the sample size when the mean is greatly different from the threshold and maximizes sample size when it is close to the threshold. This is more efficient than binomial methods presented earlier which aim for constant precision or a fixed width confidence interval, because those methods may require excessive sample sizes to terminate sampling when $\hat{p}$ may be of little interest to the investigator.[37]

VIS is normally considered to be most appropriate for situations in which spatial heterogeneity is suspected, but Nyrop and Binns[5] point out that if the field must be traversed to determine if a pest is present, VIS would be the most appropriate method, regardless of the variance structure. This suggests that for direct pests or for studies of insect-vectored pathogens, VIS would be the method of choice.

The major drawback of VIS is the difficulty in implementing the sampling program.[17] Sampling charts may be developed or a programmable calculator/notebook computer can be used to determine sample size within each segment.[37] However, the extra complexity of the sampling program will undoubtedly make user acceptance more difficult than if a classification system is used, particularly if pest control advisors are not available.

The following discussion on the development of a binomial VIS is taken primarily from Hoy.[37] First, an algorithm to calculate the sample size required to achieve the desired precision is determined as a function of $\hat{p}$. This requires that a critical lower and upper proportion of sampling units infested (θ_1 and θ_2) are defined. The upper level is set at the point at which a treatment is always required regardless of other factors (e.g., weather, population density of natural enemies, etc.) and the lower level is the point at which a treatment would never be required. When $\hat{p} \leq \theta_1$ the upper bound for the confidence interval is set at p^* where $p^* = ((\theta_1 + \theta_2)/2)$ (i.e., the average treatment threshold). When $\hat{p} \geq \theta_2$, the lower bound for the confidence interval is set at p^*. Under these circumstances the required sample size is given by

$$n = \frac{z_\alpha^2 p^*(1 - p^*)}{(p^* - \hat{p})^2} \tag{32}$$

where $\hat{p}$ is the sequential estimate of the true population value of P. When $\theta_1 \leq \hat{p} \leq \theta_2$, replace $\hat{p}$ by either θ_1 or θ_2 to obtain the maximum sample size for the entire field (n_{max}).

Operational or time constraints may also dictate the choice of n_{max}. Hoy points out that n_{max} chosen in this fashion can be used to calculate an estimated average threshold and θ_1 and θ_2 can be calculated from Equation 32. It should be noted that Equation 32 is basically a rearrangement of Equation 8 where $p^* - \hat{p} = h$ (h = width of the confidence interval desired).

The number of segments used in the sampling program is determined in a somewhat arbitrary fashion, but should take into consideration the dispersal ability of the arthropod in question.[37] The maximum number of sampling units which must be examined within a single segment is then $n_{max}/(\#\ segments)$. When $\theta_1 \leq p \leq \theta_2$, n_{max} samples are taken within a segment and when p is not within this range, then the number of samples per segment is reduced by estimating the sampling intensity (SI) required to achieve the desired precision for the current estimate of p using

$$SI = NINT\left[\frac{n - n_t}{\#\ segments\ remaining}\right] \tag{33}$$

where NINT is the nearest integer function, and n_t is the number of sampling units already examined.

In the following example, $\theta_1 = 0.25$, $\theta_2 = 0.40$, the number of segments = 10. Therefore, $p^* = 0.325$,

$$n_{max} = NINT\left[\frac{(1.96)^2(0.325)(1 - 0.325)}{(0.325 - 0.25)^2}\right] = 150$$

and the sampling intensity for the first segment is $150/10 = 15$. In the first segment, if three of the leaves are infested, then $\hat{p} = 3/15 = 0.20$,

$$n = NINT\left[\frac{(1.96)^2(0.325)(1 - 0.325)}{(0.325 - 0.20)^2}\right] = 54$$

and the sampling intensity is SI = NINT [(54 − 15)/(10 − 1)] = 4 sampling units/segment. If three of the four sampling units collected in the next segment are infested, then $\hat{p} = (3 + 3)/(15 + 4) = 0.32$ and the maximum sample size (150) would be required. The sampling intensity would be SI = NINT [(150 − 19)/8] = 16 leaves. This procedure would be followed for each of the segments. Notice that if SI < 1 because $\theta_1 \geq \hat{p} \geq \theta_2$ then SI = 1 in keeping with the design considerations of VIS. However, if $\hat{p}$ from the first segment is not representative of the field (e.g., there is a large edge effect), then the VIS can be distorted, because the first segment is sampled at maximum intensity and may indicate that further SI can be minimized. This effect may be minimized by selecting more segments. As mentioned above, the dispersal abilities of the animal in question should modify the sampling method.

In this section, $\hat{p}$ has been discussed, but the use of p_T can also be used to extend the usefulness of the VIS plan. This will be most important when n_{max} is constrained by operational and time factors; in this situation, changing the tally threshold to a higher level will move p_T^* to a lower level and require fewer samples to achieve the desired accuracy. However, simulations to test the validity of this approach must still be performed.

V. FUTURE STUDIES IN BINOMIAL SAMPLING

Binomial methods relying on an increased tally threshold may find more applicability in the future for reasons somewhat independent of the increased stability and changes in ASN discussed above. As stated earlier, when used in a decision-making context, an estimate of the typical damage the plant sustains is critical. Because the mean is the prototype nonresistant measure of the typical population value,[39] it may be preferable to switch to other measures of central tendency to adequately represent the damage sustained by the "typical" plant (e.g., a trimmed mean, the median, the bi-weight, etc.). However, if it is known that a given level of pests are required before damage occurs, then using that level as a tally threshold and using the standard SPRT or VIS for binomial sampling may allow decisions as to what proportion of the leaves have damaging population levels present rather than a mean population density level. This would allow the added variance associated with the required conversion of p_T-to-m to be eliminated, resulting in fewer samples being required. For example, two to three twospotted spider mites on a pear leaf can result in leaf burn. By setting the tally threshold to 2, the proportion of the leaves damaged can be set using θ_1 and θ_2.

The use of the proportion infested rather than the mean population density also fits in well with the imprecise nature of most economic thresholds. It makes little sense to classify or estimate the population to a greater degree of accuracy than was used in the calculation of the threshold. In particular, the sampling methods used by the individual developing the thresholds may not have provided an adequate estimate of the population density which was responsible for the damage observed.

The lack of quantitative yield loss-feeding damage functions may be viewed by some as a serious deterrent to adopting classification systems. While it is easy to predict this will improve significantly in the future, the variances associated with these functions probably always will be much higher than those for statistical sampling plans and much less predictable because of the complex interactions of the plant and its environment. However, conservative thresholds can be estimated iteratively over time to achieve a reasonable compromise and implementation of IPM programs should not be stopped until the final yield loss-feeding damage functions are developed.

Another area that must be addressed in future studies is the minimum distance between samples (i.e., geostatistics of some sort) for different pest groups. This would allow large-scale implementation of sampling programs to proceed from research-sized plots to large-scale agriculture.

The advances in binomial sampling obtained in the past few years have the potential to dramatically change pest management in systems where they have been developed. However, I expect that as the field of exploratory data analysis and robust statistics matures and becomes more widely accessible, it will probably result in an even more dramatic change in sampling design and damage assessment.

ACKNOWLEDGMENTS

I thank Jan Nyrop, Cornell University, Tom Unruh, USDA-ARS, Yakima, Washington, and Andy Taylor, University of Hawaii at Manoa for their review of earlier drafts of this manuscript. I also thank Jan Nyrop for some of the programs used to analyze binomial samples.

REFERENCES

1. Pedigo, L. P., *Entomology and Pest Management*, Macmillan, New York, 1989, 646.
2. Kono, T. and Sugino, T., On the estimation of the density of rice stem infested by the rice stem borer, *Jpn. J. Appl. Entomol. Zool.*, 2, 184, 1958.
3. Bliss, C. I. and Fisher, R. A., Fitting the negative binomial distribution to biological data, *Biometrics*, 9, 176, 1953.
4. Jones, V. P., Developing sampling plans for spider mites: those that don't remember the past may have to repeat it, *J. Econ. Entomol.*, 83, 1656, 1990.
5. Nyrop, J. P. and Binns, M. R., Quantitative methods for designing and analyzing sampling programs for use in pest management, in *Handbook of Pest Management in Agriculture*, Pimentel, D., Ed., CRC Press, Boca Raton, FL, 1990, 67.
6. Wilson, L. T., Leigh, T. F., and Maggi, V., Presence-absence sampling of spider mite densities on cotton, *Calif. Agric.*, 35, 10, 1981.
7. Cochran, W. G., *Sampling Techniques*, 3rd ed., John Wiley & Sons, New York, 1977.
8. Zar, J. H., *Biostatistical Analysis*, Prentice-Hall, Englewood Cliffs, NJ, 1974, 620.
9. Hahn, G. and Meeker, W. Q., *Statistical Intervals*, John Wiley & Sons, New York, 1991, 392.
10. Karindinos, M. G., Optimum sample size and comments on some published formulae, *Bull. Entomol. Soc. Am.*, 22, 417, 1976.
11. Kuno, E., A new method of sequential sampling to obtain the population estimates with a fixed level of precision, *Res. Popul. Ecol.*, 11, 127, 1969.
12. Gerrard, D. J. and Chiang, H. C., Density estimation of corn rootworm egg populations based upon frequency of occurrence, *Ecology*, 51, 237, 1970.
13. Binns, M. R. and Bostanian, N. J., Robustness in empirically based binomial decision rules for integrated pest management, *J. Econ. Entomol.*, 420, 1990.
14. Schaalje, G. B., Butts, R. A., and Lysyk, T. J., Simulation studies of binomial sampling: a new variance estimator and density predictor, with special reference to the Russian wheat aphid (Homoptera: Aphididae), *J. Econ. Entomol.*, 84, 140, 1991.
15. Kuno, E., Evaluation of statistical precision and design of efficient sampling for the population estimation based on frequency of occurrence, *Res. Popul. Ecol.*, 28, 305, 1986.
16. Taylor, L. R., Aggregation, variance and the mean, *Nature*, 189, 732, 1961.
17. Binns, M. R. and Nyrop, J. P., Sampling insect populations for the purpose of IPM decision making, *Annu. Rev. Entomol.*, 37, 427, 1992.
18. Schaalje, G. B. and Butts, R. A., Binomial sampling for predicting density of Russian whete aphid (Homoptera: Aphididae) on winter wheat in the fall using a measurement error model, *J. Econ. Entomol.*, 85, 1167, 1992.
19. Hepworth, G. and Macfarlane, J. R., The variance of the estimated population density from a presence-absence sample, *J. Econ. Entomol.*, 85, 1992, 2240.

20. Southwood, T. R. E., *Ecological Methods*, 2nd ed., John Wiley & Sons, New York, 1978.
21. Hurlbert, S. H., Spatial distribution of the montane unicorn, *Oikos*, 58, 257, 1990.
22. Wilson, L. T. and Room, P. M., Clumping patterns of fruit and arthropods in cotton with implications for binomial sampling, *Environ. Entomol.*, 12, 50, 1983.
23. Nyrop, J. P. and Binns, M. R., Algorithms for computing operating characteristic and average sample number functions for sequential sampling plans based on binomial count models and revised plans for European red mite (Acari: Tetranychidae) on apple, *J. Econ. Entomol.*, 85, 1253, 1992.
24. Taylor, L. R., Woiwod, I. P., and Perry, J. N., The negative binomial as a dynamic ecological model for aggregation, and the density dependence of k, *J. Anim. Ecol.*, 48, 289, 1979.
25. Taylor, L. R., Assessing and interpreting the spatial distributions of insect populations, *Annu. Rev. Entomol.*, 29, 321, 1984.
26. Binns, M. R. and Bostanian, N. J., Robust binomial decision rules for integrated pest management based on the negative binomial distribution, *Am. Entomol.*, 50, 1990.
27. Binns, M. R. and Bostanian, N. J., Binomial and censored sampling in estimation and decision making for the negative binomial distribution, *Biometrics*, 44, 473, 1988.
28. Wald, A., *Sequential Analysis*, John Wiley & Sons, New York, 1947.
29. Wetherill, G. B. and Glazebrook, K. D., *Sequential Methods in Statistics*, 3rd ed., Chapman & Hall, New York, 1986.
30. Fowler, G. W. and Lynch, A. M., Sampling plants in insect pest management based on Wald's sequential probability ratio test, *Environ. Entomol.*, 16, 345, 1987.
31. Fowler, G. W., Errors in Sampling Plans Based on Wald's Sequential Probability Ratio Test, U.S. Department of Agriculture For. Serv. Gen. Tech. Rep. NC-46, USDA, Washington, DC, 1978.
32. Nyrop, J. P., Agnello, A. M., Kovach, J., and Reissig, W. H., Binomial sequential classification sampling plans for European red mite (Acari: Tetranychidae) with special reference to performance criteria, *J. Econ. Entomol.*, 82, 482, 1989.
33. Nyrop, J. P. and Simmons, G. A., Errors incurred when using Iwao's sequential decision rule in insect sampling, *Environ. Entomol.*, 13, 1459, 1984.
34. Oi, D. H. and Barnes, M. M., Predation by the Western predatory mite (Acari: Phytoseiidae) on Pacific spider mite (Acari: Tetranychidae) in the presence of road dust, *Environ. Entomol.*, 18, 892, 1989.
35. Oi, D. H., Factors Related to Pacific Spider Mite Outbreaks in Roadside Almond Trees, unpublished Ph.D. dissertation, University of California, Riverside, 1987.
36. Wilson, L. T., Pickett, C. H., Leigh, T. F., and Carey, J. R., Spider mite (Acari: Tetranychidae) infestation foci: cotton yield reduction, *Environ. Entomol.*, 16, 614, 1987.
37. Hoy, C. W., Variable-intensity sampling for proportion of plants infested with pests, *J. Econ. Entomol.*, 84, 148, 1991.
38. Hoy, C. W., Jennison, C., Shelton, A. M., and Andaloro, J. T., Variable-intensity sampling: a new technique for decision making in cabbage pest management, *J. Econ. Entomol.*, 76, 139, 1983.
39. Mosteller, F. and Tukey, J. W., *Data Analysis and Regression*, Addison-Wesley, Reading, MA, 1977, 588.

Chapter 10

SEQUENTIAL SAMPLING TO DETERMINE POPULATION DENSITY

William D. Hutchison

TABLE OF CONTENTS

I. Introduction 208

II. Sampling Prerequisites 208

III. Sampling Objectives and Density Estimation 208

IV. Statistical Foundation and Assumptions for Density Estimation 209
    A. Probability Distributions 209
    B. Empirical Models 212

V. Multistage Sampling 215

VI. Performance and Validation of Sequential Sampling 217
    A. Field Validation 219
    B. Simulation 219

VII. Summary 221

Program Listing 222

Acknowledgments 240

References 241

I. INTRODUCTION

Sequential sampling of arthropod populations has become increasingly critical to the development and implementation of integrated pest management (IPM) programs worldwide.[1] Indeed, the absence of reliable monitoring methods for many arthropod pests was recently cited as the primary technical obstacle to the adoption of IPM programs.[2] One prerequisite for successful IPM is that economic thresholds and sampling work in tandem, where both technologies are essential components of reliable and implementable pest management decision rules. To implement multifaceted IPM, or sampling plans for applied research in the most efficient manner, sequential sampling will continue to be needed for pest and beneficial arthropods. For example, there will likely be more demand for comprehensive sequential sampling plans that incorporate multiple species (pests or beneficials), and plans that account for the differential impact of resistant varieties or biological control, as alternatives to chemical control are implemented.

As indicated in several related chapters (e.g., 8 and 9), there are many ways sequential sampling can be used to achieve these goals. In general terms, sequential sampling is advantageous because it provides an estimate of population intensity (either categorical or enumerative) with predetermined, acceptable levels of precision at minimum cost. The purpose of this chapter is to review the methods available for using sequential sampling to estimate population density and the corresponding performance of such plans.

II. SAMPLING PREREQUISITES

As with all sampling plans, several factors must be considered for the efficient development and implementation of a sampling program, specifically the why, what, where, when, and how of sampling. More formally, the following criteria require definition: objective(s) of the sampling program, appropriate sampling universe (e.g., single or multistage sampling), sample unit, spatial distribution (pattern) of the population, timing of sampling activity, and finally, selection and validation of the sequential sampling method. Many of these components have been reviewed in selected chapters of this volume and need not be repeated here in detail. However, selected criteria are discussed when necessary to illustrate the use of various sequential sampling methods and their evaluation. For now, one criterion that deserves further attention is the sampling objective.

III. SAMPLING OBJECTIVES AND DENSITY ESTIMATION

The objective of the sampling program clearly depends on the species of interest, available resources, and how the information is to be used. Although sampling plans that classify arthropod infestations into economic (e.g., outbreak) and noneconomic (endemic) categories have received considerable application in IPM (Chapter 8), the emphasis of this chapter will be sequential sampling to estimate arthropod density (i.e., point or parameter estimation) using enumerative, or count data. Presence/absence, or binomial sampling, also may be used to estimate population density; this approach is summarized in Chapter 9.

Plans to estimate density may require more sampling time in the field, but enumerative sampling methods continue to be useful for both IPM and research purposes. Specific examples include: (1) species where individuals are relatively large in relation to the sample unit and amenable to visual counts under field conditions (e.g., lepidopteran larvae in cabbage),[3] (2) species where binomial plans are not practical because of an unusually rapid increase in the proportion of sample units infested relative to mean density (e.g., both low and high densities of pea aphid, *Acyrthosiphon pisum* (Harris), in alfalfa, where the proportion infested rapidly exceeds 0.90),[4] (3) situations where density (or variance) estimates are needed over time as input to models for forecasting purposes,[5] (4) regional monitoring efforts such as the USDA-APHIS CAPS program,[6] and (5) research applications where density estimates are needed with high precision for life table analysis or applied ecological studies.[7]

IV. STATISTICAL FOUNDATION AND ASSUMPTIONS FOR DENSITY ESTIMATION

Each of the following sequential sampling methods for estimation have been derived from statistical theory for optimum sample size, based on the normal distribution. Selection of final sequential plans, however, is usually based on sample size formulas that have been modified to incorporate probability distributions or empirical regression models that more typically describe the spatial pattern of arthropod species in the field.

A. PROBABILITY DISTRIBUTIONS

Examples of the most common probability distributions include the Poisson and the Negative Binomial (NB), which characterize populations as being dispersed in random and aggregated patterns, respectively.[7] Before discussing the use of each distribution or model, the necessary formulas and definitions for precision are presented. Sequential sampling to estimate mean density can be accomplished by using at least three criteria for controlling precision about the mean estimate.

Three common definitions for precision include:

C = standard error (SE)/mean(m), i.e., the SE is within a fixed proportion of m.

d = fixed proportion of m, set to $1/2$ the width of a confidence interval, with $1 - \alpha$ probability, where $Z_{\alpha/2}$ is a standard normal deviate such that $P(Z > Z_{\alpha/2}) = \alpha/2$, and α is typically set at 0.05 for the 95% confidence interval; $Z_{\alpha/2}$ is adequately represented by $t_{\alpha/2}$ of the Student t distribution.

h = fixed positive number, set to $1/2$ the width of a confidence interval, with $1 - \alpha$ probability, based on the normal distribution with $Z_{\alpha/2}$ often set at 0.05.

Throughout this discussion, conventions for precision level and parameter estimates primarily follow those of Karandinos,[8] with the exception of d. Here, d, rather than D by Karandinos, is used because of the common use of D^9 or D_0^{10} for SE/m (equal to C above). Additionally, other symbols such as c^{11} or CV,[11] have been used for the same precision level parameter (C). Because of the variety of symbols used, there can be confusion. More importantly, C is sometimes confused with d, leading to incorrect assumptions and interpretation of desired levels of precision.

Here, m and s^2 are used to refer to the population sample, recognizing that these are estimates of the true μ (mean) and δ^2 (variance), respectively. As previously stated, the basis for each sampling plan begins with the following relationship between the m, s^2, and sample size, n. Beginning with the general case (no specific distribution assumption) precision is defined as,

$$C = (s^2/n)^{1/2}/m, (= SE/m) \qquad (1)$$

and

$$n = s^2/m^2C^2 \tag{2}$$

which is the optimum sample size (OSS). OSS formulas for other distributions have also been presented by Karandinos,[8] Ruesink,[12] and Chapter 6, and may be reviewed for additional detail. To use the OSS relationship for sequential sampling (when m is not previously known), it is necessary to replace m with Tn/n, where Tn is equal to the cumulative number of individuals counted after any given number of samples (n). Replacing m with Tn/n, and solving for Tn, yields

$$Tn = (s(n)^{1/2})/C \tag{3}$$

where s is the estimate of the standard deviation (δ) of the population, and Tn becomes the "stop-line" as a function of n for a fixed-precision sampling plan. In practice, one continues to sample until the cumulative number of individuals counted is greater than or equal to the stop-line.

If we have reason to believe that the spatial pattern of the population is dispersed at random, i.e., we cannot reject the null hypothesis that the presence of any given individual is influenced by the presence of another individual, the population's spatial pattern may be described by the Poisson distribution, where $s^2 = m$,

$$n = 1/(mC^2) \tag{4}$$

and the corresponding sequential stop-line is

$$Tn = 1/C^2 \tag{5}$$

For the Poisson, the $Tn - n$ relationship becomes a horizontal stop-line, parallel to the x-axis of n. Where there is evidence for an aggregated dispersion pattern (e.g., $s^2 > m$), we may select the NB distribution where, $s^2 = m + m^2/k$, the clumping parameter $k = m^2/s^2 - m$,

$$n = (1/m + 1/k)/C^2 \tag{6}$$

and the corresponding stop-line is

$$Tn = nk/(C^2nk - 1) \tag{7}$$

This relationship is also in agreement with the NB stop-line presented by Kuno.[10] For the NB, Binns[13] also derived a similar stop-line, where

$$Tn = g^4(nk - 0.5 - g^2) + g^2 \tag{8}$$

and g is a constant that defines the required level of precision, either as C or d (see also Nyrop and Binns[11]). In contrast to the standard NB stop-line (Equation 7), this method has the advantage of mean density (Tn/n) being independent of n. This property should minimize some of the uncertainty associated with actual C or confidence interval estimates. In addition, this relation may also be more robust than the stop-lines based on empirical regression models.

If we select d as our measure of precision, we set $1/2$ the length of the confidence interval (e.g., 95% confidence interval) equal to $d(m)$, where d = the desired proportion of m. This approach is interpreted as providing an estimate of m that is within a fixed proportion of m, with $1 - \alpha$ confidence. Assuming that the mean is normally distributed,

$$dm = (Z_{\alpha/2})(s^2/n)^{1/2} \quad \text{or} \quad dm = (Z_{\alpha/2})(\text{SE}) \tag{9}$$

$$n = (Z_{\alpha/2}^2 s^2)/(d^2 m^2) \tag{10}$$

and

$$Tn = (Z_{\alpha/2}/d)(s)(n)^{1/2} \tag{11}$$

For the Poisson distribution,

$$n = (Z_{\alpha/2}/d)^2/m \tag{12}$$

and

$$Tn = (Z_{\alpha/2}/d)^2 \tag{13}$$

For the NB distribution,

$$n = \left[(Z_{\alpha/2})^2(1/m + 1/k)\right]/d^2 \tag{14}$$

and

$$Tn = \left[k(Z_{\alpha/2}/d)^2 n\right]/\left[kn - (Z_{\alpha/2}/d)^2\right] \tag{15}$$

A useful relationship between C (= fixed ratio of SE/m) and d (= fixed proportion of m) is that, because of Equation 8,

$$d = (Z_{\alpha/2})(C) \tag{16}$$

When α is set to = 0.05, and sample size approaches $n \geq 30$, the relationship becomes $d = 1.96(C)$, or $d \approx 2(C)$. Therefore, the same stop-line formulas for C, summarized above (Equations 3, 5, and 7), can also be used to evaluate the performance of the confidence interval approach (Equations 11, 13, and 15) by setting $d \approx 2(C)$. For example, with $d = 0.25$ (and 95% confidence interval with $Z_{\alpha/2} = 1.96$), $C = 0.1276$; thus, the $d = 0.25$ plan requires a higher sample size than a fixed-precision plan of $C = 0.25$. Because of this relationship, required sample size for confidence interval (d-based) plans are always higher than fixed precision (C-based) plans where the same d and C values are used.[1,8,11]

If we define precision by a fixed positive value (h), i.e., a fixed value about m, and set h equal to 1/2 the confidence interval,

$$h = Z_{\alpha/2} * s/(n)^{1/2} \tag{17}$$

Sequential sampling formulas can be then derived for each distribution. For example, for the Poisson distribution where

$$n = (Z_{\alpha/2}/h)^2 * m \tag{18}$$

and we replace m with Tn/n and solve for Tn,

$$Tn = n^2/\left[h(Z_{\alpha/2})\right]^2 \tag{19}$$

For the NB distribution,

$$n = (Z_{\alpha/2}/h)^2 * [(km + m^2)/k] \tag{20}$$

and

$$Tn = \left[-kn - kn\left[1 + 4n/k(Z_{\alpha/2}/h)^2\right]^{1/2}\right]/2* \tag{21}$$

Plans based on h have received limited use in entomological applications. This is probably because of the lack of flexibility over a broad density range. For example, a fixed positive value of $h = 1.0$ may be useful for a low density range of 5 to 10 individuals per sample unit. However, at high densities such as $m = 40$, $h = 1.0$ (estimated m should be between 39 to 41 for true $m = 40$), begins to require a high level of precision and correspondingly large sample sizes (see also Karandinos[8]). Alternatively, with both C and d being ratios, the values of acceptable variation in the mean estimate change with m. For example, with $d = 0.15$ (estimate of sample m is within 15% of the true mean), the boundary for $m = 10$ or $m = 40$ is between 8.5 to 11.5 and 34 to 46, respectively. The practical use of h may be limited to applications with a restricted (low) density range or for quality control, where regardless of m, only a fixed (maximum) level of variability in m can be tolerated. Otherwise, a very efficient sampling method must be available to process large numbers of samples at high densities.

B. EMPIRICAL MODELS

For several reasons, the Poisson and NB distributions recently have received less attention as a basis for developing sampling plans. Two common concerns have included the fact that a series of s^2-m pairs for a given species may often fit several probability distributions (including Poisson and NB), or the dependence of k of the NB on mean density.[14,15] Subsequently, the problem with fluctuating k for sampling purposes is the difficulty in estimating a common k value.[15] Although Nyrop and Binns[11] have recently argued that a common k is less critical, particularly when sampling plans are validated by simulation, the ambiguity resulting from variable k values among s^2-m pairs has nevertheless discouraged use of the NB as a basis for sequential sampling. On the other hand, empirical models have been promoted as being virtually density-independent, with their parameter estimates reliable over a broad density range.

The first empirical model to be used for sequential sampling was Iwao's[16] regression of Lloyd's[17] mean crowding (MC) as a function of m. MC was developed as an index to describe spatial pattern, and is defined as $x^* = m + (s^2 + m - 1)$.[17] Iwao[16] then developed the MC regression model of x^* as a $f(m)$, by setting $x^* = \alpha + \beta m$, providing both statistical and biological interpretation of spatial pattern over a broad density range. With this model, $\alpha + 1$ is considered the basic unit of contagion of the population, and β is the clumping parameter that describes the spatial pattern of the

---

* Because this is a quadratic equation, there are two possible quadratic roots; however, only the relevant solution for this application is given.

population units. For example, with $\alpha = 5$, and $\alpha + 1$ individuals per unit, we have $5 + 1$ eggs per egg mass. If $\beta < 1$, the population is assumed to be dispersed in a regular pattern; for $\beta = 1$, the population is dispersed at random (cf. Poisson); and for $\beta > 1$, the population is assumed to be dispersed in an aggregated pattern.

Using the MC model, Kuno[10,18] developed one of the first sequential sampling stop-lines to estimate m. The most common has been based on C as a measure of precision, and replacing m with Tn/n,

$$Tn = (1 + \alpha)/[C^2 - (\beta - 1/n)] \tag{22}$$

When $\beta = 1$, which is indicative of the Poisson distribution,

$$Tn = (1 + \alpha)/C^2 \tag{23}$$

and for the special case of $\beta = 1$ and $\alpha = 0$, this equation is equivalent to Equation 5. This stop-line is also related to a common k of the NB where $\alpha = 0$, and $k = 1/\beta - 1$, by

$$Tn = 1/[C^2 - (1/nk)]^{18} \tag{24}$$

Examples of both Green's and Kuno's sequential sampling plans, developed for *A. pisum*[9] on alfalfa, are shown in Figure 1. These results illustrate the additional numbers of individuals to be counted to maintain a higher level of precision ($C = 0.20$),

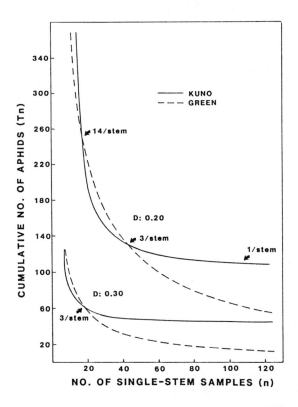

FIGURE 1. Comparison of Green's and Kuno's sequential sampling plans to estimate *A. pisum* population density for two precision levels (where $D = C$ of this chapter). Note that a stop-line for the NB distribution would have a similar pattern; the stop-line for the Poisson distribution would be a horizontal line — insensitive to changes in sample size (n) (taken from Hutchison et al.)[9].

and significant differences between the two sampling plans at low population densities (i.e., < 3 aphids/stem).

Kuno[10,18] also provided formulas to define precision by d, a fixed proportion of m; for $\beta = 1$,

$$Tn = (d^2/\alpha + 1) * n^2 \tag{25}$$

and, for $\beta > 1$,

$$Tn = \left[n * \left[((\alpha + 1)^2 + 4n((\beta - 1)d^2))^{1/2} - (\alpha + 1)\right]\right]/2(\beta - 1) \tag{26}$$

The second empirical model is Taylor's power law (TPL),[15] which has been promoted as a robust model of the s^2-m relationship and, as such, an attractive alternative to the probability distribution approach. For example, in one of Taylor's early papers,[19] he illustrated the fit of Poisson and NB distributions to selected density ranges for the European corn borer, *Ostrinia nubilalis* (Hübner). The same data set was also fit to TPL for the complete density range. TPL in fact has been successfully fit to a multitude of data sets for a variety plant and animal species.[15,20] Furthermore, as a log-log transformation it does generally provide better fits to arthropod data sets than the MC model.[15,20]

TPL is a power function of the s^2-m relation where $s^2 = am^b$. The power estimate, b, is a measure of aggregation and is typically viewed as a species-specific constant, while "a" is considered a "sampling factor", dependent on sample unit selection, or locale. Parameters are estimated by a least-squares linear regression of $\log_{10}(s^2)$ on $\log_{10}(m)$, such that the original power function becomes $\log(s^2) = \log(a) + b(\log(m))$.

Using C for precision level control, Green[21] was the first to develop a sequential sampling plan for field use based on TPL. Beginning with Equation 2, and using TPL as the predictor of s^2, OSS is defined by

$$n = am^b/C^2m^2 \tag{27}$$

Replacing Tn/n for m, the following stop-line was developed,

$$\ln(Tn) = \ln(C^2/a)/[(b - 2) + ((b - 1)/(b - 2))\ln(n)] \tag{28}$$

If we define precision as d,

$$\log_{10} Tn = \left[\ln\left(d^2/Z^2_{\alpha/2}a\right)\right]/[(b - 2) + ((b - 1)/(b - 2))\ln(n)]^{11} \tag{29}$$

Because either MC or TPL regression models can be used to fit a broad density range, there has been a virtual explosion of precision-based sequential count plans developed during the past 20 years, and particularly in the past 10 years.[22-52] Of the plans developed to date, those based on probability distributions now represent a small proportion of the total, with the TPL-based plan of Green[21] being used most often.

Ostensibly, one reason for the increased use of empirical models as the foundation for sequential sampling is the assumption that the relationship (i.e., slope coefficient) is not sensitive to changes in population density (cf. k of NB). The inference is that each model provides a consistent description of a species' spatial pattern throughout the density range, along with the expectation that sampling plans based on these parameters will always yield the predetermined "fixed-precision" level and calculated sample size. As illustrated by several authors,[9,11,53] and discussed in the following section on performance evaluation, this is not always true for either model. However, computer simulation can be used as a validation tool to detect any unexpected trends from aggregated or random assumptions of the model, and sampling plans can be adjusted accordingly.

V. MULTISTAGE SAMPLING

Thus far, all sampling methods have assumed that only simple random sampling is used, where the sampling universe is viewed as a collection of sample units occurring in a single plane. This often may be appropriate but there are several instances where this assumption does not hold and a nested or multistage sampling plan is required. The need for multistage sampling results from two different aspects of the sampling universe, but statistically, the analysis is similar.

The sampling universe may be physically defined by the cropping or livestock system of interest, or it may be arbitrarily defined based on preliminary analysis of sample data. For example, an apple orchard poses a natural, physical hierarchical sampling universe that must be considered when sampling apple arthropods infesting foliage, i.e., trees within the orchard, branches within a tree, and leaves within a branch.[54-55] A monocultured crop such as alfalfa may require the more arbitrary consideration of blocks within a field, quadrats within a block, stems within a quadrat, and leaves within a stem.[56-58] In either instance, where a multistage approach may be necessary, preliminary sample data should be collected and analyzed by nested analysis of variance (NANOVA) to document where the primary sources of variation occur within the hierarchy.

Use of NANOVA allows for the partitioning of the variation within each stratum to determine the number of strata necessary and permits optimum allocation of sampling effort within each stratum. Secondarily, optimum allocation requires an estimate of the time (or cost) necessary to locate, collect, and process samples within each stratum.

One example of a NANOVA is illustrated in Table 1 for green cloverworm, *Pathypena scabra* (F.), eggs in soybean.[56] In this example, several arbitrary classifications were evaluated to document significant sources of variability necessary for optimum allocation of sequential samples. As shown, most of the observed variability occurred between leaves within a sample. With these results, one could select leaves as the sample unit and use simple random sampling across a soybean field. However, because the variance components were well defined, Buntin and Pedigo[56] developed a two-stage sampling plan to determine the optimum number of leaves per sample and the number of samples per field. A two-stage approach generally follows the formula provided by Snedecor and Cochran,[59] where,

$$n_L = (C_S \delta_L^2 / C_L \delta_S^2)^{1/2} \tag{30}$$

For the green cloverworm example, n_L = the optimum number of leaves per sample, and C_L and C_S refer to the cost of locating and counting individuals within a leaf and a sample, respectively. The mean square error terms (variance components) from the NANOVA are used as estimates of δ_L^2 and δ_S^2. In this example, C_L and C_S required 0.67 and 0.50 person-minutes, respectively, yielding an optimum number of four leaves per sample.

TABLE 1
Nested Analysis of Variance (NANOVA) Results for Eggs of the Green Cloverworm,
***Pathypena scabra* (F.), Collected in Soybean**

Strata	df	Expected mean square error	Variance component	% of total variation
Fields[a]	30	$\delta_L^2 + \delta_S^2 + \delta_P^2 + \delta_Q^2 + \delta_F^2$	0.0001315	0.56
Quadrats/F	31	$\delta_L^2 + \delta_S^2 + \delta_P^2 + \delta_Q^2$	0.0000036	0.03
Plots/Q/F	186	$\delta_L^2 + \delta_S^2 + \delta_P^2$	0.0000011	0.01
Samples/P/Q/F	992	$\delta_L^2 + \delta_S^2$	0.0004351	3.83
Leaves/S/P/Q/F	31,868	δ_L^2	0.0108610	95.57
Total	33,107		0.0116423	100.00

[a] The field stratum actually refers to fields (F) X sample (S) dates; P = plot, Q = quadrat.

Source: (Taken from Buntin and Pedigo)[56]

Once the within-sample allocation is calculated, the number of samples to be collected can be determined using any of the sequential sampling formulas (i.e., for Tn) previously described for probability distributions or empirical models. Similar examples have been published for alfalfa pests,[57] including a three-stage sampling procedure for the leafminer, *Agromyza frontella*,[58] and for several arthropods infesting orchards.[11,54,55,60]

One potential problem with using average estimates of δ^2 components from NANOVA is the known sensitivity of the s^2-m relationship. For this reason, Nyrop and Binns[11] incorporated the TPL model as an alternative δ^2 predictor. For example, continuing with the leaf per sample example, each variance component becomes a function of TPL, and thereby changes with m,

$$s_S^2 = am^b \tag{31}$$

and

$$s_L^2 = ym^z \tag{32}$$

Substituting these s^2 estimators into Equation 30 yields the following

$$n_S = T / \left\{ T_S + \left[(T_S y / a)(i + rm) m^{z-b} \right]^{1/2} \right\} \tag{33}$$

and

$$n_L = \left\{ (T_S y m^z) / [(i + rm) a m^b] \right\}^{1/2} \tag{34}$$

In these formulas, the original C_S variable has been replaced with the time to sample (T_S). In addition, a function for T_L was added to account for the fact that the time to count individuals on a given leaf likely will be a function of the population density of the organisms on the leaf. The relationship was defined as

$$T_L = i + rm^{11} \tag{35}$$

and is so indicated in Equations 33 and 34. Also, $T = S(T_S + LT_L)$. A specific example of this approach is the hierarchical plan developed for spotted tentiform leafminer, *Phyllonorycter blancardella* (F.).[11]

Finally, a similar two-stage approach can be used to define the variance components using the model for mean crowding. Kuno[10,61] provided the following formula, which is presented here using the leaf (L) and samples (S) notation.

$$n_S = \{1/n_L[(\alpha + 1/m) + (\beta_1(\beta_2 - 1)/\beta_2)]\}/C^2 + 1/S[(\beta_1 - \beta_2)/\beta_2] \quad (36)$$

where n_S = number of samples to be taken (optimum sample size), S = the total number of samples possible in the sample universe (e.g., trees in an orchard), n_L = the number of leaves taken per sample, α is the intercept of the MC model for the within-sample distribution (i.e., among leaves), β_1 = the slope of MC for the within-sample and between-sample distributions, β_2 = the slope of MC for the between-sample distribution, and m and C are mean density and precision (SE/m), respectively, as previously defined. To use the MC model for each stratum in a sequential sampling procedure,

$$Tn = (\alpha + 1)/C^2 - \{[((\beta_1\beta_2 - \beta_1 + \beta_2)/\beta + ((\beta_1 - \beta_2)/\beta_2)n_L) - 1]/n\} \quad (37)$$

where Tn is the cumulative number of individuals counted, n_L, C, α, β_1, and β_2 are defined as in Equation 36, and n = cumulative number of leaves already sampled. Specific examples of this approach include those for mites on apple[55] and a blackfly on citrus.[60]

Another variation of multistage sampling is the variable intensity sampling (VIS), developed by Hoy et al.[3] VIS is a hybrid of traditional sequential classification methods (Chapter 8) and sequential sampling to estimate population density. VIS explicitly accounts for the need to have representative coverage of the sample universe, by designating the number of blocks or quadrats within the field, as well as the number of samples within each block. However, with VIS the number of samples within a quadrat (e.g., leaves within a sample using the previous example) can vary depending on the information collected from each successive sample. VIS is similar to classification sampling because it is based on the confidence interval of the mean, which determines the stop limits, and requires more sampling to occur as the estimated mean approaches the threshold. Nyrop and Binns[11] recently provided an excellent overview of this method, along with the development of a VIS procedure using a nested model.

VI. PERFORMANCE AND VALIDATION OF SEQUENTIAL SAMPLING

The validation of sequential sampling plans is necessary because of multiple sources of uncertainty inherent in the development and implementation of any sampling procedure. With regard to sampling plan development, uncertainty is important because of the direct effect on indices used to describe spatial pattern and subsequent stop-line criteria for estimating density. Extreme examples of the stochastic nature of spatial pattern within a species are those studies that have documented the variability of b for TPL in separate geographic locations,[62] or on different host plants within the same location.[27] Sawyer[63] also conducted extensive simulation studies to document the instability of TPL. Alternatively, Taylor[64] has shown that the density-dependent nature of k of the NB distribution can be used in a positive sense to quantify change in population pattern. Finally, as Kuno[18] recently noted, variability will occur with any of the spatial pattern models and thereby affect sampling plan criteria.

Although some controversy exists over the stability of TPL vs. MC, or TPL vs. NB, as models of spatial pattern, it is instructive to remember that in one of Taylor's early papers,[19] he notes that while b appears to be a species-specific attribute, "the index b...remains constant for the same organism in *the same environment*..." Clearly, "same environment" is a strict qualification that could preclude application of spatial pattern results among geographically distinct populations, arthropod-host plant associations, or changes in IPM management strategies. In addition, Kuno,[10] in his first paper on the MC model, suggested that because the s^2-m relationship can change in the field, sequential sampling plans are to be tested with independent data sets, and modified, if necessary, prior to implementation. Despite these qualifications, the stochastic nature of the sampling plans, based on either model, has been only recently considered.

For IPM and entomological research applications, there are three primary sources of uncertainty: biological, managerial (i.e., external perturbations to the system), and statistical (see also Chapter 4). Some may be more intuitively obvious than others but only recently have each of the stochastic elements of sequential sampling been examined in detail.

How the biology and behavior of the species' population determines its spatial pattern is clearly fundamental to the development of a sampling plan for the life stage of interest. For example, multiple mortality factors may differentially influence age-specific mortality, and subsequent spatial pattern of arthropod life stages.[65] This is particularly true for many Lepidoptera, where considerable changes in age-specific mortality and dispersal occur as larvae eclose from egg masses and mature to the late instars.[66] Age- and sex-specific differences for a diplopod, *Trigoniulus lumbricinus* (Gerst), were documented by Banerjee,[67] where $b = 1.35$ for all life stages combined (i.e., suggesting an aggregated dispersion pattern), but $b = 0.46$ for males, 0.86 for females, and 0.30 for immatures. Change in the spatial pattern of aphids may also occur as a field is colonized over time[68] (e.g., from aggregated to random as colonies are established) or in response to predation.[69]

Spatial pattern may also change as the population responds to environmental change, particularly management effects. For example, the influence of chemical control on the distribution of arthropod pests should be accounted for if the sampling plan is to be used in commercial fields treated with pesticide.[70,71] Very little work has been done to document possible changes in spatial pattern among commercial varieties of a given host crop or a comparison of resistant hybrids and current commercial hybrids. It is conceivable that potential differences might exist that would necessitate changes in a sequential sampling plan as resistant varieties are implemented.

Once the appropriate sampling data have been collected under the environmental or managerial conditions, assumptions of the statistical methods used to analyze the data and develop the plan must be considered. As with any statistical estimate of a population parameter, such as b of TPL, variability and thus uncertainty are associated with the estimate. The only question is how much variability and how this will affect sampling plan performance. It also should be emphasized that these questions be considered despite a high R^2 or low standard error for the coefficients of the regression model.

Because of the variability within a probability distribution or empirical model of the s^2-m relationship, sequential sampling plans will not yield the same precision level (C or d) every time they are used (for the same known m). As indicated by Nyrop and Binns,[11] this occurs for two reasons. First, despite a good statistical fit of the data used to quantify the relationship (e.g., TPL), the model is based on a limited set of observed data. Second, the stop-lines are based on expected values (e.g., mean density) for random observations. For example, for any particular m, the variance actually represents an average estimate of a distribution of variances (random variables). The expected outcome of a desired precision level or expected sample size is based on the use of the average variance and will only occur in the long run. An additional source of variability in sequential sampling includes a slight positive bias in the mean estimate,[11] primarily because the count for the last sample taken must be greater than zero.[72] In the long run (e.g., with > 100 sampling runs via simulation), this bias is usually negligible. However, for any given sequential sample, the count for the last sample may have considerable effects on the final mean and precision level obtained, particularly at high densities.[9]

A. FIELD VALIDATION

Field validation is the best, final test of how well a sampling plan will perform under a variety of conditions. The primary drawback, however, is that extensive field testing is expensive. The extent to which this can be done is usually limited by resources. Depending on the number of validations, the approach may also be limited by the density ranges available. In addition, depending on the personnel involved, this may become an exercise in person-person variability,[73] another aspect of how robust the overall sampling method or sample unit may be. Despite being an additional critical component of sampling plan performance, this is a different validation problem, i.e., that of the sample unit and the accuracy of a sample unit. In this chapter, the primary purpose is to present methods to assess the performance of the plan's ability to maintain precision as expected. Sampling unit issues should be evaluated prior to validation of precision (e.g., by NANOVA), given the biology of the organism and practical considerations (Chapter 4). One of the best ways to incorporate field validation is to use it in tandem with computer simulation methods. For example, use computer simulation first to modify, if necessary, sequential sampling stop-lines, then use the simulation-adjusted plans for final field validation.[74]

B. SIMULATION

The primary advantage of simulation is that it provides a systematic way to examine the behavior of sequential sampling results in general, and the performance of specific sampling plans for a given species or application.[1,9,11] Specifically, simulation is useful for studying the performance of plans under a variety of assumptions or actual data sets of arthropod counts. Given appropriate assumptions and data for validation purposes, simulation is also inexpensive relative to field validation, and savings appreciate over time as the same computer programs are applied to a number of applications.

There are two primary approaches to simulation analysis for sequential sampling, Monte Carlo[11] and bootstrap.[9,75] When using the Monte Carlo method, the population is assumed to fit a conventional probability distribution, such as the Poisson or NB.[3,11] The bootstrap approach assumes no specific underlying distribution of the data, but re-samples data files containing actual sample counts for the species of interest.[9,53,75]

Nyrop and Binns[11] developed a Monte Carlo-NB approach for precision-level estimation where the TPL parameters are used to estimate the s^2 for any given m, estimate k and a NB frequency distribution of counts. For each sampling run (or pass), a random number generator is used to take successive random samples from the NB distribution (with replacement) until the cumulative total reaches or exceeds the stop-line (Tn). Sampling then stops and actual m, s^2, n, and C values for a particular simulation run are calculated. Usually 100 to 500 such samplings will be conducted for each mean density of interest to generate distributions of actual means, precision levels, and sample sizes. This algorithm provides estimates of average m density, average n, average C, and the distribution of actual C values for a given true m. The latter output allows one to not only evaluate the relative frequency of simulation runs that yielded a C value less than or equal to the desired C, but the frequency of C values > 0.30, for example, generally considered an upper limit for IPM purposes.[7] One criterion for reliable performance of a plan is that only $< 10\%$ of the C values are greater than the desired C (e.g., 0.25 or 0.30), suggesting, with 90% confidence, that the plan is performing as expected. In addition to using the mean estimates of the TPL parameters (a and b), the program also permits the user to incorporate the variability in the TPL model by using the mean square error (mse) term of the regression and the variance of b.

At the time of this writing, the algorithm explicitly sampled only from the NB distribution. However, use of TPL input that is more indicative of a population dispersed at random (i.e., $b = 1$), should influence subsequent k values and the NB such that the results may still lead to reasonable conclusions for populations dispersed at random.

Other examples of the Monte Carlo simulation approach, where the NB was used as the underlying distribution, include validation of sampling plans for cabbage Lepidoptera,[3,62] cotton arthropods,[76] and alfalfa weevil.[74] In addition, Rudd[77] used Monte Carlo simulation to evaluate sequential sampling plans for soybean insect pests using both the Poisson and NB distributions.

Bootstrap simulation[75] differs from the Monte Carlo approach in that rather than sample from a theoretical probability distribution, samples are taken directly from actual count data sets. Such data sets should reflect the range of population densities likely encountered in the field, they should be independent of data sets used to develop the sampling plan, and their n should be higher than that of any expected sequential sample n. The only disadvantage of this approach is that there may be a limited number of data sets available that meet these criteria. However, because many data sets are usually collected in an effort to build robust sampling plans, saving some of the data for validation studies usually is not an unreasonable requirement.

The primary advantage of the bootstrap method is that there are no *a priori* assumptions about the population's spatial distribution. As previously discussed, there are many sources of biological and statistical variability that influence all empirical relationships of spatial pattern (i.e., s^2-m relation). It is therefore desirable to be able to sample from such data sets, with all of the field variability included. Recent examples of the bootstrap simulation approach include those of Hutchison et al.[9] for *A. pisum* and Jones[53] for spider mites. The algorithm of Hutchison et al., included in this chapter, allows for the evaluation of fixed-precision stop-line plans of Kuno, based on MC, and that of Green, based on TPL. Jones' program was developed to analyze Green's plan and also allows for multistage or hierarchical sampling (such as trees per orchard and leaves per tree).

A comparison of the Monte Carlo and bootstrap simulation approaches was done to illustrate similarities and differences for *A. pisum* on alfalfa,[9] using Green's plan with TPL (Table 2). In this example there are several low-density data points (< 3 aphids per stem) from two geographic locations (Wisconsin and Oklahoma) in the TPL regression; the majority of low-density data were from Oklahoma with $b = 1.12$, whereas the Wisconsin $b = 1.54$. The Oklahoma b value was not significantly different from 1.0 ($p > 0.05$), suggesting that the population was dispersed at random. Clearly, the Wisconsin aphids were aggregated. As expected with linear regression, the larger s^2-m pairs from the Wisconsin data set had considerable influence on the regression of the combined relationship, such that overall $b = 1.44$.

TABLE 2
Comparison of Monte Carlo and Bootstrap Simulation Procedures to Examine the Performance of Actual Precision-Level Control for *Acyrthosiphon pisum* (Harris) on Alfalfa, with the Desired Precision (C) Equal to 0.25[a]

	Average C for 100 sequential sampling runs			% of C values ≤ 0.30		
Mean (m)	Monte Carlo	Monte Carlo +s^2 included	Bootstrap	Monte Carlo	Monte Carlo +s^2 included	Bootstrap
0.87	0.26	0.26	0.26	78	73	92
4.10	0.25	0.26	0.17[b]	78	71	100[b]
8.10	0.24	0.27	0.18[b]	82	65	100[b]
24.50	0.22	0.25	0.22	80	79	86
41.40	0.27	0.25	0.24	72	79	86

[a] Comparison was done for Taylor's power law (TPL) with desired $C = 0.25$, using Monte Carlo algorithm of Nyrop and Binns,[11] and bootstrap algorithm of Hutchison et al.[9] presented in this chapter. Input data, taken from the TPL regression included: $a = 3.175$, $b = 1.438$, var(b) = 0.1492, mse = 0.089, and mean of ln(m) = 0.244, and $n = 63$ (taken from Hutchison et al.[9]). Performance of Green's[21] TPL-based plan was based on 100 simulated sequential sampling "runs" for both simulation methods.

[b] Population densities where actual precision-level control was better (more conservative) than expected, based on bootstrap results, indicating that the sampling plan would require more samples than necessary to maintain precision at 0.25.

Given the overall relationship, with expected tendency toward aggregation, it is reasonable to assume that a number of fixed-precision level plans should be appropriate; for example, Green's plan with $b = 1.44$, Kuno's plan with $\beta = 1.41$,[9] or the NB approach, as suggested by Nyrop and Binns.[11]

As shown in Table 2, with the desired $C = 0.25$, both simulation approaches illustrated the variability and distribution of actual precision level (C) estimates from each set of simulation runs for a given density. However, differences were observed between the two simulation methods for two of the "low" densities tested (4.1 and 8.1 aphids per stem). Specifically, the Monte Carlo approach, assuming the NB distribution for all densities tested, yielded higher C value estimates than the bootstrap approach. A similar pattern emerged when the known variability of the TPL model [mse and $s^2(b)$] was included. The bootstrap results were more conservative than necessary for IPM, with actual C values ranging from 0.17 to 0.18. Thus, because of the implicit assumption of an aggregated distribution (i.e., $b = 1.44$), the TPL plan is recommending that more samples than necessary be taken at these low densities. However, at these densities, the populations (both Oklahoma and Wisconsin) are more indicative of a random dispersion pattern, and fewer samples are necessary to maintain $C \approx 0.25$. Use of the bootstrap approach, with actual data sets, permits the illustration of this response *and* the need to rerun the simulation using higher values for C. In this example, a similar trend was observed for Kuno's fixed-precision plan; C values of 0.35 were found to provide the actual precision-level control of $C \approx 0.25$ at the low densities and final stop-line plans were adjusted accordingly.[9]

VII. SUMMARY

For any particular pest-commodity system, IPM may currently be implemented at many different levels. Ferro[78] recently summarized three phases of transition in IPM: (1) the pesticide management phase, where economic or "action" thresholds for pests are available to minimize pesticide use, (2) the cultural management phase, where knowledge of pest and crop biology is used to manipulate pest populations (e.g., planting date), and (3) the biologically intensive[79] phase, where knowledge of the biology and quantitative impact of natural enemies is used to manage pests, and further minimize, but not necessarily preclude, pesticidal control. Obviously, each additional level of integration builds on the previous level. A critical step in the process is the implementation of sampling and thresholds which are essential for minimizing pesticide use, and thereby necessary for the implementation of biological control.[11,80] Thus, decision rules based on thresholds and practical sampling plans will continue to provide the foundation for further refinement of IPM programs, particularly biologically-intensive IPM.

Although many advances have been made in the development of practical sampling plans, many plans have not been fully implemented. The recent trend toward more on-farm research and IPM demonstration[81] may help to foster more active implementation of IPM by producers. Likewise, a similar pattern of research and demonstration may be necessary for urban clientele. Finally, more work must be done to document the value of the sampling programs to consultants and producers. Placing monetary values on IPM information, compared to traditional control costs, will become increasingly important for successful implementation to continue.[82,83]

Program Listing

PROGRAM sampling:

{This program performs bootstrap sampling of actual data sets of animal populations to evaluate the performance of Kuno's[10] and Green's[21] sequential sampling plans, to estimate mean population density with a fixed level of precision, where precision C is defined as the ratio of the standard error per mean. The program was written in Turbo Pascal V.4.0, for use on DOS-based personal computers. Input to the program requires the regression coefficient parameters for either Iwao's mean crowding (MC)[16] (for Kuno's plan) or Taylor's power law (TPL)[15] (for Green's plan), the desired precision level (e.g., $C = 0.25$, for the SE to be within 25% of the mean), an ASCII (*.dat) data file containing count data in a single column (no spaces, commas, etc.), and the number of simulated sampling runs from a given data set (usually set to 100). Users should also note that the performance of d-based confidence interval plans (see Chapter) can also be evaluated for the MC and TPL models using this program. For example, if $d = 0.25$ were to be evaluated, $C = 0.1276$ would be used as input to the simulation, for a 95% confidence interval (i.e., $d = 1.96(C)$). The following algorithm was developed to illustrate the stochastic nature of the Green and Kuno sampling plans using count data for arthropods; further details can be found in Hutchison et al.[9] Interested users of this program are encouraged to contact WDH, University of Minnesota, for this program on diskette.}

```
USES Crt, TPlus, TPWindow, First,
        ksample, gsample, gvsample, paplan, PRINTER;

CONST
    Fgw  = White;
    Bgbk = Black;
    Fgy  = Yellow;
    Bgbu = Blue;
```

```
TYPE
    s_type = (kuno, green, none);

VAR
    w1 : WindowPointer;
    Item, Position : INTEGER;
    SaveArea : ARRAY [1..80] OF BYTE;
    done : BOOLEAN;
    cur_sam : s_type;                    {current sampling type}
    ans : CHAR;
```

{This procedure initializes the main parameters.}

```
PROCEDURE init;

BEGIN
    cur_sam := none;

END
```

```
PROCEDURE clear_vals (VAR fname : STRING80; VAR dprec, alpha, beta : REAL; VAR pass :
    INTEGER);

BEGIN
    fname := ' ';
    dprec := 0.0;
    alpha := 0.0;
    beta := 0.0;
    pass := 0;

END;
```

```
PROCEDURE present_vals (VAR fname : STRING80; VAR dprec, alpha, beta : REAL; VAR pass :
    INTEGER);

VAR
    w4 : WindowPointer;
    TotFields, Field, ReturnCode, CursorPos,
    I, Item, Position : INTEGER;
    SaveArea :ARRAY [1..80] OF BYTE;
    done : BOOLEAN;
    MapOption : CHAR;

BEGIN
    w4 := getscreen (4);
    ShowWindow (HorizBlinds, w4);
    CursorON;

MapOption := 'W';
    TotFields := 5;
FOR field := 1 TO PROCEDURE DO BEGIN
    CursorPos := 1;
    CASE Field OF
        1: MapString(MapOption,45,7,25,0,0,0,0,ReturnCode,CursorPos,fname,'>>>>>>>>');
        2: MapReal(MapOption,49,9,25,0,0,0,0,ReturnCode,CursorPos,deprec,'####.###');
        3: MapReal(MapOption,49,11,25,0,0,0,0,ReturnCode,CursorPos,alpha,'####.###');
        4: MapReal(MapOption,49,13,25,0,0,0,0,ReturnCode,Cursor,Pos,beta,'####.###');
        5: MapInt(MapOption,53,15,25,0,0,0,0,ReturnCode,CursorPos,pass,'####');
                    {CASE}
        END; {FOR}
```

```
Field := 1 ;
MapOption := 'L';
REPEAT
    CursorPos := 1;
    CASE Field OF
        1: MapString(MapOption,45,7,25,0,0,0,0,ReturnCode,CursorPos,fname,'>>>>>>>>');
        2: MapReal(MapOption,49,9,25,0,0,0,0,ReturnCode,CursorPos,dprec,'####.###');
        3: MapReal(MapOption,49,11,25,0,0,0,0,ReturnCode,CursorPos,alpha,'####.###');
        4: MapReal(MapOption,49,13,15,0,0,0,0,ReturnCode,CursorPos,beta,'####.###');
        5: MapInt(MapOption,53,15,25,0,0,0,0,ReturnCode,CursorPos,pass,'####');
    END;      {CASE}

    CASE ReturnCode OF
        arrowupkey : IF (Field <= 1) THEN Field := TotFields ELSE Field := Field - 1;
        arrowdnkey, enterkey:
                     IF (Field >= Totfields) THEN Field := 1 ELSE Field := Field + 1;
    END; {case}

UNTIL (ReturnCode - esckey);

CursorOff;
END;
```

{This procedure saves the current initialization values to a file.}

```
PROCEDURE save_sam (cur_sam : s_type; fname : STRING80; deprec, alpha, beta : REAL; pass :
        INTEGER);

VAR
    save_file : STRING;
    outfile : TEXT;

BEGIN
    CLRSCR;
    WRITELN; WRITELN; WRITELN;        {get file name}
    WRITELN ('Enter name of file to');
    WRITE ('save initialization data (filename.INT) :');
    READLM (save_file;
    ASSIGN (outfile, save_file;

IF    (cur_sam = kuno) THEN BEGIN
    WRITELN (outfile, 'kuno');         {put kuno data}
    WRITELN (outfile, fname);
    WRITELN (outfile, dprec);
    WRITELN (outfile, alpha);
    WRITELN (outfile, beta);
    WRITELN (outfile, pass);
    END

ELSE BEGIN
    WRITELN (outfile, 'green');        {put green data}
    WRITELN (outfile, fname);
    WRITELN (outfile, dprec);
    WRITELN (outfile, alpha);
    WRITELN (outfile, beta);
    WRITELN (outfile, pass);
    END;

CLOSE (outfile);

END;
```

{This procedure retrieves the current initialization values from a file.}

```
PROCEDURE ret_sam (VAR cur_sam : s_type; VAR fname : STRING80; VAR dprec, alpha, beta :
        REAL; VAR pas : INTEGER);

VAR
    save_file,
    temp : STRING;
    infile : TEXT;

BEGIN
    CLRSCR;
    CursorOn;
    WRITELN; WRITELN; WRITELN;              {get file name}
    WRITELN ('Enter name of file to');
    WRITE ('retrieve initialization data (filename.INT) : ');
    READLN (save_file);
    ASSIGN (infile, save_file);
    RESET (infile);

    READLN (infile, temp);                  {kuno or green sampling input}
        IF (temp = 'kuno') THEN cur_sam := kuno
        ELSE cur_sam := green;
        IF (cur_sam = kuno) THEN BEGIN
            READLN (infile, fname);         {get kuno sampling input}
            READLN (infile, dprec);
            READLN (infile, alpha);
            READLN (infile, beta);
            READLN (infile, pass);
            END
        ELSE BEGIN
            READLN (infile, fname);         {get green sampling input}
            READLN (infile, dprec);
            READLN (infile, alpha);
            READLN (infile, beta);
            READLN (infile, pass);
            END;
    CLOSE (infile);
    cursorOff;
```

{This procedure prints a hardcopy of the values in Kuno for each pass.}

```
PROCEDURE print_k;

VAR
    ifile : TEXT;           {input file}
    mean, vari, std, se, dval: REAL;
    pass, n, pi : integer;

BEGIN
    ASSIGN (ifile, 'kstemp.dat');
    RESET (ifile);

WRITELN (LST);
    WRITELN (LST,'     KUNO'S SAMPLING PLAN');
    WRITELN (LST,'     STATISTICS FOR EACH PASS');
    WRITELN (LST);
    WRITELN (LST,'PASS    MEAN    VAR STD  SE  D  PI  N');
```

```
        WRITELN (LST,'---- ---- --- --- -- - ---');
        WHILE (NOT EOF(ifile)) DO BEGIN
           READLN (ifile, pass);
           READLN (ifile, mean);
           READLN (ifile, vari);
           READLN (ifile, std);
           READLN (ifile, se);
           READLN (ifile, dval);
           READLN (ifile, pi);
           READLN (ifile, n);
           WRITELN  (LST,  pass:3,'         ',mean:622,' ',vari:6:2,' ',std:6:2,' ',se:6:2,' 'dval:5:2,' ',pi:3,'
              ,n:3);
        END;

        CLOSE (ifile);

END;
```

{This procedure prints a hardcopy of the values in Green for each pass.}

```
PROCEDURE print_g;

VAR
    ifile : TEXT;             {input file}
    mean, vari, std, se, dval: REAL;
    pass, n, pi : integer;

BEGIN
    ASSIGN (ifile, 'gstemp.dat');
    RESET (ifile);
    WRITELN (LST);
    WRITELN (LST,'        GREEN'S SAMPLING PLAN ');
    WRITELN (LST,'        STATISTICS FOR EACH PASS');
    WRITELN (LST);
    WRITELN (LST,'PASS    MEAN    VAR  STD  SE  D  PI  N');
    WRITELN (LST,'---- ---- --- --- -- - -- -');
    WHILE (NOT EOF(ifile)) NO BEGIN
    READLN (ifile, mean);
    READLN (ifile, vari);
    READLN (ifile, std);
    READLN (ifile, se);
    READLN (ifile, dval);
    READLN (ifile, pi);
    READLN (ifile, n);
    WRITELN (LST, PASS3,' ',mean:6:2,' ',vari:6:2,' ',std:6:2,' ',se:6:2,'  ',dval:5:2,' ',pi:3,' ',n:3);
    END;
    CLOSE (ifile);

END;
```

```
PROCEDURE DriveKuno ;

VAR
    w2, w6 : WindowPointer;
    pass,
    Item, Positon : INTEGER;
    SaveArea : ARRAY [1..80] OF BYTE;
    done : BOOLEAN;
    fname : STRING80;
```

```
        dprec, alpha, beta : REAA;
        ans : CHAR;
            {Stubs}
procedure helpfile; begin end;


BEGIN
    cur_sam := kuno;           {tell exoss it is using kuno}
    w2 := getscreen (2);
    ShowWindow (CurtainClose, w2);
    clear_vals (fname, dprec, alpha, beta, pass);
    done := false;
    WHILE (NOT done) DO BEGIN
        WITH w2^ DO
        MenuPick (SWx, WSy, WEx, WEy, LBracket, RBracket, RBracket, Item, Fgy, Bgbu, 5, 1);
        CASE Item OF
        0 : done := true;
        1 : BEGIN
            PushWindow (w2);
            present_vals (fname, dprec, alpha, beta, pass);
        2 : BEGIN
            PushWindow (w2);
            ret_sam (cur_sam, fname, dprec, alpha, beta, pass);
          END;
        3: BEGIN
            PushWindow (w2);
            save_sam (cur_sam, fname, dprec, alpha, beta, pass);
          END
        4: BEGIN
            PushWindow (w2);
            w6 := getscreen (5);
            ShowWindow (Normal,w6);
            kunosap (fname, dprec, alpha, beta, pass);
            ClearScn;
            WRITELN ('Do you want a print out of each pass? (Y/N) ');
            ans := READKEY;
                   IF (ans = 'Y') THEN print_k;
          END;
        5: BEGIN
            PushWindow (w2);
            helpfile;
          END
        END;        {CASE}
        Showindow (CurtainClose, w2);
    END;
```

PROCEDURE DriveGreen ;

VAR
    w3, w5 : WindowPointer;
    pass,
    Item, Position : INTEGER;
    SaveArea : ARRAY [1..80] OF BYTE;
    done : BOOLEAN;
    fname : STRING80;
    dprec, alpha, beta : REAL;
    ans : CHAR;
            {Stubs}
procedure helpfile; begin end;

```
BEGIN
    cur_sam := green;              {tell exoss it is using green}
    w3 := getscreen (3);
    ShowWindow (CurtainClose, w3);
    clear_vals (fname, dprec, alpha, beta, pass);
    done := false;
    WHILE (NOT done) DO BEGIN
        WITH w3^ DO
        MenuPick (WSx, WSy, WEx, WEy, LBracket, RBracket, Item, Fgy, BGbu, 5, 1);
        CASE Item OF
          0 : done := true;
          1 : BEGIN
          PushWindow (w3);
          present_vals (fname, dprec, alpha, beta, pass);
              END;
          2 : BEGIN
          PushWindow (w3);
          ret_sam (cur_sam, fname, dprec, alpha, beta, pass)

END;

          3 : BEGIN
          PushWindow (w3);
          save_sam (cur_same, fname, dprec, alpha, beta, pass);
              END;
          4 : BEGIN
          PushWindow (w3);
          w5 := getscreen (5);
          ShowWindow (Normal, w5);
          greensamp (fname, dprec, alpha, beta, pass);
          ClearScn;
          WRITELN ('Do you want a print out of each pass? (Y/N) ');
          ans := READKEY;
          ans := UPCASE (ans)
          PushWindow (w3);
          helpfile;
          END;
     ShowWindow (CurtainClose, w3);
     END; {WHILE}
     PopWindow (Normal);

END;
```

```
BEGIN
    init;                 {initialize main variables}
    CursorOff;
    w1 := getscreen (1);
    ShowDelay := 15;
    ClearScn;
    ShowWindow (CurtainClose, w1;
    done := FALSE;
    WHILE (NOT done) DO BEGIN
        WITH w1^ DO
        MenuPick (WSx, WSy, WEx, WEy, LBracket, RBracket, Item, Fgy, Bgby, 3, 1);
        CASE Item OF
        0 : BEGIN
        CursorOn;
        HALT;
        END;
```

```
            1 : BEGIN
        PushWindow (w1);
        DriveKuno;
            END;
        2 : BEGIN
        PushWindow (w1);
        WRITELN ('eNTER 'c' FOR gREEN'S SIMULATION USING CONSTANT a b VALUES');
        WRITE (' OR 'v' FOR gREEN'S SIMULATION USING VARIABLE a b VALUES : ');
        READLN (ANS);
        ANS := UPCASE (ANS);
        IF (ans = 'C') THEN DriveGreen
        ELSE greenvsamp;
            END
        3 : BEGIN
        PushWindow (w1);
        pap
            END;
        END;           {Case}
    ShowWindow (CurtainClose, w1);
    End;               {while}
    CursorOn;
END.


UNIT ksample;
{Kuno's Sampling Plan}
INTERFACE
    USES printer;
    PROCEDURE kunosamp (fname : STRING;
            {seed,} dprec, alpha, beta : REAL;
            pass : INTEGER);


IMPLEMENTATION

PROCEDURE kunosamp;

CONST
    max_el = 200;

TYPE
    int_array = ARRAY {1..max_el} OF INTEGER;
    real_array = ARRAY {1..max?el} OF REAL;

VAR
    nfrq : int_array;
    val : real_array;
    i,                 {loop control}
    npts,
    nct,
    n, aven,
    iele,
    nmin, nmax,
    a1, a2             {random generator variables}
    : INTEGER;
    dact, aved,
    rn, drn,
pbar,
    pvar,
    pstd,
    pse,
```

```
       tn, tn2,
       hd, hd2,
       tnsl,
       tbar, htbar, htbar2,
       tvar,
       tstd,
       tse,
       r,
       allvar, allstd, allse
        : REAL;
       sub_done : boolean;          {loop control}
       outfile : TEXT;
```

{This procedure initializes necessary values.}

```
PROCEDURE init;

VAR
    i : INTEGER;
    done : boolean;          {loop control}

BEGIN
    FOR i := 1 TO max_el DO BEGIN
    val {i} := 0.0;
      nfrq {i} := 0;
      END;
    htbar := 0;
    htbar2 := 0;
    hd := 0;
    hd2 := 0;
    aved := 0;
    aven := 0;
    allvar := 0;
    allstd := 0;
    allse := 0;

  ASSIGN (outfile, 'kstemp.dat');
  REWRITE (outfile);

END;
```

{This procedure loads the val array and counts the number of elements.}

```
PROCEDURE ger_samp_data (var count : integer; var val : real_array: fname : string);

VAR
    infile : text;
    I : INTEGER;

BEGIN
    ASSIGN (infile, fname);          {open input file}
    RESET (infile);
    count := 0;
    WHILE )NOT EOF (infile)) DO BEGIN      {loop until file empty}
        count := count + 1;
        READLN (infile, val{count});
        END;
    count := count - 1;
    CLOSE (infile);
```

END;

{This procedure calculates the initial statistical values.}

Procedure calc_stats (count : INTEGER; val : real_array; rn : REAL;
          VAR pbar, pvar, pstd, pse : REAL);

VAR
    d,
    pop, pop2, : REAL;   {temp values}
    j : INTEGER   {loop control}

BEGIN
      pop := 0; pop2 := 0;
    FOR j := 1 TO count DO BEGIN
        pop := pop + val{j};
        pop2 := pop2 = sqr(val{j});
      END;

    pbar := pop / rn;
    pvar := (pop2 - (sqr(pop)) / rn) / (rn - 1);
    pstd := sqrt (pvar);
    pse := pstd / sqrt(rn);
    d := pse / pbar;
    WRITELN (LST,'Observed Stats');
    WRITELN (LST,'mean :',pbar:6:2);
    WRITELN (LST,'var :',pvar:6:2);
    WRITELN (LST,'std :',pstd:6:2);
    WRITELN (LST,'se : ',pse:5:2);
    WRITELN (LST,'d : ',d:5:2);
    WRITELN (LST,'n : ',count:3);

END;

{This procedure inits specific values.}
PROCEDURE samp_init (VAR nct, n : INTEGER; VAR tn, tn2, rn : REAL; count : integer);

VAR i : integer;   {loop control}
    done : boolean;

BEGIN
    nct := 0;

    tn := 0.0;
    tn2 := 0.0;
    re := 1.0;

    i := 1;
    WHILE (SQR(dprec) - ((beta - 1 / i)) <= 0.0) do
        i := 1 + 1;

    IF (done) THEN n := i
    ELSE BEGIN
        WRITELN('Minimal sequential sample size is greater   than');
        WRITELN('THE NUMBER OF SAMPLES IN THE INPUT DATA FILE.');
        HALT'
    END;

END;

{This procedure inits specific values.}

```
PROCEDURE samp_init (VAR nct, n : INTEGER; VAR tn, tn2, rn : REAL; count : integer);

VAR
    i : integer;            {loop control}
    done : boolean;         {loop control}

BEGIN
    nct := 0;
    tn := 0.0;
    rn := 1.0;
    i := 1;
    WHILE (( SQR (dprec) - ((beta - 1) / 1)) < = 0.0) do
        i := i + i;

    IF (done) THEN n := i
    ELSE BEGIN
        WRITELN('Minimal sequential sample size is greater         than');
        WRITELN ('the number of samples in the input data      file.');
        HALT
    END;

END;
```

{This procedure prints out the cumulative statistics.}

```
PROCEDURE show_cummul (pass, aven : integer;
                      nfrq : int_array;
                      htbar, allvar, allstd, allse, aved : real);

VAR
                        {temp values}
    avemean,
    avedev,
    avevar,
    avestd,
    avese : real;

BEGIN
    avemean L = htbar / pass;
    avevar := allvar / pass;
    avestd := allstd / pass;
    avese := allse / pass;
    aved := aved / pass;
    aven := round (aven / pass);
    WRITELN (LST);
    WRITELN (LST,' cumulative Stats');
    WRITELN (LST,'mean :',avemean:6:2);
    WRITELN (LST,'var :',avevar:6:2);
    WRITELN (LST,'std :',avestd:6:2);
    WRITELN (LST,'se :',avese:5:2);
    WRITELN (LST,'d :',aved:5:2);
    WRITELN (LST,'n :',aven:3);

END;
```

{This function returns the minimum of two integers}

```
FUNCTION mino (first, second : INTEGER) : INTEGER;

BEGIN
    IF (first <= second) THEN mino := first
    ELSE mino := second;
END;
```

{This function returns the maximum of two integers}

```
FUNCTION maxo (first, second : INTEGER) : INTEGER;

BEGIN
    IF (first >= second) THEN maxo := first
    ELSE maxo := second;
END;
```

{This function returns a random number between 0 and 1.}

```
FUNCTION CombRandom : REAL;

VAR
    f : REAL;
    FUNCTION Ran1 : REAL;
    VAR
        t : REAL;
    BEGIN
        t := (a1 * 32749 + 3) MOD 32749;
        a1 := TRUNC (t);
        Ran1 := ABS(t/32749);
    END;       {ran1}

    FUNCTION Ran2 : REAL;
    VAR
        t : REAL;
    BEGIN
        t := (a2 * 10001 + 3) MOD 17417;
        a1 := TRUNC(t);
        Ran2 := ABS(t/17217);
    END;       {ran2}
{_____}

BEGIN   {CombRandom}
{    f := Ran2; }              {select random generator}
  { IF (f > 0.5) then} CombRandom := Random;
{    ELSE CombRandom := Ran1;}
END;     {CombRandom}

BEGIN
    init;                {initialize values}
    get_samp_data (npts, val, fname); {load val}
    rn := npts;          {load rn}
    a1 := 1; a2 := 203;   {init random variables}
    RANDOMIZE;
    {calculate initial stats}
    calc_stats (npts, val, rn, pbar, pvar, psted, pse);
```

```
FOR i := 1 TO pass DO BEGIN        {sampling loop}

   samp_init (nct, n, tn, tn2, rn, npts); {init these values}
   sub_done := false;
   WHILE (NOT sub_done) DO BEGIN
       r    := CombRandom;
       irle := round (drn * r);
       END

   tn  := tn + val[iele];           {accumulate random sample}
   tn2 := tn2 + sqr(val[iele]);
   rn  := n;

   {KUNOsampling plan}
   tnsl := (alpha + 1) / (SWR(dprec) - ((beta - 1 / rn));

   {sample exceed boundary?}
   IF (tn < tnsl)
   THEN BEGIN
       n := n + 1;
       sub_done := false;
     END
   ELSE BEGIN
       infrq[n] := nfrq[n] +1;
       nmin := mino (n, nmin);
       nmax := maxo (n, nmax);

       tbar   := tn / n;
       htbar  := htbar + tbar;
       ntbar2 := htbar2 + sqr(tbar);

       tvar   := (tn2 - sqr(tn) / rn) / (rn - 1);
       allvar := allvar + tvar;
       tstd   := sqrt(tvar);
       allstd := allsted + tsted;
       tse    := tstd / sqrt(rn);
       allse  := allse + tse;

       dact := tse / tbar;
       hd   := hd + dact;
       hd2  := hd2 + sqr(dact);

       WRITELN (outfile,i:3);
       WRITELN (outfile,tbar:6:2);
       WRITELN (outfile,tvar:6:2);
       WRITELN (outfile,tstd:6:2);
       WRITELN (outfile,tse:5:2);
       WRITELN (outfile,dact:3:2);
       WRITELN (outfile,n:3);

       aved := aved + dact;
       aven := aven + n;
       sub_done := true;
    END;         {sub_done loop}
 END;            {for loop}

 show_cummul (pass, aven, nfrq,  {print cumulative stats}
         htbar, allvar, allstd, allse, aved;

 CLOSE (outfile);
END;      {main}
{_____}
```

```
BEGIN
END.


UNIT gsample;
{Green's Sampling Plan}
INTERFACE
    USES printer;

        PROCEDURE greensamp (fname : STRING;
        {seed,} dprec, alpha, beta : REAL;
        pass : INTEGER);


IMPLEMENTATION


PROCEDURE greensamp;


CONST
    max_el  =  250;


TYPE
    int_array   =  ARRAY [1..max_el] OF INTEGER;
    real_array  =  ARRAY [1..max_el] OF REAL;


VAR
    nfrq : int_array;
    val : real_array;
    i,                  {loop control}
    npts,
    pin, pic,
    pi,
    nct,
    n, aven,
    iele,
    nmin, nmax,
    a1, a2          {random generator variables}
     : INTEGER;
    dact, aved,
    rn, drn,
    pbar,
    pvar,
    pstd,
    pse,
    tn, tn2,
    hd, hd2,
    tnsl,
    tbar, htbar, htbar2,
    tvar,
    tstd,
    tse,
    r,
    allvar, allstd, allse
     : REAL;
    sub_done : boolean          {loop control}
    outfile : TEXT;
{_____}
{This procedure initializes necessary values.}
```

PROCEDURE init;

VAR
   i : INTEGER;   {loop control}
   done : boolean;

BEGIN

   FOR i := 1 TO max_el DO BEGIN
      nfrq[i] := 0;
      END;
   htbar := 0;
   htbar2 := 0;
   hd := 0;
   hd2 := 0;
   aved := 0;
   aven := 0;
   allvar := 0;
   allstd := 0;
   allse := 0;
   pic := 0;
  ASSIGN (outfile, 'gstemp.dat');
  REWRITE (outfile);

   fname := fname + '.DAT'
END;
{_____}
{This procedure loads the val array and counts the number of elements.}

PROCEDURE get_samp_data (var count, pin : integer; var val : real_array; fname : string);

VAR
   infile : text;
   I : INTEGER;

BEGIN
   ASSIGN (infile, fname);   {open input file}
   RESET (infile);

   pin := 0;
   count := 0;
   WHILE (NOT EOF(infile)) DO BEGIN   {loop until file empty}
      count := count + 1;
      READLN (infile, val[count]);   {load val}
      IF (val[count] > 0) THEN pin := pin + 1;
      END;
   count := count − 1;

   CLOSE (infile);

END;
{_____}
{This procedure calculates the initial statistical values.}

PROCEDURE calc_stats (count, pin : INTEGER; val : real_array; rn : REAL;
     VAR pbar, pvar, pstd, pse : REAL);

```
VAR
  d,
  pi, tn,
  pop, pop2 : REAL;          {temp values}
  j : INTEGER;               {loop control}

BEGIN

    pop := 0; pop2 := 0;
    FOR j := 1 TO count DO BEGIN
        pop := pop + val[j];
        pop2 := pop2 + sqr(val[j]);
        END;

    pbar := pop / rn;
    pvar := (pop2 - (sqr(pop))) / rn / (rn - 1);
    pstd := sqrt(pvar);
    pse := pstd / sqrt(rn);
    d := pse / pbar;
    pi := (pin / count) * 100;
    tn := (pbar * count);

    WRITELN (LST,'Observed Stats');
    WRITELN (LST,'PI   :',pi:6:2);
    WRITELN (LST,'mean   :',pbar:6:2);
    WRITELN (LST,'var  :',pvar:6:2);
    WRITELN (LST,'std  :',pstd:6:2);
    WRITELN (LST,'se   : ',pse:5:2);
    WRITELN (LST,'d    : ',d:5:2);
    WRITELN (LST,'n    :  ',count:3);
    WRITELN (LST,'Tn   :',tn:6:2);

END;

{_____}
{This procedure inits specific values.}

PROCEDURE samp_init (VAR nct, n, pin : INTEGER; VAR tn, tn2, rn : REAL; count : integer);

VAR
    i : integer;            {loop control}
    done : boolean;         {loop control}

BEGIN
    nct := 0;
    tn := 0.0;
    tn2 := 0.0;
    rn := 1.0;
    n := 1;
    pin := 0;

END;

{_____}
{This procedure prints out the cumulative statistics.}

PROCEDURE show_cummul (pass, aven, pic : integer;
        nfrq : int_array;
        htbar, allvar, allstd, allse, aved : real);
```

```
VAR
                    {temp values}
    pi,
    tn,
    avenR,
    avemean,
    avedev,
    avevar,
    avestd,
    avese : real;

BEGIN
    avemean := htbar / pass;
    avevar  := allvar / pass;
    avestd  := allstd / pass;
    avese   := allse / pass;
    aved    := aved / pass;
    avenR   := ?? / pass; (can't read line)

    tn := (avemean * pass);

    WRITELN (LST);
    WRITELN (LST,'Cumulative Stats');
    WRITELN (LST,'PI    :',pi:6:2);
    WRITELN (LST,'mean  :',avemean:6:2);
    WRITELN (LST,'var   :',avevar:6:2);
    WRITELN (LST,'std   :',avestd:6:2);
    WRITELN (LST,'se   : ',avese:5:2);
    WRITELN (LST,'d    : ',aved:5:2);
    WRITELN (LST,'n    : ',avenR:5:2);
    WRITELN (LST,'Tn    :',tn:6:2);

END;

{_____}
{This function returns the minimum of two integers}

FUNCTION mino (first, second : INTEGER) : INTEGER;

BEGIN
    IF (first <= second) THEN mino := first
    ELSE mino := second;

END;

{_____}
{This function returns the maximum of two integers}

FUNCTION maxo (first, second : INTEGER) : INTEGER;

BEGIN
    IF (first <= second) THEN maxo := first
    ELSE maxo := second;

END;

{_____}
{This function returns a random number between 0 and 1.}
```

```
FUNCTION CombRandom : REAL;

VAR
    f : REAL;

{_____}
    FUNCTION Ran1 : REAL;
    VAR
        t : REAL;
    BEGIN
        t  := (a1 * 32749 + 3) MOD 32749;
        a1 := TRUNC(t);
        Ran1 := ABS(t/32749);
    END;          {ran1}

{_____}

    FUNCTION Ran2 : REAL;
    VAR
        t : REAL;
    BEGIN
        t  := (a2 * 10001 + 3 (MOD 17417);
        a2 := TRUNC(t);
        Ran2 := ABS(t/17417);
    END;          {ran2}

BEGIN {CombRandom}
{  f := Ran2; }              {select random generator}
  { IF (f > 0.5) then} CombRandom := Random;
{ ELSE CombRandom := Ran1;}
END;  {CombRandom}

{_____}
BEGIN
    init;                                      {initialize values}
    get_samp_data(npts, pin, val, fname);      {load val}
    rn := npts;                                {init rn}
    a1 := 1; a2 := 203;                        {init random variables}
    RANDOMIZE;

    {calculate initial stats}
    calc_stats (npts, pin, val, rn, pbar, pvar, pstd, pse);

    FOR i := 1 TO pass DO BEGIN                {sampling loop}
    samp_init (nct, n, pin, tn, tn2, rn, npts);    {init these values}

    sub_done := false;
    WHILE (NOT sub_done) DO BEGIN

        iele := 0; drn := npts;
        WHILE (iele <= 0) or (iele > npts) DO BEGIN
            r := combRandom;
            iele := round (drn * r);
        END;

        tn := tn + val[iele];        {accumulate random sample}
        tn2 := tn2 + sqr(val[iele]);
        rn := n;
        IF (val[iele] > 0.0) THEN pin := pin + 1;
```

```
{GREEN sampling plan}
tnsl := EXP (((LN(SQR(DPREC)/alpha)) / (beta - 2)) +
    ((beta - 1)/(beta - 2)*LN(n)));

{sample exceed boundary?}
IF (tn < tnsl)
THEN BEGIN
    n := n + 1;
    sub_done := false;
END
ELSE BEGIN
    nfrq[n] := nfrq[n] + 1;
    nmin := mino (n, nmin);
    nmax := maxo (n, nmax);

    tbar := tn / n;
    htbar := htbar + tbar;
    htbar2 := htbar2 + sqr(tbar);
    tvar :- (tn2 - sqr(tn) / rn) / (rn - 1);
    allvar := allvar + tvar;
    tstd := sqrt(tvar);
    allstd := allstd + tstd;
    tse := tstd / sqrt(rn);
    allse := allse + tse;
    dact := tse / tbar;
    hd := hd + dact;
    pi := round ((pin / n) * 100);
    pic := pic + pi;

    WRITELN (outfile,i:3);
    WRITELN (outfile,tbar:6:2);
    WRITELN (outfile,tvar:6:2);
    WRITELN (outfile,tstd:6:2);
    WRITELN (outfile,tse:5:2);
    WRITELN (outfile,dact:3:2);
    WRITELN (outfile,pi:3);
    WRITELN (outfile,n:3);
    aved := aved + dact;
    aven := aven + n;
    sub_done := true;
    END;
  END;       {sub_done loop}
 END;        {for loop}
show_cummul (pass, aven, pic, nfrq,  {print cumulative stats}
    htbar, allvar, allstd, allse, aved);

CLOSE (outfile);

END;       {main}
{_____}
BEGIN
END.
```

ACKNOWLEDGMENTS

Appreciation is extended to Andrew L. Sargent, Army High Performance Computing Research Center, University of Minnesota, for development and review of sequential sampling plan formulas based on confidence intervals or fixed positive values. I also thank David Andow, Department of Entomology, University of Minnesota, and Jan Nyrop, New York State Agricultural Experimental Station, Geneva, NY, for reviewing an earlier draft of this manuscript. This work was supported by the

College of Agriculture, University of Minnesota, and approved as publication 20,374 of the Minnesota Agriculture Experiment Station.

REFERENCES

1. Binns, M. R. and Nyrop, J. P., Sampling insect populations for the purpose of IPM decision making, *Annu. Rev. Entomol.*, 37, 427, 1992.
2. Wearing, C. H., Evaluating the IPM implementation process, *Annu. Rev. Entomol.*, 33, 17, 1988.
3. Hoy, C. W., Jennison, C., Shelton, A. M., and Andaloro, J. T., Variable intensity sampling: a new technique for decision making in cabbage pest management, *J. Econ. Entomol.*, 76, 139, 1983.
4. Hutchison, W. D., unpublished data, 1986.
5. Christensen, J. B., Gutierrez, A. P., Cothran, W. R., and Summers, C. G., The within-field spatial pattern of the larval Egyptian alfalfa weevil *Hypera brunneipennis* (Coleoptera: Curculionidae): an application of parameter estimates in simulation, *Can. Entomol.*, 109, 1599, 1977.
6. NAPIS, CAPS commentary, The Cooperative Agricultural Pest Survey Newsletter, Vol. 1.
7. Southwood, T. R. E., *Ecological Methods, with Particular Reference to the Study of Insect Populations*, Chapman & Hall, London, 1978, chap. 2.
8. Karandinos, M. G., Optimum sample size and comments on some published formulae, *Bull. Entomol. Soc. Am.*, 22, 417, 1976.
9. Hutchison, W. D., Hogg, D. B., Poswal, M. A., Berberet, R. C., and Cuperus, G. W., Implications of the stochastic nature of Kuno's and Green's fixed-precision stop lines: sampling plans for the pea aphid (Homoptera: Aphididae) in alfalfa as an example, *J. Econ. Entomol.*, 81, 749, 1988.
10. Kuno, E., A new method of sequential sampling to obtain the population estimates with a fixed level of precision, *Res. Popul. Ecol.*, 11, 127, 1969.
11. Nyrop, J. P. and Binns, M., Quantitative methods for designing and analyzing sampling programs for use in pest management, in *Handbook of Pest Management in Agriculture*, Vol. 2, Pimentel, D., Ed., CRC Press, Boca Raton, FL, 1991, 67.
12. Ruesink, W. G., Introduction to sampling theory, in *Sampling Methods in Soybean Entomology*, Kogan, M. and Herzog, D. C., Eds., Springer-Verlag, New York, 1980, 61.
13. Binns, M., Sequential estimation of the mean of a negative binomial distribution, *Biometrica*, 62, 433, 1975.
14. Myers, J. H., Selecting a measure of dispersion, *Environ. Entomol.*, 7, 619, 1978.
15. Taylor, L. R., Assessing and interpreting the spatial distributions of insect populations, *Annu. Rev. Entomol.*, 29, 321, 1984.
16. Iwao, S., A new regression method for analyzing the aggregation pattern of animal populations, *Res. Popul. Ecol.*, 10, 1, 1968.
17. Lloyd, M., Mean crowding, *J. Anim. Ecol.*, 36, 1, 1967.
18. Kuno, E., Sampling and analysis of insect populations, *Annu. Rev. Entomol.*, 36, 285, 1991.
19. Taylor, L. R., A natural law for the spatial disposition of insects, Proc. XII International Congress of Entomology, 1965, 396.
20. Taylor, L. R., Woiwod, I. P., and Perry, J. N., The density-dependence of spatial behaviour and the rarity of randomness, *J. Anim. Ecol.*, 47, 383, 1978.
21. Green, R. H., On fixed precision level sequential sampling, *Res. Popul. Ecol.*, 12, 249, 1970.
22. Allen, J., Gonzalez, D., and Gokhale, D. V., Sequential sampling plans for the bollworm, *Heliothis zea*, *Environ. Entomol.*, 1, 771, 1972.
23. Allsopp, P. G. and Bull, R. M., Spatial patterns and sequential sampling plans for melolonthine larvae (Coleoptera: Scarabeidae) in southern Queensland sugarcane, *Bull. Entomol. Res.*, 79, 251, 1989.
24. Baumgartner, J., Bieri, M., and Délucchi, V., Sampling *Acyrthosiphon pisum* (Harris) in pea fields, *Mitt. Schweiz. Ent. Ges.*, 56, 173, 1983.
25. Bechinski, E. J. and Pedigo, L. P., Population dispersion and development of sampling plans for *Orius insidiosus* and *Nabis* spp in soybeans, *Environ. Entomol.*, 10, 956, 1981.
26. Davis, P. M. and Pedigo, L. P., Analysis of spatial patterns and sequential count plans for stalk borer (Lepidoptera: Noctuidae), *Environ. Entomol.*, 18, 504, 1989.
27. Ekbom, B. S., Spatial distribution of *Rhopalosiphum padi (L.)* (Homoptera: Aphididae) in spring cereals in Sweden and its importance for sampling, *Environ. Entomol.*, 14, 312, 1985.
28. Elliot, N. C. and Kieckhefer R. W., Cereal aphid populations in winter wheat: spatial distributions and sampling with fixed levels of precision, *Environ. Entomol.*, 15, 954, 1986.

29. Finch, S., Skinner, G., and Freeman, G. H., The distribution and analysis of cabbage root fly egg populations, *Ann. Appl. Biol.*, 79, 1, 1975.
30. Finch, S., Skinner, G., and Freeman, G. H., Distribution and analysis of cabbage root fly pupal populations, *Ann. Appl. Biol.*, 88, 351, 1978.
31. Grout, T. G., Binomial and sequential sampling of *Euseius tularensis* (Acari: Phytoseiidae), a predator of citrus red mite (Acari: Tetranychidae) and citrus thrips (Thysanoptera: Thripidae), *J. Econ. Entomol.*, 78, 567, 1985.
32. Iperti, G., Lapchin, L., Ferran, A., Rabasse, J. M., and Lyon, J. P., Sequential sampling of adult *Coccinella septempunctata* in wheat fields, *Can. Entomol.*, 120, 773, 1988.
33. Jones, V. P. and Parrella, M. P., Dispersion indices and sequential sampling plans for the citrus red mite (Acari: Tetranychidae), *J. Econ. Entomol.*, 77, 75, 1984.
34. McAuslane, H. J., Ellis, C. R., and Allen, O. B., Sequential sampling of adult northern and western corn rootworms (Coleoptera: Chrysomelidae) in southern Ontario, *Can. Entomol.*, 119, 577, 1987.
35. Mollet, J. A., Trumble, J. T., Walker, G. P., and Sevacherian, V., Sampling schemes for determining population intensity of *Tetranychus cinnabarinus* (Boisduval) (Acarina: Tetranychidae) in cotton, *Environ. Entomol.*, 13, 1015, 1984.
36. Palumbo, J. C., Fargo, W. S., and Bonjour, E. L., Spatial dispersion patterns and sequential sampling plans for squash bugs (Heteroptera: Coreidae) in summer squash, *J. Econ. Entomol.*, 84, 1796, 1991.
37. Poston, F. L., Whitworth, R. J., Welch, S. M., and Loera, J., Sampling southwestern corn borer populations in postharvest corn, *Environ. Entomol.*, 12, 33, 1983.
38. Salifu, A. B. and Hodgson, C. J., Dispersion patterns and sequential sampling plans for *Megalurothrips sjostedti* (Trybom) (Thysanoptera: Thripidae) in cowpeas, *Bull. Entomol. Res.*, 77, 441, 1987.
39. Shepherd, R. F. and Otvos, I. S., Pest management of Douglas-fir tussock moth (Lepidoptera: Lymantridae): a sequential sampling method to determine egg mass density, *Can. Entomol.*, 116, 1041, 1984.
40. Smith, A. M. and Hepworth, G., Sampling statistics and a sampling plan for eggs of pea weevil (Coleoptera: Bruchidae), *J. Econ. Entomol.*, 1791, 1992.
41. Sweeney, J. D. and Miller, G. F., Distribution of *Barbara colfaxiana* (Lepidoptera: Tortricidae) eggs within and among Douglas-fir crowns and methods for estimating egg densities, *Can. Entomol.*, 121, 569, 1989.
42. Terry, L. I. and DeGrandi-Hoffman, G., Monitoring western flower thrips (Thysanoptera: Thripidae) in "Granny Smith" apple blossom clusters, *Can. Entomol.*, 120, 1003, 1988.
43. Terry, L. I., Bradley, J. R., Jr., and Van Duyn, J. W., *Heliothis zea* (Lepidoptera: Noctuidae) eggs in soybeans: within-field distribution and precision level sequential count plans, *Environ. Entomol.*, 18, 908, 1989.
44. Wada, T. and Kobayashi, M., Distribution pattern and sampling techniques of the rice leafroller, *Cnapharocrosis medinalis*, in a paddy field, *Jpn. J. Appl. Entomol. Zool.*, 29, 230, 1985.
45. Weinzierl, R. A., Berry, R. E., and Fisher, G. C., Sweep-net sampling for western spotted cucumber beetle (Coleoptera: Chrisomelidae) in snap beans: spatial distribution, economic injury level, and sequential sampling plans, *J. Econ. Entomol.*, 80, 1278, 1987.
46. Yano, E., Spatial distribution of greenhouse whitefly (*Trialeurodes vaporariorum* Westwood) and a suggested sampling plan for estimating its density in greenhouses, *Res. Popul. Ecol.*, 25, 309, 1983.
47. Zehnder, G. W. and Trumble, J. T., Sequential sampling plans with fixed levels of precision for *Liriomyza* species (Diptera: Agromyzidae) in fresh market tomatoes, *J. Econ. Entomol.*, 78, 138, 1985.
48. Boiteau, G., Bradley, J. R., Jr., Van Duyn, J. W., and Stinner, R. E., Bean leaf beetle: micro-spatial patterns and sequential sampling of field populations, *Environ. Entomol.*, 8, 1139, 1979.
49. Kolodny-Hirsch, D. M., Evaluation of methods for sampling gypsy moth (Lepidoptera: Lymantriidae) egg mass populations and development of sequential sampling plans, *Environ. Entomol.*, 15, 122, 1986.
50. Roux, O., von Arx, R., and Baumgartner, J., Estimating potato tuberworm (Lepidoptera: Gelechiidae) damage in stored potatoes in Tunisia, *J. Econ. Entomol.*, 85, 2246, 1992.
51. Smith, M. A. and Hepworth, G., Sampling statistics and a sampling plan for eggs of pea weevil (Coleoptera: Bruchidae), *J. Econ. Entomol.*, 85, 1791, 1992.
52. Allsopp, P. G., Ladd, T. L., Jr., and Klein, M. G., Sample sizes and distributions of Japanese beetles (Coleoptera: Scarabaeidae) captured in lure traps, *J. Econ. Entomol.*, 85, 1797, 1992.
53. Jones, V. P., Developing sampling plans for spider mites (Acari: Tetranychidae): those who don't remember the past may have to repeat it, *J. Econ. Entomol.*, 83, 1656, 1990.
54. Croft, B. A., Welch, S. M., and Dover, M. J., Dispersion statistics and sample size estimates for populations of the mite species *Panonychus ulmi* and *Amblyseius fallacis* on apple, *Environ. Entomol.*, 5, 227, 1976.

55. Zahner, P. H. and Baumgartner, J., Sampling statistics for *Panonychus ulmi* (Koch) (Acarina, Tetranychidae) and *Tetranychus urticae* Koch (Acarina, Tetranychidae) feeding on apple trees, *Res. Popul. Ecol.*, 26, 97, 1984.
56. Buntin, G. D. and Pedigo, L. P., Dispersion and sequential sampling of green cloverworm eggs in soybeans, *Environ. Entomol.*, 10, 980, 1981.
57. Harcourt, D. G., Mukerji, M. K., and Guppy, J. C., Estimation of egg populations of the alfalfa weevil, *Hypera postica* (Coleoptera: Curculionidae), *Can. Entomol.*, 106, 337, 1974.
58. Harcourt, D. G. and Binns, M. R., A sampling system for estimating egg and larval populations of *Agromyza frontella* (Diptera: Agromyzidae) in alfalfa, *Can. Entomol.*, 112, 375, 1980.
59. Snedecor, G. W. and Cochran, W. G., *Statistical Methods*, The Iowa State University Press, Ames, 1967.
60. Dowell, R. V. and Cherry, R. H., Detection of, and sampling procedures for, the citrus blackfly in urban southern Florida, *Res. Popul. Ecol.*, 28, 19, 1981.
61. Kuno, E., Multistage sampling for population estimation, *Res. Popul. Ecol.*, 18, 39, 1976.
62. Trumble, J. T., Brewer, M. J., Shelton, A. M., and Nyrop, J. P., Transportability of fixed precision level sampling plans, *Res. Popul. Ecol.*, 31, 325, 1989.
63. Sawyer, A. J., Inconsistency of Taylor's *b*: simulated sampling with different quadrat sizes and spatial distributions, *Res. Popul. Ecol.*, 31, 11, 1989.
64. Taylor, L. R., Woiwod, I. P., and Perry, J. N., The negative binomial as a dynamic ecological model for aggregation, and the density dependence of k, *J. Anim. Ecol.*, 48, 289, 1979.
65. Wilson, L. T., Estimating the abundance and impact of arthropod natural enemies in IPM systems, in *Biological Control in Agricultural IPM Systems*, Hoy, M. A. and Herzog, D. C., Eds., Academic Press, Orlando, FL, 1985, 303.
66. Shelton, A. M., Nyrop, J. P., and Foster, R. E., Distribution of European corn borer (Lepidoptera: Pyralidae) egg masses and larvae on sweet corn in New York, *Environ. Entomol.*, 15, 501, 1986.
67. Banerjee, B., Variance to mean ratio and the spatial distribution of animals, *Experientia*, 32, 993, 1976.
68. Trumble, J. T., Oatman, E. R., and Voth, V., Temporal variation in the spatial dispersion patterns of aphids (Homoptera: Aphididae) infesting strawberries, *Environ. Entomol.*, 12, 595, 1983.
69. Gutierrez, A. P., Summers, C. G., and Baumgartner, J., The phenology and distribution of aphids in California alfalfa as modified by ladybird beetle predation (Coleoptera: Coccinellidae), *Can. Entomol.*, 112, 489, 1980.
70. Trumble, J. T., Implications of changes in arthropod distribution following chemical application, *Res. Popul. Ecol.*, 27, 277, 1985.
71. Taylor, R. A. J., On the accuracy of insecticide efficacy reports, *Environ. Entomol.*, 16, 1, 1987.
72. Anscombe, F. J., Large sample theory of sequential estimation, *Biometrika*, 36, 455, 1949.
73. Shufran, K. A. and Raney, H. G., Influence of inter-observer variation on insect scouting observations and management decisions, *J. Econ. Entomol.*, 82, 180, 1989.
74. Barney, R. J. and Legg, D. E., Development and validation of sequential sampling plans for alfalfa weevil (Coleoptera: Curculionidae) larvae using a six-stem, 100-square-meter subsampling method, *J. Econ. Entomol.*, 81, 658, 1988.
75. Efron, B. and Tibshirani, R., Bootstrap methods for standard errors, confidence intervals, and other measures of statistical accuracy, *Stat. Sci.*, 1, 54, 1986.
76. Willson, L. J. and Young, J. H., Sequential estimation of insect population densities with a fixed coefficient of variation, *Environ. Entomol.*, 12, 669, 1983.
77. Rudd, W. G., Sequential estimation of soybean arthropod densities, in *Sampling Methods in Soybean Entomology*, Kogan, M. and Herzog, D. C., Eds., Springer-Verlag, New York, 1980, 94.
78. Ferro, D. N., Integrated pest management in vegetables in Massachusetts, in *Successful Implementation of Integrated Pest Management for Agricultural Crops*, Leslie, A. R. and Cuperus, G. W., Eds., Lewis Publishers, Boca Raton, FL, 1993, 6.
79. Frisbie, R. E. and Smith, J. W., Jr., Biologically intensive integrated pest management: the future, in *Progress and Perspectives for the 21st Century*, Menn, J. J. and Steinhauer, A. L., Eds., Entomological Society of America, Lanham, MD, 1989, 151.
80. Hutchison, W. D., Beasley, C. A., Martin, J. M., and Henneberry, T. J., Timing insecticide applications for pink bollworm (Lepidoptera: Gelechiidae) management: comparison of egg and larva treatment thresholds, *J. Econ. Entomol.*, 84, 470, 1991.
81. Steffey, K. L. and Gray, M. E., Extension-research synergism: enhancing the continuum from discovery to delivery, *Am. Entomol.*, 38, 204, 1992.
82. Nyrop, J. P., Foster, R. E., and Onstad, D. W., Value of sample information in pest control decision making, *J. Econ. Entomol.*, 79, 1421, 1986.
83. Foster, R. E., Tollefson, J. J., Nyrop, J. P., and Hein, G. L., Value of adult corn rootworm (Coleoptera: Chrysomelidae) population estimates in pest management decision making, *J. Econ. Entomol.*, 79, 303, 1986.

Chapter 11

SAMPLING TO PREDICT OR MONITOR BIOLOGICAL CONTROL

Jan P. Nyrop and Wopke van der Werf

TABLE OF CONTENTS

I. Introduction 246

II. Predicting Biological Control via Prey-Natural Enemy Ratios 246

III. Monitoring Population Density Through Time 251
    A. Cascaded Tripartite Sequential Classification 252
    B. Binomial Count Tripartite Plans 260
    C. Adaptive Frequency Classification Monitoring 262

IV. Summary and Conclusions 269

Appendix 269

References 336

I. INTRODUCTION

A primary objective of integrated pest management is to move from crop protection systems that rely on broad spectrum pesticides to more environmentally benign systems built upon target-specific chemical management tools, host plant resistance, cultural practices, and biological control. Sampling and pest control decision making are foundations of most integrated pest management systems. Well-designed pest control decision rules can reduce the need for prophylactic chemical pesticide use and thereby assist in establishing environments wherein biological control agents can survive.

When making decisions about the need for pesticide use, it often suffices to estimate or classify pest density at a single point in time. If the phenology of a population is uncertain, it may be necessary to repeat sampling; however, a decision is still sought concerning pest density over a relatively narrow time span. Most currently used pest control decision rules are designed to meet this need.

With greater reliance on biological control, desired information on population density changes. While it is still necessary to ascertain that density remains below some critical level where economic injury occurs, it usually will be necessary to either make this determination many times during some period of interest, say a growing season, or to make a prediction about it. Conventional classification or estimation sampling procedures can be used to repeatedly sample a population through time; however, this usually will not be an effective use of sampling resources. In this chapter, we present methods that can be used to either predict, based on the ratio of pest to natural enemy, that a population will remain below a critical level, or to parsimoniously sample a population through time to answer the same question.

Little work has been done in the area of sampling for predicting or monitoring biological control.[1] We describe one method that can be used for prediction and two that can be used for monitoring. The first procedure is based on using the ratio of pests to natural enemies to predict the likelihood for biological control. Sequential methods for classifying pest-natural enemy ratios have previously been described.[2] Here, we review this work and present a FORTRAN computer program that can be used to design and analyze these sampling plans. The second and third methods are based on sampling a population repeatedly through time by using information on expected population growth along with sample information on present density to schedule future sampling. FORTRAN computer programs for designing and analyzing sampling protocols founded on this approach are also provided.

We will illustrate the methods using data that describe sampling distributions for European red mite, *Panonychus ulmi* (Koch), and its predator *Typhlodromus pyri* (Scheuten) in New York apples.[1,3]

II. PREDICTING BIOLOGICAL CONTROL VIA PREY-NATURAL ENEMY RATIOS

The ratio of pest to natural enemy, R, can sometimes be used to determine whether natural enemies are sufficiently abundant to control pest population growth.[4] If R is less than some critical ratio CR, value, it is assumed biological control will ensue. In addition to the ratio R, pest density, m_N, in relation to an intervention threshold, T, may also be of interest. If m_N exceeds T, some immediate action may be required. The critical ratio might also be conditional on m_N so that if m_N exceeds T, biological control might not occur even if $R \leq$ CR.

Consideration of R in relation to CR and m_N in relation to T results in four regions into which the pest and natural enemy populations can be jointly classified

(Figure 1). In region 1, $R \leq \text{CR}$ and $m_N \leq T$. Under these conditions biological control is likely to occur. In region 2, $R \leq \text{CR}$ and $m_N > T$. As a result, the pest may be regulated by natural enemies; however, significant plant damage is likely to occur before this happens. In region 3, $R > \text{CR}$ and $m_N \leq T$. With these conditions biological control may not occur; however, intervention is not needed now and another sample should be taken at a later date. Finally, in region 4 $R > \text{CR}$ and $m_N > T$. Under these conditions immediate intervention is necessary.

A sequential procedure for simultaneously classifying R and m_N can be constructed using confidence limits about estimates of the ratio and pest density. Stop

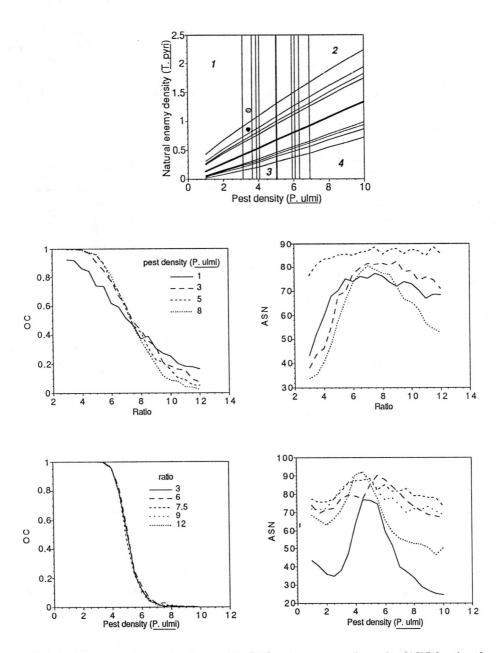

FIGURE 1. Stop lines and operating characteristic (OC) and average sample number (ASN) functions for a sequential ratio classification sampling plan.

limits are comprised of two sets, one with which the estimated prey density is compared and one with which the estimated predator density is compared. The stop limits for the prey are based on confidence limits about T. If an estimated density exceeds the upper confidence limit, sampling is terminated, and the population classified as $> T$. Conversely, if the estimated density is less than the lower confidence limit, density is classified as $\leq T$. Sampling continues until one of the stop boundaries is crossed. Binns (Chapter 8) provides a complete description of these stop lines.

Stop limits for the ratio classification work as follows: from a sample of size n_i the pest (m_N) and natural enemy (m_D) densities are estimated. The subscripts N and D denote numerator and denominator of the ratio. Both the sample size and m_N are used to determine the upper and lower ratio stop limits (ULR and LLR). Classification decisions are made by comparing the natural enemy density to ULR and LLR according to the following rules: If $m_D <$ LLR stop sampling and classify R as $>$ CR. If $m_D >$ ULR stop and classify R as $\leq$ CR. If neither of these conditions are met, take another sample of n_i observations, calculate the means (m_N and m_D) based on all the observations, and repeat the comparison. The stop limits are the densities of natural enemies (denominator of the ratio) given sample size n and prey density m_N that produce confidence limits for an estimated ratio which are greater than (LLR) and less than (ULR), the critical ratio CR. It is important to note that when both ratio and prey (pest) density are being classified, stop boundaries for both parameters must be crossed before sampling is terminated. A batch sampling procedure where n_i samples are examined before estimated means are compared with stop lines is used for two reasons. First, without batch sampling an unwieldy number of stop limits must be created because there is a new set of limits for each sample size. Second, batch sampling with $n_i \geq 20$ is currently necessary for determination of the operating characteristic (OC) and average sample number (ASN) functions. These functions must be simulated and require generation of correlated bivariate random deviates. Normal bivariate deviates are easily computed and with $n_i \geq 20$ the assumption of normality should be reasonably robust. It is not clear how correlated bivariate random variables that follow some other distribution might be simulated. The entire procedure is truncated so that after a total of n_t observations, a classification is made by comparing the estimated ratio to CR and estimated pest density to T.

The equation

$$r\left[\left(1 - z_{\alpha/2}^2 C_{ND}\right) \pm \frac{z_{\alpha/2}^2 \sqrt{(C_N + C_D - 2C_{ND}) - z_{\alpha/2}^2(C_N C_D - C_{ND}^2)}}{\left(1 - z_{\alpha/2}^2 C_D\right)}\right] \quad (1)$$

is used to compute ratio confidence limits.[5] In this equation the subscripts N and D again denote the numerator (pest) and denominator (natural enemy) of the ratio. The other variables are defined as follows: $C_N = s_N^2/(nm_N^2)$, $C_D = s_D^2/(nm_D^2)$, $C_{ND} = \rho_{ND} s_N s_D/(nm_D m_N)$, $s^2 =$ the sample variance, m_N and m_D are the estimated means for the numerator (pest) and denominator (natural enemy), ρ_{ND} is an estimated correlation coefficient between the numerator and denominator, $r = m_N/m_D$, and $z_{\alpha/2}$ is a standard normal deviate such that $P(Z \leq z_{\alpha/2}) = \alpha/2$.

Calculation of stop limits using Equation 1 requires that the correlation and variances for the numerator and denominator are known or can be written in terms of the means and the sample size. The variances usually can be modeled precisely as a function of the respective means using Taylor's power law (TPL), $s^2 = am^b$.[6] It also may be possible to model the correlation as a function of the means, or more simply, it may be a constant.

With the correlation specified and the variances modeled in terms of the means, stop lines can be determined for any value of n and m_N. To calculate ULR, values for n and m_N are chosen, and a variable x is initially set to m_N/CR. The upper confidence limit for $R = m_N/x$ is calculated and compared to CR. If CR is greater than or equal to the confidence limit, $\text{ULR} = x$; otherwise, x is increased by some value d, the confidence limits are calculated for a new R, and the comparison is made again. This procedure is repeated until CR is greater than or equal to the upper confidence interval. If the LLR are to be calculated, the procedure is identical with the exception that a lower confidence limit is calculated, the comparison criterion is that CR is less than or equal to the lower confidence limits and x is decreased by some value d. By setting d to a small value (ca., 0.01) x's can be found so that the confidence intervals are approximately equal to CR. Program "ratio" performs these calculations (see Appendix).

OC and ASN functions are constructed using simulation. Because both the pest-natural enemy ratio and the pest density are simultaneously classified, there are two sets of OC and ASN functions. For the ratio, the null hypothesis is that there are sufficient natural enemies for biological control, and hence the OC is the probability of accepting this hypothesis given any true ratio and a fixed pest density. For the pest density, the null hypothesis is that the density is less than the intervention threshold and the OC is the probability of accepting this hypothesis given any true pest density and a fixed ratio. The two sets of OC and ASN functions can be studied two ways. First, sets of OC and ASN functions can be plotted where each member of the set is indexed by a ratio or pest density (Figure 1). Alternately, the probability of making an incorrect classification (PIC) and ASN given any true ratio and pest density can be determined and plotted using an x-y-z ordinate system (Figure 2). The PIC function is a convenient way of representing the probability of an incorrect decision; however, three-dimensional figures are often difficult to interpret. To construct any of these functions, sampling is simulated from two jointly distributed populations: one representing the natural enemies and one representing the pest. Means from n_i observations are assumed to be bivariate normally distributed with variances defined as s^2/n_i and s^2 modeled using TPL. The FORTRAN program "ratio" computes stop lines, ASN, OC, and PIC functions (see Appendix).

To illustrate the method we use information on the sampling distribution of *P. ulmi* and *T. pyri*. Parameters for TPL for *P. ulmi* are $a = 4.32$ and $b = 1.2$. The same parameters for *T. pyri* are 2.38 and 1.2, respectively. Correlation between counts of this pest and natural enemy are variable; however, a conservative estimate is -0.25. Confidence intervals for a ratio become wider as the correlation decreases. Therefore, when a range of correlation coefficients can apply, use of the smallest value will produce the most conservative stop lines. The CR for this system is approximately 7.5, and one intervention threshold for *P. ulmi* is 5.0 per leaf. Stop lines were constructed using z values of 1.28 for both the ratio and prey density limits. The batch sample size (n_i) was set to 20, the maximum sample size was 100, and d was set to 0.01.

Stop lines are shown in Figure 1. Vertical lines are used with the pest density and those that run at approximately 45° from the origin are used with the natural enemy density. There are four sets of thin lines for the prey and natural enemy. Each set corresponds to a sample size of 20, 40, 60, or 80, with the interval between the lines becoming narrower as the sample size increases. The thick line in the center of each set of stop lines is used if the maximum sample size is reached by comparing the estimated mean to it.

An example will clarify use of the stop lines. Suppose a sample of 20 leaves is taken resulting in estimated pest and predator means of 3.4 and 0.85, respectively. This

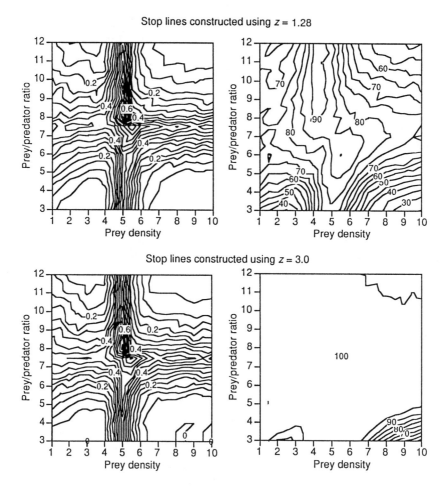

FIGURE 2. Probability of incorrect classification (PIC) and average sample number functions (ASN) for a ratio sequential classification sampling plan.

point is shown in Figure 1 as a dark circle; it lies to the right of the left-most stop limit for classifying pest density and below the uppermost limit for classifying the ratio. As a result, another sample must be taken. Suppose that after another sample of 20 leaves, the estimated density for pest and predator from the joint sample of 40 leaves was 3.5 and 1.2. This point is shown in Figure 1 as a shaded circle; it lies above the second stop line for the ratio (corresponding to a sample size of 40) and to the left of the second stop line for the pest. Sampling is terminated with the decision that the pest density is less than the intervention threshold and there are sufficient predators for biological control.

Also shown in Figure 1 are conditional OC and ASN functions for this sampling plan. Recall that the OC functions for the ratio are conditional on a particular pest density and OC functions for the pest are conditional on a ratio value. The OC functions for the ratio become steeper and improve at higher prey densities. This is because the variance of the ratio decreases with increasing prey density and because more samples are taken when the prey density is close to the intervention threshold. The OC functions for the prey are largely independent of the ratio; however, ASN values plotted in relation to prey density are greatly influenced by the ratio value. As expected, more samples are required to make a decision when the ratio is close to CR.

Shown in Figure 2 are three-dimensional plots of the ASN and PIC functions for the sampling plan described above as well as one where the z values were set to 3.0. Increasing z makes the stop limits wider. For $z = 3.0$, this results in a larger ASN and in many cases use of the maximum sample size of 100. The two sampling plans are compared to show that use of the sequential procedure with $z = 1.28$ results in approximately the same PIC as that for a plan based on $z = 3.0$ and with considerable savings in sample size.

The greatest drawback to using sequential ratio classification is that complete enumeration of counts is necessary. For many small organisms such as mites this is nearly impossible under field situations. As a result, sampling procedures for classifying or estimating densities of small, prolific organisms have been developed that substitute presence or absence of animals on sample units for enumeration. This also should be done for the ratio classification procedure and, if accomplished, would provide a tool that would be more readily used. It is straightforward to modify the stop lines so that they are expressed in terms of presence-absence counts. However, it is not clear how random variables that correctly model all sources of variation should be generated during simulations to determine performance of ratio classification schemes based on presence-absence counts.

III. MONITORING POPULATION DENSITY THROUGH TIME

The ratio of a pest to natural enemy cannot always be used as an index of the likelihood for biological control. At least three situations may result in this index being inappropriate or not usable. First, the CR, the ratio for which biological control can be expected, may not be constant with prey or predator density or other factors such as weather may influence this index. Second, more than one natural enemy species may be involved in the interaction and the relative species mix may influence the CR. For example, at least three phytoseiid predators can be found in commercial New York apple orchards (*T. pyri*, *T. longipilus*, and *A. fallacis*), as well as two stigmaeiid predators. This predator complex often provides effective biological control; however, a single CR for this complex is not tenable. Finally, it may not be practical to measure natural enemy density. Again, referring to the mite predator-prey (pest) system in apples, densities of predaceous mites can be estimated, although it is often difficult to do so in the field because phytoseiid predator mites superficially resemble their prey and are often concealed along the midrib of leaves. In other situations predators may be difficult to find or may be very mobile. When pest and natural enemy densities cannot both be estimated, ratios cannot be used, and another method for measuring the effectiveness of biological control is required.

One solution is to estimate or classify pest population density through time, and as long as the density remains below a threshold density that dictates intervention, biological control is effective. Using this scenario, it is not necessary to estimate or classify natural enemy density. However, it is often not practical or necessary to sample a pest population frequently through time. If, for example, a pest's density was much less than an intervention threshold, the population should not have to be sampled again as soon as a population whose density was only slightly less than an intervention threshold. What is needed is a sampling scheme that allows rapid classification of population density at a particular point in time, and that provides a measure of when the population should be resampled if the current density is less than the intervention threshold.

If knowledge of population growth is available, sample information can be combined with this knowledge to forecast future density. Such forecasts can be used as a basis for scheduling future samples. Wilson[7] developed such a system that used estimates of pests and natural enemies and nonlinear regression to determine when the next sample should be taken. Here, we present two methods based on sequential classification of density and simple forecasts of population growth. Each of these methods can be viewed as belonging to a family of potential monitoring protocols that do two things. First, these protocols determine whether a pest density exceeds an intervention threshold. Second, if the density is less than the threshold, the sampling protocols determine how long one can wait before sampling the pest population again and be reasonably sure that density will not have grown so that it exceeds an intervention threshold. The first method makes use of tripartite sequential classification (TSC). Binns (Chapter 8) describes this sampling method in detail. We briefly review the technique to maintain continuity in the presentation. The second method uses both sequential classification and fixed sample size estimation. Both procedures cascade individual sampling plans through time to monitor a population trajectory.

A. CASCADED TRIPARTITE SEQUENTIAL CLASSIFICATION

Tripartite sequential classification sampling plans can be used to classify population density into one of three categories that are defined by two critical densities; cd_1 and cd_2, where $cd_1 < cd_2$. The three categories defined by the two critical densities are $u \leq cd_1$, $cd_1 < u \leq cd_2$, and $u > cd_2$ where u is the population mean. Based on the two critical densities, two dichotomous sequential classification (DSC) sampling plans are constructed. The dichotomous plans could be based on one of several different sequential classification schemes. We will use Wald's[8] sequential probability ratio test (SPRT). Using this test, the two dichotomous classification plans are constructed with the following null and alternate hypotheses:

DSC$_1$
$H_{10}: u = m_{10}$
$H_{11}: u = m_{11}, m_{10} < m_{11}$

DSC$_2$
$H_{20}: u = m_{20}$
$H_{21}: u = m_{21}, m_{20} < m_{21}$

Normally $cd_1 = (m_{10} + m_{11})/2$ and $cd_2 = (m_{20} + m_{21})/2$. It is also required that $m_{20} > m_{11}$. The mean density is classified as less than cd_i if H_{i0} is accepted and greater than cd_i if H_{i0} is rejected.

When two DSC sampling plans are used simultaneously to form a tripartite sequential classification plan, composite stop lines are formed as shown in Figure 3. When using a TSC plan, sampling can be terminated with one of three possible decisions: (1) if samples fall in region 1 the decision is to accept H_{10}, which leads to the conclusion that $u \leq cd_1$; (2) if the samples fall in region 2, the decision is to accept H_{20} and concurrently reject H_{10}, leading to the conclusion that $cd_1 < u \leq cd_2$; and (3) if the samples fall into region 3, the decision is to reject H_{20}, leading to the conclusion that $u > cd_2$.

With a DSC plan there are two performance criteria: OC and ASN. The OC is defined as the probability of accepting the null hypothesis given any true mean. With a TSC plan there are three possible decisions and three corresponding probabilities of making these decisions given any true mean ($pdec_i$, $i = 1, 2, 3$). Given any two $pdec_i$, the third is one minus the sum of these two. As with a DSC plan, an ASN also exists for a TSC plan; however, it is not monotonically convex.

Tripartite sequential classification sampling schemes can be used to monitor a population through time by allowing the time interval between samples to be indexed by the less-than-threshold classification. For example, if a population was classified as

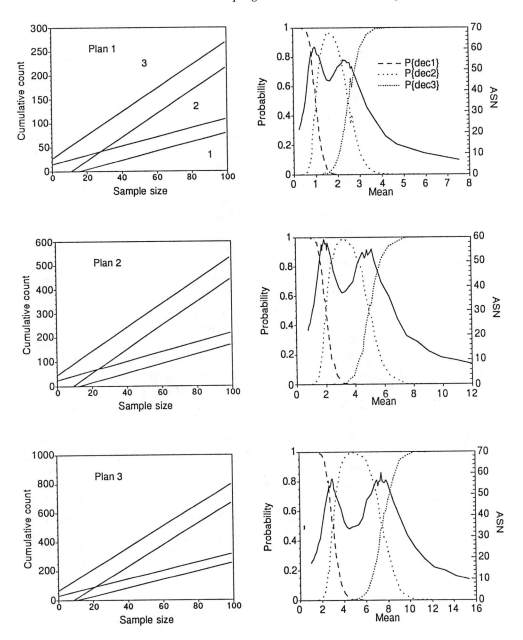

FIGURE 3. Stop limits and performance of three tripartite sequential classification sampling plans.

$\leq cd_1$ in Figure 3, the population might not be resampled for some specified time interval. If, however, it was classified as $> cd_1$ and $\leq cd_2$, it might be resampled again in a much shorter time interval. One way of determining the critical densities is to let cd_2 be the intervention threshold and to let cd_1 be a density, which, if allowed to grow unchecked for a period of time, would result in a density no greater than cd_2. The time period is the longest possible time to the next sample. If a density was classified as $> cd_1$ and $\leq cd_2$, the next sample might be taken one half of the specified time interval in the future. It is useful to call the shortest time interval between samples the sample interval (*sint*). Thus, if the mean is classified as $\leq cd_1$, the time period to the next sample will be some multiple of *sint*, and if the mean is classified as $> cd_1$ and $\leq cd_2$, the time period to the next sample is *sint*. Different TSC plans can be used at different times when sampling the same population process. When one or more TSC plans are used to monitor a population in the manner described above, we will refer to the process as cascaded tripartite sequential classification sampling. The FORTRAN program "sprt" (see Appendix) can be used to construct and analyze TSC plans based on Wald's SPRT.

The performance of tripartite classification sampling plans, cascaded through time, can be evaluated using four criteria that describe expected performance when sampling the entire population process. In addition, the outcome of each sampling bout (i.e., each time the population density is sampled) also can be examined. The four overall performance criteria are an operating characteristic, a total average sample size, the expected number of sampling bouts, and expected loss. These criteria have their basis in DSC sampling plans; however, they differ in that they refer to a trajectory of population density through time and not to specific density values. The OC is defined as the probability of not intervening over all sample bouts. The total average sample size (TASN) is the expected number of samples required to sample the population process over the time of interest. The expected number of sampling bouts (ESB) is the average number of times the population process must be sampled. Finally, the expected loss (EL) is the expected cumulative density allowed without intervention. With DSC sampling, densities allowed without intervention have a probability of occurring defined by the OC function. When tripartite plans are cascaded through time, a decision to not intervene at one point in time may still allow for intervention at the next sampling time, albeit with a different density. An expected loss function for cascaded plans must account for this. With this introduction, we will now define these performance criteria mathematically.

Because population density is being sampled through time, some variables are time dependent and for these variables, time are denoted by a subscript. To keep the notation simple, it is necessary to consider two time scales. The first and obvious one is chronological time. The second time scale that will be convenient to use is the one for sample bouts (i.e., 1, 2, 3 . . .). For example, it will be clearer to write the density at sampling bout two as d_2 instead of d_{23} where 23 specifies the chronological time when sampling bout two occurred.

The population process to be sampled is defined by d_t where d is density and t is time. During the time interval of interest there are a known set of times (st) during which sampling might take place defined by a starting point (i.e., st = 1), a fixed time interval between samples (*sint*), and an ending point (st = end). We use the variable *sb* to denote the sampling bout where *sb* belongs to the set {1, 2, 3, . . . , last}. For all combinations of density and time there exist functions that describe the outcome of sampling. These functions are the probability of decision functions (pdec*i*) and (ASN) function previously described for tripartite plans. As a review, *pdec*1 is the probability of no intervention and waiting two or more time intervals (*sint*) before sampling again, *pdec*2 is the probability of no intervention and waiting one sample time interval before sampling again, and *pdec*3 is the probability of intervening. The loss at any sampling time is defined as the cumulative density to that time point:

$$l_t = \int_0^t d_x \qquad (2)$$

At the first sampling time, the density is d_1, where the temporal subscript now refers to the sample bout, the probability of sampling at this time is one $ps_1 = 1$), the probability of each of the three decisions is given by *pdeci*$_1$ where the subscript denotes the sampling bout, and the loss is given by l_1. At the second sampling time the density is d_2, the probability of sampling at this time is $ps_2 = pdec2_1(ps_1)$, and the probability of each of the three decisions is given by $ps_2(pdeci_2)$. These relationships apply because sampling at time two can only occur if a decision was made to resample after one time interval (*sint*) during the first sample bout and because the probability of reaching a particular decision during the second sampling bout must be conditioned on the probability of sampling at that time. At the third sampling time the density is d_3, the probability of sampling at this time is $ps_3 = pdec2_2(ps_2) + pdec1_1(ps_1)$, and the probability of each of the three decisions is given by $ps_3(pdeci_3)$. Sampling at time three can only occur if a decision was made to resample after one time interval during the second sample bout or if a decision was made to resample after two time-intervals during the first sample bout. Hence, the probability of sampling at the third sampling time is the sum of these two probabilities. The probability of sampling at periods 3 through last can be generalized as:

$$p_{sb} = pdec2_{sb-1}(ps_{sb-1}) + pdec1_{sb-2}(ps_{sb-2}) \tag{3}$$

Note that this equation will apply to all sampling periods provided $ps_0 = 1$, $pdec2_0 = 1$, $ps_{-1} = 0$, and $pdec1_{-1} = 0$.

The OC, total ASN, and ESB are now calculated as:

$$OC = 1 - \sum_{1}^{last} ps_{sb}(pdec3_{sb}) \tag{4}$$

$$TASN = \sum_{1}^{last} ps_{sb}(asn_{sb}) \tag{5}$$

$$ESB = \sum_{1}^{last} ps_{sb} \tag{6}$$

The OC and TASN are weighted sums of the respective values at each sample bout with the weights determined by the probability of sampling. The ESB is simply the sum of the probability of sampling because the value being weighted is one.

The last performance measure to be computed is EL. Recall that this is the expected cumulative density allowed to occur without intervention. At any sampling bout a measure of loss would be the cumulative density to that point times the product of the probability of sampling and the probability of intervening. Summing these loss measures over all sampling bouts provides a measure of loss when sampling the entire population process. This sum is easily computed provided two endpoints are considered. First, if the last sample period is reached, then the loss at that time is simply the probability of sampling times the loss. Second, at the next to last sampling period a decision to wait two time periods to sample again is equivalent to not intervening because sampling will not be repeated. Thus, the loss that would occur at the last sampling time should apply to these cases. The following equation accounts for these endpoints:

$$EL = \sum_{1}^{last-1} [l_{sb} ps_{sb}(pdec3_{sb})] + (ps_{end} l_{end}) + (ps_{end-1} l_{end} pdec1_{end-1}) \tag{7}$$

The measure of loss we have proposed is an expected value and therefore masks extreme values and provides no information on the distribution of possible values.

Loss also should be considered in terms of extreme values and in fact, managers may be more interested in the likelihood of a particular large cumulative density occurring when using a particular sampling plan than an expected value. Such values can be computed using the cumulative probability of intervening ($spdec3_t$) and the time dependent loss.

The cumulative probability of intervening is calculated as:

$$spdec3_{sb} = \sum_{1}^{sb} ps_j\, pdec3_j \quad (8)$$

A loss can be calculated for each sb entry and, therefore, a loss value for any specific $spdec3$ can be determined using interpolation.

The FORTRAN program "cascade" (see Appendix) can be used to evaluate the performance of TSC plans cascaded through time.

We illustrate the use and investigate some of the properties of cascaded TSC plans by studying plans developed for sampling *P. ulmi* in apples. Three TSC plans were constructed for use at different times during the period June 1 to August 31. Two sequential probability ratio tests based on the negative binomial distribution formed the basis for each TSC plan.

The following thresholds were used to define the DSC plans built around cd_2: June 1 to June 30—2.5 motile mites per leaf, July 1 to July 31—5.0 motiles per leaf, and August 1 and after—7.5 motiles per leaf. The cd_1 used to develop the second set of DSC plans were calculated as the densities that would result in no more than the intervention threshold density after 14 d, assuming the population grows exponentially with a growth rate of 0.065 ($N_t = N_0 e^{0.065t}$ where $t = 14$). This growth rate was an average determined by fitting an exponential model to 14 data sets that described *P. ulmi* dynamics. Thus, we set the minimum time interval to the next sample ($sint$) to 7 d and the resample interval if the mean was classified as $\leq cd_1$ to 14 d. Parameters used to construct the sampling plans are shown in Table 1.

Stop lines and probability of decision ($pdeci$) and ASN functions for these sampling plans are shown in Figure 3. Maximum ASN values occur when the density equals cd_1 or cd_2. These are also the points where $pdeci$ functions intersect. The jaggedness of the functions in Figure 3 occurs because the functions were estimated using simulation ($n = 500$).

TABLE 1
Parameters Used to Construct Tripartite Sequential Classification Sampling Plans for Use with *P. ulmi*

Plan[a]	cd_i[b]	k[c]	H_0	H_1	Alpha	Beta
1.1	1.0	0.301	0.7	1.3	0.1	0.1
1.2	2.5	0.467	2.0	3.0	0.15	0.15
2.1	2.0	0.418	1.6	2.4	0.15	0.15
2.2	5.0	0.667	4.2	5.8	0.15	0.15
3.1	3.0	0.513	2.3	3.7	0.1	0.1
3.2	7.5	0.827	6.5	8.5	0.15	0.15

[a] For each plan there are two thresholds and two sets of SPRT parameters. The number following the decimal point signifies whether it is plan 1 or 2 of a tripartite scheme.
[b] The first value for each plan is cd_1 and the second is cd_2.
[c] Parameter k for negative binomial distribution computed using moments and based on a predicted variance calculated using the model $s^2 = am^b$.

Performance of the sampling plans cascaded through time was studied by applying the plans to a set of populations described by exponential growth ($r = 0.03$ to 0.075) and to seven actual population trajectories. The first sampling plan was applied during the time period 1 to 30, sampling plan 2 was applied during the period 31 to 60, and sampling plan 3 was used during the period 61 to end. Performance of cascaded dichotomous sampling plans was also studied. This was done to determine the saving in sampling costs that would result and what errors might result from using the tripartite plans. These dichotomous plans were constructed using the parameters for the tripartite plans based on cd_2 shown in Table 1. These plans were cascaded in time by applying the sampling plans every 7 d.

OC values obtained with the cascaded tripartite plans are shown adjacent to each population trajectory in Figure 4. Also shown in this figure are the time-dependent intervention thresholds. The OC is very steep when densities exceed the intervention threshold (cumulative density of 150 to 180). In fact, the OC is steeper than might be predicted based on the *pdec3* values for each tripartite sampling plan. This is because the OC for cascaded plans is a summation of probabilities over all sample bouts; the sum of *pdec3* values much less than one over a set of sampling bouts can still lead to large (i.e., close to 1.0) overall probability of intervention. A numerical example using a cascaded dichotomous plan will clarify this. Suppose sampling occurred three times where the densities were 2.5, 6, and 8 and the probabilities of intervening at these times (1 − OC) were 0.4, 0.5, and 0.6. The probability of sampling at the first time is 1.0, so the probability of sampling at time 2 is $1(1 - 0.4) = 0.6$. The cumulative probability of intervening at the second sampling time is $0.4 + (0.5 * 0.6) = 0.7$. The probability of sampling at the third time period is $0.6(1 - 0.5) = 0.3$ and the cumulative probability of intervening is $0.7 + (0.3 * 0.6) = 0.88$.

The four overall performance criteria for the four sampling plans are shown in Figure 5. In addition to expected loss, losses that would occur with probability 0.2 and 0.05 are also shown. The OC for tripartite and dichotomous plans are essentially equal. Slightly more total samples were required by the dichotomous plans. More importantly, the dichotomous plans required approximately twice the number of bouts

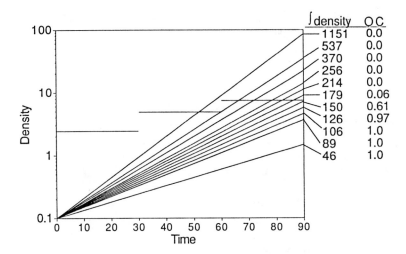

FIGURE 4. Populations with exponential growth, cumulative density, and OC values for cascaded tripartite sequential classification sampling plans applied to the populations. Short horizontal lines are intervention thresholds.

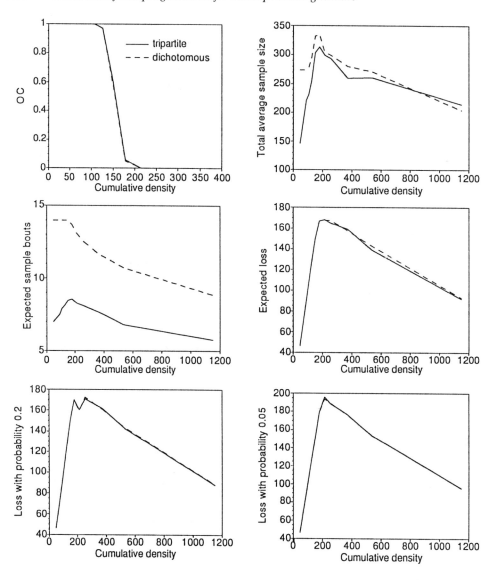

FIGURE 5. Performance of cascaded tripartite and dichotomous sampling plans.

as the tripartite plans. Because sample bouts are usually more costly than samples per bout, tripartite plans should significantly reduce overall sampling costs. Losses for the tripartite and dichotomous plans were nearly identical. Intervention thresholds for *P. ulmi* are designed to prevent a cumulative density of 500. This was always achieved with at least probability 0.95. As a result, the tripartite plans may, in fact, be overly conservative.

The seven population trajectories sampled are shown in Figure 6. Sampling was started at time 12 and the last sample was taken at time 89. Results of applying the tripartite and dichotomous sampling plans to these populations are given in Table 2.

Performance of the tripartite and dichotomous plans was similar except for the total number of samples and the number of bouts which were both larger for the dichotomous plans. Use of the tripartite plans reduced the number of bouts by one half to one third. With populations 1, 2, 4, and 6 intervention always occurred. These

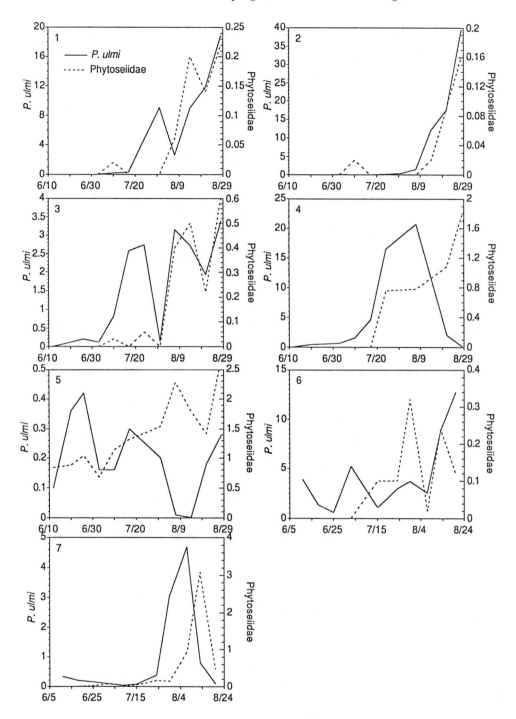

FIGURE 6. Dynamics of seven *P. ulmi*-phytoseiid mite populations.

populations all had cumulative densities at or near the level for which damage occurs (500). With population 4 biological control occurred; however, not before a high density of *P. ulmi* was present. For populations 3, 5, and 7, biological control was successful and there was no probability of intervention.

TABLE 2
Results of Cascading Tripartite and Dichotomous Complete Count and Tripartite Binomial Count Sequential Classification Sampling Plans to Sample Seven Population Trajectories

Population	∫ density	OC	ASN	Bouts	Loss	Loss with P{0.2}	Loss with P{0.05}
Tripartite							
1	329.7	0	145.8	4.8	63.4	60.0	68.0
2	363.1	0	118.8	6.0	163.1	142.1	157.9
3	103.9	1	229.9	6.6	103.9	103.9	103.9
4	462.2	0	119.9	4.1	109.5	96.5	108.0
5	15.5	1	129.5	6.0	15.5	15.5	15.5
6	363.4	0	25.2	1.0	0	0	0
7	69.5	1	180.2	6.5	69.5	69.5	69.5
Dichotomous							
1	329.7	0	199.1	7.7	60.0	57.63	67.1
2	363.1	0	190.6	10.0	58.4	48.90	56.1
3	103.9	1	236.7	12.0	103.7	103.90	103.9
4	462.2	0	161.0	6.7	92.1	91.60	106.7
5	15.5	1	235.3	12.0	15.5	15.50	15.5
6	363.4	0	24.8	1.1	0	0	0
7	69.5	1	239.4	12.0	69.5	69.50	69.5
Binomial							
1	329.7	0.08	270.3	5.4	143.0	211.9	329.7
2	363.1	0.10	223.6	6.0	188.7	155.4	363.1
3	103.9	0.98	327.4	6.4	103.1	103.9	103.9
4	462.2	0.01	219.5	4.3	132.9	110.8	233.2
5	15.5	1.00	176.2	6.0	15.4	5.4	15.4
6	363.4	0.03	163.5	2.6	65.5	159.5	319.8
7	69.5	0.96	245.2	6.3	68.4	69.5	69.5

Note: The maximum potential number of sampling bouts is 12.

B. BINOMIAL COUNT TRIPARTITE PLANS

Counting small, numerous organisms such as mites in the field is tedious and often not practical. To circumvent this problem, sampling plans have been developed that substitute presence or absence of an organism on a sample unit for complete enumeration. The shortcoming of this approach is that sampling plans based on binomial counts are less precise than plans that use enumeration. This shortcoming can be ameliorated by defining a positive binomial score as a sample unit with more than T_p organisms where T_p is called a tally point (Jones, Chapter 9). However, when T_p becomes too large the benefits of binomial counts are reduced, and our experience has been that practitioners favor binomial plans with a tally count of zero. We developed binomial count tripartite plans for *P. ulmi* based on the negative binomial distribution using the FORTRAN programs in the Appendix and discovered that when they are cascaded through time, they performed much better than expected. The reason for this is identical to the explanation given for the large cumulative probability of intervention when individual pdec3 values are modest. Here, we briefly summarize these findings because room does not permit a detailed exposition.

Parameters used to develop tally zero binomial sampling procedures are shown in Table 3. We also constructed plans using a tally count of four and found the performance, when they were used once in time, of the tally 4 sampling plans to be superior to the tally 0 sampling plans. This is not unexpected and parallels the conclusions of Nyrop and Binns.[3]

Performance of the tally 0 and tally 4 sampling plans cascaded through time was studied by applying the plans to the set of exponential populations. The OC for the tally 0 plan was not as steep as that for the tally 4 plan, and losses for the tally 0 plan were greater than for the tally 4 plans. Intervention thresholds for *P. ulmi* are designed to prevent a cumulative density of 500, which was achieved by the tally 0 plans with probability 0.95. This is an important result because it previously was suggested[3] that tally zero plans be avoided because of their poor precision. When tally 0 plans are cascaded they are quite adequate and tally 0 counts are easier to make than tally 4 counts.

Performance of the tally 0 binomial and complete enumeration sampling plans with the exponential populations is compared in Figure 7. By all accounts, the tally 0 binomial plan performed adequately and in one sense was superior to the complete enumeration plans because it is not as conservative as the complete enumeration plans. Of course, the complete count plans could be modified by using higher thresholds.

Performance of the tally 0 and complete count plans did not differ greatly, except that the ASN was greater for the binomial plans as were for the losses incurred with probability 0.2 and 0.05 (Table 2). These losses were higher for the tally 0 plans; however, they were still acceptable. With populations 1, 2, 4, and 6 intervention is practically assured. These populations all had cumulative densities at or near the level for which damage occurs (500). For populations 3, 5, and 7 biological control was successful, and there was very low or no probability of intervention. The ASN for the binomial plans probably could be reduced without sacrificing performance because much of the variability that influences the OC function is due to variation in the binomial count-mean density model which is not affected by sample size.

TABLE 3
Parameters Used to Construct Binomial Tripartite Sequential Classification Sampling Plans for Use with *P. ulmi*

Plan[a]	Tally	cd_i[b]	k[c]	Threshold[d]	H_0	H_1	Alpha	Beta
1.1	0	1.0	0.301	0.356	0.306	0.406	0.075	0.075
1.2	0	2.5	0.467	0.578	0.529	0.628	0.075	0.075
2.1	0	2.0	0.418	0.52	0.480	0.560	0.1	0.1
2.2	0	5.0	0.667	0.760	0.720	0.800	0.075	0.075
3.1	0	3.0	0.513	0.627	0.587	0.667	0.1	0.1
3.2	0	7.5	0.827	0.852	0.812	0.892	0.05	0.05

[a] For each plan there are two thresholds and two sets of SPRT parameters. The number following the decimal point signifies whether it is plan 1 or 2 of TSC.
[b] The first value for each plan is cd_1 and the second is cd_2.
[c] Parameter k for negative binomial distribution computed using moments and based on a predicted variance calculated using the model $s^2 = au^b$.
[d] cd_i expressed as the proportion of sample units with > Tally count.

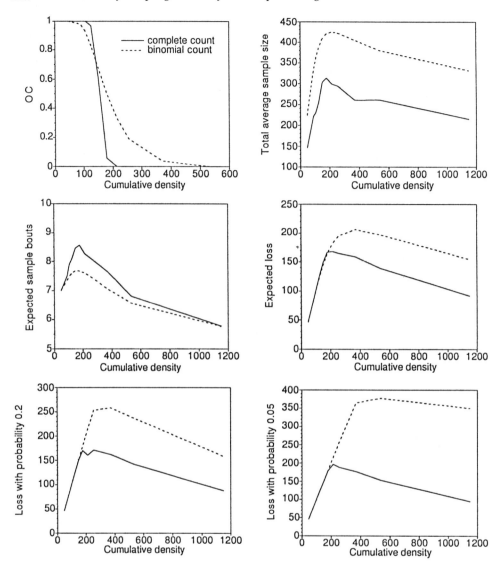

FIGURE 7. Performance of cascaded complete count and binomial tripartite sequential classification sampling plans.

C. ADAPTIVE FREQUENCY CLASSIFICATION MONITORING

If density is classified as less than an intervention threshold, tripartite classification sampling plans determine when the population should be sampled again by classifying density into one of two categories. If, for example, a decision was made to resample after the maximum waiting time, it does not matter whether the actual density was just slightly less than the critical density used to denote the maximum waiting time (cd_1) or much less than this value. However, the time sampling can be delayed depending on the actual density, might be longer if the true mean is much less than cd_1. The waiting time to the next sample bout might be more precisely determined by allowing density to be classified into four or more categories. However, such plans would require three or more dichotomous classification procedures and would likely be very cumbersome to develop. A more rational approach is to use both classification and estimation to formulate sampling plans for monitoring a population through time. We have developed such a sampling protocol and call the scheme adaptive frequency classification monitoring (AFCM). As with TSC plans, AFCM plans are cascaded through time to monitor a population trajectory. Performance criteria for cascaded AFCM plans are identical to those for cascaded TSC plans. Here, we describe AFCM sampling and illustrate its use by developing plans for monitoring *P. ulmi*. FORTRAN programs "afcm" and "cascade" perform necessary computations (see Appendix).

In AFCM, a sequential classification procedure is used to determine whether a density exceeds an intervention threshold. Unlike TSC plans though, densities are not classified as less than the intervention threshold. Instead, an estimated density is used to determine the waiting time to the next sample bout. Wald's SPRT is used to construct a classification stop line for determining whether density exceeds the intervention threshold. The mean for the alternate hypothesis is set equal to the intervention threshold. The mean for the null hypothesis is set equal to a density which, if allowed to grow at the maximum growth rate for the minimum time until the next sample bout, would not exceed the intervention threshold. Note that the densities used to construct these hypotheses are different from those used to specify stop lines for the TSC plans. With TSC plans null and alternate hypotheses are constructed around the intervention threshold while with AFCM plans, the intervention threshold is the density used for the alternate hypothesis. With TSC plans null and alternate hypotheses are also constructed around a density which, if allowed to grow at the maximum growth rate for the maximum time until the next sample is taken, would equal the intervention threshold. With AFCM plans the density for the null hypothesis is defined using the minimum time to the next sample bout.

In AFCM schemes, if a decision is to not intervene, the time to the next sample bout is based on an upper confidence limit for an estimated density obtained via a fixed sample size. We reasoned that individuals using an AFCM sampling plan would not be motivated to spend much sampling effort if the density was below an intervention threshold. Thus, we established a maximum sample size, which, if reached before a decision was made to intervene, indicated that the density was below the intervention threshold and a waiting time to the next sample should be determined. While a waiting time could be determined for each possible estimated density, this is not practical. Instead, we specify waiting times as multiples of the resample interval (*sint*; the minimum time between sample bouts) and then determine ranges of estimated densities that correspond to each waiting time. How these ranges of densities are determined is explained below.

AFCM stop lines are a composite of the upper stop limit for the sequential classification procedure and the maximum sample size. Shown in Figure 8 is an illustrative stop line. The first step in constructing an AFCM stop line is to plot the upper stop line for the sequential classification plan. Let $thr(t)$ be the intervention threshold at the current sampling time (t). As for the TSC plans, we assume that population growth can be described by a simple exponential model, although some other model could also be used. The null and alternate hypotheses for the sequential classification procedure are then:

$$H_0: \quad u = thr(t)/[\exp(r \cdot sint)]$$

$$H_1: \quad u = thr(t).$$

Alpha and beta for the SPRT are chosen to produce a desired width between the stop

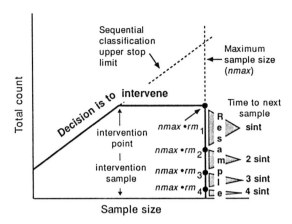

FIGURE 8. Adaptive frequency classification monitoring (AFCM) stop limit.

lines. Smaller alpha and beta result in wider stop lines that increase the average sample size and improve precision of the SPRT. With AFCM plans, this affects the proportion of sampling bouts that terminate with a decision to intervene before the maximum sample size is reached.

The second step is to plot a vertical line corresponding to the maximum sample size. Points on this line define the beginning and end of line segments which, if intersected by the total number of animals found after examining the maximum number of samples, determine the waiting time until the population is sampled again. Let ucl be an upper confidence limit for a mean estimated using the maximum sample size (n_{max}). Mean densities (rm) with upper confidence intervals ucl that solve

$$\text{ucl} = \frac{thr(t + i * sint)}{e^{(r * i * sint)}} \quad (9)$$

are the points on the maximum sample size line that define the line segments. In Equation 9, i is an integer that ranges from one to the maximum number of $sint$ time intervals in the future sampling might be delayed, $thr(t + i * sint)$ is the intervention threshold at time $t + i * sint$ and an exponential model has been used to define population growth. In Figure 8, i ranges from 1 to 4 and the mean densities with upper confidence intervals defined by Equation 9 are labeled as $rm_1 * n_{max}$ (rm = resample mean). If the total number of animals (total) in n_{max} samples satisfies

$$rm_i \cdot n_{max} > \text{total} \geq rm_{i+1} \cdot n_{max} \quad (10)$$

then sampling is to be done $i \cdot sint$ time steps in the future. If total $> rm_1 \cdot n_{max}$ a decision to intervene is made.

Equation 9 cannot be used directly to find the points (means multiplied by the maximum sample size) on the maximum sample size line that defines the line segments which correspond to specific waiting times. This is because Equation 9 is solved for an upper confidence limit; however, knowing the upper confidence limit for an estimated mean does not always allow determination of the mean itself. This problem is easily overcome on a computer by first determining ucl via Equation 9 and then performing a line search to find the mean density with this upper confidence interval. This is the strategy used in program "afcm."

It is important to note that the point on the maximum sample size line that defines the start of the line segment used to specify the shortest waiting time to the next sample bout ($n_{max} \cdot rm_1$) will usually lie below the upper sequential stop limit. This occurs because the SPRT uses a likelihood ratio test to determine whether sample data are most consistent with H_0 or with H_1, whereas the points on the maximum sample size line are defined using an upper confidence limit for the estimated density. If the point $n_{max} \cdot rm_1$ lies below the upper stop limit, a horizontal line can be drawn from this point to the upper stop limit and this line now defines part of the "intervene decision" stop limit (Figure 8). This can be done because once the total count exceeds $n_{max} \cdot rm_1$, a decision to intervene will always be made regardless of additional samples taken. If $n_{max} \cdot rm_1$ lies above the upper stop limit for the classification procedure, then the intersection of the stop limit and the maximum sample size line defines the start of the first line segment, which, if intersected by the total count, specifies a decision to sample again after *sint* time steps. The count corresponding to the intersection of the SPRT upper stop limit and $rm_1 \cdot n_{max}$ is called the intersection point and the sample size where this occurs is called the intersection sample.

A final point that must be made about AFCM stop limits, as we have formulated them, concerns the intervention threshold. In the AFCM protocol we have allowed the intervention threshold to be any nondecreasing continuous function of time and computation of waiting times to the next sample bout takes into account future threshold values. In contrast, the threshold for the TSC procedure was a discrete step function and computation of the waiting time to the next sample bout was based on the current threshold. By allowing the intervention threshold to be a nondecreasing continuous function of time and, by formulating waiting times based on future thresholds, waiting times to the next sample bout are lengthened. There is a price for this, though, because a different sampling plan must now be constructed for every possible sampling bout with a unique threshold. For example, if *sint* = 7 d and a population is to be monitored for 92 d with a different threshold for each day, 14 sets of stop lines must be determined. The large number of sampling plans required to monitor a population over several weeks or months is a possible drawback to AFCM-based monitoring plans.

AFCM plans are cascaded like TSC plans to monitor a population through time. Each AFCM plan has a set of performance characteristics consisting of an ASN function, pdec*i* functions for each of the resample decisions (1,..., *i* where *i* is the maximum *sint* interval sampling that may be delayed), and probability of a decision to intervene. Program "afcm" computes these performance characteristics for individual sampling plans. To determine how cascaded AFCM plans perform, the computations used with the cascaded TSC plans need only be slightly modified. All that is required is that, instead of sampling and decision probabilities being based on the outcome from two sample bouts in the past, *i* sample bouts in the past must now be considered. Program "cascade" does this.

To monitor *P. ulmi* we constructed AFCM plans based on a negative binomial distribution with the variance modeled using TPL. Additional parameters were *sint* = 7 d, *i* = 4, *r* for the exponential growth model = 0.065, alpha and beta for the SPRT = 0.1, n_{max} = 50, and alpha for the upper confidence limit = 0.3. Stop line parameters for protocols 1, 5, 9, and 13 from a total of 14 are shown in Table 4. Probability of decision functions (pdec*i*) and ASN functions for these plans are shown in Figure 9. Also shown in this figure are the proportion of decisions reached by (1) crossing the SPRT upper stop limit, (2) crossing the intersection point, and (3) reaching the maximum sample size. Probabilities of intervening were greater than specified via the SPRT parameters alpha and beta. SPRT alpha is the probability of

TABLE 4
Stop Line Parameters for Four AFCM Sampling Plans Used to Monitor European Red Mite

Plan	Threshold	Time	H_0	SPRT		Intersection			
				Intercept	Slope	Point	Sample size	Waiting time	Decision count
1	2.5	1	1.59	27.44	1.98	87	30	7	87
								14	65
								21	47
								28	33
5	4.91	29	3.12	36.38	3.89	149	30	7	149
								14	99
								21	65
								28	44
9	6.13	57	3.89	39.88	4.85	184	30	7	184
								14	120
								21	78
								28	51
13	7.29	85	4.63	42.9	5.77	216	30	7	216
								14	135
								21	84
								28	52

accepting the null hypothesis when the alternate hypothesis is true and for a conventional SPRT the probability of intervening approximately equals 1-alpha when the mean is equal to the alternate hypothesis. For the AFCM plans the probability of intervening was always in excess of this value because use of the intersection point as a stop limit increased the number of decisions to intervene. Similarly, when the density equals the null hypothesis, the probability of intervening is approximately equal to beta for a conventional SPRT. However, almost all decisions made when the density was close to the null hypothesis value were made by either reaching the maximum sample size or by crossing the intersection point. Both situations result in more decisions to intervene being made than if only the SPRT stop lines had been used. Overall, these factors cause the procedure to result in a high proportion of intervene decisions when the density is considerably less than the intervention threshold.

We used the AFCM protocols to sample the set of exponentially growing populations and the historical mite counts to which the cascaded TSC procedures were applied. Shown in Figure 10 are the performance criteria for cascaded TSC and AFCM plans used to monitor the exponential populations. When cascaded, the AFCM protocols resulted in more decisions to intervene than with the TSC protocol. The AFCM plans required approximately two thirds as many total samples as the TSC plans and approximately one half as many sample bouts. At low densities, this was due to the ability of AFCM plans to lengthen the waiting time beyond the 14-d maximum of the TSC plans. At high densities, the AFCM procedure had lower total average samples sizes and number of sample bouts because this procedure resulted in an intervention decision more quickly than the TSC protocols.

Shown in Table 5 are the results from applying the TSC and AFCM plans to the seven historical populations. For population 1 AFCM and TSC plans provided similar results. For population 2 the AFCM protocol did not always result in intervention (OC = 0.14). This occurred because at the third from last sample bout, 14% of the

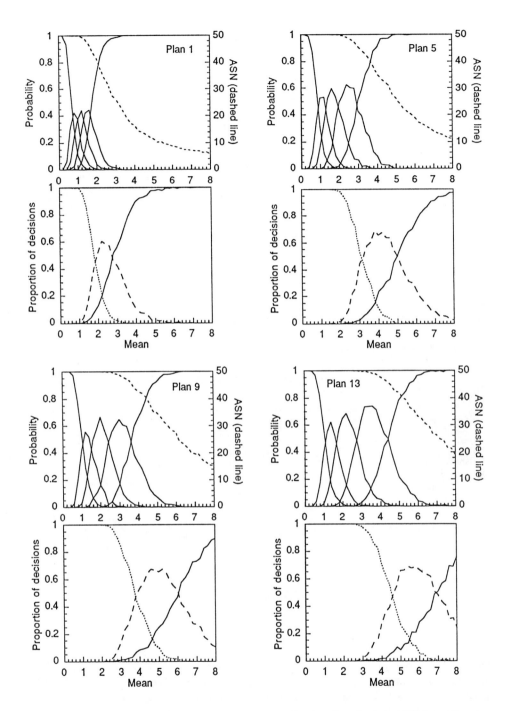

FIGURE 9. Probability of decision and average sample number functions for four AFCM sampling plans and the proportions of decisions made by (1) reaching the maximum sample size (small dashed line), (2) crossing the horizontal portion of the upper stop limit (large dashed line), and (3) crossing the SPRT upper stop limit (solid line). The probability of decision function that starts at 1.0 with mean = 0.0 is the decision to wait four time intervals until the next sample bout. This is followed by decisions to wait 3, 2, and 1 time interval to the next bout. The final function is for a treat decision.

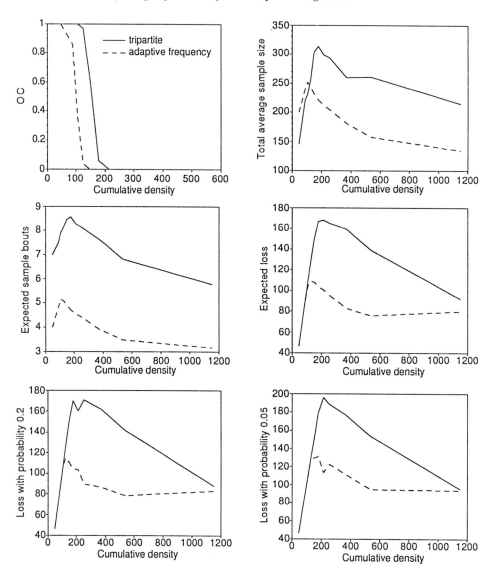

FIGURE 10. Performance of cascaded adaptive frequency classification monitoring plans and cascaded tripartite sequential classification sampling plans applied to exponentially growing populations.

decisions were to wait 4 weeks before sampling again, which was equivalent to a decision to not intervene because these populations were never sampled again. By extending the sampling period to 96 d from 92 d the OC was reduced to 0.0 and the ASN and expected number of sample bouts increased to 169.2 and 4.0, respectively. For population 3 the AFCM protocol resulted in a low rate of intervention (0.17). A higher rate of intervention occurred with population 7 where a density of 4.8 at time 75 resulted in a probability of intervention equal to 0.77. The expected number of sample bouts was always less with the AFCM protocol.

In comparison to TSC protocols, the AFCM plans resulted in a greater proportion of intervene decisions and the AFCM plans usually required fewer sampling resources (sample bouts plus number of samples per bout). Space does not permit development

TABLE 5
Results of Applying Cascaded Adaptive Frequency Classification Monitoring Plans and Tripartite Sequential Classification Sampling Plans to Seven Populations

	Adaptive frequency classification Monitoring					Tripartite sequential classification			
1	329.7	0.0	167.19	4.0	184.4	0.0	145.8	4.8	63.4
2	363.1	0.14	166.24	3.9	307.7	0.0	118.8	6.0	163.1
3	103.9	0.83	216.14	4.3	97.7	1.0	229.9	6.6	103.9
4	462.2	0.0	116.5	3.0	129.3	0.0	119.9	4.1	109.5
5	15.5	1.0	150	3.0	15.5	1.0	129.5	6.0	15.5
6	363.4	0.0	22.88	1.0	0.4	0.0	25.2	1.0	1.8
7	69.5	0.38	196.1	4.1	52.9	1.0	180.2	6.5	69.5

and analysis of binomial count plans for the AFCM protocol; however, this is a straightforward endeavor.

IV. SUMMARY AND CONCLUSIONS

Monitoring for biological control is currently not well developed and additional research is needed. Pest-natural enemy ratios sometimes can be used to predict the likelihood for biological control. If such a ratio is appropriate, the ratio classification sampling method described in this chapter can be used. However, the usefulness of this method would be enhanced if binomial counts could be substituted for complete enumeration. Cascaded tripartite plans provide a method of monitoring a population without the need for determining the abundance of natural enemies. At least one extension to this approach should prove useful: one set of plans could be used if natural enemies are present in samples and another set based on a larger pest population growth rate could be used when natural enemies were not present. In this way the time interval between samples might be made longer. Adaptive frequency classification monitoring shows even greater promise as a monitoring tool. Further work on this procedure is needed to determine how to design AFCM sampling plans that are not overly conservative in recommending intervention without increasing required sampling resources.

APPENDIX

ratio.for: sequential classification of ratios

Background: Program ratio.for is designed to construct and analyze ratio sequential classification sampling plans. These plans are designed to simultaneously classify (1) a pest density as either greater or less than an intervention threshold and (2) a pest/natural enemy ratio as either greater or less than a critical value. Stop lines are based on confidence intervals about the intervention threshold and critical ratio. Samples are processed in batches where the batch size should be > 20. Means from each batch sample are assumed to be bivariate normal distributed with the variance modeled as a function of the mean via TPL and the correlation between means constant.

The program works as follows:

1. Model parameters and data required to compute ratio stop limits are input. Required data are TPL parameters for pest and natural enemy (ya, yb, xa, xb),

correlation coefficient used to describe statistical correlation between counts of prey and natural enemy (rho), the critical ratio (*cr*), the intervention density for the pest (*t*), standard normal *z* values for computing ratio and pest density stop lines (*zcr*, *zt*) and a delta value for computing the ratio stop limits (*xdel*). Larger *z* values result in more stringent stop lines. Delta should be ≤ 0.1. These values are displayed and confirmed. Parameters for specifying the range of prey densities (numerator of ratio) and range of sample sizes for which ratio stop limits are to be calculated are then input and confirmed.

2. Ratio and pest density stop limits are computed. Stop limits can be displayed on the screen and written to a file. Stop limits for the ratio are indexed by the sample size and numerator (pest) density.

3. Performance of the sampling plan is determined via simulation. Data to be entered are parameters for specifying the range of prey densities and ratios to be used in the simulation, the correlation coefficient to be used in the simulation, and the number of Monte Carlo runs. Output consists of performance parameters indexed by the prey density (*Y*) and prey/natural enemy ratio (*R*). Performance parameters are the probabilities of making classifications one through four (dec1 = $Y \leq T, R \leq CR$, dec2 = $Y > T, R \leq CR$, dec3 = $Y \leq T, R > CR$ the ASN, and dec4 = $Y > T, R > CR$), the average sample size, the the probability of an incorrect classification (PIC), the ratio-only OC and ASN (*r*OC, *r*ASN), and the pest density-only OC and ASN (*t*OC, *t*ASN). The PIC is based on Pdec1-Pdec4. The ratio-only and pest density-only OC and ASN are determined by making decisions using only the ratio or pest density stop lines. These parameters are only written to the tab delimited output file and are not displayed on the screen.

An example run is shown below. User inputs are underlined and comments are in italics.

> ratio
Enter name for output file (≤ 12 characters)
test.out
Input model parameters
TPL parameters (var = a*m**b)
Parameters for numerator (y):
a:
4.32
b:
1.42
Parameters for denominator (x):
a:
2.38
b:
1.2
correlation coefficient:
−0.25
critical ratio:
7.5
intervention threshold:
5.0
z for ratio: *z is a standard normal deviate*
1.28
z for intervention threshold:
1.28

delta for determining x: *values much smaller than 0.001 greatly increase computation time*
0.01
    TPL parameters

ya	yb	xa	xb	rho	cr	t	zcr	zt	xdel
4.320	1.420	2.380	1.200	−0.25	7.50	5.00	1.28	1.28	0.010000

Parameters correct? (Y or N)
y
Input parameters for specifying numerator (y)
minimum:
1
maximum:
10
delta:
1
  ymin ymax ydel
    1   10   1
Parameters correct? (Y or N)
y
Input parameters for specifying sample sizes'
batch size
20
maximum sample size
100
  batch size maximum
    20   100
Parameters correct? (Y or N)
y
Display stop limits? (Y or N)
n
Write stop limits to file? (Y or N)
y
stop limits determined for following range of y
minimum = 1.00 maximum = 10.00
input parameters for specifying y during simulations   *it is not necessary to use the same range for y in the simulation as was used to determine the stop limits; however, stop limits required in the simulation for y values beyond those for which stop limits were actually determined will be linear interpolations*
minimum:
1
maximum:
10
delta:
1
  ymin ymax ydel
    1   10   1
Parameters correct? (Y or N)
y
input parameters for specifying ratio   *the same is true for the ratio as stated above for y*
minimum:
3.5
maximum:
10.5
delta:
0.5
  ymin ymax ydel
   3.5   10.5   0.5
Paramaters correct? (Y or N)
y

Correlation used to compute stop lines = −0.25
input correlation for use in simulations:
−0.25
input number of Monte Carlo runs
500
Repeat simulations (Y or N)?
n

simulations are performed and output is written to the tab delimited file 'test.out'. A portion of the output is shown below:

TPL parameters

ya	yb	xa	xb	rho	cr	t	zcr	zt	xdel
4.320	1.420	2.380	1.200	−0.25	7.50	5.00	1.28	1.28	0.010000

		Ratio		Pest	
N	Numerator	Upper_limit	Lower_limit	Upper_limit	Lower_limit
20.	1.00	0.43	0.02	6.87	3.135
20.	2.00	0.67	0.08	6.87	3.135
20.	3.00	0.89	0.15	6.87	3.135
20.	4.00	1.09	0.21	6.87	3.135
20.	5.00	1.30	0.30	6.87	3.135
20.	6.00	1.49	0.38	6.87	3.135
20.	7.00	1.68	0.45	6.87	3.135
20.	8.00	1.88	0.55	6.87	3.135
20.	9.00	2.06	0.63	6.87	3.135
20.	10.00	2.24	0.72	6.87	3.135
40.	1.00	0.31	0.04	6.32	3.681
40.	2.00	0.52	0.12	6.32	3.681
40.	3.00	0.71	0.20	6.32	3.681
40.	4.00	0.89	0.29	6.32	3.681
40.	5.00	1.08	0.39	6.32	3.681
40.	6.00	1.26	0.48	6.32	3.681
40.	7.00	1.43	0.57	6.32	3.681
40.	8.00	1.61	0.68	6.32	3.681
40.	9.00	1.77	0.77	6.32	3.681

...

Correlation used in simulations = −0.250
Monte Carlo iterations = 500

Y	R	Pdec1	Pdec2	Pdec3	Pdec4	ASN	PIC	rOC	rASN	tOC	tASN
1.00	3.50	0.910	0.000	0.090	0.000	52.44	0.090	0.910	52.44	1.000	20.00
2.00	3.50	0.964	0.000	0.036	0.000	43.20	0.036	0.958	42.64	1.000	21.24
3.00	3.50	0.994	0.000	0.006	0.000	42.52	0.006	0.980	37.36	1.000	33.16
4.00	3.50	0.966	0.032	0.000	0.002	63.80	0.034	0.990	36.64	0.954	60.76
5.00	3.50	0.544	0.452	0.000	0.004	81.80	0.456	0.992	32.32	0.528	74.44
6.00	3.50	0.088	0.910	0.000	0.002	64.76	0.090	0.996	29.48	0.086	56.96
7.00	3.50	0.034	0.964	0.000	0.002	46.40	0.036	0.998	29.68	0.034	37.12
8.00	3.50	0.012	0.984	0.000	0.004	37.08	0.016	0.996	28.36	0.012	28.64
9.00	3.50	0.000	1.000	0.000	0.000	32.40	0.000	1.000	27.76	0.000	24.44
10.00	3.50	0.000	1.000	0.000	0.000	26.92	0.000	1.000	25.28	0.000	21.64
1.00	4.00	0.848	0.000	0.152	0.000	61.56	0.152	0.848	61.56	1.000	20.00
2.00	4.00	0.948	0.000	0.052	0.000	55.00	0.052	0.938	54.04	1.000	21.60
3.00	4.00	0.990	0.000	0.010	0.000	50.04	0.010	0.968	45.96	1.000	32.76
4.00	4.00	0.952	0.044	0.002	0.002	66.56	0.048	0.968	43.12	0.946	62.56

...

program ratio

```
c   Program written in Fortran 77 compiled using Microsoft FORTRAN
c   compiler v5.1.
c   This program requires a uniform {0,1} random number generator. This
c   version makes use of the RANDOM subroutine provided in Microsoft
c   Fortran. Small sections of code must be changed in the subroutines
c   if a similar is not provided by the compiler.
c   Program written by Jan Nyrop, Department of Entomology, NYSAES,
c   Cornell University, Geneva, NY 14456
c   Last modification date: 7 October, 1992


c ********************************************************************
c ***************************** Variable definitions ****************************
c ********************************************************************

c asn - average sample size for complete plan determined via simulation
c batch - batch sample size (real)
c cr - critical ratio of pest (y) to natural enemy (x); y/x
c file - name of output file
c i - integer index
c j - integer index
c k - integer index
c l - integer index
c llr - lower limit for ratio determined via interpolation during simulation
c m - integer index
c mci - Monte Carlo iterations (integer)
c mcruns - Monte Carlo iterations (real)
c n(25) - sample size; n(i) is the total number of samples after i batches
c ncount - number of samples taken during a Monte Carlo bout
c ni - batch sample size (integer)
c nsteps - maximum number of batch samples to take during a Monte Carlo bout
c nt - maximum sample size
c pdec1 - probability of making classification 1
c pdec2 - probability of making classification 2
c pdec3 - probability of making classification 3
c pdec4 - probability of making classification 4
c pic - probability of making an incorrect classification
c r1dec - classification 1 counter
c r2dec - classification 2 counter
c r3dec - classification 3 counter
c r4dec - classification 4 counter
c rasn - asn based only on classifying the ratio
c rdel - delta for determining the ratio during simulation
c respond - character keyboard input
c rflag - flag used to indicate when a ratio classification has been made
c rho - correlation between pest and natural enemy used to build stop limits
c rll(100,25) - ratio lower stop limits indexed by pest density (100) and sample size
c   (25)
c rmax - maximum ratio used in simulations
c rmin - minimum ratio used in simulations
c roc - OC (P{r ≤ cr}) based only on classifying the ratio
c roccnt - counter for computing roc
c rsamp - counter for computing rasn
c rsteps - number of ratios used in simulation
c rul(100,25) - ratio upper stop limits indexed by pest density (100) and sample
c   size (25)
c samples - number of samples taken during a Monte Carlo bout
c simrat - pest/natural enemy ratios used in simulations
c simrho - pest/natural enemy correlation used in simulations
c simx - natural enemy density used in simulations
c simy - pest density used in simulations
```

c stdxm - standard deviation of the mean for natural enemy
c stdym - - standard deviation of the mean for pest
c t - pest threshold density
c tab - character tab
c tasn - average sample number based only on classifying pest density
c tflag - flag used to indicate when a pest classification has been made
c tll(25) - pest lower stop limits indexed by sample size (25)
c toc - OC (P{pest $\leq$ t}) based only on classifying the pest density
c toccnt - counter for computing toc
c totsamp - total samples taken during a simulation for specified pest density and ratio
c tsamp - counter for computing tasn
c tul(25) - pest upper stop limits indexed by sample size (25)
c ulr - upper limit for ratio determined via interpolation during simulation
c xdel - delta used when determining ratio stop limits
c xm - bivariate normal deviate generated by subroutine binorm used to represent natural enemy density
c xmean - estimated natural enemy mean density during a Monte Carlo bout
c xtpla - TPL parameter a for natural enemy
c xtplb - TPL parameter b for natural enemy
c y(100) - pest densities used to compute ratio stop limits
c ydel - delta used to compute y(i) when calculating ratio stop limits
c ym - bivariate normal deviate generated by subroutine binorm used to represent pest density
c ymax - maximum y used to compute ratio limits
c ymean - estimated pest mean density during a Monte Carlo bout
c ymin - minimum y used to compute ratio limits
c ysimstp - number of pest density values used in simulation
c ysteps - number of pest density values used to compute ratio limits
c ytpla - TPL parameter a for pest
c ytplb - TPL parameter b for pest
c zcr - standard normal z used with ratio limits
c zt - standard normal z used with pest limits
c****************** Variable declarations ******************
```
      real ymin,ymax,ydel,ytpla,ytplb,xtpla,xtplb,rho,cr,t,zcr,zt,xdel,
     + n(25),tul(25),tll(25),y(100),rul(100,25),rll(100,25)
      real rmin,rmax,rdel,rsteps,simrat,simy,simx,batch,r1dec,r2dec,
     +r3dec,r4dec,stdym,stdxm,mcruns,ym,xm,xmean,ymean,ulr,llr,
     +totsamp,pdec1,pdec2,pdec3,pdec4,samples,asn,roccnt,toccnt,
     +roc,toc,rsamp,tsamp,rasn,tasn,pic,simrho
      integer ni,nt,nsteps,ysteps,i,j,ncount,k,l,m,mci,ysimstp,rflag,tflag
      character*1 respond,tab
      character*12 file
      common/params/ytpla,ytplb,xtpla,xtplb,rho,cr,t,zcr,zt,xdel
      tab  =  char(9)
```

c open file for output
```
      write (*,'(a)') ' Enter name for output file ( ≤ 12 characters)'
      read (*,'(a)') file
      open (unit  =  10,file  =  file,status  =  'unknown')
```

c ***
c ************************* Define models and sampling plan *************************
c ***

```
1     write (*,*) 'Input model parameters'
      write (*,*) 'TPL parameters (var  =  a*m**b)'
      write (*,*) 'Parameters for numerator (y):'
      write (*,*) 'a:'
      read (*,*) ytpla
      write (*,*) 'b:'
      read (*,*) ytplb
```

```
      write (*,*) 'Parameters for denominator (x):'
      write (*,*) 'a:'
      read (*,*)xtpla
      write (*,*) 'b:'
      read (*,*) xtplb
      write (*,*) 'correlation coefficient:'
      read (*,*) rho
      write (*,*) 'critical ratio:'
      read (*,*) cr
      write (*,*) 'intervention threshold:'
      read (*,*) t
      write (*,*) 'z for ratio:'
      read (*,*) zcr
      write (*,*) 'z for intervention threshold:'
      read (*,*) zt
      write (*,*) 'delta for determining x:'
      read (*,*) xdel

c display and confirm
      write (*,10)
10    format ('   TPL parameters')
      write (*,11)
11    format ('   ya   yb   xa   xb   rho   cr   t   zcr   zt   xdel
     +')
      write (*,20) ytpla,ytplb,xtpla,xtplb,rho,cr,t,zcr,zt,xdel
20    format (1x,f5.3,1x,f5.3,1x,f5.3,1x,f5.3,1x,f5.2,1x,f4.2,1x,f5.2,
     + 1x,f4.2,1x,f4.2,1x,f8.6)
30    write (*,*) 'Parameters correct? (Y or N)'
      read (*,'(a)') respond
      if (respond.ne.'N'.and.respond.ne.'n') then
         if (respond.ne.'Y'.and.respond.ne.'y') goto 30
      end if
      if (respond .eq. 'N' .or. respond .eq. 'n') goto 1
      write (10,40)
40    format ('        TPL parameters')
      write (10,41)
41    format ('1  ya   yb   xa   xb   rho   cr   t   zcr   zt   xdel
     +')
      write (10,50) ytpla,ytplb,xtpla,xtplb,rho,cr,t,zcr,zt,xdel
50    format (1x,f5.3,1x,f5.3,1x,f5.3,1x,f5.3,1x,f5.2,1x,f4.2,1x,f5.2,
     + 1x,f4.2,1x,f4.2,1x,f8.6)
51    write (°,*) 'Input parameters for specifying numerator (y)'
      write (*,*) 'minimum:'
      read (*,*) ymin
      write (*,*) 'maximum:'
      read (*,*) ymax
      write (*,*) 'delta:'
      read (*,*)ydel
      ysteps = aint((ymax - ymin)/ydel) + 1
      ymax = ymin + ydel*(ysteps-1)
      write (*,60)
60    format ('  ymin   ymax   ydel')
      write (*,70) ymin,ymax,ydel
70    format (1x,f5.2,1x,f6.2,1x,f6.3)
80    write (*,*) 'Parameters correct? (Y or N)'
      read (*,'(a)') respond
      if (respond.ne.'N'.and.respond.ne.'n') then
         if (respond.ne.'Y'.and.respond.ne.'y') goto 80
      end if
      if (respond .eq. 'N' .or. respond .eq. 'n') goto 51
81    write (*,*) 'Input parameters for specifying sample sizes'
```

```
        write (*,*) 'batch size'
        read (*,*) ni
        write (*,*) 'maximum sample size'
        read (*,*) nt
        nsteps = aint(nt/ni)
        no = ni*steps
        write (*,90)
90      format ('  batch size   maximum')
        write (*,100) ni, nt
100     format (7x,i4,3x,i6)
110     write (*,*) 'Parameters correct? (Y or N)'
        read (*,'(a)') respond
        if (respond.ne.'N'and.respond.ne.'n') then
           if (respond.ne.'Y'.and.respond.ne.'y') goto 110
        end if
        if (respond .eq. 'N' .or. respond .eq. 'n') goto 81

c ****************************** Compute stop limits ********************************
c ****************************************************************************
c ****************************** sequential limits ********************************

        do 220 i = 1,nsteps − 1
          n(i) = float(ni*i)
          call tlim (n(i),tul(i),tll(i))
          do 230 j = 1,ysteps
            y(j) = ymin +(ydel*(j − 1))
            call ratlim (n(i),y(j),rul(j,i),rll(j,i))
230       continue
220     continue

c ***** terminal stop limits *****

        n(nsteps) = float(nt)
        tul(nsteps) = t
        tll(nsteps) = t
        do 240 j = 1,ysteps
          rul(j,nsteps) = y(j)/cr
          rll(j,nsteps) = y(j)/cr
240     continue

c ***** output stop lines *****

250     write (*,*) 'Display stop limits? (Y or N)'
        read (*,'(a)') respond
        if (respond.ne.'N'.and.respond.ne.'n') then
          if (respond.ne.'Y'.and.respond.ne.'y') goto 250
        end if
        if (respond .eq. 'Y' .or. respond .eq. 'y') then
          write (*,251)
251       format ('      Ratio limits        Pest limits
     +')
          write (*,252)
252       format ('       N Numerator Upper_limit Lower_limit Upper_limit
     +Lower_limit')
          ncount = 0
          do 253 i = 1,nsteps
            do 254 j = 1,ysteps
              write (*,255) n(i),y(j),rul(j,i),rll(j,i),tul(i),tll(i)
255           format (1x,f6.0,3x,f6.2,6x,f6.2,6x,f6.2,6x,f6.2,5x,f6.2)
              ncount = ncount + 1
              if (ncount.gt.20) then
```

```
              pause 'input blank line to continue display'
              ncount = 0
              write (*,256)
256           format ('       N Numerator Upper_limit Lower_limit',1x,
     +        'Upper_limit Lower_limit')
              end if
254     continue
253     continue
        endif
260     write (*,*) 'Write stop limits to file? (Y or N)'
        read (*,'(a)') respond
        if (respond.ne.'N'.and.respond.ne.'n') then
          if (respond.ne.'Y'.and.respond.ne.'y') goto 260
        end if
        if (respond .eq. 'Y' .or. respond .eq. 'y') then
          write (10,261) tab,tab,tab,tab
261       format (1x,a1,a1,' Ratio',a1,a1,'Pest')
          write (10,262) tab,tab,tab,tab,tab
262       format (' N',a1,'Numerator',a1,'Upper_limit',a1,'Lower_limit',
     +    a1,'Upper_limit',a1,'Lower_limit')
          do 263 i = 1,nsteps
            do 264 j = 1,ysteps
              write (10,265) n(i),tab,y(j),tab,rul(j,i),tab,rll(j,i),
     +    tab,tul(i),tab,tll(i)
265           format (1x,f6.0,a1,f6.2,a1,f6.2,a1,f6.2,a1,f6.2,a1,f6.3)
264         continue
263       continue
        endif


c ******************************************************************************
c ***************************** Simulate performance ***************************
c ******************************************************************************


c *************** input parameters ***************

295     write (*,*) 'stop limits determined for following range of y'
        write (*,300) ymin,ymax
300     format (' minimum = ',f5.2,' maximum = ',f5.2)
305     write (*,*) 'input parameters for specifying y during simulations'
        write (*,*) 'minimum:'
        read (*,*) ymin
        write (*,*) 'maximum:'
        read (*,*) ymax
        write (*,*) 'delta:'
        read (*,*) ydel
        ysimstp = aint((ymax - ymin)/ydel) + 1
        ymax = ymin + ydel*(ysimstp - 1)
        write (*,310)
310     format ('  ymin   ymax   ydel')
        write (*,320) ymin,ymax,ydel
320     format (1x,f5.2,1x,f6.2,1x,f6.3)
330     write (*,*) 'Parameters correct? (Y or N)'
        read (*,'(a)') respond
        if (respond.ne.'N/.and.respond.ne.'n') then
          if (respond.ne.'Y'.and.respond.ne.'y') goto 330
        end if
        if (respond .eq. 'N' .or. respond .eq. 'n') goto 305
335     write (*,*) 'input parameters for specifying ratio'
        write (*,*) 'minimum:'
        read (*,*) rmin
```

```
            write (*,*) 'maximum:'
            read (*,*) rmax
            write (*,*) 'delta:'
            read (*,*) rdel
            rsteps = aint((rmax − rmin)/rdel) + 1
            rmax = rmin + rdel*(rsteps − 1)
            write (*,340)
340     format (' rmin   rmax   rdel')
            write (*,350) rmin,rmax,rdel
350     format (1x,f5.2,1x,f6.2,1x,f6.3)
360     write (*,*) 'Parameters correct? (Y or N)'
            read (*,'(a)') respond
            if (respond.ne.'N'.and.respond.ne.'n') then
               if (respond.ne.'Y'.and.respond.ne.'y') goto 360
            end if
            if (respond .eq. 'N' .or. respond .eq. 'n') goto 335
            write (*,365) rho
365     format (' Correlation used to compute stop lines = ',f6.3)
            write (*,*) 'Input correlation for use in simulations:'
            read (*,*) simrho
            write (*,*) 'input number of Monte Carlo runs'
            read (*,*) mci
            mcruns = float(mci)
            batch = float(ni)
            write (10,366) simrho
366     format (' Correlation used in simulations = ',f6.3)
            write (10,367) mci
367     format (' Monte carlo iterations = ',i4)
            write (*,370)
370     format ('Y    R    Pdec1 Pdec2 Pdec3 Pdec4 ASN   PIC')
            write (10,380) tab,tab,tab,tab,tab,tab,tab,tab,tab,tab
380     format (' Y',a1,'R',a1,'Pdec1',a1,'Pdec2',a1,'Pdec3',a1,'Pdec4',
          + a1,'ASN',a1,'PIC',a1,'rOC',a1,'rASN',a1,'tOC',a1,'tASN')


c ****************** Computations ********************

c      ***** loop for ratio *****

            do 400 k = 1,rsteps
               simrat = rmin + rdel*(k − 1)

c      ***** loop for numerator (y) *****

               do 410 l = 1,ysimstp
                  simy = ymin + ydel*(l − 1)
                  simx = simy/simrat
                  r1dec = 0.0
                  r2dec = 0.0
                  r3dec = 0.0
                  r4dec = 0.0
                  totsamp = 0.0
                  roccnt = 0.0
                  toccnt = 0.0
                  rsamp = 0.0
                  tsamp = 0.0
                  stdym = sqrt((ytpla*simy**ytplb)/batch)
                  stdsm = sqrt((xtpla*simx**xtplb)/batch)

c      ***** loop for Monte Carlo runs *****
```

```
              do 420 m  =  1,mci
                xmean  =  0.0
                ymean  =  0.0
                rflag  =  0
                tflag  =  0
                do 430 i  =  1,nsteps
                  samples  =  float(i*ni)
                  call binorm (symy,simx,stdym,stdxm,simrho,ym,xm)
                  xmean  =  ((xmean*float((i−1)*ni))+(xm*batch))/samples
                  ymean  =  ((ymean*float((i−1)*ni))+(ym*batch))/samples
                  call interp (rul(1,i),y,ymean,ysteps,0,ulr)
                  call interp (rll(1,i),y,ymean,ysteps,0,llr)
                  if (rflag.eq.0) then
                    if (xmean.ge.ulr) then
                      rflag  =  1
                      roccnt  =  roccnt + 1.0
                      rsamp  =  rsamp + samples
                    end if
                    if (xmean.lt.llr) then
                      rflat  =  1
                      rsamp  =  rsamp + samples
                    end if
                  end if
                  if (tflag.eq.0) then
                    if (ymean.le.tll(i)) then
                      tflag  =  1
                      toccnt  =  toccnt + 1.0
                      tsamp  =  tsamp + samples
                    end if
                    if (ymean.gt.tul(i)) then
                      tflag  =  1
                      tsamp  =  tsamp + samples
                    end if
                  end if
                  if (ymean.le.tll(i).and.xmean.ge.ulr) then
                    r1dec  =  r1dec + 1.0
                    totsamp  =  totsamp + samples
                    goto 500
                  endif
                  if (ymean.gt.tul(i).and.xmean.ge.ulr) then
                    r2dec  =  r2dec + 1.0
                    totsamp  =  totsamp + samples
                    goto 500
                  endif
                  if (ymean.le.tll(i).and.xmean.t.llr) then
                    r3dec  =  r3dec + 1.0
                    totsamp  =  totsamp + samples
                    goto 500
                  endif
                  if (ymean.gt.tul(i).and.xmean.lt.llr) then
                    r4dec  =  r4dec + 1.0
                    totsamp  =  totsamp + samples
                    goto 500
                  endif
430           continue
500           continue
420         continue

c *****************************************************************************
c ******************************* Output results *****************************
c *****************************************************************************
```

```
              pdec1 = r1dec/mcruns
              pdec2 = r2dec/mcruns
              pdec3 = r3dec/mcruns
              pdec4 = r4dec/mcruns
              asn = totsamp/mcruns
              if (simy.le.t.and.simrat.le.cr) pic = 1-pdec1
              if (symy.gy.t.and.simrat.le.cr) pic = 1 − pdec2
              if (simy.le.t.and.simrat.gt.cr) pic = 1 − pdec3
              if (symy.gt.t.and.simrat.gt.cr) pic = 1 − pdec4
              roc = roccnt/mcruns
              toc = toccnt/mcruns
              rasn = rsamp/mcruns
              tasn = tsamp/mcruns
              write (*,510) simy,simrat,pdec1,pdec2,pdec3,pdec4,asn,pic
510           format (1x,f6.2,2x,f6.2,2x,4(f5.3,2x),f7.2,2x,f5.3)
              write (10,520) simy,tab,simrat,tab,pdec1,tab,pdec2,tab,pdec3,
       +tab,pdec4,tab,asn,tab,pic,tab,roc,tab,rasn,tab,toc,tab,tasn,
       +tab
520           format (1x,f6.2,a1,f6.2,a1,4(f5.3,a1),f7.2,a1,f5.3,a1,2(f5.3,a1,f7.2,a1))
410      continue
400      continue

600      write (*,*) 'Repeat simulations (Y or N)?'
         read (*,'(a)') respond
         if (respond.ne.'N'.and.respond.ne.'n') then
            if (respond.ne.'Y'.and.respond.ne.'y') goto 600
         end if
         if (respond .eq. 'Y' .or. respond .eq. 'y') goto 295
         close (10)
         stop
         end


c  *******************************************************************************
c  ******************************** Subroutines *********************************
c  *******************************************************************************

         subroutine tlim (n,ul,ll)

c  *******************************************************************************
c  Subroutine computes stop limits for comparing pest density to
c  *******************************************************************************

         real n,ul,ll,var,ytpla,ytplb,xtpla,xtplb,rho,cr,t,zcr,zt,xdel
         common/params/ytpla,ytplb,xtpla,xtplb,rho,cr,t,zcr,zt,xdel
         var = ytpla*t**ytplb
         ul = t +zt*sqrt(var/n)
         ll = t −zt*sqrt(var/n)
         return
         end


         subroutine ratlim(n,y,ul,ll)

c  *******************************************************************************
c  Subroutine computes stop limits for comparing natural enemy density
c  to in order to classify the pest-natural enemy ratio
c  *******************************************************************************

         real n,ul,ll,ytpla,ytplb,xtpla,xtplb,rho,cr,t,zcr,zt,xdel,
```

```
      +x,vy,vx,cyy,cxx,cyx,var1,var2,var3,var4,rhat,rhigh,rlow
       integer j
       common/params/ytpla,ytplb,xtpla,xtplb,rho,cr,t,zcr,zt,xdel
       do 10 j = 1,2

c    *** note: j = 1 for upper interval and j = 2 for lower interval ***

       x = y/cr
20     vy = ytpla*y**ytplb
       vx = xtpla*x**xtplb
       cyy = vy/(n*y**2)
       cxx = vx/(n*x**2)
       cyx = (rho*sqrt(vx)*sqrt(vy))/(n*x*y)
       var1 = (cyy+cxx-(2*cyx))-(zcr**2*(cyy*cxx-cyx**2))
       rhat = y/x

c    ***** test for imaginary roots of var1 *****

       if (var1.lt.0.0.and.j.eq.1) then
         x = x+xdel
         goto 20
       end if
       if (var1.t.0.0.and.j.eq.2) then
         x = x-xdel
         if (x.lt.0.0) then
           ll = -999.0
           goto 10
         endif
         goto 20
       end if

c    ***** compute confidence intervals *****

       var2 = 1-(zcr**2*cyx)
       var3 = 1-(zcr**2*cxx)
       var4 = zcr*sqrt(var1)
       if (j.eq.1) then
         rhigh = rhat*(var2+var4)/var3
       else
         rlow = rhat*(var2-var4)/var3
       endif

     test if conditions have been met for confidence limits to be stop limits

       if (j.eq.1) then
         if (rhigh.lt.cr.and.rhigh.gt.0.0) then
           ul = x
         else
           x = x+xdel
           goto 20
         endif
       else
         if (rlow.gt.cr) then
           ll = x
         else
           x = x-xdel
           if (x.lt.0.0) then
             ll = -999.0
             goto 10
```

```fortran
              endif
              goto 20
           endif
         endif
10    continue
      return
      end


      subroutine binorm (m1,m2,stdv1,stdv2,rho,x1,x2)

c     ****************************************************************
c     Subroutine computes bivariate normally distributed random variables
c     using the method described by Johnson and Kotz in Distributions in
c     Statistics: Continuous Multivariate Distributions. John Wiley & Sons,
c     New York. The subroutine use a uniform {0,1} random variate which
c     must be provided if the FORTRAN compiler does not support
c     it [call random(r1) below].
c     ****************************************************************

      real m1,m2,stdv1,stdv2,rho,r1,r2,u1,u2,y1,y2,x1,x2

c     generate first uniform normal variate

      call random(r1)
      if (r1.le.0.0) r1 = 0.0000001
      call random(r2)
      u1 = sqrt(-2.*log(r1))*(cos(6.283*r2))

c     generate second uniform normal variate

      call random(r1)
      if (r1.le.0.0) r1 = 0.0000001
      call random(r2)
      u2 = sqrt(-2.*log(r1))*(cos(6.283*r2))
      y1 = u1
      y2 = rho*u1 + sqrt(1 - rho**2)*u2
      x1 = m1 + y1*stdv1
      x2 = m2 + y2*stdv2
      return
      end


      subroutine interp (y,x,dummyx,k,trunc,yvalue)

c     ****************************************************************
c     This subroutine performs a linear interpolation of a function y = f(x).
c     Vectors of y values (y) and x values (x) are passed along with a dummy
c     x value (dummyx) for which a y value (yvalue) is sought. The dimension
c     of the y and x arrays (k) must also be passed. In addition a variable
c     trunc is passed that when = 1 causes truncation of the function so that
c     when dummyx < x(1) or dummyx > x(k), yvalue is constrained to be y(1) or
c     y(k). When trunc is not equal to 1, yvalue is interpolated beyond y(1)
c     or y(k).
c     ****************************************************************

      real y,x,dummyx,yvalue
      integer k,trunc,j
      dimension y(k),x(k)

      if (dummyx.le.x(1)) then
```

```
        if (trunc.eq.1) then
           yvalue = y(1)
           goto 20
        else
           yvalue = y(1)−((y(2)−y(1))/(x(2)−x(1))*(x(1)−dummyx))
           goto 20
        endif
     endif

     do 10 j = 2,k
        if (dummyx.le.x(j)) then
           yvalue = y(j − 1)+((y(j)−y(j−1))/(x(j)−x(j − 1))*(dummyx−x(j − 1)))
           goto 20
        endif
10   continue

     if (trunc.eq.1) then
        yvalue = y(k)
        goto 20
     else
        yvalue = y(k)+((y(k)−y(k−1))/(x(k)−x(k − 1))*(dummyx−x(k)))
        goto 20
     endif

20   continue
     return
     end
```

sprt: sequential classification

Background: Program sprt.for analyzes the performance of dichotomous and tripartite sequential classification sampling plans based on Wald's sequential probability ratio test (SPRT). Tripartite plans are based on two dichotomous plans. Five models can be used to describe sample observations; binomial, Poisson, and negative binomial distributions and two models where binomial counts are substituted for complete enumeration. One binomial count model is the empirical relationship $\ln(-\ln(1 - p_T)) = \gamma + \delta n(m)$ and the other is based on the negative binomial distribution

$$P_T = 1 - \sum_{v=0}^{T} \Pr\{x = v\}$$

In both of these equations P_T is the proportion of sample units with more than T organisms. T is referred to as a tally threshold.

OC and ASN functions for dichotomous plans based on the three probability distribution functions are computed using an approximate formula. When the negative binomial distribution is specified, the parameter k can either be a function of the mean by modeling the variance via TPL ($s^2 = \alpha m^\beta$) or k can be a constant. For the negative binomial distribution OC and ASN functions are first computed using a nominal value of k. When $k = f(\text{TPL})$ this is the k that corresponds to a mean equal to the average of the means used to specify the null and alternate hypotheses. OC and ASN functions for the negative binomial distribution can also be simulated. For $k = f(\text{TPL})$, k in the simulation is computed using TPL. When k is a constant a new k for the simulations is input. OC and ASN functions for binomial count plans are the expected values described by Nyrop and Binns.[3]

The program functions as follows:

1. A sample model is chosen and data required to describe the model are input. Required data for each model are listed below:

    (i) Binomial: none
    (ii) Poisson: none
    (iii) Negative binomial: if $k = f(TPL)$, intercept and slope for TPL model.
    (iv) Empirical p − m: intercept, slope, variance of the slope, mean square error, number of data points, and mean of independent variable used in regression
    (v) NGB based p − m: intercept, slope variance of the slope, mean square error, number of data points, and mean of independent variable used in TPL regression

2. Parameters for the first SPRT are input. If a tripartite plan is to be constructed the first SPRT is the one used to classify the density into the lowest category. If binomial count models are used, a threshold density is first specified and then the proportion of occupied sample units is computed and displayed. The SPRT is based on this proportion.

3. OC and ASN functions for the specified plan are computed and displayed. Displayed values are interpolated from a more complete set of OC and ASN values. Interpolated values are shown to facilitate comparison of plans. If desired, the complete set of OC and ASN values can be written to the output file. If the OC and ASN values are not acceptable, new SPRT parameters can be entered.

4. Steps 2 and 3 are repeated for the second SPRT of a tripartite plan. If only a dichotomous SPRT is desired, program execution can be halted.

5. Both SPRT plans are displayed. If desired, parameters for either plan can be changed and OC and ASN functions are recomputed.

6. The two SPRT plans are used to construct a tripartite plan and the performance of this plan is simulated. To start the simulations the number of Monte Carlo runs as well as the maximum sample size are input. If the maximum sample size is reached, sampling is terminated by comparing the sample mean to the midpoints of the means used to form the null and alternate hypotheses. The means of populations for which sampling is simulated are automatically determined. This is done as follows: for each SPRT, OC and ASN values are determined for 29 means. The means used in the tripartite simulation are determined as:

    i) The first 18 means from the first SPRT
    ii) Means 19–29 from the first SPRT unless a mean from the first SPRT > the first mean from the second SPRT
    iii) All the means from the second SPRT

When two SPRTs are combined to form a tripartite plan it is possible for the lower stop limit of the uppermost set of limits to lie below the lower limit of the lowermost set of limits for some means. It is also possible for the upper stop limit for the lowermost set of limits to lie above the upper limit of the uppermost set of limits. When either of these situations occur in the simulation, a decision to stop sampling is not made until sample data lie beyond both sets of stop limits.

An example run using a binomial count model based on the negative binomial distribution is shown below. User inputs are underlined and comments are in italics.

```
> sprt
Enter name for output file (≤ 12 characters)
```

ngb.txt
Choose a sampling model
1) Binomial
2) Poisson
3) Negative binomial
4) Empirical p-m
5) NGB based p-m
enter a number 1-5-m
5
Input TPL parameters
intercept;
4.32
slope;
1.42
variance of the slope;
0.004
mse from the regression;
0.278
number of data points for regression;
147
mean of the independent variable {lnm};
0.728

NGB p-m model with TPL used

intercept	slope	var(slope)	mse	N	mean of lnm
4.320	1.420	0.0040	0.278	147.0	0.728

Parameters correct? (Y or N)
y
Enter plan number; 0   *plan number is initially set to 0*
1
threshold as density; 0.0   *a threshold density is required to compute P_T it is initially set to 0.0*
1.0
tally threshold; 0.0
0
Intervention proportion = 0.3564
mean for null hypothesis;
0.306
mean for alternative hypothesis;
0.406
alpha;
0.075
beta;
0.075

H0 mean	H1 mean	k	alpha	beta	
0.306	0.406	0.301	0.075	0.075	*k is computed using TPL*

Parameters correct? (Y or N)
y

OC	Mean	ASN
0.99	0.39	72.15
0.90	0.55	82.85
0.80	0.67	86.14
0.70	0.78	87.56
0.60	0.88	88.12
0.50	0.99	87.82
0.40	1.11	82.89
0.30	1.28	74.20
0.20	1.50	64.32
0.10	1.90	51.22
0.01	3.32	22.30

Redo OC and ASN? (Y or N)
<u>n</u>
Write OC and ASN functions to file? (Y or N)
<u>n</u>
Stop now? (Y or N)
<u>n</u>
Enter plan number; 1
<u>2</u>
threshold as density; 0.0
<u>2.5</u>
tally threshold; 0.0   *tally threshold for plan 2 = that for plan 1*
intervention proportion = 0.5785
mean for null hypothesis;
<u>0.529</u>
mean for alternative hypothesis;
<u>0.628</u>
alpha;
<u>0.075</u>
beta;
<u>0.075</u>

H0 mean	H1 mean	k	alpha	beta	
0.529	0.628	0.467	0.075	0.075	*H0 and H1 means for plan 2 are larger than those for plan 1*

Parameters correct? (Y or N)
<u>y</u>

OC	Mean	ASN
0.99	0.85	31.45
0.90	1.37	60.06
0.80	1.68	71.46
0.70	1.95	77.19
0.60	2.22	81.48
0.50	2.50	82.67
0.40	2.81	79.14
0.30	3.26	72.70
0.20	3.85	64.11
0.10	4.88	49.46
0.01	8.71	21.82

Redo OC and ASN? (Y or N)
<u>n</u>
Write OC and ASN functions to file? (Y or N)
<u>n</u>

Stop now? (Y or N)
n

NEG p-m model with TPL used

intercept	slope	var(slope)	mse	N	mean of lnm
4.320	1.420	0.0040	0.278	147.0	0.728

H0 mean	H1 mean	k	alpha	beta	H0 mean	H1 mean	k	alpha	beta
0.306	0.406	0.301	0.075	0.075	0.529	0.628	0.467	0.075	0.075

Low intercept	High intercept	Slope	Low intercept	High intercept	Slope
−5.73	5.73	0.355	−6.16	6.16	0.579

intercepts and slopes are those for stop lines

OC	Mean	ASN	OC	Mean	ASN
0.99	0.39	72.15	0.99	0.85	31.45
0.90	0.55	82.85	0.90	1.37	60.06
0.80	0.67	86.14	0.80	1.68	71.46
0.70	0.78	87.56	0.70	1.95	77.19
0.60	0.88	88.12	0.60	2.22	81.48
0.50	0.99	87.82	0.50	2.50	82.67
0.40	1.11	82.89	0.40	2.81	79.14
0.30	1.28	74.20	0.30	3.26	72.70
0.20	1.50	64.32	0.20	3.85	64.11
0.10	1.90	51.22	0.10	4.88	49.46
0.01	3.32	22.30	0.01	8.71	21.82

Redo a plan? (Y or N)?
n
starting simulations...
Maximum sample size?
100
Monte Carlo iterations?
500
Simulations are performed and output is written to the tab delimited file 'ngb.txt'.
A portion of the output is shown below:

Max. sample size = 100   iterations = 500

number of means = 45

Mean	P{dec1}	P{dec2}	P{dec3}	ASN
0.357	0.993	0.007	0.000	37.60
0.396	0.985	0.015	0.000	41.09
0.437	0.968	0.032	0.000	44.92
0.480	0.946	0.054	0.000	48.75
0.527	0.916	0.084	0.000	52.52
0.576	0.880	0.120	0.000	56.22
0.628	0.839	0.161	0.000	59.68
0.682	0.793	0.206	0.001	62.99
0.739	0.735	0.263	0.002	66.04
0.797	0.680	0.316	0.004	68.51
0.809	0.670	0.326	0.005	68.99
0.821	0.657	0.337	0.005	69.45

P{dec1} is the probability of classifying the density as ≤ 1.0, P{dec2} is the probability of classifying the density as > 1.0 and ≤ 2.5, and P{dec3} is the probability of classifying the density as > 2.5

```
      program sprt

c     Program written in Fortran 77: IBM version compiled using Microsoft
c     FORTRAN compiler v5.1.
c     This program requires a uniform {0,1} random number generator. This
c     version makes use of the RANDOM subroutine provided in Microsoft
c     Fortran. If the FORTRAN compiler used does not provide a similar
c     function, one must be provided and small sections of code in some
c     subroutines must be changed.
c     Program written by Jan Nyrop, Department of Entomology, NYSAES,
c     Cornell University, Geneva, NY 14456
c     Last modification date: 3 August, 1992

c ******************************************************************************
c *****************************   Variable definitions   **********************************
c ******************************************************************************

c     a: intermediate SPRT variable
c     alpha(2): alpha for SPRT plans 1 and 2
c     asn(29,2): ASN corresponding to OC(29,plan(2))
c     asnprt(11,2): displayed ASN values corresponding to mean(11,sampling
c        plan(2))
c     b: intermediate SPRT variable
c     beta(2): beta for SPRT plans 1 and 2
c     bvar1: variable used to compute binomial SPRT
c     c: intercept for empirical p-m model (ln(-ln(p)) = c + d(ln(m))
c     confirm: integer used to flag whether sampling model has been confirmed
c     count: number of samples generated during one simulated sampling bout
c     d: slope for empirical p-m model (ln(-ln(p)) = c + d(ln(m))
c     distr: distribution index
c     dummy: x value passed to interpolation routine for which a y value is
c        returned
c     easn: expected average sample size when binomial counts are substituted for
c        complete counts
c     edec: expected probability of a decision (e.g., OC) when binomial counts are
c        substituted for complete count
c     file: name for output file
c     h(14): variable used in computation of dichotomous SPRT OC
c     highi(2): stop limit parameter for SPRT plans 1 & 2
c     i: plan number (1 or 2)
c     i2: do loop index
c     icount: index for simmean
```

c   initun: initial file unit number
c   init(2): used to determine if plans 1 & 2 have been initialized
c   ivalues: number of values passed to interpolation routine
c   j: index
c   jj: index
c   k(2): NGB k used to construct dichotomous SPRT
c   ksim: NGB k used in tripartite SPRT simulations
c   lna: ln(a) - intermediate SPRT variable
c   lnb: ln(b) - intermediate SPRT variable
c   lowi(2): stop limit parameter for SPRT plans 1 & 2
c   m(29,2): mean corresponding to an OC(29,plan(2))
c   m0(2): null hypothesis mean for SPRT plan 1 & 2
c   m1(2): alternate hypothesis mean for SPRT plans 1 & 2
c   mci: Monte Carlo iterations
c   mean(11,2): means corresponding to ocprint(11,plan(2))
c   mlnm: mean of ln(m) used in empirical p-m or TPL model
c   mse: mean square error for either empirical p-m or TPL model
c   mt: threshold density used to compute a threshold proportion for binomial
            count sampling
c   nmax: maximum sample size for simulating tripartite SPRT OC and ASN
c   ngbk: flag to indicate whether constant k or k = f(TPL) is to be used
c   oc(29,2): dichotomous SPRT OC function for m(29, plans(2))
c   ocprint(11): SPRT OC values for which means are displayed on screen
c   p: NGB p corresponding to an OC value
c   pdec1(58): P{decision m < m0(1)}
c   pdec2(58): P{decision m1(1) < m < m0(2)}
c   pdec3(58): P{decision m > m1(2)}
c   pi(2): threshold density expressed as a proportion of binomial counts
c   pmbin(29,2): mean density corresponding to threshold proportions used to
            compute OC and ASN functions for dichotomous SPRT
c   pnb0: NGB parameters for m1
c   pnb1: NGB parameters for m0
c   pxgtt: probability that x > t when x is distributed as NGB
c   q: NGB q
c   qnb0: NGB parameter for m1
c   qnb1: NGB parameter for m0
c   q0: binomial parameter for m0
c   q1: binomial parameter for m1
c   refm: a reference set of means for which pxgtt is calculated
c   respond: used to record keyboard input to questions (Y, N)
c   result: y value returned from interpolation routine
c   revoc(29,2): 1 - OC; used in interpolation routine
c   rn: number of data points in either TPL or empirical p-m model
c   seq3: number of > m1(2) sequential decisions
c   seq2: number of > m1(1) and < m0(2) sequential decisions
c   seq2dec: P{m1(1) < m < m0(2)|m} for sequential decisions
c   seq1: number of < m0(1) sequential decisions
c   seq1dec: P{m < m0(1)|m} for sequential decisions
c   simasn(58): ASN for simulated means
c   simmean(58): means for which P{dec} and ASN functions are to be simulated
c   sll(2): stop line
c   slope(2): stop limit parameters for SPRT plans 1 & 2
c   slu(2): stop line
c   spmbin(58): means for which P{dec} and ASN functions are to be simulated;
c   means based on binomial count proportions
c   sum: sum of NGB variates during a simulated sampling bout
c   sumpdec: sum of P{deci} functions used to standardize so sum = 1.0
c   t: tab character
c   tally: integer used to define a positive binomial count
c   temp:
c   ter3: number of > m1(2) terminal decisions

```
      c       ter2: number of  > m1(1) and  < m0(2) terminal decisions
      c       ter2dec: P{m1(1) < m < m2(0)|m} for terminal decisions
      c       ter1: number of  < m0(1) terminal decisions
      c       ter1dec: P{m < m0(1)|m} for terminal decisions
      c       thresh(2): threshold for SPRT plans 1 & 2
      c       totcnt: total samples generated for simulated sampling from a mean
      c       tpla: a for var  =  a(mean)^b
      c       tplb: b for var  =  a(mean)^b
      c       unit: file connection number
      c       var: computed variance
      c       vb: variance of the slope b
      c       vd: variance of the slope d
      c       vnb1: intermediate variable for computing stop limits
      c       vnb2: intermediate variable for computing stop limits
      c       weight: weights for computing expected asn and P{dec} functions for binomial count sampling
      c       x: random variate generated during simulations
      c       xbar: average of x values

              character*1 respond,t
              character*12 file
              integer i,i2,icount,init(2),ivalues,j,jj,mci,nmax,unit,initun,distr,ngbk,confirm
              real a,alpha(2),asn(29,2),asnprt(11,2),b,beta(2),count,
             +dummy,h(14),highi(2),k(2),ksim,lna,lnb,lowi(2),m(29,2),m0(2),
             +m1(2),mean(11,2),oc(29,2),ocprint(11),p,pdec1(58),pdec2(58),
             +pdec3(58),pnb0,pnb1,q,qnb0,qnb1,result,revoc(29,2),seq3,seq2,
             +seq2dec,seq1,seq1dec,simasn(58),q0,q1,bvar1,sumpdec,temp
              real simmean(58),sll(2),slope(2),slu(2),sum,ter3,ter2,ter2dec,
             +ter1,ter1dec,thresh(2),totcnt,tpla,tplb,vnb1,vnb2,x,xbar
              real c,d,vd,mse,rn,mlnm,weight(11),z(11),edec(58),easn(58),
             +mt(2),pi(2),pmbin(29,2),spmbin(58),temp1(58),temp2(58),temp3(58),
             +vb,tally,var,pxsum,refm(101),pxgtt(101)

              data h/5,4.5,4.,3.5,3.,2.5,2.,1.5,1.,.5,.4,.3,.2,.1/
              data ocprint/0.99,0.9,0.8,0.7,0.6,0.5,0.4,0.3,0.2,0.1,0.01/
              data init/0,0/
              data weight /0.1974,0.1856,0.1638,0.1358,0.1062,0.0776, +0.0534,0.0346,0.0212,
              0.012,0.0064/
              data z /0.0,0.375,0.625,0.875,1.125,1.375,1.625,1.875,2.125,2.375,2.625/

              t  =  char(9)
              tpla  =  0.
              tplb  =  0.
              i  =  0.

c open file for output

              write (*,'(a)') ' Enter name for output file (≤ 12 characters)'
              read (*,'(a)') file
              open (unit = 10,file = file,status = 'unknown')

c the initialized value of unit is the screen

              initun  =  0
              unit  =  initun

c ************************************************************************
c *********************** Model selection and parameterization *************************
c ************************************************************************

              confirm  =  0
```

```
500     write (*,*) 'Choose a sampling model'
        write (*,*) '1) binomial'
        write (*,*) '2) Poisson'
        write (*,*) '3) Negative binomial'
        write (*,*) '4) Empirical p-m'
        write (*,*) '5) NGB based p-m'
        write (*,*) 'enter a number 1-5'
        read (*,*) distr
        if (distr.lt.1.or.distr.gt.5) goto 500
505     continue

        if (distr.eq.1) then
           write (unit,510)
510        format (' Binomial distribution used')
        endif

        if (distr.eq.2) then
           write (unit,520)
520        format (' Poisson distribution used')
        endif
        if (distr.eq.3) then
           if (confirm.eq.0) then
530           write (*,*) 'use 1) constant k or 2) k  =  f(TPL)'
              read (*,*) ngbk
              if (ngbk.ne.1.and.ngbk.ne.2) goto 530
           end if
           if (ngbk.eq.1) then
              write (unit,540)
540           format (' NGB model with constant k used')
           else
              if (confirm,eq.0) then
                 write (*,*) 'Input TPL parameters'
                 write (*,*) 'intercept;'
                 read (*,*) tpla
                 write (*,*) 'slope;'
                 read (*,*) tplb
              end if
              write (unit,550) tpla,tplb
550           format (' NGB model with k  =  f(TPL) used',/,
       +              'a  =  ',f5.3,1x,'b  =  ',f5.3)
           end if
        end if

        if (distr.eq.4) then
           if (confirm.eq.0) then
              write (*,*) 'Input empirical p-m model parameters'
              write (*,*) 'intercept;'
              read (*,*) c
              write (*,*) 'slope;'
              read (*,*) d
              write (*,*) 'variance of the slope;'
              read (*,*) vd
              write (*,*) 'mse from the p-m regression;'
              read (*,*) mse
              write (*,*) 'number of data points for p-m regression;'
              read (*,*) rn
              write (*,*) 'mean of the independent variable {lnm};'
              read (*,*) mlnm
           end if
           write (unit,560)
```

```
560         format(' Empirical p-m model used',/,
       +          ' intercept  slope  var(slope)   mse    N     mean of lnm'}
            write (unit,565) c,d,vd,mse,rn,mlnm
565         format (3x,f6.3,1x,f6.3,5x,f7.4,1x,f6.3,1x,f4.0,3x,f8.3)
          endif

          if (distr.eq.5) then
            if (confirm.eq.0) then
              write (*,*) 'Input TPL parameters'
              write (*,*) 'intercept;'
              read (*,*) tpla
              write (*,*) 'slope;'
              read (*,*) tplb
              write (*,*) 'variance of the slope;'
              read (*,*) vb
              write (*,*) 'mse from the regression;'
              read (*,*) mse
              write (*,*) 'number of data points for regression;'
              read (*,*) rn
              write (*,*) 'mean of the independent variable {lnm};'
              read (*,*) mlnm
            endif
            write (unit,570)
570         format(' NGB p-m model with TPL used',/,
       +          ' intercept  slope  var(slope)   mse    N     mean of lnm')
            write (unit,575) tpla,tplb,vb,mse,rn,mlnm
575         format (3x,f6.3,1x,f6.3,5x,f7.4,1x,f6.3,1x,f4.0,3x,f8.3)
          end if

          if (confirm.eq.0) then
580         write (*,*) 'Parameters correct? (Y or N)'
            read (*,'(a)') respond
            if (respond.ne.'N'.and.respond.ne.'n') then
              if (respond.ne.'Y'.and.respond.ne.'y') goto 580
            end if
            if (respond .eq. 'N' .or. respond .eq. 'n') then
              goto 500
            else
              unit = 10
              confirm = 1
              goto 505
            end if
          end if

          unit = initun

c ******************************************************************************
c ***************************** SPRT parameterization **************************
c ******************************************************************************

7         continue
          write (*,8) i
8         format (' Enter plan number; ',i3)
          read (*,*) i
          if (i.ne.1.and.i.ne.2) goto 7
          if (i.eq.1.and.init(1).eq.0) init(1) = 1
          if (i.eq.2.and.init(2).eq.0) init(2) = 1

9         if (distr.eq.4.or.distr.eq.5) then
            write (*,10) mt(i)
10          format (' threshold as density; ',f6.3)
            read (*,*) mt(i)
```

```
              if (distr.eq.4) then
                pi(i) = 1−exp(−exp(c+d*log(mt(i))))
              else
                write (*,11) tally
11              format (' tally threshold; ',f4.0)
                if (i.eq.1) then
                   read (*,*) tally
                endif
                var = tpla*mt(i)**tplb
                k(i) = mt(i)**2/(var−mt(i))
                call ngbcdf(mt(i),k(i),tally,pxsum)
                pi(i) = 1.0-pxsum
              endif
              write (*,12) pi(i),k(i)
12            format (' intervention proportion = ',f6.4,
     +        ' k based on TPL = ',f6.4)
            endif

            write (*,13) m0(i)
13          format ('mean for null hypothesis; ', f6.3)
            read (*,*) m0(i)
            write (*,14) m1(i)
14          format (' mean for alternative hypothesis; ',f6.3)
            read (*,*) m1(i)
            write (*,15) alpha(i)
15          format (' alpha; ',f5.3)
            read (*,*) alpha(i)
            write (*,16) beta(i)
16          format (' beta; ',f5.3)
            read (*,*) beta(i)
            thresh(i) = (m0(i)+m1(i))/2.
            if (distr.eq.3) then
              if (ngbk.eq.1) then
                 write (*,17) k(i)
17               format (' NGB k; ',f5.3)
                 read (*,*) k(i)
              else
                 k(i) = thresh(i)**2/(tpla*thresh(i)**tplb-thresh(i))
              end if
            end if

            write (*,18)
18          format (' H0 mean   H1 mean        alpha   beta   NGB k')
            write (*,19) m0(i),m1(i),alpha(i),beta(i),k(i)
19          format(f7.3,3x,f7.3,11x,f4.3,4x,f4.3,4x,f5.3)
20          write (*,*) 'Parameters correct? (Y or N)'
            read (*,'(a)') respond
            if (respond.ne.'N'.and.respond.ne.'n') then
               if (respond.ne.'Y'.and.respond.ne.'y') goto 20
            end if
            if (respond .eq. 'N' .or. respond .eq. 'n') goto 9

c ******************************************************************************
c ******************************* SPRT computations ****************************
c ******************************************************************************

c variables used for all distributions

            a = (1−beta(i))/alpha(i)
            lna = log(a)
            b = beta(i)/(1−alpha(i))
            lnb = log(b)
```

c variables used for binomial and negative binomial

```
      if (distr.eq.1.or.distr.eq.4.or.distr.eq.5) then
        q1    = 1−m1(i)
        q0    = 1−m0(i)
        bvar1 = log((m1(i)*q0)/(m0(i)*q1))
      end if
      if (distr.eq.3) then
        pnb0 = m0(i)/k(i)
        pnb1 = m1(i)/k(i)
        qnb0 = 1 + pnb0
        qnb1 = 1 + pnb1
        vnb1 = (pnb1*qnb0)/(pnb0*qnb1)
        vnb2 = qnb0/qnb1
      end if
```

c calculate stop limit parameters

```
      if (distr.eq.1.or.distr.eq.4.or.distr.eq.5) then
        lowi(i)  = lnb/bvar1
        highi(i) = lna/bvar1
        slope(i) = log(q0/q1)/bvar1
      else if (distr.eq.2) then
        lowi(i)  = lnb/log(m1(i)/m0(i))
        highi(i) = lna/log(m1(i)/m0(i))
        slope(i) = (m1(i) − m0(i))/log(m1(i)/m0(i)))
      else
        lowi(i)  = lnb/log(vnb1)
        highi(i) = lna/log(vnb1)
        slope(i) = k(i)*(log(qnb1/qnb0)/log(vnb1))
      end if
```

c OC and ASN computations

```
      do 30 j = 1,14
        oc(j,i)    = (a**h(j) − 1)/(a**h(j) − b**h(j))
        revoc(j,i) = 1 − oc(j,i)
        if (distr.eq.1.or.distr.eq.4.or.distr.eq.5) then
          m(j,i)   = (1 −(q1/q0)**h(j))/((m1(i)/m0(i))**h(j) − (q1/q0)
     +              **h(j))
          asn(j,i) = (lnb*oc(j,i) + (1 −oc(j,i))*lna)/m(j,i)*bvar1
     +              + log(q1/q0))
        else if (distr.eq.2) then
          m(j,i)   = ((m1(i) − m0(i))*h(j))/((m1(i)/m0(i))**h(j) − 1)
          asn(j,i) = (lnb*oc(j,i) + lna*(1 −oc(j,i)))/(m(j,i)*
     +              + log(m1(i)/m0(i)) + m0(i) − m1(i))
        else
          p        = (1 − vnb2**h(j))/(vnb1**h(j) − 1)
          m(j,i)   = p*k(i)
          asn(j,i) = (lnb*oc(j,i) + lna*(1 −oc(j,i)))/(k(i)*log(vnb2)
     +              + k(i)*p*log(vnb1))
        end if
30    continue
      oc(15,i)    = lna/(lna − lnb)
      revoc(15,i) = 1 − oc(13,i)
      if (distr.eq.1.or.distr.eq.4.or.distr.eq.5) then
        m(15,i)   = −log(q1/q0)/bvar1
        asn(15,i) = −lna*lnb/(log(m1(i)/m0(i))*log(q0/q1))
      else if (distr.eq.2) then
        m(15,i)   = (m1(i) − m0(i))/log(m1(i)/m0(i))
        asn(15,i) = −(lna*lnb)/(m(j,i)*(log(m1(i)/m0(i)))**2)
```

```
          else
            p   = log(qnb1/qnb0)/log(vnb1)
            q   = 1+p
            m(15,i) = p*k(i)
            asn(15,i) = -(lna*lnb)/(k(i)*p*q*(log(vnb1))**2)
          end if
          do 35 index = 1,14
            j  = 15 - index
            jj = 15 + index
            oc(jj,i) = ((1/a**h(j))-1)/((1/a**h(j))-(1/b**h(j)))
            revoc(jj,i) = 1-oc(jj,i)
            if (distr.eq.1.or.distr.eq.4.or.distr.eq.5) then
              m(jj,i) = (1-(1/(q1/q0)**h(j)))/((1/(m1(i)/m0(i))**h(j)) -
       +(1/(q1/q0)**h(j)))
              asn(jj,i) = (lnb*oc(jj,i)+(1-oc(jj,i))*lna)/(m(jj,i)*bvar1
       +log(q1/q0))
            else if (distr.eq.2) then
              m(jj,i) = ((m1(i) - m0(i))*(-h(j)))/(1/(m1(i)/m0(i))**h(j)-1)
              asn(jj,i) = (lnb*oc(jj,i)+lna*(1-oc(jj,i)))/(m(jj,i)*
       +log(m1(i)/m0(i))+m0(i)-m1(i))
            else
              p  = (1 - 1/vnb2**h(j))/(1/vnb1**h(j)-1)
              m(jj,i) = p*k(i)
              asn(jj,i) = (lnb*oc(jj,i)+lna*(1-oc(jj,i)))/(k(i)*log(vnb2)
       +k(i)*p*log(vnb1))
            end if
35        continue

c *****************************************************************************
c ******************* Incorporation of p-m model variablity in OC and ASN *******************
c *****************************************************************************

c If distr = 4 (empirical binomial) or distr = 5 (NGB binomial)
c then do the following:
c 1) Convert m(j,i) which is expressed as a proportion to
c a density (pmbin(j,i)).
c 2) Compute effect of variation in p-m model or variation in
c NGB k on OC and ASN
c 3) Set revoc(j,i) and asn(j,i) to result from binemp and binngb
c subroutines

          if (distr.eq.4) then
            do 36 j = 1,29
              pmbin(j,i) = exp((log(-log(1-m(j,i)))-c)/d)
36          continue
            call binemp(29,pmbin(1,i),d,vd,mse,rn,mlnm,weight,z,
       + oc(1,i),asn(1,i),edec,easn)
          endif
          if (distr.eq.5) then
            do 37 j = 1,101
              refm(j) = float(j-1)
              call ngbcdf(refm(j),k(i),tally,pxsum)
              pxgtt(j) = 1 - pxsum
37          continue
            do 38 j = 1,29
              call tabex(refm,pxgtt,m(j,i),101,pmbin(j,i))
38          continue
            call binngb(29,pmbin(1,i),m(1,i),tpla,tplb,mse,rn,vb,mlnm,
       + oc(1,i),asn(1,i),tally,z,weight,edec,easn)
          endif
          if (distr.eq.4.or.distr.eq.5) then
            do 39 j = 1,29
```

```
              revoc(j,i) = 1−edec(j)
              asn(j,i) = easn(j)
39         continue
           endif

c ****************************************************************************
c ******************************** display results ******************************
c ****************************************************************************

           write (*,40)
40         format (' OC   Mean   ASN')
           do 45 j = 1,11
              ivalues = 29
              dummy = 1−ocprint(j)
              if (distr.eq.4.or.distr.eq.5) then
                 call tabex(pmbin(1,i),revoc(1,i),dummy,ivalues,result)
                 if (result.le. 0.0) then
                    result = 0.000
                 endif
              else
                 call tabex(m(1,i),revoc(1,i),dummy,ivalues,result)
              endif
              mean(j,i) = result
              if (distr.eq.4) then
                 dummy = 1−exp(−exp(c +d*log(result)))
              elseif (distr.eq.5) then
                 call ngbcdf(result,k(i),tally,pxsum)
                 dummy = 1−pxsum
              else
                 dummy = result
              endif
              call tabex(asn(1,i),m(1,i),dummy,ivalues,result)
              asnprt(j,i) = result
              write (*,50) ocprint(j),mean(j,i),asnprt(j,i)
50         format (1x,f4.2,1x,f6.2,1x,f7.2)
45         continue

60         write (*,*) 'Redo OC and ASN? (Y or N)'
           read (*,'(a)') respond
           if (respond.ne.'N'.and.respond.ne.'n') then
              if (respond.ne.'Y'.and.respond.ne.'y') goto 60
           end if
           if (respond .eq. 'Y' .or. respond .eq. 'y') goto 9
61         write (*,*) 'Write OC and ASN functions to file? (Y or N)'
           read (*,'(a)') respond
           if (respond.ne.'N'.and.respond.ne.'n') then
              if (respond.ne.'Y'.and.respond.ne.'y') goto 61
           end if
           if (respond .eq. 'Y' .or. respond .eq. 'y') then
              write (10,62) i
62         format (' Complete OC and ASN functions for plan ',i2)
              write (10,63)
63         format (' H0 mean   H1 mean   k   alpha   beta')
              write (10,64) m0(i),m1(i),k(i),alpha(i),beta(i)
64         format(f7.3,3x,f7.3,3x,f5.3,3x,f4.3,4x,f4.3)
              write (10,65)
65         format (' Low intercept High intercept Slope')
              write (10,66) lowi(i), highi(i), slope(i)
66         format (6x,f8.2,9x,f7.2,1x,f6.3)
              if (distr.eq.4.or.distr.eq.5) then
                 write (10,67)
```

```
   67          format (' threshold proportion')
               write (10,68) mt(i),pi(i)
   68          format (2(3x,f6.3,7x,f6.4,3x))
               if (distr.eq.5) then
                  write (10,69) tally
   69             format(' tally threshold  =  ',f5.1)
               end if
            end if
            write (10,70) t,t
   70       format (' Mean',a1,'OC',a1,'ASN')
            if (distr.eq.4.or.distr.eq.5) then
               do 71 j  =  1,29
                  write (10,72) pmbin(j,i),t,edec(j),t,asn(j,i)
   72             format (1x,f6.2,a1,f4.2,a1,f7.2)
   71          continue
            else
               do 73 j  =  1,29
                  write (10,74) m(j,i),t,oc(j,i),t,asn(j,i)
   74             format (1x,f6.2,a1,f4.2,a1,f7.2)
   73          continue
            end if
         end if

c NGB model simulation

         if (distr.eq.3) then
   75       write (*,*) 'Simulate OC and ASN? (Y or N)'
            read (*,'(a)') respond
            if (respond.ne.'N'.and.respond.ne.'n') then
               if (respond.ne.'Y'.and.respond.ne.'y') goto 75
            end if
            if (respond .eq. 'Y' .or. respond .eq. 'y') then
   76          format (' Simulated OC and ASN functions for plan ',i2)
               write (10,77)
   77          format (' H0 mean    H1 mean    k    alpha    beta')
               write (10,78) m0(i),m1(i),k(i),alpha(i),beta(i)
   78          format(f7.3,3x,f7.3,3x,f5.3,3x,f4.3,4x,f4.3)
               write (10,79)
   79          format (' Low intercept    High intercept    Slope')
               write (10,80) lowi(i), highi(i), slope(i)
   80          format (6x,f8.2,9x,f7.2,1x,f6.3)
               call ngbsim (m(1,i), thresh(i),tpla,tplb,k(i),ngbk,lowi,highi,
     +         slope,oc(1,i),asn(1,i),29)
            end if
         end if

   90    write (*,*) 'Stop now? (Y or N)'
         read (*,'(a)') respond
         if (respond.ne.'N'.and.respond.ne.'n') then
            if (respond.ne.'Y'.and.respond.ne.'y') goto 90
         end if
         if (respond .eq. 'Y' .or. respond .eq. 'y') goto 1000

         if (init(1).eq.0.or.init(2).eq.0) goto 7

c ****************************************************************************
c *********************** display results for both plans and verify ***********************
c ****************************************************************************

         call print (unit,m0,m1,k,alpha,beta,lowi,highi,slope,
     +   ocprint,mean,asnprt,mt,pi,distr,tally)
```

```
          if (unit .eq. initun) then
100         write (*,*) 'Redo a plan? (Y or N)?'
            read (*,'(a)') respond
            if (respond.ne.'N'.and.respond.ne.'n') then
              if (respond.ne.'Y'.and.respond.ne.'y') goto 100
            end if
            if (respond .eq. 'Y' .or. respond .eq. 'y') then
              goto 7
            else
              unit = 10
              call print (unit,m0,m1,k,alpha,beta,lowi,highi,slope,
     +        ocprint,mean,asnprt,mt,pi,distr,tally)
            end if
          end if

c ******************************************************************************
c ************************ Simulation of tripartite plan performance ************************
c ******************************************************************************

          write (*,*) 'starting simulations...'
          write (*,*) 'Maximum sample size? '
          read (*,*) nmax
          write (*,*) 'Monte Carlo iterations?'
          read (*,*) mci
150       format (' Max. sample size = ',i4,' iterations = ',i4)
          icount = 0
          do 200 j = 1,18
            icount = icount + 1
            simmean(icount) = m(j,1)
200       continue
          do 120 j = 19,29
            if (m(j,1).gt.m(1,2)) then
              temp = m(j−1,1)
              goto 220
            endif
            icount = icount + 1
            simmean(icount) = m(j,1)
210       continue
220       do 230 j = 1,29
            if (temp.lt.m(j,2)) then
              icount = icount + 1
              simmean(icount) = m(j,2)
            endif
230       continue
          write (*,240) icount
          write (10,240) icount
240       format (' number of means = ',i3)

          do 300 j = 1,icount
          write (*,250) j,icount
250       format (' simulating from mean ',i2,' of ',i2)

c set decision counters to zero

                  seq1 = 0.
                  seq3 = 0.
                  seq2 = 0.
                  ter1 = 0.
                  ter3 = 0.
                  ter2 = 0.
                  totcnt = 0.
```

```
           if (distr.eq.3) then
             if (ngbk.eq.1) then
               if (simmean(j).le.((m1(1)+m0(2))/2.)) then
                 ksim  =  k(1)
               else
                 ksim  =  k(2)
               end if
             else
               ksim  =  simmean(j)**2/(tpla*simmean(j)**tplb – simmean(j))
             end if
           endif

c Monte Carlo loop

           do 400 i2  =  1,mci
             count  =  0.
             sum  =  0.
             if (distr.eq.1.or.distr.eq.4.or.distr.eq.5) then
               call binomial(simmean(j),x)
             else if (distr.eq.2) then
               call poisson(simmean(j),x)
             else
               call ngb(simmean(j),ksim,x)
             end if
             sum  =  sum + x
             count  =  count + 1.

c compute stop lines and compare samples to them

450        continue
           slu(1)  =  highi(1) + slope(1)*count
           sll(1)  =  lowi(1) + slope(1)*count
           slu(2)  =  highi(2) + slope(2)*count
           sll(2)  =  lowi(2) + slope(2)*count
           if ((sum.gt.slu(2).and.sum.gt.slu(1)).or.(sum.lt.sll(1).and.
      +    sum.lt.sll(2))) then
             if (sum.lt.sll(1).and.sum.lt.sll(2)) then
               totcnt  =  totcnt + count
               seq1  =  seq1 + 1.
             else
               totcnt  =  totcnt + count
               seq3  =  seq3 + 1.
             end if
           else if (sum.gt.slu(1) .and. sum.lt.sll(2)) then
             seq2  =  seq2 + 1.
             totcnt  =  totcnt + count
           else if (count .ge. nmax) then
  xbar  =  sum/float(count)
             if (xbar.le.thresh(1)) then
               ter1  =  ter1 + 1.
               totcnt  =  totcnt + count
             else if (xbar.gt.thresh(1).and.xbar.le.thresh(2)) then
               ter2  =  ter2 + 1.
               totcnt  =  totcnt + count
             else
               ter3  =  ter3 + 1.
             totcnt  =  totcnt + count
             end if
           else
             if (distr.eq.1.or.distr.eq.4.or.distr.eq.5) then
               call binomial(simmean(j),x)
```

```
              else if (distr.eq.2) then
                call poisson(simmean(j),x)
              else
                call ngb(simmean(j),ksim,x)
              end if
              sum   =  sum + x
              count =  count + 1.
              goto 450
            end if
400       continue


c compute OC values and probabilities of decision

          seq1dec  =  seq1/float(mci)
          ter1dec  =  ter1/float(mci)
          seq2dec  =  seq2/float(mci)
          ter2dec  =  ter2/float(mci)
          pdec1(j) =  seq1dec + ter1dec
          pdec2(j) =  seq2dec + ter2dec
          pdec3(j) =  1 - (pdec1(j) + pdec2(j))


c compute average sample size

          simasn(j) = totcnt/float(mci)
300     continue


c ******************************************************************************
c ***************** Incorporation of p-m model variability in P{dec} and ASN *****************
c ******************************************************************************


c If distr = 4 (empirical binomial) or distr = 5 (NGB binomial)
c then do the following:
c 1) convert simmean(j), currently a proportion, to a density
c [spmbin(j)]
c 2) compute effect of variation in p-m model on pdec and asn

        if (distr.eq.4) then
          do 460 j = 1,icount
            spmbin(j) = exp((log(-log(1-simmean(j)))-c)/d)
460       continue
          call binemp(icount,spmbin,d,vd,mse,rn,mlnm,weight,z,pdec1,
     +      simasn,temp1,easn)
          call binemp(icount,spmbin,d,vd,mse,rn,mlnm,weight,z,pdec2,
     +      simasn,temp2,easn)
          call binemp(icount,spmbin,d,vd,mse,rn,mlnm,weight,z,pdec3,
     +      simasn,temp3,easn)
        endif
        if (distr.eq.5) then
          do 463 j = 1,icount
            call tabex(refm,pxgtt,simmean(j),icount,spmbin(j))
463       continue
          call binngb(icount,spmbin,simmean,tpla,tplb,mse,rn,vb,mlnm,
     +      pdec1,simasn,tally,z,weight,temp1,easn)
          call binngb(icount,spmbin,simmean,tpla,tplb,mse,rn,vb,mlnm,
     +      pdec2,simasn,tally,z,weight,temp2,easn)
          call binngb(icount,spmbin,simmean,tpla,tplb,mse,rn,vb,mlnm,
     +      pdec3,simasn,tally,z,weight,temp3,easn)
        endif
```

```
          if (distr.eq.4.or.distr.eq.5) then
             do 465 j = 1,icount
                pdec1(j) = temp1(j)
                pdec2(j) = temp2(j)
                pdec3(j) = temp3(j)
                simasn(j) = easn(j)
                simmean(j – = spmbin(j)
465          continue
          endif

c adjust P{dec} functions so that they sum to 1.0

          do 466 j = 1,count
             sumpdec = pdec1(j) + pdec2(j) + pdec3(j)
466       continue
          do 467 j = 1,count
             pdec1(j) = pdec1(j)/sumpdec
             pdec2(j) = pdec2(j)/sumpdec
             pdec3(j) = pdec3(j)/sumpdec
467       continue

          write(*,470)
470       format(' Mean  P{dec1}  P{dec2}  P{dec3}  ASN')
          write(10,471) t,t,t,t
471       format(' Mean',a1,'P{dec1}',a1,'P{dec2}',a1,'P{dec3}',a1,
         +'ASN')
          do 480 j = 1,icount
             write(*,481) simmean(j),pdec1(j),pdec2(j),pdec3(j),simasn(j)
481          format(1x,f7.3,3x,f7.3,3x,f7.3,3x,f7.3,3x,f7.2)
             write(10,482) simmean(j),t,pdec1(j),t,pdec2(j),t,pdec3(j),t,
         +simasn(j)
482          format(1x,f7.3,a1,f7.3,a1,f7.3,a1,f7.3,a1,f7.2)
480       continue
          pause

1000      continue
          close (10)
          end


          subroutine tabex(val,arg,dummy,k,value)

c *********************************************************************
c This subroutine uses linear interpolation to compute y = f(x).
c Sets of y values (val) and x arguments (arg) are passed along
c with a dummy x value (dummy) for which y (value) is to be
c determined. k is the size of the val and arg arrays. Values
c in the argument (arg) array must be in ascending order.
c *********************************************************************
          real val,arg,dummy,value
          integer k,j
          dimension val(k),arg(k)
          do 1 j = 2,k
          if (dummy.gt.arg(j)) go to 1
2         value = (dummy – arg(j –1))*(val(j) – val(j –1))/
         +(arg(j) – arg(j –1)) + val(j –1)
          return
1         continue
          j = k
          go go 2
          end


          subroutine ngb(mean,k,x)
```

```
c ****************************************************************************
c This subroutine generates negative binomial distributed random
c variables using a rejection method. It requires uniform {0,1}
c random variates be generated as the variable r. If the Fortran
c compiler does not support a 'random' function, one must be provided.
c ****************************************************************************
      real x,mean,k,px,ppx,pxsum,r
      call random(r)
      x = 0.0
      px = 1/((1+mean/k)**k)
      ppx = px
      pxsum = px
      if (r .le. px) goto 4
5     x = x + 1
      px = ((k+x -1.)/x)*(mean/(mean+k))*ppx
      ppx = px
      pxsum = pxsum + px
      if (r .le. pxsum) goto 4
      if (x. gt. 200.) goto 4
      goto 5
4     continue
      return
      end

      subroutine binomial(p,x)

c ****************************************************************************
c This subroutine generates binomial distributed random
c variables x {x = 0 or 1} using a rejection method.
c The subroutine requires uniform {0,1} random variates be generated
c as the variable r. If the Fortran compiler does not support a
c 'random' function, one must be provided.
c ****************************************************************************
      real p,x,r
      call random(r)
      if (r.lt.p) then
         x = 1.
      else
         x = 0.0
      end if
      return
      end

      subroutine poisson(mean,poivar)

c ****************************************************************************
c Generates pseudo random Poisson distributed random variables using a
c rejection method. A uniform random variate (r) is required. If
c the FORTRAN compiler does not support a function (random in this routine)
c for generating uniform deviates, one must be supplied.
c ****************************************************************************
      real mean,poivar,r,px,pxsum

      call random(r)
      poivar = 0.
      px = exp(-mean)
      pxsum = px
      if (r.le.pxdum) return

10    continue
         poivar = poivar + 1.
         px = px*mean/poivar
         pxsum = pxsum + px
```

```
            if (r.le.pxsum) return
            if (poivar.ge.500) return
       goto 10

       return
       end

       subroutine binemp (entries,m,d,vd,mse,rn,mlnm,weight,z,novdec,
      + novasn,edec,easn)

c ********************************************************************************
c Subroutine computes expected P{dec} and ASN for sequential binomial sampling
c programs based on the empirical model ln(−ln(pT)) = a + bln(m) where pT is
c the proportion of samples with less than T animals and m is the mean. A
c maximum of 58 nominal values can be passed.
c ********************************************************************************

c variable definitions

c c,d; intercept and slope of p0 − m relationship
c dec(k,j), asn(k,j); P{dec} and asn values indexed by the means (m(j))
c     and the bias weights k
c edec(j), easn(j); expected P{dec} and asn functions
c m(j) mean density corresponding to 1 −p(j) where p(j) is the proportion of
c     samples with ≤ tally threshold
c mlnm; mean of independent variable of p −m relationship
c mse; mean square error of p0 − m relationship
c negbm(j) means biased by P{x ≤ tally threshold} being less than nominal value
c novdec(j), novasn(j): P{dec} and asn values exclusive of variance in the p −m
c     relationship
c posbdec(j),negbdec(j),posbasn(j),negbasn(j): oc and asn values for biased p
c     values calculated for the set of means = m(j)
c posbm(j) means biased by P{x ≤ tally threshold} being greater than nominal value
c vd, rn; variance of d and data points used in p −m relationship
c vlnlnp(j); variance of ln(−ln(1 −pT)
c weight(k) set of values for weighting biased P{dec} and ASN curves
c z(k): standard normal values associated with weight(k)
c declarations

       integer entries,j,k

       real d,vd,mse,rn,mlnm,novdec,novasn,m,weight(11),z(11),
      + vlnlnp(58),temp,posbm(58),negbm(58),posbdec(58),negbdec(58),
      + posbasn(58),negbasn(58),bdec(11,58),basn(11,58),edec,easn,
      + dummy,result

       dimension novdec(entries),novasn(entries),m(entries),
      + edec(entries),easn(entries)

c computations

       do 30 j = 1,entries
          vlnlnp(j) = mse/rn + (log(m(j)) − mlnm)**2*vd + mse
          bdec(1,j) = novdec(j)
          basn(1,j) = novasn(j)
30     continue
       do 40 k = 2,11
          do 50 j = 1,entries
             temp = exp(z(k)*sqrt(vlnlnp(j))/d)
             posbm(j) = m(j)/temp
             negbm(j) = m(j)*temp
50        continue
```

```
c     compute P{dec} and ASN values for biased means that have the same p value
c     as the nominal means

              do 60 j = 1,entries
                 dummy = m(j)
                 call tabex(novdec,posbm,dummy,entries,result)
                 posbdec(j) = result
                 call tabex(novdec,negbm,dummy,entries,result)
                 negbdec(j) = result
                 call tabex(novasn,posbm,dummy,entries,result)
                 posbasn(j) = result
                 call tabex(novasn,negbm,dummy,entries,result)
                 negbasn(j) = result
                 bdec(k,j) = (posbdec(j) + negbdec(j))/2.
                 basn(k,j) = (posbasn(j) + negbasn(j))/2.
60            continue
40         continue

c     compute expected P{dec} and ASN values

           do 70 j = 1,entries
              edec(j) = 0.0
              easn(j) = 0.0
              do 80 k = 1,11
                 edec(j) = edec(j) + bdec(k,j) * weight(k)
                 easn(j) = easn(j) + basn(k,j) * weight(k)
80            continue
70         continue
           return
           end
           subroutine binngb(entries,m,p,a,b,mse,rn,vb,mlnm,novdec,
         + novasn,tally,z,weight,edec,easn)

c *******************************************************************************
c     Subroutine computes expected P{dec} and ASN for sequential binomial sampling
c     programs on a negative binomial model. Expected values are based on
c     variation in k modeled using variation about TPL ( residuals about TPL are
c     assumed to be normally distributed. A maximum of 58 nominal values can be
c     passed.
c *******************************************************************************

c     variable definitions

c     a        TPL parameter
c     b        TPL parameter
c     dummy    temporary variable
c     easn     expected ASN
c     edec     expected P{dec}
c     entries  number of P{dec} and ASN values passed
c     i        integer index
c     j        integer index
c     ktemp    temporary variable
c     lnm      ln(mean)
c     lnv      ln(variance)
c     m        mean
c     mlnm     mean of ln(mean) used in TPL
c     mse      means square error for TPL regression
c     negasn   asn corresponding to negative deviation about TPL
c     negdec   P{dec} corresponding to negative deviation about TPL
c     novasn   nominal ASN
c     novdec   nominal P{dec}
c     p        proportion of samples with more than T organisms (corresponds to m)
c     posasn   asn corresponding to positive deviation about TPL
```

```
c posdec   P{dec} corresponding to positive deviation about TPL
c pxsum    NGB cumulative distribution function value
c rn       data points in TPL regression
c tally    T value or p-m model
c v        variance
c vb       variance of TPL b
c vlnv     variance of ln(variance predicted via TPL)
c weight   normal probability for z
c z        standard normal deviate

      integer entries,j,i
      real m,p,a,b,mse,rn,vb,mlnm,novdec,novasn,tally,z,weight,edec,
     +easn,v,lnv,lnm,vlnv,dummy,ktemp,pxsum,posdec,negdec,
     +posasn,negasn
      dimension m(entries),p (entries),novdec (entries),novasn(entries),
     +z(11),weight(11),edec (entries),easn(entries)

c computations

      do 1 j = 1,entries
        v = a*m(j)**b
        lnv = log(v)
        lnm = log(m(j))
        vlnv = mse/rn + (lnm-mlnm)**2*vb + mse
        edec(j) = 0.0
        easn(j) = 0.0
        do 35 i = 1,11
          dummy = lnv + z(i)*sqrt(vlnv)
          dummy = exp(dummy)
          ktemp = (dummy - m(j))/m(j)**2
          if (ktemp.lt.0.01) ktemp = 0.01
          ktemp = 1./ktemp
          call ngbcdf (m(j),ktemp,tally,pxsum)
          dummy = 1.0-pxsum
          call tabex(novdec,p,dummy,entries,posdec)
          if (posdec.gt.1.) posdec = 1.0
          if (posdec.lt.0.) posdec = 0.
          call tabex(novasn,p,dummy,entries,posasn)
          if (posasn.lt.1.0) posasn = 1.0

c repeat for negative deviation

          dummy = lnv - z(i)*sqrt(vlnv)
          dummy = exp(dummy)
          ktemp = (dummy-m(j))/m(j)**2
          if (ktemp.lt.0.01) ktemp = 0.01
          ktemp = 1./ktemp
          call ngbcdf (m(j),ktemp,tally,pxsum)
          dummy = 1.0-pxsum
          call tabex(novdec,p,dummy,entries,negdec)
          if (negdec.gt.1.) negdec = 1.0
          if (negdec.lt.0.) negdec = 0.
          call tabex(novasn,p,dummy,entries,negasn)
          if (negasn.lt.1.0) negasn = 1.0
          edec(j) = edec(j) + ((posdec + negdec)/2)*weight(i)
          easn(j) = easn(j) + ((posasn + negasn)/2)*weight(i)
35      continue
1     continue
      return
      end

      subroutine ngbcdf (m,k,t,pxsum)
```

```
c ******************************************************************************
c This routine computes the probability that x is ≤ T when x is
c distributed as a negative binomial random variable with m
c (the mean) and k (the dispersion parameter).
c The routine uses a recursive relationship for computing the
c probability mass function for the negative binomial distribution.
c ******************************************************************************
      real m,p,q,k,px,pxp,pxsum,x,t
10    continue
      p = m/k
      q = 1+p
      x = 0.
      pxsum = 0.
      px = 1/q**k
      pxsum = pxsum + px
      pxp = px
15    continue
      if (x.lt.t) then
         x = x+1
         px = ((k+x-1)*p/(x*q))*pxp
         pxp = px
         pxsum = pxsum+px
         goto 15
      end if
      return
      end

      subroutine print (unit,m0,m1,k,alpha,beta,lowi,highi,slope,
     + ocprint,mean,asnprt,mt,pi,distr,tally)
      real m0(2),m1(2),k(2),alpha(2),beta(2),lowi(2),highi(2),slope(2),
     + ocprint(11),mean(11,2),asnprt(11,2),mt(2),pi(2),tally
      integer unit,j,distr
      write (unit,1)
1     format ('H0 mean  H1 mean   k    alpha   beta',
     +' H0 mean  H1 mean   k    alpha   beta')
      write (unit,2) m0(1),m1(1),k(1),alpha(1),beta(1),
     +m0(2),m1(2),k(2),alpha(2),beta(2)
2     format(2(f7.3,3x,f7.3,3x,f5.3,3x,f4.3,4x,f4.3))
      write (unit,3)
3     format (' Low intercept   High intercept   Slope',
     +' Low intercept   High intercept   Slope')
      write (unit,4) lowi(1), highi(1), slope(1),
     +lowi(2), highi(2), slope(2)
4     format (2(6x,f8.2,9x,f7.2,1x,f6.3))
      write (unit,5)
5     format ('   OC   Mean   ASN',1x,'   OC   Mean   ASN')
      do 10 j = 1,11
         write (unit,11) ocprint(j),mean(j,1),asnprt(j,1),
     +   ocprint(j),mean(j,2),asnprt(j,2)
11       format (2(1x,f4.2,1x,f6.2,1x,f7.2))
10    continue
      if (distr.eq.4.or.distr.eq.5) then
         write (unit,15)
15       format (' threshold  proportion   threshold  proportion')
         write (unit,16) mt(1),pi(1),mt(2),pi(2)
16       format (2(3x,f6.3,7x,f6.4,3x))
         if (distr.eq.5) then
            write (unit,17) tally
17          format(' tally threshold = ',f5.1)
         endif
```

```
          endif
          return
          end

          subroutine ngbsim (m,thresh,tpla,tplb,nomk,ngbk,lowi,highi,
         +slope,nomoc,nomasn,values)

c ****************************************************************************
c This subroutine simulates the performance of an SPRT when a negative
c binomial model is used and k is either 1) a constant, or 2) a function
c of TPL.
c ****************************************************************************
          real m,thresh,tpla,tplb,lowi,highi,slope,nomoc,nomasn,ngbsoc,
         +ngbsasn,seql,seqh,terl,terh,totcnt,k,sum,slu,sll,xbar,var,x,nomk
          integer values,i,j,nmax,mci,count,ngbk
          character*1 respond,tab
          dimension m(values),nomoc(values),nomasn(values)
          tab = char(9)

c variables unique to subroutine:

c count, sum; number of samples, sum of animals in total samples from one run
c k; k from negative binomial distribution used in sensitivity analysis
c mci; number of Monte Carlo iterations
c mci; number of monte carlo iterations
c ngbsoc, ngbsasn; oc and asn values determined via simulation
c nmax; maximum number of samples to be taken
c nmax; maximum number of samples to be taken
c seql, seqh, terl, terh; number of above and below threshold
c sequential decisions, number of above and below terminal decisions
c slu, sll; upper and lower stop lines
c totcnt; total number of samples drawn
c tpla, tplb; parameters for Taylors power law
c xbar; mean of sample observations
c
c variables passed to subroutine

c lowi, highi, slope; intercepts and slopes for stop lines
c m; means for which oc and asn are to be computed
c ngbk; indicates whether 1) constant k or 2) k = f(TPL)
c nomk; nominal value of k used to construct stop lines
c nomoc, nomasn; nominal oc and asn values
c thresh; average of means for null and alternate hypothesis
c pla, tplb; a and b for Taylor's power law

c obtain necessary data

1         write (*,*) 'input maximum sample size'
          read (*,*) nmax
          write (*,*) 'input monte carlo iterations'
          read (*,*) mci
          k = 0.0
          if (ngbk.eq.1) then
             write (*,5) nomk
5            format (' Nominal k = ',f5.3,/,
         +   ' input k value for simulations')
             read (*,*) k
          end if
          write (*,10) nmax, mci,k
10        format (' maximum samples = ',i4,' monte carlo iterations = ',i5
         +' k* = ',f5.3,/,'*if k = 0, k = f(TPL)')
```

```
11      write (*,*) 'parameters correct? (y or n)'
        read (*,'(a)') respond
          if (respond.ne.'N'and.respond.ne.'n') then
            if (respond.ne.'Y'.and.respond.ne.'y') goto 11
  end if
        if (respond.eq. 'N' .or. respond.eq. 'n') goto 1
        write (10,*) 'Simulation of SPRT performance with variable k'
        write (10,12) nmax,mci, k
12      format (' maximum samples  =  ',i4,' monte carlo iterations  =  ',i5,
       +'  k*  =  ',f5.3,/, *if k  =  0, k  =  f(TPL)')

        write (*,20)
20         format (' Mean   Nom_OC   Sim_OC   Nom_ASN   Sim_ASN')
        write (10,21) tab,tab,tab,tab
21      format (' Mean',a1,'Nom_OC',a1,'Sim_OC',a1,'Nom_ASN',a1,
       +'Sim_ASN')

            do 100 i  =  1,values

c set decision counters to zero and compute k

            seql  =  0.
            seqh  =  0.
            terl  =  0.
            terh  =  0.
            totcnt  =  0.
            if (ngbk.eq.2) then
              var  =  tpla*m(i)**tplb
              k  =  (m(i)**2)/(var − m(i))
            end if

c Monte Carlo loop

            do 110,j = 1,mci
              count  =  0.
              sum  =  0.
        call ngb(m(i),k,x)
              sum  =  sum + x
              count  =  count + 1

c compute stop lines and compare samples to them

120           continue
              slu  =  highi + slope*float(count)
              sll  =  lowi + slope*float(count)
              if (sum.gt.slu.or.sum.lt.sll) then
                if (sum.lt.sll) then
                  ttcnt  =  totcnt + count
                  seql  =  seql + 1.
                else
                  totcnt  =  totcnt + count
                  seqh  =  seqh + 1.
                end if
              else if (count .ge. nmax) then
                xbar  =  sum/float(count)
                if (xbar.lt.thresh) then
                  terl  =  terl + 1.
                  totcnt  =  totcnt + count
                else
                  terh  =  terh + 1.
                  totcnt  =  totcnt + count
                end if
```

```
              else
                call ngb(m(i),k,x)
                sum = sum + x
                count = count + 1
                goto 120
              end if
110     continue

c compute OC values
        ngbsoc = (seql + terl)/float(mci)

c compute average sample size

        ngbsasn = totcnt/float(mci)

c write results

        write (*,130) m(i),nomoc(i),ngbsoc,nomasn(i),ngbsasn
130     format (1x,f6.2,2x,f5.3,3x,f5.3,3x,f7.2,2x,f7.2)
        write (10,131) m(i),tab,nomoc(i),tab,ngbsoc,tab,nomasn(i),tab,
       +  ngbsasn
131     format (1x,f6.2,a1,f5.3,a1,f5.3,a1,f7.2,a1,f7.2)

100     continue

        if (ngbk.eq.1) then
           write (*,*) 'repeat simulations for new k (y or n)?'
           read (*,'(a)') respond
200     if (respond.ne.'N'.and.respond.ne.'n') then
            if (respond.ne.'Y'.and.respond.ne.'y') goto 200
        end if
        if (respond.eq. 'Y' .or. respond.eq. 'y') goto 1
        end if

        return
        end
```

cascade: cascading of sampling decisions to monitor density through time

Background: This program evaluates the performance of sampling plans cascaded through time. Each sample plan independently results in decisions to either resample the population i time intervals in the future or to intervene. A maximum of 20 sampling plans can be cascaded. The program works by first building a table of sampling statistics indexed by sampling time and then by computing summaries of these tables. Extended output yields the time-indexed tables. Two input data files are required; one provides the sampling statistics (P{dec} and ASN) for each sampling plan used and time schedule for using each plan, and the other data file describes the population trajectories to be sampled. At each sampling bout (time) a particular sampling plan is used that can result in one of ndel + 1 decisions; intervene, or resample after $i = 1,\ldots,$ ndel time periods. A maximum of five resample time intervals can be used. Probabilities for these decisions and the average sample number are determined via interpolation from an input data file. Each sample plan can have a maximum of 100 data points that describe the probabilities for these decisions [pdec(i, j, k); $i = 1,\ldots, 100, j = 1,\ldots, 5$ (plan number), $k = 1,\ldots, 6$ (decision)]. A table is built for each possible sampling bout that consists of the time [determined by the beginning sampling time (begtime) and sample interval (deltime)], the population density (dent) which is interpolated from an input file, the probability of sampling at that time (psamp), the six possible decision probabilities, the cumulative probability of making a decision to intervene (sumptrt) and average sample number, and the loss (losst) which is defined as the cumulative population density.

Summary statistics [indexed by the population (pindex)] consist of the cumulative density for the population, the OC (oc), the average total sample size (avgtotn), the expected number of sampling bouts (expbout), the expected loss (exploss), and losses with probability 0.5, 0.2, and 0.05. At the end of the summary table is the statement "if P() loss = Cum.den, SUM P(dec3) < 1 − P()." This means that if the loss with probability x equals the cumulative density, then the cumulative probability of a decision to intervene is less than x.

The program functions as follows:

1. Files that contain the P{dec} and ASN data and the population trajectories to be sampled are specified along with a file to which output is written. Data are read from the respective files which have the following formats. P{dec} and ASN data file. The first line contains two integers separated by a tab, space, or comma. The first number specifies the number of sampling plans and the second the number of time intervals (≤ 5) sampling might be delayed. The second line contains two integers delimited by a tab, space, or comma that specify the number of data points (n) for the first sampling plan and the time when the sampling plan should first be used. The third through $n + 2$ lines contain the mean, P{dec(1)}, P{dec(2)}, ..., P{dec(ndel + 1)}, and ASN values for the first sampling plan delimited by a tab, space, or comma. The $n +$ third line contains the number of data points for the second sampling plan and when the second plan should first be used. The data file then repeats to conclusion. An example is shown below:

3	2			
46	0			
0.245	1.000	0.000	0.000	21.59
0.276	1.000	0.000	0.000	22.82
0.312	1.000	0.000	0.000	23.89
0.353	1.000	0.000	0.000	25.39
0.402	0.998	0.002	0.000	28.34
0.459	0.994	0.006	0.000	30.21
0.526	0.994	0.006	0.000	34.66
0.606	0.980	0.020	0.000	40.71
0.700	0.936	0.064	0.000	50.40
0.812	0.862	0.138	0.000	58.41
0.837	0.772	0.228	0.000	56.41
0.863	0.720	0.280	0.000	58.13
0.890	0.676	0.324	0.000	56.68
...				
49	30			
0.789	1.000	0.000	0.000	21.93
0.856	1.000	0.000	0.000	23.54
...				

Population trajectory data file: the first line contains the number of populations in the file (maximum of 15). Subsequent lines have the time and densities for each population delimited by a tab, space, or comma. An example with seven populations and time values of 12, 20, ..., 89 is shown below:

7							
12	0	0	0	0	0.1	3.88	0.34
20	0	0	0.12	0.42	0.36	1.36	0.2
26	0	0.02	0.2	0.52	0.42	0.58	0.16
33	0	0.1	0.12	0.64	0.16	5.22	0.09
40	0.22	0.02	0.82	1.58	0.16	1.9	0.04
47	0.32	0.06	2.58	4.58	0.3	1.1	0.08
61	9.05	0.38	0.14	16.53	0.2	3.7	0.38
68	2.6	1.44	3.16	20.74	0.01	2.58	3.04
75	8.98	12.22	2.72	8.4	0	8.9	4.68
82	11.82	17.68	1.94	2.02	0.18	12.7	0.8
89	18.84	39.46	3.38	0.16	0.28	18.6	0.1

The file containing the P{dec} and ASN data can be quickly constructed from the output file generated by programs 'sprt' or 'afcm.' It is important that the P{dec} functions in these files be convex at the tails; otherwise, interpolated values may be seriously in error. For example, if the last three values for the P{dec(3)} function were 1.0, 1.0, 0.99 when ndel = 2; these should be changed to 1.0, 1.0, 1.0.

2. Whether or not extended output is desired is specified.
3. When samples are to be taken is specified by providing a starting time, an ending time, and a time interval between samples. If the time interval lies outside the time horizon for which population trajectory data are available, densities are truncated to either the first or last value. Note that while sampling plans constructed for cascading usually are designed with a specific time interval between samples in mind, this feature allows other time intervals to be used for sensitivity analysis.
4. A population scale factor can be applied if desired. This factor scales the densities in the population trajectory file by the multiple provided. A value of 1.0 must be entered if no scaling is desired.
5. Computations are made and results are written to the output file. If desired computations can be repeated for a new population scale factor or for a new sampling time schedule.

An example is shown below. Ulser inputs are underlined and comments are in italics.

```
> cascade
Enter name of file with P{dec} & ASN data
ngb.in
Enter name of file with trajectory data
histpop.in
P{dec} & ASN file = ngb.in Population data file = histpop.in
Is this correct? (Y or N)
y
```

Enter name of file for output
ngb.out
Select extended(1) or summary(2) output
1
Write P{dec} & ASN data to file? (Y or N)
n
Starting time for plan 1 = 0
Starting time for population trajectories = 12
Ending time for population trajectories = 89
Input time for first sample:
12
Input time interval between samples:
7
Input time for last sample:
89
Sampling schedule:

	Start	Interval	Last sample at
	12	7	89

Is this correct? (Y or N)
y
Input population scale factor: 0.0
1.0
computations are performed and displayed on the screen
Repeat for new population scale? (Y or N)
n
Repeat for new time model? (Y or N)
n
a portion of the output file is shown below:
  P{dec} & ASN file = ngb.in   Population data file = histpop.in
  Results for population # 1

Time	Density
12.00	0.00
20.00	0.00
26.00	0.00
33.00	0.00
40.00	0.22
47.00	0.32
61.00	9.05
68.00	2.60
75.00	8.98
82.00	11.82
89.00	18.84

Time	Density	P{samp}	Pd	Pd2	Pd3	Pd4	Pd5	Pd6	p{trt}	ASN	Loss
12.00	0.00	1.000	1.000	0.000	0.000	0.000	0.000	0.000	0.000	21.59	0.00
19.00	0.00	0.000	1.000	0.000	0.000	0.000	0.000	0.000	0.000	21.59	0.00
26.00	0.00	1.000	1.000	0.000	0.000	0.000	0.000	0.000	0.000	21.59	0.00
33.00	0.00	0.000	1.000	0.000	0.000	0.000	0.000	0.000	0.000	21.93	0.00
40.00	0.22	1.000	0.000	0.000	0.000	0.000	0.000	0.000	0.000	21.93	0.77
47.00	0.32	0.000	0.000	0.000	0.000	0.000	0.000	0.000	0.000	21.93	2.66
54.00	4.68	1.000	0.000	0.655	0.345	0.000	0.000	0.000	0.345	51.48	35.46
61.00	9.05	0.655	0.000	0.069	0.931	0.000	0.000	0.000	0.955	39.45	68.25
68.00	2.60	0.045	0.800	0.200	0.000	0.000	0.000	0.000	0.955	51.74	109.03
75.00	8.98	0.009	0.000	0.084	0.916	0.000	0.000	0.000	0.963	40.36	149.56
82.00	11.82	0.037	0.000	0.002	0.998	0.000	0.000	0.000	1.000	17.19	222.36
89.00	18.84	0.000	0.000	0.000	1.000	0.000	0.000	0.000	1.000	9.72	329.67

Note: Pd1–Pd6 values are output even when the number of delays are < 6. The last Pdi values that are not all zeroes are the probabilities for an intervention decision.
Results for population # 2

Time	Density
12.00	0.00
20.00	0.00
26.00	0.02

...

Start	Interval	Last sample at	Possible bouts
12.0	7.0	89.0	12.0

Population scale factor = 1.000

Pop.	Cum.den.	OC	ASN	Bouts	Exp.loss	P(0.5)loss	P(0.2)loss	P(0.05)loss
1	329.67	0.00	145.79	4.75	63.35	43.79	59.94	68.01
2	363.08	0.00	118.75	6.00	163.09	110.76	142.16	157.86
3	103.85	1.00	229.89	6.64	103.85	103.85	103.85	103.85
4	462.20	0.00	119.94	4.08	109.47	73.64	96.52	107.96
5	15.45	1.00	129.46	6.00	15.45	15.45	15.45	15.45
6	363.41	0.00	25.19	1.05	1.78	0.00	0.00	0.00
7	69.53	1.00	180.16	6.48	69.49	69.53	69.53	69.53

** if P()loss = Cum.den, SUM(pdec3) < 1 − P()

program cascade

```
c      Program written in Fortran 77: IBM version compiled using Microsoft
c      FORTRAN compiler v5.1.
c      Program written by Jan Nyrop, Department of Entomology, NYSAES,
c      Cornell University, Geneva, NY 14456
c      Last modification date: 2 February, 1993


c ******************************************************************************
c ****************************** Variable descriptions ****************************
c ******************************************************************************

c asn(j,i): ASN values for mean j and sampling plan i
c avgden: average of two densities used to compute cumulative density
c avgtotn(k): average total number of samples taken for each of k
c     trajectories
c begtime: time for first sample
c bout: sample bout index
c cplan: sampling plan currently being used
c delta: time interval used to compute cumulative density
c deltime: sampling time interval
c dent: population density at current sampling time
c endtime: last sampling time
c expbout(k): expect number of sampling bouts for each of k trajectories
c exploss(k): expected loss for each of k population trajectories
c file1: file name
c file2: file name
c file3: file name
c i: integer index
c iend: integer index
c ii: integer index
```

c iin1: file unit number
c iin2: file unit number
c input: character to record keyboard input
c intasn: interpolated asn value
c intpd(i): interpolated P{dec(i)} value
c iout: file unit number
c ipoint: integer index
c ipops: integer index
c j: integer index
c losst: loss (cumulative density) at sampling time t
c losssb(sb): loss at sampling bout sb
c lossp1(k): interpolated loss for probability p1 and population k
c lossp2(k): interpolated loss for probability p2 and population k
c lossp3(k): interpolated loss for probability p3 and population k
c lsttime: last sampling time
c mean(j,i): mean for P{dec} and ASN functions for tripartite plan i
c mult: multiplier for scaling population density
c ndel: number of time intervals resampling might be delayed
c oc(k): OC value for each of k population trajectories
c oldpopsc: old population scale value
c pdec(j,k,i): probability of making decision i with plan k for mean j
c pindex: population index
c plans: number of sampling plans used
c points(i): number of P{dec} and ASN data points in sampling plan i
c popscale: population scale factor
c potbouts: number of potential sampling bouts
c prpdec(j,i): probability of making decision j i time intervals previous
c prpsamp(i): probability of sampling i time intervals previous
c psamp: probability of sampling at a particular time
c respond: keyboard input
c respond2: keyboard input
c sched(i): time to begin using sampling plan i
c sum1: summed values
c sum2: summed values
c sum3: summed values
c sumden(j,i): cumulative density through time for population i
c sumptrt(sb): probability of intervention summed through sample bout sb
c time: sampling time
c trcnt: number of population trajectories to be sampled
c trden(j,i): population density for trajectory i
c trtime(j): times corresponding to trden(j,i)

c **
c ********************************* Declarations *************************************
c **

```
      real mean(100,20),pdec(100,20,6),
     +asn(100,20),oc(15),expbout(15),avgtotn(15),exploss(15),sched(20),
     +trtime(100),trden(100,15),sumden(100,15),losssb(50),sumptrt(50),
     +lossp1(15),lossp2(15),lossp3(15),prpsamp(5),prpdec(5,5),intpd(6)
      real time,dent,psamp,intasn,sum1,sum2,temp,
     +sum3,deltime,endtime,losst,begtime,popscale,mult,oldpopsc,
     +lsttime,delta,avgden,potbouts


      integer plans,cplan,respond,pindex,trcnt,ipops,k,
     +points(20),iin1,iin2,iout,i,iend,j,ii,ipoint,bout,ndel


      character*12 file1, file2, file3
      character*1 t,input
```

```
            t = char(9)
            oldpopsc = 1.0
            popscale = 0.0

            do 500 k = 1,6
              do 510 j = 1,20
                do 530 i = 1,100
                  pdec(i,j,k) = 0.0
520         continue
510         continue
            intpd(k) = 0.0
500     continue

c ****************************************************************************
c ***************************** Model parameterization **************************
c ****************************************************************************

            iin1 = 20
            iin2 = 21
            iout = 15

1       write (*,'(a)') ' Enter name of file with P{dec} & ASN data'
            read (*,'(a)') file1
            write (*,'(a)') ' Enter name of file with trajectory data'
            read (*,'(a)') file2
            write (*,2) file1,file2
2       format (' P{dec} & ASN file = ',a12,' Population data file = ',
            +a12)
3       write (*,*) 'Is this correct? (Y or N)'
            read (*,'(a)') input
                if (input.ne.'N'.and.input.ne.'n') then
                  if (input.ne.'Y'.and.input.ne.'y') goto 3
                end if
                if (input .eq. 'N' .or. input .eq. 'n') goto 1
            write (*,'(a)') ' Enter name of file for output'
            read (*,'(a)') file3
            open (unit = iin1,file = file1,status = 'old')
            open (unit = iin2,file = file2,status = 'old')
            open (unit = iout,file = file3,status = 'unknown')
            write (iout,4) file1,file2
4       format (' P{dec} & ASN file = ',a12,' Population data file = ',
            +a12)

c determine whether extended or summary output is desired

10      write (*,*) 'Select extended(1) or summary (2) output'
            read (*,*) respond
            if ((respond.ne.1).and.(respond.ne.2)) goto 10

c read P{dec} and ASN data

            read (iin1,*) plans,ndel
            do 15 i = 1,plans
              read (iin1,*) points(i),sched(i)
              iend = points(i)
              do 20 j = 1,iend
                read (iin1,*) mean(j,i),(pdec(j,i,k),k = 1,ndel + 1),asn(j,i)
20          continue
15          continue
```

c read population trajectory data

```
          read (iin2,*) ipops
          trcnt - 0
22        if (.not. eof(iin2)) then
            trcnt = trcnt + 1
            read (iin2,*) trtime(trcnt), (trcnt), (trden(trcnt,j), j = 1,ipops)
            goto 22
          endif
          close (iin2)
```

c if extended results are requested determine whether P{dec} and ASN functions
c should be written to file:

```
          if (respond.eq.1) then
25          write (*,*) 'Write P{dec} & ASN data to file? (Y or N)'
            read (*,'(a)') input
            if (input.ne.'N'.and.input.ne.'n') then
              if (input.ne.'Y'.and.input.ne.'y') goto 25
            end if
            if (input .eq. 'Y' .or. input .eq. 'y') then
              do 30 i = 1,plans
                write (iout,31) i
31              format (' P{dec} and ASN functions for sampling plan ',i2)
                write (iout,33) t,t,t,t,t,t,t
33              format (' Mean',a1,'Pd1',a1,'Pd2',a1,'Pd3',a1,'Pd4',a1,
     +                  'Pd5',a1,'Pd6',a1,'ASN')
                iend = points(i)
                do 35 j = 1,iend
                  write (iout,36) mean(j,i),t,pdec(j,i,1),t,pdec(j,i,2),t,
     +                            pdec(j,i,3),t,pdec(j,i,4),t,
     +                            pdec(j,i,5),t,pdec(j,i,6),t,asn(j,i)
36                format (1x,f6.2,a1,6(f5.3,a1),f7.2)
35              continue
30            continue
            end if
          end if
```

c determine time interval between samples and last sampling time

```
39        continue
          write (*,40) sched(1)
40        format (' Starting time for plan 1  = ',f6.1)
          write (*,41) trtime(1)
41        format (' Starting time for population trajectories = ',f6.1)
          write (*,42) trtime(trcnt)
42        format (' Ending time for population trajectories = ',f6.1)
          write (*,*) 'Input time for first sample:'
          read (*,*) begtime
          write (*,*) 'Input time interval between samples:'
          read (*,*) deltime
          write (*,*) 'Input time for last sample:'
          read (*,*) endtime
          potbouts = aint((endtime − begtime)/deltime) + 1
          lsttime = begtime + ((potbouts − 1)*deltime)
          write (*,*) 'Sampling schedule:'
          write (*,43)
43        format ('Start   Interval   Last sample at')
          write (*,44) begtime,deltime,lsttime
44        format (1x,f6.1,2x,f6.1,5x,f6.1)
```

```
45      write (*,*) 'Is this correct? (Y or N)'
        read (*,'(a)') input
          if (input.ne.'N'.and.input.ne.'n') then
            if (input.ne.'Y'.and.input.ne.'y') goto 45
          end if
          if (input .eq. 'N' .or. input .eq. 'n') goto 39

c apply population scale factor

46      write (*,47) popscale
47      format ('Input population scale factor: ',f7.3)
        read (*,*) popscale
        mult  =  popscale/oldpopsc
        oldpopsc  =  popscale
        do 48 i  =  1,trcnt
          do 49 j  =  1,ipops
            trden(i,j)  =  trden(i,j)*mult
49        continue
48      continue

c ******************************************************************************
c ********************************** Calculations ******************************
c ******************************************************************************

c compute the area beneath each population trajectory curve (sumden)

        do 50 j  =  1,ipops
          sumden(1,j)  =  0.0
          do 51 i  =  1,trcnt − 1
            delta  =  trtime(i + 1) − trtime(i)
            avgden  =  (trden(i + 1,j) + trden(i,j))/2.0
            sumden(i + 1,j)  =  sumden(i,j) + (delta*avgden)
51        continue
50      continue

c loop based on populations in trajectory file
        pindex  =  1
60      continue

c if detailed output is required, print header

        if (respond.eq.1) then
          write (iout,61) pindex
61        format ( 'Results for population # ',i2)
          write (iout,62) t
62        format (/,'Time',a1,'Density')
          do 63 ii  =  1,trcnt
            write (iout,64) trtime(ii),t,trden(ii,pindex)
64          format(1x,f7.2,a1,f7.2)
63        continue
          write (iout,65)t,t,t,t,t,t,t,t,t,t,t
65        format (/'Time',a1,'Plan',a1,'Density',a1,'P{samp}',a1,'Pd1'
     +    a1,'Pd2',a1,'Pd3',a1,'Pd4',a1'Pd5',a1,'Pd6',a1,'P{trt}',a1,
     +    'ASN',a1,'Loss')
        end if

c initialize probabilities at time 0, set time to 0, set sums to 0

        do 66 i  =  1,5
          do 67 j  =  1,5
            prpdec(i,j)  =  0.0
```

```
       67      continue
               prpsamp(i) = 0.0
       66      continue

               prpdec(ndel,1) = 1.0
               prpsamp(1) = 1.0
               time = begtime
               bout = 0
               sum1 = 0.
               sum2 = 0.
               sum3 = 0.
               exploss(pindex) = 0.

        c loop based on time

       70      continue

               bout = bout + 1

        c determine which sampling plan to use

               i = plans
       75      if ((time + 0.001).ge.sched(i)) then
                 cplan = i
               else
                 i = i-1
                 go to 75
               end if

        c compute the time dependent density

               call interp(trden(1,pindex),trtime,time,trcnt,1,dent)

        c compute the time dependent loss function

               call interp(sumden(1,pindex),trtime,time,trcnt,1,losst)
               losssb(bout) = losst

        c compute the interpolated probabilities for each decision and
        c the interpolated asn

               ipoint = points(cplan)
               do 78 j = 1,ndel + 1
                 call interp(pdec(1,cplan,j),mean(1,cplan),dent,ipoint,0,
             +               intpd(j))
                 if (intpd(j).lt.0.0) intpd(j) = 0.0
                 if (intpd(j).gt.1.0) intpd(j) = 1.0
       78      continue
               call interp(asn(1,cplan),mean(1,cplan),dent,ipoint,1,intasn)

        c calculate the probability of sampling in the current time period
               psamp = 0
               do 90 j = 1,ndel
                 psamp = psamp + (prpsamp(j)*prpdec(ndel + 1 − j,j))
       90      continue
```

```
c calculate sums

      sum1 = sum1 + (psamp*intpd(ndel + 1))
      sum2 = sum2 + (psamp*intasn)
      sum3 = sum3 + psamp
      sumptrt(bout) = sum1

c determine if it is the last sampling time

      if ((time + deltime).lt.(endtime + 0.0001)) then
         exploss(pindex) = exploss(pindex) + losst*psamp*intpd(ndel + 1)

c delay probabilities of decision 1 to ndel; prpdec(decision,delay)

         do 95 j = 1,ndel
            do 100 k = ndel,2,-1
               prpdec(j,k) = prpdec(j,k - 1)
100         continue
            prpdec(j,1) = intpd(j)
95       continue

c delay probabilities of sampling; prpsamp(delay)

         do 110 j = ndel,2,-1
            prpsamp(j) = prpsamp(j - 1)
110      continue
         prpsamp(1) = psamp
         if (respond.eq.1) then
            write (iout,120) time,t,cplan,t,dent,t,psamp,t,intpd(1),t,
     +         intpd(2),t,intpd(3),t,intpd(4),t,intpd(5),t,intpd(6),t,sum1,
     +         t,intasn,t,losst
120         format(1x,f7.2,a1,i2,a1,f7.2,a1,7(f5.3,a1),f7.3,a1,f7.2,a1,
     +         f7.2)
         end if
         time = time + deltime
         go to 70
      else
         temp = 0.0
         do 125 j = 2,ndel
            intpd(1) = intpd(1) + intpd(j)
            intpd(j) = 0.0
125      continue
         temp = 0.0
         do 130 j = ndel - 1,1,-1
            do 135 i = 1,ndel - j
               temp = temp + prpsamp(j)*prpdec(i,j)
135         continue
130      continue
         exploss(pindex) = exploss(pindex) + losst*psamp + temp*losst
         if (respond.eq.1) then
            write (iout,140) time,t,cplan,t,dent,t,psamp,t,intpd(1),t,
     +         intpd(2),t,intpd(3),t,intpd(4),t,intpd(5),t,intpd(6),t,sum1,
     +         t,intasn,t,losst
140         format(1x,f7.2,a1,i2,a1,f7.2,a1,7(f5.3,a1),f7.3,a1,f7.2,a1,
     +         f7.2)
         end if
      end if

c end of time loop
```

c compute OC, bouts, average total sample size, losses

```
        oc(pindex) = 1 –sum1
        avgtotn(pindex) = sum2
        expbout(pindex) = sum3
        if (sumptrt(bout).lt.0.5) then
           lossp1(pindex) = losst
        else
           call interp(losssb,sumptrt,0.5,bout,1,lossp1(pindex))
        endif
        if (sumptrt(bout).lt.0.8) then
           lossp2(pindex) = losst
        else
           call interp(losssb,sumptrt,0.8,bout,1,lossp2(pindex))
        endif
        if (sumptrt(bout).lt.0.95) then
           lossp3(pindex) = losst
        else
           call interp(losssb,sumptrt,0.95,bout,1,lossp3(pindex))
        endif
```

c increment population index

```
        pindex = pindex + 1
        if (pindex.le.ipops) go to 60
        pindex = pindex – 1
```

c write summary statistics to output file and to the screen

```
        write (iout,200)
200     format (' Start   Interval   Last sample at   Possible bouts')
        write (iout,205) begtime,deltime,lsttime,potbouts
205     format (f6.1,2x,f6.1,7x,f6.1,12x,f6.1)
        write (*,210)
210     format (' Start   Interval   Last sample at   Possible bouts')
        write (*,215) begtime,deltime,lsttime,potbouts
215     format (f6.1,2x,f6.1,7x,f6.1,12x,f6.1)
        write (iout,220) popscale
220     format (' Population scale factor  = ',f6.3)
        write (*,225) popscale
225     format (' Population scale factor  = ',f6.3)
        write (*,230)
230     format (' Pop  Sum.den  OC      ASN       Bouts  Exp.loss  P0.5loss
       +  P0.2loss  P0.05loss')
        write (iout,235) t,t,t,t,t,t,t
235     format (' Pop.',a1,'Cum.den.',a1,'OC',a1,'ASN',
       +a1,'Bouts',a1,'Exp.loss',a1,'P(0.5)loss',a1,'P(0.2)loss',a1,
       +'P(0.05)loss')
        do 240 j = 1,pindex
           write (iout,245) j,t,sumden(trtcnt,j),t,oc(j),t,
       +    avgtotn(j),t,expbout(j),t,exploss(j),t,lossp1(j),t,
       +    lossp2(j),t,lossp3(j)
           write (*,246) j,sumden(trtcnt,j),oc(j),avgtotn(j),expbout(j),
           exploss(j),lossp1(j),lossp2(j),lossp3(j)
245     format (1x,i2,a1,f7.2,a1,f4.2,a1,f7.2,a1,f5.2,a1,f7.2,
       +  a1,f7.2,a1,f7.2,a1,f7.2)
246     format (1x,i2,3x,f7.2,2x,f4.2,2x,f7.2,2x,f5.2,2x,f7.2,
       +  3x,f7.2,3x,f7.2,3x,f7.2)
240     continue
        write (iout,250)
```

```
250   format (' ** if P()loss  =  Cum.den, SUM(ptrt)  <  1 − P()')

260   write (*,*) 'Repeat for new population scale? (Y or N)'
      read (*,'(a)') input
      if (input.ne.'N'.and.input.ne.'n') then
        if (input.ne.'Y'.and.input.ne.'y') goto 260
      end if
      if (input .eq. 'Y' .or. input .eq. 'y') goto 46

270   write (*,*) 'Repeat for new time model? (Y or N)'
      read (*,'(a)') input
      if (input.ne.'N'.and.input.ne.'n') then
        if (input.ne.'Y'.and.input.ne.'y') goto 270
      end if
      if (input .eq. 'Y' .or. input .eq. 'y') goto 39

      close (iout)
      close (iin1)
      close (iin2)
      end

      subroutine interp (y,x,dummyx,k,trunc,yvalue)

c ******************************************************************************
c This subroutine performs a linear interpolation of a function y = f(x).
c Vectors of y values (y) and x values (x) are passed along with a dummy
c x value (dummyx) for which a y value (yvalue) is sought. The dimension
c of the y and x arrays (k) must also be passed. In addition a variable
c trunc is passed that when = 1 causes trunction of the function so that
c when dummyx  <  x(1) or dummyx  >  x(k), yvalue is constrained to be y(1) or
c y(k). When trunc is not equal to 1, yvalue is interpolated beyond y(1)
c or y(k).
c ******************************************************************************

      real y,x,dummyx,yvalue
      integer k,trunc,j
      dimension y(k),x(k)

      if (dummyx.le.x(1)) then
        if (trunc.eq.1) then
          yvalue  =  y(1)
          goto 20
        else
          yvalue  =  y(1) − ((y(2) − y(1))/(x(2) − x(1))*(x(1) − dummyx))
          goto 20
        endif
      endif

      do 10 j  =  2,k
      if (dummyx.le.x(j)) then
        yvalue  =  y(j −1) + ((y(j) − y(j −1))/(x(j) − x(j −1))*(dummyx − x(j −1)))
        goto 20
      endif
10    continue

      if (trunc.eq.1) then
        yvalue  =  y(k)
        goto 20
      else
        yvalue  =  y(k) + ((y(k) − y(k −1))/(x(k) − x(k −1))*(dummyx − x(k)))
        goto 20
      endif
```

```
20      continue
        return
        end
```

afcm: adaptive frequency classification monitoring

Background: This program constructs stop lines for and evaluates the performance of a set of adaptive frequency classification monitoring (AFCM) sampling plans. AFCM sampling plans are designed to be used as a cascaded set to monitor a population through time. Decisions made via a single AFCM sampling plan are to sample the population again after some time interval or to intervene. Time intervals between sample bouts are specified as a multiple of some minimum time. The program computes descriptions of stop lines for each plan as well as probabilities of making various decisions, average sample number functions, and probabilities of terminating sampling one of three ways. All performance criteria are determined by simulating sampling from a specified distribution and applying the stop lines to the generated random variables. Sampling decisions for which probabilities are determined include resampling the population $1, 2, \ldots, n$ minimum time intervals in the future and the probability of intervening. Sampling can be terminated by the sample path crossing the upper stop line for the SPRT, by crossing the intersection point, or by reaching the maximum sample size.

The programs work as follows:

1. Parameters used to describe the sampling plans are specified in a file read by the program. An example file is shown below. Labels read by the program and used to describe parameters occupy columns 1 to 30. Data values begin in column 31; the range of columns the data may be in are indicated in parentheses. The first two lines of the file are headers and are ignored; however, they must be present in the file. Currently only a negative binomial distribution can be used to describe sample counts and this distribution is specified by the integer 1. It is straightforward to modify the program to allow other distributions such as a Poisson or a binomial count model. Methods used in the program 'sprt' can be adapted for this purpose. In the remainder of this section text enclosed by single quotes denotes the label for a parameter. The 'growth rate' is the parameter r for the exponential model used to describe population growth. 'SPRT alpha and beta' are parameters for the sequential probability ratio test. The 'maximum sample size' is the maximum number of samples that will be taken before reaching a decision. The 'confidence interval alpha' is used to specify the confidence interval about estimated means used to compute the waiting time until the population is sampled again. The 'resample time interval' is the minimum time until the next sample will be taken. The 'resample time delays' specifies the number of resample time intervals sampling might be delayed. The maximum for this parameter is 5. For example, if the 'resample time interval' is 3 days and the 'resample time delays' is 3, plans would be constructed that would result in decisions to resample after 3, 6, and 9 days or to intervene. The 'number of sampling plans' specifies the number of plans to be constructed; this parameter must be ≤ 20. Each sampling plan is tied to s specific time and intervention threshold. The first plan is constructed for sampling at the 'sampling start time'. The second plan is to be used at the

'sampling start time' + the 'resample time interval' and the last plan is constructed for sampling at the 'sampling start time'+('number of sampling plans' − 1)*('resample time interval'). 'TPL alpha and beta' are the parameters for Taylor's power law. When the negative binomial distribution is used, k is assumed to be a function of the mean. Intervention thresholds are interpolated from a set of values specified in the parameter data file. The 'threshold data points' specifies the number of values. The 'monte carlo iterations' specifies the number of simulated sampling bouts used to estimate performance criteria. Mean densities during the simulations are specified by the 'density start'ing value, 'stop'ping value, and 'incr'ement.

2. The program functions by first constructing stop lines for all sampling plans. These stop lines can then be displayed on the screen and are written to an output file. If the intersection of the SPRT upper stop limit on the maximum sample size line exceeds the first decision count, program execution is terminated. This situation signifies that inappropriate parameters (probably SPRT alpha and beta or the confidence interval alpha) have been used.
3. After generation of the stop lines, performance of each plan is simulated.

file for use with afcm.for

```
           123456789 123456789 123456789    123456789 123456789    1234567890
           distribution                     1                      (31-33)
           growth rate                      0.065                  (31-35)
           SPRT alpha and beta              0.2     0.2            (31-35,36-40)
           maximum sample size              50                     (31-33)
           confidence interval alpha        0.30                   (31-35)
           resample time interval           7.0                    (31-33)
           resample time delays             4                      (31)
           sampling start time              1                      (31-34)
           number of sampling plans         14                     (31-32)
           TPL alpha and beta               4.32    1.42           (31-35,36-40)
           threshold data points            4                      (31-33)
           threshold; time, value
           1.0     2.5                                             (1-5,6-10)
           30.0    5.0
           90.0    7.5
           100.0   7.5
           monte carlo iteration            500                    (31-36)
           density start, stop,incr.        0.2   15.0   0.2       (31-37,38-44,45-50)
```

An example run is shown below. user inputs are underlined and comments are in italics.

```
 > afcm
Enter name of file with parameters
afcm.in
Parameter file = afcm.in
Is this correct (Y or N)?
y
Enter name of file for output
afcm.out
```

Input parameters are displayed and confirmed

distribution	1
growth rate	0.065
SPRT alpha and beta	0.2 0.2
maximum sample size	50
confidence interval alpha	0.30
resample time interval	7.0
resample time delays	4
sampling start time	1
number of sampling plans	14
TPL alpha and beta	4.32 1.42
threshold data points	4
threshold; time, value	
monte carlo iteration	500
density start, stop, incr.	0.2 15.0 0.2

Is this correct?
<u>y</u>
Stop lines computed for all plans
Display results for each plan (Y or N)?
<u>n</u>
Performance criteria are then simulated; density and sampling plan being used are displayed on the screen. A portion of the output file is shown below. Parameters usd to construct the sampling plans are written to the output file; these are not shown.

Results for plan 1

Time = 1.0   Threshold = 2.50
SPRT H0 = 1.59   intercept = 27.44   slope = 1.98

intersection =   87 n at intersection =   30

Waiting time	Decision count	Threshold
7.0	87	3.10
14.0	65	3.71
21.0	47	4.31
28.0	33	4.91

Results for plan 2

Time = 8.0   Threshold = 3.10
SPRT H0 = 1.97   intercept = 30.03   slope = 2.46

intersection =   104 n at intersection =   30

Waiting time	Decision count	Threshold
7.0	104	3.71
14.0	76	4.31
21.0	54	4.91
28.0	36	5.25

Results for plan   14

                    Time = 92.0   Threshold = 7.50
                    SPRT H0 = 4.76   intercept = 43.41   slope = 5.93

intersection =   216   n at intersection =   5.93

Waiting time	Decision count	Threshold
7.0	216	7.50
14.0	135	7.50
21.0	84	7.50
28.0	52	7.50

14   4   *number of plans and delay intervals; used by 'cascade'*
    40   1.0   *number of data points and starting time for use; used by cascade*
performance criteria for the first plan:

density	pd1	pd2	pd3	pd4	pd5	asn	pseq1	psq2	pterminal
0.20	0.998	0.002	0.000	0.000	0.000	50.00	0.000	0.000	1.000
0.40	0.936	0.062	0.002	0.000	0.000	50.00	0.000	0.000	1.000
0.60	0.630	0.304	0.058	0.008	0.000	50.00	0.000	0.000	1.000
0.80	0.306	0.422	0.240	0.028	0.004	49.91	0.002	0.002	0.996
1.00	0.106	0.378	0.374	0.140	0.002	50.00	0.000	0.000	1.000
1.20	0.040	0.160	0.440	0.296	0.064	49.59	0.004	0.040	0.956
1.40	0.010	0.088	0.316	0.422	0.164	48.66	0.018	0.122	0.860
1.60	0.006	0.034	0.178	0.446	0.336	46.66	0.042	0.266	0.692
1.80	0.000	0.012	0.112	0.318	0.558	43.78	0.088	0.428	0.484
2.00	0.000	0.006	0.040	0.246	0.708	41.47	0.130	0.522	0.348
2.20	0.000	0.000	0.014	0.140	0.846	37.54	0.204	0.604	0.192
2.40	0.000	0.000	0.010	0.070	0.920	34.21	0.302	0.584	0.114
2.60	0.000	0.000	0.008	0.032	0.960	31.79	0.376	0.570	0.054
2.80	0.000	0.000	0.002	0.020	0.978	27.69	0.516	0.456	0.028
3.00	0.000	0.000	0.000	0.014	0.986	26.10	0.560	0.424	0.016
3.20	0.000	0.000	0.000	0.008	0.992	24.31	0.626	0.366	0.008
3.40	0.000	0.000	0.000	0.000	1.000	21.73	0.706	0.290	0.004

.
.
.

    40   8.0   *results for the second plan*

0.20	1.000	0.000	0.000	0.000	0.000	50.00	0.000	0.000	1.000
0.40	0.952	0.048	0.000	0.000	0.000	50.00	0.000	0.000	1.000
0.60	0.736	0.246	0.018	0.000	0.000	50.00	0.000	0.000	1.000
0.80	0.420	0.460	0.114	0.006	0.000	50.00	0.000	0.000	1.000
1.00	0.148	0.466	0.334	0.052	0.000	50.00	0.000	0.000	1.000
1.20	0.058	0.298	0.466	0.172	0.006	49.98	0.000	0.004	0.996
1.40	0.014	0.202	0.448	0.294	0.042	49.77	0.002	0.032	0.966
1.60	0.004	0.100	0.312	0.454	0.130	49.05	0.010	0.090	0.900

    program afcm

c   Program written in Fortran 77: IBM version compiled using Microsoft
c   FORTRAN compiler v5.1.
c   This program requires a uniform {0,1} random number generator. This

```
c       version makes use of the RANDOM subroutine provided in Microsoft
c       Fortran. If the FORTRAN compiler used does not provide a similar
c       function, one must be provided and small sections of code in some
c       subroutines must be changed.
c       Program written by Jan Nyrop and Wopke van der Werf
c       Department of Entomology, NYSAES,
c       Cornell University, Geneva, NY 14456
c       Last modification date: 2 February, 1993


c *******************************************************************************
c ****************************** Variable descriptions ******************************
c *******************************************************************************

      c asn, average sample size
      c avg, average of h0 and h1
      c cialpha, alpha used to construct confidence limits
      c cid(i), critical initial density; density at time that will not
      c exceed a threshold at time t +i*sint
      c counter, integer counter
      c dec(i), number of sample bouts that end with dec(i),
      c i = 1 is the longest waiting time, i = ndelay + 1 = intervene
      c deccnt(nplans,i), count of organisms that dictate specific time delays
      c (sint*i) to the next sample bout
      c deltime(i), time delays (i*sint) between sample bouts
      c density, mean density used in simulations
      c device, output device
      c distr, distribution of sample counts
      c dstart,dstop,dincr, starting and stopping density,density increment
      c for simulations
      c fthresh(nplan,ndelay), future threshold values used to calculate cid
      c ftime, future time; used to compute the future threshold
      c when determining deccnt
      c growth, function for calculating growth rate
      c h0(nplans),h1(nplans), hypotheses for SPRT for each sampling plan
      c i, integer index
      c ii, integer index
      c intsct(nplans), total count of organisms where upper stop
      c line for SPRT = count corresponding to decision to wait 1 time interval
      c intsctn(nplans), sample size that corresponds to intsct on SPRT stop line
      c j, integer index
      c k, ngb k
      c lowi(nplans),highi(nplans),slope(nplans), low and high
      c intercept, slope for SPRT
      c mci, monte carlo iterations
      c ndelays, number of time delays (sint) sampling may be delayed; maximum of 5
      c nmax, maximum sample size
      c nplans, number of plans to be developed; maximum of 20
      c nthresh, number of values threshold is to be interpolated
      c from pdec(ndelay + 1), probability of a decision indexed by mean density
      c pseq1, probability of reaching a decision by crossing the SPRT stop line
      c pseq2, probability of reaching a decision by crossing intsct before nmax
      c pterm, probability of reaching a decision after taking
      c the maximum number of samples
      c r, growth rate for exponential model
      c samples, number of samples taken during a monte carlo iteration
      c sem, estimated standard error of the mean
      c seq1, number of sampling decisions made by crossing upper SPRT limit
      c seq2, number of sampling decisions reached by crossing
      c intsct before the maximum sample size is reached
      c sint, sample time interval; minimum time delay between samples
```

```
c     sprta, sprtb, alpha and beta for sprt
c     start, starting time
c     terminal, number of sampling decisions made with the maximum
c     number of samples
c     thresh(nthresh), threshold values corresponding to tthresh
c     time(nplans), time each sampling plan is first used
c     total, total of x over samples in a monte carlo iteration
c     tpla, tplb, a and b for Taylor's power law
c     tsamps, total number of samples taken during simulation for a mean
c     tthresh(nthresh), times corresponding to threshold values
c     ucl, upper confidence level (based on cialpha) for xbar
c     usl, upper stop limit for SPRT
c     uslnmax, upper stop limit for SPRT at nmax
c     x, random variable
c     xbar, estimated mean density
c     znormal, function that computes a standard normal deviate for alpha

      integer distr,nmax,start,nplans,nthresh,j,ndelays,i,
     +uslnmax,counter,ii,intsct(20),intsctn(20),deccnt(20,5),
     +mci,device,dec(6),tsamps,seq1,seq2,terminal,samples,n,
     +dindex,points
      real r,sprta,sprtb,cialpha,sint,tpla,tplb,growth,time(20),
     +h0(20),h1(20),lowi(20),highi(20),slope(20),k,avg,sem,ucl,
     +xbar,ftime,znormal,dstart,dstop,dincr,density,x,total,usl
      real tthresh(100),thresh(100),cid(5),fthresh(20,5),
     +deltime(5),pdec(6),asn,pseq1,pseq2,pterm
      character*30 label(14)
      character*12 file1,file2
      character*1 input
      character*45 frmt
      character*2 rep

      frmt = ' (1x,f7.2,2x,  (f5.3,2x),f7.2,3(2x,f5.3))'

1     write ((a)') ' Enter name of file with parameters'
      read (*,'(a)') file1
      write (*,2) file1
2     format (' Parameter file = ',a12)
3     write (*,*) 'Is this correct? (Y or N)'
      read (*,'(a)') input
      if (input.ne.'N'.and.input.ne.'n') then
         if (input.ne.'Y'.and.input.ne.'y') goto 3
      end if
      if (input .eq. 'N' .or. input .eq. 'n') goto 1
      write (*,'(a)') ' Enter name of file for output'
      read (*,'(a)') file2
      open (unit = 1,file = file1,status = 'old')
      open (unit = 2,file = file2,status = 'unknown')

      read (1,10)
10    format (/)
      read (1,11) label(1),distr
11    format (a30,i3)
      read (1,15) label(2),r
15    format (a30,f6.3)
      read (1,20) label(3),sprta,sprtb
20    format (a30,f5.3,1x,f5.3)
      read (1,25) label(4),nmax
25    format (a30,i4)
      read (1,30) label(5),cialpha
```

```
30      format (a30,f5.3)
        read (1,35) label(6),sint
35      format (a30,f6.1)
        read (1,36) label(7),ndelays
36      format (a30,i1)
        read (1,40) label(8),start
40      format (a30,i3)
        read (1,45) label(9),nplans
45      format (a30,i2)
        read (1,50) label(10),tpla,tplb
50      format (a30,f5.3,1x,f5.3)
        read (1,55) label(11),nthresh
55      format (a30,i3)
        read (1,60) label(12)
60      format (a30)
        do 70 j = 1,nthresh
          read (1,65) tthresh(j),thresh(j)
65        format (f5.1,1x,f5.1)
70      continue
        read (1,80) label(13),mci
80      format (a30,i6)
        read (1,85) label(14),dstart,dstop,dincr
85      format (a30,f7.2,f7.2,f6.3)

        write (2,100) label(1),distr
        write (*,100) label(1),distr
100     format (' ',a30,i3)
        write (2,105) label(2),r
        write (*,105) label(2),r
105     format (1 ',a30,f6.3)
        write (2,120) label(3),sprta,sprtb
        write (*,120) label(3),sprta,sprtb
120     format (' ',a30,f5.3,1x,f5.3)
        write (2,125) label(4),nmax
        write (*,125) label(4),nmax
125     format (' ',a30,i4)
        write (2,130) label(5),calpha
        write (*,130) label(5),cialpha
130     format (' ',a30,f5.3)
        write (2,135) label(6),sint
        write (*,135) label(6),sint
135     format (' ',a30,f6.1)
        write (2,136) label(7),ndelays
        write (*,136) label(7),ndelays
136     format (' ',a30,i1)
        write (2,140) label(8),start
        write (*,140) label(8),start
140     format (' ',a30,i3)
        write (2,145) label(9),nplans
        write (*,145) label(9),nplans
145     format (' ',a30,i2)
        write (2,150) label(10),tpla,tplb
        write (*,150) label(10),tpla,tplb
150     format (' ',a30,f5.3,1x,f5.3)
        write (2,155) label911),nthresh
        write (*,155) label(11),nthresh
155     format (' ',a30,i3)
        write (2,160) label(13),mci
        write (*,160) label(13),mci
160     format (' ',a30,i6)
```

```
      write (2,165) label(14),dstart,dstop,dincr
      write (*,165) label(14),dstart,dstop,dincr
165   format (' ',a30,f7.2,f7.2,f6.3)

180   write (*,*) 'Is this correct? (Y or N)'
      read (*,'(a)') input
      if (input.ne.'N'.and.input.ne.'n') then
        if (input.ne.'Y'.and.input.ne.'y') goto 180
      end if
      if (input .eq. 'N' .or. input .eq. 'n') goto 5000

      n = ndelays + 1
      write (rep,190) n
190   format (i2)
      frmt(13:14) = rep

c ********************** Stop lines *************************

c ****** time loop ******

      do 200 j = 1,nplans
        time(j) = float(start) + float(j-1)*sint

c ****** determine threshold ******

      call interp (thresh,tthresh,time(j),nthresh,2,h1(j))

c ******* compute sprt stop lines *******

      h0(j) = h1(j)/growth(r,sint)
      if (distr.eq.1) then
        avg = (h1(j) + h0(j))/2.0
        k = avg**2/(tpla*avg**tplb - avg)
      endif
      call sprtlim (distr,h1(j),h0(j),sprta,sprtb,k,lowi(j),
     +              highi(j),slope(j))

c ****** compute fixed sample size stop line ******

c     compute each critical initial density (cid)

      do 220 i = 1,ndelays
        deltime(i) = float(i)*sint
        ftime = time(j) + deltime(i)
        call interp (thresh,tthresh,ftime,nthresh,2,fthresh(j,i))
        cid(i) = fthresh(j,i)/growth(r,deltime(i))
220   continue

c     find means with upper confidence limit = cid

      uslnmax = nint( highi(j) + slope(j)*float(nmax))
      do 230 i = 1,ulsnmax
        xbar = float(i)/float(nmax)
        sem = sqrt(trpla*xbar**tplb/nmax)
        counter = 0
        ucl = xbar + znormal(cialpha)*sem
```

```fortran
              if (ucl*float(nmax).gt.uslnmax) then
                write (*,*) 'UCL  >  max sprt upper stop limit'
                write (2,*) 'UCL  >  max sprt upper stop limit'
                goto 5000
              endif
              do 240 ii = 1,ndelays
                if (ucl.lt.cid(ii)) then
                  deccnt(j,ii) = i
                  counter = counter + 1
                endif
240           continue
              if (counter.eq.0) goto 250
230         continue
250       continue


c ****** integrate sprt and fixed sample size stop line ******

          intsct(j)  = deccnt(j,1)
          intsctn(j) = nint((float(intsct(j)) – highi(j))/slope(j))


c ****** output stop lines for each sampling plan ******

          device = 2
          call display1 (j,device,time,h1,h0,highi,slope,intsct,intsctn,
     +        deltime,deccnt,fthresh,ndelays)

200     continue

        write (*,*) 'Stop lines computed for all plans'
        write (*,*) 'Display results for each plan (Y or N)?'
360     read (*,'(a)') input
        if (input.ne.'N'.and.input.ne.'n') then
          if (input.ne.'Y'.and.input.ne.'y') goto 360
        end if
        if (input .eq. 'Y' .or. input .eq. 'y') then
          device = 0
          do 370 j = 1,nplans
            call display1 (j,device,time,h1,h0,highi,slope,intsct,
     +          intsctn,deltime,deccnt,fthresh,ndelays)
            pause 'Enter to continue'
370       continue
        endif


c ****** write number of plans and delays to output file ******

        write (2,380) nplans,ndelays
380     format (1x,i2,2x,i2)


c ***************** Pdec and ASN functions *****************

c ****** determine number of data points for each function ******

        points = anint((dstop – dstart)/dincr) + 1

c ****** loop for sampling plans ******
```

```
      do 400 j = 1,nplans
        write (*,401) j,nplans
401     format (' ','sampling from plan ',i2,' of ',i2,' total')
        write (2,402) points, time(j)
402     format (1x,i3,2x,f6.1)


c ****** loop for density ******

        do 405 dindex = 1,points
          density = dstart + dincr*float(dindex – 1)
          k = density**2/(tpla*density**tplb – density)
          do 410 i = 1,6
            pdec(i) = 0.0
            dec(i) = 0
410       continue
          tsamps = 0
          seq1 = 0
          seq2 = 0
          terminal = 0


c ****** loop for Carlo iterations ******

            do 420 i = 1,mci
              samples = 0
              total = 0.0
c generate random variable
425           call ngb(density,k,x)
              total = total + x
              samples = samples + 1
c # of samples < intersection number
              if (samples.lt.intsctn(j)) then
                usl = highi(j) + slope(j)*float(samples)
                if (total.gt.usl) then
                  dec(ndelays + 1) = dec(ndelays + 1) + 1
                  tsamps = tsamps + samples
                  seq1 = seq1 + 1
                  togo 420
                else
                  goto 425
                endif
c # of sample >= intersection number and < maximum samples
              elseif (samples.lt.nmax) then
                if (total.gt.intsct(j)) then
                  dec(ndelays + 1) = dec(ndelays + 1) + 1
                  tsamps = tsamps + samples
                  seq2 = seq2 + 1
                  goto 420
                else
                  goto 425
                endif
c # of samples = maximum samples
              elseif (samples.eq.nmax) then
                do 430 ii = ndelays,1, – 1
                  if (total.lt.deccnt(j,ii)) then
                    dec(ndelays – ii + 1) = dec(ndelays – ii + 1) + 1
                    tsamps = tsamps + samples
                    terminal = terminal + 1
                    goto 420
```

```
                          endif
430              continue
                 dec(ndelays + 1)  =  dec(ndelays + 1) + 1
                 tsamps  =  tsamps + samples
                 terminal  =  terminal + 1
              endif
420           continue

c ****** end of loop for Monte Carlo iterations ******

c ****** compute statistics and write results ******

              do 450 i  =  1,ndelays + 1
                 pdec(i)  =  float(dec(i))/float(mci)
450           continue
              asn  =  float(tsamps)/float(mci)
              pseq1  =  float(seq1)/float(mci)
              pseq2  =  float(seq2)/float(mci)
              pterm  =  float(terminal)/float(mci)
              write (2,frmt) density,(pdec(i), i  =  1,ndelays + 1),asn,
     +                   pseq1,pseq2,pterm
              write (*,frmt) density,(pdec(i), i  =  1,ndelays + 1),asn,
     +                   pseq1,pseq2,pterm
405           continue

c ****** end of loop for density ******

400    continue

c ****** end loop for sampling plans ******

5000  continue
       close (1)
       close (2)
       stop
       end

c ********************************************************************
c *************************** Functions and Subroutines ***************************
c ********************************************************************

         function growth(r,t)
         real r,t
         growth  =  exp(r*t)
         return
         end

         subroutine sprtlim (distr,h1,h0,alpha,beta,k,lowi,highi,
     + slope)

c ********************************************************************
c This subroutine computes stop limit parameters for an SPRT
c Parameters passed are; distr − distribution, 1  =  ngb, 2  =  Poisson,
c 3  =  binomial; h1 and h0 null and alternate hypotheses
c ********************************************************************
         real h1,h0,alpha,beta,lowi,highi,slope,k,a,lna,b,lnb,q1,q0,
     + bvar1,pnb0,pnb1,qnb0,qnb1,vnb1,vnb2
         integer distr
```

```
c variables used for all distributions

      a    = (1-beta)/alpha
      lna  = log(a)
      b    = beta/(1-alpha)
      lnb  = log(b)

c variables used for binomial and negative binomial

      if (distr.eq.3) then
         q1   = 1-h1
         a0   = 1-h0
         bvar1 = log((h1*q0)/(h0*q1))
      end if
      if (distr.eq.1) then
         pnb0 = h0/k
         pnb1 = h1/k
         qnb0 = 1 + pnb0
         qnb1 = 1 + pnb1
         vnb1 = (pnb1*qnb0)/pnb0*qnb1)
         vnb2 = qnb0/qnb1
      end if

c calculate stop limit parameters

      if (distr.eq.3) then
         lowi  = lnb/bvar1
         highi = lna/bvar1
         slope = log(q0/q1)/bvar1
      else if (distr.eq.2) then
         lowi  = lnb/log(h1/h0)
         highi = lna/log(h1/h0)
         slope = (h1 - h0)/log(h1/h0)
      else
         lowi  = lnb/log(vnb1)
         highi = lna/log(vnb1)
         slope = k*(log(qnb1/qnb0)/log(vnb1))
      end if
      return
      end

      subroutine display1 (j,device,time,h1,h0,highi,slope,intsct,
     +intsctn,deltime,deccnt,fthresh,ndelays)
      integer device,j,intsctn(20),intsct(20),deccnt(20,10),i,ndelays
      real time(20),h1(20),h0(20),highi(20),slope(20),deltime(10),
     +fthresh(20,10)
      write (device,10) j
10    format (' Results for plan ',i3)
      write (device,20) time(j),h1(j)
20    format (' Time = ',f6.1,' Threshold = ',f6.2)
      write (device,30) h0(j),highi(j),slope(j)
30    format (' SPRT H0 = ',f6.2,' intercept = ',f7.2,
     +' slope = ',f6.2)
      write (device,40) intsct(j),intsctn(j)
40    format (' intersection = ',i7,' n at intersection = ',i4)
      write (device,50)
50    format (' Waiting time  Decision count  threshold')
      do 55 i = 1,ndelays
         write (device,60) deltime(i),deccnt(j,i),fthresh(j,i)
```

```
      60    format (' ',f6.1,8x,i5,11x,f7.2)
      55    continue
            return
            end

            subroutine interp (y,x,dummyx,k,trunc,yvalue)

c     *****************************************************************************
c     This subroutine performs a linear interpolation of a function y = f(x).
c     Vectors of y values (y) and x values (x) are passed along with a dummy
c     x value (dummyx) for which a y value (yvalue) is sought. The dimension
c     of the y and x arrays (k) must also be passed. In addition a variable
c     trunc is passed that when = 1 causes truncation of the function so that
c     when dummyx < x(1) or dummyx > x(k), yvalue is constrained to be
c     y(1) or y(k). When trunc is not equal to 1, yvalue is interpolated
c     beyond y(1) or y(k).
c     *****************************************************************************

            real y,x,dummyx,yvalue
            integer k,trunc,j
            dimension y(k),x(k)

            if (dummyx.le.x(1)) then
              if (trunc.eq.1) then
                yvalue = y(1)
                goto 20
              else
                yvalue = y(1) – ((y(2) – y(1))/(x(2) – x(1))*(x(1) – dummyx))
                goto 20
              endif
            endif

            do 10 j = 2,k
              if (dummyx.le.x(j)) then
                yvalue = y(j –1) + ((y(j) – y(j –1))/(x(j) – x(j –1))*(dummyx – x(j –1)))
                goto 20
              endif
      10    continue

            if (trunc.eq.1) then
              yvalue = y(k)
              goto 20
            else
              yvalue = y(k) + ((y(k) – y(k –1))/(x(k) – x(k –1))*(dummyx – x(k)))
              goto 20
            endif

      20    continue
            return
            end

****************************************************************************
*     The function ZNORMAL returns the x-value at which the cumulative      *
*     normal probability density function reaches the y-value p.            *
*     The algorithm was taken from Abramowitch & Stegun (1972):             *
*     Handbook of mathematical functions, 9th ed. Dover, New York,          *
*     1046 pp. p. 933, equation 26.2.23                                     *
*     error < 4.5E-4                                                        *
*                                                                           *
*     Wopke van der Werf, 4/30/92                                           *
****************************************************************************
```

```
      function znormal(a)
      real c0,c1,c2,d1,d2,d3,cdf,p,t,x,a
      cdf = 1 - a
      c0  = 2.515517
      c1  = 0.802853
      c2  = 0.010328
      d1  = 1.432788
      d2  = 0.189269
      d3  = 0.001308

      if ( cdf .le. 0. ) then
         znormal = -5.
         return
      else if ( ( cdf .gt. 0. ) .and. (cdf . lt . 0.5 ) ) then
         p = cdf
         t = sqrt( log( p**(-2.) ) )
         x = t - ( c0 + c1 * t + c2 * t**2. ) / ( 1 + d1 * t + d2 * t**2.
     $                                            + d3 * t**3. )
         znormal = -x
         return
      else if ( ( cdf .ge. 0.5 ) .and. ( cdf .lt. 1. ) ) then
         p = 1 - cdf
         t = sqrt( log( p**(-2.) ) )
         x = t - ( c0 + c1 * t + c2 * t**2. ) / ( 1 + d1 * t + d2 * t**2.
     $                                            + d3 * t**3. )
         znormal = x
         return
      else if ( cdf .ge. 1. ) then
         znormal = 5.
         return
      endif

      return
      end


      subroutine ngb(mean,k,x)

c *******************************************************************
c This subroutine generates negative binomial distributed random
c variables using a rejection method. It requires uniform {0,1}
c random variates be generated as the variable r. If the Fortran
c compiler does not support a 'random' function, one must be provided.
c *******************************************************************
      real x,mean,k,px,ppx,pxsum,r
      call random(r)
      x = 0.0
      px = 1/((1+mean/k)**k)
      ppx = px
      pxsum = px
      if (r. .le. px) goto 4
 5    x = x+1
      px = ((k+x-1.)/x)*(mean/(mean+k))*ppx
      ppx = px
      pxsum = pxsum+px
      if (r .le. pxsum) goto 4
      if (x .gt. 200.) goto 4
      goto 5
 4    continue
      return
      end
```

REFERENCES

1. Binns, M. R. and Nyrop, J. P., Sampling insect populations for the purpose of IPM decision making, *Annu. Rev. Entomol.*, 37, 427, 1992.
2. Nyrop, J. P., Sequential classification of prey-predator ratios with application to European red mite (Acari: Tetranychidae) and *Typlodromus pyri* (Acari: Phytoseiidae) in New York apple orchards, *J. Econ. Entomol.*, 80, 14, 1988.
3. Nyrop, J. P. and Binns, M. R., Algorithms for computing operating characteristic and average sample number functions for sequential sampling plans based on binomial count models and revised plans for European red mite (Acari: Tetranychidae) on apple, *J. Econ. Entomol.*, 85, 1253, 1992.
4. Croft, B. A. and Nelson, E. E., An index to predict efficient interaction of *Typhlodromus occidentalis* in control of *Tetranychus mcdanieli* in Southern California apple trees, *J. Econ. Enomol.*, 64, 845, 1972.
5. Cochran, W. G., *Sampling Techniques*, 3rd ed., John Wiley & Sons, New York, 1977, chap. 6.
6. Taylor, L. R., Aggregation, variance, and the mean, *Nature (London)*, 189, 732, 1961.
7. Wald, A., *Sequential Analysis*, John Wiley & Sons, New York, 1947.
8. Wilson, L. T., Estimating the abundance and impact of arthropod natural enemies in IPM systems, in *Biological Control in Agricultural IPM Systems*, Hoy, M. A. and Herzog, D. C., Eds., Academic Press, New York, 1985, pp. 303.

Chapter 12

TIME-SEQUENTIAL SAMPLING FOR TAKING TACTICAL ACTION

Larry P. Pedigo

TABLE OF CONTENTS

I. Concept of Time-Sequential Sampling 338

II. Development of a Time-Sequential Sampling Plan 338
    A. Plan Requirements 338
    B. General Calculations 339
    C. Specific Calculations 340
        1. Formulas for Calculation of h_1 and h_2 340
        2. Formulas for Calculation of b_t 340
            a. Negative Binomial Distribution 340
            b. Poisson Distribution 340
            c. Binomial Distribution 340
        3. Formulas for Calculation of w_i 342
            a. Negative Binomial and Binomial Distributions 342
            b. Poisson Distribution 342

III. Using a Time-Sequential Sampling Plan 342

IV. Applications of Time-Sequential Sampling 342
    A. Development of a Time-Sequential Sampling Plan for the Nondamaging Stage of a Pest 342
        1. Sampling Protocols and Adult Population Configurations 343
        2. Calculations for the Time-Sequential Sampling Plan 344
        3. Time-Sequential Sampling Design and Use 344
    B. Development of a Time-Sequential Sampling Plan for the Damaging Stage of a Pest 347

V. Advantages and Limitations of Time-Sequential Sampling 349

VI. Code Listing of a Computer Program for Calculation of a Time-Sequential Sampling Plan 349

Acknowledgments 352

References 353

I. CONCEPT OF TIME-SEQUENTIAL SAMPLING

A primary objective of most established IPM programs is to cure an acute or chronic crop disorder caused by pests. This curative approach, referred to as therapy, is based on pest assessment followed by timely action if economic damage is occurring or is imminent.[1] Because of its efficiency and convenience, therapy is one of the most widely used strategies in IPM, and usually the first strategy developed in dealing with a new pest problem.

By its very nature, IPM therapy relies on an efficient sampling program, sound economic thresholds, and effective tactical agents, such as pesticides. A major impediment to the development of therapy in the past has been the design of efficient sampling programs. Of the attempts to improve sampling design for IPM decision making none has met with more success than sequential sampling.

Particularly, classifying pest status at a point in time using Wald's sequential probability ratio test (SPRT)[2] has been a major element in improving efficiency. As discussed in Chapter 8, SPRT is used without *a priori* knowledge of numbers of sampling units to be taken in reaching a decision. Instead, sampling units are taken, and status of the population assessed after each succeeding sampling unit. SPRT sampling plans have been developed for many agricultural arthropod pests[3] and development continues for many others. Although intermediate population levels may still require considerable sampling, average savings of 50% may be realized from successful SPRT plans.[4]

Although efficiency gains have been made with SPRT plans, conventional sequential sampling does not address decision making in a time sense. To make decisions, independent classifications, one on each sample date, are made throughout the time a pest is potentially damaging. Even then, conventional sequential sampling may fail to produce the information required to adequately deal with a pest population.[5] Indeed, a major requirement in management is knowing when to take action, which in reality is determining the economic threshold.[6] This timing element was not formally addressed in SPRT plans until recently.[7]

In a study with green cloverworm (*Plathypena scabra* (F.)) adults in soybean, Pedigo and van Schaik[8] developed an application of Wald's SPRT method to classify populations in a time sense. These authors reasoned that insect numbers have characteristic distributions in time, just as they do in space; therefore, methods for determining pest status at one time could be used, through time, to classify ultimate status of the population. They proposed that the SPRT, with modification, could be used to classify populations as to outbreak or endemic configuration, calling the approach *time-sequential sampling*.

II. DEVELOPMENT OF A TIME-SEQUENTIAL SAMPLING PLAN

A. PLAN REQUIREMENTS

Time-sequential sampling (TSS) requires inputs similar to those of conventional or spatial SPRT plans. Therefore, to construct a TSS plan, it is necessary to establish the following:

1. A mathematical model of data distribution through time
2. Classification limits of outbreak vs. nonoutbreak populations
3. Levels of acceptable risk in making classification errors

These requirements are the same as in SPRT plans, except the first two, density and classification, require procedures that reflect a time element.

Distribution models frequently used in SPRT plans are the Poisson, negative binomial, and binomial.[9] These statistical models also can be used in a time sense for TSS by pooling sampling-unit means over all sample dates and then testing these means for goodness-of-fit with the models using chi-square analysis.[10] For example, counts from ten sampling units may be taken weekly to establish a weekly mean in each of three locations. Subsequently, ten weekly means may be gathered at each location for each of five growing seasons (150 mean values). The 150 means then are pooled and tested in a single analysis to determine distribution of the data. Once the type of distribution is determined, appropriate formulas can be selected for calculating decision lines for the TSS plan.

A time element is included in the establishment of classification limits of TSS by selecting multiple limits over the entire sampling period. Whereas in conventional SPRT plans a single set of class limits is chosen to describe an endemic (m_0) and an outbreak (m_1) population, class limits are set for each sample date in the TSS plan. These changing m_0 and m_1 values reflect differences in growth rates or trajectories between an endemic and an outbreak population as a growing season progresses. Values of m_0 and m_1 are selected from actual means observed in endemic and outbreak populations through time or from curves fitted to such means. Characteristically, the difference between values of m_0 and m_1 is small at first, grows to a maximum at population peak, and then declines as the populations subside.

Levels chosen for acceptable risk, α (probability of calling a population endemic when it is outbreak) and β (probability of calling a population outbreak when it is endemic), are selected similarly to conventional SPRT procedures. A level of 0.1 is often chosen for both α and β, i.e., decisions from the plan are expected to be correct nine times out of ten.

B. GENERAL CALCULATIONS

In TSS plans the general formulas for decision lines (boundaries of decision zones) appear similar to those of conventional SPRT plans but are defined in a time sense. After the tth sample they are given as:

$$d_{1,t} = h_1 + b_t \text{ (lower)} \tag{1}$$

$$d_{2,t} = h_2 + b_t \text{ (upper)} \tag{2}$$

where d_t is a weighted cumulative number of arthropods observed on the tth date (as opposed to the conventional d which is the unweighted cumulative number at a location); h_1 and h_2 are intercepts; and b_t is a slope-like parameter that varies from sampling date to sampling date. In establishing the overall TSS plan, it is necessary to calculate class boundaries for every potential sample date because m_0 and m_1 values are designated and vary with each date.

A weighted cumulative number is necessary in TSS plans for the analysis to distinguish more precisely between endemic and outbreak populations. The weighting coefficient (w) is specific for a date (i) and is a device that reflects change in the difference between the two means. Basically, w values increase as the difference between m_0 and m_1 increases, and they decrease as the difference decreases. Consequently, w gives greatest weight to sampling dates when the differences between population types are greatest and easiest to distinguish. Although w is present in conventional SPRT, it is indistinguishable there because it is a constant, being divided out of the left-hand side and appearing as the denominator in the right-hand side of the general boundary formulas. Like b_t, calculation of w_i depends on dispersion pattern of the data.

C. SPECIFIC CALCULATIONS

A summary of the specific steps and calculations of a TSS plan is as follows:

1. Calculate mean arthropod density separately for each location on each sample date and determine the mathematical model (e.g., negative binomial, Poisson) that best fits distribution of these means. For this determination, pool all available sampling data, regardless of outbreak or endemic status. If data from many locations are available on a date, a distribution model should be determined separately for each date. However, lacking such an amount of data, field means from all sampling dates usually are pooled for analysis of the distribution.
2. Classify each available set of sampling data as outbreak (population ultimately exceeded the economic injury level) or endemic for each year or period of interest.
3. For each population type (outbreak or endemic) calculate mean density over all locations separately for each sample date.
4. Set class limits, m_0 and m_1 values, for each sample date in the sampling period of interest. These values reflect the density of a typical endemic population (m_0) and outbreak population (m_1) on a specific sample date. (For the binomial distribution, probabilities of species occurrence in a sample are set in place of class-limit densities.)
5. Choose levels of acceptable risk of making classification errors, α and β.
6. Calculate boundaries of the decision zones, using Equations 1 and 2, for each sample date (e.g., every 2 to 3 d) over the entire sampling period. The entire TSS scheme would normally cover the length of time required for most individuals in a generation to pass through the target stage.
7. Calculate a weighting coefficient (w) for each sampling time in the program as specified for the type of data distribution.
8. Present the plan in graphical or tabular form as with spatial sequential sampling, with instructions on use of the weighting coefficient.

1. Formulas for Calculation of h_1 and h_2

Formulas for h values are calculated for all distributions as:

$$h_1 = \log_e \left[\frac{\beta}{1 - \alpha} \right] \qquad (3)$$

$$h_2 = \log_e \left[\frac{1 - \beta}{\alpha} \right] \qquad (4)$$

where α and β are acceptable risk values as discussed earlier.

2. Formulas for Calculation of b_t

Formulas for b_t are specific to the type of distribution pattern, as shown under the following headings.

a. Negative Binomial Distribution

$$b_t = -k \sum_{i=1}^{t} \log_e \left[\frac{q_{0,i}}{q_{1,i}} \right] \quad (5)$$

where k = clumping factor, a parameter of the negative binomial distribution, as estimated from the data (see Southwood[9] for estimation procedures) and

$$q_{0,i} = 1 + p_{0,i} \quad (6)$$

and

$$q_{1,i} = 1 + p_{1,i} \quad (7)$$

Further,

$$p_{0,i} = m_{0,i}/k \quad (8)$$

= a parameter of the negative binomial distribution associated with the arthropod population in the ith sample of an endemic (low) population ($m_{0,i}$ is the mean number of arthropods expected in the ith sample of an endemic population) and

$$p_{1,i} = m_{1,i}/k \quad (9)$$

= a parameter of the negative binomial distribution associated with the insect population in the ith sample of an outbreak (high) population ($m_{1,i}$ is the mean number of insects expected in the ith sample of an outbreak population).

b. Poisson Distribution

$$b_t = \sum_{i=1}^{t} (m_{1,i} - m_{0,i}) \quad (10)$$

where $m_{0,i}$ = mean number of arthropods expected in the ith sample of an endemic population and $m_{1,i}$ = mean number of arthropods expected in the ith sample of an outbreak population.

c. Binomial Distribution

$$b_t = \sum_{i=1}^{t} n_i \log_e \left[\frac{q_{0,i}}{q_{1,i}} \right] \quad (11)$$

where n_i = number of samples taken in the ith sampling period, and

$$q_{0,i} = 1 - p_{0,i} \quad (12)$$

and

$$q_{1,i} = 1 - p_{1,i} \quad (13)$$

and $p_{0,i}$ = probability of observing the species in the ith period in an endemic population (value estimated from previous years' data and is analogous to the $m_{0,i}$ in the previous two distributions), and $p_{1,i}$ = probability of observing the species in the

ith period in an outbreak population (value estimated from previous years' data and is analogous to $m_{1,i}$ in the previous two distributions).

3. Formulas for Calculation of w_i

Formulas used for the calculation of w_i depend on the type of data distribution.

a. Negative Binomial and Binomial Distributions

The same formula is used for both distributions as follows:

$$w_i = \log_e \left[\frac{p_{1,i} q_{0,i}}{p_{0,i} q_{1,i}} \right] \tag{14}$$

where q and p values for the negative binomial distribution are derived from Equations 6 through 8 inclusive, and those for the binomial are derived from Equations 9 through 12 inclusive.

b. Poisson Distribution

$$w_i = \log_e \left[\frac{m_{1,i}}{m_{0,i}} \right] \tag{15}$$

where m values are as defined previously.

III. USING A TIME-SEQUENTIAL SAMPLING PLAN

Time-sequential sampling plans are used similarly to those of conventional (spatial) sequential sampling. Differences are primarily in the way data are accumulated. With TSS plans, estimates of population density are summed over successive sampling dates for a location rather than place to place in that location. In addition, the mean density from a field on a given sampling date is multiplied by the date-specific weighting coefficient (w) before it is summed. Specifically, this procedure is as follows:

$$d_t = \sum_{i=1}^{t} w_i r_i \tag{16}$$

where d_t is the weighted cumulative number of arthropods observed; w_i is the weighting coefficient; and r_i for negative binomial and Poisson distributions is the number of arthropods in the ith sample. For the binomial distribution, r_i is the number of samples observed to have the arthropod species present in the ith sample.

Because of the weighting coefficient, a tabular presentation of time-sequential sampling is the most convenient form for use under field conditions.

IV. APPLICATIONS OF TIME-SEQUENTIAL SAMPLING

A. DEVELOPMENT OF A TIME-SEQUENTIAL SAMPLING PLAN FOR THE NONDAMAGING STAGE OF A PEST

Either the damaging stage or the nondamaging stage of a pest can be sampled for use in IPM decision making. When used to classify the population of a nondamaging pest stage, TSS can allow decisions on the need for advanced tactical action or indicate whether or not the damaging stage should be sampled and, if so, allow predictions on when to take samples. TSS has been used to classify populations of nondamaging stages with the sorghum shoot fly, *Atherigona soccata* Rondani, on sorghum (egg stage)[11] and the green cloverworm on soybean (adult stage).[8] The use of TSS to classify the green cloverworm adult population is a useful example to show development and use of this concept.

The green cloverworm (GCW) is distributed over the eastern one half of North America from the Florida peninsula and into Canada. The species overwinters in the southern U.S., migrating to northern locations and establishing annual populations there.[12] Outbreaks of the pest are believed to be caused by unusually large emigrations northward, causing economic populations of larvae in the first generation. Therefore, it is feasible to determine the probability of damaging larval populations by sampling immigrating adults. By classifying adult populations, years and fields within years with potential problems can be identified, and larval samples can be taken in suspected outbreak fields to confirm the need for insecticides. In years and fields showing low outbreak potential, sampling larvae is unnecessary, and some larval sampling costs can therefore be avoided.

1. Sampling Protocols and Adult Population Configurations

Samples of adult moths were taken by using a flushing technique. The technique consisted of walking between two soybean rows and beating plants with aluminum sweepnet handles as walking progressed. By striking the plants, moths were disturbed and flew up and down wind of the sampler. A sampling unit consisted of counting moth numbers flushed in a walked area of 0.1 ha; ten sites were included in the area walked to obtain the count.

Flushing samples of the immigrant GCW adults were taken every 2 to 3 d in each of three fields early in the growing seasons from 1977 through 1981 near Ames, IA. Data from years were grouped according to whether the GCW larval population that developed later in the season was of an outbreak type or endemic type[12] for that year. Subsequently, adult means were calculated across years and fields in sampling sequence for each population type. Means for each type were plotted through time according to sampling sequence. These plots showed two distinct configurations for the adults, with strongly diverging population growth rates on the third sampling date (Figure 1). Furthermore, it was evident that a total of ca. nine samples (each 2 to 3 d) was adequate to classify the type of population configuration in any one year.

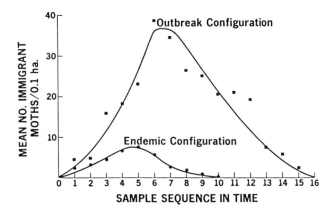

FIGURE 1. Outbreak and endemic configurations of green cloverworm adults. Data show mean number at a sampling time.

2. Calculations for the Time-Sequential Sampling Plan

To analyze distribution of the data through time, counts of moth number per 0.1 ha from each of the three fields at each sample time were pooled from 1977, 1979, and 1981. These pooled data were used to determine the most appropriate mathematical model representing data distribution, and the model was used to select appropriate formulas for calculation of TSS decision lines. Data from 1978 and 1980 were reserved for later testing of the TSS program. A test of the Poisson series was conducted in the analysis first, but the model was rejected because of a significant chi-square value. The null hypothesis for the negative binomial model was accepted because $\chi^2 = 62.3$, df = 49, and $p > 0.05$. A common k of 1.31 was calculated for the distribution.

Class limits were set for endemic (m_0) and outbreak (m_1) populations for each sample time for nine times (Table 1). These critical densities were obtained from the means in Figure 1. The level of acceptable error was set at 0.1 for both α and β in the plan.

3. Time-Sequential Sampling Design and Use

The design of the TSS plan for adult GCW is shown in Figure 2. A definite curvature in the lines is obvious, as opposed to the straight lines of a conventional SPRT plan. This characteristic is the result of the varying differentials between m_0 and m_1 on different sampling periods and iterative calculations.

The TSS plan is presented in a more useful form in Table 2. The tabular form is most convenient because weighting factors are listed and spaces provided for making calculations and accumulating weighted counts.

To use the TSS plan for GCW adults, sampling is begun with soybean seedlings and continued weekly until the first moth is found. On the date of detection and on following sampling dates, ten sampling units covering a total area of 0.1 ha are taken and summed. The summed counts are multiplied by the weighting coefficient to produce a weighted count. A running total of weighted counts is kept by sampling every 2 to 3 d after detecting the first moth until a decision can be made.

The place of the TSS plan within the entire IPM recommendation algorithm is given in Figure 3. If the decision indicates an outbreak population with destructive

TABLE 1
Critical Green Cloverworm Moth Densities Describing Endemic and Outbreak Population Configurations

Sample period	Critical density[a]	
	m_0	m_1
1	2.3	4.3
2	3.0	4.6
3	4.3	15.9
4	6.6	18.2
5	7.8	23.0
6	5.5	38.5
7	2.8	34.5
8	1.7	26.3
9	0.8	25.2

[a] Number of moths/0.1 ha. m_0 = endemic population, m_1 = outbreak population.

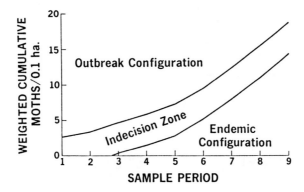

FIGURE 2. Time-sequential sampling graph for green cloverworm adults.

potential of the larvae, the algorithm calls for a return to the field in 335 degree-days (°D) (52°F [ca. 11.1°C] base) to sample larvae, when numbers of middle-sized individuals should be peaking.[12,13] If an endemic configuration is indicated, directions call for cessation of sampling activities, with no larval sample being necessary. In the instance of no decision after the ninth sampling period, larval sampling is recommended 335°D after the date of highest adult count.

The TSS program for adult GCW on soybean was validated using data from three fields each in 1978 and 1980. Results of the validation indicated correct decisions in

TABLE 2
Time-Sequential Sampling Form for Green Cloverworm Adults

Grower: _____ Beginning date: _____
Field: _____ Decision: _____
Scout: _____

| | a | b | c | d | | Decision criteria | | |
| | | | | | | e | | |
Sample number	Date	Number counted	Weighting factor	Weighted count		Lower limit	Running total of weighted count	Upper limit
1	___	___	×0.1845 =	___		0	___	2.77
2	___	___	×0.1115 =	___	E	0	___	3.19 O
3	___	___	×0.1864 =	___	N	0.26	___	4.65 U
4	___	___	×0.1113 =	___	D	1.44	___	5.83 T
5	___	___	×0.0996 =	___	E	2.72	___	7.11 B
6	___	___	×0.1798 =	___	M	5.03	___	9.42 R
7	___	___	×0.3458 =	___	I	7.86	___	12.25 E
8	___	___	×0.5217 =	___	C	10.75	___	15.15 A
9	___	___	×0.9177 =	___		14.06	___	18.46 K

Directions: Sample field weekly beginning at stage V1 to V2. Take 10 flush samples that together cover 0.1 ha. When the first moth is sighted, enter the total number counted in column b and the date in column a.

Multiply the number in b by the number in c and record in d. Continue sampling every 2 to 3 days. Keep a running total of the numbers from d in column e. If the number in e exceeds the upper limit, the population is outbreak, and a larval sample is recommended after 335° D (52° F [11.1° C] base). If the number is below the lower limit, it is endemic and sampling activities are discontinued. If no decision can be reached after nine sample periods, return to the field after 335° D from the date with the highest count to obtain a larval sample.

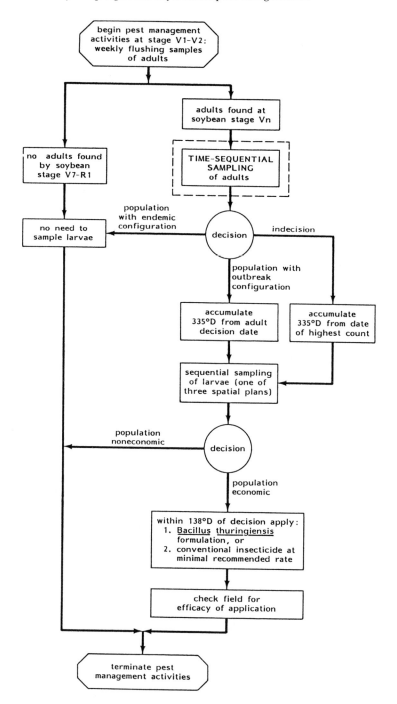

FIGURE 3. Recommendation algorithm of a pest management system for the green cloverworm in soybean, showing the role of the adult time-sequential sampling plan.

all fields but one (Table 3). A peculiar immigration pattern in that field did not allow a decision; therefore, a larval sample would have been automatically recommended. In fields where decisions were made, savings of 44 to 67% of sampling effort were realized, compared to a fixed program of nine sampling periods. These results suggest very significant savings in resources for managing this agricultural pest, because a number of visits to a field are eliminated.

TABLE 3
Results of Using the Green Cloverworm Time-Sequential Sampling Plan with Data from an Outbreak Year (1978) and an Endemic Year (1980)

Year	Field	No. of samples	Decision	Decision disposition	Savings[a] %
1978	1	5	Outbreak	Correct	44
	2	4	Outbreak	Correct	56
	3	3	Outbreak	Correct	67
1980	1	9	Indecision	None	0
	2	4	Endemic	Correct	56
	3	5	Endemic	Correct	44

[a] Based on a fixed sampling program with nine sample dates.

B. DEVELOPMENT OF A TIME-SEQUENTIAL SAMPLING PLAN FOR THE DAMAGING STAGE OF A PEST

In some instances it may be advantageous or even necessary to delay pest management decisions until the damaging stage of a pest is present, rather than sampling the nondamaging stage. In this regard, TSS plans have been developed for both arthropod injury[11] and the arthropods themselves.[14]

When the TSS plan is used to make decisions on action to be taken directly against the damaging stage, the approach can actually describe the economic threshold (ET).[14] As such, the TSS plan has been referred to as a *dichotomous ET*.[14]

The GCW provides an example of developing a TSS plan that subsequently serves as a dichotomous ET for the pest on soybean. To develop the plan, data on weekly larval counts from ground-cloth sampling units of 60 cm of soybean row[4] were summarized from the years 1977 through 1980. These counts represented economically important populations near Ames, IA in the years 1978 and 1979 and subeconomic populations in 1977 and 1980. Economically important populations exceed calculated economic injury levels (EIL) of 3.4, 8.1, and 12.3/60 cm of row at soybean stages V2, V6, and R2.[15] For each EIL, the cost divided by value ratio was 0.5 bushels/acre. Economic and subeconomic population types were described through a 7-week period (beginning the week of June 20), similarly to that of the adult populations described in the previous section (Figure 4). Seven sets of m_0 and m_1 values (one set each week) were chosen based on these curves; the values represented limits of the subeconomic and economic populations, respectively. These values, expressed in number of larvae per 60 cm of row, consecutively, each week, from week 1, were as follows: m_0 0.1, 0.6, 1.5, 2.0, 2.2, 2.3, and 2.5; m_1 1.0, 2.0, 4.5, 9.0, 12.5, 10.0, and 5.5.

Poisson and negative binomial models were fitted to pooled data, and the negative binomial was chosen as the best fit ($\chi^2 = 12.9$, $p > 0.05$, $k = 1.164$). Using appropriate m_0 and m_1 values, k value, and setting α and β both at 0.1, a TSS plan was developed according to the Pedigo and van Schaik procedure.[8] This scheme is presented as Table 4.

To validate this TSS plan for GCW larvae, 12 computer simulations, four each of low-, medium-, and high-density populations, were run. Sampling unit numbers generated randomly from the simulations were entered into the TSS sampling plan for population classification. Correct decisions were made in all instances of high- and low-density populations, giving average savings in sampling time of ca. 25% over a fixed sample number. The four medium-density simulations resulted in two instances

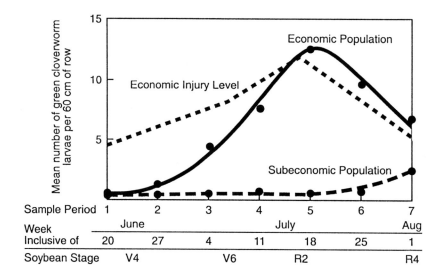

FIGURE 4. Economic and subeconomic configurations of green cloverworm larval populations during the first generation.

TABLE 4
Time-Sequential Sampling Form for Green Cloverworm Larvae in Iowa Soybean

Grower: _____ Beginning date: _____
Field: _____ Decision: _____
Scout: _____

	a	b	c	d		Decision criteria			
							e		
							Running total of		
Sample number	Date	Number counted	Weighting factor	Weighted count		Lower limit	weighted count	Upper limit	
1	____	____	×1.7649 =	____	S	0	____	2.82	E
2	____	____	×0.6197 =	____	U	0	____	3.50	C
3	____	____	×0.3443 =	____	B	0	____	4.38	O
4	____	____	×0.3371 =	____	E	1.35	____	5.74	N
5	____	____	×0.3356 =	____	C	2.98	____	7.37	O
6	____	____	×0.2994 =	____	O	4.34	____	8.73	M
7	____	____	×0.1903 =	____	N	5.04	____	9.43	I
					O				C
					M				
					I				
					C				

Directions: Sample field weekly beginning week of June 20. Take five 60-cm ground-cloth samples in an average-sized field, and calculate a mean. Enter mean in column b and the date in column a. Multiply mean in b by number in c and record in d. Continue sampling once per week. Keep a running total of the numbers from d in column e. If the number in e exceeds upper limit, the population is economic, and insecticide treatment is recommended. If the number is below lower limit, it is subeconomic, and sampling activities can be discontinued. If no decision can be reached after seven sample periods, treat to reduce risk of loss.

where no decision could be made, as well as two incorrect decisions. In both instances of incorrect decisions, insecticide treatment was recommended when treatment was unnecessary. Here, simulated populations peaked just below the EIL.

V. ADVANTAGES AND LIMITATIONS OF TIME-SEQUENTIAL SAMPLING

When used to make IPM decisions either with the nondamaging or damaging arthropod stages, the TSS approach has several advantages. In particular, it is objective, easy to calculate, and simple to use. A major disadvantage of the approach is that several years of population data at various densities are required to characterize pest population types. Additionally, some pests may not have distinct economic and noneconomic configurations from year to year. However, for many arthropod species, TSS deserves strong consideration as a decision-making tool, potentially giving substantial savings in sampling, and, therefore, management costs.

CODE LISTING OF A COMPUTER PROGRAM FOR CALCULATION OF A TIME-SEQUENTIAL SAMPLING PLAN

A computer program, written by J. W. van Schaik and modified by M. J. Wallendorf, performs necessary calculations for the development of a TSS plan. The code listing was written in BASIC language and prompts the user for all information required for calculations.

```
1000  CLS
1020  REM              PROGRAM TSS
1040  REM         SEQUENTIAL SAMPLING PROGRAM
1060  REM         WRITTEN SEPT. 1984 BY JAN VAN SCHAIK
1080  REM         revised July 1991 by Mike Wallendorf
1100  REM
1120  REM   variables:
1140  REM     A   alpha  (probability of incorrectly deciding that a population
                         is outbreak)
1160  REM     B   beta   (probability of incorrectly deciding that a population is
                         endemic)
1180  REM     K   clumping factor (used with negative binomial distribution)
1200  REM     N   number of periods
1220  REM     NS(10) array of sample sizes (used with binomial distribution)
1240  REM     E(10)  array of endemic thresholds
1260  REM     O(10)  array of outbreak thresholds
1280  REM     DC$ distribution code (PO:  poisson,
1300  REM                            NB:  negative binomial,
1320  REM                            BI:  binomial)
1340  REM
1360  PRINT
1380  PRINT "      Time Sequential Sampling Teaching Program"
1400  PRINT : PRINT
1420  DIM NS(10), E(10), O(10)
1440  maxper = 10
1460  PRINT "This computer program calculates the decision limits"
1480  PRINT "   of a Time Sequential Sampling Program (TSS)."
1500  PRINT
```

```
1520    PRINT "Three pieces of information are required:"
1540    PRINT "    (1) the underlying distribution for the TSS program,"
1560    PRINT "    (2) the number of time periods the program will span, and"
1580    PRINT "    (3) the economic threshold points for both endemic and"
1600    PRINT "        outbreak populations in each time period."
1620    PRINT
1640    PRINT "This computer program returns three values for each time period:"
1660    PRINT "    (1) a weighting factor used to weight the observed count"
1680    PRINT "        of a particular time period,"
1700    PRINT "    (2) the upper decision limit above which the population"
1720    PRINT "        is classified as OUTBREAK, and"
1740    PRINT "    (3) the lower decision limit below which the population"
1760    PRINT "        is classified as ENDEMIC."
1780    PRINT : PRINT
1800    INPUT "Do you want to continue? (Y/N) ", junk$: REM 1800
1820    IF junk$ = "Y" OR junk$ = "y" GOTO 1860
1840    IF junk$ = "N" OR junk$ = "n" GOTO 4640 ELSE GOTO 1800
1860    CLS : REM 1860
1880    PRINT "What is the underlying distribution for the TSS program"
1900    PRINT "    (1) Poisson,"
1920    PRINT "    (2) Negative Binomial, or"
1940    PRINT "    (3) Binomial?"
1960    PRINT
1980    INPUT "Choose 1, 2, or 3 : ", DCN: REM 1980
2000    PRINT
2020    IF (DCN < 1) OR (DCN > 3) GOTO 1980
2060    IF DCN = 1 THEN DC$ = "PO"
2980    IF DCN = 2 THEN DC$ = "NB"
2100    IF DCN = 3 THEN DC$ = "BI"
2120    PRINT : REM 2120
2140    PRINT "Please enter the number of time periods (between 1 and"; maxper;
        ")"
2160    INPUT "    that the TSS program is to cover: ", N
2180    PRINT
2200    IF (N < 1) OR (N > maxper) GOTO 2120
2220    F (DC$ < > "NB") GOTO 2340
2240      PRINT
2260      PRINT "The Negative Binomial distribution requires a "
2280      INPUT "clumping factor (greater than zero): ", K
2300      PRINT
2320      IF (K < = 0) GOTO 2240 ELSE GOTO 3040
2340    IF (DC$ < > "BI") GOTO 3040: REM 2340
2360      PRINT "The Binomial distribution requires the number of samples"
2380      PRINT "    taken in any one period. Is this number the same for"
2400      INPUT "    all time periods? (Y/N) ", junk$
2420      PRINT
2440      IF (junk$ = "y") OR (junk$ = "Y") GOTO 2480
2460      IF (junk$ = "n") OR (junk$ = "N") GOTO 2660 ELSE GOTO 2260
2480      PRINT "Please enter the number of samples taken "
2490      INPUT "    in any one time period ( > 0): ", NS1
2520      PRINT
2540      IF (NS1 > 0) GOTO 2555
2550      PRINT "    The number of samples should exceed 0.": PRINT : GOTO
                2480
2555      PRINT
2560      FOR T = 1 TO N
2580         NS(T) = NS1
2600      NEXT T
2620      GOTO 3040
2640    REM ************************* for differing n's
```

```
2660                         PRINT "          Number of"
2680                         PRINT "Period   samples"
2700     FOR T = 1 TO N: REM 2700
2720       IF T = 1 THEN INPUT " 1        ", NS(T)
2740       IF T = 2 THEN INPUT " 2        ", NS(T)
2760       IF T = 3 THEN INPUT " 3        ", NS(T)
2780       IF T = 4 THEN INPUT " 4        ", NS(T)
2800       IF T = 5 THEN INPUT " 5        ", NS(T)
2820       IF T = 6 THEN INPUT " 6        ", NS(T)
2840       IF T = 7 THEN INPUT " 7        ", NS(T)
2860       IF T = 8 THEN INPUT " 8        ", NS(T)
2880       IF T = 9 THEN INPUT " 9        ", NS(T)
2900       IF T = 10 THEN INPUT " 10       ", NS(T)
2920       IF (NS(T) > 0) GOTO 2940
2930       PRINT "The number of samples should exceed 0.": GOTO 2700
2940     NEXT T
2960  PRINT
2980  INPUT "Is this OK (Y/N) ? ", junk$
3000  IF junk$ = "Y" OR junk$ = "y" GOTO 3040
3020  IF junk$ = "N" OR junk$ = "n" GOTO 2640 ELSE GOTO 2700
3040  CLS : REM 3040
3060  PRINT "FOR EACH TIME PERIOD, ENTER THE ENDEMIC AND
            OUTBREAK THRESHOLD VALUES."
3080  PRINT
3100  IF (DC$ < > "BI") THEN GOSUB 3140 ELSE GOSUB 3220
3120  GOTO 3320
3140  PRINT "With the POISSON and NEGATIVE BINOMIAL distributions, "
3160  PRINT "      the THRESHOLD values entered are interpreted as mean "
3180  PRINT "      numbers of insects."
3200  RETURN
3220  PRINT "With the BINOMIAL distribution, the THRESHOLD values are"
3240  PRINT "      interpreted as the probabilities of observing an "
3260  PRINT "      insect in a sample. Here the THRESHOLD should be "
3280  PRINT "      between 0 and 1."
3300  RETURN
3320  FOR T = 1 TO N
3340    PRINT : PRINT
3360    PRINT "FOR TIME PERIOD "; T
3380    INPUT "      Please enter the ENDEMIC threshold: ", E(T)
3400    IF (E(T) < = 0) THEN PRINT "THRESHOLD should exceed 0.":
          GOTO 3380
3420    IF (DC$ < > "BI") OR (E(T) < 1) GOTO 3480
3440      PRINT "          For the BINOMIAL distribution, ";
3450      PRINT "          the threshold should lie between 0 and 1."
3460      GOTO 3380
3480    PRINT
3520    INPUT "      Please enter the OUTBREAK threshold: ", O(T)
3540    IF (O(T) > E(T)) GOTO 3560
3550    PRINT "      OUTBREAK should exceed ENDEMIC.": GOTO 3480
3560    IF (DC$ < > "BI") OR (O(T) < 1) GOTO 3620
3580      PRINT "          For the BINOMIAL distribution, ";
3581      PRINT "          the threshold should lie between 0 and 1."
3600      GOTO 3480
3620    NEXT T
3640  CLS
3670  PRINT
3680  PRINT "ALPHA is the acceptable level of risk for incorrectly"
3700  PRINT "      calling a population OUTBREAK when it is actually EN-
              DEMIC."
3720  PRINT
```

```
3740 INPUT "     Please enter the ALPHA probability level (0 to 1):", A
3760 IF (A ≤ 0) OR (A ≥ 1) THEN PRINT "Between 0 and 1 please.":
     GOTO 3740
3780 PRINT
3800 PRINT "BETA is the acceptable level of risk for incorrectly"
3820 PRINT "     calling a population ENDEMIC when it is actually OUTBREAK."
3840 PRINT
3860 INPUT "     Please enter the BETA probability level (0 to 1):", b
3880 IF (b ≤ 0) OR (b ≥ 1) THEN PRINT "Between 0 and 1 please.":
     GOTO 3860
3885 CLS : PRINT "Alpha = "; A; "   Beta = "; b
3890 PRINT
3900 REM
3920 REM   The numerical guts of the program. NO DOCUMENTATION
         AVAILABLE.
3940 REM
3960 H1 = LOG(b / (1 − A))
3980 H2 = LOG((1 − b) / A)
4000 A = 0: b = 0
4005 PRINT "Computations:"
4010 PRINT "                    Upper      Lower"
4015 PRINT "Period    Weight    Limit      Limit"
4020 FOR T = 1 TO N
4040   IF (DC$ < > "NB") GOTO 4200
4060     P0 = E(T) / K
4080     Q0 = 1 + P0
4100     P1 = 0(T) / K
4120     Q1 = 1 + P1
4140     A = A − K * LOG(Q0 / Q1)
4160     w = LOG((P1 * Q0) / (P0 * Q1))
4180     GOTO 4420
4200   IF (DC$ < > "PO") GOTO 4280
4220     A = A + 0(T) − E(T)
4240     w = LOG(0(T) / E(T))
4260     GOTO 4420
4280 REM CALCULATION FOR BI
4300     P0 = E(T)
4320     Q0 = 1 − P0
4340     P1 = 0(T)
4360     Q1 = 1 − P1
4380     A = A + NS(T) * LOG(Q0 / Q1)
4400     w = LOG((P1 * Q0) / (P0 * Q1))
4420   Upper = H2 + A
4440   lower = H1 + A: IF lower < 0 THEN lower = 0
4460   PRINT
4544   PRINT T; "    "; w; "    "; Upper; "    "; lower
4580 NEXT T
4600 PRINT
4640 END
4660 REM ******************************************
```

ACKNOWLEDGMENTS

I thank D. Buntin of the University of Georgia, and my graduate students J. Browde, T. DeGooyer, T. Klubertanz, and M. Zeiss for their many suggestions in improving this presentation. Special thanks are also due M. Wallendorf of the Department of Statistics at Iowa State University for verifying formulas and modifying computer code. Journal Paper J-15139 of the Iowa Agriculture and Home Economics Experiment Station, Ames, Iowa; Project No. 2580 and 2903.

REFERENCES

1. Pedigo, L. P., Integrating preventive and therapeutic tactics in soybean insect management, in *Pest Management of Soybean*, Copping, L. G., Green, M. B., and Rees, R. T., Eds., Elsevier, London, 1992, 10.
2. Wald, A., *Sequential Analysis*, John Wiley & Sons, New York, 1947.
3. Fowler, G. W. and Lynch, A. M., Bibliography of sequential sampling plans in insect pest management based on Wald's sequential probability ratio test. *Great Lakes Entomol.*, 20, 165, 1987.
4. Pedigo, L. P., *Entomology and Pest Management*, Macmillan, New York, 1989.
5. Strayer, J. M., Shepard, M. A., and Turnipseed, S. G., Sequential sampling for management decisions on the velvetbean caterpillar on soybeans, *J. Ga. Entomol. Soc.*, 12, 220, 1977.
6. Pedigo, L. P., Hutchins, S. H., and Higley, L. G., Economic injury levels in theory and practice, *Annu. Rev. Entomol.*, 31, 341, 1986.
7. Binns, M. R. and Nyrop, J. P., Sampling insect populations for the purpose of IPM decision making, *Annu. Rev. Entomol.*, 37, 427, 1992.
8. Pedigo, L. P. and van Schaik, J. W., Time-sequential sampling: a new use of the sequential probability ratio test for pest management decisions, *Bull. Entomol. Soc. Am.*, 30, 32, 1984.
9. Southwood, T. R. E., *Ecological Methods*, Chapman & Hall, London, 1978.
10. Andrewartha, H. G., *Introduction to the Study of Animal Populations*, University of Chicago Press, Chicago, 1961.
11. Zongo, J. O., Vincent, C., and Stewart, R. K., Time-sequential sampling of sorghum shoot fly, *Atherigona soccata* Rondani (Dipters: Muscidae), in Burkina Faso, *Trop. Pest Manage.*, 1993, in press.
12. Pedigo, L. P., Bechinski, E. J., and Higgins, R. A., Partial life tables of the green cloverworm (Lepidoptera: Noctuidae) in soybean and a hypothesis of population dynamics in Iowa, *Environ. Entomol.*, 12, 186, 1983.
13. Hammond, R. B., Poston, F. L., and Pedigo, L. P., Growth of the green cloverworm and a thermal-unit system for development, *Environ. Entomol.*, 8, 639, 1979.
14. Pedigo, L. P., Higley, L. G., and Davis, P. M., Concepts and advances in economic thresholds for soybean entomology, in *Proc. World Soybean Res. Conference IV*, Vol. 3, Pascale, A. J., Ed., 1989, 1487.
15. Fehr, W. R. and Caviness, C. E., *Stages of Soybean Development*, Special Rep. 80, Iowa State University Cooperative Extension Service, Ames, 1977.

Section IV: Sampling Programs

Chapter 13

SAMPLING METHODS FOR INSECT MANAGEMENT IN ALFALFA

Richard C. Berberet and William D. Hutchison

TABLE OF CONTENTS

I. Alfalfa Forage Production 358

II. Alfalfa Seed Production 358

III. Overview of Insect Damage in Alfalfa Forage Production 359
    A. Damage Resulting in Quantitative Losses 359
    B. Damage Resulting in Qualitative Losses 360

IV. Sampling and Decision Making for Insect Control in Alfalfa 361
    A. Defoliators...362
        1. Alfalfa Weevil Complex 362
        2. Variegated Cutworm...............................365
        3. Army Cutworm 365
        4. Lepidopteran Foliage-Feeding Complex 366
        5. Blister Beetles 367
    B. Fluid Feeders..368
        1. Potato Leafhopper 368
        2. Pea Aphid 369
        3. Blue Alfalfa Aphid 372
        4. Spotted Alfalfa Aphid...............................373
    C. Seed Production Pests..................................374
        1. *Lygus* Bugs 374

V. Summary 376

References 376

ISBN 0-8493-2923-X/94/$0.00 + .50
© 1994 by CRC Press, Inc.

I. ALFALFA FORAGE PRODUCTION

Alfalfa, *Medicago sativa* L., is the most highly valued forage crop grown in North America, as evidenced by the fact that over 40% of the world hectarage is grown on this continent.[1] Large areas are devoted to alfalfa production in both the U.S. (10 to 11 million ha)[2] and Canada (4 to 5 million ha).[3] The largest areas devoted to alfalfa production are found in the northcentral region of the US. and the prairie provinces (Alberta, Saskatchewan, and Manitoba) of Canada. Among the attributes that have contributed to the success of this remarkable crop are its wide range of adaptation (it is grown virtually worldwide), value as a source of protein for livestock, and capability for nitrogen fixation.

Another important attribute of this perennial legume is its capability to maintain profitable forage production for 4 to 5 years or longer on deep, well-drained soils, with favorable climates, and care in crop management. However, an analysis of profitability in alfalfa forage production by Cuperus et al.[4] showed that high establishment costs for alfalfa result in limited net income until the third year of stand life. With each additional year that a full, productive stand can be maintained, net profits in excess of $200/ha may be realized.

Within *M. sativa* and the closely related species *M. falcata* L., alfalfa breeders have found a tremendously wide range of climatic adaptation. Survival in diverse climatic conditions is dependent, in part, on varying degrees of dormancy among cultivars. The most dormant cultivars typically possess winter hardiness that allows them to survive extremely cold conditions prevailing in some areas of the northern U.S. and Canada.[5] By comparison, maximum productivity is attained under the hot, dry conditions of the southwestern states with nondormant cultivars.

Harvest schedules vary greatly over the geographic range of alfalfa production in North America. As few as one or two harvests are taken in the prairie provinces of Canada,[3] while irrigated stands in Arizona and California may be cut from 6 to 10 times/year. According to Sheaffer et al.,[6] one or more of the following criteria are used to schedule harvests: (1) stage of growth (cutting at early bloom stage is frequently recommended), (2) fixed intervals normally ranging between 30 and 45 d, and (3) emergence of new crown shoots. In some instances, timing of harvests may be adjusted to assist in control of an insect pest species. For example, early harvesting may be recommended as an alternative to insecticide application for control of the alfalfa weevil, *Hypera postica* (Gyllenhal), in the northern U.S.[7,8] and Canada.[3] Although amounts of forage harvested annually are quite variable within any state or province, depending on factors such as temperature, rainfall, and pest infestations, the great production potential of alfalfa has been demonstrated by yields up to 24 Mg/ha without irrigation in Michigan.[9]

II. ALFALFA SEED PRODUCTION

During the years from 1945 to 1955, a great transition occurred as the majority of alfalfa seed production in the U.S. moved from the central plains to areas west of the Rocky Mountains. Before this period of transition, most seed was harvested from broadcast stands grown under nonirrigated conditions and planted primarily for forage production in states ranging from Oklahoma to South Dakota. Since the early 1950s, seed production has been centered in the western states of California, Idaho, Nevada, Oregon, and Washington. These states provide about 75% of the total production in the U.S., primarily from irrigated stands planted in rows to maximize alfalfa seed yields.[10] A relatively small amount of seed production occurs in Canada, primarily in the provinces of Alberta, Manitoba, and Saskatchewan.

As the alfalfa seed industry moved to the western states, a system of production was established that has resulted in impressive yields. While average yields were usually below 100 kg/ha in the Plains states before 1945,[11] specialized methods resulted in yields averaging over 700 kg/ha in California by 1984.[10] The principal components of the alfalfa seed production system include: (1) use of row plantings established with low seeding rates; (2) pollination by honey bees, *Apis mellifera* L. (California and Nevada) or the alkali bee, *Nomia melanderi* Cockerell, and the alfalfa leafcutting bee, *Megachile rotundata* F. (Idaho, Oregon, Washington, and in Canada); (3) development of integrated pest management (IPM) programs that provide effective regulation of damaging species without destroying pollinator populations.[10,11]

Marble[12] summarized several advantages of row plantings over broadcast stands for seed production that are relevant to management of pollinators and pest, including: (1) greater light penetration and access to flowers by pollinators as the plant canopy is less dense, (2) lower incidence of foliar diseases due to reduced lodging and lower humidity in the canopy, and (3) better penetration of the canopy with pesticides used for insect and weed control. Fewer insecticide applications are being made during seed production due to integration of chemical and biological control of a key pest group, the *Lygus* spp. For example, control provided by predators including bigeyed bugs, *Geocoris* spp., and damsel bugs, *Nabis* spp., in combination with one or two carefully timed prebloom sprays greatly reduces the need for insecticides during seed-set in the northwestern U.S. and Canada.[10,11] Improved sampling and decision-making procedures have reduced the numbers of insecticide applications made in all areas where seed is produced.

III. OVERVIEW OF INSECT DAMAGE IN ALFALFA FORAGE PRODUCTION

In the following discussion, it becomes evident that alfalfa forage production must be viewed somewhat differently than production of many other field crops. The concept of *direct* vs. *indirect* damage is useful in describing the potential for economic losses (losses exceeding the cost of available controls) resulting from insect pest infestations in field crops. *Direct* damage is described as that inflicted on yield-forming structures of the plants, usually the fruiting structures, as in the heads of grain sorghum, *Sorghum bicolor* (L.) Moench, the pods of peanut, *Arachis hypogaea* L., or the blossoms and pods of alfalfa grown for seed. Typically, the potential for economic loss is great when pests cause *direct* damage. By comparison, *indirect* damage is that inflicted on plant parts other than those harvested by humans such as the leaves of peanuts or soybeans. Damage to these plant parts usually does not result in economic loss unless capability of plants to produce fruit is diminished.

In alfalfa forage production, the importance of foliar growth of plants must be emphasized as it is highly valued by humans for use in feeding livestock. Any pest that limits forage production either by consuming foliage or reducing plant growth causes *direct* damage to alfalfa grown for hay. The feeding of these species causes quantitative losses in alfalfa production. Additionally, the importance of foliage-feeding pests in reducing protein and digestible nutrient content of alfalfa is considered in the section on qualitative losses.

A. DAMAGE RESULTING IN QUANTITATIVE LOSSES

The alfalfa weevil, variegated cutworm, *Peridroma saucia* (Hübner), fall armyworm, *Spodoptera frugiperda* (J. E. Smith), alfalfa caterpillar, *Colias eurytheme* Boisduval, and other insects with chewing mouthparts reduce yields in an obvious manner by removing foliage, particularly the leaves. Damage to growing terminals may result in shorter stems and delayed maturity.[13] Newly formed crown buds and stems may also be destroyed when these pests infest stands in early spring as in the case of alfalfa weevil larvae in the southern U.S.,[13] or when the timing of the infestation coincides with harvesting such that larvae feed extensively on regrowth, e.g., the alfalfa weevil in the northern U.S.[14] and the variegated cutworm.[15] Defoliators have caused yield reductions ranging up to 2 to 3 Mg/ha in a single crop of alfalfa.[16,17] Several studies have shown that residual effects of extensive defoliation by the alfalfa weevil may result in reduced plant vigor and productivity in subsequent growth cycles.[18-20]

The potato leafhopper, *Empoasca fabae* (Harris), pea aphid, *Acyrthosiphon pisum* (Harris), and spotted alfalfa aphid, *Therioaphis maculata* (Buckton), are among the more common fluid-feeding insects, with potential to cause reduced growth and productivity of alfalfa. Obviously, the symptoms resulting from infestation by these species, having piercing-sucking mouthparts, are quite different than those associated with feeding of the defoliators. Characteristic patterns of chlorosis often occur in association with feeding of each species. For example, the potato leafhopper causes a V-shaped pattern of chlorosis on the tips of leaves that is called 'hopperburn'. Plants infested by the spotted alfalfa aphid often exhibit a symptom known as 'veinbanding', or clearing of veins in newly formed leaves near growing terminals. If infestations persist, extensive chlorosis followed by necrosis of tissues may lead to loss of leaves. Severe infestations of the spotted alfalfa aphid, in particular, can result in death of plants.[21]

It has been hypothesized that patterns of chlorosis accompanying infestations of fluid feeding insects result from injection of toxic substances with salivary secretions.[22,23] However, existence of toxins in these secretions has not been proven for any of the fluid feeders that infest alfalfa. Although chlorosis is an indicator of infestation and injury to plants, it is usually stunting of plant growth by pests such as leafhoppers and aphids that leads to yield reductions. The severity of losses has been demonstrated in susceptible cultivars by Hower and Flinn,[24] who reported that infestation by eight potato leafhopper nymphs per plant reduced growth by 54% and yield (dry weight) by 57%. In a study by Kindler et al.,[25] yield reductions caused by the pea aphid and spotted alfalfa aphid were 48% and 38%, respectively.

B. DAMAGE RESULTING IN QUALITATIVE LOSSES

As the value of alfalfa hay for livestock feed is based to a great extent on protein and digestible nutrient content, concern about insect damage is not limited to yield reductions alone. However, a review of the effects of plant growth and maturation on alfalfa forage quality is critical to interpreting possible pest effects.

Throughout crop growth, leaves are consistently the primary contributors to crude protein content and digestibility. In early stages of vegetative growth, alfalfa foliage may be comprised of up to 60% leaves (by weight) having a crude protein content of about 30% (Table 1) and *in vitro* true digestibility (IVTD) of over 90%.[26] By comparison, the crude protein content of stems is typically <20% and IVTD averages about 70%.[26] At the early bloom stage, a time at which many growers cut their alfalfa, leaf content of forage may be just 35 to 37%. Crude protein content of these leaves is about 25 to 27%, which represents a small reduction compared to the early vegetative stage (Table 1). The most critical reduction in quality as alfalfa matures occurs in the stem component, which at early bloom may have a crude protein content of 10% or less and IVTD of about 50%.[26]

Because the leaves are such important contributors to forage quality, defoliators such as lepidopteran larvae and the alfalfa weevil have great potential to reduce quality. Hintz et al.[27] reported significant reductions in crude protein and *in vitro* dry

TABLE 1
Proportions of Leaf and Stem Components and Forage Quality at Varied Alfalfa Growth Stages

			% Crude protein		
Stage	Height (cm)	% Leaves	Leaf	Stem	Plant
Vegetative	21	60	29	18	24
Vegetative	37	43	27	14	20
Bud	61	38	27	12	18
First bloom	77	38	27	11	17
Full bloom	86	34	26	10	16

Note: Data from a field study using first growth alfalfa, Stillwater, OK.

matter digestibility (IVDMD) due to infestations of alfalfa weevil larvae. However, no consistent reductions were observed by Berberet et al.,[18] and Berberet and McNew,[13] or Liu and Fick.[26] The extent of quality reduction may be influenced by the timing of damaging infestations relative to growth of the alfalfa. Berberet and McNew[13] theorized that with heavy damage and stunting of plants in the early vegetative stage, the relatively high crude protein content of stems tended to mask reductions from leaf loss. Defoliation closer to harvest seems to result in greater quality reductions.

Although fluid-feeding insects do not reduce crude protein content and digestibility by removal of leaf tissues, some of these species can cause quality losses through interference with normal plant physiological processes. Through blockage of vascular tissues by salivary secretions such as those used in forming feeding sheaths, or perhaps by induction of a toxic reaction resulting from components of salivary secretions, the potato leafhopper alters plant physiology in several ways. For example, four and eight potato leafhopper adult females per plant reduced photosynthetic rates by 60 and 80%, respectively, within 21 d.[28] In a period of just 7 d, these levels of infestation reduced leaf crude protein by 22 to 35%. Other studies have shown similar reductions in crude protein content of forage and have illustrated that these reductions result in 300 to 500 kg/ha less protein at harvest time.[29,30] An added concern is the reduction of root carbohydrate reserves observed in association with potato leafhopper infestation which may lead to greater winter kill and reduced yield at first cut the following year. Results of several studies have shown no consistent reduction in forage quality caused by infestations of the pea aphid as estimated by decreased crude protein content, IVDMD, or increased fiber content.[20,25,31,32] Summers and Coviello[33] observed no decreases in crude protein or percentage of total digestible nutrients (TDN) after infestation of blue alfalfa aphid, *Acyrthosiphon kondoi* Shingi.

IV. SAMPLING AND DECISION MAKING FOR INSECT CONTROL IN ALFALFA

It is not the intent of this chapter to include all insect pests that may occur in alfalfa grown for forage or seed production, but only those key pests for which at least rudimentary sampling procedures and economic thresholds have been developed. We have discussed seasonal life histories of pests, and the sampling methods included are primarily those intended for field scouting and decision making. We have not attempted to include detailed descriptions of sampling methods used for research purposes. Several of the species have broad geographic distributions and some variation often exists from region to region in recommended sampling and decision-making procedures. Although this chapter provides a review of these procedures, detailed information should be sought from extension specialists in respective states or provinces. These are also the sources from which information can be obtained regarding insecticides recommended for control of the pests.

A. DEFOLIATORS
1. Alfalfa Weevil Complex

Three populations of the alfalfa weevil have resulted from what apparently were three separate introductions of this pest into the U.S. The first of these was discovered in Utah during 1904,[34] and has since been regarded as the western strain of the weevil; it is widespread over the region. A second introduction, detected in California during 1939,[35] was identified as the Egyptian alfalfa weevil, *Hypera brunneipennis* (Bohman), and has been confined to alfalfa production areas of California and Arizona. The third alfalfa weevil population was discovered in Maryland during 1951,[36] and what has since been called the eastern strain has colonized all of the eastern and central U.S. and southern areas of Canada. Manglitz and Ratcliffe[37] cited evidence from recent genetic studies indicating that there is but one alfalfa weevil species in North America. Partial sterility in crosses of the western strain with the eastern strain (and Egyptian weevil) appear to result from the presence of a rickettsial organism infecting the western strain.[38,39]

Although the alfalfa weevil typically completes just one generation/year throughout its range in North America, climatic variations from northern areas of the U.S. and Canada to the southern states result in quite different phenological events in its seasonal life history. The following description provides comparisons of alfalfa weevil life history in northern vs. southern portions of its range.

Following emergence from the pupal stage, the aestivation period of adult weevils lasts for at least several weeks during summer and fall.[40,41] This period is normally spent in field borders, creek bottoms, and similar areas sheltered by dense vegetation. Although timing varies over years and locations, most adults return to alfalfa fields in October and November and, if temperatures remain above the ovipositional threshold of 1.7°C,[42] deposition of eggs begins within 3 to 4 weeks.

Suitability of temperature conditions for oviposition and survival of eggs through the months from December to March is the primary determinant for timing of alfalfa weevil larval populations. Cold temperatures prevent virtually all oviposition in fall and winter in the northernmost states such as Michigan[43] and Wisconsin,[44] and in western states at higher elevations such as in Utah.[45] The usual time period for most oviposition in these states is from April to June. An area that is intermediate in terms of timing for oviposition extends through states such as Illinois,[46] Ohio,[47] and Pennsylvania.[48] Numbers of eggs laid in fall and winter in these states vary greatly, depending on prevailing temperatures, as does the percentage of eggs that survive to hatch in spring. Morrison and Pass[49] reported that temperatures below −15°C may be lethal to eggs, however, the percent mortality is dependent not only upon how low temperatures fall but also the duration of exposure.

The percentage of eggs laid and viability throughout winter increases greatly in southern states. The peak in egg numbers typically occurs in January and February in states ranging from Alabama,[50] to Oklahoma,[16] and in southern California.[51] Although viability of overwintering eggs remains relatively high during most years in these states, there are occasions when lethal temperatures cause extensive mortality[52] and delay the onset of larval infestations.

In northern states and Canada, larval populations eclose primarily from eggs that are laid in spring, and they typically peak during May and June.[8,26,53] In general, the highest larval numbers coincide with maturation of the first crop of alfalfa. In areas that are intermediate for timing of oviposition and survival of overwintering eggs, there is corresponding variation in timing of peak larval populations among years and locations. The extent of infestation and damage in spring when the alfalfa plants are in the early vegetative stage depends on numbers of eggs that survive through the winter. The peak in larval numbers in the southernmost states occurs as early as February in southern California[51] and extends to mid-April in Alabama.[50]

Predictive models often serve a valuable function in alerting growers to the probable timing of damaging pest infestations. With the relative consistency of the seasonal life history of the alfalfa weevil in Ontario, Canada, Harcourt[53] has developed a highly effective model for predicting events that are critical to its management. These include peak hatch and peak prevalence of each instar. In one study, predictions were accurate within 1 d of the actual events over a 4-year period. The primary input for the model is degree-day accumulation from April 1.

Development of a similar model has been attempted in Oklahoma. However, the extent of variation in the seasonal life history of the weevil in this locality has caused great difficulty in the modeling effort. Whereas the timing of instar peaks varied by little more than 1 week over the years in Ontario,[53] in Oklahoma timing has varied by 4 to 6 weeks.[16] The model developed by Stark[52] using data collected over an 18-year period requires five inputs; total egg number on January 1, viable eggs on February 1, ovipositional (threshold = $1.7°C$) and developmental (threshold = $90°C$) degree-days (CDD) accumulated from January 1 to February 15, and the number of days on which the minimum temperature was at or below $-12°C$ for this period. This model has been effective for predicting timing of larval populations above the economic threshold within 2-week time intervals specified as late February, early March, late March, and early April.

The primary purpose of the two models described above and those that have been developed in other states is to provide growers and consultants a time frame when scouting efforts must be intensified to assure maximum profitability of insecticide applications for control of weevil larvae. For example, the early warning system for alfalfa weevil in Ontario is activated at peak hatch which occurs with accumulation of about 155 CDD from April 1.[53] The prescribed time for beginning field scouting in Oklahoma has been 85 CDD (150 FDD) from January 1.[54] Although this recommendation has never resulted in losses occurring before scouting was begun, in some years field checks have started well before the threat of damage exists. The newly developed model will improve accuracy of the recommended starting date. In Minnesota and Wisconsin, it is recommended that field scouting begin with accumulation of about 175 CDD (300 FDD) from January 1.[8]

Two methods for assessment of alfalfa weevil larval population densities for decision-making purposes that have been used for many years are visual estimates of the percentage of damaged plant terminals and sweep-net sampling.[55] Visual estimates typically involve recommendations for spraying when 30 to 40% of the plant terminals begin to show feeding damage.[7,8] When sweep-net sampling is used, spraying is typically recommended when 20 or more weevil larvae are collected per sweep.[7,51]

Because of concerns such as those expressed by Cothran and Summers[56] regarding application of sprays that are not warranted, sampling procedures adopted in several states now involve removal of foliage from measured areas or stem sampling as suggested by Armbrust et al.[57] One of the first stem sampling plans described was that of Guppy et al.[58] in which the sample unit was a six-stem bouquet collected from each of 16 randomly selected 0.9 m^2 quadrats per field. The authors estimated the time requirement for this method at 2 h per field. Another sampling plan was developed by Ruesink[59] using a dynamic set of economic thresholds based on degree-day accumulation, average stem height, and number of larvae in a 30-stem sample. This procedure was incorporated into an IPM program for alfalfa by Wedberg et al.[60] and has since been adopted by many states.

As used in most states that have adopted the 'shake-bucket' procedure, degree-day totals from January 1 are divided into intervals for treatment recommendations as shown in Table 2. This information is combined with results of a 30-stem field sample when a decision is made regarding insecticide application for weevil larvae. Stems are pulled at random over the field and placed in a container. The average height of alfalfa is estimated by measuring at least 10 stems, and the number of large larvae in 30 stems is determined after vigorous shaking in the container to remove them from the foliage. These three inputs are combined in the recommendation table for decisions regarding the need for spraying. Although the decision-making format has remained the same, numbers of larvae needed for a decision to spray does vary considerably, depending upon the region where the procedure is being used (Table 2). In southern states where the seasonal occurrence of weevil larvae coincides closely with the onset of alfalfa growth in February and March, spraying may be recommended with lower numbers than in the North where the larval peak occurs near the time of first cut.

Legg et al.[61] and Barney and Legg[62] have expressed concerns about the accuracy of the 30-stem sample for decision-making related to alfalfa weevil control. They have proposed that a minimum of five samples per field, each comprised of six stems taken in a 100-m^2 area, provides a more reliable basis for a treatment decision. They have also developed and validated a sequential sampling plan using six-stem per 100 m^2 subsamples.[63] Field validation has shown that required numbers of samples for estimation of weevil population density with a 95% confidence interval ranges from 2.1 to 9.7, depending upon the degree of aggregation (clumping) assumed in the spatial pattern.

TABLE 2
Portions of the Alfalfa Weevil Control Recommendation Chart from Illinois and Oklahoma

(FDD) from January 1	Alfalfa height (in.)							
	11	12	13	14	15	16	17	18 or more
Illinois								
390–510								
SPRAY	52	52	58	64	68	72	76	80
Resample 50 DD	8–51	8–51	8–57	14–63	14–67	14–71	18–75	18–79
Resample 100 DD	0–7	0–7	0–7	0–13	0–13	0–13	0–17	0–17
Oklahoma								
390–530								
SPRAY	20	20	25	30	30	35	35	40
Resample 3–5 d	8–19	8–19	8–24	14–29	14–29	14–34	18–34	18–39
Resample 5–7 d	0–7	0–7	0–7	0–13	0–13	0–13	0–17	0–17

Note: This table shows alfalfa weevil larval numbers required for profitable insecticide application at comparable DD accumulations in Illinois and Oklahoma.

Although the sampling plans described above do consider some of the variables that affect loss in alfalfa production due to the weevil, they are not based on true dynamic economic thresholds because changing values of alfalfa forage and control costs are not included for decision making. A computer program called 'ALFWEEV' has been developed in Oklahoma that takes these factors into consideration along with plant height and numbers of larvae collected in sampling.[64]

When peak larval numbers occur near the time of first cut, it is possible that regrowth could be damaged by larvae and/or adult weevils. This is particularly true if a decision was made to use harvesting as a control measure, rather than spraying, to eliminate an infestation before first cut. Sampling methods described above are not designed to deal with infestations in the regrowth. However, Buntin and Pedigo[65] found that a stubble-spray is warranted if regrowth is delayed more than 3 to 5 d after first cut.

2. Variegated Cutworm

The variegated cutworm is an occasional pest of alfalfa throughout most of the U.S. and southern Canada.[66] Though outbreaks of this species occur at irregular intervals, it has great damage potential. This species completes several generations/year, however, infestations in alfalfa typically build up in the first crop, and the final stages of larval development coincide with first cut. Whereas the presence of even large numbers of cutworms may escape notice in the dense foliage of the mature first crop, the potential for damage to the new growth after cutting may be great.

In a comprehensive study of damage by the variegated cutworm on regrowth after the first cut of alfalfa, Buntin and Pedigo[15,17] found that six or more larvae per 0.1 m^2 completely suppressed growth and caused significant delays in development of the second crop. If the second cut is to be taken at first bloom, economic injury levels range from 2.8/0.1 m^2 [treatment cost = $16/ha ($7/acre) and value of hay = $60/ton] to 4.8 larvae per 0.1 m^2 [cost = $22/ha ($10/acre) and value = $77/ton].[17] Another aspect of stubble defoliation relates to the duration over which damage occurs. When large numbers of cutworms are present, growth of the second crop may be delayed from 2 to 3 weeks.[67] In an extreme case, complete destruction of regrowth was observed for 24 d after first cut.[68] Whether stubble defoliation is due to the variegated cutworm or, as mentioned earlier, the alfalfa weevil, insecticide application may be justified if damage persists for more than 3 to 5 d.[65]

No sampling procedures have been published for use in scouting fields to evaluate the need for sprays to control the variegated cutworm. However, there are several aspects of its biology that should be considered when assessing population density of this pest. The cutworm feeds primarily at night, and during daytime hours seeks shelter beneath windrows, if hay has not been baled, or in soil around plant crowns after hay has been removed. As decisions regarding insecticide applications often must be made before hay has been removed, checking beneath windrows may be required if it is suspected that a damaging infestation may exist. Preparations can then be completed for spraying immediately after hay has been removed from the field. Although studies have not been completed to determine required numbers of samples, it is unlikely that a proper decision could be made unless at least one 0.1 m^2 area is checked in a minimum of 10 locations per field.

3. Army Cutworm

The geographic range of the army cutworm, *Euxoa auxiliaris* (Grote), includes most of the central and western U.S. and the prairie provinces of Canada.[69] The single generation completed each year begins with deposition of eggs in October. Favored sites for oviposition are areas with little ground cover such as newly planted fields of alfalfa or fields that have recently been harvested or grazed. Larvae hatch in fall, and early instars overwinter. Damage due to feeding by this cutworm begins as early as February in the Southern Plains and later in areas where winter weather prevails until March or April. A period of 1 to 2 months is required in early spring for completion of larval development, after which the potential for damage declines as they tunnel into the soil for pupation. Typically, years in which outbreaks occur are preceded by 1 to 2 years with gradually increasing populations.[69]

It is important that fields be scouted for this pest in spring as soon as temperatures begin to warm sufficiently to allow plant growth. This cutworm often causes heavy damage, particularly in new stands, before producers become aware of infestations. As with the variegated cutworm, detailed scouting procedures have not been developed. Also, research has not been conducted to determine economic threshold levels to guide treatment decisions. As the army cutworm is a nocturnal insect, soil around crowns must be thoroughly searched when sampling. In the absence of specific guidelines, use population densities given for the variegated cutworm to determine when spray applications are warranted.

4. Lepidopteran Foliage-Feeding Complex

The phrase 'foliage-feeding complex' refers to a variety of lepidopteran caterpillars that infest alfalfa fields throughout the summer months wherever the crop is grown in North America. The species that comprise this complex vary from one region to another as do those that predominate. Perhaps the most widely distributed is the alfalfa caterpillar. However, this species rarely causes serious damage except in the southwestern U.S.[70] Other widely distributed species include the alfalfa looper, *Autographa californica* (Speyer), and the green cloverworm, *Plathypena scabra* (F.).[37] Additional species that are common in the central and southern U.S. include the fall armyworm, corn earworm, *Helicoverpa zea* (Boddie), and webworms, *Achyra* spp.[71,72] The beet armyworm, *S. exigua* (Hübner), and western yellowstriped armyworm, *S. praefica* (Grote), are among the common foliage feeders in the western U.S.[45,51]

Even though all of the species comprising the foliage-feeding complex complete more than one generation/year, occurrence of larvae is often in synchrony with harvest cycles for alfalfa during summer. The usual pattern is as described for the alfalfa caterpillar by Summers et al.,[51] in which adults oviposit in fields where alfalfa growth is short following a recent harvest. Larvae hatch and increase in size as the alfalfa grows such that the greatest potential for defoliation often occurs as the time for the next harvest nears. Alfalfa crops in which the highest numbers of foliage-feeders occur are typically those produced in midsummer. Cutting alfalfa can often serve as an effective alternative to spraying to control these caterpillars. Spraying is recommended only if numbers exceed the economic threshold at least 1 to 2 weeks in advance of harvest.

As two or more species in this foliage-feeding complex often infest fields at the same time and cause similar damage, it usually is not necessary to distinguish species in scouting and decision making. There have been no comprehensive scouting procedures developed for the foliage-feeding complex, however, recommendations in several states involve taking 5 to 20 sweeps in at least five locations per field. The point at which insecticide application is recommended varies from five to eight larvae per sweep.[51,73]

With regard to sweepnet sampling, there is no agreement among entomologists on an appropriate sample unit size. Of the two most common interpretations, the 'pendulum' sweep involves swinging the net in a 90° arc through the foliage in front of the body perpendicular to the direction the sampler is walking through the field. This type of sweep results in a sample covering an area of about 0.5 m² (4 to 5 ft²). A second type of sweep described in many extension bulletins involves drawing the net through the foliage in a 180° arc from one side of the body to the other as the sampler walks the field. This sweep is estimated to cover about twice the area of the 'pendulum' sweep and would be expected to result in twice the number of insects collected. Descriptions of sweepnet sampling and economic thresholds expressed in insects per sweep in this chapter have been standardized to 'pendulum' sweeps. The standard diameter for a sweepnet used in alfalfa field scouting is 37 cm (15 in.).

5. Blister Beetles

The blister beetle family (Meloidae) is a relatively large group of beetles, with many species having fairly wide distribution over the alfalfa-growing areas of North America. Surveys in Colorado[74] and Kansas[75] indicate that at least seven species are common in alfalfa grown in the central plains of the U.S. Perhaps the species most commonly observed is the black blister beetle, *Epicauta pennsylvanica* (De Geer). In addition, there are several species that are termed "striped blister beetles" including *E. lemniscata* (F.) and *E. occidentalis* Werner.[75] Although blister beetles are apparently somewhat more common in the Southern Plains states than in the northern states and Canada, concerns about blister beetle infestations in alfalfa extend into areas such as Idaho,[7] Minnesota, and Wisconsin.[8]

Blister beetles have a rather unusual seasonal life history in that the immature stages of those species commonly found in alfalfa typically do not occur within alfalfa fields, but in border areas and grasslands that are infested by grasshoppers. Usually, one generation is completed each year, beginning with egg deposition, during late summer in the soil in areas that are favored by grasshoppers as egg-laying sites. The food source for larvae of these blister beetles is the eggs of grasshoppers, laid in clusters (egg pods) in the soil. Immatures overwinter in the soil, and adult emergence occurs in late spring or early summer, depending upon the species in question. It is adults that invade alfalfa fields in search of foliage, and particularly blossoms, on which they feed. There is much greater likelihood of blister beetle infestations in alfalfa that is blooming than that in the vegetative stage of growth.[74,76]

The amount of feeding done by blister beetles is rarely cause for concern by alfalfa growers. The major problem created by the beetles is the presence of a chemical within their bodies called cantharidin. If beetles are killed in the harvesting process and their bodies are present in forage fed to livestock, severe poisoning can result. The danger of a fatal reaction is particularly acute for horses, and it has been estimated that bodies of as few as 25 beetles could be lethal to a small horse.[74] Two species cited by Blodgett and Higgins[75] as having great potential to cause poisoning in horses fed alfalfa hay are the striped blister beetles, *E. lemniscata* and *E. occidentalis*. Adults of these species have been collected for a period of over 3 months during the summer in Kansas, and they have the tendency to form large aggregations or "swarms". If beetles are killed in the harvesting process, the presence of swarms in fields can result in concentrations of beetle remains within one or a few bales of hay.

Common recommendations for avoiding the presence of blister beetle remains in forage include cutting before alfalfa is in bloom to reduce the likelihood of infestation and cutting without a crimper so that beetles are not crushed as the hay passes through the windrower. Another precaution involves checking fields before cutting to determine if swarms are present. As it is unlikely that sufficient resources would be available to permit careful scouting of entire fields, the following observations regarding the behavior of these beetles will assist in identifying areas for close attention. First, it has been found that the probability of blister beetle infestation is increased when alfalfa is located adjacent to rangeland[74] or areas of weedy vegetation.[76] Blodgett and Higgins[76] suggest that scouting efforts be concentrated in areas close to the field borders to optimize blister beetle detection. Scouting should be done by taking frequent samples with a sweepnet while walking the field within 5 to 10 m of the border. If more than one to two beetles are collected per sweep in a particular part of the field, insecticide application in that area may be warranted. An insecticide having a 1 or 0 d preharvest interval (e.g., malathion) can be applied to infested areas to kill beetles before cutting. The chance of blister beetle contamination is greatest in forage cut during midsummer, and infestations are much less likely before June 1 or after September 15 in most regions.

B. FLUID FEEDERS
1. Potato Leafhopper

The potato leafhopper has a large number of host plants and is considered by many entomologists to be the most consistently damaging insect pest of alfalfa in the northcentral and northeastern U.S.[77,78] As the leafhopper has no diapausing stage in which to survive subfreezing temperatures, it overwinters only in states bordering the Gulf of Mexico and migrates into the northern U.S. and Canada in spring of each year. Timing of its arrival and severity of infestations is dependent on the passage of weather fronts moving from southwestern to northern states that serve to carry the insects northward.[79] In addition, the numbers of leafhoppers available to be carried northward is dependent upon the quality of host plants and occurrence of lethal temperatures in overwintering sites.

Typically, potato leafhoppers arrive in the northern U.S. between mid-May and early June. Migrating populations usually consist of a preponderance of females, which may outnumber males 10:1. Eggs are inserted into stem tissues of alfalfa plants and, after eclosing, nymphs feed on the stems with piercing-sucking mouthparts. Nymphal development requires 15 to 30 d, depending upon temperature, or a total of 330 to 360 CDD.[80] Females lay an average of 104 eggs over the adult life span of 30 to 60 d.[80]

The potato leafhopper feeds by making repeated insertions of stylets into cells of alfalfa stems. This probing and injection of salivary secretions results in clogging of phloem tissues. The most obvious symptom of feeding damage by the leafhopper is the V-shaped chlorotic area, called 'hopperburn,' produced at the tips of host leaves. Feeding results in stunting of plant growth and reduced yields with populations as low as 0.3 to 2.0 leafhoppers per sweep in plants < 30 cm tall. Crude protein content of forage may be reduced if numbers approach 5 to 6 per sweep over a 20- to 30-d period.[81] In addition, delayed maturity of infested alfalfa may limit the number of harvests that can be taken in a season.[82]

Nominal thresholds used in many states for decision making regarding insecticide applications for control of the potato leafhopper are as follows: alfalfa height < 8 cm (3 in.), 0.3 leafhoppers per sweep; 8 to 18 cm (3 to 7 in.), 0.5 per sweep; 20 to 28 cm (8 to 11 in.), 1.0 per sweep; > 30 cm (12 in.), 2.0 per sweep.[81] Although both nymphs and adults are counted in most instances, for timely decisions made soon after cutting and removal of forage, only adults are present.

Dynamic economic thresholds, published by Gessell et al.[83] in Pennsylvania, include the value of forage and control cost. The sampling protocol for these requires five sets of 20 sweep per field, counting all potato leafhoppers, and calculating the average number per sweep. Three categories of plant heights are used in decision making; e.g., 0 to 10 cm (4 in.), 10 to 20 cm (4 to 8 in.), and 20 to 30 cm (8 to 12 in.). For each category, a table based on ranges of values for alfalfa forage and control costs is used to determine if the average number of leafhoppers collected warrants spraying.

Because migration of the potato leafhopper into the northcentral and northeastern U.S. usually occurs between mid-May and mid-June, sampling is initiated after the first alfalfa crop is cut. Numbers typically decline rapidly in late August or September; thus, sampling is not necessary in fall. It is critical that sampling be conducted soon after each harvest during June, July, and August in areas where alfalfa is prone to damage by this pest.

Lamp and Smith[84] recently reviewed sampling methods for potato leafhopper published through 1987. They concluded that sweepnet sampling continues to be the most efficient way to monitor populations for decision-making purposes. A variety of sample-unit sizes have been recommended, ranging from 5 to 50 or 100 sweep. Shields and Specker[85] found no significant differences in precision among 10-, 20-, and 25-sweep units for estimating low population densities (< 0.2 per sweep), however, they concluded that the 10-sweep unit was the most efficient when leafhopper numbers exceeded 0.2 per sweep.

Sequential sampling plans have been developed for decision making regarding insecticide applications for potato leafhopper control in New York[85] and Virginia.[86] Researchers developing these plans have agreed that the spatial pattern for the leafhopper in fields is usually random and conforms to the Poisson distribution, particularly when the population is comprised only of adults. Consequently, this distribution was selected when developing leafhopper sequential sampling stop-lines using Wald's sequential probability ratio test.[87,88]

The primary difference between the plans developed in New York and Virginia is in the class limits (m_1 and m_2) or threshold bracketing values selected to calculate the stop lines. In both instances, researchers conducted extensive computer simulations to evaluate the effect of a range of bracketing values. However, Luna et al.[86] concluded that no differences existed among plant height/economic threshold combinations while Shields and Specker[85] found clear relationships between class limit values and performance of their plan and incorporated adjusted class limits (Table 3). To use this plan, a minimum of three sets of ten sweeps must be taken. Compare the total number of potato leafhoppers collected with values in the table for the appropriate crop height to decide if insecticide application is needed. For example, if height is < 3 in., no treatment is required if two or less leafhoppers are collected. If nine or more are collected, spraying is recommended. If the total is between these limits, additional sets of ten sweeps must be taken until a decision is reached (Table 3).

Over 3 years, Luna et al.[86] observed time savings ranging from 28 to 55% using their plan in comparison with a conventional plan requiring 100 sweeps per field. Shields and Specker[85] calculated average time savings of 68% in a similar comparison. When their plan was evaluated in 79 fields in Wisconsin, an average of just 3.3 samples (10 sweeps per sample) was required to reach a management decision. The decisions agreed with those attained using the conventional 100 sweep (5 sets of 20 sweeps) sample in 96% of the comparisons.

An expert system designated 'PLEXUS' is currently being developed and validated in Pennsylvania to assist with decision making for potato leafhopper control in alfalfa.[89] This is a dynamic system that incorporates forage values, control costs, plant heights, and other parameters for treatment recommendations.

2. Pea Aphid

The pea aphid is distributed virtually worldwide as a pest of alfalfa and peas, *Pisum sativum* L.[90] This aphid is known to exhibit facultative diapause, overwintering as eggs in the northern latitudes of North America,[91] while surviving the winter as nymphs and adults in areas such as coastal California.[51] Pea aphids complete four nymphal

TABLE 3
Sequential Table for Sampling *E. fabae* Adults on Alfalfa in New York[a]

Crop height (in.)	Sample (site) no.	Cumulative number of *E. fabae* adults		
		No treatment	Continue sampling	Treatment
< 3	3	≤ 2	3–8	≥ 9
	4	≤ 4	5–10	≥ 11
	5	≤ 5	6–12	≥ 13
	6	≤ 7	8–14	≥ 15
	7	≤ 9	10–15	≥ 16
	8	≤ 11	12–17	≥ 18
	9	≤ 13	14–19	≥ 20
	10	≤ 15	16–21	≥ 22
3–6	3	≤ 9	10–19	≥ 20
	4	≤ 14	15–24	≥ 25
	5	≤ 18	19–29	≥ 30
	6	≤ 23	24–34	≥ 35
	7	≤ 28	29–39	≥ 40
	8	≤ 33	34–44	≥ 45
	9	≤ 38	39–48	≥ 49
	10	≤ 43	44–53	≥ 54
7–10	3	≤ 19	20–40	≥ 41
	4	≤ 29	30–49	≥ 50
	5	≤ 39	40–59	≥ 60
	6	≤ 49	50–69	≥ 70
	7	≤ 59	60–79	≥ 80
	8	≤ 69	70–89	≥ 90
	9	≤ 79	80–99	≥ 100
	10	≤ 89	90–109	≥ 110
> 10	3	≤ 44	45–74	≥ 75
	4	≤ 64	65–94	≥ 95
	5	≤ 84	85–114	≥ 115
	6	≤ 104	105–134	≥ 135
	7	≤ 124	125–154	≥ 155
	8	≤ 144	145–174	≥ 175
	9	≤ 164	165–194	≥ 195
	10	≤ 184	185–214	≥ 215

[a] Sample unit = 10 sweeps.

stages, with developmental times ranging from 6.2 d at 25°C (77°F) to 19 d at 10°C (50°F).[92] Rapid development in combination with parthenogenetic reproduction results in population growth for the pea aphid that is typically exponential in the absence of most biotic and abiotic mortality factors. Cool weather conditions of late winter and spring are favorable for increasing numbers of pea aphids, while temperatures above 25°C are detrimental to population growth and survival.[93] Consequently, damage by the pea aphid typically occurs in alfalfa crops produced before mid-June across the southern and central U.S.[20,51,94] Infestations may persist through the summer in the northern U.S. and Canada.

Feeding of the pea aphid results in stunting of plants, with obvious chlorosis and wilting in the presence of high numbers. Yield reductions caused by the pea aphid are highly variable, in part because of difficulty in comparing sweep samples and stem samples. Hobbs et al.[95] observed no yield loss in irrigated alfalfa with infestations of 1400 to 1800 aphids per sweep. However, Wilson and Quisenberry[20] found a significant reduction at a population density of 90 pea aphids per stem when soil moisture was not limiting. Cuperus et al.[31] reported an economic threshold of 1.2 aphids per stem 2 weeks before cutting in water-stressed alfalfa. Currently, population densities at which insecticide application is recommended range from 100 to 300 per sweep for alfalfa heights up to 25 to 30 cm (10 to 12 in.).[7,8,45,96] Spraying is recommended when numbers reach 40 to 50 per stem for plants up to 25 cm and 70 to 80 per stem when taller.[51,96] In California, it is recommended that a sweepnet sample be taken to determine whether the population density of lady beetles is sufficient to eliminate the aphid infestation without need of an insecticide application. A field should not be sprayed if it is estimated that there is one lady beetle adult per 5 to 10 aphids or 3 lady beetle larvae per 40 aphids.[51]

In addition to use of sweep nets and stem sampling, a variety of sampling devices have been used for the pea aphid including yellow-pan and suction traps[97] and D-Vac® suction samplers.[98] Stem sampling has been used extensively in research on aphids in alfalfa and is becoming more widely used in field scouting. Cuperus et al.[31] and Buntin and Isenhour[99] found stem sampling to be an accurate and efficient alternative to sweep sampling. Not only were CVs (coefficients of variation) lower for stem samples in the study of Cuperus et al.,[31] but substantial time savings were accrued as well. Whereas sweep sampling may be an efficient means of determining if aphids are present in low numbers, at high population densities, the sheet volume collected may require extensive time for counting. At population densities of 0 to 15 per stem (0 to 80 per sweep), Buntin and Isenhour[99] found no difference in RNP (relative net precision) between 30-stem and 20-sweep samples. They calculated RNP, as follows:

$$\text{RNP} = [1/(\text{Cost} * \text{RV})] * 100 \tag{1}$$

where RV = SEM (standard error of mean)/Mean density * 100.

Buntin and Isenhour[99] also developed a regression equation for converting sweep-net counts to stem sample estimates, as follows:

$$\text{Aphids/stem} = -9.9873 + 0.308(\text{SW}) - 0.00008(\text{SW}^2) \tag{2}$$

where SW = number of aphids per sweep (using a 20-sweep sample).

Hutchison[100] compared three-stem with one-stem sample units and found no significant differences in mean aphids per stem at high population densities (e.g., 80 aphids per stem or higher). Because pea aphids prefer top portions of stems, he found that taking only the distal two thirds of stems did not reduce accuracy. Taking distal portions of single stems resulted in less dislodging of aphids and considerable time savings in field scouting.

Over a wide range of population densities, up to 90 insects per stem, the pea aphid exhibits an aggregated spatial pattern in alfalfa fields that conforms to the negative binomial distribution.[101-103] It is only at low densities (< 5 aphids per stem) that random spatial patterns have been observed. These observations on statistical distributions for the pea aphid were used by Hutchison et al.[103] in developing a stop-line sequential sampling plan for estimating population density using the procedure of Kuno.[104] This plan involves collecting single-stem sample units and accumulating totals until the required number of aphids for a given number of samples is attained

(Table 4). At this point the mean number of aphids per stem is calculated by dividing total aphids by the number of samples. The plan has two stop-lines, one for mean densities ranging from 3 to 15 aphids per stem and one for densities either below or above this range.

3. Blue Alfalfa Aphid

The blue alfalfa aphid was first detected in North America (California) during 1974, and by 1980 had spread into several central plains states including Nebraska, Kansas, and Oklahoma.[105] Recently, the presence of this species in Maryland has been confirmed.[106] Seasonal occurrence of the blue alfalfa aphid through spring and

TABLE 4
Stop-Line Sampling Plans for Estimating *A. pisum* Population Density in Alfalfa Using Kuno's Procedure[a]

Cumulative[b] no. sample units (stems)	Cumulative aphid count (T_n) needed to stop[c]	
	Low / high density (≤ 3 or > 15.1) ($D = 0.30$)	Intermediate density (3.1–15) ($D = 0.35$)
5	451	116
6	178	83
7	124	70
8	101	62
9	89	57
10	81	54
11	75	51
12	71	49
13	68	48
14	65	46
15	63	45
16	62	45
17	60	44
18	59	43
19	58	43
20	57	42
22	56	41
24	55	41
26	54	40
28	53	40
30	52	39
32	51	39
34	51	39
36	50	39
38	50	38
40	50	38
42	49	38
44	49	38
46	49	38
48	49	38
50	49	37

[a] Sampling (T_n) criteria based on computer simulation results designed to maintain the precision level $D = 0.25$; for selected density ranges using Kuno's stop-line plan. (see also Chapter 10)
[b] Minimum sample size, 5 stems; maximum, 50 stems.
[c] Density ranges refer to total number of aphids per stem.

early summer is quite similar to that of the pea aphid, and fields are often infested by combined populations of these species.[51,94] However, the blue alfalfa aphid is rarely collected in fall and is not known to produce an overwintering egg stage. This species apparently makes annual northward migrations into the Plains from overwintering sites in the extreme southern U.S.

In contrast to the observation by Stern et al.[105] regarding suppression of pea aphid populations by blue alfalfa aphid, Poswal and Berberet[107] found that the two species tend to increase in concert without evidence of competitive displacement. Lower and upper thresholds for development for the blue alfalfa aphid have been determined to be 3.4° and 27.1°C, respectively, and adults do not reproduce at temperatures above 29°C.[108]

Both Stern et al.[105] and Poswal et al.[109] observed that infestations of the blue alfalfa aphid occur at a time in early spring when there are few natural enemies available to limit population increases. The blue alfalfa aphid prefers to feed on new growth of alfalfa plants, and large numbers often occur in plant terminals. Damage symptoms include stunting of growth and deformation of leaves, with chlorosis and leaf drop evident in heavily infested fields. Stern et al.[105] reported that as few as 10 to 12 aphids per stem on regrowth after cutting may result in sufficient population increase to cause losses in excess of control costs. These authors state that damage symptoms become evident when numbers reach 20 per stem on new growth, however, alfalfa that is 25 to 30 cm in height can withstand 40 to 50 blue alfalfa aphids per stem. Bishop et al.[110] developed a series of regression analyses for relating aphid-days per stem (number of aphids per stem X days infested) with 10, 20, and 30% yield loss estimates in tolerant (cv. = 'CUF 101') and susceptible (cv. = 'Hunter River' and 'Condura 73') alfalfas. Economic thresholds published in Oklahoma range from 10 aphids per stem on alfalfa up to 25 cm (10 in.) in height to 30 per stem on taller plants.[96] In California, control is recommended if numbers reach 40 to 50 per stem in alfalfa over 25 cm tall.[51]

Rohitha and Penman[111] conducted a study comparing stem sampling, suction sampling, and removal of foliage from measured areas for determination of blue alfalfa aphid population densities. They concluded that sampling efficiency, based on cost per unit and variability among samples, is greatest when removing foliage from measured areas of 700 cm^2 and 80 cm^2 when alfalfa is 3.5 and 12 cm tall, respectively. When plant height exceeds 60 cm, stem sampling is more efficient. They also acknowledged that while useful for research studies, time required for area removal sampling limits its practicality for field scouting.

Single stems were used as sample units in studies of Poswal et al.,[109] with 6 to 12 units having standard errors within 5 to 13% of the means. Bishop and McKenzie[112] devised a method for improving recovery of aphids from stem samples in the field. It involves covering a container with hardware cloth (1.5-cm mesh) onto which stems can be beaten for removal of the insects. This method allows 83 to 95% recovery of aphids from stems. Recommended scouting procedures in California involve counting blue alfalfa aphids on five to six single-stem units at each of five locations per field.[51] In Oklahoma, a sample of 30 stems is collected at random per 20 acres of alfalfa, and aphids are shaken into a container for counting.[96] Spraying is recommended when aphid numbers per stem exceed the threshold levels cited above.

4. Spotted Alfalfa Aphid

In contrast to the aphids previously discussed, the spotted alfalfa aphid is well adapted to relatively high temperatures and dry conditions. Graham[113] reported that the most favorable temperature for reproduction and growth of this species is 30°C

(86°F), with relative humidity at 25 to 30%. Thus, warm, arid regions of the southwestern U.S. are ideally suited for damaging infestations of this species. Nielson and Barnes[114] observed that population densities may exceed economic threshold levels at any time during the growing season (March to October) in Arizona. They found that the primary limiting factor on population increase is rainfall. As a result of northward migrations, the aphid often infests alfalfa throughout the western U.S.[7,11,45,51] and into the central Plains.[96] Although a holocyclic (egg-laying) strain of the spotted alfalfa aphid is known to occur in the central U.S.,[115] it is believed that infestations in this area result primarily from adult migrations or survival of nymphs and adults through winter when the weather is relatively mild.[94,116]

The spotted alfalfa aphid can cause severe stunting of alfalfa, with chlorosis and necrosis of leaves, followed by leaf drop and death of plants within a 2- to 3-week period of high population densities. Early symptoms associated with the toxic effect induced by this aphid include discoloration or clearing of tissues along veins of newly formed leaves (veinbanding). In Oklahoma, stands of susceptible alfalfa cultivars have been destroyed in mid-winter by infestations on crown bud growth that is typically present.

The economic threshold used for the spotted alfalfa aphid soon after its discovery in the U.S. was five aphids per leaf.[114] Nielson[117] concluded from a study of several sampling methods including counts from leaves, stems, whole plants, and sweeping that the leaf counts provided the greatest accuracy and efficiency. In most states, however, stem sampling procedures like those already described for pea aphid and blue alfalfa aphid are used, and economic thresholds are expressed as aphids per stem. Population densities at which insecticide application is recommended range from 10 to 20 per stem up to 40 to 50 per stem in established alfalfa, depending upon plant height and time of year.[7,45,51,96] The spotted alfalfa aphid often infests new seedlings in fall. Because of limited tolerance of seedlings, the economic threshold is typically one aphid per plant.[11,96]

C. SEED PRODUCTION PESTS

Many of the species that have been discussed as pests in alfalfa forage production can also cause losses in seed production. Damage by these insects is primarily indirect for seed production in that fruiting structures are often not injured. Although these pests are not discussed again in this section, it is important to note that the defoliation, stunting, and reduced vigor that occur as a result of feeding by alfalfa weevil, cutworms, aphids, and other species can result in reduced seed production, whether in broadcast stands intended primarily for harvest of forage or those planted in rows specifically for seed. In addition, populations of key pests in seed production, such as *Lygus* spp., often increase during vegetative stages of alfalfa growth which may lead to severe damage to fruiting structures as the crop matures. It is generally recommended that fields from which seed is to be harvested be scouted 7 to 10 d before alfalfa begins blooming to determine if sprays are needed to control insect pests before pollinator activity begins.[11,118] More detailed information on the life history and sampling of seed production pests is given below.

1. *Lygus* Bugs

Of over 30 species of *Lygus* known to occur in North America, relatively few are common in alfalfa. The principal species found in seed production areas of the western U.S. are *L. hesperus* Knight and *L. elisus* Van Duzee.[11,118] Those present in greatest numbers in fields of alfalfa seed in Saskatchewan, Canada, are *L. borealis* (Kelton) and *L. unctuosus* (Kelton).[119] Another species that is common in alfalfa fields in many areas of North America is *L. lineolaris* (Palisot de Beauvois).[119-121] *Lygus* spp. feed on a wide variety of plants including many weed species such as mustards, Russian thistle, dandelion, and lambsquarters. Crop species that serve as hosts include beans, sugarbeets, cotton, and alfalfa.[11] Adults of *Lygus* spp. are highly mobile and frequently migrate among fields of cultivated crops or from weeds to crops. Damaging infestations in alfalfa seed production are often the result of migrations from nearby alfalfa fields being cut for hay.[118]

In most regions of Canada, *Lygus* spp. apparently complete one to two generations per year, although Khattat and Stewart[120] reported three generations of *L. lineolaris* in Quebec. Three to four generations are completed in the alfalfa seed-producing areas of the U.S.[11] *Lygus* spp. overwinter as adults in plant debris and litter throughout most of their range in North America. In early spring they feed on various weed species, where one generation may be completed before they migrate to crops such as alfalfa.[118,120] Eggs are inserted into leaf or stem tissues of host plants. After eclosion, nymphs begin feeding on plant fluids with sucking mouthparts.

Jensen et al.[121] reported that *L. lineolaris* has the potential to reduce alfalfa forage production. However, *Lygus* spp. listed earlier have much greater importance in limiting productivity of alfalfa being grown for seed. Although both nymphs and adults feed on buds, blossoms, and immature seeds, nymphal feeding is most destructive.[118] Losses are greatest due to feeding on buds. Injured buds become discolored (light brown or tan), die, and drop from plants. Also of importance is feeding by *Lygus* spp. on immature seeds, which causes them to turn brown and shrivel. Damaged seeds do not germinate.[11,118]

Johansen et al.[11] have stated a strong case of adoption of dynamic economic thresholds for *Lygus* spp. in alfalfa seed production. Among the important considerations are the maturity of the alfalfa (prebloom vs. bloom and seed set), activity of pollinators, and the presence of beneficial predators and parasites. It is commonly recommended that numbers must reach two bugs per "pendulum" sweep during the prebloom period before spraying is warranted.[118] When alfalfa is in full bloom and setting seed, the numbers of *Lygus* spp. at which spraying is recommended ranges from two to five per sweep.[10,11,118,122] According to Davis et al.,[118] spraying is warranted during the postbloom period of seed maturation only if numbers of bugs reach eight per sweep. Other authors state that sprays are usually not profitable during this period.[11,122] Due to their greater potential for causing damage, Davis et al.[118] have recommended that each nymph be counted as two bugs when taking sweep samples.

A pest management program developed for alfalfa seed production in Washington emphasizes preservation of beneficial predators, parasites, and pollinators.[11,123] In this program, spraying is not recommended if there is a 2:1 ratio of predators such as big-eyed bugs, *Geocoris* spp., and damsel bugs, *Nabis* spp., to *Lygus* spp. Subsequent research by Gupta et al.[124] resulted in development of a model used to calculate the total numbers of big-eyed bugs and damsel bugs needed to reduce pest populations comprised of *Lygus* spp. and pea aphid to a level below the economic threshold. This model is based on the premise that a 1:1 ratio of predators to pests is sufficient for effective regulation of these pests.

As is the case for many insects infesting field crops, *Lygus* spp. exhibit an aggregated or "clumped" spatial pattern.[125] This pattern requires that several samples must be taken in each field for accurate estimation of population density. Johansen et al.[11] recommend five sweeps as the basic sample unit for use in estimating numbers of *Lygus* spp. In a field of 15 to 25 h (30 to 50 acres), at least five samples of five sweeps should be taken with one about 10 m from each side and the fifth sample in the center of the field. For larger fields, numbers of samples should be increased, with about one taken for every 5 ha (10 acres). Particular attention must be given to portions of the field adjacent to other fields from which *Lygus* spp. may migrate.

For greatest accuracy in estimating numbers of *Lygus* spp., sampling should be conducted when temperatures range from 15 to 27°C (60 to 80°F) and winds do not exceed 8 to 10 mph. Cool, rainy weather, excessively hot conditions, and high winds create difficulty in making accurate estimates, and fields sampled under such conditions should be sampled again as soon as more optimal conditions occur. The usual sampling interval is one week, however, if there is a trend for rapid population increase, samples should be taken every 3 d.

V. SUMMARY

In this chapter, we have discussed information currently available on seasonal life histories, economic thresholds, and sampling protocols as they relate to management of key insect pests of alfalfa. Extensive research has been conducted for development of decision-making criteria for some species such as the alfalfa weevil, potato leafhopper, and *Lygus* bugs. Also, at least nominal thresholds and some sampling plans have been developed for cutworms, foliage-feeding lepidopterans, and aphids. However, much additional research is needed to refine the decision-making processes for these insects and provide the basis for effective management of species about which little is known, such as plant bugs and the three-cornered alfalfa hopper, *Spissistilis festinus* (Say). Little work has been done on multiple-species interactions in alfalfa to learn how sampling and decision-making can be made more effective when alfalfa fields are infested by two or more pests at the same time.

Despite the fact that sampling plans and economic thresholds have been developed for several key pests in alfalfa, adoption of these methods by producers has been limited. For example, a recent survey of extension entomologists in several states has indicated that the reluctance of producers to scout their fields has been the primary obstacle to implementation of management programs for potato leafhopper.[126] Similar experiences have occurred with other key pests throughout the U.S. As is discussed in Chapter 22, a concerted educational effort by extension personnel and consultants must continue for greater adoption of improved decision-making processes. It is also critical that research be continued on multiple-pest interactions and development of dynamic economic thresholds and sampling plans that integrate well with other farming operations.

REFERENCES

1. Michaud, R., Lehman, W. F., and Rumbaugh, M. D., World distribution and historical development, in *Alfalfa and Alfalfa Improvement*, Hanson, A. A., Barnes, D. K., and Hill, R. R., Jr., Eds., American Society of Agronomy, Madison, WI, 1988, chap. 2.
2. Melton, W., Moutray, J. B., and Bouton, J. H., Geographic adaptation and cultivar selection, in *Alfalfa and Alfalfa Improvement*, Hanson, A. A., Barnes, D. K., and Hill, R. R., Eds., American Society of Agronomy, Madison, WI, 1988, chap. 20.
3. Goplen, B. P., Baenziger, H., Bailey, L. D., Gross, A. T. H., Hanna, M. R., Michaud, R., Richards, K. W., and Waddington, J., Growing and Managing Alfalfa in Canada, *Agriculture Canada Publ.* 1705, Ottawa, 1980.
4. Cuperus, G. W., Berberet, R. C., and Ward, C., Integrating IPM into crop management, in *Proc. Natl. IPM Symp./Workshop*, Glass, E. H., Ed., Communications Services, N.Y. State Agric. Exp. Stn., Geneva, NY, 1989, 176.

5. McKenzie, J. S., Paquin, R., and Duke, S. H., Cold and heat tolerance, in *Alfalfa and Alfalfa Improvement*, Hanson, A. A., Barnes, D. K., and Hill, R. R., Jr., Eds., American Society of Agronomy, Inc., Madison, WI, 1988, chap. 8.
6. Scheaffer, C. C., Lacefield, G. D., and Marble, V. L., Cutting schedules and stands, in *Alfalfa and Alfalfa Improvement*, Hanson, A. A., Barnes, D. K., and Hill, R. R., Jr., Eds., American Society of Agronomy, Madison, WI, 1988, chap. 12.
7. Homan, H. W. and Baird, C. R., Control of alfalfa hay insects in Idaho, *Idaho Coop. Ext. Curr. Info. Ser.*, 1992, 908.
8. Undersander, D., Martin, N., Cosgrove, D., Kelling, K., Schmitt, M., Wedberg, J., Becker, R., Grau, C., and Doll, J., *Alfalfa Management Guide*, American Society of Agronomy, Madison, WI, 1991.
9. Tesar, M. B., Fertilization and management for a yield of ten tons of alfalfa without irrigation, in *Proc. Forage and Grassland Conf.*, Balas, M. A., Ed., American Forage and Grassland Council, Hershey, PA, 1985, 327.
10. Rincker, C. M., Marble, V. M., Brown, D. E., and Johansen, C. A., Seed production practices, in *Alfalfa and Alfalfa Improvement*, Hanson, A. A., Barnes, D. K., and Hill, R. R., Jr., Eds., American Society of Agronomy, Madison, WI, 1988, chap. 32.
11. Johansen, C., Baird, C., Bitner, R., Fisher, G., Undurraga, J., and Lauderdale, R., *Alfalfa Seed Insect Pest Management*, Western Regional Ext. Publ. 0012, 1979.
12. Marble, V. M., Producing Alfalfa Seed in California, Calif. Div. Agric. Sci. Leafl., 2383, 1970.
13. Berberet, R. C. and McNew, R. W., Reduction in yield and quality of leaf and stem components of alfalfa forage due to damage by larvae of *Hypera postica* (Coleoptera: Curculionidae), *J. Econ. Entomol.*, 79, 212, 1986.
14. Fick, G. W., Alfalfa weevil on regrowth of alfalfa, *Agron. J.*, 68, 809, 1976.
15. Buntin, G. D. and Pedigo, L. P., Dry-matter accumulation, partitioning, and development of alfalfa regrowth after stubble defoliation by the variegated cutworm (Lepidoptera: Noctuidae), *J. Econ. Entomol.*, 78, 371, 1985.
16. Berberet, R. C., Senst, K. M., Nuss, K. E., and Gibson, W. P., Alfalfa weevil in Oklahoma: the first ten years, *Oklahoma Agric. Exp. Stn. Bull.*, B751, 1980.
17. Buntin, G. D. and Pedigo, L. P., Development of economic injury levels for last-stage variegated cutworm (Lepidoptera; Noctuidae) larvae in alfalfa stubble, *J. Econ. Entomol.*, 78, 1341, 1985.
18. Berberet, R. C., Morrison, R. D., and Senst, K. M., Impact of the alfalfa weevil, *Hypera postica* (Gyllenhal) (Coleoptera: Curculionidae), on forage production in nonirrigated alfalfa in the southern plains, *J. Kan. Entomol. Soc.*, 54, 312, 1981.
19. Wilson, M. C., Stewart, J. K., and Vail, H. D., Full season impact of the alfalfa weevil, meadow spittlebug, and potato leafhopper in an alfalfa field, *J. Econ. Entomol.*, 72, 830, 1979.
20. Wilson, H. K. and Quisenberry, S. S., Impact of feeding by alfalfa weevil larvae (Coleoptera: Curculionidae) and pea aphid (Homoptera: Aphididae) on yield and quality of first and second cuttings of alfalfa, *J. Econ. Entomol.*, 79, 785, 1986.
21. Howe, W. L. and Pesho, G. R., Spotted alfalfa aphid resistance in mature growth of alfalfa varieties, *J. Econ. Entomol.*, 53, 234, 1960.
22. Medler, J. T., The nature of injury to alfalfa by *Empoasca fabae* (Harris), *Ann. Entomol. Soc. Am.*, 34, 439, 1941.
23. Nickel, J. L. and Sylvester, E. S., Influence of feeding time, stylet penetration, and developmental instar on the toxic effect of the spotted alfalfa aphid, *J. Econ. Entomol.*, 52, 249, 1959.
24. Hower, A. A. and Flinn, P. W., Effects of feeding by potato leafhopper nymphs (Homoptera: Cicadellidae) on growth and quality of established stand alfalfa, *J. Econ. Entomol.*, 79, 779, 1986.
25. Kindler, S. D., Kehr, W. R., and Ogden, R. L., Influence of pea aphids and spotted alfalfa aphids on the stand, yield of dry matter, and chemical composition of resistant and susceptible varieties of alfalfa, *J. Econ. Entomol.*, 64, 653, 1971.
26. Liu, B. W. Y. and Fick, G. W., Yield and quality losses due to alfalfa weevil, *Agron. J.*, 67, 828, 1975.
27. Hintz, T. R., Wilson, M. C., and Armbrust, E. J., Impact of alfalfa weevil larval feeding on the quality and yield of first cutting alfalfa, *J. Econ. Entomol.*, 69, 749, 1976.
28. Flinn, P. W., Hower, A. A., and Knievel, D. P., Physiological response of alfalfa to injury by *Empoasca fabae* (Homoptera: Cicadellidae), *Environ. Entomol.*, 19, 176, 1990.
29. Shaw, M. C. and Wilson, M. C., The potato leafhopper: scourge of leaf protein!—and root carbohydrates too?, in *Proc. 16th Natl. Alfalfa Symp.*, Wilson, M. C., Ed., Certified Alfalfa Seed Council, Inc., Woodland, CA, 1986.
30. Hutchins, S. H. and Pedigo, L. P., Phenological disruption and economic consequence of injury to alfalfa induced by potato leafhopper (Homoptera: Cicadellidae), *J. Econ. Entomol.*, 83, 1587, 1990.
31. Cuperus, G. W., Radcliffe, E. B., Barnes, D. K., and Marten, G. C., Economic injury levels and economic thresholds for pea aphid, *Acyrthosiphon pisum* (Harris), on alfalfa, *Crop Prot.*, 1, 453, 1982.

32. Stucker, D. S., Establishment of Economic Injury Levels and Economic Thresholds for Pea Aphid on Alfalfa, M.S. thesis, University of Minnesota, St. Paul, 1986.
33. Summers, C. G. and Coviello, R. L., Impact of *Acyrthosiphon kondoi* (Homoptera: Aphididae) on alfalfa: field and greenhouse studies, *J. Econ. Entomol.*, 77, 1052, 1984.
34. Titus, E. G., The alfalfa leaf weevil, *Utah Agric. Exp. Stn. Bull.*, 110, 1910.
35. Wehrle, L. P., A new insect introduction, *Bull. Brooklyn Entomol. Soc.*, 34, 170, 1939.
36. Poos, F. W. and Bissell, T. L., The alfalfa weevil in Maryland, *J. Econ. Entomol.*, 46, 178, 1953.
37. Manglitz, G. R. and Ratcliffe, R. H., Insects and mites, in *Alfalfa and Alfalfa Improvement*, Hanson, A. A., Barnes, D. K., and Hill, R. R., Jr., Eds., American Society of Agronomy, Madison, WI, 1988, chap. 22.
38. Hsiao, C. and Hsiao, T. H., Rickettsia as the cause of cytoplasmic incompatibility in the alfalfa weevil, *Hypera postica* (Gyllenhal), *J. Invertebr. Pathol.*, 45, 244, 1985.
39. Hsiao, C. and Hsiao, T. H., Hybridization and cytoplasmic incompatibility among alfalfa weevil strains, *Entomol. Exp. Appl.*, 37, 155, 1985.
40. Manglitz, G. R., Aestivation of the alfalfa weevil, *J. Econ. Entomol.*, 51, 506, 1958.
41. Litsinger, J. A. and Apple, J. W., Estival diapause of the alfalfa weevil in Wisconsin, *Ann. Entomol. Soc. Am.*, 66, 11, 1973.
42. Day, W. H., Reproductive status and survival of the alfalfa weevil adults: effects of certain foods and temperatures, *Ann. Entomol. Soc. Am.*, 64, 208, 1971.
43. Casagrande, R. A. and Stehr, F. W., Evaluating the effects of harvesting alfalfa on alfalfa weevil (Coleoptera: Curculionidae) and parasite populations in Michigan, *Can. Entomol.*, 105, 1119, 1973.
44. Litsinger, J. A. and Apple, J. W., Oviposition of the alfalfa weevil in Wisconsin, *Ann. Entomol. Soc. Am.*, 66, 17, 1973.
45. Davis, D. W. and Knowlton, G. F., Foliage pests of alfalfa, in *Insects and Nematodes Associated with Alfalfa in Utah*, Davis, D. W., Ed., *Utah Agric. Exp. Stn. Bull.*, 494, 1976, chap. 1.
46. Hsieh, F. and Armbrust, E. J., Temperature limits of alfalfa weevil oviposition and egg density in Illinois, *J. Econ. Entomol.*, 67, 203, 1974.
47. Niemczyk, H. D. and Flessel, J. K., Population dynamics of alfalfa weevil eggs in Ohio, *J. Econ. Entomol.*, 63, 242, 1970.
48. Townsend, H. G. and Yendol, W. G., Survival of overwintering alfalfa weevil eggs in Pennsylvania, *J. Econ. Entomol.*, 61, 916, 1968.
49. Morrison, W. P. and Pass, B. C., The effect of subthreshold temperatures on eggs of the alfalfa weevil, *Environ. Entomol.*, 3, 353, 1974.
50. Bass, M. H., Notes on the biology of the alfalfa weevil, *Hypera postica*, in Alabama, *Ann. Entomol. Soc. Am.*, 60, 295, 1967.
51. Summers, C. G., Gilchrist, D. G., and Norris, R. F., *Integrated Pest Management for Alfalfa Hay*, University of California, Richmond, 1981.
52. Stark, J. A., Life History of the Alfalfa Weevil, *Hypera postica*, and Temporal Predictions for Larval Populations Exceeding the Economic Threshold in Oklahoma, Ph.D. thesis, Oklahoma State University, Stillwater, 1991.
53. Harcourt, D. G., A thermal summation model for predicting seasonal occurrence of the alfalfa weevil, *Hypera postica* (Coleoptera: Curculionidae), in southern Ontario, *Can. Entomol.*, 113, 601, 1981.
54. Berberet, R. and Coppock, S., Scouting for the alfalfa weevil in Oklahoma, *Okla. Coop. Ext. Serv. Curr. Rep.*, 7177, 1985.
55. Blickenstaff, C. C., Standard survey procedures for the alfalfa weevil, *Bull. Entomol. Soc. Am.*, 12, 29, 1966.
56. Cothran, W. R. and Summers, C. G., Visual economic thresholds and potential pesticide abuse: alfalfa weevils, an example, *Environ. Entomol.*, 3, 891, 1974.
57. Armbrust, E. J., Niemczyk, H. D., Pass, B. C., and Wilson, M. C., Standardized procedures adopted for cooperative Ohio Valley states alfalfa weevil research, *J. Econ. Entomol.*, 62, 250, 1969.
58. Guppy, J. C., Harcourt, D. G., and Mukerji, M. K., Population assessment during the larval stage of the alfalfa weevil, *Hypera postica* (Coleoptera: Curculionidae), *Can. Entomol.*, 107, 785, 1975.
59. Ruesink, W. G., Modeling of pest populations in the alfalfa ecosystem with special reference to the alfalfa weevil, in *Modeling for Pest Management*, Tummala, R. L., Haynes, D. L., and Croft, B. A., Eds., Michigan State University, East Lansing, 1976, 80.
60. Wedberg, J. L., Ruesink, W. G., Armbrust, E. J., and Bartell, D. P., Alfalfa weevil pest management program, *Ill. Coop. Ext. Serv. Circ.*, 1136, 1977.
61. Legg, D. E., Shufran, K. A., and Yeargan, K. V., Evaluation of two sampling methods for their influence on the population statistics of alfalfa weevil (Coleoptera: Curculionidae) larva infestations in alfalfa, *J. Econ. Entomol.*, 78, 1468, 1985.

62. Barney, R. J. and Legg, D. E., Accuracy of a single 30-stem sample for detecting alfalfa weevil larvae (Coleoptera: Curculionidae) and making management decisions, *J. Econ. Entomol.*, 80, 512, 1987.
63. Barney, R. J. and Legg, D. E., Development and validation of sequential sampling plans for alfalfa weevil (Coleoptera: Curculionidae) larvae using a six-stem, 100-square-meter subsampling method, *J. Econ. Entomol.*, 81, 658, 1988.
64. Sabella, V. R., Stark, J. A., Berberet, R., Mulder, P., and Cuperus, G. W., ALFWEEV: alfalfa insect management expert system, *Okla. Coop. Ext. Serv. Comput. Software Ser.*, 37, 1989.
65. Buntin, G. D. and Pedigo, L. P., Management of alfalfa stubble defoliators causing a complete suppression of regrowth, *J. Econ. Entomol.*, 79, 769, 1986.
66. Frosheiser, F. I., Munson, R. D., and Wilson, M. C., *Alfalfa Analyst*, Certified Alfalfa Seed Council, Bakersfield, CA, 1972.
67. Soteres, K. M., Berberet, R. C., and McNew, R. W., Parasites of larval *Euxoa auxiliaris* (Grote) and *Peridroma saucia* (Hubner) (Lepidoptera: Noctuidae) in alfalfa fields of Oklahoma, *J. Kan. Entomol. Soc.*, 57, 63, 1984.
68. Berberet, R. C., unpublished data, 1976.
69. Burton, R. L., Starks, K. J., and Peters, D. C., The army cutworm, *Okla. Agric. Exp. Stn. Bull.*, B749, 1980.
70. Anon, Control of the Alfalfa Caterpillar, *U.S. Department of Agriculture Leaflet* Washington, DC, 1963, 325.
71. Brandenburg, R. L., Control of the variegated cutworm and fall armyworm in alfalfa, *Mo. Coop. Ext. Div.*, 1982.
72. Soteres, K. M., Berberet, R. C., and McNew, R. W., Parasitic insects associated with lepidopterous herbivores on alfalfa in Oklahoma, *Environ. Entomol.*, 13, 787, 1984.
73. Mulder, P., Insect management, in *Alfalfa Integrated Management in Oklahoma*, Diel, T., Ed., Oklahoma State University, Stillwater, 1991, 21.
74. Capinera, J. L., Gardner, D. R., and Stermitz, F. R., Cantharidin levels in blister beetles (Coleoptera: Meloidae) associated with alfalfa in Colorado, *J. Econ. Entomol.*, 78, 1052, 1985.
75. Blodgett, S. L. and Higgins, R. A., Blister beetles (Coleoptera: Meloidae) in Kansas: historical perspective and results of an intensive alfalfa survey, *J. Econ. Entomol.*, 81, 1456, 1988.
76. Blodgett, S. L. and Higgins, R. A., Blister beetles (Coleoptera: Meloidae) in Kansas alfalfa: influence of plant phenology and proximity to field edge, *J. Econ. Entomol.*, 83, 1042, 1990.
77. Gyrisco, G. G., Landman, D., York, A. C., Irwin, B. J., and Armbrust, E. J., The literature of arthropods associated with alfalfa. IV. A bibliography of the potato leafhopper, *Ill. Agric. Exp. Stn. Spec. Publ.*, 51, 1978.
78. Steffey, K. L. and Armbrust, E. J., Pest management systems for alfalfa insects, in *Handbook of Pest Management*, Vol. 3, Pimentel, D., Ed., CRC Press, Boca Raton, FL, 1991, 475.
79. Carlson, J. D., Whalon, M. E., Landis, D. A., and Gage, S. H., Evidence for long-range transport of potato leafhopper into Michigan, in *Proc. 10th Conf. on Biometeorology and Aerobiology and the Special Session on Hydrometeorology*, American Meteorological Society, Boston, 1991, 123.
80. Hogg, D. B., Potato leafhopper (Homoptera: Cicadellidae) immature development, life tables, and population dynamics under fluctuating temperature regimes, *Environ. Entomol.*, 14, 349, 1985.
81. Cuperus, G. W., Radcliffe, E. B., Barnes, D. K., and Marten, G. C., Economic injury levels and economic thresholds for potato leafhopper (Homoptera: Cicadellidae) on alfalfa in Minnesota, *J. Econ. Entomol.*, 76, 1341, 1983.
82. Hutchins, S. H. and Pedigo, L. P., Phenological disruption and economic consequence of injury to alfalfa induced by potato leafhopper (Homoptera: Cicadellidae), *J. Econ. Entomol.*, 83, 1527, 1990.
83. Gesell, S. G., Hartwig, N. L., Hower, A. A., Jr., Leath, K. T., and Stringer, W. C., A pest management program for alfalfa in Pennsylvania, *Pa. Coop. Ext. Circ.*, 284, 1982.
84. Lamp, W. O. and Smith, L. M., Objectives and problems of potato leafhopper sampling, in *Potato Leafhopper, Empoasca fabae (Harris) (Homoptera: Cicadellidae) Research: Current and Historical Perspectives*, Armbrust, E. J. and Lamp, W. O., Eds., Misc. Publ., Entomology Society of America, Lanham, MD, 1989, 3.
85. Shields, E. J. and Specker, D. R., Sampling for potato leafhopper (Homoptera: Cicadellidae) on alfalfa in New York: relative efficiency of three sampling methods and development of a sequential sampling plan, *J. Econ. Entomol.*, 82, 1091, 1989.
86. Luna, J. M., Fleischer, S. J., and Allen, W. A., Development and evaluation of sequential sampling plans for potato leafhopper (Homoptera: Cicadellidae) in alfalfa, *Environ. Entomol.*, 12, 1690, 1983.
87. Wald, A., *Sequential Analysis*, John Wiley & Sons, New York, 1947.
88. Waters, W. E., Sequential sampling in forest insect surveys, *For. Sci.*, 1, 68, 1955.

89. Hower, A., Calvin, D., Alexander, S., McCuer, J., and Lazaros, J., Managing potato leafhopper on alfalfa: computer based vs. paper based systems, in *Proc. 33rd N. Am. Alfalfa Imp. Conf.*, McCaslin, M., Ed., N. Am. Alfalfa Imp. Conf., Beltsville, MD, 1992, 27.
90. Harper, A. M., Miska, J. P., Manglitz, G. R., Irwin, B. J., and Armbrust, E. J., The literature of arthropods associated with alfalfa. III. A bibliography of the pea aphid *Acyrthosiphon pisum* (Harris) (Homoptera: Aphididae), *Ill. Agric. Exp. Stn. Spec. Publ.*, 50, 1978.
91. Hutchison, W. D. and Hogg, D. B., Time-specific life tables for the pea aphid, *Acyrthosiphon pisum* (Harris), on alfalfa, *Res. Popul. Ecol.*, 27, 231, 1985.
92. Hutchison, W. D. and Hogg, D. B., Demographic statistics for the pea aphid (Homoptera: Aphididae) in Wisconsin and a comparison with other populations, *Environ. Entomol.*, 13, 1173, 1984.
93. Campbell, A. and Mackauer, M., Reproduction and population growth of the pea aphid (Homoptera: Aphididae) under laboratory and field conditions, *Can. Entomol.*, 109, 277, 1977.
94. Berberet, R. C., Arnold, D. C., and Soteres, K. M., Geographical occurrence of *Acyrthosiphon kondoi* Shinji in Oklahoma and its seasonal incidence in relation to *A. pisum* (Harris) and *Therioaphis maculata* (Buckton) (Homoptera: Aphididae), *J. Econ. Entomol.*, 76, 1064, 1983.
95. Hobbs, G. A., Holmes, N. D., Swailes, G. E., and Church, N. S., Effect of the pea aphid, *Acyrthosiphon pisum* (Harris) (Homoptera: Aphididae), on yields of alfalfa hay on irrigated land, *Can. Entomol.*, 93, 801, 1961.
96. Mulder, P. and Berberet, R., Alfalfa aphids in Oklahoma, *Okla. Coop. Ext. Facts*, 7184, 1990.
97. Medlar, J. T. and Ghosh, A. K., Apterous aphids in water, wind, and suction traps, *J. Econ. Entomol.*, 61, 267, 1968.
98. Radcliffe, E. B., Weires, R. W., Stucker, R. E., and Barnes, D. K., Influence of cultivars and pesticides on pea aphid, spotted alfalfa aphid, and associated arthropod taxa in a Minnesota alfalfa ecosystem, *Environ. Entomol.*, 5, 1195, 1976.
99. Buntin, G. D. and Isenhour, D. J., Comparison of sweep-net and stem-count techniques for sampling pea aphids in alfalfa, *J. Entomol. Sci.*, 24, 344, 1989.
100. Hutchison, W. D., unpublished data, 1983.
101. Forsythe, H. Y. and Gyrisco, G. G., Jr., The spatial pattern of the pea aphid in alfalfa fields, *J. Econ. Entomol.*, 56, 104, 1963.
102. Gutierrez, A. P., Summers, C. G., and Baumgartner, J., The phenology and distribution of aphids in California alfalfa as modified by ladybird beetle predation, *Can. Entomol.*, 112, 489, 1980.
103. Hutchison, W. D., Hogg, D. B., Poswal, M. A., Berberet, R. C., and Cuperus, G. W., Implications of the stochastic nature of Kuno's and Green's fixed-precision stop lines: sampling plans for the pea aphid (Homoptera: Aphididae) in alfalfa as an example, *J. Econ. Entomol.*, 81, 749, 1988.
104. Kuno, E., A new method of sequential sampling to obtain population estimates with a fixed level of precision, *Res. Popul. Ecol.*, 11, 127, 1969.
105. Stern, V. M., Sharma, R., and Summers, C., Alfalfa damage from *Acyrthosiphon kondoi* and economic threshold studies in southern California, *J. Econ. Entomol.*, 73, 145, 1980.
106. Lamp, W. O., personal communication, 1992.
107. Poswal, M. A. and Berberet, R. C., Host-dependent association and its implications on the relative abundance of blue alfalfa aphid and pea aphid (Homoptera: Aphididae) on alfalfa in Oklahoma, *Res. Popul. Ecol.*, 31, 275, 1989.
108. Summers, C. G., Coviello, R. L., and Gutierrez, A. P., Influence of constant temperatures on the development and reproduction of *Acyrthosiphon kondoi* (Homoptera: Aphididae), *Environ. Entomol.*, 13, 236, 1984.
109. Poswal, M. A., Berberet, R. C., and Young, L. J., Time-specific life tables for *Acyrthosiphon kondoi* (Homoptera: Aphididae) in first crop alfalfa in Oklahoma, *Environ. Entomol.*, 19, 1001, 1990.
110. Bishop, A. L., Walters, P. J., Holtkamp, R. H., and Dominiak, B. C., Relationships between *Acyrthosiphon kondoi* and damage in three varieties of alfalfa, *J. Econ. Entomol.*, 75, 118, 1982.
111. Rohitha, B. H. and Penman, D. R., A comparison of sampling methods for bluegreen lucerne aphid (*Acyrthosiphon kondoi*) in Canterbury, New Zealand, *N. Z. J. Zool.*, 8, 539, 1981.
112. Bishop, A. L. and McKenzie, H. J., Sampling *Acyrthosiphon kondoi* Shinji in the field, *J. Aust. Entomol. Soc.*, 21, 312, 1982.
113. Graham, H. M., Effects of temperature and humidity on the biology of *Therioaphis maculata* (Buckton), *Univ. Calif. Publ. Entomol.*, 16, 47, 1959.
114. Nielson, M. W. and Barnes, O. L., Population studies of the spotted alfalfa aphid in Arizona in relation to temperature and rainfall, *Ann. Entomol. Soc. Am.*, 54, 441, 1961.
115. Manglitz, G. R., Calkins, C. O., Walstrom, R. J., Hintz, S. D., Kindler, S. D., and Peters, L. L., Holocyclic strain of the spotted alfalfa aphid in Nebraska and adjacent states, *J. Econ. Entomol.*, 59, 636, 1966.

116. Simpson, R. G. and Burkhardt, C. C., A three-year overwintering study of the spotted alfalfa aphid, *J. Econ. Entomol.*, 53, 220, 1960.
117. Nielson, M. W., Sampling technique studies on the spotted alfalfa aphid, *J. Econ. Entomol.*, 50, 385, 1957.
118. Davis, D. W., Haws, B. A., and Knowlton, G. F., Insects injurious to seed production, in *Insects and Nematodes Associated with Alfalfa in Utah*, Davis, D. W., Ed., *Utah Agric. Exp. Stn. Bull.*, 494, 1976, chap. 3.
119. Craig, C. H., Seasonal occurrence of *Lygus* spp. (Heteroptera: Miridae) on alfalfa in Saskatchewan, *Can. Entomol.*, 115, 329, 1983.
120. Khattat, A. R. and Stewart, R. K., Population fluctuations and interplant movements of *Lygus lineolaris*, *Ann. Entomol. Soc. Am.*, 73, 282, 1980.
121. Jensen, B. M., Wedburg, J. L., and Hogg, D. B., Assessment of damage caused by tarnished plant bug and alfalfa plant bug (Hemiptera: Miridae) on alfalfa grown for forage in Wisconsin, *J. Econ. Entomol.*, 84, 1024, 1991.
122. Baird, C. R. and Homan, H. W., Idaho insect control recommendations for alfalfa seed production, *Idaho Coop. Ext. Curr. Info. Ser.*, 231, 1991.
123. Johansen, C. A. and Eves, J. D., Development of a pest management program on alfalfa grown for seed, *Environ. Entomol.*, 2, 515, 1973.
124. Gupta, R. K., Tamaki, G., and Johansen, C. A., *Lygus* bug damage, predator-prey interaction, and pest management implications on alfalfa grown for seed, *Wash. State Univ. Agric. Exp. Stn. Bull.*, 0992, 1980.
125. Sevacherian, V., Spatial distribution of *Lygus* in host plants, in *Proc. Lygus Bug: Host Plant Interactions Workshop*, Scott, D. R. and O'Keeffe, L. E., Eds., University Press of Idaho, Moscow, 1976, 3.
126. Luna, J. M., Implementation of pest management programs for potato leafhopper in alfalfa, in *Potato Leafhopper, Empoasca fabae (Harris), (Homoptera: Cicadellidae) Research: Current and Historical Perspectives*, Armbrust, E. J. and Lamp, W. O., Eds., Misc. Publ., Entomology Society of America, Lanham, MD, 1989.

Chapter 14

SAMPLING PEST AND BENEFICIAL ARTHROPODS OF APPLE

Elizabeth H. Beers, Larry A. Hull, and Vincent P. Jones

TABLE OF CONTENTS

I.	Introduction	384
II.	Sampling Techniques for Apple Orchards	385
III.	Codling Moth (*Cydia pomonella* (L.)) (Lepidoptera: Tortricidae)	386
	A. Phenological Model	387
	B. Thresholds	388
IV.	Obliquebanded Leafroller (*Choristoneura rosaceana* (Harris)) (Lepidoptera: Tortricidae)	388
V.	Tufted Apple Bud Moth (*Platynota idaeusalis* (Wlk.)) (Lepidoptera: Tortricidae)	389
VI.	Pandemis Leafroller (*Pandemis pyrusana* Kearfott) (Lepidoptera: Tortricidae)	392
VII.	Apple Maggot (*Rhagoletis pomonella* (Walsh)) (Diptera: Tephritidae)	392
VIII.	San Jose Scale (*Quadraspidiotus perniciosus* (Comstock)) (Homoptera: Diaspididae)	394
IX.	Plum Curculio (*Conotrachelus nenuphar* (Herbst)) (Coleoptera: Curculionidae)	395
X.	Tarnished Plant Bug (*Lygus lineolaris* = *pratensis* (Palisot de Beauvois)) (Hemiptera: Miridae)	396
XI.	Tetranychid Mites (*Panonychus ulmi* (Koch); *Tetranychus urticae* (Koch); *Tetranychus mcdanieli* (McGregor)) (Acari: Tetranychidae)	398
	A. TSSM and MDSM Thresholds	399
	B. ERM Thresholds	401
XII.	Apple Rust Mite (*Aculus schlechtendali* (Nalepa)) (Acari: Eriophyidae)	402
	A. ARM Thresholds	403
XIII.	Phytoseiid Predatory Mites (*Typhlodromus occidentalis* (Nesbitt), Western Predatory Mite; *Amblyseius fallacis* (Garman); *Typhlodromus pyri* (Scheuten)) (Acari: Phytoseiidae)	403

ISBN 0-8493-2923-X/94/$0.00 + .50
© 1994 by CRC Press, Inc.

XIV. Stethorus punctum (*Stethorus punctum* (Leconte))
(Coleoptera: Coccinellidae) 404

XV. Rosy Apple Aphid (*Dysaphis plantaginea* (Passerini))
(Homoptera: Aphididae) 405
    A. RAA Threshold 405

XVI. Green Apple Aphid (*Aphis pomi* (De Geer)) and Spiraea Aphid
(*Aphis spiraecola* (Patch) (Homoptera: Aphididae) 406

XVII. Tentiform Leafminers (*Phyllonorycter blancardella*, Spotted Tentiform Leafminer; *Phyllonorycter crataegella*, Apple Blotch Leafminer; *Phyllonorycter elmaella*, Western Tentiform Leafminer) (Lepidoptera: Gracillariidae) ...407

XVIII. White Apple Leafhopper (*Typhlocyba pomaria* (McAtee))
(Homoptera: Cicadellidae) 409

XIX. Crop Loss Assessment 410

References 410

I. INTRODUCTION

Interest in sampling and monitoring in apple pest management has been intense for the past 2 decades. The great diversity of pests and natural enemies and the physical structure of the trees has presented a considerable challenge to researchers in this area. Apple is one of the most intensive per-acre consumers of pesticides,[1] and this statistic has stimulated efforts to develop sampling schemes and economic thresholds in an attempt to reduce pesticide use. Many successes were realized. However, with the relatively high value of the crop, and the very direct link between quality, cosmetic appearance, and price, there is considerable pressure to maintain low pest populations. The relatively low cost of agrichemicals in relation to crop value has been a significant factor in the development of integrated pest management (IPM) programs. The emphasis has shifted in recent years to developing alternative control measures to further reduce the use of pesticides.

Like most crops, apple is the host for both direct (fruit-feeding) and indirect (root, shoot, wood, or foliage-feeding) pests. Some of the most highly developed sampling/economic threshold programs have been directed at the latter group, because they have a greater probability of being present at nondamaging levels. Although thresholds have also been developed for some direct pests, much of the effort in this arena has been directed toward maximizing the efficiency of pesticides through optimal timing (e.g., the use of phenological models), application technology, or targeting the site of applications. There is also a number of apple pests that are of quarantine concern, for which monitoring methods are geared to detect the presence or absence of the pest and where mere presence is sufficient to trigger a pesticide application. Although the original impetus for many apple IPM programs was pesticide resistance, the subsequent development and implementation of these programs has thus been driven largely by production and marketing economics and, to a lesser extent, regulatory issues.

Apple trees are perennials that are relatively large and structurally complex, both of which contribute to support a great diversity of pests. Trees begin significant commercial production 2 to 6 years after planting, and the life of the orchard frequently exceeds 20 to 30 years. In addition to the trees, the orchard floor may support a variety of herbaceous plants which can both support or defeat pest management efforts. Lack of vegetation is associated with poor habitat for many beneficial arthropod species; however, certain plant species can harbor insect or disease pests. The surrounding habitat (woods, pastures, other crops, sprayed or abandoned orchards) can have a profound influence on both the diversity and the spatial or temporal distribution of pests and natural enemies. Newer orchards are often planted with size-controlling rootstocks and/or artificial support systems (trellises or posts), while many older plantings exceed 4 m in height. In addition, it is not feasible to climb or uproot mature trees in order to sample; thus, in many situations, large areas of the canopy and the entire root system are virtually inaccessible for sampling in a decision-making mode. The tremendous differences in tree size and structure among orchards of different ages, cultivars, or training systems also present substantial barriers to developing sampling schemes that are universally applicable.

If certain features of the apple production system have made sampling difficult, they have operated even more strongly on the development of economic thresholds. The per-acre value of the crop varies tremendously among years, cultivars, and individual orchards, often driven by national or international supplies or freezing episodes in winter or spring. Even totally external factors such as media coverage can drastically influence crop value. The marketing channel (fresh, processing, or juice) is also a major determinant of the crop value. Although generalizations can be made from regional averages, each grower must make the calculations independently for his/her own operation, and even for individual blocks within that operation. This requires not only a fairly rigorous structure but also extensive inputs of many types (biological, economic, climatological) for decision making. The detailed level of recordkeeping and sampling has been and will continue to be a barrier to fully implementing IPM programs.

II. SAMPLING TECHNIQUES FOR APPLE ORCHARDS

Apple researchers and pest managers have explored and exploited a number of sampling techniques. Many would-be sampling schemes have been discarded because of the laboriousness of the technique or the lack of an accompanying decision-making rule. Traps of all sorts have gained widespread popularity, since the onus is on the insect to come to the trap, rather than the sampler to seek out and count the insect. Nonselective traps, such as the blacklight or the bait trap, are no longer used much because separating the target pest from the multitude of other organisms captured is very time-consuming. The development of selective traps, often to the species level, has been a tremendous boost to their use in orchard IPM. The main types have been those that are visually attractive (shape or color), those that are baited with food odors, or those that are baited with sex pheromones (or some combination of the above). Much of the literature on monitoring via traps has been devoted to the efficiency of the trap design, pheromone chemistry, lure longevity, trap placement, or density. The identification and synthesis of insect pheromones was a tremendous leap forward in the ability to monitor many lepidopteran species and has recently been extended to some Hemiptera. Visual traps have been developed for several other pests, replacing the more laborious direct examination or limb tapping methods.

Size has been a barrier for rapid sampling of smaller arthropods. The development of a leaf-brushing machine was a major step in evaluating tetranychid and eriophyid mite populations, along with their phytoseiid or stigmaeid predators. Although this technique represents a significant step forward from visual examination of the leaves in terms of efficiency and accuracy, it is still considered by many too laborious for commercial IPM programs. The trend is being reversed again; in the newest protocols, direct examination used in a presence-absence scheme reduces the amount of time necessary for sampling.

Limb tapping is commonly recommended for certain hemipteran and coleopteran pests and is also used as a generalized sampling technique for an array of beneficials. Direct visual examination is still used in many instances, but is usually the least preferred if a range of options is available. On certain pests (i.e., those that are fairly large and are easily seen and located, such as leafhopper nymphs, aphid colonies, or leafminer mines) examination of plant parts may indeed be the easiest and most straightforward method. Any time the sample must be brought back to the laboratory, labeled, and processed before counting, there is an increase in time, which is justifiable only with a substantial increase in precision or reliability.

For many types of sampling methods, statistical techniques such as binomial and/or sequential sampling protocols have increased the efficiency of some of the more labor-intensive sampling techniques, but ease of use still constitutes a major barrier to field implementation. Sequential sampling plans tied to decision-making thresholds may be too regional to gain widespread acceptance and will be obsolete if new thresholds are established.

Because of the diversity of pests, many of which overlap temporally, the development of multispecies sampling plans would provide additional impetus toward their use. Sampling protocols published in the literature almost always target a single pest and establish the most efficient way to sample for it. For decision makers, however, techniques that reduce their overall sampling time may be easier to promote as long as they do not seriously impair precision for the insects involved.

The timing of sampling (and pest control) on apple is frequently tied to tree phenology. Tree phenology often provides a highly visible (although approximate) marker for insect phenology. In the early part of the season, apple leaves and flower parts go through more or less distinct stages. A key to these stages (e.g., silver tip, green tip, half-inch green, tight cluster, prepink, pink, bloom, and petal fall) can be found in most state university pest control recommendations for apples. Timing after these stages are some combination of calendar date and phenological stage, or some other defined point in the season (e.g., harvest).

III. CODLING MOTH (*CYDIA POMONELLA* (L.)) (LEPIDOPTERA: TORTRICIDAE)

Codling moth (CM) is a pest of apples almost worldwide and is considered a major direct fruit-feeding pest in many of those areas. It is the key pest in the irrigated fruit-growing regions of the western U.S., but due to factors such as climate, habitat, cultivars, and differing spray programs, it is of lesser importance in the eastern U.S. However, if left uncontrolled, CM can infest up to 90% of the fruits in many areas. Codling moth usually has 2 to 3 generations/year in most fruit-growing regions, but this can vary from 1 to 5 depending on the area. They overwinter as fully grown larvae and pupate as the weather warms in the spring. The first adults emerge at about the time apple trees bloom. Eggs are small, flattened, and translucent and are laid singly on leaves or fruit. Neonate larvae find a fruit and bore into it. They tunnel and feed extensively in the fruit cortex, core, and seeds, completing 5 instars (Note: keys to last-instars of some of the tortricid species can be found in Chapman and Lienk[2]). Mature larvae crawl down the trunk, seeking a sheltered spot to pupate in crevices in the bark or in the ground cover. A second generation of larvae occurs in July, with occasionally a partial third generation in September.

The difficulty in detecting CM eggs in commercial orchards has precluded the use of this stage for sampling. Presence of the larva in the fruit is readily detectable because the hole is large and typically plugged with a mound of frass. The percentage of infested fruit, although a "postfacto" sampling method, may be an indication of the potential damage by the succeeding generation. Pupae may be detected by placing cardboard bands around the trunk, where migrating larvae encounter and use them for pupation sites. This method is also postfacto in terms of preventing damage by that generation of larvae but may have some specific applications. Banding has been used as a form of nonchemical control because the pupae can be removed and destroyed.

Far and away, the most accessible and best-studied method of sampling is the sex pheromone trap for adult males. The literature on codling moth pheromones and their uses is voluminous, and only a summary of the North American literature and current recommendations is covered here. Butt[3] provides a bibliography for general reading, and Riedl et al.[4] provide an excellent summary of the status of monitoring with pheromone traps. Pheromone trapping forms one of the key components (viz., biofix) for the heat unit model[5] which has been its major use in IPM programs. The biofix, or biological fixation point, is used to synchronize the model with field populations in order to increase the accuracy of subsequent predictions. In the CM model, biofix is the first capture of males in pheromone traps in the spring. The other major use for pheromone traps in an IPM mode is the use of moth capture as a treatment threshold.

One drawback to pheromone monitoring is that movement or activity of females is not tracked. Bait traps and blacklights can be used to detect both males and females, but these methods are considered too laborious for real-time decision making. A second drawback to using pheromones for monitoring is its limitations for use in mating disruption plots, where trap shutdown is considered an essential indicator of the success of the treatment. In this situation, banding for pupae may be useful for following long-term population trends in the orchard.[6]

A. PHENOLOGICAL MODEL

The CM model as a tool for timing spray applications has been widely adopted. The collection of pheromone trap data and maximum-minimum temperatures has replaced the much more laborious search for eggs or fruit entries. Batiste et al.[7] demonstrated the usefulness of egg hatch data in timing sprays, but it was the development of the phenology model[5,8] that made this approach feasible for field use. Brunner et al.[9] provide a simplified description of the application of this method for IPM. Beers and Brunner[10] demonstrated that the historical accuracy of the model in predicting first egg hatch is substantially superior to the former calendar method (21 d after full bloom).

The target of control measures is the newly eclosed larva before it has had time to sting (chew a shallow depression) or enter the fruit. Optimum timing for control is ca. 3% egg hatch (as predicted by the model) or 250 cumulative degree-days after biofix.[8] Riedl and Croft[11] provide a similar value, i.e., 243 degree-days between biofix and first egg hatch. Timing of subsequent sprays is based not so much on insect development as it is on residual properties of the insecticide used. Brunner et al.[9] recommend a 21-d interval between the first and second codling moth spray based on the length of effective residues of azinphosmethyl, the most commonly used material in Washington. Hogmire et al.[12] recommend repeat applications (if a threshold of 5 months per trap is exceeded) every 12 to 14 d. Both the amount of precipitation and the type of insecticide used will influence the interval.

B. THRESHOLDS

A number of researchers have established a relationship between codling moth trap capture and fruit damage.[13-15] This relationship has been further used to develop action thresholds for insecticides based on a count (usually cumulative moth capture per trap). Madsen and Vakenti[16] developed a system where spray applications were made if interior traps caught ≥ 2 moths per trap per week and fruit entries were found, or if border traps caught > 2 moths per trap per week and fruit entries were found. Brunner[17] has successfully used a threshold of five cumulative moths per trap since the last spray application (or, in the case of the first spray, since the beginning of flight). This method also advises that only the area associated with the trap exceeding the threshold should be sprayed, not the whole orchard. Using these thresholds, the percentage of culled fruit due to codling moth has generally been below 1%.

Trap design, placement (both intratree and intraorchard) density, lure type, and maintenance regimes can profoundly affect the moth catch and the interpretation that may be made from it. Riedl et al.[4] summarizes the research in this area, which forms the basis of the current recommendations. The recommended trap type is a wing trap with a restricted opening (e.g., the Pherocon® 1CP). The adhesive lower surface of the trap should be replaced whenever it loses effectiveness (due to moth removal, contamination with dirt or insect parts), or about every 20 to 30 moths removed. A rubber septa baited with 1 mg of the codling moth pheromone should be placed in the adhesive in the trap bottom, and replaced every 4 weeks. In mating disruption trials, septa loaded with 10 mg have been used to track moth flight.[154] Traps should be placed in a grid evenly spaced throughout the orchard, and the design should include both interior and border traps. A trap density of one trap per hectare (2.47 acres) must be used if the decision-making rules are to apply accurately; the accuracy at other trap densities is unknown. The traps should be hung in the tree canopy about 2 m above the ground, and the surrounding foliage and fruit removed. Placement in the orchard should occur before the first moth flight; the pink stage of flower bud development is used to time trap placement, because first flight usually occurs during bloom. If other moth species are being trapped, care must be used not to cross-contaminate the traps by using the same instrument to handle the pheromone caps.

IV. OBLIQUEBANDED LEAFROLLER (*CHORISTONEURA ROSACEANA* (HARRIS)) (LEPIDOPTERA: TORTRICIDAE)

The obliquebanded leafroller (OBLR) is native to and widely distributed throughout the temperate fruit-growing regions of North America. During the last 15 years, this insect has increased in importance in many regions because it had developed resistance to a number of organophosphate insecticides.[18,19] OBLR has 1 to 2 generations/year, depending on locality. It overwinters as half-grown larvae in cocoons in sheltered areas on the tree. In early spring, activity resumes, and larvae feed on developing buds and young fruit. Adults begin to emerge in late May to early June and again in early August.

The larvae can injure the leaves and the fruit during three distinct feeding periods; however, only the feeding injury to fruit is of economic consequence. The first fruit-feeding period occurs around petal fall, when the overwintering larvae feed on the developing fruit. Fruit injured at this time abscise or exhibit deep corky scars. Larvae of the first summer brood, which feed on the developing fruit in late July and early August, usually cause the most serious injury because damaged fruit remain on the tree until harvest. Larvae of the second generation can feed on the fruit just prior to harvest before seeking overwintering shelters.

The pheromone of OBLR was identified by Roelofs and Tette.[20] Because OBLR has over 75 host plants, sex pheromone traps are used primarily for determining adult flight, subsequent peaks of larval activity, and possible timing of insecticide applications, but not for population assessment. For trapping adult OBLR, Vincent et al.[21] and Knodel and Agnello[22] found only a few differences in total trap captures, first date of captures, and population fluctuations for sticky (Pherocon® 1C and Pherocon II) vs. nonsticky (Multi-Pher® I and III) traps. Both groups concluded that the nonsticky traps offer considerable promise for monitoring adult OBLR.

At present, direct visual examination of the foliage for larvae is the only method useful for predicting damaging levels of OBLR. In order to determine the optimum time to sample for larvae, a number of researchers conducted developmental rate studies. Reissig[18] was the first to determine the effect of temperature on the developmental rate of the eggs. Onstad et al.[23] later developed a phenological model for management and monitoring purposes using developmental data on eggs, larvae, and pupae from a number of different hosts. They found this model not only accurately predicted the best time to sample for larvae of the first summer generation, but also when treatment was necessary.

For sampling larvae, Agnello et al.[24] present a comprehensive monitoring scheme. They recommend two periods for sampling OBLR, one at the bloom stage to assess the density of the overwintering larvae and one during early July to assess the status of the first summer generation. They promote the use of a sequential sampling plan for both periods (Figure 1).[24] With this method, a grower samples a minimum of ten randomly chosen fruit clusters (overwintering generation) or ten expanding terminals (summer generation) per tree and continues sampling until a decision line is reached. For the overwintering larvae they use an economic threshold of 3% of the fruit clusters containing live larvae. For the summer generation they advise growers to use a degree-day model based on first adult catch in a pheromone trap as a biofix point, then wait 600 degree-days (base: 6.1°C; 43°F) before using a similar sampling scheme as described above. Because the fruit injured by this generation remains on the tree, the grower is advised to choose between two economic thresholds which are based upon the designated market of the fruit. For fresh fruit they use a 3% infestation rate of live larvae, and for processing they use a 10% infestation rate.

V. TUFTED APPLE BUD MOTH (*PLATYNOTA IDAEUSALIS* (WLK.)) (LEPIDOPTERA: TORTRICIDAE)

The tufted apple bud moth (TABM) is a serious direct pest of apples in the mid-Atlantic region of the U.S. and is of lesser importance throughout other eastern and midwestern states. Although this pest can be found in most commercial orchards, those orchards located in intensive production areas are usually more seriously affected. This is primarily because of the increased insecticide selection pressure,

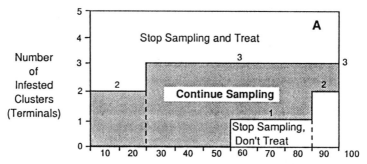

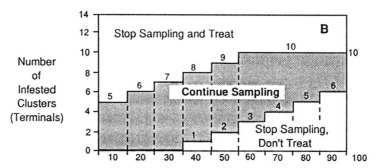

FIGURE 1. Sequential sampling plan for OBLR. The 3% infestation rate threshold (A) is recommended for overwintering larvae, and for summer generation if the fruit is destined for the fresh market; the 10% infestation rate (B) is recommended for the summer generation if the fruit is destined for the processing market. Examine at least 10 leaf, fruit, or bud clusters per tree (overwintering generation) or 10 expanding terminals (summer generation), selecting trees from throughout the block. If the trees are > 3 m (10 ft), include clusters (terminals) from the upper half of the canopy and the watersprouts. Two or more larvae per cluster (terminal) still count as one infested cluster (terminal). Continue sampling until the number of infested clusters (terminals) fall outside the shaded area. If the intersection of the two lines is reached by the 100th cluster, do not treat for OBLR. If a "Don't Treat" decision is reached, sample again in 3 to 5 d (100 degree-days); if the second sample also indicates that no treatment is necessary, this generation need not be resampled. Redrawn from Agnello et al.[24]

which has resulted in the development of resistance to the organophosphate insecticides.[25]

TABM is bivoltine, with larvae overwintering as second through fifth instars in larval shelters such as rolled leaves and decaying fruit. These shelters are located beneath trees in apple, cherry, peach, and pear orchards. The larvae become active in early spring and complete development on a variety of host plants (e.g., root suckers, dandelion, dock, and wild strawberry).[26] Adult emergence commences in early May, followed by oviposition several weeks later. Early instars feed along the leaf midrib, and beginning with third instars, create shelters by rolling and/or tying leaves to other leaves or fruit and by building shelters within fruit clusters. Second-brood adult emergence commences in late July with oviposition starting early in August and continuing until early September. Late season second-brood larvae drop to the ground during fruit harvest or with leaf fall to overwinter.

Although TABM is a leafroller, the leafrolling activity has little physiological or economic impact on the tree. It is when the larva webs a leaf onto the apple fruit and feeds directly on the fruit that it becomes a pest. This injury appears as tiny holes, as irregular scarring or channeling on the apple surface, or as an area of rot, usually found around the stem.[27] Fresh market apples injured by TABM are usually reduced to processing grades.[28] Generally, TABM injury does not reduce the grade of processing apples, but it can affect the rate of fruit drop and storability of those apples by promoting decay, both of which can have an economic impact on the grower and processor.

Meagher and Hull[29] evaluated four sampling methods using summer egg masses, larvae, or first brood fruit injury to predict apple injury at harvest. They found only high percentages of first brood fruit injury or high numbers of larval shelters adequately predicted fruit injury at harvest. They further developed a sampling plan that utilized three consecutive weekly 5-min timed counts of larval shelters taken during late July or early August. They also presented sample-size requirements for optimum numbers of trees to sample based on selected percentages of mean density. In a second study conducted by Hull and Meagher[30] on the cultivars 'Delicious' and 'Yorking', they found that the amount of fruit injury at harvest could again be accurately estimated from the amount of first brood fruit injury during late July to early August. This relationship is represented by the following equation:

$$\text{Fruit injury at harvest} = 2.22 + 1.47(x) \tag{1}$$

where x = % first brood injury. They suggested sampling 10 trees per orchard and examining 100 apples per tree (50 apples top and 50 bottom) during early August.

Later work by Knight and Hull[31] revealed a relationship between the cumulative number of adult males caught in sex pheromone traps during the first 3 to 4 weeks following initial emergence in the spring and apple injury at harvest for the cultivars 'Delicious' and 'Yorking', but not for 'Golden Delicious' (Figure 2). They also found that if a measure of tree size was incorporated with cumulative trap catch, the prediction of total fruit injury at harvest improved significantly. This relationship was developed using the wing-style pheromone trap and from orchards treated with insecticides applied for each brood (June and August). They recommend cleaning the

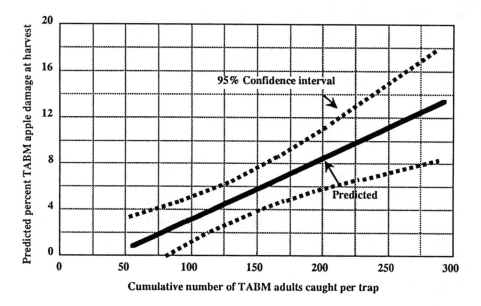

FIGURE 2. Predicted percent apple damage caused by TABM larvae, based on the total number of TABM moths (males) caught in sex pheromone traps during the first 3 weeks after initial adult catch in the spring. (From Travis et al.[107] With permission.)

trap after the capture of 20 adults. This method of sampling is much simpler and less time consuming than any of the methods described above. Also, this method can be used to determine the amount of spraying needed for both broods of TABM rather than just the second brood, if using fruit or leaf shelter sampling. Economic thresholds should be based upon the intended markets for the fruit.

VI. PANDEMIS LEAFROLLER (*PANDEMIS PYRUSANA* KEARFOTT) (LEPIDOPTERA: TORTRICIDAE)

This leafroller is a pest native to North America and is primarily restricted to the western regions of the U.S. and Canada. It overwinters as either first, second, or third instars within hibernacula in protected areas on the tree. The larvae become active in the spring and leave the hibernacula at the half-inch green stage to feed on the developing leaves. There are 2 generations/year, with peak adult flight occurring usually in June and August-September. Although larvae are primarily foliage feeders (which has no impact on mature trees) their economic impact occurs when the larvae feed on the fruit. This can occur three times yearly: mid-May, late June through early July, and just before harvest in September and October.

Monitoring the activity of *Pandemis* can be accomplished in two ways. Sex pheromone traps can be used to monitor the flight of adult males. The pheromone currently used to monitor *Pandemis* is that of the three-lined leafroller, *Pandemis limitata*. Brunner and Beers[32] reported that moth capture has not been a reliable indicator of potential leafroller damage. However, they speculate that trap captures of 15 or fewer adults per week are an indication of low leafroller pressure.

Sampling the larvae is much more labor intensive. Brunner and Beers[32] recommend the examination of 100 fruit buds/orchard at the half-inch green stage. The larvae can be sampled again in late July as they migrate to growing points (water sprouts and shoots) to feed. Although it is too late to treat for *Pandemis* at this time, sampling may provide an indication of the potential to treat for the next generation.

Information on economic thresholds is minimal. California recommends treatment if one *Pandemis* larva is found in 100 fruit cluster samples taken during the green tip to pink stage of development.

VII. APPLE MAGGOT (*RHAGOLETIS POMONELLA* (WALSH)) (DIPTERA: TEPHRITIDAE)

The apple maggot is indigenous to eastern North America, where its original host is thought to be various species of hawthorn (*Crataegus* spp.). It was first reported attacking apple in 1866 in Vermont, but the extent of the problem at that time suggests that the development of the apple-feeding race occurred much earlier.[33] Prior to the late 1970s its distribution included the northeastern U.S. and adjacent areas in southeastern Canada, all of the Atlantic states south to Florida, west to the Dakotas and southern Manitoba, and south to Arkansas. Isolated populations were also reported in Texas and Colorado.[33] In 1979, it was discovered in Portland, OR,[34] and, by 1987, it was collected in most of the western fruit-growing areas, including Washington (west of the Cascades and Spokane), Oregon, California, Utah, Idaho, and Colorado. However, in Utah and Colorado the flies were primarily restricted to hawthorn (both states) or tart cherry (Utah).[35,36]

In all areas, the apple maggot overwinters in the pupal stage and emerges from the soil in mid-summer. The adults pass through a 7- to 10-d preoviposition period and begin to insert eggs singly under the skin of the fruit once the fruit begins to soften. The three larval stages occur within the host fruit, and just prior to pupation, the last instar bores out of the host fruit, drops to the ground, and digs down through cracks in the soil to pupate. Jones et al.[37] compiled 152 orchard-years of information from the literature and found that in the East the average first emergence date was June 20 ($N = 132$), and in the West it occurred approximately 3 to 4 weeks later (July 9, $N = 20$).

Sampling for apple maggot is mostly performed using adult traps. This is because eggs and young larvae are often difficult to see in the fruit and because the primary aim of management programs is to prevent fruit damage. Paired emergence cage vs. aerial sticky trap studies have demonstrated that sticky traps capture flies within ± 1 to 2 d of observing emergence in emergence cages.[38,39] Because the flies have a 7- to 10-d preoviposition period, and may also emerge before host fruits are suitable for oviposition,[40] the accuracy of sticky traps is generally sufficient for pest management programs. Traps should be placed in the orchards either before degree-day accumulations indicate emergence is imminent or before historical data from the area suggest emergence will occur. In the western U.S., where soil moisture may be quite low over much of the summer, rain has been shown to rapidly increase emergence in both field and laboratory tests.[41,42] In New York, where rain is much more frequent during the growing season, consistent evidence for this effect is lacking.[33]

Traps for adult apple maggot fall into two broad types, yellow panels and red spheres. Both types are coated with a sticky material and placed within the tree canopy. The yellow panels are generally baited with ammonium acetate and casein hydrolysate and are theorized to represent a supernormal foliage and honeydew stimulus.[43,44] The red spheres are used either plain or in combination with apple volatiles and are thought to be a supernormal fruit stimulus.[44,45] There are two apple volatile baits commonly used, one consisting of a mixture of hexyl acetate, butyl 2-methylbutanoate, propyl hexanoate, butyl hexanoate, and hexyl butanoate in a 36 : 7 : 12 : 5 : 29 : 11 ratio (the Fein blend)[46] and the single component butylhexanoate.[47,48] The efficiency of the two types of traps seems to vary markedly from study to study, even sometimes when they are performed by the same investigator. Combination traps, such as a red sphere on a yellow rectangular background, are effective, but rarely used.[49] The cost of the commercially available traps using this design is generally excessive compared to the normal spheres or yellow panels, and the combination traps are cumbersome to handle and virtually impossible to use without the user becoming entangled in sticky material.

Longevity studies of commercial and experimental baits have been performed under field and laboratory conditions.[50] In the laboratory, Pherocon® AM panels (Trécé, Salinas, CA) lost ca. 50% of their weight within 24 h and 90% after 5 d. In the field, Pherocon® AM panels aged 9 d before use caught significantly fewer flies than new traps. The large weight loss is probably because panel-type traps have a large surface area over which a small amount of ammonium acetate is released. The same studies also indicated that some apple volatile baits suffer from longevity problems. The rubber septum used by Ladd (Ladd Research Industries, Burlington, VT) lost 58% of its weight in 17 h and ca. 87% in 2.7 d. Averill and Reissig[48] indicated they obtained similar results with this product in New York. Baits provided by Consep Membranes (Bend, OR) (butyl hexanoate) and Great Lakes IPM (Vestaburg, MI) (Fein blend) lasted at least 40 d under both laboratory or field conditions and are excellent choices in population monitoring.[50]

The disparity in bait longevity suggests that these data could be at least partially responsible for the confusion between the relative effectiveness of spheres (baited or unbaited) and yellow panels. To remove this factor, in comparisons between the two trap types, yellow panels should be replaced at weekly intervals. Baits added externally to the adhesive also can be used to augment the longevity of yellow panels.[50]

Placement of the traps in the trees has been shown to be important in trap catch.[51,52] Studies in New York[51] found that both yellow panels and red spheres were more effective when placed in the middle of the canopy radius between 2.1 (yellow panels) and 3.0 m (red spheres) from the ground. Studies in Massachusetts showed that the area around the trap within a radius of 0.25 to 0.5 m should be cleared of leaves and fruit for optimal trap efficiency.[52]

Baited spheres have become the trap of choice in New England, New York, and Pennsylvania. Studies in New York have demonstrated the superiority of these traps and suggest that a threshold of five flies per trap allows reduction of spray programs from an average of 4 sprays to 1.5 sprays per season.[53,54] The threshold is used by waiting for the cumulative catch to surpass the threshold and then applying the pesticide (in this case, azinphosmethyl). Flies caught for the next 10 d are not included in the threshold calculation because the pesticide is considered effective for at least that period of time. The next pesticide application is, therefore, applied based on the number of flies caught after the 10-d pesticide activity period.

In the western U.S., thresholds are quarantine mediated. Although they vary from location to location, they are typically based on the capture of a single fly. Until the quarantine status is lifted, sampling and thresholds cannot be used in a pest management context. In parts of Washington and Oregon where the the closely related snowberry maggot (*Rhagoletis zephyria* Snow) occurs, Brunner[55] found that the red sphere traps selectively attracted apple maggot, but not snowberry maggot. However, where snowberry maggot was not present, studies have found that the yellow panels performed as well or better than baited or unbaited sphere traps.[36,56]

Sampling for all life stages of apple maggot has been described by Cameron and Morrison.[57] This information is primarily useful for research purposes such as life table analysis because of the time required and the precise timing necessary for adequate estimation of the different stages. In particular, they note that sampling the fruit too early results in increased larval mortality, while late removal leads to inaccuracy caused by larval emergence.

In their study, Cameron and Morrison[57] state that the egg stage is uniformly distributed, but, according to their tables, it appears aggregated using the conventional interpretation of Taylor's power law β.[58] Other studies[59] indicate the egg distribution is uniform, but Averill and Prokopy[60] found that on hawthorn, eggs were clumped early in the season, later became random, and finally showed a uniform distribution. They theorized that this could result from fruit suitability changes over time and the possible effects of the oviposition-deterring pheromone.[61] These factors could also apply to apples and should be considered if egg sampling is required.

Larval distribution has been found to be uniform by two separate studies.[57,59] Cameron and Morrison[57] also found that pupal distribution is related to the distribution of fallen apples and to predation. They found that sampling directly under the fallen fruit minimized variation and provided the most efficient method for population estimation.

VIII. SAN JOSE SCALE (*QUADRASPIDIOTUS PERNICIOSUS* (COMSTOCK)) (HOMOPTERA: DIASPIDIDAE)

The San Jose scale (SJS) has been considered a key pest of apple worldwide; however, the use of dormant oils (often combined with an organophosphate insecticide) keep the scales suppressed to subeconomic levels in most fruit-growing areas. SJS overwinters as a blackcap (first instar), or as a mature, mated female. Females are flightless, but the winged males emerge and fly in spring, roughly during the blossom period of apple. First-generation crawlers (mobile stages of the nymphs) begin hatching about 1 month later, with a second generation crawler period later in the summer. Crawler movement occurs over long periods (6 to 8 weeks), thus making sampling and control of this stage difficult.

Because of its small size, it is fairly laborious to sample in all stages. Both a pheromone[62,63] and a heat-unit phenology model[64] are available for monitoring SJS. The pheromone is commercially available and is used with a tent-style trap to set a biofix (first trap capture of males in the spring) for the model. Examining traps for adult male scale is often done with a microscope because of their small size. Flight periods of the males can be fairly erratic, because trap capture is greatly influenced by wind speed.[62] In certain areas, constantly windy conditions prevent scale pheromone trapping from being a feasible method.

In general, the pheromone/heat unit phenology model is not much used in apple because of the ease of control with oil/organophosphate programs. In Washington, crawler emergence usually coincides with the second cover spray for CM, and growers alter their choice of CM insecticide to one that is effective for both SJS and CM. The window for optimum timing for control of SJS on peach is 600 to 700 degree-days after biofix, which covers the peak period of crawler activity,[65] so slight variations in second cover timing appear to be reasonable. The upper and lower developmental thresholds for SJS are very similar to CM (10.5 and 32.2°C [51 and 90°F] for SJS vs. 10 and 31.1°C [50 and 88°F] for CM); thus, it is feasible to use tables for CM heat units for SJS.

Westigard and Calvin[66] developed a sampling scheme for SJS on pear which would probably be equally useful on apple. The authors found significant correlations between percentage-infested fruit at harvest and either the total number of live scales per 10-spur sample or the percentage-infested spurs. The latter was both easier and less time consuming to sample and was better correlated to fruit damage. Spurs taken from the upper part of the tree were better correlated to fruit damage than those from the lower portion. On untreated trees, the sampling periods best related to fruit damage were late April (first instar of overwintering form), late June (first instar of the first summer generation crawlers), and early August (50% mature females, 50% immatures). Sampling during peaks of crawler activity provided the poorest correlations, although these correlations were still significant. An economic injury level of 2% infested spurs is estimated from the cost of control and the value of the crop (with an allowance for coming out better than the "break even" point). The relationship between scale population and damage also existed for trees treated with a dormant oil spray. Thus, the population estimates provide a means of determining whether further controls (i.e., against crawlers) are necessary. The regression indicated that for a late-April sample, the threshold is 1% infested spurs (for need to control first generation crawlers), and for a late-June to early-July sample, the threshold is 4% infested spurs (for need to control second generation crawlers).

Crawler movement may also be monitored using sticky bands around tree limbs, which are usually placed close to known scale infestations on the bark. Either double-sided sticky tape, or tape coated with Vaseline® is effective in trapping crawlers. Bands are removed and examined under microscope to determine crawler numbers. Dark-colored tape provides a better contrast to detect crawlers.

IX. PLUM CURCULIO (*CONOTRACHELUS NENUPHAR* (HERBST)) (COLEOPTERA: CURCULIONIDAE)

The plum curculio (PC) is a native and direct pest of apple and can be found throughout the apple-growing regions of midwestern and eastern North America; it was recently discovered in Utah. It has only 1 generation/year in the northern portion of its range and 2 generations in some southern states. PC overwinters in the adult stage in trashy areas along fence rows and woods and begins to emerge when temperatures increase above 20°C, just prior, during, or immediately following bloom on apple.[67] Adult females begin to oviposit in apples starting at petal fall. Larvae feed on the flesh of the developing fruit, causing the fruit to abscise. Adults emerge from the fallen apples during the summer and may feed briefly on apple before seeking overwintering sites.

Because the PC moves into the orchard from the surrounding environment, monitoring the movement and activity of this pest is extremely important if control measures other than routine applications of pesticides are to be successful. Unfortunately, very few reliable and practical monitoring techniques exist for this insect. There are only four published techniques for sampling PC adult abundance, all of which are done from late bloom through mid-June. The first method involves limb jarring which is usually accomplished by holding a catching frame below a limb and either shaking or tapping the limb with a rubber-covered mallet.[68] Adults will drop to the catching frame where they can be counted. There are a number of variables precluding the standardization of this method such as size of the jarred limb, size of drop cloth, strength of the blow, weather conditions, and time of day.[69] The second proposed method and one still presently recommended[67,70] involves the examination of fruit for oviposition and feeding scars.[71,72] Growers are encouraged to examine five to ten developing fruit for fresh oviposition scars on each of several trees along rows that border fence rows or woods.[67,70] However, this method does not allow for pesticidal control before the onset of egg laying. The third method involves the exploitation of PCs natural dropping behavior through the use of funnel traps.[73,74] Unfortunately, Owens et al.[74] found that this behavior was positively correlated with temperature and could only be used on warmer nights, which usually occurred after the decision-making period for PC control in Massachusetts.

In a recent study, Le Blanc et al.[69] evaluated the use of "scout-apples" to monitor adult PC activity in the spring. They found that the PC would preferentially oviposit on these fruit before attempting to oviposit on naturally-occurring fruitlets. Mature 'Granny Smith' apples from storage are suspended on a wire in contact with the "V" portion of major branch junctions. The apples are placed in trees along the border of the early pink stage and monitored for oviposition and feeding activity. The advantage of this technique is that the activity and location of the PC can be more precisely detected before control measures are required.

Work on economic thresholds for the PC has been minimal, primarily due to the inability of researchers to successfully monitor this insect as well as the low tolerance level ($\leq 1\%$) for damaged fruit at harvest. Because PC can cause 100% fruit injury, economic thresholds are based on the first detection of oviposition scars on fruit. Spraying may be limited initially to the border rows, depending upon weather conditions and population density.

X. TARNISHED PLANT BUG (*LYGUS LINEOLARIS* = *PRATENSIS* (PALISOT DE BEAUVOIS)) (HEMIPTERA: MIRIDAE)

The tarnished plant bug (TPB) is a widely distributed species in North America and is found in most areas where fruit is grown. It is a pest of major importance in the northeastern U.S. and southeastern Canada but only a sporadic pest in other growing regions. It attacks other deciduous tree fruits (e.g., pear, peach, plum), but is not a year-round orchard resident. It overwinters in the adult stage in herbaceous crops and migrates into orchards, usually during the prebloom through early postbloom period. Successive generations are produced on herbaceous hosts, both in orchard ground cover and other crops.[75] Only the adults are known to attack fruits; injury occurs from either feeding or oviposition activities. Injury may occur to the shoots, leaves, fruit blossoms and buds, or the fruit, although the latter is currently the primary concern. Attack on fruit buds causes abscission, and attack on fruitlets typically causes a characteristic deep dimple. Buds are susceptible to injury during the period from silver tip to tight cluster, and fruit malformation can occur from tight cluster to ca. 2 weeks after petal fall.[76]

Because the adult is the only injurious stage on apple, it is the usual target for sampling. Hauschild and Parker[77] used a sweep net to sample the ground cover, but the objective of their study was to study the seasonal development, especially nymphal population peaks. Boivin et al.[78] used a D-Vac® vacuum insect net for collection of phytophagous mirids in the ground cover; however, this method is more appropriate for faunal studies than for decision making in IPM programs. Direct visual examination of the tree is possible, but difficult and inefficient for detecting the highly mobile adults. The techniques that have been studied and used for IPM are limb tapping and visual sticky traps.

Prokopy et al.[79] studied the visual response of adult TPB to various hues and reflective properties of sticky-coated rectangles (15 × 20 cm). They concluded that although several aspects of hue and reflectance probably play a role in response, the most effective trap was either a non-UV reflecting white or Zoecon® yellow sticky trap, hung vertically ca. 0.7 m above the ground. A comparison of the efficiency of four sampling methods was made in a subsequent article.[80] Efficiency of the method is defined as the degree of correlation between TPB populations as measured by the sampling method and percentage fruit injury. Of the methods tested over a 2- to 5-year period, only the visual trapping method was significantly positively correlated with percentage fruit injury (with one exception); for the silver tip through tight cluster period, the correlation was significant in 4 out of 5 years. Also, the visual traps detected the highest numbers of TPB. Although there is no direct way of comparing the different methods in terms of numbers of individuals, they were set up to represent about the same amount of effort. Prokopy et al.[80] regressed cumulative trap capture vs. percentage fruit injury, dividing the sampling period into two tree phenology-based sections: (1) silver tip through tight cluster and (2) silver tip through late pink, in order to determine an action threshold. Using a predicted value of 2% fruit injury, the regression equations are solved for x values of 3 adults per trap (silver tip-tight cluster) or 4.4 adults/trap (silver tip-late pink); these were recommended as action thresholds for prebloom sprays. The economic injury level of 2% fruit damage was considered to be synonymous with the action threshold and was based on the percentage of injury resulting even when insecticidal sprays are applied for TPB in commercial orchards; in this sense, it represented a realistic goal for minimizing TPB injury.

Boivin et al.[78] evaluated the sticky trap developed by Prokopy for captures of several species of phytophagous mirids, including TPB. Traps hung 0.5 m above the ground were suitable for capturing TPB. The seasonal population trend indicated a strong pulse of adults early in the season (April and May), with only a few captures throughout the rest of the summer and fall. The data concerning the contrast between traps placed in the center vs. the orchard border indicated that border traps are important for detecting the first immigrants into the orchard. Michaud et al.[81] also compared the efficiency of sticky traps vs. limb tapping in predicting the percentage of damaged fruit at harvest. They found that although limb tapping was a better predictor, it is less user-friendly and more subject to error. Michaud et al.[81] also calculated cost of control measures and value of fruit to derive an economic threshold, using the method of Stone and Pedigo:[82]

$$ED = CC/WV \qquad (2)$$

where ED is economic damage (kg/ha), CC is the cost of chemical control ($/ha), and WV is the wholesale value of the fruit ($/kg). Four insecticides and three sampling intervals provided EIL values of 0.5 to 2.8 TPB (cumulative) per trap and 0.8 to 4.5 TPB/50 limb taps. However, the authors point out that extrapolation to other years, regions, or production systems should be made with caution because the cost of chemical control and particularly the value of the fruit tends to vary greatly. These calculations use the assumption that chemical control is 100% effective in preventing further damage.

Current field-level recommendations for monitoring TPB in New England[67] are the above-mentioned white sticky trap 15 × 20 cm (6 × 8 in.), at a density of 1 trap per 1.2 to 2 ha (3 to 5 acres) or not less than three per block. Traps should be located near the block periphery, as they are a better indicator of immigrating adults. Traps are hung vertically just below knee height, and blossoms and foliage should be removed within 30 cm (12 in.) of the trap. Traps are placed in the orchard at silver tip, checked weekly, and removed at petal fall. Treatment is recommended if cumulative average trap capture is ≥ 2.4 adults per trap (silver tip through tight cluster) or 4.1 adults per trap (silver tip through late pink stage). A second New England publication[70] also recommends the above trapping methods; however, they mention that trap results have been inconsistent in some areas, and sampling the ground cover with a sweep net may also be useful. Fifty sweeps per sample, at a rate of one sample per 5 to 6 acres is recommended. Sampling should be done when weather is dry, between 9.9 and 15.5°C (50 and 60°F), and biased toward areas where ground cover is diverse.

XI. TETRANYCHID MITES (*PANONYCHUS ULMI* (KOCH); *TETRANYCHUS URTICAE* (KOCH); *TETRANYCHUS MCDANIELI* (McGREGOR)) (ACARI: TETRANYCHIDAE)

Spider mites are among the most ubiquitous pests of apples worldwide. The primary species causing problems in North America include the European red mite (ERM), *Panonychus ulmi* (Koch), the twospotted spider mite (TSSM), *Tetranychus urticae* Koch and the McDaniel spider mite (MDSM), *Tetranychus mcdanieli* (McGregor). Problems with the ERM occur throughout North America, while problems with TSSM and MDSM are more prevalent in the drier western U.S. and Canada. There are occasionally problems reported with the brown mite, *Bryobia rubrioculus* (Scheuten), and the yellow spider mite, *Eotetranychus carpini borealis* (Ewing), but these tend to be restricted to small geographic areas or orchards which do not use dormant sprays.

TSSM and MDSM have similar life histories and distributions in the trees and so are discussed together. The mites overwinter as adult females in debris at the base of the tree or under bark scales on the trunk and scaffold limbs. The mites break diapause in spring in synchrony with the development of new shoots and buds. TSSM tend to remain in the weeds at the base of the tree before moving up into the tree, whereas MDSM tends to move directly up in the trunk. The spring populations are, therefore, most common on water sprout leaves and interior canopy leaves. As the season progresses, the mites move toward the top of the tree and by early summer they tend to be well distributed on the outer canopy.

Despite the problems with these mites in the western U.S., there is surprisingly little information on sampling them. Early studies recommended the use of leaf-brushing machines,[83] which brush the mites off the leaves and onto a rotating glass plate. Usually samples of five leaves are taken from ten trees and run through the machine and then the numbers on the plate (or if the density was high, part of the plate) were counted.[84] Optimal sample size (using constant precision as the criterion) have been calculated for TSSM using a multistage sampling program by Zahner and Baumgaertner.[85] They reported that low densities made studying its distribution pattern difficult, a phenomenon also noted by other authors.[86,87] Constant precision techniques are extremely time consuming for spider mites and are probably rarely ever used outside of the research community. Information on the dispersion of MDSM on seedling trees in a walk-in growth chamber was reported,[88] but the applicability of this information to field sampling is dubious.

Binomial sampling of TSSM and MDSM on apple has only been reported by Jones.[87] This paper was primarily concerned with developing sampling plans which have a broad crop and pest applicability, but Jones does report information on sampling both mite species on apple. Using Taylor's power law (TPL), he found that the generic values of $\alpha = 3.30$ and $\beta = 1.49$ entered into Wilson and Room's[89] binomial sampling formulas (see Chapter 9) provided excellent fit to the MDSM data set. For TSSM, the low densities found on apple precluded extensive analysis. However, the Taylor coefficients observed by Zahner and Baumgaertner[85] ($\alpha = 4.68$, $\beta = 1.73$) were within the range of α and β observed in Jones' *Tetranychus* data set, and the binomial samples developed in his paper could be used until further studies are performed.

A. TSSM AND MDSM THRESHOLDS

Most of the economic thresholds for mite damage were developed for ERM, and only Hoyt et al.[155] have done extensive studies with MDSM. Although these authors developed separate cumulative mite day-peak population relationships for the three primary mite species, there is no evidence to suggest that injury caused by the different species is fundamentally different in its effect on the tree. The higher threshold suggested by Hoyt et al.[15] (2000 to 2500 mite-days, or 80 to 100 mites per leaf) is probably associated with differences in climate and irrigation practices in the western vs. the eastern U.S. Water stress, which can occur regularly in unirrigated eastern orchards, may be a key factor in determining the tree's tolerance to mite damage.[101,155]

The European red mite overwinters as eggs, with the highest numbers laid on the outer periphery of the tree. Eggs begin hatching at about the tight cluster bud development stage, and the distribution of the mite follows the same general pattern as the eggs early in the season. Depending on the location, there can be 4 to 9 overlapping generations/year. During the summer, a generation may be completed in less than 2 weeks, requiring frequent sampling of the orchard to ensure populations have not surpassed threshold levels. During mid- to late-August and September, adult females begin laying overwintering eggs.

In contrast to MDSM and TSSM, many sampling plans for ERM have been developed. Bliss and Fisher[90] report an example of using the negative binomial to fit the counts of ERM observed on apple leaves; however, this study was performed on only 150 leaves and should be viewed strictly as an example. Chant and Muir[91] compared an imprint method, where infested leaves were placed between sheets of absorbent paper and passed between rubber rollers, with the leaf-brushing machine. They found that the number recorded using the imprint method was significantly less than the number recorded by the brushing method. This discrepancy probably occurs because the resulting dots are hard to separate when the populations are high.

Modern sampling plans for ERM started with the work of Pielou.[92] Pielou found that the ERM was distributed contagiously on apple foliage and developed a method that allowed the indexing of the population by recording the presence or absence of mites on the leaves. Pielou also fit the negative binomial distribution to 5 of 418 trees and found satisfactory fit in only one instance. Further work was not performed, primarily because his data set was so large (418 trees, 100 leaves per tree) that without computers, such a task was unreasonable.

Sekita and Yamada[93] used Iwao's patchiness regression[94] and Kuno's implementation of a constant precision sample[95] to develop two-stage sampling plans for eggs, first-generation adults, and summer adults. The primary use of their sample was conducting life table analysis or ecological studies that require a known level of precision.

Croft et al.[96] calculated dispersion statistics and sample sizes for ERM in Michigan. They evaluated several data sets, the largest varying from a mean of 1.2 to 177.5 mites per leaf. Their results showed that, using a common k value, the data departed from the negative binomial distribution ca. 50% of the time. They indicated that the true distribution for ERM was significantly less clumped than the negative binomial, but that by using the negative binomial distribution, a conservative estimate of the sample size would be generated. The sample sizes required to obtain a given level of precision are presented. Their inability to obtain a common k value is not surprising. The k values are known to be variable over a comparatively short range of means[97] and the means of the Croft et al. data set varies by over two orders of magnitude. The usefulness of this paper and estimates of sample sizes would be much greater if the authors had given the coefficients for both the among-tree and between-tree power law regressions (they are provided for the between-tree regression, but only the β value is provided for the among-tree regressions).

Mowery et al.[98] provided two sampling plans for ERM based on a combination of the negative binomial and Taylor's power law. They developed the among-tree mean-variance relationship using the negative binomial with a common k and calculated the between-tree variance component using TPL. They provide a two-stage sampling program using total enumeration similar to that presented by Croft et al.[96] Perhaps the most significant advance in sampling ERM occurs in the Mowery et al. paper. The authors took the idea of Pielou[92] and provided a presence-absence sampling plan to allow the rapid estimation of mite populations so that the sampling programs actually could be used easily by consultants and growers. The sampling plans presented here are widely used in Pennsylvania today.

A question that needs to be answered is why did the negative binomial with a common k work for Mowery et al.[98] and not for Pielou[92] or Croft et al.[96] The problems with Pielou's data are somewhat difficult to address because the data set was not completely analyzed. However, it should be noted that by digitizing the presence-absence graph provided in Pielou's original paper and using Wilson and Room's[89] equations (i.e., a constrained negative binomial), estimates of α and β values for TPL using nonlinear regressions fall within the normal range of values for α and β observed for ERM.[87] While this does not prove Pielou's data would fit a negative binomial distribution, it suggests that a binomial sampling program could have been developed if the data were analyzed using systems currently available. The k values obtained by Mowery et al.[98] were much lower than those observed by Croft et al.[96] (0.87 vs. 1.442). This is probably related to the lower range of mean values (1.3

to 26.5 mites per leaf) of the data fit by Mowery et al.[98] Because the range of means examined was much smaller and more representative of the levels experienced in commercial situations, a common k value functioned adequately (22 of 26 data sets found no difference between the predicted and observed using the common k, where 24 of 26 found no difference using maximum likelihood estimates of k). Limitations on this presence-absence program are similar to those on any such program (see Chapter 9), and estimation of population levels over approximately 5 mites per leaf requires an expansion of the sampling program to a higher tally threshold (e.g., > 1, > 2, etc. mites per leaf).

Zahner and Baumgaertner[85] presented constant precision sampling programs for ERM in Switzerland. They used both TPL and Iwao's patchiness regression to measure dispersion of the populations studied and found values close to those reported by Croft et al.[96] and Mowery et al.[98] for Taylor's coefficients. Values for Iwao's patchiness regression are similar to those reported by Sekita and Yamada[93] for within-trees, but the between-tree component is close only to the first generation adults reported by Sekita and Yamada. Values for common k were lower than those reported by both Croft et al.[96] and Mowery et al.[98] (0.46, 1.44, 0.87, respectively).

Nyrop et al.[99] presented dispersion indices (TPL) and binomial sequential sampling plans aimed at classifying a population of ERM to thresholds. This binomial procedure differs from the plans mentioned above because its primary aim is decision making, rather than estimation of the population level (see Chapter 9 for more information). The dispersion analysis provided values of α and β that were within the range of information presented by the above studies and those reported by Jones.[87] The application of the sampling programs to field situations showed that the sequential program was able to correctly classify the mite population levels in a rapid and parsimonious fashion. This program has been implemented in New York, with slight variations (using a different tally thresholds) and variable economic threshold levels depending on the time of year.[24,100]

B. ERM THRESHOLDS

Research conducted in Pennsylvania integrated a number of other factors important toward establishing meaningful economic thresholds. Hull and Beers[101] validated a series of injury thresholds for two major apple cultivars, 'Yorking' and 'Delicious'. They established four target thresholds of 0, 250, 750, and 1250 cumulative mite-days per leaf and allowed the majority of injury to occur during July and August, the normal period for mites to increase and cause injury. They found only the target threshold of 1250 cumulative mite-days caused any reduction in yield variables. They coupled this research with their previous work on the timing of mite injury[102] and the interaction between mite injury and crop load[103] to establish dynamic economic thresholds (Figure 3). Pennsylvania[104] advocates that the thresholds be reduced to one half the figure value if the growers are using the alternate row middle system of applying acaricides since only one half of the tree is treated. These thresholds are further coupled with a predator:prey ratio to optimize decision making.

Sampling for ERM eggs has been developed by several authors. Sekita and Yamada[93] and Herbert and Butler[105,106] provide egg sampling plans that can be used effectively for monitoring eggs for research purposes. Sampling eggs for pest management programs is rarely used because, if timed correctly, dormant sprays can effectively control the egg stage. In addition, eggs often are laid on each other and make accurate population estimates difficult to obtain for the cost in time associated with them.

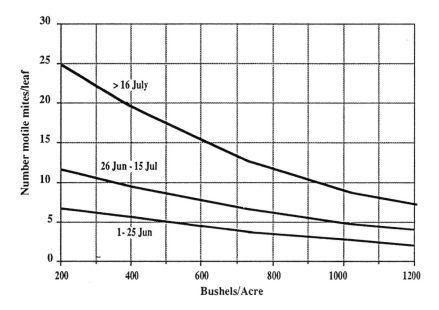

FIGURE 3. Mite action thresholds for various crop loads at different time of the season. These levels apply to healthy, vigorous trees with mite damage occurring only after June 5. (From Travis et al.[107] With permission.)

XII. APPLE RUST MITE (*ACULUS SCHLECHTENDALI* (NALEPA)) (ACARI: ERIOPHYIDAE)

Apple rust mite (ARM) is studied in the context of both a pest (feeding on fruit and leaves) and a beneficial (an alternate prey source for predatory mites, especially phytoseiids). ARM feeding also has been studied as a means of reducing leaf quality to discourage buildup of the ultimately more damaging ERM.[107] Its life history has not been much studied in North America, but Easterbrook[108] has made detailed observations. ARM overwinters as an adult in or near buds, and moves to green tissue as it expands in the prebloom period. Fruit feeding occurs during the early postbloom or fruit set period, which causes apple skin russetting. Injury is more apparent on light-colored varieties such as 'Golden Delicious'. In Washington, this situation is one of the few where control of ARM is recommended.[109] Foliar feeding causes leaf bronzing, rolling, and premature terminal bud set (reduction in the vegetative growth of shoots); in severe cases, reduction in fruit growth rate can occur.[110] Treatments for rust mites as indirect feeders are infrequent, however, because the advantage of their presence as alternate prey is felt to outweigh the possible influence on fruit size or vegetative growth.

Typhlodromus occidentalis feeds and reproduces readily on ARM[111] and is considered a necessary component of the predator-prey system of mites. *T. occidentalis* can maintain itself on ARM, and keep tetranychid mites at low levels throughout the season. Because ARM populations can reach very high levels (> 2000 per leaf) in the presence of *T. occidentalis*, it apparently does not effectively regulate ARM populations.[111]

Wardlow and Jackson[112] have done one of the few papers specifically on sampling ARM. This study is methodological rather than statistical in nature; the same number of sampling units (50 buds per block, or 5 buds from 10 shoots) were held constant and were processed using three techniques. The first was examination of buds with a binocular scope (×30), which was deemed laborious and time consuming; the second was dissection of buds, with extraction of mites in methanol; the third was similar to the second, except that buds were only sliced vertically rather than dissected. The third method was the least time consuming, but the equipment and materials involved (methanol, sieves, shaker, ×200 magnification binocular scope, Fenwick can) may not be accessible to many field-level practitioners.

Despite the lack of published work devoted to sampling, rust mites are regularly sampled in the course of many studies. ARM is removed from apple leaves with commercially available leaf-brushing machines and can be counted using the same techniques used for tetranychid mites. Hoyt[111] notes a slightly higher density in the center or periphery of the tree, whereas tetranychid mites are sometimes concentrated in low centers. Because of their small size and frequently high numbers, it is easier to do a second examination of the plate for ARM, using a smaller subsection of the plate (e.g., $\frac{1}{20}$) at higher magnification (×40). This technique is available only after the leaves are fully expended and is not useful for assessing early season populations for control decisions.

A. ARM THRESHOLDS

Tentative economic thresholds have been set for ARM feeding, although only Hoyt (as cited in Hull et al.[110]) has made a rigorous study of the effect of foliar damage on fruit size and quality parameters in Washington. This work indicated that up to 4800 mite-days could be tolerated without any loss in fruit size. Current Washington recommendations indicate that populations exceeding 50 ARM per leaf (May) or 250 ARM per leaf (late June) may require treatment, but otherwise, moderate levels of ARM should be maintained to encourage *T. occidentalis*.[109] For early season fruit damage, Agricultural Development and Advisory Service (as cited in Wardlow and Jackson[112]) has set an action threshold of 10 ARM per bud. Croft[113] indicated that 200 ARM per leaf for 10 to 14 d will not cause economic damage under Michigan conditions.

XIII. PHYTOSEIID PREDATORY MITES (*TYPHLODROMUS OCCIDENTALIS* (NESBITT), WESTERN PREDATORY MITE; *AMBLYSEIUS FALLACIS* (GARMAN); *TYPHLODROMUS PYRI* (SCHEUTEN)) (ACARI: PHYTOSEIIDAE)

These three predacious mites are the most important predators of tetranychid mites feeding on apples in North America. *T. occidentalis* is found primarily in the western U.S. and is most effective against web-spinning mites on a number of orchard systems such as tart cherries,[114] almonds,[115,116] and peaches.[117] Hoyt and Caltagirone[118] indicate that *T. occidentalis* has a lower level of effectiveness in controlling ERM early in the season in apples because the distributions of ERM (outer canopy) and *T. occidentalis* (inner portion of the tree) do not overlap. However, later in the season, the distributions do overlap, providing control of ERM. *A. fallacis* is important throughout the New England area, the Atlantic states, and Michigan. In some of the areas, such as the Mid-Atlantic states, *A. fallacis* is less important than *Stethorus punctum*, primarily because some of the pesticides used for leafroller control are detrimental to *A. fallacis*. *T. pyri* is of lesser importance (in terms of area) than the other predators, but in New York and parts of Canada it can effectively regulate populations of ERM.

Sampling programs for *T. occidentalis* on apple are not reported directly. Jones[114] reported dispersion analysis of *T. occidentalis* on tart cherry and unpublished information on apple. In both instances, TPL yielded virtually identical coefficients ($\alpha = 2.39$, $\beta = 1.23$ and $\alpha = 2.07$, $\beta = 1.23$ on tart cherry and apple, respectively). On almonds, Wilson et al.[116] found similar coefficients ($\alpha = 2.28$, $\beta = 1.24$) and Jones[117] found only slightly more clumped estimates from peach ($\alpha = 3.81$, $\beta = 1.40$).

Binomial sampling plans on apple have not been developed for *T. occidentalis*. However, by combining the information from apple, tart cherry, and peach, Jones[117] found no significant differences between the crops. Until further information is available for *T. occidentalis*, we suggest the binomial sampling program devised for tart cherries by Jones[114] or that of Wilson et al.[116] on almonds be used on apples.

The only sampling plan developed for *A. fallacis* is reported by Croft et al.[96] This study utilizes the negative binomial with a common k (0.735) to develop constant precision sample sizes. Unfortunately, like the data set used for *P. ulmi*, the negative binomial is an inadequate model for the data set (47% of the distributions depart from the negative binomial) and the α coefficient for the among-tree TPL were not reported. The widespread importance of this predator in apple production means that this should be a priority in future research on apples.

Sampling for *T. pyri* is provided by Nyrop.[119] This study utilizes a sequential classification of predator : prey ratios relative to a critical ratio (critical in the respect that biological control would be assumed to occur when the ratio is high and not occur when it is low). Because this is covered extensively elsewhere in this volume, the reader is referred to Chapter 11 for further information.

XIV. STETHORUS PUNCTUM (*STETHORUS PUNCTUM* (LECONTE)) (COLEOPTERA: COCCINELLIDAE)

This small, black lady beetle is the most important mite predator throughout Pennsylvania, New Jersey, and most of the Mid-Atlantic states. *Stethorus* adults overwinter in the leaf litter under apple trees and in nearby fence rows and woods.[120] They become active in the spring when the trees are beginning to develop and rapidly move into the trees when the mites begin to hatch at the pink to bloom stage. Adult females begin to lay eggs when a concentration of mites is found. The larvae pass through four stadia in approximately 12 to 15 d, and there can be up to 3 generations/year. The adults and larger stage larvae can consume from 50 to 100 motile mites per day.[121]

The only accepted method for counting *Stethorus* is for observers to make a 3-min observation around the periphery of a tree, recording all *Stethorus* adults and larvae that they see.[122] Asquith and Hull[123] reported on two experiments in Pennsylvania to relate the numbers of *Stethorus* counted in a 3-min observation to the number per 1000 leaves on an apple tree and to investigate the accuracy of the 3-min observation. The first experiment was conducted because the 3-min observation does not estimate the absolute predator population on a tree which was required by Mowery et al.,[124] who were attempting to construct a computer-based predictive system for determining the success of *Stethorus* in providing biological control of ERM. The r^2 values for this relationship between sampling methods for *Stethorus* adults and larvae were 63 and 89%, respectively. The second experiment was conducted to investigate the accuracy of the 3-min observation. They tested two hypotheses: (1) there is no real difference between human observers, and (2) the number of *Stethorus* in the 3-min period is directly proportional to the number per tree. No statistical difference was found between observers once they were properly trained and the number of *Stethorus* counted in a 3-min observation was directly proportional to the number per tree.

XV. ROSY APPLE APHID (*DYSAPHIS PLANTAGINEA* (PASSERINI)) (HOMOPTERA: APHIDIDAE)

The rosy apple aphid (RAA) is the most serious pest of the aphid complex attacking apple throughout North America. These aphids remove plant juices from the leaves, causing severe curling and abscission, and they secrete large quantities of honeydew, which provides a substrate for a black, sooty fungus. However, the most serious effect results from the translocation of saliva from the leaves to the fruit. This causes the apples to remain small and deformed and renders them unmarketable.[125] RAA overwinters on apple as shiny, dark green to black eggs that are deposited in the fall on the bark, especially at the base of the buds. The eggs begin to hatch in the spring when the trees are at the silver tip stage and continue until the half-inch green stage.[126] Usually 3 generations are produced on apple before the aphids begin leaving apple in late June and early July to fly to their alternate host, narrow-leaf plantain. In the fall, the winged females and males return to apple, mate, and lay overwintering eggs.

Because no method exists for distinguishing the overwintering eggs of RAA from that of the other three species overwintering on apple, egg sampling is rarely recommended. Only California recommends sampling 25 to 100 fruit spurs during the dormant season to prevent aphid colonies from quickly spreading over the tree once the eggs hatch.[127] Because the most critical time to treat for RAA is at the half-inch green stage,[126] both Pennsylvania and West Virginia recommend treatment at this time.[12,128] If no treatment is made at the half-inch green stage or control was not achieved, an assessment can be made at the early pink to pink stage to determine the necessity for spraying. New York specialists[24] recommend sampling ten fruit clusters from the interior canopy of each of ten trees per block at this time. Specialists in Pennsylvania[128] and West Virginia[12] also recommend sampling five to ten trees per block at this stage by counting the number of fruit spurs showing curled leaves and containing live aphids during a 5-min examination per tree. Growers in Ontario are encouraged to check weekly for the presence of RAA beginning at tight cluster and continuing until late June by examining 10 fruit clusters from each of 20 trees. A cluster is considered infested if more than 20 aphids are present.[129] During the postbloom period, Mowery et al. (unpublished data as cited by Hoyt et al.[130]) recommend using the proportion of aphid-infested leaves per rolled leaf cluster as an estimate of aphid density. The advantage of this technique is its brevity and economy. They developed a table which incorporated both the variance among leaf clusters from the same tree and from different trees so an individual could choose the optimal combination of clusters per tree and numbers of trees based on time and cost estimates. The only major disadvantage to this method is that a majority of the fruit injury caused by RAA is already completed at the time of sampling, and only further leaf and shoot damage can be estimated.[126]

A. RAA THRESHOLD

Economic thresholds for RAA are based on a number of indices. For example, Hull et al.[110] in Pennsylvania established a relationship between the number of infested fruit spurs per tree in a 5-min search at the early pink to pink stage and the number of damaged apples per tree at harvest. This relationship assumes that the aphids attack the blossom clusters at random, and one aphid-infested cluster yields one injured apple. From this relationship, they proposed an economic threshold of one infested cluster per tree at the pink stage. New York also uses a threshold of 1% infested clusters at the pink stage.[24] Ontario recommends treatment if 10% or more of the fruit clusters become infested at any time.[129] Nova Scotia uses an economic threshold of 1.5 colonies per meter of tree height.[131]

XVI. GREEN APPLE APHID (*APHIS POMI* (DE GEER)) AND SPIRAEA APHID (*APHIS SPIRAECOLA* (PATCH)) (HOMOPTERA: APHIDIDAE)

The green apple aphid (GAA) is a secondary indirect pest of apple and is widely distributed throughout North America and the world. Recently, close inspection of presumed GAA infestations has revealed that a substantial proportion of these infestations may belong to a different species, the spiraea aphid (SA).[132] Differentiating between species without the aid of a microscope is difficult for apterae; Halbert and Voegtlin[156] provide a description of morphological differences of all forms. The status of the overwintering host of SA is unclear, but it has a wide range of secondary hosts, including apple.[156] Therefore, until additional research is conducted on the biology and behavior of SA on apple, we consider the spatial distribution and injury effects of the two species similar and treat them as such in this section. In all fruit-growing areas the aphids overwinter as eggs, which are shiny and black and cannot be differentiated from the other two species (rosy apple aphid and apple grain aphid, *Rhopalosiphum fitchii* (Sanderson)) overwintering on apple. The eggs are laid mostly on the bark or on the buds in the fall. Hatch occurs in the spring between the silver tip and half-inch green growth stage.[126] The winged forms disperse to other parts of the tree or to other trees. Both species breed continuously during the summer. In August and during the autumn months, they are found almost exclusively on water sprouts or terminals of young trees that are still growing, and it is at such locations that the male and female sexual forms are usually produced.

Both nymphs and adults suck sap from apple trees. They prefer to feed on the succulent, young tissue found at the ends of terminal shoots and water sprouts. They can curl the foliage if high populations develop and stunt tree growth, especially on young trees. Aphids can also secrete large amounts of honeydew on which a black fungus grows that discolors both fruit and leaves; this is particularly noticeable on early[133] or light-colored apples.

For most areas in North America sampling for GAA and SA usually begins in early- to mid-June when terminal shoots are rapidly expanding. Madsen et al.[134] were the first to recommend sampling plans for GAA by counting the aphids on the third, seventh, and fifteenth leaf. Both Jokinen[135] and Hull and Grimm[136] studied the spatial distribution of GAA on bearing apple trees and derived sample size requirements for the number of shoots per tree and number of trees at various levels of precision. Hull and Grimm[136] developed two sampling schemes for use in IPM programs that would estimate the population size of GAA in an apple tree. The first method and the most accurate one for predicting aphid density utilized the number of aphids on the most heavily infested leaf per shoot. The second method depended on counting only the number of infested leaves per shoot; a leaf was considered infested if it contained one or more apterous aphids. The latter method resulted in a 10 to 20% decrease in accuracy, but also a four- to eight-fold decrease in the time required to sample. Either method can reliably predict the aphid population over the entire tree by only counting the aphids on the most infested leaf per shoot or the number of infested leaves per shoot on the lower shoots.

A number of other states recommend a slightly different method of sampling for GAA and SA. For example, Massachusetts, New York, and Michigan, as well as the Canadian province of Nova Scotia, recommend monitoring ten growing terminals on ten trees and determining the percentage of terminals infested.[67,70,130,131] Hull and Grimm[136] also examined the accuracy and speed of this sampling method. This method is by far the most rapid, but it is the least reliable for estimating population density.

The first recommended economic threshold for GAA was developed by Madsen et al.[134] They advised treatment when the aphid population exceeded 5.0 aphids on the seventh leaf from the apex. This threshold was later revised to 50% of the terminals infested.[72,137] This level is still in use today for a number of states (Massachusetts,[67] New York,[70] and Michigan[130]). Both Pennsylvania and West Virginia use an economic threshold of four infested leaves per shoot.[12,128] This threshold was derived from the work of Hull and Grimm,[136] where they related their findings on the spatial distribution and density of GAA to that of Madsen et al.[134] who found that an infestation level of 250 aphids per shoot would cause honeydew to appear on the fruit. Nova Scotia recommends treatment when the aphid population infests 10% of the terminals.[131] A very detailed study by Hamilton et al.[138] on the effect of GAA populations on shoot growth, fruit quality, and yield found very few significant relationships between aphid-days and fruit yield and vegetative growth parameters on the cultivars 'Golden Delicious' and 'Delicious'.

XVII. TENTIFORM LEAFMINERS (*PHYLLONORYCTER BLANCARDELLA*, SPOTTED TENTIFORM LEAFMINER; *PHYLLONORYCTER CRATEGELLA*, APPLE BLOTCH LEAFMINER; *PHYLLONORYCTER ELMAELLA*, WESTERN TENTIFORM LEAFMINER) (LEPIDOPTERA: GRACILLARIIDAE)

Although there is a large complex of leafminers attacking apples in North America, the spotted tentiform leafminer (STLM), apple blotch leafminer (ABLM), and the western tentiform leafminer (WTLM) are considered the most severe. These three species can be considered broadly to be ecological homologues in terms of general phenology, niche in the ecosystem, and type of injury.[130] The first generation adults emerge from the overwintering pupae around the green tip to half-inch green stage of apple development, and eggs are laid on the underside of the young leaves. The larvae eclose and bore into the leaf where the first three stages are usually confined to the leaf's spongy mesophyll and are referred to as "sap-feeders". Feeding by the last two larval stages is much more damaging because the larvae are larger and because feeding is extended to the upper palisade cells. The latter two stages are referred to as "tissue-feeders". Pupae are formed in the leaf mine, with one end extending out of the mine. There are 3 to 5 generations per year depending on location. Monitoring methods have been developed for all but the pupal stage.

An egg sampling program for *P. blancardella* before bloom has been developed in New York by Nyrop et al.[139] This sampling program relies on sampling the second, third, and fourth leaves on fruit clusters, which contain approximately 50% of the eggs found in the entire cluster. Nyrop et al.[139] provides variable intensity, sequential, and optimal fixed sampling programs. For field implementation, the sequential program is recommended, and a sampling chart (set for a threshold of 2 eggs per leaf) is available.[24] The protocol calls for sampling three fruit clusters from around the canopy of a tree and after sampling two trees, the total number of eggs is compared to a sampling chart. If the number of eggs is between 59 and 13, another tree is sampled. Sampling continues until a maximum of seven trees is sampled or until the cumulative number of eggs is greater than or less than indicated on the sampling chart. In Washington,[153] sampling for the first generation WTLM eggs is accomplished by a fixed sample of 50 leaves per block, taken from the larger cluster leaves. If three or more eggs per leaf are found, control of this generation is recommended. Because of rapid canopy expansion during this period, three eggs per leaf will result in one mine per leaf at the end of the generation.

Larval (mine) monitoring programs have been developed in New York for STLM[24] and in Utah for WTLM.[140] The New York program uses the same leaf sampling methods as described above for STLM egg sampling. They recommend taking the leaf sample at about 630 to 700 degree-days (43°F base) after the start of the second generation. Their sampling plan is geared to providing information as to whether the population is over a threshold of two mines per leaf. A sampling chart and detailed instructions are available.[24] Washington recommendations[153] for WTLM emphasizes sampling and decision-making for the second generation. One spur leaf and one shoot leaf on 25 trees per block (50 leaves) is sampled during the sapfeeding stage. If two or more mines per leaf are detected, control of the second generation is recommended. Using the same sampling procedure, a threshold of five mines per leaf is recommended for the third generation. These thresholds are further modified by the presence of parasitism. If $> 35\%$ of the tissue feeders of the previous generation were parasitized, biological control of the current generation is probable, and no chemical control measures are recommended.

Investigations on the dispersion and sampling of WTLM were performed in Utah.[140] These studies did not use the cluster sampling method described above, but rather used a random selection of leaves from the outer canopy of the tree. Sampling was performed during the second through fourth leafminer generations, and results indicate that the populations were only slightly clumped, according to TPL.[140] Binomial sampling plans which could estimate population levels accurately up to three mines per leaf were provided and validated. This same study suggested that ten trees needed to be sampled for an accurate estimate of the orchard population of three mines per leaf and four trees at two mines per leaf. However, 13 trees at three mines per leaf and eight trees at two mines per leaf are the correct numbers. Sampling for pest management is performed during the second generation at approximately the same time (start of the tissue feeding stage), as suggested in New York.

Adults of *P. blancardella* and *P. elmaella* both respond to the pheromone of the STLM.[141] The pheromone is so effective and the population levels can become so high that normal sticky traps can be completely saturated in < 1 h.[140] Nonsaturating-type pheromone traps, with a capacity of 500,000 moths, have been tested, and graduated cylinders were calibrated to provide a rapid method of estimating population density.[142] Other methods such as sticky panels without pheromone caps[143] and sweeping foliage with an insect net have been used.[144] For apple blotch leafminer, studies have shown that a 20×30 cm rectangle painted "tartar" red and hung horizontally 0.5 m high and within 0.5 m of the trunk was most efficient at trapping ABLM.[145] However, Green and Prokopy[145] noted that the superiority of the horizontally hung trap may be mostly mechanical, because they observed ABLM adults striking vertical traps and then falling away without being captured. With subsequent advances in sticky materials,[54] perhaps the easier to use and more convenient vertical orientation may be substituted for the horizontal orientation.

XVIII. WHITE APPLE LEAFHOPPER (*TYPHLOCYBA POMARIA* (McATEE)) (HOMOPTERA: CICADELLIDAE)

White apple leafhopper (WALH) feeds on apple foliage, although its excrement droplets on the fruit are sometimes of concern. WALH overwinters in the egg stage, which hatch from about tight cluster to petal fall. Nymphs pass through five instars in April and May, and adults of the first generation fly during June and July. Summer eggs are laid in the leaves (primarily veins and midribs), and a second generation of nymphs is produced in July and August. Adults of this generation peak in mid-September.

Very little formal work on sampling has been done on WALH, although there are a number of provisional sampling methods.[130] Both the adult and nymphal stages are easy to sample. Nymphs may be sampled in the field by direct visual examination of the leaves. Nymphs tend to be somewhat negatively phototactic, and will move to the underside of the leaf in strong sunlight; thus, both sides of the leaf should be examined. Some examples of sampling protocols and economic thresholds from different states are: Michigan, 10 leaves per tree, 10 trees per block, 2 to 3 active stages per leaf;[130] Massachusetts, 30 leaves per tree, 0.25 active stages per leaf;[130] New York, 50 leaves per tree, 10 trees per block, 0.25 nymphs per leaf first generation, > 1 nymph per leaf second generation;[130] California, 4 leaves per tree from 25 trees per block, 0.5 nymphs per leaf or 30% of the leaves are infested.[127]

Adults may be monitored with an unbaited apple maggot panel, which is commercially available.[146] Beers and Elsner[147] examined the response to various trap colors, and both white and yellow caught the highest numbers of adult leafhoppers.

Seyedoleslami and Croft[148] studied the spatial dispersion of WALH eggs, along with its parasite *Anagrus epos*. Eggs were found on 1- to 5-year-old wood and can be sampled any time after deposition is complete in the fall and through the following spring. Estimates of egg density are laborious and, for control purposes, there is no reason not to wait and sample the nymphs. However, overwintering eggs may be dissected for assessment of parasitism by *A. epos*. Parasitized eggs can be distinguished from normal eggs after February, but the distinction is easier to make as the adult parasite approaches emergence (early to mid-April).[149] The parasite can also be monitored with the same yellow sticky panel used for adult leafhoppers,[146] although its extremely small size makes this somewhat laborious.

Although most fruit-growing states provide chemical control recommendations for WALH, almost no data exist on the effect of WALH on fruit yield or quality. Beers and Elsner[147] found no reduction in fruit size, quality, external color, or return bloom after one season of injury by high leafhopper populations under Washington conditions. Nothing is known about multiple years of damage, however. Observations by the senior author indicate that typical packing house fruit-washing operations remove leafhopper excrement spots except in the stem and calyx cavities of the fruit. Fruit destined for direct marketing to consumers would not pass through this washing process, and tolerance for fruit contaminated by excrement may be lower. Annoyance to fruit harvesters by the adults has also been mentioned as a reason to reduce leafhopper population just prior to harvest. Beers and Elsner[147] found that pest control consultants would recommend sprays for blocks where adult leafhopper populations have corresponding sticky trap catches of ≥ 130 adults per trap per day.

XIX. CROP LOSS ASSESSMENT

Sampling methods to determine crop loss caused by the feeding of insects and mites during the season on fruit yield and quality just before packing and storage are as crucial as methods to determine the size of the given pest population for management considerations.[150] Such methods would allow pest managers to identify not only the causal organisms involved but also the extent of downgrades in fruit. Also, other orchard management factors such as bruises, limb rubs, color, etc., can be identified that affect fruit packout for a given orchard. Such information would also facilitate better storage and marketing decisions for the grower.[12] However, very little research has been done to aid the decision-making process in this area.

Hogmire et al.[151] developed a sampling plan in conjunction with a modified grading scheme constructed by Russo and Rajotte[152] to predict apple fruit quality after harvest but before any packing. The optimum sample size per orchard block consisted of 100 apple samples from each of five bins. The apples are then graded according to a modification of the Russo and Rajotte[152] grading scheme, which utilizes a 1 to 10 quantitative rating scale and determines the cause, defect, and color of each apple. If an apple has multiple defects, the two most obvious are recorded. The sampling scheme was precise enough to predict fruit defects to within 5% and packout to within 10%.

REFERENCES

1. Pimentel, D., Krammel, J., Gallahan, D., Hough, J., Merrill, A., Schreiner, I., Vittum, P., Koziol, F., Black, E., Yen, D., and Fiance, S., Benefits and costs of pesticide use in U.S. food production, *BioScience*, 28, 772, 1978.
2. Chapman, P. J. and Lienk, S. E., *Tortricid Fauna of Apple in New York*, Cornell University Press, Ithaca, NY, 1971, 28.
3. Butt, B. A., *Bibliography of the Codling Moth*, ARS W-31, U.S. Department of Agriculture, Agricultural Research Service, Washington, DC, 1975.
4. Riedl, H., Howell, J. F., McNally, P. S., and Westigard, P. H., Codling moth management: use and standardization of pheromone trapping systems, *Univ. Calif. Agric. Exp. Stn. Bull.*, 1918, 1986.
5. Riedl, H., Croft, B. A. and Howitt, A. J., Forecasting codling moth phenology based on pheromone trap catches and physiological-time models, *Can. Entomol.*, 108, 449, 1976.
6. Charmillot, P. J., Mating disruption technique to control codling moth in western Switzerland, in *Behavior-Modifying Chemicals for Insect Management*, Ridgway, R. L., Silverstein, R. M., and Inscoe, M. N., Eds., Marcel Dekker, New York, 1990, chap. 11.
7. Batiste, W. C., Berlowitz, A., Olson, W. H., DeTar, J. E., and Joos, J. L., Codling moth: estimating time of first egg hatch in the field—a supplement to sex-attractant traps in integrated control, *Environ. Entomol.*, 2, 387, 1973.
8. Welch, S. M., Croft, B. A., Brunner, J. F., and Michels, M. F., PETE: an extension phenology modeling system for management of multi-species pest complex, *Environ. Entomol.*, 7, 482, 1978.
9. Brunner, J. F., Hoyt, S. C., and Wright, A. M., Codling moth control—a new tool for timing sprays, *Wash. St. Univ. Coop Ext. Bull.*, 1072, 1982.
10. Beers, E. H. and Brunner, J. F., Implementation of the codling moth phenology model on apples in Washington State, U.S.A., *Acta Phytopathol. Entomol. Hungarica*, 27(1–4), 97, 1992.
11. Riedl, H. and Croft, B. A., *Management of the Codling Moth in Michigan*, Mich., Res. Rep. 337, Michigan State University Agric. Exp. Stn., East Lansing, 1978.
12. Hogmire, H. W., Jr., Baugher, T. A., Barrat, J. G., and Kotcon, J. B., *Integrated Orchard Management in West Virginia*, OM104, West Virginia University Coop. Ext. Serv., Wheeling, 1986, p. 26.
13. Riedl, H. and Croft, B. A., A study of phermone trap catches in relation to codling moth (Lepidoptera: Olethreutidae) damage, *Can. Entomol.*, 106(5), 525, 1974.

14. Paradis, R. O. and Comeau, A., Piégeage de la pyrale de la pomme, *Laspeyresia pomonella* (L.), dans les vergers du sud-ouest du Quebec au moyen d'une pheromone sexuelle synthetique, *Ann. Soc. Entomol. Quebec*, 17, 7, 1972.
15. Madsen, H. F. and Vakenti, J. M., Codling moth: female-baited and synthetic pheromone traps as population indicators, *Environ. Entomol.*, 1(5), 554, 1972.
16. Madsen, H. F. and Vakenti, J. M., Codling moth: use of Codlemone-baited traps and visual detection of entries to determine need of sprays, *Environ. Entomol.*, 2(6), 677, 1973.
17. Brunner, J. F., unpublished data, 1991.
18. Reissig, W. H., Biology and control of the obliquebanded leafroller on apples, *J. Econ. Entomol.*, 71, 804, 1978.
19. Reissig, W. H., Stanley, B. H., and Hebding, H. E., Azinphosmethyl resistance and weight-related response of obliquebanded leafroller (Lepidoptera: Tortricidae) larvae to insecticides, *J. Econ. Entomol.*, 79, 329, 1986.
20. Roelofs, W. L. and Tette, J. P., Sex pheromone of the obliquebanded leafroller moth, *Nature*, 226, 1172, 1970.
21. Vincent, C., Mailloux, M., Hagley, E. A. C., Reissig, W. H., Coli, W. M., and Hosmer, T. A., Monitoring the codling moth (Lepidoptera: Olethreutidae) and the obliquebanded leafroller (Lepidoptera: Tortricidae) with sticky and nonsticky traps, *J. Econ. Entomol.*, 83, 434, 1990.
22. Knodel, J. J. and Agnello, A. M., Field comparison of nonsticky and sticky pheromone traps for monitoring fruit pests in western New York, *J. Econ. Entomol.*, 83, 197, 1990.
23. Onstad, D. W., Reissig, W. H., and Shoemaker, C. A., Phenology and management of the obliquebanded leafroller (Lepidoptera: Tortricidae) in apple orchards, *J. Econ. Entomol.*, 78, 1455, 1985.
24. Agnello, A., Kovach, J., Nyrop, J., Reissig, H., and Wilcox, W., *Simplified Integrated Management Program*, IPM No. 201c, Cornell Coop. Ext., Ithaca, NY, 1991.
25. Knight, A. L., Hull, L., Rajotte, E., Hogmire, H., Horton, D., Polk, D., Walgenbach, J., Weires, R., and Whalen, J., Monitoring azinphosmethyl resistance in adult male *Platynota idaeusalis* (Lepidoptera: Tortricidae) in apple from Georgia to New York, *J. Econ. Entomol.*, 83, 329, 1990.
26. Knight, A. L. and Hull, L. A., Area-wide populations dynamics of *Platynota idaeusalis* (Lepidoptera: Tortricidae) in southcentral Pennsylvania pome and stone fruits, *Environ. Entomol.*, 17, 1000, 1988.
27. Meagher, R. L., Jr. and Hull, L. A., Site selection for oviposition and larval feeding by the tufted apple bud moth (Lepidoptera: Tortricidae) on apple in Pennsylvania, *J. Entomol. Sci.*, 26, 149, 1991.
28. Hull, L. A. and Rajotte, E. G., Effects of tufted apple bud moth (Lepidoptera: Tortricidae) injury on quality and storageability of processing apples, *J. Econ. Entomol.*, 82, 1721, 1988.
29. Meagher, R. L., Jr. and Hull, L. A., Predicting apple injury caused by *Platynota idaeusalis* (Lepidoptera: Tortricidae) from summer brood sampling, *J. Econ. Entomol.*, 79, 620, 1986.
30. Hull, L. A. and Meagher, R. L., Jr., Can we predict TABM damage at harvest?, *Mountaineer Grower*, 473, 14, 1987.
31. Knight, A. L. and Hull, L. A., Predicting seasonal apple injury by tufted apple bud moth (Lepidoptera: Tortricidae) with early-season sex pheromone trap catches and brood 1 fruit injury, *Environ. Entomol.*, 18, 939, 1989.
32. Brunner, J. F. and Beers, E. H., Pandemis and obliquebanded leafrollers, *Wash St. Univ. Ext. Bull.*, 1582, 1990.
33. Dean, R. W. and Chapman, P. J., Bionomics of the apple maggot in Eastern New York, *Search Agric.*, 3, 1, 1973.
34. AliNiazee, M. T. and Penrose, R. L., Apple maggot in Oregon: a possible new threat to the northwest apple industry, *Bull. Entomol. Soc. Am.*, 27, 245, 1981.
35. Kroening, M. K., Kondratieff, B. C., and Nelson, E. E., Host status of the apple maggot (Diptera: Tephritidae) in Colorado, *J. Econ. Entomol.*, 82, 866, 1989.
36. Jones, V. P. and Davis, D. W., Evaluation of traps for apple maggot populations associated with cherry and hawthorn in Utah, *Environ. Entomol.*, 18, 521, 1989.
37. Jones, V. P., Smith, S. L., and Davis, D. W., Comparing apple maggot adult phenology in eastern and western North America, in *Apple Maggot in the West: History, Biology and Control*, Dowell, R. V., Wilson, L. T., and Jones, V. P., Eds., University of California Press, Oakland, 1990, chap. 6.
38. Laing, J. E. and Heraty, J. M., The use of degree-days to predict emergence of the apple maggot, *Rhagoletis pomonella* (Diptera: Tephritidae), in Ontario, *Can. Entomol.*, 116, 1123, 1984.
39. Trottier, R., Rivard, I., and Neilson, W. T. A., Bait traps for monitoring apple maggot activity and their use for timing control sprays, *J. Econ. Entomol.*, 62, 211, 1975.
40. Messina, F. J. and Jones, V. P., Relationship between fruit phenology and infestation by the apple maggot (Diptera: Tephritidae) in Utah, *Ann. Entomol. Soc. Am.*, 83, 742, 1990.

41. Jones, V. P., Davis, D. W., Smith, S. L., and Allred, D. B., Phenology of apple maggot associated with cherry and hawthorn in Utah, *J. Econ. Entomol.*, 82, 788, 1989.
42. Smith, S. L. and Jones, V. P., Alteration of apple maggot (Diptera: Tephritidae) emergence by cold period duration and rain, *Environ. Entomol.*, 20, 44, 1991.
43. Prokopy, R. J., Visual responses of apple maggot flies, *Rhagoletis pomonella*: orchard studies, *Entomol. Exp. Appl.*, 11, 403, 1968.
44. Prokopy, R. J., Response of apple maggot flies to rectangles of different colors and shades, *Environ. Entomol.*, 1, 720, 1972.
45. Prokopy, R. J., Attraction of *Rhagoletis* flies to red spheres of different sizes, *Can. Entomol.*, 109, 593, 1977.
46. Fein, B. L., Reissig, W. H., and Roelofs, W. L., Identification of apple volatiles attractive to the apple maggot, *J. Chem. Ecol.*, 8, 1473, 1982.
47. Carle, S. A., Averill, A. L., Rule, G. S., Reissig, W. H., and Roelofs, W. L., Variation in host fruit volatiles attractive to apple maggot fly, *Rhagoletis pomonella*, *J. Chem. Ecol.*, 13, 795, 1986.
48. Averill, A. L. and Reissig, W. H., Development and use of volatile-baited traps in the east, in *Apple Maggot in the West: History, Biology and Control*, Dowell, R. V., Wilson, L. T., and Jones, V. P., Eds., University of California Press, Oakland, 1990, chap. 4.
49. Kring, J. B., Red spheres and yellow panels combined to attract apple maggot flies, *J. Econ. Entomol.*, 63, 466, 1970.
50. Jones, V. P., Longevity of apple maggot (Diptera: Tephritidae) lures under laboratory and field conditions in Utah, *Environ. Entomol.*, 17, 704, 1988.
51. Reissig, W. H., Performance of apple maggot traps in various apple tree canopy positions, *J. Econ. Entomol.*, 68, 534, 1975.
52. Drummond, F., Groden, E., and Prokopy, R. J., Comparative efficacy and optimal positioning of traps for monitoring apple maggot flies, *Environ. Entomol.*, 13, 232, 1984.
53. Stanley, B. H., Reissig, W. H., Roelofs, W. L., Schwarz, M. R., and Shoemaker, C. A., Timing treatments for apple maggot control using sticky sphere traps baited with synthetic apple volatiles, *J. Econ. Entomol.*, 80, 1057, 1987.
54. Agnello, A. M., Spangler, S. M., and Reissig, W. H., Development and evaluation of a more efficient monitoring system for apple maggot (Diptera: Tephritidae), *J. Econ. Entomol.*, 83, 539, 1990.
55. Brunner, J. F., personal communication, 1992.
56. Brunner, J. F., Apple maggot in Washington state: a review with special reference to its status in other western states, *Melanderia*, 45, 33, 1987.
57. Cameron, P. J. and Morrison, F. O., Sampling methods for estimating the abundance and distribution of all life stages of the apple maggot, *Rhagoletis pomonella*, *Can. Entomol.*, 106, 1025, 1974.
58. Taylor, L. R., Aggregation, variance and the mean, *Nature (London)*, 189, 732, 1961.
50. Leroux, E. J. and Mukerji, M. K., Notes on the distribution of the immature stages of the apple maggot, *Rhagoletis pomonella* (Walsh) (Diptera: Trypetidae) on apple in Quebec, *Ann. Entomol. Soc. Quebec*, 8, 60, 1963.
60. Averill, A. L. and Prokopy, R. J., Distribution patterns of *Rhagoletis pomonella* (Diptera: Tephritidae) eggs in hawthorn, *Ann. Entomol. Soc. Am.*, 82, 38, 1989.
61. Averill, A. L. and Prokopy, R. J., Residual activity of oviposition-deterring pheromone in *Rhagoletis pomonella* (Diptera: Tephritidae) and female response to infested fruit, *J. Chem. Ecol.*, 13, 167, 1987.
62. Rice, R. E. and Hoyt, S. C., Response of San Jose scale to natural and synthetic sex pheromones, *Environ. Entomol.*, 9, 190, 1980.
63. Hoyt, S. C., Westigard, P. H., and Rice, R. E., Development of pheromone trapping techniques for male San Jose scale (Homoptera: Diaspididae), *Environ. Entomol.*, 12, 371, 1983.
64. Jorgensen, C. D., Rice, R. E., Hoyt, S. C., and Westigard, P. H., Phenology of the San Jose scale (Homoptera: Diaspididae), *Can. Entomol.*, 113, 149, 1981.
65. Rice, R. E. and Jones, R. A., Timing postbloom sprays for peach twig borer (Lepidoptera: Gelichiidae) and San Jose scale (Homoptera: Diaspididae), *J. Econ. Entomol.*, 81, 293, 1988.
66. Westigard, P. H. and Calvin, L. D., Sampling San Jose scale in a pest management program on pear in southern Oregon, *J. Econ. Entomol.*, 70, 138, 1977.
67. Coli, W. M., Morin, G. E., Prokopy, R. J., Becker, C. M., Cooley, D., Manning, W. J., Lord, W. J., and Ladd, E. R., Integrated management of apple pests in Massachusetts and New England, *Coop. Ext. Serv.*, C-169, 1984.
68. Wylie, W. D., Technique in jarring for plum curculio, *J. Econ. Entomol.*, 44, 818, 1951.
69. Le Blanc, J. P. R., Hill, S. B., and Paradis, R. O., Oviposition in scout-apples by plum curculio, *Conotrachelus nenuphar* (Herbst) (Coleoptera: Curculionidae), and its relationship to subsequent damage, *Environ. Entomol.*, 13, 286, 1984.

70. Anon., Management guide for low-input sustainable apple production, *USDA Northeast LISA Apple Production Project*, Cornell University, Rodale Research Center, Rutgers University, University of Massachusetts, University of Vermont, 1990.
71. Leroux, E. J., Variation between samples of fruit, and of fruit damages mainly from insect pests, on apple in Quebec, *Can. Entomol.*, 93, 680, 1961.
72. Prokopy, R. J., Coli, W. M., Hislop, R. J., and Hauschild, K. I., Integrated management of insect and mite pests in commercial apple orchards in Massachusetts, *Environ. Entomol.*, 9, 529, 1980.
73. Le Blanc, J. P. R., Hill, S. B., and Paradis, R. O., Essais de piégeage du charancon de la prune, *Conotrachelus nenuphar* (Hbst.) (Coleoptera: Curculionidae), dans une pommeraie du sud-ouest du Quebec, *Ann. Entomol. Soc. Quebec*, 26, 182, 1981.
74. Owens, E. D., Hauschild, K. I., Hubbel, G. L., and Prokopy, R. J., Diurnal behavior of plum curculio (Coleoptera: Curculionidae) adults within host trees in nature, *Ann. Entomol. Soc. Am.*, 75, 357, 1982.
75. Parker, B. L. and Hauschild, K. I., A bibliography of the tarnished plant bug, *Lygus lineolaris* (Hemiptera: Miridae), on apple, *Bull. Entomol. Soc. Am.*, 21(2), 119, 1975.
76. Prokopy, R. J. and Hubbell, G. L., Susceptibility of apple to injury by tarnished plant bug adults, *Environ. Entomol.*, 10, 977, 1981.
77. Hauschild, K. I. and Parker, B. L., Seasonal development of the tarnished plant bug on apple in Vermont, *Environ. Entomol.*, 5(4), 675, 1976.
78. Boivin, G., Stewart, R. K., and Rivard, I., Sticky traps for monitoring phytophagous mirids (Hemiptera: Miridae) in apple orchards in southwestern Quebec, *Environ. Entomol.*, 11, 1067, 1982.
79. Prokopy, R. J., Adams, R. G., and Hauschild, K. I., Visual responses of tarnished plant bug *Lygus lineolaris* adults on apple, *Environ. Entomol.*, 8, 202, 1979.
80. Prokopy, R. J., Hubbell, G. L., Adams, R. G., and Hauschild, K. I., Visual monitoring trap for tarnished plant bug adults on apple, *Environ. Entomol.*, 11, 200, 1982.
81. Michaud, O. D., Boivin, G., and Stewart, R. K., Economic threshold for tarnished plant bug (Hemiptera: Miridae) in apple orchards, *J. Econ. Entomol.*, 82(6), 1722, 1989.
82. Stone, J. D. and Pedigo, L. P., Development and economic injury level of the green cloverworm on soybean in Iowa, *J. Econ. Entomol.*, 65, 197, 1972.
83. Henderson, C. F. and McBurnie, H. V., Sampling Technique for Determining Populations of the Citrus Red Mite and Its Predators, USDA Circ. 671, USDA, Washington, DC, 1943, 1.
84. Tanigoshi, L. K., Hoyt, S. C., and Croft, B. A., Basic biology and management components for mite pests and their natural enemies, in *Integrated Management of Insect Pests of Pome and Stone Fruits*, Croft, B. A. and Hoyt, S. C., Eds., John Wiley & Sons, New York, 1983, chap. 6.
85. Zahner, P. and Baumgaertner, J., Sampling statistics for *Panonychus ulmi* (Koch) (Acarina, Tetranychidae) and *Tetranychus urticae* Koch (Acarina, Tetranychidae) feeding on apple trees, *Res. Popul. Ecol.*, 26, 97, 1984.
86. Collyer, E., Phytophagous mites and their predators in New Zealand orchards, *N.Z. J. Agric. Res.*, 7, 551, 1964.
87. Jones, V. P., Developing sampling plans for spider mites: those that don't remember the past may have to repeat it, *J. Econ. Entomol.*, 83, 1656, 1990.
88. Tanigoshi, L. K., Browne, R. W., and Hoyt, S. C., A study on the dispersion pattern and foliage injury by *Tetranychus mcdanieli* in simple apple ecosystems, *Can. Entomol.*, 107, 439, 1975.
89. Wilson, L. T. and Room, P. M., Clumping patterns of fruit and arthropods in cotton with implications for binomial sampling, *Environ. Entomol.*, 12, 50, 1983.
90. Bliss, C. I. and Fisher, R. A., Fitting the negative binomial distribution to biological data, *Biometrics*, 9, 176, 1953.
91. Chant, D. A. and Muir, R. C., A comparison of the imprint and brushing machine methods for estimating the numbers of the fruit tree red spider mite, *Metatetranychus ulmi* (Koch) on apple leaves, *Rep. E. Malling Res. Stn. (A)*, 1954, 141, 1955.
92. Pielou, D. P., Contagious distribution in the European red mite, *Panonychus ulmi*, and a method of grading population densities from a count of mite-free leaves, *Can. J. Zool.*, 38, 645, 1960.
93. Sekita, N. and Yamada, M., Applicability of a new sequential sampling method in the field population surveys, *Appl. Entomol. Zool.*, 8, 8, 1973.
94. Iwao, S., A new regression method for analyzing the aggregation pattern of animal populations, *Res. Popul. Ecol.*, 10, 1, 1968.
95. Kuno, E., A new method of sequential sampling to obtain the population estimates with a fixed level of precision, *Res. Popul. Ecol.*, 11, 127, 1969.
96. Croft, B. A., Welch, S. M., and Dover, M. J., Dispersion statistics and sample size estimates for populations of the mite species *Panonychus ulmi* and *Amblyseius fallacis* on apple, *Environ. Entomol.*, 5, 227, 1976.

97. Taylor, L. R., Assessing and interpreting the spatial distributions of insect populations, *Annu. Rev. Entomol.*, 29, 321, 1984.
98. Mowery, P. D., Hull, L. A., and Asquith, D. L., Two new sampling plans for European red mite surveys on apple utilizing the negative binomial distribution, *Environ. Entomol.*, 9, 159, 1980.
99. Nyrop, J. P., Agnello, A. M., Kovach, J., and Reissig, W. H., Binomial sequential classification sampling plans for European red mite (Acari: Tetranychidae) with special reference to performance criteria, *J. Econ. Entomol.*, 82, 482, 1989.
100. Nyrop, J. P. and Binns, M. R., Algorithms for computing operating characteristic and average sample number functions for sequential sampling plans based on binomial count models and revised plans for European red mite (Acari: Tetranychidae) on apple, *J. Econ. Entomol.*, 85, 1253, 1992.
101. Hull, L. A. and Beers, E. H., Validation of injury thresholds for European red mite (Acari: Tetranychidae) on 'Yorking' and 'Delicious' apple, *J. Econ. Entomol.*, 83(5), 2026, 1990.
102. Beers, E. H. and Hull, L. A., Timing of mite injury affects the bloom and fruit development of apple, *J. Econ. Entomol.*, 83(2), 547, 1990.
103. Beers, E. H., Hull, L. A., and Grimm, J. W., Relationships between leaf:fruit ratio and varying levels of European red mite stress on fruit size and return bloom of apple, *J. Am. Soc. Hort. Sci.*, 112(4), 608, 1987.
104. Travis, J. W., Crassweller, R. M., Greene, G. M., Rajotte, E. G., Hull, L. A., Halbrendt, J., Brittingham, M. C., Kelly, J., Hock, W. K., Heinemann, P., Daum, D. R., Clarke, G., Harper, J. K., Becker, J. C., Serotkin, N., Chambers, R. D., and Jung, C., *Tree Fruit Production Guide, 1992–1993*, Pennsylvania State University, College of Agricultural Science, Philadelphia, 1992.
105. Herbert, H. J. and Butler, K. P., Distribution of phytophagous and predacious mites on apple trees in Nova Scotia., *Can. Entomol.*, 105, 271, 1973.
106. Herbert, H. J. and Butler, K. P., Sampling systems for European red mite, *Panonychus ulmi*, eggs on apple in Nova Scotia, *Can. Entomol.*, 105, 1519, 1973.
107. Croft, B. A. and Hoying, S. A., Competitive displacement of *Panonychus ulmi* (Acarina: Tetranychidae) by *Aculus schlechtendali* (Acarina: Eriophydae) in apple orchards, *Can. Entomol.*, 109, 1025, 1977.
108. Easterbrook, M. A., The life-history of the eriophyid mite *Aculus schlechtendali* on apples in south-east England, *Ann. Appl. Biol.*, 91, 287, 1979.
109. Beers, E. H., Grove, G. G., Williams, K. M., Parker, R., Askham, L. R., and King, E. P., *1991 Crop Protection Guide for Tree Fruits in Washington*, EB0419, Washington State University Coop. Ext., Spokane, 1991.
110. Hull, L. A., Beers, E. H., and Grimm, J. W., Action thresholds for arthropod pests of apple, in *Integrated Pest Management on Major Agricultural Systems*, Texas Agric. Exp. Stn. MP-1616, Frisbie, R. E. and Adkisson, P. L., Eds., 1986, 274.
111. Hoyt, S. C., Population studies of five mite species on apple in Washington, in *Proceedings of the 2nd International Congress of Acarology, Sutton Bonington, England, 19th–25th July, 1967*, Evans, G. O., Ed., Akadémiai Kiadó, Budapest, 1969, 117.
112. Wardlow, L. R. and Jackson, A. W., Comparison of laboratory methods for assessing numbers of apple rust mite (*Aculus schlechtendali*) overwintering on apple, *Plant Pathol.*, 33, 57, 1984.
113. Croft, B. A., Integrated Control of Apple Mites, *Mich. St. Univ. Coop. Ext. Ser. Bull.*, E-825, 1975.
114. Jones, V. P., Sampling and dispersion of the twospotted spider mite (Acari: Tetranychidae) and the western orchard predatory mite (Acari: Phytoseiidae) on tart cherry, *J. Econ. Entomol.*, 83, 1376, 1990.
115. Zalom, F. G., Hoy, M. A., Wilson, L. T., and Barnett, W., Sampling mites in almonds. II. Presence-absence sequential sampling for *Tetranychus* mite species, *Hilgardia*, 52, 14, 1984.
116. Wilson, L. T., Hoy, M. A., Zalom, F. G., and Smilanick, J. M., Sampling mites in almonds. I. Within-tree distribution and clumping pattern of mites with comments on predator-prey interactions, *Hilgardia*, 52, 1, 1984.
117. Jones, V. P., unpublished data, 1992.
118. Hoyt, S. C. and Caltagirone, L. E., The developing programs of integrated control of pests of apples in Washington and peaches in California, in *Biological Control*, Huffaker, C. B., Ed., Plenum Press, New York, 1971, chap. 18.
119. Nyrop, J. P., Sequential classification of prey:predator ratios with application to European red mite (Acari: Tetranychidae) and *Typhlodromus pyri* (Acari: Phytoseiidae) in New York apple orchards, *J. Econ. Entomol.*, 81, 14, 1988.
120. Colburn, R. B. and Asquith, D., Observations on the morphology and biology of the ladybird beetle *Stethorus punctum*, *Ann. Entomol. Soc. Am.*, 64, 1217, 1971.
121. Hull, L. A., Asquith, D., and Mowery, P. D., The functional responses of *Stethorus punctum* to densities of the European red mite, *Environ. Entomol.*, 6, 85, 1977.

122. Asquith, D. and Colburn, R. B., Integrated pest management in Pennsylvania apple orchards, *Bull. Entomol. Soc. Am.*, 17, 89, 1971.
123. Asquith, D. and Hull, L. A., Integrated pest management systems in Pennsylvania apple orchards, in *Pest Management Programs for Deciduous Tree Fruits and Nuts*, Boethel, D. J. and Eikenbary, R. D., Eds., Plenum Press, New York, 1979, chap. 6.
124. Mowery, P. D., Asquith, D., and Bode, W. M., Computer simulation for predicting the number of *Stethorus punctum* needed to control the European red mite in Pennsylvania apple trees, *J. Econ. Entomol.*, 68, 250, 1975.
125. Brunner, J. F. and Howitt, A., *Tree Fruit Insects*, Publ. 63, Cooperative Extension Service, Michigan State University, East Lansing, 1981.
126. Hull, L. A. and Starner, V. R., Effectiveness of insecticide applications timed to correspond with the development of rosy apple aphid on apple, *J. Econ. Entomol.*, 76, 594, 1983.
127. Pickel, C., Bethell, R. S., Teviotdale, B. L., and Gubler, D., *Apple Pest Management Guidelines*, Univ. Calif. Pest Manage. Guidelines Pub. 12, 1990, 24.
128. Hull, L. A. and Barrett, B. A., Economic injury levels and action thresholds for apple insect and mite pests, *Penn Fruit News*, 72, 60, 1992.
129. Anon., *Rosy Apple Aphid*, Factsheet 91-048, Ministry of Agriculture and Food, Ontario, 1991.
130. Hoyt, S. C., Leeper, J. R., Brown, G. C., and Croft, B. A., Basic biology and management components for insect IPM, in *Integrated Management of Insect Pests of Pome and Stone Fruits*, Croft, B. A. and Hoyt, S. C., Eds., John Wiley & Sons, New York, 1983, chap. 5.
131. Anon., *Orchard Outlook Newsletter*, *Rosy Apple Aphid*, Nova Scotia Department of Agriculture Market, Halifax, 1991, 8.
132. Pfeiffer, D. G., Brown, M. W., and Varn, M. W., Incidence of spirea aphid (Homoptera: Aphididae) in apple orchards in Virginia, West Virginia, and Maryland, *J. Entomol. Sci.*, 24, 145, 1989.
133. Oatman, E. R. and Legner, E. F., Bionomics of the apple aphid, *Aphis pomi*, on young nonbearing apple trees, *J. Econ. Entomol.*, 54, 1034, 1961.
134. Madsen, H. F., Westigard, P. H., and Falcon, L. A., Evaluation of insecticides and sampling methods against the apple aphid, *Aphis pomi*, *J. Econ. Entomol.*, 54, 892, 1961.
135. Jokinen, D. P., Spatial Distribution of *Aphis pomi* (De Geer) and the Predator *Aphidoletes aphidimyza* (Rondani) Relative to Growth in the Apple Tree, M.S. thesis, Michigan State University, East Lansing, 1980.
136. Hull, L. A. and Grimm, J. W., Sampling schemes for estimating populations of the apple aphid, *Aphis pomi* (Homoptera: Aphididae), on apple, *Environ. Entomol.*, 12, 1581, 1983.
137. Madsen, H. F., Peters, H. F., and Vakenti, J. M., Pest management experiences in six British Columbia apple orchards, *Can. Entomol.*, 107, 873, 1975.
138. Hamilton, G. C., Swift, F. C., and Marini, R., Effect of *Aphis pomi* (Homoptera: Aphididae) density on apples, *J. Econ. Entomol.*, 79, 471, 1986.
139. Nyrop, J. P., Reissig, W. H., Agnello, A. M., and Kovach, J., Development and evaluation of a control decision rule for first-generation spotted tentiform leafminer (Leidoptera: Gracillariidae) in New York apple orchards, *Environ. Entomol.*, 19, 1624, 1990.
140. Jones, V. P., Binomial sampling plans for tentiform leafminer (Lepidoptera: Gracillariidae) on apple in Utah, *J. Econ. Entomol.*, 84, 484, 1991.
141. Roelofs, W. L., Reissig, W. H., and Weires, R. W., Sex attractant for the spotted tentiform leaf miner moth, *Lithocolletis blancardella*, *Environ. Entomol.*, 6, 373, 1977.
142. Vincent, C., Mailloux, M., and Hagley, E. A. C., Nonsticky pheromone-baited traps for monitoring the spotted tentiform leafminer (Lepidoptera: Gracillariidae), *J. Econ. Entomol.*, 79, 1666, 1986.
143. Dutcher, J. D. and Howitt, A. J., Bionomics and control of *Lithocolletis blancardella* in Michigan, *J. Econ. Entomol.*, 71, 736, 1978.
144. Maier, C. T., Seasonal occurrence, abundance, and leaf damage of the apple blotch leafminer, *Phyllonorycter crataegella*, in Connecticut apple orchards, *Environ. Entomol.*, 10, 645, 1981.
145. Green, T. A. and Prokopy, R. J., Visual monitoring trap for the apple blotch leafminer moth, *Phyllonorycter crataegella* (Lepidoptera: Gracillariidae), *Environ. Entomol.*, 15, 562, 1986.
146. Elsner, E. A. and Beers, E. H., Yellow sticky board trapping of *Typhlocyba pomaria*, *Edwardsiana rosae*, and *Anagrus* sp. in central Washington, *Melanderia*, 46, 58, 1988.
147. Beers, E. H. and Elsner, E. A., unpublished data, 1988.
148. Seyedoleslami, H. and Croft, B. A., Spatial distribution of overwintering eggs of the white apple leafhopper, *Typhlocyba pomaria*, and parasitism by *Anagras* [sic] *epos*, *Environ. Entomol.*, 9, 624, 1980.
149. Beers, E. H. and Elsner, E. A., Phenology of white apple leafhopper eggs and a mymarid egg parasitoid in central Washington, 1987–1988, *Melanderia*, 46, 49, 1988.

150. Whalon, M. E. and Croft, B. A., Apple IPM implementation in North America, *Annu. Rev. Entomol.*, 29, 435, 1984.
151. Hogmire, H. W., Baugher, T. A., and Ingle, M., Development of a sampling plan and application of a grading scheme for determining apple packout losses, *Hortic. Sci.*, 24, 628, 1989.
152. Russo, J. M. and Rajotte, E. G., A theoretical grading scheme for production decision making: an application to fresh apples, *Pennsylvania Agric. Exp. Stn. Bull.*, 844, 1983.
153. Beers, E. H., Control strategies for leafminers and leafhoppers revisited, in *New Directions in Tree Fruit Pest Management*, Williams, K. W., Beers, E. H., and Grove, G. G., Eds., Good Fruit Grower, Yakima, WA, 1991, chap. 15.
154. Gut, L., unpublished data, 1992.
155. Hoyt, S. C., Tanigoshi, L. K., and Browne, R. W., Economic injury level studies in relation to mites on apple, *Recent Adv. Acarol.*, 1, 3, 1979.
156. Halbert, S. E. and Voegtlin, D. J., Morphological differentiation between *Aphis spiraecola* and *Aphis pomi* (Homoptera: Aphididae), *Great Lakes Entomol.*, 25(1), 1, 1992.

Chapter 15

SAMPLING ARTHROPOD PESTS IN CITRUS

J. Daniel Hare

TABLE OF CONTENTS

I. General Considerations in the Management of Citrus Pests 418

II. General Consideration of Sampling Programs in Citrus 419

III. Interrelationships between the Development of Treatment Thresholds and the Design of Sampling Programs 420
    A. Direct, Cosmetic Pests 420
    B. Indirect Pests 421
    C. Secondary Induced Pests 422

IV. Representative Programs 422
    A. California Red Scale 422
    B. Citrus Thrips 424
    C. Citrus Rust Mite 425
    D. Natural Enemies 426
        1. *Euseius tularensis* 426
        2. Parasitoids of California Red Scale 427

V. Future Needs 427

Acknowledgments 428

References 428

I. GENERAL CONSIDERATIONS IN THE MANAGEMENT OF CITRUS PESTS

Citrus is grown worldwide in tropical and subtropical regions, with both humid and relatively arid climates. The absence of killing temperatures in winter plays the most important role in limiting the areas in which citrus can be commercially grown.[1] Worldwide, about 60% of all citrus is produced for the fresh market, with the remainder sold for processing, mainly as juice or juice concentrate.[2] The proportion of fruit grown for the fresh market, however, varies among growing regions. Less than 10% of Florida oranges are sold on the fresh market, while more than 68% of California oranges are marketed as fresh fruit.[2,3] Brazil is the largest producer of oranges in the world at present, and nearly two thirds of this crop is processed.[2] In contrast, more than 80% of citrus grown in the Mediterranean basin is grown for the fresh market.[2]

Market value of fresh citrus is determined largely by fruit size and external appearance. As long as minimal ripeness criteria are reached, internal fruit composition, flavor, or nutritional quality have relatively little influence on crop value. Total yield can be a poor predictor of the value of fresh market citrus when reductions in total yield are accompanied by increases in fruit size, as is often the case when trees are treated by fruit-thinning agents[4] or are subjected to mild stresses.[5] Thus, pest management of fresh market citrus focuses more upon preserving the appearance of the fruit rather than upon maximizing total production.

The most important pests of fresh-market citrus are a variety of homopteran insects including several species of both armored and soft scales, mealybugs, and whiteflies.[6-8] Various eryiophyid and tetranychid mite species may be important in some areas. A few citrus thrips species are important pests of citrus, especially in California's San Joaquin Valley and South Africa.

In nearly all instances, the major economic impact of these species is to downgrade the surface appearance of the fruit and reduce or eliminate its value on the fresh fruit market. For example, at average prices over the past 5 years, a reduction in one grade of California oranges would reduce their value by 19% in the smaller size categories and by more than 40% in the larger size classes.[9] Further surface damage can result in culling and sale for products. The economic return to the California grower for fruit culled for external defects barely covers the cost of harvesting and grading and leaves only a few cents per box to cover other costs of production.

Unlike the cosmetic pests of other crops, the feeding activity of these arthropods causes no substantial degradation of the tissues consumed by humans, either directly or indirectly via the introduction of pathogenic or saprophytic microorganisms. In fact, California 'navel' oranges that were scarred by the citrus thrips, *Scirtothrips citri* (Moulton), matured sooner and were sweeter than unscarred fruit.[10] These results, coupled with the fact that the rind must be discarded before consumption by humans, has raised serious questions as to whether the high cosmetic standards that create the need for much of current citrus pest management are biologically rational.[10,11]

Several additional factors may complicate determinations of the actual economic status of many of these pests. Perhaps the most important is that the level of damage tolerated by consumers and packers varies with the volume of fruit available for sale, with moderately damaged fruits escaping downgrading in periods of short supply. Such periods can result from unanticipated, weather-related reductions in production from other citrus-growing regions, or from more predictable variation in supply vs. demand at various times through the harvest season. The result of the application of variable grading standards obviously complicates pest management decisions in citrus, for the pest density that may cause substantial economic loss if the crop were harvested in mid-season may have trivial economic impact if harvests were earlier or later.

External appearance of citrus fruit obviously contributes substantially less to the value of citrus grown for processing than for the fresh market. For processing, the most important factors contributing to crop value are total yield, total soluble solid (mostly sugars) content, acid content, and total juice content.[12] Pest management concerns of citrus for processing therefore focus on the effect of pests on citrus trees' overall productivity.

It was commonly believed that severe infestations of the cosmetic citrus pests could debilitate the tree sufficiently that total fruit production also would be reduced, but sound, experimental evidence supporting such assertions actually is quite meager. Such evidence is more difficult and expensive to obtain than for annual crops because, in perennial crops, the damage caused by potential pests may take several seasons of injury before it is realized.[13-18] Nevertheless, what data are available do not demonstrate that the various cosmetic pests of citrus have any effect beyond damaging marketable fruits.

Historically, pest management guidelines for citrus have tended to be conservative, because a potential 20 to 40% loss in crop value from fruit scarring was substantially more than the cost of a pesticide application to prevent it. Pesticide applications were perceived to be a form of cheap insurance; the potential losses that might accrue from a pesticide application withheld were expected to be substantially greater than the cost of making an unneeded application. Treatment thresholds were set at conservative levels largely to avoid making a costly mistake.

Recently, however, several changes in the economics of pest management have resulted in a reconsideration of the cost/benefit ratios of several pesticide applications. Among these have been the loss of a number of insecticides due to the development of insect resistance and/or nonrenewal of registrations. Additionally, growers are being held relatively more accountable both for the safety of their produce and for any deleterious consequences to nontarget organisms, including humans, within the vicinity of their groves. Finally, a considerable increase in entomologists' knowledge of the biology of citrus trees, their herbivores, and the herbivores' natural enemies has provided the needed background to permit economically acceptable biological control of many of the key pests of citrus.

All of these recent changes have altered several conventional sampling programs for citrus pests as the need for new programs, optimized for different, often higher, pest densities, became evident. Simultaneously, the development of newer, more precise sampling procedures allowed for more precise measurements of pest density and their actual impact on tree fruit production and crop value.

II. GENERAL CONSIDERATION OF SAMPLING PROGRAMS IN CITRUS

Sampling programs for large trees are inherently more complex than for annual, herbaceous plants because of the wider variety of habitats that trees offer. Metapopulations of a pest species in different parts of the tree may contribute to that species' overall pest status in complex ways. California red scale (*Aonidiella aurantii* (Maskell)), for example, utilizes all above-ground parts of the citrus tree, including leaves, green stems, and woody bark, in addition to fruit. It is only on the fruit that California red scale causes economic loss, however. The metapopulations on the bark substrates, however, contribute substantially not only to the recruitment of scale on fruits, but also to the persistence of the interaction between California red scale and its most important biological control agent, the parasitic wasp, *Aphytis melinus* DeBach.[19-21] Refinements of treatment thresholds for California red scale may require consideration not only of the density of scale on fruits, but also estimates of densities on other substrates and assessments of the natural enemy populations.

Substantial heterogeneity in pest populations can exist even within different parts of the leaf canopy. Pest densities may vary with height and exposure as well as orientation. Citrus thrips populations, for example, begin to develop on the portions of the leaf canopy facing the sun, presumably because such leaves receive more insulation and are warmer early in the season relative to leaves in the shaded portion of the canopy. Populations may reach highest densities on the shaded quadrants, however, because new leaves and stems have a longer period of growth in the absence of exposure to direct sun.[22] Sampling programs either must be structured to assess such heterogeneity, or the pattern of heterogeneity must be well enough understood, so that assessments taken in only a single part of the tree can provide accurate information for pest management decisions. These points are illustrated in the various case studies discussed in the following section.

III. INTERRELATIONSHIPS BETWEEN THE DEVELOPMENT OF TREATMENT THRESHOLDS AND THE DESIGN OF SAMPLING PROGRAMS

Sampling procedures have two general functions in pest management programs. The first is largely a research function to assess accurately the pest population densities in order to obtain quantitative data on the relationship between pest density and crop loss. After such research has identified the critical densities (i.e., the "treatment threshold") for pest management decisions, then sampling programs can be refined and simplified to determine whether the pest population is at the critical density or not. Unlike the needs of a research program, it is less important to quantify the exact population density in a pest management program than to rapidly and economically determine where the pest population density is relative to its treatment threshold. The needs of sampling programs for research therefore differ fundamentally from those for the implementation of that research in the need for accurate quantitative data over a wide range of pest densities.

A. DIRECT, COSMETIC PESTS

Treatment thresholds for the various cosmetic pests are determined by the packing house or marketing cooperative and reflect the perception that consumers demand unblemished fruit. External appearance is of fundamental importance when fruit is exported to foreign fresh markets, thus the tolerance for insect damage on exported fruit is near zero.

For example, a 1% culling rate from citrus thrips scarring is the highest that growers will accept from *Scirtothrips aurantii* Faure in South Africa.[23] In California, 1% downgrading of lemons or 5% of oranges severely scarred by *S. citri* is considered economically significant.[24,25]

A similarly low economic tolerance exists for California red scale, although fruit quality usually is not downgraded until the density exceeds ten scales per fruit at harvest.[7,26] Again, because of the long delay between the time that fruit are vulnerable to infestation and the time that they are harvested, treatment based only upon visual searches are often recommended when as few as 10% of fruit are infested with at least one live scale.[7,27] Sampling programs for these and other pests causing such cosmetic injury obviously must be optimized for early, reliable detection of a low infestation of susceptible fruit.

B. INDIRECT PESTS

A number of other insect and mite species have been considered to be indirect pests, in that by feeding on foliage, they were presumed to reduce the tree's photosynthetic capacity, thus the quantity of photosynthate available for fruit production and growth. Primary among these in California has been the citrus red mite *Panonychus citri* (McGregor). At one time, this mite was thought to be the most important pest of citrus in California,[6] although California red scale and citrus thrips usually have been given higher priority. Like most other tetranychid mites, *P. citri* feeds on leaves, kills leaf cells, and removes their contents. In the absence of detailed information on the quantitative relationship between mite damage and tree yield, treatment guidelines were based upon minimizing the visible injury to leaves.

The nominal threshold of two adult females per leaf was most widely used,[28] although some pest control advisors recommended treatments at densities as low as one adult female per two leaves.[27] These thresholds were developed in southern California, where mite injury in late summer and fall is of most concern, but they had been implemented throughout the state without validation.

The first step toward assessing the actual impact of feeding by *P. citri* on citrus production required a rapid estimate of mite populations in the field. This was facilitated by research showing that *P. citri* is uniformly distributed within the tree canopy,[28,29] and that the total mite population can be predicted accurately from the density of the adult females, the largest active stage.[30] These two findings simplify population monitoring by minimizing concerns of location (height, interior vs. exterior location, compass direction) on selection of leaves for mite counts, and by compressing the sampling time by ignoring the males and immatures.

With the use of these monitoring tools, we found that *P. citri* populations with peak densities of over 14 adult females per leaf caused no discernable reduction in rates of leaf photosynthesis,[31,32] and populations approaching 10 adult females per leaf caused no consistent economic loss on orange. While these high mite populations caused reductions in total yield of up to 10%, yield reductions were accompanied by an increase in the size of remaining fruit. Because larger fruit are, on average, more valuable than smaller fruit,[9] the increase in fruit size more than compensated for the reduction in total yield. Often, the crop from untreated trees was more valuable than that from acaricide-treated trees even before the cost of acaricide treatments were subtracted.[5,9] The increases in fruit size probably results from a slightly increased rate of fruit abortion during the early summer "June drop" period, causing photosynthate to be allocated to a smaller number of fruit, thereby allowing those fruit to grow to a larger size. In short, rather than being an economic pest, *P. citri* has offsetting beneficial roles as a fruit thinning agent and as an alternate food source for some of the predaceous mite species beneficial in suppression of citrus thrips.[5]

A relationship between the percent of leaves infested with mites and the absolute numbers of mites per leaf also was developed to facilitate scouting on a commercial scale.[28,30] Simple presence-absence sampling obviously is more rapid than counting all adult females. These relationships may have somewhat less utility than was anticipated when they were developed, because 100% of all leaves become infested at a density of about eight adult female mites per leaf.[28] This value is below the density that causes consistent economic loss, thus the occurrence of uninfested leaves in this presence-absence sampling scheme is useful only in determining that the resident mite population is still too sparse to be treated.

Treatment thresholds are defined for relatively few of the other indirect pests of citrus, and the lack of this knowledge has hampered the further development of successful biological control programs for the more important cosmetic pests. For example, several lepidopterans commonly feed on citrus in the spring in California, and many of these are thought to cause economic damage. Unfortunately, there are no data relating density to loss in crop value for any of them at the present time. Treatment guidelines have been promulgated for these insects, however, and they are based upon the number of "worms" or caterpillars seen per fixed interval of sampling time.[7]

These thresholds are inadequate not only because their economic reality has yet to be demonstrated, but also because the ability to see and count caterpillars increases with experience. Therefore, it is far from clear how many insecticide applications directed at these lepidopteran pests are justified economically at the present time. Perhaps as important an economic consequence of these treatments against caterpillar populations is the deposition of insecticide residues toxic to effective biological control agents of California red scale. Current research presently is in progress to put treatments of the various lepidopteran pests of citrus in California on a more rational basis.

C. SECONDARY INDUCED PESTS

Citrus has its complement of relatively innocuous arthropod species that occasionally become pests under unusual circumstances. The citrus red mite, for example, is one species that tends to increase following insecticide applications directed at other insect pests.[33,34] In areas where *P. citri* is under good biological control by insect predators, insecticide applications can be used as a research tool in crop loss studies to boost *P. citri* populations to densities higher than they would normally reach.[35]

Of more economic importance than *P. citri* are the various honeydew-secreting homopterans that are induced to pest status by the presence of ants. These include many of the mealybugs and soft scale species, such as the brown soft scale, *Coccus hesperidum* L., the black scale, *Saissetia olea* (Bernard), the wooly whitefly, *Aleurothrixus floccosus* (Maskell), and the citrus mealybug, *Planococcus citri* (Risso). Several ant species, most notably the Argentine ant, *Iridomyrmex humulis* (Mayr), tend these homopterans for their honeydew and protect them from their biological control agents.[36,37] Ants apparently also interfere with the predators and parasitoids of other citrus pests that do not produce honeydew but are within the same tree as the honeydew producers. Thus, increases in densities of the California red scale and citrus red mite also occur in trees tended by ants.[38,39] The need to suppress ant populations in order to keep them from disrupting biological control programs continues to be one of the more difficult problems yet to be solved in citrus pest management.

IV. REPRESENTATIVE PROGRAMS

A. CALIFORNIA RED SCALE

Although California red scales utilize all above-ground tissues of citrus trees, the exterior substrates (fruit, leaves, and green twigs) are most suitable for scale survival and growth. Only the first-instar crawlers and adult males of California red scale are motile; females remain at the site of settling for the duration of their lives. After the third molt, females begin emitting a sex pheromone. Winged male scales emerge coincident with the onset of pheromone production and mate shortly after emergence. Adult males do not live for more than 1 d.[40,41] Crawler production begins several days after the female has mated.

California red scale has from two to four generations per year in California and is active throughout the year. Harsh winter temperatures may kill the younger stages,[42] and differential mortality of the younger stages may promote synchrony of scale populations in the interior citrus-growing regions of California,[43] where winters are more severe than in the coastal citrus districts.

The homogeneity of the scale's distribution within trees may be temperature dependent. In hot climates, e.g., California's San Joaquin Valley, scale densities on the peripheral substrates (e.g., leaves and fruit) are reduced in the most exposed sectors and the upper regions of the leaf canopy.[44,45] In less extreme climates, however, such heterogeneous distributions are not found.[46]

Contemporary sampling programs for California red scale in the San Joaquin Valley utilize traps baited with synthetic pheromone to monitor the seasonal flight activity of male California red scale. Initially, California red scale pheromone traps were white, because male scales were most attracted to white or yellow in the absence of pheromone; color had no effect in the presence of pheromone, however.[47] Some workers are now using yellow pheromone traps to monitor those pests attracted to yellow, e.g., citrus thrips, in addition to California red scale, using the same trap.[48,49] Flight activity is organized into four flight periods,[26] and these flights are used to schedule pesticide applications. Optimal timing occurs during the second half of the first flight or within 2 to 4 weeks after the onset of the second flight period.[50] These time periods coincide with the onset of production of crawlers by females mated during the first and second male flight periods, respectively.

In addition to predicting when pesticide applications should be made, the density of males caught in pheromone traps can be used to predict not only when but if an insecticide application is justified. In research completed before widespread releases of *A. melinus* in the San Joaquin Valley, the percentage of fruit infested at the end of the season was directly related to the density of males caught per trap during the previous summer.[26] This information has been used to determine if the scale population in mid-summer, as indexed by the number of males trapped, is expected to cause a level of fruit infestation at harvest high enough to warrant treating. Unfortunately, this relationship may only hold true in the absence of *A. melinus* and other effective natural enemies. Because *A. melinus* kills more female than male scales, the regression equation relating fruit infestation at harvest to the capture of males would overestimate fruit infestation if *A. melinus* was present or was being released.

In theory, pheromone trapping should also aid in scheduling releases of commercially reared *A. melinus*. Female scales are both the most attractive and the most suitable for utilization by *A. melinus* when they are virgin third-instars.[51-53] The onset of this life stage coincides with emergence and flight of male scales, thus the optimal time to release *A. melinus* is at or shortly before the onset of a flight of male California red scale.[54] While developmental degree-day models[53,55] can be useful in predicting optimal release times using the previous male flight as a starting point ("biofix" for degree-day accumulations), the logistical problems in obtaining and releasing sufficient quantities of commercially reared wasps at exactly the right time remain considerable. Therefore, current recommendations call for multiple *Aphytis* releases, made about 2 weeks apart during the scale's critical developmental period.[7]

The relationships between California red scale trap catches and fruit infestation may not apply to cultivars other than 'navel' orange, even in the absence of *A. melinus*. First 'navel' orange fruits are initiated in the spring and harvested in the following fall or winter, so there are no mature fruit available when summer-generation California red scale crawlers are settling. Second, young orange fruits cannot become infested until mid-July in the San Joaquin Valley because of their small size prior to this time. The same situation does not apply to 'Valencia' oranges even in the same area, because harvest occurs about 15 months after fruit set (i.e., the following summer), thus mature fruit are indeed available to colonization by crawlers. While pheromone traps still can be used to monitor and quantify the resident scale population, lower treatment thresholds must be used because of the shorter time interval between crawler production and harvest of 'Valencia' compared to 'navel' oranges.

Lemon fruits in California's coastal regions may become infested almost any time by California red scale due to the more indeterminate nature of flower and fruit production of coastal lemons. Lemon fruit may be initiated at any time during the year, and lemon trees are harvested three to five times annually at irregular intervals. This, plus the activity of established *A. melinus* populations precludes relating density of male scales caught in traps with infestation rates during any particular harvest. Visual inspection of the fruit is used in this locale, and a treatment threshold of 5% of inspected fruit with more than ten live scales commonly is used.[7] Fortunately, biological control of California red scale is substantially better in the California coastal districts than in the inland valleys; thus, the need to monitor and treat California red scale populations is substantially reduced in the coastal regions.

Pheromone traps are also used to monitor populations of California red scale in South Africa,[48,49] but no relationship is yet proven between the density of males caught in traps and fruit infestation levels. In South Africa, variability of infestation levels is introduced in part by the activity of resident *Aphytis* parasitoids, and parasitoid activity is itself differentially affected by insecticide applications targeted against the South African citrus thrips, *S. aurantii*. Thus, California red scale pheromone traps may be best used simply as a population monitoring tool in South Africa.[49] These results also reinforce the idea that as *A. melinus* becomes better established in California's San Joaquin Valley, pheromone traps may require recalibration to accurately predict fruit infestation rates from catches of male scale.

B. CITRUS THRIPS

S. citri is a species indigenous to California and exploited citrus as trees were planted in California. *S. citri* overwinters in the egg stage within green host plant tissues (leaves, green twigs, fruits). First-generation larvae eclose at about the time that citrus trees flush in the spring. The larvae then pass through two feeding stadia followed by two nonfeeding stadia before emerging as adults. These first-generation adults reproduce at about the same time that trees are initiating new fruits (i.e., at "petal-fall"), and as many as five more thrips generations can occur annually. Thrips larvae cause economic damage by feeding beneath the sepals of young fruits. This causes a circular or semicircular surface lesion that continues to grow as the fruit expands, often leaving a circular ring around the sepal end of the mature fruit. Fruit are susceptible to damage by thrips only until they reach about 4 cm in size. Larvae of later generations feed on expanding leaves and leaf buds, causing some disfigurement, but this damage is not of economic importance.[7]

In an extensive study of *S. citri* in California, Grout et al.[22] showed that populations are initiated in the most exposed (southwest) quadrants, and thrips densities in this quadrant peaked first, but that these differences did not persist beyond the second generation. Overall, adult thrips densities were greatest in the northeast quadrant, perhaps because tree shoot growth persisted for a longer time on the shadiest side of the tree.

Several traps have been developed to facilitate the monitoring of citrus thrips. The first was simply a box with a piece of tangle-foot-treated paper at the bottom and a screen at the top. Leaf terminals were shaken against the screen, and all insects dislodged fell to the bottom and were immobilized for later enumeration.[56]

The second was based upon the premise that citrus thrips dropped from the tree and pupated in the soil and litter below the leaf canopy. A two-sided sticky plate was placed on a frame beneath a citrus tree. As thrips larvae dropped from the canopy, a portion of them would be trapped on the upper sticky surface of the plate. The supporting frame enclosed a sample of soil and litter, and emerging adults would be trapped on the lower surface of the plate as they flew upward.[57] Thus, pre- and postpupation individuals could be monitored simultaneously. Simplified versions of this trap have been developed with no loss of effectiveness.[58]

The utility of emergence traps on the ground was called into question by the realization that the majority of thrips pupate in protected locations within the tree,[59] and this led to the design of a new trap that was based upon providing pupation sites within the tree and was constructed from a paper towel.[22]

These traps, and other methods of population assessment, including D-Vac® vacuum sampling, yellow sticky cards suspended within the leaf canopy, and counting thrips directly on immature fruit, were evaluated for their ability or predict levels of fruit scarring at harvest. Unfortunately, none of these methods of population assessment provided consistent correlations with fruit scarring; thus, their value may be limited to providing relative abundance data on different life stages for the monitoring of phenology.[22,25]

Current monitoring programs for commercial citrus in California therefore rely upon direct examination of newly-formed fruit until they reach approximately 4 cm in diam. this occurs approximately six to eight weeks after petal fall. Twenty-five fruits are removed, and the area under the sepals is examined with a hand lens. Treatments are usually warranted if 5% of 'navel' fruit are infested with at least one thrips, if 10% of 'Valencia' fruit are infested, or if 15% of lemon fruit are infested with thrips.[7] These thresholds can be doubled if mites predaceous on thrips are present at densities ≥ 0.2 mites/leaf.[7] Practically, such thresholds relate largely to presence or absence, which may be appropriate, given the current low tolerances of packers for thrips-damaged fruit.

In South Africa, researchers have been somewhat more successful in relating catches of their native citrus thrips, *S. aurantii*, to scarring at harvest using yellow traps of various designs.[23,60] the treatment thresholds are so low, however (e.g., 9 to 20 thrips per set of 3 traps, depending upon the design[23,60]) that the economic benefits of implementing a trapping program may be minimal.

C. CITRUS RUST MITE

The Citrus rust mite, *Phyllocoptruta oleivora* (Ashmead) is the most serious pest of citrus in Florida. It infests all above-ground parts of commercial citrus cultivars, but the infestations on fruit are most important economically.[8,61] Feeding injury to the fruit may take different forms depending upon the cultivar and maturity of the fruit when attacked, which causes the formation of silvered or russeted lesions on the fruit surface.[62] Such scarring has minimal effects upon internal fruit characteristics until more than half of the fruit surface is injured, at which time preharvest fruit drop increases and fruit growth is reduced.[63,64] High citrus rust mite populations feeding upon bark or leaves are thought to reduce tree growth and vitality, although quantitative data are lacking. Therefore, the citrus rust mite is a cosmetic pest for Florida citrus grown for the fresh market,[2] but is a potential indirect pest of citrus grown for processing.

Current recommendations for sampling citrus rust mite in Florida exploit a number of important relationships. First, the percent of fruit surface damaged is linearly related to the density and duration of mite populations ("mite-days") per cm^2 on individual fruits.[65] Because fruit injury becomes visible shortly after being attacked by mites, growers can monitor and treat mites as injury occurs in order to maintain accumulated mite-days below levels necessary to cause economic injury.[8] The calculation of accumulated mite-days per cm^2 is facilitated by the existence of a predictive relationship between the number of hand-lens fields (fruit surface area observed with a 10 × hand lens) infested with citrus rust mites and their overall density.[8] Thus, the expected percent area of fruit surfaced damaged ultimately can be predicted by the proportion of lens fields with at least one citrus rust mite. An alternative way to predict mite density employs the "HB scale" from plant pathology.[66] In this method, the density of citrus rust mites seen in a hand-lens field of view is estimated and assigned a scale value based upon a preexisting frequency distribution of mite densities. Thus, the scout need not determine exactly how many mites show up within the field of view, but rather determine if the number is between 1 and 3, 4 and 6, 7 and 12, and so on. The class intervals of the frequency distribution become wider with increasing mite density. The scaled values then are converted back to mite densities for final population estimation.

Treatment thresholds for the citrus rust mite are substantially higher for fruit destined for processing than for the fresh market.[8] The parasitic fungus *Hirsutella thompsonii* Fisher is an effective natural enemy of the citrus rust mite, and treatment thresholds can be raised based upon the number of infected mites observed.

Current sampling plans recommend sampling 50 fruit per 10-acre block, with little regard to the number of fruit per tree.[8] Sampling plans in the future may be improved by incorporating recent research on the pattern of dispersion of mites within and among fruits. Mites are clumped near the equatorial, shaded parts of the fruit[67] and are more abundant in the shady tree quadrants.[68] Both of these observations are consistent with behavioral observations that mites avoid direct sunlight.[69] Mites are also more abundant in the middle portion of the tree canopy[67] and are aggregated among fruit.[68] Knowledge of these patterns of distribution have been used to develop more precise estimates as to the number of samples to be taken, and how fruit should be chosen to estimate mite density at various levels of precision.[67,68]

D. NATURAL ENEMIES

Despite the low tolerances for cosmetic damage by most citrus pests, control by natural enemies has been economically effective in many cases. In many of the earlier cases, natural enemies of foreign origin were identified and released, and the pest reached an equilibrium density acceptably below economic injury levels. Although this scenario remains the most desired, current approaches toward natural control are likely to emphasize conservation and exploitation of resident natural enemies rather than the discovery and importation of new ones. To determine if a resident natural enemy population has the capacity to suppress pest populations, growers and scouts will require the ability to sample populations of natural enemies as precisely as they do the pests. The following reviews some results of research addressing the need to develop sampling procedures for natural enemies of important citrus pests.

1. *Euseius tularensis*

Euseius tularensis Congdon apparently is an omnivorous mite but will feed on both the citrus red mite and citrus thrips,[70] and standardized ampling plans for *E. tularensis* on 'Washington navel' orange have been developed.[28,71] The plan is based upon sampling units of 20 leaves taken from the interior of citrus trees, at random, around the periphery and determining if they are infested with *E. tularensis*. The density of predaceous mites per leaf is a linear function of the proportion of leaves infested with mites. Thus densities, up to two predaceous mites per leaf, can be predicted from the proportion of leaves infested.

Although the density of predaceous mites relative to that of thrips necessary to provide economically effective thrips suppression is not yet known, a number of pest control advisors double the treatment threshold for citrus thrips if predaceous mites are present at a density ≥ 0.2 predaceous mites per leaf. Similarly, the presence of predaceous mites at about one half the density of adult female citrus red mites early in the season usually precludes the need for an acaricide treatment.[7]

2. Parasitoids of California Red Scale

Several parasitoids of California red scale are attracted to fluorescent yellow and will be captured in the same yellow traps that are used to monitor thrips and other citrus pests.[72] Such traps have been used to discern general population trends in parasitoid abundance and distribution of parasitoids within orchards.[19,73,74] Currently, trap catches cannot be used to determine if the density of resident female parasitoids is sufficient to provide economically effective biological control of California red scale, however. Because of the difficulty in sexing minute wasps caught on sticky traps, it is not known if both sexes are equally attracted to these traps, or if the numbers caught actually reflect the densities of adult female wasps within the grove.

For absolute density determinations, two procedures have been widely used. The first is simply to collect samples of plant tissue with scale, then bring them back to the laboratory, lift scale covers under a microscope, and determine parasitization rates.[44,49,75] The second is to hang detached fruit with fixed numbers of same-aged scales from a laboratory colony in trees for a defined time, then collect them, and, again, examine scales for the presence of parasitoids under a microscope.[20,73] These procedures were used to determine aspects of the distribution and host selection behavior of *Aphytis* species within citrus trees but are too tedious to be used routinely by scouts in pest management programs.

In South Africa, *Aphytis* spp. are collected in fluorescent yellow California red scale pheromone traps.[48,49] Data presently at hand do not indicate that the presence of the scale's pheromone markedly increases the catch of *Aphytis* spp., i.e., the scale's sex pheromone seemingly is not also a kairomone for its natural enemy. Nevertheless, because yellow traps are used to monitor citrus thrips and citrus psylla (*Trioza erytreae* (Del Guercio)), yellow pheromone traps may permit the monitoring of several pests and some of their natural enemies simultaneously. A major disadvantage of fluorescent yellow traps is that many other innocuous insect species also positively respond to them, and the advantage of specificity of traps based upon sex pheromones is lost.

V. FUTURE NEEDS

A great deal is known of the population dynamics and distribution of most of the major citrus pests, and relatively sophisticated sampling programs have been developed for them. Their widespread commercial implementation has not been rapid, largely because the treatment thresholds to which those sampling programs are focused are variable and often are not known early enough in order to make a rational pest management decision.

Much of this has to do with the cosmetic nature of the injury caused by these pests and its time-varying relationship with fruit supply and demand in the marketplace. Clearly, citrus pest management would be simplified if grading standards were uniformly applied within and between seasons so that more precise relationships between pest density and economic loss could be developed. Citrus pest management would be simplified even more if consumers became educated to recognize cosmetic damage to fruit rinds as merely imperfections of appearance, and not a predictor of reduced nutritional quality.[11] Although sensible, current marketing conditions are unlikely to bring this about in the foreseeable future.[10] Therefore, it does not seem likely that citrus pest management programs can take full advantage of current sampling methodology without a more refined understanding of the economic losses associated with pest populations at particular densities and time intervals.

ACKNOWLEDGMENTS

I thank J. G. Morse and R. F. Luck for their thoughtful reviews of a previous draft of the chapter. My research on citrus pest management has been supported through various grants from the California Citrus Research Board and the University of California Statewide Integrated Pest Management Project.

REFERENCES

1. Burke, J. H., The commercial citrus regions of the world, in *The Citrus Industry*, Vol. 1, Reuther, W., Webber, H. J., and Batchelor, L. D., Eds., University of California, Berkeley, 1967, 40.
2. Wardowski, W. F., Nagy, S., and Grierson, W., *Fresh Citrus Fruits*, AVI Publishing, Westport, CT, 1986.
3. Behr, R., Fairchild, G., Brown, M., and Bedigian, K., 1991–1992 Florida orange outlook, *Citrus Ind.*, 72(12), 7, 1991.
4. Hield, H. Z., Burns, R. M., and Coggins, C. W., Jr., Some fruit thinning effects of napthaleneacetic acid on Wilking mandarin, *Proc. Am. Soc. Hortic. Sci.*, 81, 218, 1962.
5. Hare, J. D., Pehrson, J. E., Clemens, T., Menge, J. L., Coggins, C. W., Embleton, T. W., and Meyer, J. L., Effects of managing citrus red mite (Acari: Tetranychidae) and cultural practices on total yield, fruit size, and crop value of 'Navel' orange, *J. Econ. Entomol.*, 83, 976, 1990.
6. Ebeling, W., *Subtropical Fruit Pests*, University of California, Berkeley, 1959.
7. Pehrson, J. E., Flaherty, D. L., O'Connell, N. V., Phillips, P. A., and Morse, J. G., *Integrated Pest Management for Citrus*, 2nd ed., Publ. 3303, University of California, Berkeley, 1991.
8. Knapp, J. L. and Fasulo, T. R., Citrus rust mite, in *Florida Citrus Integrated Pest and Crop Management Handbook, Section IV*, Knapp, J. L., Ed., University of Florida, Gainesville, 1983.
9. Hare, J. D., Pehrson, J. E., Clemens, T., Menge, J. A., Coggins, C. W., Embleton, T. W., and Meyer, J. L., Effects of citrus red mite (Acari: Tetranychidae) and cultural practices on total yield, fruit size, and crop value of 'Navel' orange: years 3 and 4, *J. Econ. Entomol.*, 85, 486, 1992.
10. Arpaia, M. L. and Morse, J. G., Citrus thrips *Scirtothrips citri* (Moulton) (Thys., Thripidae) scarring and navel orange fruit quality in California, *J. Appl. Entomol.*, 111, 28, 1991.
11. Brown, M., An orange is an orange, *Environment*, 17, 6, 1975.
12. Jackson, L., *Citrus Growing in Florida*, 3rd ed., University of Florida Press, Gainesville, 1991.
13. Barnes, M. M. and Andrews, K. L., Effects of spider mites on almond tree growth and productivity, *J. Econ. Entomol.*, 71, 555, 1978.
14. Barnes, M. M. and Moffitt, H. R., A five-year study of the effects of the walnut aphid and the European red mite on Persian walnut productivity in coastal orchards, *J. Econ. Entomol.*, 71, 71, 1978.
15. Beers, E. H. and Hull, L. A., Timing of mite injury affects the bloom and fruit development of apple, *J. Econ. Entomol.*, 83, 547, 1990.
16. Welter, S. C., Barnes, M. M., Ting, I. P., and Hayashi, J. T., Impact of various levels of late-season spider mite (Acari: Tetranychidae) feeding damage on almond growth and yield, *Environ. Entomol.*, 13, 52, 1984.
17. Welter, S. C., McNally, P. S., and Farham, D. S., Willamette mite (Acari: Tetranychidae) impact on grape productivity and quality: a re-appraisal, *Environ. Entomol.*, 18, 408, 1989.
18. Welter, S. C., Freeman, R., and Farham, D. S., Recovery of 'Zinfandel' grapevines from feeding damage by Willamette spider mite (Acari: Tetranychidae): implications for economic injury level studies in perennial crops, *Environ. Entomol.*, 20, 104, 1991.

19. Reeve, J. D. and Murdoch, W. W., Biological control by the parasitoid *Aphytis melinus*, and population stability of the California red scale, *J. Anim. Ecol.*, 55, 1069, 1986.
20. Murdoch, W. W., Luck, R. F., Walde, S. J., and Yu, D. S., A refuge for red scale under control by *Aphytis*: structural aspects, *Ecology*, 70, 1707, 1989.
21. Yu, D. S., Luck, R. F., and Murdoch, W. W., Competition, resource partitioning, and coexistence of an endoparasitoid *Encarsia perniciosi* and an ectoparasitoid *Aphytis melinus* of the California red scale, *Ecol. Entomol.*, 15, 469, 1990.
22. Grout, T. G., Morse, J. G., O'Connell, N. V., Flaherty, D. L., Goodell, P. B., Freeman, M. W., and Coviello, R. L., Citrus thrips (Thysanoptera: Thripidae) phenology and sampling in the San Joaquin Valley, *J. Econ. Entomol.*, 79, 1516, 1986.
23. Samways, M. J., Spatial distribution of *Scirtothrips aurantii* Faure (Thysanoptera: Thripidae) and threshold level for one per cent damage on citrus fruit based on trapping with fluorescent yellow sticky traps, *Bull. Entomol. Res.*, 76, 649, 1986.
24. Phillips, P. A., Thrips in coastal lemons: is control economical?, *Citrograph*, 66, 111, 1981.
25. Rhodes, A. A. and Morse, J. G., *Scirtothrips citri* sampling and damage prediction on California navel oranges, *Agric. Ecosyst. Environ.*, 26, 117, 1989.
26. Moreno, D. S. and Kennett, C. E., Predictive year-end California red scale (Homoptera: Diaspididae) orange fruit infestations based on catches of males in the San Joaquin Valley, *J. Econ. Entomol.*, 78, 1, 1985.
27. Luck, R. F., Integrated pest management in California Citrus, in *Proceedings of the International Society of Citriculture*, Vol. 2, Matsumoto, K., Ed., International Society of Citriculture, Japan Chapter, Shimizu, 1981, 630.
28. Zalom, F. G., Kennett, C. E., O'Connell, N. V., Flaherty, D., Morse, J. G., and Wilson, L. T., Distribution of *Panonychus citri* (McGregor) and *Euseius tularensis* Congdon on central California orange trees with implications for binomial sampling, *Agric. Ecosyst. Environ.*, 14, 119, 1985.
29. Jones, V. P. and Parrella, M. P., Dispersion indices and sequential sampling plans for the citrus red mite (Acari: Tetranychidae), *J. Econ. Entomol.*, 77, 75, 1984.
30. Jones, V. P. and Parrella, M. P., Intratree regression sampling plans for the citrus red mite (Aacari: Tetranychidae) on lemons in southern California, *J. Econ. Entomol.*, 77, 810, 1984.
31. Hare, J. D. and Youngman, R. R., Gas exchange of orange (*Citrus sinensis*) leaves in response to feeding injury by the citrus red mite (Acari: Tetranichidae), *J. Econ. Entomol.*, 80, 1249, 1987.
32. Hare, J. D., Pehrson, J. E., Clemens, T., and Youngman, R. R., Combined effects of differential irrigation and feeding injury by the citrus red mite (Acari: Tetranychidae) on gas exchange of orange leaves, *J. Econ. Entomol.*, 82, 204, 1989.
33. Lewis, H. C., Three year's use of DDT on citrus in central California, *Citrograph*, 33, 546, 1948.
34. Phillips, P. A., Machlitt, D., and Mead, M., Non-target effects of dimethoate and acephate against *Euseius tularensis* Congdon and *Aphytis melinus* DeBach on lemons in California, *Crop Prot.*, 6, 388, 1987.
35. Hare, J. D. and Phillips, P. A., Economic impact of the citrus red mite on southern California coastal lemons, *J. Econ. Entomol.*, 85, 1926, 1992.
36. Flanders, S. E., The role of the ant in the biological control of homopterous insects, *Can. Entomol.*, 83, 93, 1951.
37. Bartlett, B. R., The influence of ants upon parasites, predators, and scale insects, *Ann. Entomol. Soc. Am.*, 54, 543, 1961.
38. Haney, P. B., Luck, R. F., and Moreno, D. S., Increases in densities of the citrus red mite, *Panonychus citri* (Acarina: Tetranychidae), in association with the Argentine ant, *Iridomyrmex humulis* (Hymenoptera: Formicidae), in southern California citrus, *Entomophaga*, 32, 49, 1987.
39. Samways, M. J., Ant assemblage structure and ecological management in citrus and subtropical fruit orchards in southern Africa, in *Applied Myrmecology: A World Perspective*, Vander Meer, R. K., Jaffe, K., and Cedeno, A., Eds., Westview Press, Boulder, CO, 1990, 570.
40. Tashiro, H. and Beavers, J. B., Growth and development of the California red scale, *Aonidiella aurantii*, *Ann. Entomol. Soc. Am.*, 61, 1009, 1968.
41. Yan, J.-Y. and Isman, M. B., Environmental factors limiting emergence and longevity of male California red scale (Homoptera: Diaspididae), *Environ. Entomol.*, 15, 971, 1986.
42. Abdelrahman, I., Growth, development and innate capacity for increase in *Aphytis chrysomphali* Mercet and *A. melinus* DeBach, parasites of California red scale, *Aonidiella aurantii* (Mask.), in relation to temperature. *Aust. J. Zool.*, 22, 213, 1974.
43. Carroll, D. P. and Luck, R. F., Within-tree distribution of California red scale, *Aonidiella aurantii* (Maskell) (Homoptera: Diaspididae), and its parasitoid *Comperiella bifasciata* Howard (Hymenoptera: Encyrtidae) on orange trees in the San Joaquin Valley, *Environ. Entomol.*, 13, 179, 1984.

44. Carroll, D. P. and Luck, R. F., Bionomics of California red scale, *Aonidiella aurantii* (Maskell) (Homoptera: Diaspididae), on orange fruits, leaves, and wood in California's San Joaquin Valley, *Environ. Entomol.*, 13, 847, 1984.
45. Samways, M. J., Relationship between red scale, *Aonidiella aurantii* (Maskell) (Hemiptera: Diaspididae), and its natural enemies in the upper and lower parts of citrus trees in South Africa, *Bull. Entomol. Res.*, 75, 379, 1985.
46. Bodenheimer, F. S., *Citrus Entomology*, Dr. W. Junk, The Hague, The Netherlands, 1951.
47. Rice, R. E. and Moreno, D. S., Flight of male California red scale, *Ann. Entomol. Soc. Am.*, 63, 91, 1970.
48. Samways, M. E., Comparative monitoring of red scale, *Aonidiella aurantii* (Mask.) (Hom., Diaspididae) and its *Aphytis* spp. (Hym., Aphelinidae) parasitoids, *J. Appl. Entomol.*, 105, 483, 1988.
49. Grout, T. G. and Richards, G. I., Value of pheromone traps for predicting infestations of red scale, *Aonidiella aurantii* (Maskell) (Hom., Diaspididae), limited by natural enemy activity and insecticides used to control citrus thrips, *Scirtothrips aurantii* Faure (Thys., Thripidae), *J. Appl. Entomol.*, 111, 20, 1991.
50. Walker, G. P., Aitken, D. C. G., O'Connell, N. V., and Smith, D., Using phenology to time insecticide applications for control of California red scale (Homoptera: Diaspididae) on citrus, *J. Econ. Entomol.*, 83, 189, 1990.
51. Luck, R. R. and Podoler, H., Competitive exclusion of *Aphytis lingnanensis* by *A. melinus*: potential role of host size, *Ecology*, 66, 904, 1985.
52. Opp, S. B. and Luck, R. F., Effects of host size on selected fitness components of *A. melinus* and *A. lingnanensis* (Hymenoptera: Aphelinidae), *Ann. Entomol. Soc. Am.*, 79, 700, 1986.
53. Yu, D. S. and Luck, R. F., Temperature-dependent size and development of California red scale (Homoptera: Diaspididae) and its effect on host availability for the ectoparasitoid, *Aphytis melinus* DeBach (Hymenoptera: Aphelinidae). *Environ. Entomol.*, 17, 154, 1988.
54. Phillips, P. A., Timing *Aphytis* release in coastal citrus, *Citrograph*, 72, 128, 1987.
55. Kennett, C. E. and Hoffman, R. W., Seasonal development of the California red scale (Homoptera: Diaspididae) in San Joaquin Valley citrus based on degree-day accumulation, *J. Econ. Entomol.*, 78, 73, 1985.
56. McGregor, E. A., A device for determining the relative degree of insect occurrence, *Pan-Pac. Entomol.*, 3, 29, 1926.
57. Reed, D. K. and Rich, J. R., A new survey technique for citrus thrips, *J. Econ. Entomol.*, 68, 739, 1975.
58. Tanigoshi, L. K. and Moreno, D. S., Traps for monitoring populations of the citrus thrips, *Scirtothrips citri* (Thysanoptera: Thripidae), *Can. Entomol.*, 113, 9, 1981.
59. Grout, T. G., Morse, J. G., and Brawner, O. L., Location of citrus thrips (Thysanoptera: Thripidae) pupation: tree or ground, *J. Econ. Entomol.*, 79, 59, 1986.
60. Grout, T. G. and Richards, G. I., Monitoring citrus thrips, *Scirtothrips aurantii* Faure (Thysanoptera, Thripidae), with yellow card traps and the effect of latitude on treatment thresholds, *J. Appl. Entomol.*, 109, 385, 1990.
61. Yother, W. W. and Mason, A. C., *The Citrus Rust Mite and its Control*, USDA Tech. Bull. 176, USDA, Washington, DC, 1930.
62. McCoy, C. W. and Albrigo, L. G., Feeding injury to the orange caused by the citrus rust mite, *Phyllocoptruta oleivora* (Prostigmata: Eriophyoidae), *Ann. Entomol. Soc. Am.*, 68, 289, 1975.
63. Allen, J. C., The effect of citrus rust mite damage on citrus fruit drop, *J. Econ. Entomol.*, 71, 746, 1978.
64. Allen, J. C., Effect of citrus rust mite damage on citrus fruit growth, *J. Econ. Entomol.*, 72, 195, 1979.
65. Allen, J. C., A model for predicting citrus rust mite damage on Valencia orange fruit, *Environ. Entomol.*, 5, 1083, 1976.
66. Rogers, J. S., Estimating citrus rust mite population densities with the HB system, *Citrus Ind.*, 73(1), 60, 1992.
67. Peña, J. E. and Baranowski, R. M., Dispersion indices and sampling plans for the broad mite (Acari: Tarsonemidae) and the citrus rust mite (Acari: Eriophydae) on limes, *Environ. Entomol.*, 19, 378, 1990.
68. Hall, D. G., Childers, C. C., and Eger, J. E., Estimating citrus rust mite (Acari: Eriophyidae) levels on fruit in individual citrus trees, *Environ. Entomol.*, 20, 382, 1991.
69. Albrigo, L. G. and McCoy, C. W., Characteristic injury by citrus rust mite to orange leaves and fruit, *Proc. Fla. State Hortic. Soc.*, 87, 48, 1974.
70. Tanigoshi, L. K., Nishio-Wong, J. Y., and Fargerlund, J., Greenhouse- and laboratory-rearing studies of *Euseius hibisci* (Chant) (Acari: Phytoseiidae) a natural enemy of the citrus thrips, *Scirtothrips citri* (Moulton) (Thysanoptera: Thripidae), *Environ. Entomol.*, 12, 1298, 1983.

71. Grout, T. G., Binomial and sequential sampling of *Euseius tularensis* (Acari: Phytoseiidae), a predator of citrus red mite (Acari: Tetranychidae) and citrus thrips (Thysanoptera: Thripidae), *J. Econ. Entomol.*, 78, 567, 1985.
72. Moreno, D. S., Gregory, W. A., and Tanigoshi, L. K., Flight response of *Aphytis melinus* (Hymenoptera: Aphelinidae) and *Scirtothrips citri* (Thysanoptera: Thripidae) to trap color, size, and shape, *Environ. Entomol.*, 13, 935, 1984.
73. Reeve, J. D. and Murdoch, W. W., Aggregation by parasitoids in the successful control of the California red scale: a test of theory, *J. Anim. Ecol.*, 54, 797, 1985.
74. Samways, M. J., Spatial and temporal population patterns of *Aonidiella aurantii* (Maskell) (Hemiptera: Diaspididae) parasitoids (Hymenoptera: Aphelinidae and Encyrtidae) caught on yellow sticky traps in citrus, *Bull. Entomol. Res.*, 76, 265, 1986.
75. Smith, A. D. and Maelzer, D. A., Aggregation of parasitoids and density-independence of parasitism in field populations of the wasp *Aphytis melinus* and its host, the red scale, *Aonidiella aurantii*, *Ecol. Entomol.*, 11, 425, 1986.

Chapter 16

SAMPLING ARTHROPOD PESTS IN FIELD CORN

Jon J. Tollefson and Dennis D. Calvin

TABLE OF CONTENTS

I. Introduction 434

II. Sampling to Manage Corn Rootworm 435
    A. Larval Sampling 436
    B. Egg Sampling 438
    C. Adult Sampling..439
        1. Counting Beetles on Corn Plants 441
        2. Sticky Traps 442
        3. Emergence Traps 444
        4. Sequential Sampling 445
    D. The Economics of Corn Rootworm Sampling and Management 445

III. Sampling to Manage European Corn Borer 447
    A. Variation in Seasonal Development and Occurrence 448
        1. Expression of Voltinism at a Geographic Site 448
        2. Coexistence of European Corn Borer Races 451
    B. Relationship between Crop Phenology and ECB Egg Deposition 453
    C. Relationship between Plant Phenology and Susceptibility to
       European Corn Borer Feeding Injury 457
        1. Plant Phenology and Seed Yield Reduction Potential............457
        2. Economic Injury Levels and Economic Thresholds 458
    D. Sampling to Manage European Corn Borer 462
        1. First Generation 462
        2. Second Generation..................................464
        3. Larval Sampling to Initialize the European Corn Borer
           Phenology Model..................................465
    E. The Economics of ECB Sampling and Management................466

References 470

I. INTRODUCTION

There is a good possibility that corn (maize), *Zea mays*, originated in southern Mexico. It appears to have been domesticated in the Tehuacan Valley and its cultivation soon spread throughout the Americas, where it became a mainstay food. Presently corn is cultivated worldwide, but half of the world's crop is produced in the U.S.

In the U.S., much of the corn never leaves the farm. Nearly 90% is field corn that is fed to cattle and hogs. There are three field corns: flint corns (kernels of hard starch), flour corns (kernels of soft starch), and dent corns (kernels of hard starch capped by soft starch). The dent corns are the most widely grown field corn because they are especially high yielding and are well adapted to the "Corn Belt" of the northcentral U.S. In addition to field corns, there are sweet corns and popcorns. Sweet corn has a higher sugar content and is harvested green for human food. Popcorn, with kernels that burst open upon heating, is used mainly as a confectionery.

Corn is basically a long-season, moisture-loving crop. Planting with the first warm weather, during early May in the Corn Belt, results in shorter plants that do not lodge as easily. Early emerging plants shade weeds more quickly and pollinate and form ears before moisture becomes scarce in late summer. Timely planting helps corn mature before fall so that it may be allowed to stand in the field and dry naturally.

Because corn is monoecious and the female and male flowers (ears and tassels, respectively) are separate on the plant, pollination is relatively easy to control. This ability, combined with corn's demonstrated strong heterotic response (increase in yield due to hybrid vigor), has resulted in a large hybrid seed corn industry that produces all the seed farmers plant in the U.S. Most of the hybrids planted today are single crosses, a single crossing of two inbred parents. These single crosses are homogeneously heterozygous. The homogeneity facilitates mechanized production because fields are very uniform in stature and phenology. However, because some inbreds are favored as parents, the resulting genetic similarity in corn across the Corn Belt increases the risk of losses to infestations of pests and diseases. The heterozygosity means the progeny of the corn planted by the farmer will segregate and not breed true. To maximize commercial field corn's genetic potential, farmers must buy new hybrid seed each year.

The most important insect pests of field corn grown in the Corn Belt are the corn rootworms (CRW) and the European corn borer. It is currently estimated that the CRW alone costs U.S. farmers $1 billion annually in insecticide treatments and crop losses (Metcalf[1]). While both pests can cause direct losses (feeding on the silks by rootworm beetles interferes with pollination and tunneling through the ear shank by corn borer larvae causes ear droppage) most of the damage caused by both pests is indirect. The feeding on corn roots by rootworm larvae and the tunneling in the stalk by corn borer larvae (1) disrupts the physiology of the plant, causing a reduction in grain production, and (2) increases root and stalk lodging, decreasing harvested yield further. Pest injury that indirectly leads to yield reduction is more sensitive to environmental variation. Therefore, predicting its magnitude, even with reliable pest population estimates, is more unreliable and requires more complex economic injury levels. The value of sampling these pests will vary greatly because of the environmental interaction.

The genetic plasticity of corn and the strong seed corn industry that supplies all of the farmers' seed have encouraged improvements in corn's ability to tolerate CRW and corn borers. Lines of corn with larger root systems have been produced to tolerate feeding by CRW larvae. Other lines are immune to the establishment of the first generation corn borer larvae. However, host resistance alone has not been sufficient to manage the pests. Corn roots are only moderately tolerant of CRW feeding, and lines that resist corn borer invasion do not yield as well as some lines that are more suspectible.

Control tactics that are more commonly relied on by farmers are cultural techniques that help avoid infestations and applications of insecticides to eliminate the pests if they become established. Over much of the Corn Belt, annual crop rotation will completely avoid CRW infestations. Cultural techniques that help reduce corn borer infestations include: destroying the residue from the previous corn crop that contains overwintering larvae, timely planting so the crop is not unusually attractive to females for egg laying, and early harvest to avoid excessive stalk lodging and ear droppage. Even with these tools, insecticides must be used because almost half of the field corn is not rotated, and corn borer infestations cannot be completely avoided by cultural practices. Because insecticides must be used, the principles of pest management cannot be applied unless sampling methods are provided that allow farmers or their consultants to apply insecticides to susceptible fields only when a pest infestation is expected.

II. SAMPLING TO MANAGE CORN ROOTWORM

Diabrotica, a New World genus of chrysomelids, has its greatest diversity in the tropics, but its most severe pest status is north of Mexico. Of the 338 tropical species, six are pests; of the seven species (including two subspecies) that occur in the U.S., four are pests (Krysan[2]). Of these four pest species, three belong to the *virgifera* group: the western CRW, *Diabrotica virgifera virgifera* LeConte; the Mexican CRW, *Diabrotica virgifera zeae* Krysan and Smith; and the northern CRW, *Diabrotica barberi* Smith and Lawrence. The fourth, the southern corn rootworm, *Diabrotica undecimpunctata howardi* Barber, is a member of the *fucata* group. Differences in the biologies of the two groups result in differences in their severity as pests of field corn in the U.S. Corn Belt. The *virgifera* group is univoltine, overwinters as eggs, and, probably because of the hardiness of the egg, can survive cold winter temperatures (Krysan[2]). Because they can overwinter throughout the Corn Belt, the *virgifera* group is the more serious pest of field corn. The *fucata* group is multivoltine, overwinters as adults, and cannot survive subfreezing temperatures. Populations of the southern CRW seldom reach densities that require control. Because the *virgifera* group includes the species that growers must manage most frequently, the CRW sampling techniques discussed are those used for this group.

Adult CRW can cause yield losses by feeding on the female flowers before they are pollinated. Injury results in poor pollination and reduced seed set. Usually controls to prevent silk feeding in field corn are not necessary because the flowers are pollinated too early in the adult emergence period for populations to reach economic densities. Larval feeding on the roots of corn causes substantially greater yield losses, and growers will invest more resources (time and money) to predict and avoid larval injury.

The best correlation between estimated pest densities and anticipated damage should be produced by sampling the stage causing injury. While sampling earlier stages may be more efficient and provide greater lead time, the environment acting on both pest and host will reduce the precision of damage prediction. In the case of a univoltine insect that overwinters as an egg such as the CRW, the prior adult stage whose progeny will produce the damage being predicted is a complete growing season removed. Because it is of greatest interest to predict larval injury, this discussion of

CRW sampling begins with larval sampling and move back to the previous season's adult stage.

A. LARVAL SAMPLING

Northern and western CRW larvae are intimately associated with corn roots. Chiang et al.[3] reported that greater than 60% of all northern CRW larvae were within a 20-cm cube of soil surrounding the corn plant. Sechriest[4] found that greater than 90% of CRW larvae were contained in a 20 × 20 × 10-cm soil cube collected at the base of the corn plant. Fisher and Bergman[5] concluded that the differences between the two studies could be due to different species compositions or to sampling at different times. Bergman et al.[6] had reported that first-stage larvae are generally scattered and that second and third instars are mostly concentrated in the upper 10 cm of the root zone. Regardless of the reason, Fisher and Bergman[5] concluded the larval sampling unit should include a corn plant.

The concentration of CRW larvae at the plant base is probably a behavioral consequence. The female lays eggs in clusters that are scattered in the upper soil layer. Newly hatched larvae locate roots that grow into the vicinity of where they have hatched. Small larvae begin feeding on root hairs. "Successful" larvae—those that survive—move toward the base of the plant as they grow and require larger roots.

Fisher and Bergman[5] classified larval-sampling techniques into three categories: (1) plant-area soil samples, (2) soil cubes or blocks, and (3) soil cores. Plant-area soil samples are the simplest technique. It involves collecting soil from the root zone with a shovel or trowel. Sampling precision is low because the dimensions of the sample unit are poorly controlled, but attempts have been made to use the sampling technique for documenting the need for applying a granular larvicide to the soil when the corn is cultivated. The technique may be used to detect the presence of larvae and to estimate the stage of larval development. The crude economic threshold of one to two larvae per root mass and larval damage present has been used when prescribing curative insecticide treatments.

A research-based sampling program has not been defined for the plant-area soil sampling technique. The way the first author uses the sampling technique for survey purposes is to walk 50 to 100 paces into a cornfield, dig up several plants, place each root mass with the surrounding soil into a separate pail, and carry the pails to the edge of the field. There the soil can be shaken off the roots onto a cloth spread on the ground and the loose soil from the pail added to the pile. Water can be added to the pail, the soil remaining on the root system washed off in the pail, and the plant immersed in the water to soak while the soil on the cloth is inspected for larvae. The number of larvae that float to the surface of the water and that are found by carefully examining the root system (the examination includes breaking open roots that show feeding injury) are added to the number found in the soil.

Sampling by removing soil cubes or blocks reduces variation by controlling the dimension of the sample unit. The size of a soil sample can best be controlled by using a template. In practical field sampling, however, a surrogate template is created by removing soil with a shovel of specified width that has been marked to standardize the depth to which it is inserted. Bergman et al.[6] reported that an 18-cm cube surrounding a corn plant will contain most of the larvae feeding on the plant.

Collecting soil cores is a common way of sampling soil arthropods. A number of custom-made and commercially marketed sampling devices have been used to sample CRW larvae. The device most commonly used is the golf-cup cutter which cuts a core 10 cm in diameter to a depth of 21 cm. It is equipped with a lever that facilitates the ejection of the soil core. The core sampler offers two advantages over removing soil blocks: the dimensions of the sample unit are more precisely controlled, and the smaller volume of soil collected reduces processing costs.

Bergman et al.[6,7] compared the CRW larval sampling efficiencies of the soil-cube and core-sampling techniques. They demonstrated that the larvae are distributed in an aggregated manner (Bergman et al.[7]) and that a stratified-random design was most appropriate to avoid systematic sampling errors (Bergman et al.[6]). The optimum allocation of sampling resources was calculated to be one cube and two cores at each location (Bergman et al.[6]). The sample sizes required to achieve specified sampling precisions are shown in Table 1. The overall precision is increased by sampling at more locations per field. Because the core sampler produced smaller volumes of soil, it was cheaper (requiring less time) to process. The core-sampling technique was significantly more variable, however, and required a sufficiently larger sample size to achieve similar precision, so that the technique is more costly overall. The most efficient CRW larval-sampling technique, therefore, is the cube-sampling technique.

The techniques for removing larvae from soil samples have been grouped by Fisher and Bergman[5] into (1) hand searching, (2) behavioral methods, and (3) mechanical methods, including dry sieving and wet sieving with flotation. During hand searching, the soil particles are broken apart and the sample visually inspected for CRW larvae. To aid in processing, a screen may be used to sift the soil. Because very little equipment is needed, the hand searching technique is most commonly used in the field for on-site determination of larval presence and phenology. This is the least efficient and most variable extraction technique because many small larvae are overlooked. Weiss and Mayo[8] recovered less than 22% of the known number of CRW larvae in a soil sample; the recovery was inconsistent, varying from 0 to 21.2%.

Berlese-Tullgren funnels may be used to extract CRW larvae from soil samples. Edwards and Fletcher[9] reported it took 3 d to process a soil sample. This behavioral technique is not used for CRW management because the handling time requires a large number of funnels if more than a few samples are processed and there is a delay before management recommendations can be offered.

While a large-mesh screen may be used to break up soil particles during visual searching, the larvae are too fragile to separate from dry soil by forcing the sample through a series of sieves. Larvae may be removed from soil by passing it through increasingly finer sieves, however, if large particles are broken up and the soil suspended in water by agitation. Wet sieving is the most efficient way of recovering larvae, and researchers have automated the sieving process by motorizing stacked sieves.

TABLE 1
Allocation of Sampling Resources for Estimating Corn Rootworm Larval Populations (modified from Bergman et al.[6])

Precision (%)	Core sampling technique		Cube sampling technique	
	No. of sites	Cost (human-h)	No. of sites	Cost (human-h)
10	2525	707	865	294
15	1123	314	385	131
20	632	177	217	74
25	404	113	139	47
30	281	79	97	33
40	158	43	55	19
50	101	28	35	12

Flotation is relatively quick and may be used in the field by pest scouts. It involves dispersing the soil in one or more water baths and collecting the larvae that float to the surface. Recovery efficiency may be improved by adding salt or sugar to enhance the buoyancy of small larvae.

Bergman et al.[6] reported that recovery efficiency can be maximized by combining wet sieving and flotation. They placed samples in a bucket and dispersed the soil with a pressurized stream of water. The liquid was poured off the soil into a second pail. The flotation process was repeated two more times, and all three liquid portions were decanted through 30- and 80-mesh sieves. The residue trapped by the finer sieve was transferred to a saturated salt solution. Larvae that floated to the surface of the salt solution were retrieved and counted. The researchers reported that it required 15 min to process a single sample.

Corn rootworm larval samples may either be processed in the field or returned to the laboratory. When a large number of samples are taken and larvae are extracted in the laboratory, samples may be stored. Piedrahita[10] stored larval samples for a short time at 8°C, but Bergman et al.[6] found that larvae in samples that were refrigerated but not frozen began to deteriorate within 2 to 3 d. Fisher[11] froze larval samples for up to 60 d without the efficiency of larval recovery declining.

The recommended sampling plan for CRW larvae consists of extracting larvae from a soil unit no smaller than 18×18 cm by at least 10 cm in depth. The sample should contain a root system. Because most of the larvae are found at the plant base, the density of larvae can be approximated by multiplying the number of larvae by the density of corn plants per unit of area. The selection of the larval-extraction technique will depend on the reason for sampling. If a larval-management decision is needed, the larvae should be extracted in the field using visual searching, and if possible, flotation. For research, a wet-sieving/flotation method will produce a more accurate estimate of larval density.

B. EGG SAMPLING

Corn rootworms oviposit in the soil. From his review of the literature, Ruesink[12] concluded the factors that most greatly influence the location of CRW eggs in the soil are soil moisture, soil type, compaction, surface debris, and, perhaps, proximity to the host plant. Because they are not accomplished burrowers, females use existing openings such as soil cracks, crevices at the plant/soil interface, earthworm burrows, and cavities left by decaying plant material to gain access to the soil. The depth at which the eggs are laid is determined by the availability of moisture, which is a factor of rainfall or irrigation and soil type. Several authors have reported finding 95% of the northern CRW eggs in the top 10 cm of soil (Sisson and Chiang,[13] Foster et al.,[14] and Patel and Apple[15]). Variation in soil moisture seems to have a greater impact on the western CRW. Gustin[16] reported that as little as 32% of the western CRW's eggs were found in the upper 7.5 cm of soil under dry conditions. Sampling to the depth of 20.3 cm, Hein et al.[17] found 11 to 50% of the northern and 16 to 65% of the western CRW eggs deeper than 10 cm. They suggested that sampling should be conducted to a depth of at least 20 cm, and it might be necessary to sample deeper if deep cracks are present from drought. Gray et al.[18] confirmed CRW eggs could occur deeper in the soil profile. They reported 60% of the western CRW eggs were found 20 to 30 cm deep and they recommended sampling to the greater depth when intensive ecological studies are conducted.

When used in a CRW management program, egg sampling offers the advantages of providing a longer period of time over which to sample—fall and spring—and it leaves open more control options. Soil samples may be collected in the fall and processed throughout the winter. If the need for control can be determined before the corn crop is planted, a preventive management tactic such as rotation to a nonhost crop may be used rather than applying an insecticide.

Egg sampling suffers several disadvantages, however. Being a step removed from the injurious stage (larval) causes the prediction of larval injury to be more dramatically influenced by environmental variation. The eggs are laid in clutches, resulting in a highly aggregated distribution and large sample variation. The variability forces large sample sizes for precise population estimation. Because the small eggs are imbedded in the soil, specialized equipment is required for extraction and counting the eggs. The large sample sizes and specialized sample processing equipment causes egg sampling to be expensive.

During the 1960s the most common method of extracting soil samples for sampling CRW eggs was through the use of a core sampler. Various devices were used, including a 2-cm auger, a 5-cm bulb setter, and a 10-cm golf-cup cutter. In 1964 Gunderson[19] proposed a standardized sampling program for the golf-cup cutter. In 1979 Foster et al.[14] added the "spade method", which used a small, collapsible shovel, and the "frame method".

The frame method consisted of driving a 10-cm-wide metal frame into the soil perpendicular to the direction of the corn rows. Soil was removed from a length of the frame equivalent to the row spacing to a depth of 10 cm. Species of CRW have shown a horizontal ovipositional preference with respect to the host plant, and this preference may be influenced by edaphic and environmental conditions and agronomic practices (Sisson and Chiang,[13] Pruess et al.,[20] Chiang et al.,[3] and Weiss et al.[21]). The frame method is independent of these biases because it samples a consistent subunit of the sampling universe, mid-row to mid-row. Ruesink and Shaw[22] reported mechanizing the frame method by excavating a trench with a gasoline-powered trencher.

All of the methods published for removing eggs from soil involve some combination of washing, sieving, and flotation. Probably the fastest and most efficient mechanized extraction procedure was developed at the Illinois Natural History Survey (Shaw et al.[23]) based upon a design described by Chandler et al.[24] The extraction procedure was reported to reach efficiencies as high as 97%, and Foster[25] reported processing as many as 40 samples per day.

Hein et al.[17] optimized the allocation of sampling resources using the frame soil-sampling method and the Illinois extraction device. Multiple samples were taken at sites within quadrats within fields. Sample variability was affected by egg density. The optimum allocation of sampling resources at low densities was two samples per site. At higher densities, which include the economic threshold, taking one soil sample per site was most efficient. The costs of sampling CRW eggs with specified levels of precision are presented in Table 2.

Because of the large amount of variability in egg sampling, sample sizes are large for reasonable levels of precision. The large sample sizes and the specialized equipment needed to extract and count CRW eggs causes egg sampling to be used primarily in research to confirm the presence of an infestation. There has been speculation that egg sampling could be used to supplement more cost efficient sampling techniques for making the final management decisions in fields that are very close to the economic threshold.

C. ADULT SAMPLING

All three types of sampling estimates—absolute, relative, and population indices—have been used with adult CRW sampling, depending on the management objective. For example, the number of beetles feeding on silks that cause economic

TABLE 2
Optimum Sampling Programs and Costs for Corn Rootworm Egg Sampling Using the Frame Method (modified from Hein et al.[17])

Sampling constraints	Precision (SE / m)	Sampling scheme Q / F[a]	S / Q[b]	Field costs (h)	Total costs (h)
Total cost minimized					
Mean < 2.5[c]	0.10	199	2	43	109
	0.25	32	2	7	18
Mean > 2.5	0.10	85	1	18	35
	0.25	14	1	3	6
Field cost minimized					
Mean < 2.5	0.10	106	9	29	188
	0.25	17	9	5	30
Mean > 2.5	0.10	50	5	12	62
	0.25	8	5	2	10

[a] Quadrants per field.
[b] Sites per quadrant.
[c] Expressed as eggs per 0.5 l soil.

reduction in seed set is highly variable. The relationship is strongly influenced by the vigor with which the line produces silk, by how well pollen shed and silk emergence are synchronized, and by environmental conditions, e.g., moisture stress delays silk emergence. Rather than basing a control decision on beetle numbers, using a population index has been suggested. If beetles eat unpollinated silks faster than they are emerging from the husks and if pollen is present, insecticidal sprays are recommended to control the beetles.

A second management objective might be to suppress oviposition so that there will not be an economic larval population the following year. In this case the purpose of sampling is to predict the onset of egg laying as well as beetle density. If adulticides are applied too early, they will lose their effectiveness before adult emergence and immigration are completed. If applications are delayed too long in an effort to extend their effectiveness, economic densities of eggs may be laid before the beetles are killed. Collecting beetles over a specified time interval is a relative sampling method that can be used to provide valuable information concerning the sex ratio and stage of ovarian development. Small, wide-mouthed jars containing 70% alcohol have been used to collect the beetles. The sampler holds the open jar under likely sources of beetles such as fresh silks, and disturbs them. In an attempt to escape, the excited beetles drop from the silks, falling into the jar. The sample size is a specified time interval; catching as many beetles as possible in 10 min has been most commonly used. A collecting label is added, the jar is closed, and later the beetles can be counted, a sex ratio calculated, and the females dissected to determine if egg laying has begun.

Chiang and Flaskerd[26] were critical of the beetle-collecting sampling technique because it does not provide an estimate of density per unit area. They also suggested that the technique might be influenced by the growth stage of the crop, differences in agility of the insect, and differences in the experience and ability of the collector. Tollefson et al.[27] demonstrated that the amount of sampling experience influenced the number of beetles they collected from the same field and that there was a significant interaction with crop phenology. During pollination, when nearly all the silks are fresh, succulent, and attractive to the beetles, the samplers caught similar numbers of beetles. As the fields matured and the presence of green silks was more sporadic, experienced samplers caught significantly more beetles than inexperienced samplers.

While collecting beetles for an interval of time does not produce a reliable density estimate, it is favored for making species- and sex-ratio determinations and estimating when egg laying begins. The two sampling techniques most commonly used to collect density estimates for making management decisions are counting the beetles on the host plant and deploying sticky traps to capture beetles.

1. Counting Beetles on Corn Plants

Visually counting adult CRW on corn plants, proposed by Chiang and Flaskerd,[26] is a population-intensity estimate that produces a type of absolute-population estimate. The host density is known or easily counted, so as long as the proportion of the insect population found on the host remains relatively uniform, a density per unit area can be calculated by multiplying the average number of beetles per plant by the number of hosts per unit area. Counting the number of beetles on corn plants is the most commonly used sampling technique for making CRW management decisions. It numerically defines the pest density and may be quantitatively related to the likelihood of subsequent larval damage.

Morris[28] suggested that smaller sampling units, carefully selected, may yield the most efficient sampling design. When sampling began to be applied commercially in CRW management, some practitioners chose a subunit of the whole plant, the ear zone, as their sampling unit.[70] The ear zone includes the area from, and including, the upper surface of the leaf below the ear to and including the lower surface of the leaf above the ear.

The visual beetle-count sampling method begins by selecting a plant at random. The selection should be made before the sampler is close enough to the plant to disturb beetles wandering about on the leaves. As the sampler slowly approaches the chosen plant, beetles exposed on the leaf surfaces and stalk are counted. When within reach of the plant, the silks on the tip of the primary ear should be grasped in one hand and held closed to confine the beetles within the ear tip. The sampler then starts from the bottom of the plant and works up to the tassel, visually counting beetles on the whole plant. The free hand is used to pull leaves away from the stalk to locate beetles behind leaf sheaths. When the plant has been thoroughly examined, the ear tip is released. One hand should be held open under the ear tip as the husks and silks are stripped back with the other hand. The beetles feeding on silks and kernels will attempt to escape by dropping from the ear tip. As they fall, they will land on the open hand, where they can be identified and counted before they take flight. The final total count for the whole plant is recorded. It is easy to separate species as they are counted, so totals can be recorded by species if desired.

Steffey et al.[29] used cost and variance components for visual beetle counts to design an optimum sampling program. Fifty-nine Iowa cornfields were divided into quadrants and the quadrants subdivided into four equal plots to create 16 equal-size plots. Sample sites were assigned systematically to the center of each plot. The amount of time to count and record beetle numbers was recorded as "human-h". Variance components were calculated using a nested analysis of variance with fields, quadrants within fields, sites within quadrants, and plants within sites as the sources of variation.

Steffey et al.[29] demonstrated that the whole-plant count is less variable than the subunit beetle counts from the ear zone. The smaller error results in a sample size

TABLE 3
Allocation of Sample Resources for Desired Precision of Population Intensity Estimates Obtained by Counting Adult Corn Rootworms on Two Randomly Selected Corn Plants at Each Sample Site (modified from Steffey et al.[29])

		Cost (human-h)	
No. of sites	Precision (%)	16 ha	65 ha
220	15	6.8	8.9
80	20	2.5	3.2
40	26	1.2	1.6
27	30	0.8	1.1
14	40	0.4	0.6
9	50	0.3	0.4

that is sufficiently smaller, 54 plants vs. 160 ear zones for 30% precision (Tables 3 and 4, respectively), to offset the lower cost of sampling a subunit. The study also showed that field size did not significantly influence sampling variance components, so the same sampling plan can be used in fields of any size. For estimating beetle populations to make management decisions, the sampling plan that provides 30% precision offers the best balance between accuracy and cost. The most efficient allocation of sampling resources is to count beetles on two randomly selected plants at each of 27 sites in the field.

To achieve 30% precision when sampling 54 plants, it is necessary to sample all quadrants. Coverage of the quadrants will be guaranteed by using a stratified-random sampling plan to spread the sample sites across the field. A realistic pattern is an inverted U, where samplers move into the field through the center of two quadrants until they are two thirds of the way through the field. The samplers then move across rows until they are into the other half of the field and walk out, sampling along the way. The sample sites can be evenly spaced along the path by counting paces between sites.

2. Sticky Traps

When adult CRW are sampled to make management decisions, sampling is conducted over very different environmental conditions by persons with a wide range of experience. The variability raises questions. Would farmers scouting their own

TABLE 4
Allocation of Sample Resources for Desired Precision of Population Intensity Estimates Obtained by Counting Adult Corn Rootworms on Five Randomly Selected Ear Zones at Each Sample Site (Modified from Steffey et al.[29])

		Cost (human-h)	
No. of sites	Precision (%)	16 ha	65 ha
115	20	3.5	4.6
32	30	1.0	1.3
15	40	0.5	0.6
9	50	0.3	0.4

fields be confident that they were able to find and accurately count the active beetles? Would population estimates produced by commercial scouts, who must sample all day and nearly every day to service their contracted acres, be affected by diurnal or meteorological changes? To help minimize these sources of variation, passive traps have been used to monitor adult populations. Traps aid inexperienced samplers by capturing the beetles for them, providing an opportunity to leisurely inspect the trappings, and to seek assistance in identifying a specimen if needed. Traps that are positioned in fields for extended periods of time tend to average out daily and environmental fluctuations.

The investigation of the possibility of using passive, sticky traps was begun by Tollefson et al.[30] By using different sized paper cartons painted different colors, they identified 0.95-l cartons, painted yellow, as being the most efficient tool. Hein and Tollefson[31] compared commercially available sticky traps with the yellow carton and selected the unbaited Pherocon® AM trap as the most reliable and readily available sampling device.

When Steffey et al.[29] optimized visual counts, they also included the cylindrical traps used by Tollefson et al.[30] (Table 5). As with the visual counts, the greatest source of variation was between individual samples. The most efficient allocation of sampling resources was to use a single trap at each sample site. The variation between field quadrants was relatively small but large enough that it had to be considered to produce acceptable levels of precision. To guarantee traps would be placed in all quadrants of a field, a stratified-random sampling plan was suggested as the best way to deploy traps.

Sticky traps are relative sampling techniques, and a factor that influences catch is adhesive effectiveness. The adhesive loses its ability to capture insects as it weathers and becomes fouled over time. Karr and Tollefson[32] determined the temporal efficiency of the Pherocon® AM trap. The catch efficiency of traps that had been left in a cornfield for up to 21 d was compared to those that had been in the field only 2 d. Averaged across fields, trap efficiency did not decline significantly for 10 d. In one field with a low beetle density, the traps caught significantly fewer beetles after only 6 d. The resulting recommendation was that unbaited Pherocon® AM sticky traps could be left in a cornfield for at least 1 week, and probably for as much 10 d, without a significant reduction in adult CRW catch.

The number of sticky traps needed to produce catches with prescribed precision levels was calculated by Hein and Tollefson.[33] The Pherocon® traps were used to sample CRW beetles in 34 fields over 2 years. The following year, larval damage to corn roots was determined in strips of the field, where no soil insecticide had been applied and in companion areas treated with insecticide to prevent larval damage.

TABLE 5
Allocation of Sample Resources for Desired Precision of Relative Population Estimates Using Pherocon® AM Sticky Traps to Capture Adult Rootworms (modified from Steffey et al.[29])

		Cost (human-h)	
No. of sites	Precision (%)	16 ha	65 ha
98	20	13.9	15.8
26	30	3.7	4.2
13	40	1.8	2.1
8	49	1.1	1.3

Twelve traps, arranged systematically across the field, will produce a relative beetle-density estimate with 10% precision. If an average of more than six beetles per trap per day are caught, the field should have the potential for economic larval damage to corn the following season.

The recommended method of field sampling adult CRW with the Pherocon® AM sticky trap is to use 12 traps in a systematic arrangement. It is suggested the traps be placed into the field in an inverted "U" pattern and spaced far enough apart to cover all quadrants of the field. To avoid border effects, traps should not be placed close to the edges of fields. If the traps along the arms of the "U" are placed in the same row and the row is clearly marked, they will be easier to find when they are recovered.

The sticky card is positioned on the plant by wrapping it around the plant with the sticky side out and clipping the corners with paper clips or using the slot-and-tab if provided. The trap is held on the plant by pushing a nail through the holes punched in the top of the trap. Hein and Tollefson[31] recommended placing the trap at ear height because they found that beetle captures at approximately 4 ft agreed best with the visual intensity estimates.

The sticky traps should be removed from the field in 7 to 10 d. If sampling is to continue, they can be replaced with new traps at the same time. The trap is recovered by removing it from the plant and folding it shut so that the adhesive is not exposed. The traps may be opened and beetles counted outside the field. If traps are to be stored, they should be refrigerated or, preferably, frozen to to retard insect decomposition. If the sex of the beetles is to be determined, they can be scraped from the trap and soaked in a solvent to remove some of the adhesive, but a timed beetle collection will produce a cleaner sample. Mechanic's hand cleaner works well for removing adhesive from hands, tools, and work areas.

3. Emergence Traps

To understand how environmental, edaphic, and agronomic factors affect CRW population dynamics, an absolute population estimate is necessary. Extracting eggs and larvae from a standard unit of soil will provide an absolute density estimate, but this sampling is expensive. Capturing the beetles as they emerge from the soil is relatively easy and offers an inexpensive way of obtaining an absolute estimate of the number of CRW that completed development within a cornfield. Hein et al.[34] designed, field evaluated, and optimized sample-resource allocation for an emergence cage that adjusted to fit a standard subunit of the sampling universe.

The conical screen cage has a rectangular wooden frame that can be adjusted so that its length matches common row spacings. The trap is positioned in the cornfield by cutting a corn plant off approximately 50 cm above the soil surface and placing the trap over the stump, perpendicular to the row direction. If soil has been cultivated into the corn row to form a ridge, furrows are dug across the ridge so that the frame will sit flush with the soil surface and a tight seal can be maintained. A half-liter carton, with slits in the bottom to accommodate the stalk and coated on the inside with adhesive, is slid over the cut-off stalk. Subsequently, the cage is placed over the carton, and the soil is mounded around the frame to prevent beetle escape. When beetles emerge from the soil, they crawl to the top of the cage, fall into the trap, and are snared by the adhesive. Beetle emergence may be monitored by raising the trap, replacing the carton with a fresh one, and returning the old carton to the laboratory for counting.

Because the intent of using the emergence cage is to obtain an absolute population estimate, it is important to know if the cage modifies the microenvironment sufficiently to influence density estimates. Fisher[35] compared beetle emergence from the area around cut corn plants with emergence from the area around intact plants. The total number of beetles that emerged did not differ, but emergence was accelerated where tops had been removed. He attributed the accelerated emergence to increased temperatures and humidities within cages where the plant canopy had been opened by removal of the plant tops. The root systems appeared sound for several weeks after removing the tops, but potential impact can be reduced by randomly selecting new plants periodically and moving the cages throughout the beetle emergence period.

To optimize the allocation of sample resources using the emergence cage, Hein et al.[34] sampled beetle emergence in 14 fields over a 5-year period. A three-stage sampling design was used with multiple traps at sites and several sites within each of the field's quadrants. The data were analyzed by using a nested analysis of variance, and the variance components and sampling costs were used to calculate the optimum allocation of sampling resources within each stratum. The optimum sampling design was a single cage at a sample site. Sampling-unit numbers necessary to estimate average beetle emergence within a standard error of 10% were calculated at 80 and 222 traps for western and northern CRW, respectively. The sampling error for the northern CRW was larger because several fields had low populations of this species. When these fields were dropped from the analysis, sample size for estimating northern CRW emergence with 10% accuracy was 126 traps.

4. Sequential Sampling

When CRW are sampled for commercial reasons, i.e., to make decisions concerning managing the pest, it is often adequate to merely classify the population as "economic" or "noneconomic". Waters[36] applied sequential analysis techniques to classify forest insect abundance at a savings in sampling costs of up to 50%. The report generated interest in using sequential sampling programs to help reduce sampling costs when it is necessary to simply classify insect abundance.

Steffey et al.[29] reported that beetles should be counted on two randomly selected plants at each of 27 sites in a cornfield to estimate beetle density on a per-plant basis with 30% precision. Because it is necessary to space the sample sites across the complete field, sampling at 30% precision requires about one human-h per field. The cost may not be justified if a field is obviously above or below the threshold.

Foster et al.[37] developed a scheme and evaluated the economics of using sequential sampling to sample adult CRW for predicting the need for taking management actions. The decision lines calculated are presented in Table 6. Using cost and regression analysis, Foster et al.[37] calculated that if Waters' sequential sampling method had been used in a scouting program conducted in 43 Iowa cornfields, a 36% savings in time would have been realized. The savings would lose value if the decisions reached were not correct, but the management decision reached using sequential sampling agreed with the fixed sample-size program in all 43 fields. When each of 225 visits to the fields was considered separately, the agreement between sequential and fixed sample-size sampling exceeded 96%.

D. THE ECONOMICS OF CORN ROOTWORM SAMPLING AND MANAGEMENT

Tollefson participated in a study (Foster et al.[38]) that used Bayesian analysis to estimate the value of sampling adult CRW populations for predicting the need to apply insecticides the following season for the protection of corn roots from larvae. Over a 3-year period in three areas in Iowa, adult populations were monitored in 77 cornfields. Insecticide-treated and untreated areas were left in the fields the following season to determine the amount of yield loss that was prevented by the application of insecticides. At the time the study was conducted, 1979 through 1982, it was estimated

TABLE 6
Sequential-Sampling Decision Table for Classifying the Likelihood of Larval Damage the Following Season Based on Counting Corn Rootworm Beetles on Corn Plants[37]

No. of plants		Cumulative no. of beetles		
10		3		17
12		5		19
14		7		21
16		9		23
18		11		25
20	No	13		27
24		17	Continue	31	Control
28	Control	21		35	
32		25	Sampling	39	Necessary
36	Necessary	29		43	
40		33		48	
44		37		52	
48		41		56	
52		45		60	
54		47		62	

it cost $2.50/acre to scout the fields, insecticides cost $12.00/acre, and field corn was worth $3.00/bushel. The value of the sample information was calculated to be zero. As a result, the most efficient strategy was not to sample and always apply an insecticide. The conclusion supports the prophylactic use of insecticides, a concept that is at odds with the pest management philosophy.

A number of changes have improved the value of the sample information. The price of insecticides has increased by $2 to $3/acre while the value of corn has declined to around $2.00/bushel. These changes have increased the losses that occur if an insecticide is used when it is not needed, hence making it more desirable to sample.

The authors also raised some reservations concerning the data. Of the 77 fields sampled, only 4 (5%) had adult CRW populations that were below the economic threshold of 1 beetle per plant. A manager of a pest management program reported that about 40% of his fields were below the economic threshold. If this sample likelihood of 40% of the fields being below the economic threshold is substituted into the algorithm, the value of sampling increases to slightly less than $6.00/acre. It is now clearly better to base insecticide use on sample information.

There are reasons the sample of fields could have been biased. The fields were not chosen at random. Cooperating farmers were recruited through the assistance of the county agriculturist. These farmers constituted a subset of all farmers, i.e., those that relied on public information in their farming operation. The farmers were told what the objectives of the research were when they were recruited. They were then asked to volunteer a field—again a deviation from random field selection. There is some reason to suspect the growers may have biased the sample toward greater CRW infestation because one farmer stated, "I have always had problems with that field; maybe you can figure out what is wrong." The Bayesian analysis indicated that if only two of ten fields would have been below the economic threshold, it would have paid to base insecticide treatments on sample information.

The Foster et al.[38] study illustrated a difficulty in predicting yield loss resulting from a pest that causes indirect damage. Beetle numbers correctly predicted economic root damage 83% of the time but only predicted economic yield loss 49% of the time. This is because the injury, root damage, only agreed with yield loss 53% of the time. In other words, while the number of beetles present is a pretty good predictor of the amount of damage the subsequent larval stage will cause, the injury to the roots is not well correlated with yield loss. Agronomic, edaphic, and environmental factors greatly influence the impact root loss will have on yield. Regression analysis was used to determine that beetle number explained only 17% of the variation in yield loss. When agronomic (phosphorous required) and environmental (June rainfall) factors were added to the linear regression equation, it explained 42% of the variation in yield loss. An economic threshold that is reliable and valuable for an indirect pest such as the CRW will have to be a comprehensive threshold that incorporates understanding environmental and agronomic interactions.

III. SAMPLING TO MANAGE EUROPEAN CORN BORER

The European corn borer (ECB), *Ostrinia nubilalis* (Hubner), was first introduced into North America from Europe in the early 1900s. Since its introduction, the ECB has expanded its distribution by adapting to a wide range of environmental conditions. In addition to attacking corn, the pest is known to attack snap beans, tomatoes, potatoes, peppers, apples, small grains, and cotton. Hodgson[39] lists over 200 crop and noncrop plants that the ECB is known to use as a host; these plants represent 131 genera of 40 plant families.

Beside having a wide host range and geographic distribution, ECB populations are quite variable in their voltinism expression (generations/year) and other developmental characteristics. It is believed that two races of the borer were introduced into North America, which differed in voltinism and host plant preference.[40] The original eastern race, established near Boston, had 2 generations/year with facultative diapause and a polyphagous feeding habit. The original western race, established near Lake Erie in Ontario, was univoltine and fed mostly on corn. As the pest expanded its range, the number of generations per season increased to two across most of the corn Belt, with a partial third generation occurring in the southern Corn Belt during warm summers. In the southern U.S., the pest can express up to 5 generations/year (Figure 1).

Another developmental characteristic that varies between ECB populations is the sex pheromone communication system. Males of the ECB can be separated by their attraction to blends of the isomeric sex pheromone 11-tetradecenyl acetate.[41] Males that are attracted to a mixture of 97% (Z)-11-tetradecenyl acetate and 3% (E)-11-tetradecenyl acetate are designated as Z-type borers, while males that are attracted to a mixture of 1% (Z)-11-tetradecenyl acetate and 99% (E)-11-tetradecenyl acetate are designated as E-type borers. Hudon et al.[40] suggested the existence of three ECB races: E-bivoltine, Z-bivoltine, and Z-univoltine. The existence of these three races and the coexistence of races at a geographic location influence both timing and length of insect activities. This affects when sampling should occur and whether or not it is economically justified. Sampling ECB populations requires an understanding of the insect's biology as it influences the timing and occurrence of the pest.

This section is not intended as a complete literature review on ECB biology and management but rather a synthesis of knowledge about the insect's biology and its influence on sampling methodology. Much of the material contained in the section is drawn from the second author's personal files and field experience with the pest. To

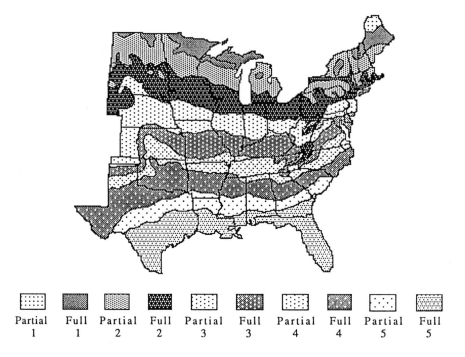

FIGURE 1. Predicted number of European corn borer generations and flights at each geographic location in the U.S. using 30-year climatology. Spatial interpolation of temperature data was provided by Zedx Inc., Boalsburg, PA.

help IPM practitioners understand the complex nature of ECB management, information is included on the effects of climate on seasonal occurrence of ECB life stages, the influence of corn plant phenology on attractiveness of cornfields to attack by ECB, and the influence of corn plant phenology on potential yield reductions from ECB feeding injury. This information is discussed in the context of ECB sampling. Finally, the economics of ECB scouting, in relation to local climate, are addressed.

A. VARIATION IN SEASONAL DEVELOPMENT AND OCCURRENCE

Management of *O. nubilalis* is not simple. The appropriate management decision is a function of geographic site as it affects climatic conditions which, in turn, affect other factors such as developmental rates, race mixture, postdiapause development and diapause induction, the complex of hosts, and host phenology. Climatic conditions also affect the host's ability to deal with stresses caused by the insect's feeding. The economics of management are influenced by crop market value, crop yield potential, control tactics and their efficacy, and the efficiency and effectiveness of sampling procedures. The seasonal developmental pattern of ECB at a geographic site determines whether there is any hope of economically controlling the pest in field corn. Therefore, before implementing a scouting program, one needs to know when ECB are active and whether economic control is possible at their site.

1. Expression of Voltinism at a Geographic Site

The driving variable that explains seasonal development of the ECB at a geographic site is weather, particularly seasonal degree-day (DD) accumulations. Table 7 shows the estimated number of degree-days required for an *O. nubilalis* population to reach initiation, peak, and termination of each life stage during a given generation

TABLE 7
Predicted Degree-Days (Developmental Threshold = 12.5°C) Required to Reach Initiation, Peak, and Termination of each Z-Multivoltine and Z-Univoltine European Corn Borer Developmental Stage

	Voltine type					
	Multivoltine			Univoltine		
Life stage	Init.	Peak	Termin.	Init.	Peak	Termin.
Overwintering						
Postdiapause dev.	75	200	350	200	425	650
Adult emergence	172	297	447	297	522	747
Preoviposition dev.	260	385	535	385	610	835
	(First Flight Period)					
First generation						
Egg stage	310	435	585	435	660	885
First instar	358	483	633	483	708	933
Second instar	407	532	682	532	757	982
Third instar	461	586	736	586	811	1036
Fourth instar	522	647	797	647	872	1097
Fifth instar	626	751	901			
Pupal stage	723	848	998			
Preoviposition dev.	811	936	1086			
	(Second Flight Period)					
Second generation						
Egg stage	861	986	1136			
First instar	909	1034	1184			
Second instar	958	1083	1233			
Third instar	1012	1137	1287			
Fourth instar	1073	1198	1348			
Fifth instar	1177	1302	1452			
Pupal stage	1274	1399	1549			
Preoviposition dev.	1362	1487	1637			
	(Third Flight Period)					
Third generation						
Egg stage	1412	1537	1687			
First instar	1460	1585	1735			
Second instar	1509	1634	1784			
Third instar	1563	1688	1838			
Fourth instar	1624	1749	1899			
Fifth instar	1728	1853	2003			
Pupal state	1825	1950	2100			
Preoviposition dev.	1913	2038	2188			
	(Fourth Flight Period)					
Fourth generation						
Egg stage	1963	2088	2238			
First instar	2011	2136	2286			
Second instar	2060	2185	2335			
Third instar	2114	2239	2389			
Fourth instar	2175	2300	2450			
Fifth instar	2279	2404	2554			
Pupal stage	2376	2501	2651			
Preoviposition dev.	2464	2589	2739			
	(Fifth Flight Period)					

TABLE 7—(*Continued*)

	Voltine type					
	Multivoltine			Univoltine		
Life stage	Init.	Peak	Termin.	Init.	Peak	Termin.
Fifth generation						
Egg stage	2514	2639	2789			
First instar	2562	2687	2837			
Second instar	2611	2736	2886			
Third instar	2665	2790	2940			
Fourth instar	2726	2851	3001			
Fifth instar	2830	2955	3105			
Pupal stage	2927	3052	3202			
Preoviposition dev.	3015	3140	3290			
	(Sixth Flight Period)					

(the accuracy of these predictions may be questionable beyond two generations). The seasonal DD (beginning January 1) needed to reach each life stage is dependent on whether the population is Z-multivoltine (bivoltine in much of its range) or Z-univoltine. The degree-day requirements of the E-multivoltine race have not been studied sufficiently; however, the spring flight resulting from overwintering individuals is known to occur earlier than that of the Z-multivoltine population and the flight of first generation adults (summer flight) occurs later.

Based on research conducted by Song[42] it was estimated that spring pupation begins at approximately 75 DD for a Z-multivoltine population, peaks at around 200 DD, and terminates at 350 DD, using a developmental threshold of 12.5°C.[43] A Z-univoltine population initiates spring pupation at about 200 DD, peaks at 425 DD, and terminates at 650 DD. The shape of the spring pupation curve at a site where both Z-multivoltine and Z-univoltine races coexist is dependent on the relative abundance of each race in the population but begins at approximately 75 DD and terminates at 650 DD. Figure 2 shows the variation in the spring pupation period of five geographic locations in Pennsylvania. At locations where the bivoltine race is dominant over the univoltine race (Mercer, Lancaster, Centre, and Lycoming Co.), the spring pupation curve is highly skewed toward early pupation (majority of pupation between 75 and 350 DD), while at locations where the univoltine race is dominant (Bradford Co.), the spring pupation curve is highly skewed toward later pupation (majority of pupation between 200 and 650 DD).

The shape of the spring pupation period at a site sets the pattern for phenological events in the life cycle of the pest population for the remainder of the season. The phenological events in the life cycle of an ECB population of concern in sampling and management are the adult flight period, the egg deposition period, and the first-instar eclosion period. Methods to monitor the pest during these periods are discussed in detail in later subsections.

Table 7 can be used to estimate when adult ECB should be actively depositing eggs at a given geographic location. The overwintering population initiates the adult flight at approximately 260 DD and terminates at 535 DD, if the population is 100% Z-multivoltine. If the ECB population is 100% univoltine at a site, then the flight begins at approximately 385 DD and terminates at 835 DD. At sites where mixed voltinism occurs, the flight begins at about 260 DD and continues through 835 DD. The second flight period begins at approximately 811 DD and terminates at 1086 DD. In locations where both a Z-multivoltine and Z-univoltine population coexist, the

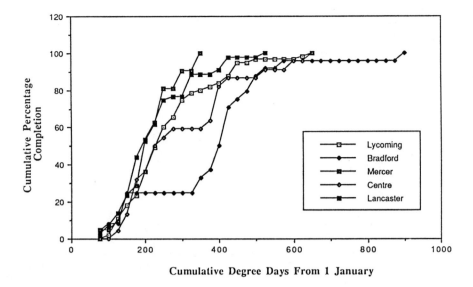

FIGURE 2. Spring pupation period of European corn borer populations collected at five geographic locations in Pennsylvania. (Adapted from Song.[42])

Z-univoltine and second flight of the Z-multivoltine populations may overlap and appear as one extended flight. The adult flight corresponds closely to the egg deposition period.

The degree-days shown are for the termination of a developmental stage, not the initiation. Therefore, the degree-day values shown for the first generation egg stage represent the period of first-instar eclosion (egg hatch). Using Table 7, a scout can estimate when to initiate and terminate scouting activities and the optimal period for implementation of control tactics. The degree-day predictions provided in Table 7 should be used only as a guideline for the occurrence of key phenological events and should be supplemented with field monitoring and/or a blacklight trap to monitor the flight.

2. Coexistence of European Corn Borer Races

Eckenrode et al.[44] studied the flight patterns of *O. nubilalis* in New York and observed three distinct flight peaks near King Ferry. They concluded that both univoltine and bivoltine races coexisted throughout much of the state. Glover et al.[45] verified this finding using electrophoretic techniques to separate the races.

In Pennsylvania, flight patterns vary greatly between geographic locations and across years. In the southeast region of Pennsylvania, two flights per year are common and are separated by about 1 month. In this area, however, each flight is typically bimodal (two humps) and longer than would be anticipated for a Z-multivoltine race (Figure 3). Dr. Charles Mason[46] has shown in Delaware that when Z-multivoltine and E-multivoltine races coexist, there is a bimodalism in the flight pattern. He also indicated that the E-multivoltine race flight activity is earlier during the spring flight than the Z-multivoltine race and later during the summer flight period. The outcome of this coexistence is an extended spring and summer adult flight period, egg deposition period, and first instar eclosion period. From a scouting perspective, this means that field monitoring must continue over a longer time period. From a

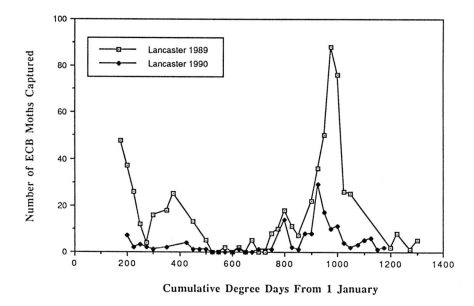

FIGURE 3. Summer adult flight activity for European corn borer collected using a blacklight trap in Lancaster Co., PA, in 1989 and 1990. (Adapted from Song.[42])

management perspective, this means that control efforts must take place over a longer time period and in some cases may not be cost effective.

In central Pennsylvania, the flight period varies greatly from year to year. In some years, the flight shows the typical Z-bivoltine pattern but never completely ends after the spring flight (Figure 4). In other years, there is a small spring flight (late May to mid-June), a large mid-summer flight (late June to mid-July), and a late-summer flight

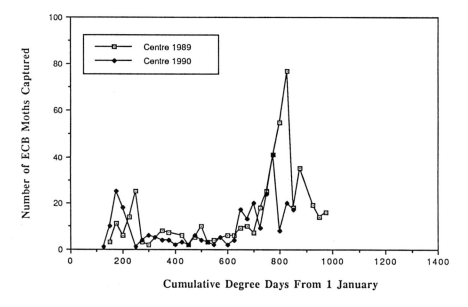

FIGURE 4. Summer adult flight activity for European corn borer collected using a blacklight trap in Centre Co., PA, in 1989 and 1990. (Adapted from Song.[42])

(late July to early September). In cool years, the flight appears to be continuous over the entire growing season. This variation in flight pattern is caused by changes in race dominance (multivoltine vs. univoltine) between years.

In the western regions of Pennsylvania, the ECB has a typical Z-bivoltine flight pattern. The spring flight occurs in late May to mid-June and the summer flight occurs from late July to late August.

Finally, in the northern highlands of Pennsylvania, the flight pattern is more typical of a univoltine population, with a much reduced bivoltine component. The main flight occurs during mid-summer in July. It is predicted that races will coexist in geographic locations where year-to-year variation in seasonal degree-days is adequate for bivoltine and univoltine genotypes to successfully exploit the environment and become fifth instars in time to overwinter after the last generation. Currently, no one has provided a suitable explanation of why the E-multivoltine race occurs sympatrically with the Z-multivoltine race.

Figures 5a to c are stylized representations of typical Z-multivoltine, E-multivoltine, and Z-univoltine flight periods. Figure 5a represents the most common flight pattern across the ECBs geographic range (central and western Corn Belt states). In areas where this flight pattern is common, European corn borer management has been promoted to control the pest. However, as are shown later in this chapter, as the sites become progressively cooler within this region, the ability to economically control the pest in field corn declines. Figure 5b shows a flight pattern that might occur in some coastal regions in the Mid-Atlantic states where the E-multivoltine race is known to exist. A univoltine flight is represented in Figure 5c and is typical in northern fringes of the ECBs geographic range and mountainous areas where the seasonal degree-days are reduced because of elevation. Figures 6a to c illustrate three possible flight patterns that develop when combinations of the three races coexist at a geographic location. When races coexist, management of the pest in field corn becomes more difficult using conventional insecticides because of the extended period of egg deposition and first-instar eclosion.

The relative magnitude of each race at a geographic location is determined by the frequency of seasonal degree-day accumulations at climatic extremes.[42] A genetic polymorphism of univoltine and bivoltine individuals maintains itself in the ECB population at sites where the shortest growing seasons will not allow second-generation insects to become fifth instars in time to overwinter; but during most years, the growing season is long enough for at least some of the individuals in a bivoltine population to become the fifth instars (from either the first or second generation). Figure 7 shows the areas in Pennsylvania where a univoltine, bivoltine, and mixed voltine population are predicted based on spatially interpolated 30-year climatic seasonal degree-day accumulations.

B. RELATIONSHIP BETWEEN CROP PHENOLOGY AND ECB EGG DEPOSITION

When sampling ECB populations it is important to understand the relationship between plant growth stage and the pest. Like all insects, the ECB has specific habitat requirements. The larval stage needs a host that will provide shelter from abiotic (weather, accidents, etc.) and biotic (parasites, diseases, predators, etc.) mortality factors and nutrients needed for proper growth and development. Because the larval stage does not have the long-distance mobility to seek out a suitable habitat, the adult stage determines where their offspring will feed.

Adults emerge in most geographic areas synchronized to the appropriate corn growth stage in the spring. In areas where the E-multivoltine race occurs, some ECB adults may emerge before the corn is in the ideal stage for successful colonization.

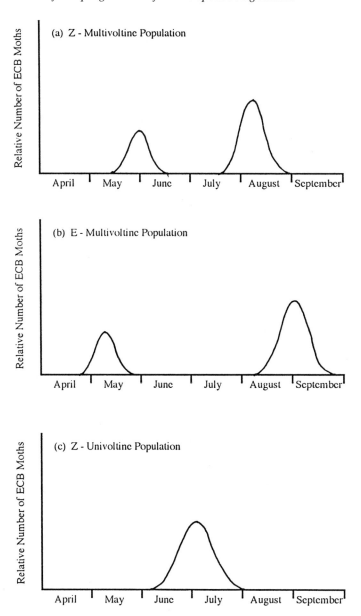

FIGURE 5. Stylized representation of a typical (a) Z-multivoltine, (b) E-multivoltine, and (c) Z-univoltine European corn borer flight period.

Therefore, they tend to exploit noncrop and crop species that begin growth and development before corn is planted or is adequate size for colonization.

To illustrate the interaction between ECB egg deposition by females and corn plant development, a typical Corn Belt growing season and Z-multivoltine race are used because of the simplicity of this system. In most areas of the Corn Belt, field corn is planted between mid-April and mid-June. ECB adult emergence occurs in the spring sometime between early May and mid-June. These adults move first to weedy areas of high humidity for mating and egg development. Once the eggs are ready to be laid, the females seek out cornfields that are tall enough to block wind movement

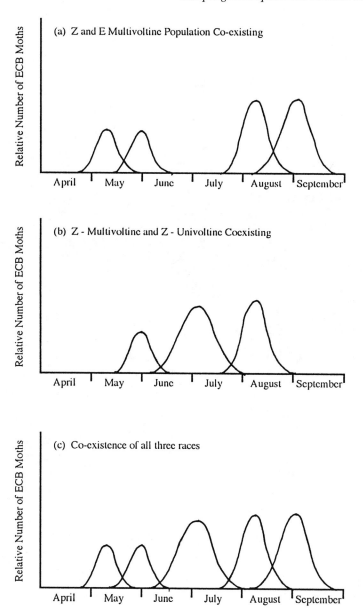

FIGURE 6. Stylized representation of a European corn borer flight period where (a) Z-multivoltine and E-multivoltine races coexist, (b) Z-multivoltine and Z-univoltine races coexist, and (c) all three races coexist.

(the area between rows is closed by leaves), which provide a high humidity environment and the available water needed for proper egg development. Field corn 46 cm (18 in.) or taller does not have adequate concentrations of 2, 4- dihydroxy-7-methoxy-1,4-benzoxazin-3-one (DIMBOA), a natural toxin to ECB to cause high mortality in neonate larvae;[47] therefore, larvae can colonize the crop.

When scouting field corn in the spring for ECB eggs and larvae, a scout should look across the landscape and pick the tallest fields in the area. Fields less than 46 cm tall are not attractive to female ECB for oviposition. Only in years when corn is

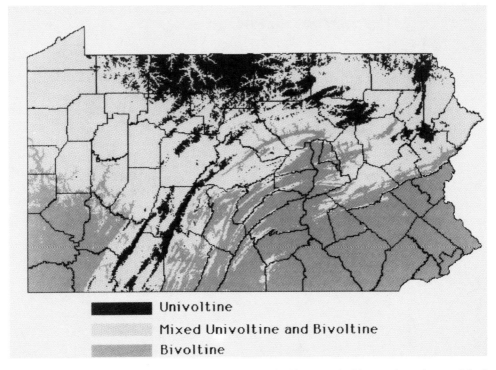

FIGURE 7. Predicted areas in Pennsylvania where Z-bivoltine, Z-univoltine, and a mixture of both European corn borer races coexist. Spatial interpolation of temperature data was provided by Zedx Inc., Boalsburg, PA.

planted late because of wet conditions or when all fields in the areas are of similar height will ECB lay large numbers of eggs in field corn less than 46 cm tall. In fact, if alternative hosts such as potatoes, apples, or noncrop species are abundant in the area, spring-emerging female ECB will lay their eggs on these hosts, rather than on short corn.

Once field corn has exceeded 46 cm, there does not appear to be any significant difference in attractiveness based on corn height.[48] Therefore, in areas where a univoltine population dominates or coexists with a multivoltine race, ECB females may continue ovipositing in a field throughout the vegetative growth period and move between adjacent fields.

As the season progresses, the corn initiates reproductive development, indicated by flowering (tasseling and silking) and pollination. It is during the green-silk stage that field corn is most attractive to female ECB. The reason for this attraction is not well understood but probably has to do with food quality and stalk succulence. It is possible that the species has evolved to lay eggs according to the availability of pollen, a high protein source, allowing neonate larvae to use the pollen as a quick energy source and then tunnel into the stalk before the plant lignifies. Regardless of the reason, the attractiveness of this plant growth stage is well documented.[47]

Once the silks have turned brown in a cornfield, the amount of egg laying declines rapidly. Only when all cornfields in area are the brown-silk stage by the time the summer generation of ECB adults emerge will these fields experience heavy egg deposition. Typically any fields in the green-silk state, when most fields in the immediate area are in the brown-silk state, are subject to heavy attack by ECB. Females will move to crops, such as peppers, if all cornfields in the area have progressed past the brown-silk stage.

In summary, spring-emerging ECB adults are attracted to the earliest planted cornfields, while summer-emerging ECB adults are attracted to the latest planted corn in the area. If the univoltine race occurs in the area, they may attack corn from 46 cm tall to the end of vegetative growth and into the reproductive stages of plant growth. The E-multivoltine race typically emerges before field corn is available for colonization and moves to alternative hosts. This is also true of the Z-multivoltine race in years when wet conditions prevent early planting, but temperatures are warm enough for normal ECB development. By understanding the relationship between corn phenology and corn's attractiveness to ECB females, a scout can limit which fields need to be sampled and reduce the time required for sampling activities.

C. RELATIONSHIP BETWEEN PLANT PHENOLOGY AND SUSCEPTIBILITY TO EUROPEAN CORN BORER FEEDING INJURY

A corn plant proceeds through a series of phenological stages during its growth and development.[49] During each of these stages, the plant allocates its resources to areas of rapid growth and for general plant maintenance. Because the plant varies in its ability to withstand stress during various stages of plant growth, the stage or stages when it is attacked by ECB influences its ability to deal with the insect's feeding injury. The period in the plant's development that ECB can attack and cause injury is from approximately 46 cm tall to physiological maturity. In general, during vegetative growth the majority of photosynthate produced by the plant and minerals are used for leaf and root growth and stem elongation. During the reproductive growth period the majority of photosynthate and nutrients are used for tassel and ear formation and grain fill.

1. Plant Phenology and Seed Yield Reduction Potential

When sampling to determine the potential impact of ECB on a cornfield, the synchrony of attack relative to plant growth stage is very important. Numerous researchers have conducted studies to assess the impact of ECB feeding on corn seed yield;[50-63] but their results varied, depending on plant growth stage when ECB stem feeding occurred and environmental conditions. Because the synchrony of ECB attack varies greatly from location to location, it is important to understand the relationship between timing of injury relative to crop phenology and the plant's response to injury.

Calvin et al.[64] published the first paper that attempted to mathematically integrate plant growth stage with the timing of second generation ECB stalk feeding in Kansas to predict potential yield reductions. This relationship was reasonable for the southern Midwest, but not other corn production regions. Bode and Calvin[65] conducted a study to determine the effect of ECB larval feeding on corn yields in Pennsylvania. Because both bivoltine and univoltine races of ECB occurred in the study area, they artificially infested four plant growth stages (10th leaf, 16th leaf, blister, and dough stage) of a common hybrid with ECB third instars. Table 8 shows the average percentage yield reduction per larva per plant by plant growth stage. Stalk tunneling during the vegetative growth stage is more damaging than during the reproductive stages. In addition, the work of Calvin et al.[64] and Bode and Calvin[65] show that from initiation of the reproductive period to physiological maturity, the impact of ECB stalk feeding declines. During the grain filling period, yield loss from ECB feeding is the result of stem tunneling which interferes with photosynthate and water movement in the plant. The later in the grain filling period that ECB stem tunneling is initiated, the lower the impact on yield. Injury during the vegetative growth period can influence plant height, leaf number, leaf size, and grain weight.[51]

Understanding the relationship between host developmental stage and ECB feeding injury can be used by a scout to assess the value of monitoring individual fields. As

TABLE 8
Percentage Reduction in Yield per *O. nubilalis* Larva per Plant
($\pm$ Standard Error) when Larval Feeding is Initiated during Each
Plant Growth Stage[a]

Plant stage	% Yield reduction per *O. nubilalis* per plant		
	1986	1987	Average
10th leaf	5.26 ± 1.05	6.61 ± 1.30	5.94 ± 1.17
16th leaf	4.33 ± 1.31	5.68 ± 0.93	5.01 ± 1.12
Blister	3.81 ± 0.96	2.45 ± 0.98	3.13 ± 0.97
Dough	3.01 ± 1.07	1.81 ± 1.00	2.41 ± 1.44

[a] Adapted from Bode and Calvin.[65]

discussed in the subsection on the economics of scouting, the returns to scouting are partially dependent on potential yield reductions caused by ECB feeding. The degree-days required of a hybrid to reach maturity, planting date, projected synchrony of ECB attack, and other additional plant stresses determine the potential yield loss caused by ECB feeding.

2. Economic Injury Levels and Economic Thresholds

The first economic injury levels (EIL) were developed from yield loss relationships established by Deay et al.[54] and Patch et al.[62] Both authors reported between 1 and 3% yield reduction per ECB larva per plant. Using this information and the experience of research and extension personnel, economic thresholds (ET) were established for both first and second generation ECB. Because management decisions are made during the egg stage and damage is the result of larval feeding, the ET used for ECB management are expressed as number of egg masses per plant. The original ET for first generation was 50% of the whorl-stage plants showing leaf feeding and/or egg masses with live larvae in the whorl region. For the second generation, the threshold was 50% of the plants with egg masses present. These ET served as guidelines for pest management decision makers, but were limited in their flexibility to account for changing economic and biological conditions.

Calvin[66] evaluated the utility of the ET used for second generation ECB management and found that they were inadequate for ECB management in Kansas. By comparing a computer-based ECB management model that accounts for changes in biological and economic factors with the traditional ET used at the time, he found that a corn producer would lose on average $2.40/ha ($0.96/acre) using the traditional ET and would gain on average $22.60/ha ($9.04/acre) using the computer-generated ET. It was determined that simple, single-value ET are of little utility in managing the ECB.

Showers et al.[47] provided a paper-based description of the ECB management model for development of ET. Average yield loss percentages for corn attacked by ECB during any of five plant phenological stages provided in the paper were modified from Lynch[60] and are representative of Midwest growing conditions. Bode and Calvin[65] provided yield loss levels for four corn growth stages (Table 8) and EIL (Table 9) that are representative of growing conditions in the northeast U.S. Bode and Calvin[65] and Showers et al.[47] used Equation 1 to calculate EIL:

$$EIL = TC/(CV * PL * PC) \qquad (1)$$

where TC = the total cost of a control tactic, including the cost of an insecticide and the cost of application, CV = expected crop value when marketed, PL = expected

TABLE 9
Economic Injury Level Values for *O. nubilalis* Larval Populations Initiating Tunneling During One of Four Plant Growth Stages when Control Cost and Crop Value are Varied and 100% Control is Assumed, Expressed as Larvae per Plant[a,b]

Crop value $/ha ($/acre)	Control cost, $/ha ($/acre)				
	30 (12)	35 (14)	40 (16)	45 (18)	50 (20)
10th leaf					
500 (200)	1.01	1.18	1.35	1.52	1.68
625 (250)	0.81	0.94	1.08	1.21	1.35
750 (300)	0.67	0.79	0.90	1.01	1.12
875 (350)	0.58	0.67	0.77	0.87	0.96
16th leaf					
500 (200)	1.20	1.40	1.60	1.80	2.00
625 (250)	0.96	1.12	1.28	1.44	1.60
750 (300)	0.80	0.93	1.06	1.20	1.33
875 (350)	0.68	0.80	0.91	1.03	1.14
Blister					
500 (200)	1.92	2.24	2.56	2.88	3.19
625 (250)	1.53	1.79	2.04	2.30	2.56
750 (300)	1.28	1.49	1.70	1.92	2.13
875 (350)	1.10	1.28	1.46	1.64	1.83
Dough					
500 (200)	2.49	2.90	3.32	3.73	4.15
625 (250)	1.99	2.32	2.66	2.99	3.32
750 (300)	1.66	1.94	2.12	2.49	2.77
875 (350)	1.42	1.66	1.90	2.13	2.37

[a] Crop value is calculated by multiplying the crop market value ($/kg) times the yield (kg/ha).
[b] Adapted from bode and Calvin.[65]

average proportional yield reduction per larva per plant, and PC = the expected proportional control from an insecticide application or other control tactic. Crop value is calculated by multiplying the expected yield [kg/ha (bushels/acre)] by the expected market value of the crop [$/ha ($/acre)]. Table 9 lists the EIL calculated for each crop growth stage assuming 100% control of the population. Note that the EIL is influenced by the effectiveness of the control tactic. As the effectiveness of the control tactic declines, the EIL increases. Therefore, to determine the true EIL, the EIL listed in the table must be divided by the expected proportional control. If the proportional control is unknown, 0.67 can be assumed. The subsection on the economics of sampling and management discusses the effect of the ECB flight period and insecticide kill function on proportional control.

Because a control decision cannot be made when larvae are present in the plant, the ET must be derived. In the case of ECB, the ET is expressed as the number of egg masses per plant. To calculate the ET for a cornfield, one must know the length of the appropriate generation's egg deposition period (equivalent to flight period).

Figure 8 shows the predicted length of the second generation flight period for all geographic locations in the U.S.. For most areas, with the exception of areas where races coexist, the first generation flight period is about 21 d. Once the appropriate flight period is determined, Equation 2 can be used to calculate the ET:

$$ET = EIL * PO/4.6 \tag{2}$$

where EIL = the economic injury level expressed as larvae per plant and PO = the proportion of the egg deposition period completed when the egg mass sample was collected. The value 4.6 is the expected average proportion of the population that will survive from the egg stage to the third instar (0.2) times the expected average number of ECB eggs per egg mass (23 eggs per egg mass). Equations 3 through 5 can be used to calculate the proportion of egg deposition completed by the last egg mass sample date. When the egg mass sample date is earlier than, or equal to, the peak in the egg deposition period and after initiation of the egg deposition period:

$$PO = x^2/a(a + b) \tag{3}$$

When the egg mass sample date is later than the peak of the egg deposition period but before the termination of the egg deposition period:

$$PO = 1 - (a + b - x)^2/b(a + b) \tag{4}$$

And when the sample date is later than or equal to the termination of oviposition:

$$PO = 1.0 \tag{5}$$

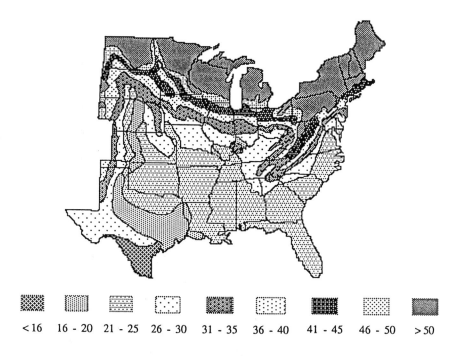

FIGURE 8. Predicted number of days from initiation to termination of the first generation European corn borer flight (second generation egg deposition) period for all geographic locations in the U.S. based on 30-year climatology. Spatial interpolation of temperature data was provided by Zedx Inc., Boalsburg, PA.

where x = (sample date) − (days from initiation of the egg deposition period), a = days from the initiation of the egg deposition period to the peak of the egg deposition period, and b = days from peak to termination of the egg deposition period. Figure 9 illustrates the relationship of each variable in the equations to a triangular egg deposition period.

The following example illustrates how an ET is calculated using the information provided in Figure 8 and Table 9.

A farm is located in central Iowa where the second generation egg deposition period is between 31 and 35 d (Figure 8). For the example we will use 33 d for the egg deposition period. The blacklight trap indicates that the second flight began on July 29 (initiation of egg deposition period). On August 6, the scout sampled for egg masses and noted that the corn was in the blister stage. He estimates that the cost to control will be $40.00/ha ($16.00/acre) and the value of the crop will be $625/ha ($250/acre). From Table 9 he finds that the EIL is 2.04 larvae per plant if he were to obtain 100% control of the infestation. However, he knows from experience that the control tactic will only kill 67% of the ECB population in the field. Using the information provided, he calculates the ET for the field to be scouted.

Step 1. Calculate the EIL when 67% control is expected.

$$EIL = 2.04/0.67 = 3.04 \text{ larvae/plant}$$

Step 2. Calculate the proportion of the egg deposition period complete (*PO*).

$$PO = (8)^2/16.5(16.5 + 16.5) = 0.12 \quad \text{or} \quad 12\%$$

Step 3. Calculate the ET

$$ET = 3.04 * 0.12/4.6 = 0.08 \text{ egg masses/plant}$$

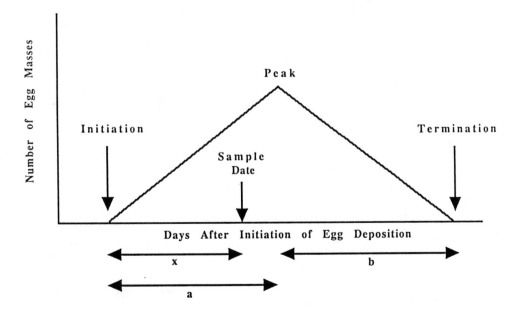

FIGURE 9. Stylized European corn borer egg deposition period and information to calculate the proportion of the egg deposition period completed by the sample date. The variable "x" is the days from initiation of the egg deposition period to the sample date. The variable "a" is the days from initiation to peak of the egg-deposition period. The variable "b" is the days from peak to termination of the egg deposition period.

D. SAMPLING TO MANAGE EUROPEAN CORN BORER

Earlier sections in this chapter discussed factors that influence ECB seasonal occurrence, the relationship between crop phenology and ECB egg deposition, the relationship between plant phenology and susceptibility to ECB feeding injury, and the relationship between the EIL and ET. In this section, sampling procedures are presented that utilize the knowledge discussed in previous subsections.

The value of a pest management scouting program is the site specific information provided to assist with decision making. To adequately manage the ECB, information is needed to determine if the insect is active at the site, whether there is a likelihood that the numbers of the pest will exceed the ET, when a control tactic should be timed for maximum effectiveness, and whether the control tactic provided acceptable control. Without this information, a farmer must rely on past experience to determine the need for control and to time control efforts. In the case of ECB management, this is a poor approach because the timing of key events in the insect's life cycle and the magnitude of population levels vary greatly between years. The degree to which pest management information can be obtained, however, depends on the human and time resources available to a farmer. The following subsections describe a ECB scouting program designed to provide all the above information for both first and second generation and possible shortcuts that can be taken to minimize scouting effort without increasing the risk of a wrong decision.

1. First Generation

The first step in a first generation ECB scouting program is to determine whether the insect is active in the area. This can be done in one of four ways. First, a blacklight trap can be set up to collect adult ECB; the numbers collected in the trap correspond to the relative number of egg masses being deposited in the surrounding cornfields (i.e., when high numbers are captured in the trap, egg laying activity is high in the area). Second, if the field scout can identify adult ECB moths, walking through grassy areas around cornfields will "scare up" moths if they are present. Third, corn stubble from last year can be split to see if the overwintering individuals are in the larval or pupal stage or if empty pupal cases are present in the stubble. If no empty pupal cases are found, adults have not begun to emerge. However, if all the stalks contain empty pupal cases, the entire adult emergence period is over and it is probably too late to properly time management activities. The ideal situation is when only a few of the stalks contain empty pupal cases and most contain live larvae and pupae; this indicates that the spring adult emergence period has just begun. Fourth, if neither a blacklight trap nor the expertise to identify adult ECB is available, scouting can be initiated when the corn reaches 46 cm. Prior to reaching this height, corn plants contain a toxin that prevents establishment of small ECB larvae in the plant. In fact, there is no need to scout fields that have not reached this height.

Once it is established that ECB adults are active in the area and/or 200 DD have accumulated since the cornfield reached 46 cm, then it is time to initiate a full-scale scouting program. The following scouting information was extracted from Showers et al.[47]

When scouting for first generation egg masses and larvae in whorl-stage corn plants, it is important to know where the adult females deposit their eggs. During the vegetative stages of growth, the majority of egg masses are deposited on the underside of fully emerged leaves, although some egg masses will be deposited on the stem and upper surface of the leaf. As a rule-of-thumb, fresh egg masses are deposited (on the third of the leaf nearest the stem) on the leaves that have just fully expanded out of the whorl. Larvae emerging from these egg masses move directly into the whorl area for shelter and to feed. Once in the whorl, the larvae tend to stay between the unexpanded leaves until the tassel begins to emerge from the whorl. At this time, the larvae are forced to move to other locations on the plant.

First and second instar ECB feed on the mesophyl of leaves, leaving areas on the leaves that appear transparent (windowpane effect) because the mesophyl has been removed, leaving one layer of epidermis. Early feeding can be detected by looking down into the whorl area and observing the unfurling leaves. At this time, if small larvae are present in the whorl, the outside leaves in the whorl will have small scarred areas. As the larvae grow, they begin to eat through the furled leaves in the whorl. This injury can either look like many odd-shaped holes in the newly emerging leaves or as a row of nice circular holes across the leaf.

When sampling for ECB, it is important not to sample along the edge of fields. As with many insects, ECB numbers tend to be higher along the edge near grassy weeds or wind breaks. To avoid an edge effect and obtain a representative sample from the field, do not sample the outside ten rows; if the field is large enough, walk 30 m or more into the field before initiating sampling. Within a field pay close attention to topographical, environmental, and cultural features that influence plant height, plant maturity, and plant density. In general, ECB adults resulting from overwintering larvae will lay more eggs in areas of the field with high plant populations or taller plants. If more than one variety is planted in the field, each variety should be sampled separately to avoid differences in attraction to adult females and larval survival.

In fields from 15 to 40 acres, sample 5 sets of 20 consecutive plants. In larger fields, take the same number of plants from subsets of 40 to 50 acres. In fields less than 15 acres, take 5 sets of 10 consecutive plants. Although less precise, as a practical compromise for very large fields (80 acres or more), surveys could be limited to 5 sets of 25 consecutive plants per field. Samples should be collected so that all areas of the field are represented. Figure 10 shows the estimated relationship between sampling cost per acre per visit and field size (assuming 0.67 h per field per visit and $9.00/h in

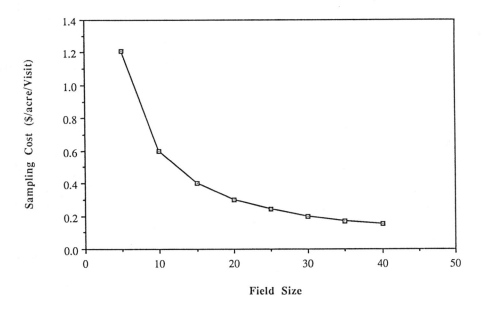

FIGURE 10. Estimated costs in dollars/ha ($/acre) per visit to scout for European corn borer by field size. The cost does not include time between fields and farms or the cost of operating a vehicle.

labor costs). These costs, however, do not include the cost of moving between fields on a farm and between farms or the costs of operating a vehicle. To calculate these costs, a person must value their time, determine the time required to move between fields on a farm and between farms, and the operating costs of a vehicle. Treatment guidelines and additional sampling details are available in Showers et al.[47]

2. Second Generation

The sampling process for second generation ECB is similar to the first generation, but management is much more complicated because of the small window available in the insect's life cycle for successful control. As with the first generation, a blacklight trap can be used to determine the initiation of adult activity in the area.

At this point in the chapter, a short discussion of the insect's behavior is needed. When ECB females move into a cornfield to deposit their eggs, they select the leaves near the ear. Larvae that emerge from these eggs move rapidly to protected sites, such as the ear below the husk, the leaf collar region, and accumulations of pollen and debris that collect at the base of the leaf at the stem interface. When the small larvae become second and third instars, some will tunnel into the leaf midrib until it is no longer large enough for their body. By the third and fourth stages, the majority of larvae tunnel into the stem, although some will continue to feed on or within the ear and in the leaf collar region. Because all insecticides used for ECB control work on contact and microbials, such as *Bacillus thuringiensis* Kurstaki, act as stomach toxins, the insect must come in contact with the poison before moving to a sheltered area on the plant. Therefore, effective control of the pest requires that the material be on the plant when the first instars eclose. Other limitations to management are discussed in the following subsection.

The main difference between sampling for first and second generation is the size of the corn plant. Typically, corn has entered the reproductive stages of development when second generation adults are actively seeking sites for egg deposition. As mentioned earlier, the females prefer fields in the green-silk stage of corn development. Because of the females' attraction to chemical volatiles given off by corn silks, the vertical egg deposition pattern on the plant follows a bell-shaped curve (Calvin et al.[67]), with about 91% of the egg masses being deposited on the three leaves above and below the ear leaf. However, this distribution may be skewed toward the top of the plant in more mature cornfields (fields in brown-silk stage when egg masses were deposited).

Knowledge of the dispersion pattern of ECB egg masses in the field and the vertical distribution on the plant can be used to increase sampling efficiency. Calvin et al.[67] conducted field research to estimate the number of sampling units needed per field to obtain a statistically valid estimate of the mean number of ECB egg masses per field. Their results indicated that when sampling at 32% completion of egg deposition in the field (optimal time to scout), 160 and 65 five-plant sampling units per field were needed to adequately sample ECB egg mass populations when mean density was 0.40 and 1.35 egg masses per five-plant sampling unit, respectively (this range roughly represents the range of egg mass numbers per five-plant sample occurring at 32% completion of oviposition). If 160 and 65 five-plant sampling units per field were taken and all leaves on the plant were examined, the time required per field would be 9.3 and 3.8 h, respectively, to scout a single field. Obviously, this was not a workable sampling program.

In an attempt to develop a sampling approach that minimized the time in a field for decision making, Calvin et al.[67] evaluated a sampling program that used 20 five-plant sampling units. Although this sampling unit number was not adequate at the 95% confidence and 25% precision levels, it proved to be quite adequate when the economics of decision making were included in the analysis. The proper management decision was reached in 83% of the fields tested. The cost of reducing the sampling unit number size to 20 five-plant samples and increasing the risk of wrong decisions was very low compared to the benefits that resulted from time saved in the field. The 20 five-plant sampling unit required only 1.16 h.

Calvin et al.[67] also evaluated reducing the number of corn leaves sampled. They found that by sampling only the three leaves above and below the ear leaf and the ear leaf that the time for scouting was reduced to 0.67 h per field without increasing the number of improper management decisions. Figure 10 shows the relationship between field size and scouting costs for ECB management. Clearly, the cost of scouting to manage ECB when farm and field size is small is considerably higher than for larger farms and fields. For this reason, field size must be considered before implementing a scouting program. Second generation ECB management and sampling procedures are discussed in detail by Showers et al.[47]

3. Larval Sampling to Initialize the European Corn Borer Phenology Model

Larval scouting can be used to collect larvae to initialize the ECB phenology model.[68] Sampling to initialize the ECB phenology model typically occurs between mid-June and mid-July; although in cooler areas of the insect's geographic range, sampling may be necessary in late July.

To manage the ECB second generation, a ECB phenology model was developed by Calvin.[69] Higgins et al.[68] further refined the model and developed computer software and a user's guide. The model predicts the time period of the second generation egg deposition period to optimize the timing of egg mass scouting and control tactic implementation. To initialize the model, information is needed on the age structure of the first generation and historic and real time maximum and minimum daily temperatures. The larval age structure is determined by sampling the first generation when the majority of the population is in the larval stage. For best results, the most abundant larval stage should be the fourth at the time the sample is collected. Higgins et al.[68] provide information on classification of larvae into the appropriate age class.

Unlike egg mass and larval sampling to assess potential or estimated damage, larval sampling to initialize the model requires a landscape approach to sampling rather than sampling within an individual field. A landscape approach is preferred because corn plant phenology during the first generation flight period influences the timing of egg deposition in the field. For example, early in the flight period many cornfields may not be tall enough to attract females to deposit their eggs. However, as the flight proceeds, these cornfields may reach a suitable stage of development to attract female ECB that are searching for sites to deposit their eggs. This changing of cornfield phenology over the first generation flight period can result in a segregation of the ECB age structure. If a sample were drawn to initialize the model from only one field, the sample would not include the entire ECB age structure in the general region and lead to an improper prediction of the second generation egg-deposition period. For this reason, several fields of varying phenological age should be sampled to estimate the entire range of the local ECB population age structure.

To collect a sample of the first generation larval age structure, locate a field in the area that is more mature than the majority of fields. Dissect the corn plants, collect the larvae, and save them in a vial of 70% alcohol for later determination of the population's age structure. With experience, the age class of a larvae can be determined in the field with a high degree of accuracy. It is not important to take a random sample. Locate corn plants that show leaf injury in the whorl or on the expanded leaves to reduce the sampling effort. This may cause a slight bias in the sample toward old larvae, but will have only a minor effect on the prediction. If possible, collect 20 to 40 larvae from the field. Next, locate a field that is medium in maturity for the area. Again, only dissect plants that show leaf injury and try to collect another 20 to 40 larvae. Finally, locate a field that is relatively young in maturity compared to other fields in the immediate area. Dissect plants and collect as many larvae as possible. Most likely, however, few larvae will be found in this field. An ideal sample size to initialize the model is 60 to 100 larvae, although a reasonable prediction can be made with 30 to 40 larvae.

E. THE ECONOMICS OF ECB SAMPLING AND MANAGEMENT

Sampling to manage the ECB should only be attempted in geographic areas where there is an opportunity to effectively control the pest. As shown earlier, the cost of scouting ECB is determined by field size and the number of visits to a field. In areas where field size is large, the cost of scouting is relatively low, whereas the cost of scouting a small field is relatively high. A key variable that determines average field size is the local topography. In relatively flat areas fields tend to be larger, while in hilly and mountainous areas fields tend to be smaller. Topography is also involved in controlling climate at a given location. A key climatic variable that influences the economics of ECB management is temperature and its effect on the rate of growing degree-days accumulated over the growing season. In cooler areas, the generation time is much longer than in warmer areas. This relationship affects the length of time that the life stage susceptible to control measures is active at a geographic location and the time span over which a control tactic must protect the crop.

In the case of ECB, the life stage of importance for control is the first instar. Figure 11 shows the variation in the time from initiation to termination of the second generation flight period for Pennsylvania. The period of second generation first instar eclosion (egg hatch) is similar in length but shifted about 50 DD (base threshold = 12.5°C) later than egg deposition.

To illustrate the effect of the first instar eclosion period length on ECB management, a 21- and 42-d period were selected (Figure 12). Figure 13 shows an insecticide kill function for ECB that is similar to a commonly used product that provides the longest period of protection. Using functions in Figures 12 and 13, the maximum level of control and optimal date for application can be predicted for the product. Figure 14a shows the predicted maximum level of control using the product for a 21-d first instar eclosion period. Applying the product 5 d after initiation of first-instar eclosion provides the maximum level of control, with 70.2% of the infestation controlled. In contrast, the same product only provides 48.2% control of the ECB population when the first-instar eclosion period is 42 d long (Figure 14b). To obtain this level of control, the product was applied 15 d after initiation of the first-instar eclosion period. From this example, it can be seen that the length of the first-instar eclosion period influences the ability of a control tactic to protect the crop.

Table 10 shows the benefit, cost, and average expected net gain for one to three insecticide applications aimed at controlling a potential ECB population of three larvae per plant when the first-instar eclosion period is 21- and 42-d long. When one application was applied optimally during the 21-d eclosion period, the population was decreased by 2.11 larvae per plant (70.2%). The benefit of this action was $19.81/acre and cost of control was $16.00/acre for a net gain of $3.81/acre. If a scout visited a

Sampling Arthropod Pests in Field Corn   **467**

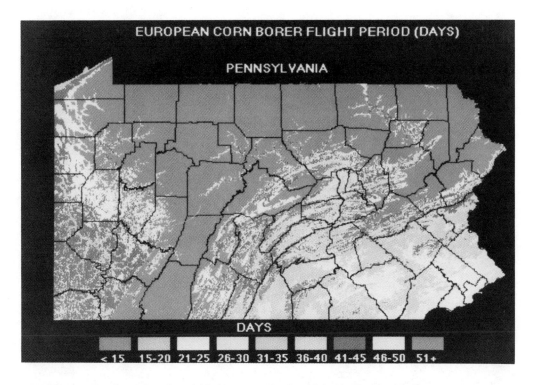

FIGURE 11. Predicted days from initiation to termination of the first generation European corn borer flight (second generation egg deposition) period for all geographic locations in Pennsylvania. Spatial interpolation of temperature data was provided by Zedx Inc., Boalsburg, PA.

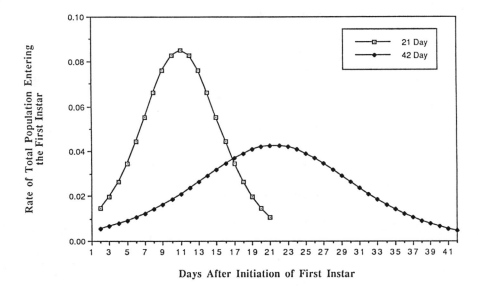

FIGURE 12. Rate of European corn borer population entering the first instar relative to days after initiation of first-instar eclosion.

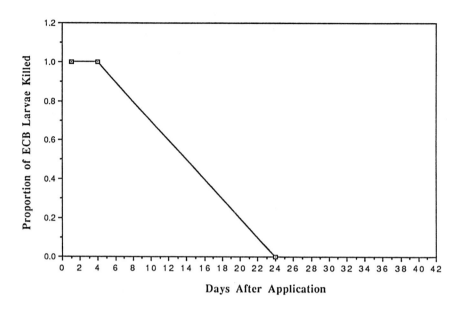

FIGURE 13. Proportion of European corn borer first-instar population killed by a representative insecticide by days after application.

TABLE 10
Benefit, Cost, and Net Gain of Insecticide Applications to Control Three European Corn Borer Larvae per Plant, when the First-Instar Eclosion Period is 21 and 42 d long[a]

No. of applications	No. of ECB larvae	Benefit	Cost	Net gain
\multicolumn{5}{c}{21-Day First-Instar Eclosion Period}				
0	3.0	0	0	0
1	0.89	$19.81	$16.00	$3.81
2	0.27	$5.92	$32.00	−$6.27
3	0.08	$1.75	$48.00	−$20.52
\multicolumn{5}{c}{42-Day First-Instar Eclosion Period}				
0	3.0	0	0	0
1	1.45	$14.55	$16.00	−$1.45
2	0.15	$12.30	$32.00	−$5.15
3	0.02	$1.23	$48.00	−$19.92

[a] Assumptions: crop value = $300/acre; control cost = $16.00/acre; yield reduction/ECB = 3.13%.

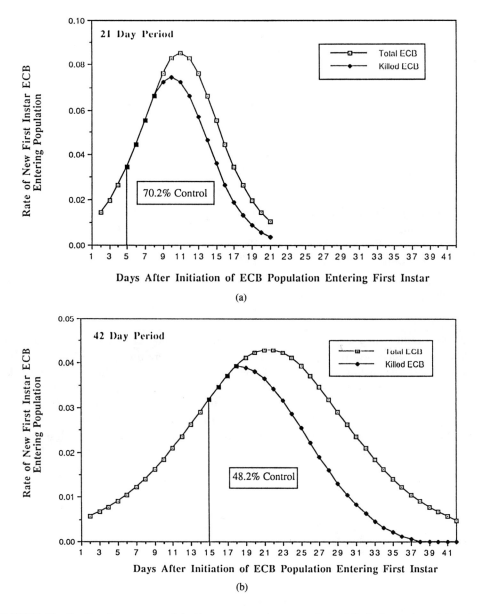

FIGURE 14. Predicted percentage of the European corn borer population killed when an insecticide with the kill function shown in Figure 13 is applied to maximize the level of control for (a) a 21-d first-instar eclosion period and (b) a 42-d first-instar eclosion period.

field two times to collect the information necessary to make a control decision and to time the insecticide application, then the cost of scouting would be approximately $3.10/acre for a 5-acre field and $0.70/acre for a 20-acre field. It can be seen that the activities associated with ECB management, when the flight period is 21 d, barely result in an economic return when the field is only 5 acres, whereas the net gain when the cost of scouting a 20-acre field is included is $3.11/acre. Table 10 also shows that two or more applications of an insecticide result in a negative net gain, because the marginal change in benefit is less than the marginal change in cost (i.e., the cost

TABLE 11
Net Gain from One Insecticide Application to Control Several Infestation Levels of European Corn Borer Larvae

No. of ECB larvae	Length of ECB first-instar eclosion period	
	21 days	42 days
3	$3.81	−$1.45
4	$10.40	$2.10
5	$16.99	$6.63
6	$23.55	$11.16

increased by $16.00/acre for each additional application, while the benefit increased at a decreasing rate).

In contrast to the 21-d first-instar eclosion period, there is no positive economic solution to controlling a potential population of 3 ECB per plant for the 42-d first-instar eclosion period using the insecticide. Regardless of the number of applications applied, a producer would always end up with a negative net gain to management. Table 11 shows the net gain for a 21- and 42-d first-instar eclosion period when various infestation levels of ECB are expected.

The economics of ECB management is site specific and dependent on a number of variables. A decision to implement a scouting program must consider field size, historic ECB infestation levels, potential yield reductions for the site as influenced by host and pest synchrony, and the effects of climate on the egg-deposition period. Climate regulates the developmental rate of the host, the complex of host plants, the developmental rate of the ECB, and the mix of races at a geographic site. Topography influences the climate and average field size in a geographic area. Although it has not been stated previously, implementation of a scouting program to manage ECB is not always economically feasible. The economic returns to the producer and crop consultant vary greatly across the ECBs geographic range. Before implementing a scouting program for ECB, all the variables discussed in this section should be considered relative to conditions at a geographic location.

REFERENCES

1. Metcalf, R. L., Foreword, in *Methods for the Study of Pest Diabrotica*, Krysan, J. L. and Miller, T. A., Eds., Springer-Verlag, New York, 1986, vii.
2. Krysan, J. L., Introduction: biology, distribution, and identification of pest *Diabrotica*, in *Methods for the Study of Pest Diabrotica*, Krysan, J. L. and Miller, T. A., Eds., Springer-Verlag, New York, 1986, chap. 1.
3. Chiang, H. C., Sisson, V., and Rasmussen, D. E., Conversion of results of concentrated samples to density estimates of egg and larval populations of the northern corn rootworm, *J. Econ. Entomol.*, 62, 578, 1969.
4. Sechriest, R. E., Observations of the biology and behavior of corn rootworms, *Proc. North Cent. Branch Entomol. Soc. Am.*, 24, 129, 1969.
5. Fisher, J. and Bergman, M., Field sampling of larvae and pupae, in *Methods for the Study of Pest Diabrotica*, Krysan, J. L. and Miller, T. A., Eds., Springer-Verlag, New York, 1986, chap. 6.
6. Bergman, M. K., Tollefson, J. J., and Hinz, P. N., Sampling scheme for estimating populations of corn rootworm larvae, *Environ. Entomol.*, 10, 986, 1981.

7. Bergman, M. K., Tollefson, J. J., and Hinz, P. N., Spatial dispersion of corn rootworm larvae (Coleoptera: Chrysomelidae) in Iowa cornfields, *Environ. Entomol.*, 12, 1443, 1983.
8. Weiss, M. J. and Mayo, Z B, Potential of corn rootworm (Coleoptera: Chrysomelidae) larval counts to estimate larval populations to make control decisions, *J. Econ. Entomol.*, 76, 158, 1983.
9. Edwards, C. H. and Fletcher, K. E., A comparison of extraction methods for terrestrial anthropods, in *Methods of Study in Quantitative Soil Ecology*: Population, Production, and Energy Flow, Blackwell Scientific, Oxford, U.K., 1971.
10. Piedrahita, O. H., Electrophoretic Identification of Larvae of Two *Diabratica* (Coleoptera: Chrysomelidae) and Their Competition and Movement, M.S. thesis, University of Guelph, Ontario, Canada, 1984.
11. Fisher, J. R., System for extracting corn rootworm larvae from soil samples, *J. Econ. Entomol.*, 74, 103, 1981.
12. Ruesink, W. G., Egg sampling techniques, in *Methods for the Study of Pest Diabrotica*, Krysan, J. L. and Miller, T. A., Eds., Springer-Verlag, New York, 1986, chap. 5.
13. Sisson, V. E. and Chiang, H. C., The distribution of northern corn rootworm eggs within a field, *Proc. North Cent. Branch Entomol. Soc. Am.*, 19, 93, 1964.
14. Foster, R. E., Ruesink, W. G., and Luckmann, W. H., Northern corn rootworm egg sampling, *J. Econ. Entomol.*, 72, 659, 1979.
15. Patel, K. K. and Apple, J. W., Ecological studies on the eggs of the northern corn rootworm, *J. Econ. Entomol.*, 60, 496, 1967.
16. Gustin, R. D., Effect of two moisture and population levels on oviposition of the western corn rootworm, *Environ. Entomol.*, 8, 406, 1979.
17. Hein, G. L., Tollefson, J. J., and Hinz, P. N., Design and cost considerations in the sampling of northern and western corn rootworm (Coleoptera: Crysomelidae) eggs, *J. Econ. Entomol.*, 78, 1495, 1985.
18. Gray, M. E., Hein, G. L., Boetel, M. A., and Walgenbach, D. D., Western and northern corn rootworm (Coleoptera: Chrysomelidae) egg densities at three soil depths: implications for future ecological studies, *J. Kan. Entomol. Soc.*, 65, 354, 1992.
19. Gunderson, H., Proposal on uniform sampling technique for rootworm eggs, *Proc. North Cent. Branch Entomol. Soc. Am.*, 19, 97, 1964.
20. Pruess, K. P., Weekman, G. T., and Somerhalder, B. R., Western corn rootworm egg distribution and adult emergence under two corn tillage systems, *J. Econ. Entomol.*, 61, 1424, 1968.
21. Weiss, M. J., Mayo, Z B, and Newton, J. P., Influence of irrigation practices on the spatial distribution of corn rootworm (Coleoptera: Chrysomelidae) eggs in the soil, *Environ. Entomol.*, 12, 1293, 1983.
22. Ruesink, W. G. and Shaw, J. T., Evaluation of a trench method for sampling eggs of the northern and western corn rootworms (Coleoptera: Chrysomelidae), *J. Econ. Entomol.*, 76, 1195, 1983.
23. Shaw, J. T., Ellis, R. O., and Luckmann, W. H., Apparatus and procedure for extracting corn rootworm eggs from soil, *Ill. Nat. Hist. Surv. Biol. Notes*, 96, 1976.
24. Chandler, J. H., Musick, G. J., and Fairchild, M. L., Apparatus and procedure for separation of corn rootworm eggs from soil, *J. Econ. Entomol.*, 59, 1409, 1966.
25. Foster, R. E., Corn Rootworm Egg Sampling, M.S. thesis, University of Illinois, Champaign, 1977.
26. Chiang, H. C. and Flaskerd, R. G., Sampling methods of adult populations of the corn rootworms, *Proc. North Cent. Branch Entomol. Soc. Am.*, 20, 67, 1965.
27. Tollefson, J. J., Witkowski, J. F., Owens, J. C., and Hinz, P. N., Influence of sampler variation on adult corn rootworm population estimates, *Environ. Entomol.*, 8, 215, 1979.
28. Morris, R. R., Sampling insect populations, *Annu. Rev. Entomol.*, 5, 243, 1960.
29. Steffey, K. L., Tollefson, J. J., and Hinz, P. N., Sampling plan for population estimation of northern and western corn rootworm adults in Iowa cornfields, *Environ. Entomol.*, 11, 287, 1982.
30. Tollefson, J. J., Owens, J. C., and Witkowski, J. F., Influence of sticky trap color and size on adult corn rootworm population estimates, *Proc. North Cent. Branch Entomol. Soc. Am.*, 30, 83, 1975.
31. Hein, G. L. and Tollefson, J. J., Comparison of adult corn rootworm (Coleoptera: Chrysomelidae) trapping techniques as population estimators, *Environ. Entomol.*, 13, 266, 1984.
32. Karr, L. L. and Tollefson, J. J., Durability of the Pherocon AM trap for adult western and northern corn rootworm (Coleoptera: Chrysomelidae) sampling, *J. Econ. Entomol.*, 80, 891, 1987.
33. Hein, G. L. and Tollefson, J. J., Use of the Pherocon trap as a scouting tool for predicting subsequent corn rootworm (Coleoptera: Chrysomelidae) larval damage, *J. Econ. Entomol.*, 78, 200, 1985.
34. Hein, G. L., Bergman, M. K., Bruss, R. G., and Tollefson, J. J., Absolute sampling technique for corn rootworm (Coleoptera: Chrysomelidae) adult emergence that adjusts to fit common-row spacing, *Environ. Entomol.*, 78, 1503, 1985.
35. Fisher, J. R., Comparison of emergence of *Diabrotica virgifera virgifera* (Coleoptera: Chrysomelidae) from cut and uncut corn plants in artificial and natural infestations, *J. Kan. Entomol. Soc.*, 57, 405, 1984.

36. Waters, W. E., Sequential sampling in forest insect surveys, *For. Sci.*, 1, 68, 1955.
37. Foster, R. E., Tollefson, J. J., and Steffey, K. L., Sequential sampling plans for adult corn rootworms (Coleoptera: Chrysomelidae), *J. Econ. Entomol.*, 75, 791, 1982.
38. Foster, R. E., Tollefson, J. J., Nyrop, J. P., and Hein, G. L., Value of adult corn rootworm (Coleoptera: Chrysomelidae) population estimates in pest management decision making, *J. Econ. Entomol.*, 79, 303, 1986.
39. Hodgson, B. E., The Host Plants of the European Corn Borer in New England, USDA. Tech. Bull. No. 77, 1928, USDA, Washington, DC, 63.
40. Hudon, M. E., LeRoux, E. J., and Harcourt, D. G., Seventy years of European corn borer (*Ostrinia nubilalis*) research in North America, in *Biology and Population Dynamics of Invertebrate Crop Pests*, Russel, G. E., Ed., Intercept, Andover, NH, 1989, 1.
41. Glover, T. J., Robbins, P. S., Eckenrode, C. J., and Roelofs, W. L., Genetic control of voltinism characteristics in European corn borer races assessed with a marker gene, *Arch. Insect Biochem. Physiol.*, 20, 107, 1992.
42. Song, P. Z., Seasonal Developmental Differences in European Corn Borer, *Ostrinia nubilalis* Hubner, (Lepidoptera: Pyralidae), Populations from Two Geographic Locations in Pennsylvania, M.S. thesis, The Pennsylvania State University, University Park, 1992.
43. Calvin, D. D., Higgins, R. A., Knapp, M. C., Poston, F. L., Welch, S. M., Showers, W. B., Witkowski, J. F., Mason, C. E., Chiang, H. C., and Keaster, A. J., Similarities in developmental rates of geographically separate European corn borer (Lepidoptera: Pyralidae) populations, *Environ. Entomol.*, 20, 441, 1991.
44. Eckenrode, C. J., Robbins, P. S., and Andaloro, J. T., Variations in flight patterns of European corn borer (Lepidoptera: Pyralidae) in New York, *Environ. Entomol.*, 12, 393, 1983.
45. Glover, T. J., Knodel, J. J., Robbins, P. S., Eckenrode, C. J., and Roelofs, W. L., Gene flow among three races of European corn borers (Lepidoptera: Pyralidae) in New York State, *Environ. Entomol.*, 20, 1356, 1991.
46. Mason, C. E., personal communication, 1992.
47. Showers, W. B., Witkowski, J. F., Mason, C. E., Calvin, D. D., Higgins, R. A., and Dively, G. P., European corn borer development and management, North Central Regional Extension Publication No. 327, 1989.
48. Patch, L. H., Height of corn as a factor in egg laying by the European corn borer moth in the one-generation area, *J. Agric. Res.*, 64, 503, 1942.
49. Richie, S. W. and Hanway, J. J., How a Corn Plant Develops, Cooperative Extension Service Special Report 48, Iowa State University, Ames, 1982.
50. Chiang, H. C., Cutcomp, L. K., and Hodson, A. C., The effects of the second generation European corn borer on field corn, *J. Econ. Entomol.*, 47, 1015, 1954.
51. Chiang, H. C. and Holdaway, F. G., Effect of *Pyrausta nubilalis* (Hbn.) on the growth of leaves and internodes of field corn, *J. Econ. Entomol.*, 52, 567, 1959.
52. Chiang, H. C., Holdaway, F. G., Brindley, T. A., and Neiswander, C. R., European corn borer populations in relation to the estimation of crop loss, *J. Econ. Entomol.*, 53, 517, 1960.
53. Chiang, H. C. and Hudon, M., Effect of European corn borer infestation on yield of two borer resistant inbred lines of maize, *Phytoprotection*, 58, 72, 1977.
54. Deay, H. O., Patch, L. O., and Snelling, R. O., Loss in yield of dent corn infested with August generation of the European corn borer, *J. Econ. Entomol.*, 42, 81, 1949.
55. Everett, T. R., Chiang, H. C., and Hibbs, E. T., Some factors influencing populations of European corn borer in the north central states, *Minn. Agric. Exp. Stn. Tech. Bull. 229*, 1958.
56. Hudon, M., Boivin, G., and Boily, R., Grain corn responses to five tillage methods under different European corn borer infestations, *Agric. Ecosyst. Environ.*, 30, 27, 1990.
57. Jarvis, J. L., Everett, T. R., Brindley, T. A., and Dicke, F. F., Evaluating the effect of European corn borer populations on corn yield, *Iowa State J. Sci.*, 36, 115, 1961.
58. Keller, N. P., Bergstrom, G. C., and Carruthers, R. I., Potential yield reductions in maize associated with an anthracnose/European corn borer pest complex in New York, *Phytopathology*, 76, 586, 1986.
59. Kwolek, W. V. and Brindley, T. A., The effects of the European corn borer, *Pyrausta nubilalis* (Hbn.), on corn yield, *Iowa State J. Sci.*, 33, 293, 1959.
60. Lynch, R. E., European corn borer: yield losses in relationship to hybrid and stage of corn development, *J. Econ. Entomol.*, 73, 159, 1980.
61. Patch, L. H., Still, G. W., Schlosberg, M., App, B. A., and Crooks, C. A., Comparative injury by the European corn borer to open-pollinated and hybrid field corn, *J. Agric. Res.*, 63, 473, 1941.
62. Patch, L. H., Deay, H. O., and Snelling, R. O., Stalk breakage of dent corn infested with August generation of the European corn borer, *J. Econ. Entomol.*, 44, 534, 1951.

63. Raemisch, D. R. and Walgenbach, D. D., Assessment of European corn borer (Lepidoptera: Pyralidae) impact on grain and silage yield in three areas of eastern South Dakota, *J. Kan. Entomol. Soc.*, 57, 79, 1984.
64. Calvin, D. D., Knapp, M. C., Xingquan, K., Poston, F. L., and Welch, S. M., Influence of European corn borer (Lepidoptera: Pyralidae) feeding on various stages of field corn in Kansas, *J. Econ. Entomol.*, 81, 1203, 1988.
65. Bode, W. M. and Calvin, D. D., Yield-loss relationships and economic injury levels for European corn borer (Lepidoptera: Pyralidae) populations infesting Pennsylvania field corn, *J. Econ. Entomol.*, 83, 1595, 1990.
66. Calvin, D. D., Welch, S. M., and Poston, F. L., Evaluation of a management model for second-generation European corn borer (Lepidoptera: Pyralidae) for use in Kansas, *J. Econ. Entomol*, 81, 335, 1988.
67. Calvin, D. D., Knapp, M. C., Xingquan, K., Poston, F. L., and Welch, S. M., Using a decision model to optimize European corn borer (Lepidoptera: Pyralidae) egg-mass sampling, *Environ. Entomol.*, 15, 1212, 1986.
68. Higgins, R. A., Poston, F. L., Welch, S. M., Calvin, D. D., Droste, B., Pontius, J., Brandsberg, G., Fischer, K., and Wolf, R., European Corn Borer Phenology and Management Models (User's Guide), Kansas Cooperative Extension Service, Manhattan, 1986.
69. Calvin, D. D., Evaluation and Revision of a European Corn Borer, *Ostrinia nubilalis*, Decision Model, Ph.D. thesis, Kansas State University, Manhattan, 1985.
70. Raun, E. S., personal communication.

Chapter 17

ESTIMATING ABUNDANCE, IMPACT, AND INTERACTIONS AMONG ARTHROPODS IN COTTON AGROECOSYSTEMS

L. T. Wilson

TABLE OF CONTENTS

I. Introduction 476

II. The Cotton Agroecosystem 477
   A. Cotton Growth and Development 477
   B. Arthropod Species Richness 478
   C. Damage Potential of Herbivorous Arthropods 478

III. Spatial Patterns of Distribution 480
   A. Intrinsic Factors Affecting Spatial Aggregation 484
   B. Extrinsic Factors Affecting Spatial Aggregation 484

IV. Quantifying Spatial Aggregation 486
   A. Single Point Estimates 486
   B. Aggregation as a Function of Density 487
   C. Impact of Sample Unit Size on Perceived Aggregation 488
   D. Impact of Sampler Efficiency on Perceived Aggregation 490
   E. Probability Distribution Functions 492
   F. Statistical Hierarchy 495

V. Sample Size Estimation 497
   A. Defining the Objectives of Sampling 497
   B. Population Sampling (Parameter Estimation) 498
   C. Commercial Monitoring (Decision Estimation) 500

VI. Optimizing Sampling Cost Reliability 504

VII. Binomial vs. Enumerative Sampling 505

VIII. Integrating Natural Enemies into the Decision Process 506

IX. Future Directions 507

Acknowledgments 508

References 508

ISBN 0-8493-2923-X/94/$0.00 + .50
© 1994 by CRC Press, Inc.

I. INTRODUCTION

My intent in writing this chapter is to provide readers with an overview of basic quantitative sampling principles as applied to cotton agroecosystems. This topic is broadly based and integrates aspects of several disciplines including ecology, economics, plant physiology, sociology, statistics, and, of course, entomology. Collectively these disciplines constitute a component of *biological systems analysis*, or if applied to the management of agricultural systems, *integrated cropping systems management*. Cropping systems by nature are complex and involve a multitude of interactions that are poorly understood at a quantitative level. A change in one component of a cropping system inevitably results in perturbations throughout that system. For example, a shift in soil structure from sandy-loam at one end of a field to clay-loam at the opposite end will have an effect on the soil's water holding capacity. This will, in turn, impact the soil biota and the growth and development of the crop. To a varying degree, depending on the species in question, the abundance and distribution of arthropods will also be impacted.

Other factors, including those intrinsic to the species being sampled, those due to interaction of a species with other biotic agents, and those that are attributable to sampling method or sampler, can greatly affect our perception of spatial patterns and processes. The combined impact of these factors results in complex spatial processes and species interactions. If estimation procedures are to provide accurate and reliable estimates of the abundance, impact, and interactions among arthropods, they must account for the effects of these variables. The integration of cropping systems management with sampling methodology continues to develop at an escalating rate. Although the key to understanding a biological system will not be found in this or any book, if coupled with an inquisitive and open mind, and a healthy enthusiasm for agriculture and field research, this book will hopefully encourage readers to address this challenging and exciting area.

As a prelude to the more statistical sections, the first section focuses on general crop growth and development, and the temporal assemblage of herbivorous arthropods found in cotton. My intent is to provide the readers a "biological" point of reference for later sections. Readers who wish to obtain a more detailed understanding of cotton production and management are encouraged to read *Integrated Pest Management Systems and Cotton Production* by Frisbie et al.[1] The second section focuses on how spatial patterns or our perception of spatial patterns are affected by sampler error, biotic, and physical factors. Subsequent sections focus on several areas which collectively encompass much of the current applied sampling literature. Included are sections on sample size estimation, economics, and decision criteria, binomial and enumerative sampling, and the role of natural enemies in the decision process.

As a final point of introduction, this is primarily a principles chapter and does not attempt to catalog and discuss the wide range of sampling methods that have been developed for cotton. How you swing a sweepnet or visually examine a cotton plant is, without question, important. However, this type of information can only be obtained by hands-on field experience. With few exceptions the principles which I discuss can be applied to any sampling method. Relevant references on how to sample arthropods and other organisms that inhabit cotton are provided in Wilson et al.[2] A thorough review of general sampling methods is provided by Southwood.[3]

II. THE COTTON AGROECOSYSTEM

A. COTTON GROWTH AND DEVELOPMENT

Cotton is produced on more than 32 million ha in over 40 countries,[4] and is grown as far north as 47°N lat. in the Ukraine and as far south as 32°S lat. in South America and Australia.[5] Approximately 90% of world production involves the species *Gossypium hirsutum* (L.), with an additional 8% involving the species *G. barbadense* (L.).[6] Cotton is produced in 17 states within the U.S., from California to Florida.

Cotton has an indeterminate growth pattern that makes it well suited for both dryland and irrigated production systems. Average fiber production ranges from 167 kg/ha, where rainfall is limited, to greater than 2400 kg/ha in the irrigated deserts of Arizona and California. In most of the U.S. Cotton Belt, a single fruiting cycle is completed each season. However, in the irrigated deserts of Arizona and California, season length, water, and nutrients are sufficient for a second cycle of fruiting. Although fiber production is the primary goal of the cotton industry, cottonseed accounts for about 17.3% of the total world oilseed production.[7]

The seasonal pattern of cotton growth and development can be divided into five stages (Figure 1), with each stage differing in its metabolic supply/demand balance.[8] During the *pre-squaring stage* (prefruit bud production), the cotton plant progresses from a condition of stress at germination, through early seedling development, to the establishment of sufficient leaf area to enable maximum potential growth of existing leaf, root, and stem tissue, and a maximum rate of production of new structures. The *early squaring stage* is characterized by rapid vegetative growth, resulting in the establishment of much of the cotton's canopy structure and the initiation of fruit bud production. The *peak squaring-early boll stage* is characterized by a period of rapid flux in metabolite availability. Many of the earliest produced squares reach anthesis (flowering) and begin to accumulate biomass at a rapid rate. As a result, the crop progresses from a favorable metabolic supply/demand balance to a metabolite deficit. The result is a near cessation of vegetative growth and the initiation of leaf senescence and fruit bud abortion. During the *boll maturation stage* the supply/demand balance continues to drop as demand by a greater number of bolls increases. Production of new fruit buds approaches zero, while the mortality of existing fruit buds and small bolls rapidly increases. Lower leaves in particular begin to senesce as

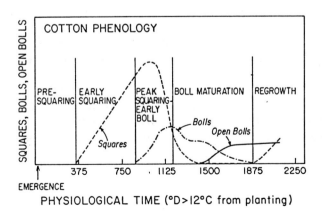

FIGURE 1. Phenological stages of cotton growth and development. (From Wilson et al.[124])

they export nonstructural carbohydrates and nitrogen to developing bolls. The *regrowth stage* is reached when the crop has matured a sufficient number of bolls to reduce the demand for metabolites and allow return to a favorable metabolic supply/demand balance. The regrowth stage is analogous to the early squaring stage and is again characterized by a favorable metabolic supply/demand balance, and correspondingly rapid vegetative growth and rapid production of fruiting buds.

The seasonal pattern of metabolic supply/demand balance has a major impact on the response of cotton to arthropod damage, with each crop growth stage differing in its sensitivity. Cotton is most responsive to leaf and fruit injury during the boll maturation stage and least responsive very early or very late in the season.[8] Moderate levels of injury to fruiting structures during early squaring can result in a yield increase.[8-10] Injury that might appear extremely severe during early stages of crop growth may have no effect on yield, quality, or maturity. In contrast, low levels of injury by fruit feeding herbivores during the boll maturation stages can have a major negative impact on yield and in some cases quality.[9]

B. ARTHROPOD SPECIES RICHNESS

A cotton agroecosystem can possess a diverse complex of arthropod species. Ellington et al.[11] identified over 50 species of herbivorous arthropods that feed on cotton in the western U.S. However, few are of major economic importance. Only nine groups are responsible for the majority of damage by arthropods to cotton across the U.S., with the majority of these normally prevented from reaching economically damaging levels by a complex of natural enemies.[12-14] Table 1 summarizes the importance of each herbivore group, ranked by their estimated damage for each cotton producing state. As seen in this table, *Lygus hesperus* Knight and spider mites, *Tetranychus* spp., account for an estimated 97% of arthropod-induced yield loss in California. In contrast these herbivores accounted for an estimated 4.5% of arthropod-induced yield loss in Texas. The complex of herbivorous arthropods inhabiting cotton in each geographic region is quite varied and is dependent upon season length, rainfall, temperature, surrounding vegetation, and agronomic as well as pest management practices. Each of these factors varies from one cotton growing region to the next.

C. DAMAGE POTENTIAL OF HERBIVOROUS ARTHROPODS

Although it is difficult to generalize about the seasonal abundance of each arthropod for a wide geographical area, herbivores can be grouped by the manner in which they feed upon cotton (Table 2). This grouping is relevant when assessing the impact of a species on crop damage and yield. Insects that feed on fruiting structures (direct pests) typically have lower economic injury levels than those that feed on leaves or terminal meristem tissue (indirect pests). For example, the economic injury level for the cotton bollworm, *Heliocoverpa* (= *Heliothis*) *zea* (Boddie) on irrigated cotton is approximately 0.1 third instars per plant during the boll maturation stage[8] In contrast, the beet armyworm, *Spodoptera exigua* Hübner, a leaf feeder and occasional fruit feeder, will not cause economic loss even at densities of 6 to 11 third instars per plant at this same crop growth stage.[15] Smaller arthropods require even higher densities before causing economic loss. Goodell and Roberts[16] and Wilson[17] present data suggesting that 80% of leaves with spider mites at the location within the canopy that is most likely to have mites[18] is not sufficient to cause economic loss. This is equal to approximately 150 spider mites per plant. Detailed analyses examining the time at which spider mites invade a cotton field, the rate with which they increase in population size, and the resulting yield loss suggests that 80% infested leaves is a

TABLE 1
Estimated Yield Loss to Cotton by Herbivorous Arthropods as a Percent of Each State's Losses.[154]

State	Aphids and whiteflies	Armyworms and loopers	Boll weevil	Fleahoppers	Heliothis spp.	Lygus spp.	Pink bollworm	Spider mites	Thrips	Others
Alabama	1.96	10.91	35.98	0.00	32.39	18.35	0.00	0.04	0.04	0.33
Arizona	8.97	5.13	0.43	0.00	8.55	17.52	56.41	1.71	0.00	1.28
Arkansas	10.17	0.00	36.94	0.00	32.37	1.72	0.00	0.00	18.80	0.00
California	0.00	0.95	0.00	0.00	1.97	60.33	0.47	36.28	0.00	0.00
Florida	2.50	25.67	15.75	0.00	50.97	0.09	0.00	0.00	0.37	4.65
Georgia	6.40	34.34	0.00	0.00	55.06	1.05	0.00	0.00	0.00	3.15
Louisiana	5.24	0.00	43.75	2.77	43.75	2.92	0.00	0.45	1.12	0.00
Mississippi	6.31	0.27	52.51	0.30	26.13	7.40	0.00	0.54	6.25	0.29
Missouri	31.62	0.00	63.37	0.00	0.00	0.00	0.00	2.51	0.00	2.50
New Mexico	5.77	3.22	0.00	7.11	33.69	5.64	0.40	1.61	42.28	0.28
N. Carolina	0.00	0.00	0.00	0.00	90.05	0.00	0.00	0.00	9.42	0.53
Oklahoma	0.55	0.00	85.08	0.00	13.26	0.00	0.00	0.00	0.00	1.11
S. Carolina	3.25	0.62	0.00	0.00	32.67	0.38	0.00	0.88	51.69	10.51
Tennessee	0.00	0.00	87.55	0.00	0.00	0.88	0.00	0.00	6.91	4.66
Texas	16.45	0.00	45.32	4.68	23.87	0.16	0.00	4.35	1.94	3.23
Virginia	0.00	0.00	0.00	0.00	100.00	0.00	0.00	0.00	0.00	0.00
All states	5.97	1.95	29.82	1.19	20.28	22.23	1.52	12.04	3.90	1.10

TABLE 2
Feeding Preference for Different Plant Parts for the Major Cotton Herbivores

Herbivore	Fruiting structures	Leaves	Terminal meristem tissue
Aphids and whiteflies		●	
Armyworms and loopers	•	●	●
Boll weevil	●		
Fleahoppers spp.	●		
Heliothis spp.	●		•
Lygus spp.	●		•
Pink bollworm	●		
Spider mites		●	
Thrips		●	•

● = primary form of damage to cotton; • = secondary form of damage.

conservative threshold and that economic loss will not occur until considerably higher densities are reached.[19]

III. SPATIAL PATTERNS OF DISTRIBUTION

Central to a sound management or research program is one or more sampling procedures that address the spatial dynamics of the species being studied. Taylor et al.,[20] in analyzing 156 sets of field data comprising 102 species, presented strong evidence that most arthropods are spatially aggregated. Wilson and Room[21] in their analysis of 23 arthropod species or age classes and three categories of fruiting structures, reached a similar conclusion for the cotton agroecosystem.

Aggregation implies a degree of nonrandom spatial association between individuals within a species or between individuals of differing species (see Reilly and Sterling[22] and Schoenig and Wilson[23]). The degree of spatial association and resultant aggregation varies dramatically among different species of arthropods. As an example, the between-plant spatial pattern of the egg stage of *Heliocoverpa* (= *Heliothis*) *armigera* (Hübner), tends to be slightly aggregated, while tetranychid mites are highly aggregated. Data for *H. zea* suggest that an average of 1.3 to 1.89 eggs are oviposited during each oviposition visit to cotton.[24,25] Upon hatching, neonate *Heliocoverpa* larvae exhibit only limited plant-to-plant movement[26] and at low levels of mortality largely maintain the spatial aggregation that was imposed by the ovipositing females. Similarly, once a female tetranychid mite begins to oviposit, often on a leaf surface, she will likely remain on the leaf unless consumed by a predator or agitated by predator feeding. Similarly, a large proportion of her progeny will remain on the substrate where the eggs were laid. Plant-to-plant movement of spider mites is most common in female mites prior to oviposition, although for most of the season plant-to-plant movement is rare. Interplant movement commonly occurs when host quality decreases rapidly, such as during the fruit maturation phase,[18] and during periods of rapid spider mite population increase.[27] When host quality is low, spider mites respond by moving to plant terminals, which facilitates wind dispersal.

Each species and often each age class within a species has a preference for a particular part of the crop canopy, and although the within-plant distributions of species overlap, they are nevertheless often very distinct.[14,18,25,29-36] Figure 2 derived from data presented by Pickett et al.,[37] Wilson and Gutierrez,[34] and Wilson et al.[18,25,35,38] approximates the vertical within-plant distribution of five species of generalist predators (*Chrysoperla* (= *Chrysopa*) *carnea* Stephens, *Frankliniella occidentalis* Pergande, *Geocoris* spp., *Orius tristicolor* White, and *Nabis americoferus* Carayon) and five species of herbivores (*F. occidentalis*, *H. zea*, *Lygus hesperus*, *Tetranychus* spp., and *Trichoplusia ni* (Hübner)). *Frankliniella occidentalis* is listed as both an entomophage and herbivore because it exhibits a predilection to feeding on spider mite eggs.[39-41]

The vertical within-plant distribution for most of these species is proportional to the number of nodes on the plant. For example, *H. zea* eggs are on average found ca. 27% of the nodal distance from the terminal. Fairly early in a season, when the plants have an average of ten mainstem nodes, the seventh mainstem node above the cotyledonary node will on average have the greatest number of eggs. Similarly, later in the season when 20 mainstem nodes are present, the 15th mainstem node will on average have the greatest egg density. Because internode elongation is much less in the upper compared to the middle or lower part of the plant, the large majority of *Heliocoverpa* eggs are located near the plant terminal.

Not all cotton arthropods have a vertical distribution pattern which is proportional to the number of mainstem nodes. *Tetranychus* spp. and, to a lesser degree, *Frankliniella occidentalis* are attuned to the nutrient status of the host plant. For example, when spider mite population density increases rapidly and begins to deplete plant nutrients, the mites respond by dispersing up the plant to feed on less damaged leaves. An explosive spider mite population will ultimately reach the terminal of the plant from which they will then begin to initiate aerial dispersal in search of more favorable habitats. As a result of this response to plant quality, spider mites will be found low on the plant, very early in a season. During the middle of a season they rapidly disperse up the plant, and very late in the season they will be found in the upper terminal nodes.[18]

For the nine arthropods shown in Figure 2, 19 vertical distributions (by age class and time of season) are shown. The closer the average mainstem node of distribution for the different species, the greater the degree of overlap in their vertical distributions. The degree of overlap comparing the distribution of any two of these can be calculated using Figure 3. For example, the average relative node of distribution from the cotton terminal for nymphal *O. tristicolor* and eggs of *T. ni* is 0.438 and 0.420, respectively. Therefore, the difference between their average relative nodes is 0.018. From Figure 3 this value corresponds to an overlap of 94%, suggesting a relatively high probability of encounter. In comparison, the average relative node of distribution of adult *L. hesperus* and immature *N. americoferus* is 0.258 and 0.582, respectively. The difference of 0.324 corresponds to an overlap of only 11%, suggesting that even if adult *Lygus* were suitable prey for *Nabis*, encounters between these stages of these species would be exceedingly rare except when one or both species are at high densities.

Although vertical synchrony is an important component that should be used when estimating the relative value of different entomophagous arthropods, it is insufficient in determining the ability of an entomophage to control one or more herbivorous species. At each node of a cotton plant there are several structures which a species might occupy. Table 3, modified from Wilson and Gutierrez[34] and Pickett et al.,[37] suggests considerable variation in the structural distribution of different predators on cotton. Data presented by these authors also suggest that seasonal changes in the number of each structure type (e.g., leaves, stems, fruit) directly affect the distribution of each species, by providing a dynamic set of resources upon which each species (or age class) uniquely distributes itself based on its individual biological requisites.

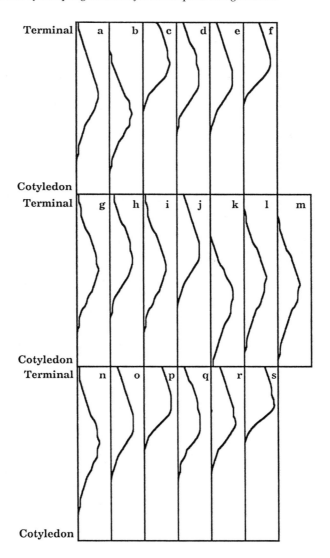

FIGURE 2. Relative within-plant mainstem node distribution of cotton arthropods: (a) *Trichoplusia ni* eggs, (b) *Trichoplusia ni* larvae, (c) *Heliocoverpa zea* eggs, (d) *Heliocoverpa zea* larvae, (e) *Lygus hesperus* nymphs, (f) *Lygus hesperus* adults, (g) *Orius tristicolor* nymphs, (h) *Orius tristicolor* adults, (i) *Geocoris* spp. nymphs, (j) *Geocoris* spp. adults, (k) *Chrysoperla carnea* larvae, (l) *Nabis americoferus* nymphs, (m) *Nabis americoferus* adults, (n) *Frankliniella occidentalis* (immatures + adults) at 500°D (> 12°C from planting), (o) *Frankliniella occidentalis* at 1000°D, (p) *Frankliniella occidentalis* at 1500°D, (q) *Tetranychus* spp. (mobile stages) at 500°D, (r) *Tetranychus* spp. at 1000°D, and (s) *Tetranychus* spp. at 1500°D. (From Wilson.[61])

Studies by Jones and Gutierrez[42] show that the structural distribution of *C. carnea* and *O. tristicolor* switches from leaves when spider mites are abundant to flowers when adult thrips are abundant. In contrast, *Geocoris* spp. and *Collops vitatus* Say does not switch in this way; they search primarily on leaves. Although two species may occupy the same location they may be asynchronous. *Orius tristicolor* is a voracious consumer of spider mites, but it typically does not reach high abundance until mid-season, long after spider mites have colonized the cotton fields. Other factors, such as time of day (probably a temperature effect), have also been shown to

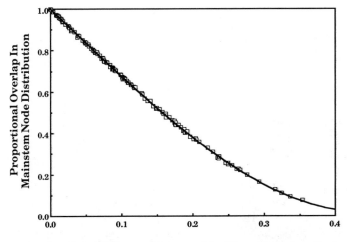

FIGURE 3. Proportional within-plant mainstem node distribution overlap of cotton arthropods as a function of the mean mainstem node location. (From Wilson.[61])

TABLE 3
Proportion of Predators upon Different Parts of the Cotton Plant (Averaged for Season)[34,37]

Predator	Fruit	Leaves	Branches
Orius			
Adults	0.764	0.228	0.008
Nymphs	0.550	0.440	0.010
Geocoris			
Adults	0.716	0.243	0.041
Nymphs	0.626	0.345	0.029
Eggs	0.279	0.706	0.015
Chrysoperla			
Larvae	0.532	0.457	0.011
Eggs	0.115	0.878	0.007
Nabis			
Adults	0.625	0.375	0.000
Nymphs	0.273	0.727	0.000
Frankliniella			
Adults	0.338	0.662	—
Larvae	0.166	0.834	—

influence the efficiency of sampling methods for entomophagous arthropods in cotton, suggesting that they also influence vertical and/or structural distributions.[43,44]

Although the majority if not all cotton arthropods have spatial patterns that can be classified as aggregated, it would be erroneous to conclude that their spatial patterns are always aggregated. While some species tend to be more highly aggregated than others, any single species might have a uniform, random, or aggregated spatial pattern, depending on the conditions under which it is sampled. The degree to which a species exhibits spatial aggregation is a function of several factors, including those intrinsic to the species being studied, those specific to the biotic and physical conditions present at the time of study, and those that are a function of the estimation procedures used in the study (Table 4).

A. INTRINSIC FACTORS AFFECTING SPATIAL AGGREGATION

Factors that are intrinsic to a species, such as oviposition pattern, life history strategies (e.g., polyembryony), immature dispersal characteristics, or the presence or absence of aggregation pheromones, sets bounds on the aggregation characteristics of each arthropod species. The response of a species to a kairomone produced by one or more of its hosts similarly results in a spatial pattern related to that of the host species. As a general rule, species that do not produce an aggregating pheromone become progressively less aggregated with age. This effect is largely influenced by dispersal and mortality. Even species that oviposit in masses, such as armyworms and pentatomids, exhibit only limited aggregation by the time they enter pupation or molt to the adult stage. A species spatial pattern can progress from being aggregated, to random, to uniform as density decreases.[45] At densities approaching zero, an increasing number of sample units must be examined to distinguish an aggregated spatial pattern from either a random or a uniform spatial pattern.

B. EXTRINSIC FACTORS AFFECTING SPATIAL AGGREGATION

Factors that are extrinsic to a species, such as natural enemy-induced mortality, reduce the degree of spatial aggregation. *Chrysoperla carnea* and *O. tristicolor* aggregate in areas having the greatest density of prey and thereby reduce prey aggregation

TABLE 4
Factors Affecting the Spatial Pattern of Biological Organisms

Intrinsic to the species being sampled	Examples
Oviposition pattern	*Helicoverpa* (= *Heliothis*) *zea* (Boddie) oviposits eggs individually while *Spodoptera exigua* Hübner oviposits in masses of several hundred eggs.
Polyembryonic life strategies	*Copidosoma truncatellum* (Dalm.) inserts a single egg into the egg of *Trichoplusia ni* Hübner, which results in the development of several hundred parasitoids.
Dispersal characteristics of immatures	Nymphal *Lygus hesperus* Knight rapidly disperse upon emerging from their egg and do not demonstrate a propensity to aggregate with other *Lygus* nymphs. *Estigmene acrea* Drury remain aggregated during the first three larval instars and only begin to disperse during the last two instars.
Aggregation pheromones	Female *Tectocoris diapthalmus* (Thunberg) produce an aggregation pheromone that attracts males of the species.
Host aggregations kairomones	*Hyposoter exiguae* (Vier.) uses kairomones produced by *Spodoptera exigua* larvae to locate its prey. This results in the spatial pattern of this parasitoid being similar to that of its host.
Species density	A species spatial pattern is dependent upon its density. The variance/mean ratio will progress from < 1.0 (uniform spatial pattern) at low densities, to = 1.0 (random spatial pattern), to > 1.0 (aggregated spatial pattern) at higher densities. The density range at which the spatial pattern changes from uniform to random to aggregated will be lower for a species that exhibits a high propensity to aggregate, such as *Aphis gossypii* Glover, and higher for a species that exhibits a low propensity to aggregate, such as *Anthonomus grandis* Boheman.

TABLE 4—(Continued)

Extrinsic biotic factors	Examples
Natural enemy-induced mortality	Natural enemy mortality reduces the degree of spatial aggregation. Natural enemy species, such as *Chrysoperla carnea* (Stephens) and *Orius tristicolor* White, aggregate in areas having the greatest density of tetranychid mite hosts and thereby reduce host aggregation to a greater degree than do natural enemies, such as *Geocoris* spp., that aggregate less to areas of high host density.
Host quality	As host quality decreases with cotton crop maturation, the rate of dispersal of the arthropod herbivores increases, resulting in reduced spatial aggregation. This is particularly apparent with the more aggregated species such as the *Tetranychus* spp. Highly heterogeneous host quality, such as that due to sand streaks and uneven topology, can result in increased aggregation.
Host refugia	Alternative host refugia, such as an almond orchard upwind to a cotton field, can result in higher densities of *Tetranychus urticae* Koch on the side of the cotton field closest to the almonds. Infestations of green amaranth, *A. hybridus*, can result in clumped infestations of *Spodoptera exigua* Hübner, particularly in the adjacent cotton plants.
Abiotic-induced mortality	Rainfall, extreme temperatures, low humidity, and wind-induced mortality reduce the degree of spatial aggregation by essentially randomly thinning the population.

Extrinsic abiotic factors	Examples
Environmental toxicants	Some insecticides, particularly pyrethroids, have been observed to increase the spatial aggregation of *Heliothis* spp. eggs. When ovipositing females experience sublethal doses of pyrethroids, the distance flown between oviposition events can decrease drastically and result in single leaves or single plant terminals receiving up to several dozen eggs.
Prevailing winds	Prevailing winds can result in direction gradients in densities for a range of cotton arthropods.

Estimation procedures	Examples
Sampler bias	Procedures which have a lower catch efficiency under estimate the degree of spatial aggregation. This problem is exacerbated at low densities.
Sample unit size	As the sample unit size is increasingly reduced or expanded in comparison to a "biologically relevant" sample unit (species and age class specific), the apparent spatial pattern will approach random.
Sample size	Except when the total population is sampled, samples provide estimates of parameters and not the actual parameter value. The lower the sample size, the more difficult it is to statistically separate an observed spatial pattern from what would be expected from either a uniform, random, or aggregated spatial pattern.
Species density	As density decreases, sample size must increase to enable statistical separation of an observed spatial pattern from what would be expected from either a uniform, random, or aggregated spatial pattern.

to a greater degree than do natural enemies, such as *Geocoris*, which appear not to aggregate in response to prey. As host quality decreases with cotton crop maturation, the rate of dispersal of herbivorous arthropods increases, resulting in reduced spatial aggregation. This is particularly apparent with the more aggregated species such as the *Tetranychus* spp. Highly heterogeneous host quality, such as that due to sand streaks, hard pans, and uneven topology, can result in increased aggregation. Alternative host refugia, such as an almond orchard upwind to a cotton field, can result in higher densities of *Tetranychus urticae* Koch on the side of the cotton field closest to the almonds. Infestations of *Spodoptera exigua* (Hübner) on green amaranth, *Amaranthus hybridus*, can result in clumped infestations on adjacent cotton plants. Abiotic-induced mortality, such as that caused by rainfall, extreme temperatures, low humidity, and wind reduce the degree of spatial aggregation by randomly thinning the population. Environmental toxicants, particularly pyrethroid insecticides, have been observed to increase the spatial aggregation of *Heliocoverpa* eggs. Following exposure to sublethal doses of pyrethroids, female *Heliocoverpa* often oviposit several eggs on a single leaf or terminal before flying to a subsequent plant. *Heliocoverpa* females place eggs further down the plant in fields that are water stressed. It has been postulated that this is a survival mechanism which affords protection to the eggs from extremely high temperatures in the upper canopy of water-stressed plants. Also, under water stress, plant terminals are first to be affected. Because neonate *Heliocoverpa* larvae usually first feeds on the structure upon which they were oviposited (after eating their eggshell), it would not be to the advantage of a moths genes to oviposit on a plant part that is of poor food quality. Prevailing winds can result in spatial gradients in densities for a range of cotton-arthropods. Most cotton inhabiting arthropods are aggregated at the edges of fields that are proximate to the prevailing wind direction.

IV. QUANTIFYING SPATIAL AGGREGATION

A. SINGLE POINT ESTIMATES

The variance/mean ratio for individual samples can be used to estimate whether a population's spatial pattern is uniform ($\sigma^2/\mu < 1$), random ($\sigma^2/\mu = 1$), or aggregated ($\sigma^2/\mu > 1$). An appropriate statistic for testing whether a sample is significantly more or less aggregated than that found for a randomly distributed variate of equal mean density, is the ratio of the sample variance (S_s^2) to the variance expected with a randomly distributed population (σ_r^2).[46] This ratio is approximated by an F test with the numerator and denominator degrees of freedom both equal to the sample size minus one (Equation 1a).

$$F = S_s^2 / \sigma_r^2 \tag{1a}$$

Because the variance for a randomly distributed variate is equal to its mean, the estimated F statistic simplifies to the sample variance/mean ratio (Equation 1b).

$$F = S_s^2 / \bar{x} \tag{1b}$$

The degree of aggregation exhibited by a species is density dependent. For the majority, if not all cotton arthropods, as the density increases, the degree of aggregation also increases. As a result, it is easier to statistically separate an aggregated spatial pattern from either a random or uniform pattern at higher densities (Figure 4). Similarly, the greater the degree of intrinsic spatial aggregation, the easier it is to discriminate an observed spatial pattern from either a random or uniform pattern.

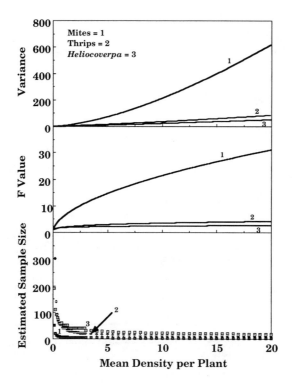

FIGURE 4. Effect of density on variance, estimated F statistic, and estimated sample size for spider mites (highly aggregated), Western flower thrips (moderately aggregated), and *Heliocoverpa* eggs (slightly aggregated). (From Wilson.[61])

B. AGGREGATION AS A FUNCTION OF DENSITY

Statistics which focus on single sample estimates of spatial aggregation, although of some utility, do not provide a dynamic representation of how aggregation changes with density. Several authors have developed equations which to varying degrees quantify this change: most noticeably Iwao's patchiness regression index,[47-49] Lloyd's mean crowding,[50] Morisita's index,[51-53] and Taylor's equation[20,54-57] (also see Chapter 3). The simplest and most straightforward of these, and from my experience the most consistently reliable, although with qualifications (see *Estimation Procedures and Spatial Aggregation*), is Taylor's equation.

First used by Fracker and Brischle[54] to study spatial aggregation in a biological system, but independently proposed by Taylor,[56] Taylor's equation estimates the relationship between the variance and mean density of a sample using a power function:

$$S^2 = a\bar{x}^b \tag{2}$$

Taylor's a and b coefficients are estimated using $\log_e$-$\log_e$ transformations or nonlinear regression techniques.[56,58] Both estimation procedures are subject to some error,[45,58,59] and preliminary analyses should employ each technique before a decision is made as to which is most appropriate for a particular data set. For cotton-inhabiting arthropods, Taylor's a coefficient typically ranges from about 1 to 20, or higher if counting all stages of highly aggregated species such as *Aphis gossypii* Glover and *Tetranychus* spp., while the b coefficient ranges from about 1 to 1.6. Wilson and

Room[21] present *a* and *b* coefficients for cotton arthropods with values that are in some cases less than 1.0. My opinion is that with most arthropod sampling in cotton, coefficients with values less than 1.0 are an artifact of curve fitting or random sample variability.

C. IMPACT OF SAMPLE UNIT SIZE ON PERCEIVED AGGREGATION

The concept of a *biologically relevant sample unit*, proposed by Morisita,[60] has direct bearing on this issue. In essence, a *biologically relevant sample unit* is the sample unit size that captures a species intrinsic spatial aggregation. As the sample unit size increases above or below this size, the perceived aggregation pattern approaches random and then uniform. The *biologically relevant sample unit size* is a species characteristic that tends to be small for small-sized species.[2,45] For example, when sampling a foliar pathogen, the *biologically relevant sample unit size* is likely a fraction of a square centimeter. In comparison, adult *Heliocoverpa* likely orient to whole cotton plants or small groups of cotton plants (ca. 0.03 to 0.25 m^2).

At extremely small sample unit sizes that approach the perceptive range of an arthropod, Taylor's coefficients may have values less than 1.0. It is likely that an individual insect or mite exhibits a degree of behaviorally mediated habitat partitioning when crowded by other individuals of its own species. For example, with individual cotton aphids within aphid colonies, where the sample unit size is in the range of a square millimeter, the aphid spatial pattern would be uniform and Taylor's coefficients would have values less than 1.0. However, using a leaf or a plant as the sample unit size, the aphid population would have an aggregated spatial pattern and the resulting estimates of Taylor's coefficients both would be greater than 1.0.

The perceived degree of spatial aggregation for a sample is not always influenced by sample unit size. An exception occurs when sample units are the result of collapsing randomly collected, spatially distinct subunits. This approach is commonly used to sample many arthropods, nematodes, and pathogens of cotton. For example, when recording immature flower thrips and some predators, many samplers will walk through a field, in the process collecting leaves in a random manner. The leaves are examined individually, but a total is recorded for the group of leaves. As a result, a group of leaves and not individual leaves is a sample unit. Table 5 shows the frequency distribution for individual leaves, and the frequency distribution that would be obtained, were the sampler to have collapsed the leaves into groups of three

TABLE 5
The Variance and Mean for Individually Sampled Leaves and That Obtained by Collapsing the Leaves in Groups of Three

Frequency class	Number of sample units in each frequency class	
	1 Leaf/sample unit	3 Leaves/sample unit
0	900	729
1	50	122
2	50	128
3	0	14
4	0	7
5	0	0
Mean	0.15	0.45
Variance	0.228	0.683
Variance/mean	1.517	1.517

leaves. Both the mean and the variance per sample unit are proportional to the number of leaves in each sample unit. As a result, the $S^2/\bar{x}$ ratio is the same for the one- and three-leaf sample unit size.

If repeated samples are taken at different population densities, the relationship between the variance and mean is distinct for each sample unit size. Expressing the variance as a function of both Taylor's equation and the number of subunits per sample unit, we have the following:[61]

$$S_c^2 = c^{1-b} a \bar{x}_c^b \qquad (3a)$$

Or, if expressed in terms of density per subunit $(\bar{x})$,[61]

$$S_c^2 = ca\bar{x}^b \qquad (3b)$$

where c = the number of subunits per sample unit, a and b are Taylor's coefficients estimated when the subunit is equal to the sample unit, and $\bar{x}_c$ = the density per sample unit ($= c\bar{x}$).

The implication of this relationship is that the random grouping of spatially distinct sample subunits changes the estimated variance-mean relationship and, as a result, the estimates for Taylor's a and b coefficients.

When cluster sampling, such as would occur when collapsing data from consecutive plants along a row of cotton, estimates for both coefficients would decrease with an increase in the number of plants sampled as a sample unit. An exception would occur when the *biologically relevant sample unit* is larger than the number of plants per sample unit; in which case, the values for the coefficients would increase up to the point where the sample unit size is equal to the size of the *biologically relevant sample unit*.

When Taylor's coefficients are both equal to 1.0, the variance and mean are equal and the spatial pattern is random at all densities. When both coefficients are greater than 1.0, the estimated variance is greater than the mean for a wide range of densities. However, ignoring for the moment that an estimated variance/mean ratio may be greater than 1.0, but not significantly so, it is possible for both coefficients to be greater than 1.0, but for a species to have a uniform spatial pattern for a range of densities, a random spatial pattern for a single density, and an aggregated spatial pattern at still another range of densities. This is demonstrated by the following. Dividing both sides of Equation 2 by the mean density,

$$S^2/\bar{x} = a\bar{x}^{b-1} \qquad (4a)$$

substituting 1 for the variance/mean ratio for a random spatial pattern,

$$1 = a\bar{x}^{b-1} \qquad (4b)$$

taking the $\log_e$ of both sides,

$$0 = \log_e(a) + (b-1)\log_e(\bar{x}) \qquad (4c)$$

rearranging,

$$\log_e(\bar{x}) = -\log_e(a)/(b-1) \qquad (4d)$$

and taking the antilog$_e$ of both sides gives Equation 5 for estimating the density at which a species spatial pattern will be random. The equal sign in Equation 5 is replaced by a greater than sign (>) for a uniformly distributed spatial pattern and by a less than sign (<) for one that is aggregated:[61]

$$\bar{x} = e^{-\log_e(a)/(b-1)} \tag{5}$$

When solved for a range of a and b coefficients, a three-dimensional surface is generated, representing the density at which a species has an aggregated spatial pattern for each combination of a nd b (Figure 5). For a specific combination of a and b with values that are greater than 1, a random spatial pattern occurs at a single density that is less than one organism per sample unit. For many species this is a significant point. The density at which direct pests, such as cotton bollworm, tobacco budworm, boll weevil, and *Lygus* bug, cause economic loss, is less than one feeding-stage per plant or one per terminal region, using standard sample units for these species. The density at which many entomophagous species are capable of suppressing most herbivorous arthropods is less than one per leaf, a common sample unit for these arthropods. Likewise, for most of a growing season, the majority of arthropods will be at a density of less than one per sample unit (for most commonly used sample unit sizes). Combined with a limited sample size normally used to assess a species' abundance or management status, the above considerations make it difficult to separate a species' spatial pattern, when at low densities, from what would be expected for a random spatial pattern.

D. IMPACT OF SAMPLER EFFICIENCY ON PERCEIVED AGGREGATION

As indicated earlier in this chapter, the physical size of a sample unit, relative to the biologically relevant sample unit, affects the perceived degree of spatial aggregation. Sampler bias or catch efficiency can also have a profound impact on the estimated degree of spatial aggregation. The lower the catch efficiency, i.e., the

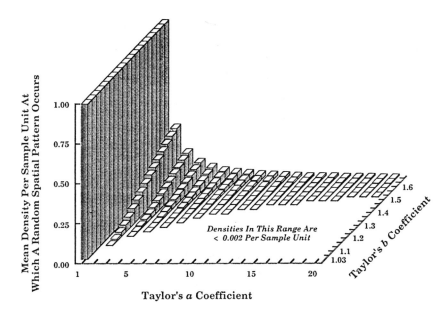

FIGURE 5. Densities at which spatial patterns are random for combinations of Taylor's a and b coefficients. (From Wilson.[61])

proportion of individuals that are recorded during sampling for the unit area immediately represented by the sample, the lower the resulting estimate of the degree of spatial aggregation. As the efficiency of the sampler decreases, the curve describing the relationship between the variance and the mean approaches the variance-mean curve for a random or Poisson distributed variate ($S^2 = \bar{x}$). The greater the estimated degree of aggregation, say comparing *Tetranychus* spp. with *H. zea*, the easier it is to distinguish a species' spatial pattern from that based on random expectation, except for extremely inefficient sampling methods or samplers.

The efficiency of different sampling methods varies with the method used, the species being sampled, and with stage of crop growth. Smith et al.[62] reported that the sweepnet has an efficiency of ca. 10% when sampling predators in cotton. Wilson and Gutierrez[34] similarly reported that the sweepnet has an approximately 12% efficiency when estimating total predator abundance in cotton. However, Wilson and Gutierrez[34] also reported that immature predators were captured less efficiently with the sweepnet than were adult predators, with the efficiency varying from 3.3 to 22.8% for the immatures compared with 13.9 to 26.5% for the adults. Wilson and Gutierrez[34] also reported that immature and adult predators were captured with greatest efficiency during the period of rapid fruiting. Field observations suggest that this is likely due to the greater abundance of suitable prey (predominantly flower thrips) in the more exposed parts of the plant during this stage of growth.

Variability in sampler efficiency is in some cases due to the sampler and not the method used when sampling. Lincoln[63] reported that one scout counted 207% more squares (flower buds) on the cotton plants than did a second scout. Wilson et al.[18] similarly reported one sampler recording 42 and 65% as many cotton leaves infested with mites and predators, respectively, as found by three other samplers. Inexperienced samplers or samplers who are in a hurry seem most prone to sampler bias.

Analyses suggest that Taylor's coefficients both approach a minimum value of 1.0 as sampler efficiency approaches 0 (0%), and asymptotically approach a maximum value for a species as the efficiency of the sampling method increases. The form of the equation describing Taylor's coefficients as a function of density is approximated in Equation 6:[61]

$$a_{i,e} = 1 + (a_i - 1)(1 - e^{-\alpha_a \lambda^{\beta_a}}) \quad (6a)$$

$$b_{i,e} = 1 + (b_i - 1)(1 - e^{-\alpha_b \lambda^{\beta_b}}) \quad (6b)$$

where $a_{i,e}$ and $b_{i,e}$ = Taylor's coefficients for the ith species sampled with λ efficiency; a_i and b_i = Taylor's coefficients for the ith species sampled with absolute efficiency ($\lambda = 1.0$); α and β = constants describing changes in $a_{i,e}$ and $b_{i,e}$ with changes in sampling efficiency, and differ for both a and b. Also, they should not be confused with error rates used in sample size estimation; and λ = sampling efficiency (values ranging from > 0 to 1.0).

α and β can be estimated through rearranging and double $\log_e$ transformation of Equations 6a and 6b and regressing the transformed y values against $\log_e(\lambda)$ as follows:[61]

$$\log_e(-\log_e(1 - (a_{i,e} - 1)/(a_i - 1))) = \log_e(\alpha_a) + \beta_a \log_e(\lambda) \quad (7a)$$

$$\log_e(-\log_e(1 - (b_{i,e} - 1)/(b_i - 1))) = \log_e(\alpha_b) + \beta_b \log_e(\lambda) \quad (7b)$$

As a result, the variance estimated as a function of sampler efficiency, and intrinsic aggregation characteristics of a species has the following form:[61]

$$S^2 = \left[1 + (a_i - 1)(1 - e^{-\alpha_a \lambda^{\beta_a}})\right] \bar{x}^{[1 + (b_i - 1)(1 - e^{-\alpha_b \lambda^{\beta_b}})]} \quad (8)$$

Figure 6 shows the pattern of Taylor's coefficients changing with density, at different sampler efficiencies, for *Heliocoverpa* eggs and the mobile stages of *Tetranychus* spp., while Figure 7a and b shows the effect of changing sampler efficiency on spatial aggregation.

Taylor et al.[20] considered the coefficient b constant for a species, with only a affected by sample unit size. Banerjee,[64] Wilson,[45] and data presented in this chapter show that both coefficients are subject to change due to changes in age-specific mortality, plant quality, age-specific dispersal, sample unit size, sampler bias, sampler efficiency, and spatial heterogeneity (see Table 4). However, the coefficients for an age class of a species are often fairly constant over a wide range of conditions (for comparable sample unit sizes). The uncritical acceptance by many authors of Taylor's interpretation of the static nature of the b coefficient is likely due to many authors not having tested the dynamic nature of Taylor's coefficients with respect to the above variables. Or, it may be due to the nature of the b coefficient in Taylor's equation. While the variance is directly proportional to the a coefficient, the b coefficient as an exponent has an exponential effect on the variance. As a result, b is much less sensitive to change than is a, and therefore measuring changes in b is statistically more difficult.

E. PROBABILITY DISTRIBUTION FUNCTIONS

The simple but robust nature of Taylor's equation is of considerable use when categorizing a species' spatial pattern. However, a drawback for some is that it cannot be used to estimate the frequency distribution of arthropods as a function of density. Fisher[65] wrote one of the first papers which compares the distribution of an arthropod (a tick) on different sample units (sheep) to what one would expect using a discrete probability distribution function (also see Haldane[66]). Subsequent authors have fit arthropod data to numerous discrete probability distribution functions, ranging from the simple models such as the Poisson to more complicated models such as the Pascal type H (five parameters),[67] in an attempt to better describe their

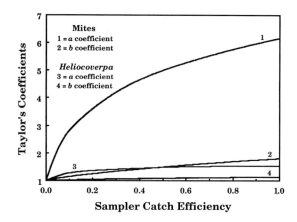

FIGURE 6. Effect of sampler efficiency on Taylor's coefficients for *Heliocoverpa zea* eggs, and mobile stages of spider mites. (From Wilson.[61])

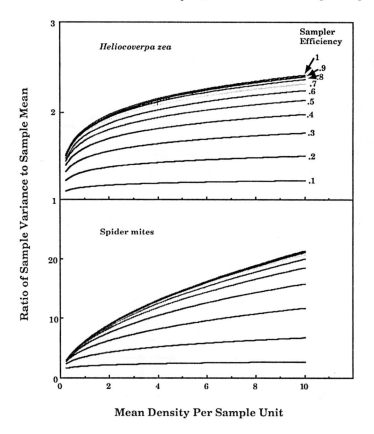

FIGURE 7a–b. Effect of sampler efficiency on the spatial aggregation estimated by the variance/mean ratio, for *Heliocoverpa zea* eggs, and mobile stages of spider mites. (From Wilson.[61])

aggregation characteristics. Gates and Ethridge[68] developed a computer program that enabled data to be fit to eight of the most common probability distribution functions (Poisson, binomial, negative binomial, Poisson with zeros, Neyman type A, logarithmic with zeros, Poisson-binomial, and Thomas double Poisson). Wilson et al.[69] modified the gates and Ethridge[68] program by incorporating a method of moments estimator for the Thomas double Poisson. This approach provides lower critical values than does the maximum likelihood estimator, enabling greater statistical separation between cotton arthropod field data and this distribution.[69] Wilson et al.[69] also replaced the χ^2 goodness of fit test statistic with a discrete Kolmogorov-Smirnov test, which is generally more powerful because the χ^2 test groups frequency classes with few observations, thereby losing information (see Sokal and Rohlf[70] and Zar[71]).

The most widely used of the probability distribution functions is the negative binomial distribution. Although the negative binomial distribution has been criticized as having several ecological constraints,[20,72-75] others have found this model useful in fitting biological frequency distribution data.[65,69,76-84] Wilson et al.[69] found that the negative binomial distribution provides the best fit to both visual whole plant and sweepnet sample data for a range of cotton-inhabiting arthropods and plant parts. Of the eight probability distributions tested, the negative binomial distribution was least affected by changes in density, species, and season for the visual data. However, it fits sweepnet data less well at low densities, because the estimated variance is often less than the mean for this sampling method over the range of densities tested.

The negative binomial distribution parameter k describes the relationship between the variance and mean. Anscombe[76] and Bliss and Owen[80] show the dynamic relationship between k and mean density. Although numerous authors have attempted to estimate a common k (an average k), because k changes dynamically with density as does aggregation, derivation of a common k makes little biological or statistical sense, except possibly when the negative binomial distribution is to be applied over a limited range of densities. Taylor et al.[20,73,74] contend that the change in k with density is an undesirable attribute for an index of aggregation. One can argue that because variance changes with density in a predictable fashion, and because k is a function of both the variance and density, that k should be dynamic as well.

Various procedures for estimating k have been developed (see Southwood[3] and Chapter 3). The most common of these is probably the method of moments estimator.[77] The moment estimate of k is calculated as

$$k = x^2/(S^2 - \bar{x}) \tag{9a}$$

Equation 9a does not readily allow an estimate of how k changes dynamically with density. However, this limitation is overcome by substituting Taylor's equation for the variance.

$$k = x^2/(a\bar{x}^b - x) \tag{9b}$$

Factoring x from the numerator and denominator on the right side of the equation produces the following:

$$k = x/(a\bar{x}^{b-1} - 1) \tag{9c}$$

The integration of Taylor's equation into Equation 9c allows for k, or the reciprocal of k as it is often presented, to be estimated dynamically for any density, where the variance is greater than the mean. The value of $1/k$ decreases (k increases) with increasing density and is greater for the more aggregated species (Figure 8). As density increases, the frequency distribution shifts to the higher frequency classes (Figure 9a to t). The more highly aggregated species exhibit greater

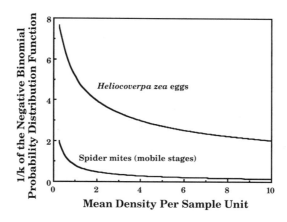

FIGURE 8. Effect of density on $1/k$ of the negative binomial probability distribution function. (From Wilson.[61])

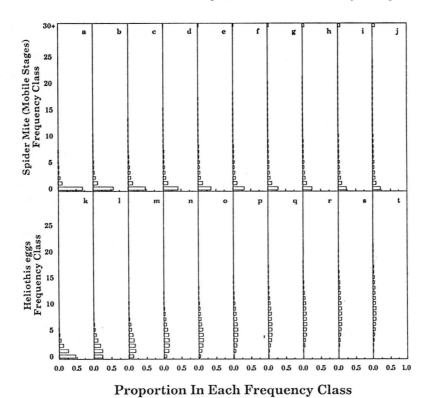

FIGURE 9a–t. Effect of density on a highly aggregated species (mobile stages of spider mites), and a slightly aggregated species (egg stage of *Heliocoverpa zea*). (a–j) 1, 2, 3, ..., 8, 9, 10 spider mites per leaf, (k–t) 1, 2, 3, ..., 8, 9, 10 *Heliocoverpa* eggs per plant. (From Wilson.[61])

variability in their frequency distribution, which is responsible for the greater sampling effort required to reliably estimate their density (see Section V).

F. STATISTICAL HIERARCHY

If the $S^2/\bar{x}$ ratio can be called an α spatial statistic, it is appropriate to classify equations such as Taylor's equation as β statistics, and probability distribution functions as χ statistics. Each statistical level provides greater spatial information. Collectively, all three encompass the extent of spatial statistics used by the majority of entomologists. Although of considerable use, statistics at these levels summarize spatial information in a point (α statistic) or a one-dimensional (β and χ statistics) context. The spatial extent of individuals comprising a population is collapsed in the summarization and as a result is lost.

Biologists have largely followed the lead of geologists and geographers in the use of two- and three-dimensional spatial statistics, commonly referred to as geostatistics, but for the sake of continuity, herein referred to as δ statistics. δ statistics, such as variograms, cross-variograms, semi-variograms, kriging, and cokriging describe the spatial association between pairs of data increasing distances apart (see Brooker,[85] Cressie and Hawkins,[86] Edwards and Penney,[87] and Isaaks and Srivastava[88,89]). Each of these techniques has broad applicability for describing the spatial association between individuals within a species or between individuals of different species. In cotton, these techniques have been used to describe the establishment and dispersal of bollworm within a cotton agroecosystem.[90] As might be expected, the further the distance between data pairs, the lesser the correlation between these points. δ statistics should also be ideally suited to estimating the size of biologically relevant sample units and for estimating the spatial association between arthropod species. However, an overriding problem is the lack of adequate spatial data. Only a limited number of entomological research efforts contain a sufficient number of sample units to enable comprehensive analysis using δ statistical approaches.

In contrast to δ statistics, which focus on dispersion statistics, a wide range of simulation approaches have been used to predict the dispersal as well as the spatial extent of arthropod, other animal, and plant populations; see Okubo[91] for additional examples. Culin et al.[92] developed a simple model that predicts the dispersal of the cotton boll weevil, *Anthonomus grandis* Boheman, across the southeastern U.S. A diffusion model presented by McKibben and Smith[93] and McKibben et al.[94] suggests that a small percentage of migrating boll weevils may disperse in excess of 300 km/year. McKibben et al.[94] attribute long-range dispersal to a small percentage of the weevils flying at sufficient elevations to experience high wind speeds that can keep the weevils aloft for extended periods of time. Corbett and Plant,[95] using classical diffusion modeling, predicted the movement of spider mites and an introduced phytoseiid predator, *Metaseiulus* (= *Typhlodromus* = *Galendromus*) *occidentalis* (Nesbitt) from interplanted strips of alfalfa into adjoining cotton. Although Corbett's model contains only limited biological detail, it provides considerable insight into how crop interplanting might be used in conjunction with introduced entomophages to regulate a key herbivorous arthropod.

An aspect of the author's research addresses the dispersal of the cotton boll weevil between individual cotton fields and overwintering habitats. During the fall, diapausing weevils migrate from cotton to overwintering habitats.[96,97] The following spring, weevils that survive winter conditions migrate from the overwintering habitats to adjacent cotton fields. Weevils that find cotton in the squaring stage will survive and oviposit. The movement of weevils from cotton to overwintering habitats and from overwintering habitats to cotton is estimated using a diffusion model that accounts for the distance between each habitat (ca. 10,000) and field (ca. 3800), wind speed and wind direction, the size of each habitat and field, the effect of the attractiveness of each habitat to the weevils, and the vigor of the weevils, measured as percent fat body content.[98] The physical size and longitude and latitude of each field and overwintering habitat is exported to the diffusion model via a geographic information system (GIS) database.[99] The estimate weevil densities for each cotton field are then exported to a mechanistic crop-pest simulation model, which predicts the regional and local seasonal dynamics of both the cotton crop and the boll weevil, including the impact of feeding injury upon the growth and development of individual cotton fields.[98] For the interested reader, Aronoff[100] and Star and Estes[101] provide a thorough review of the development and application of GIS databases.

The use of δ statistics and spatial modeling in cotton, as briefly discussed above, is very much in its infancy, and it is premature to speculate as to how broadly these approaches will be adopted. However, there is ample evidence to suggest that both have utility in structuring sustainable intercropping systems, such as proposed by Corbett and Plant;[95] and in the management of highly dispersive arthropods such as the cotton bollworm, boll weevil, and *Lygus* bug. Little more than a decade ago, Phillips et al.[102] described the development of a regional management program for the cotton bollworm in Arkansas. At that time, each field in the management region was individually scouted and a management decision reached, based on the mean density across fields for the whole region. Geostatistical and modeling approaches have the potential to greatly extend the scope and utility of regional crop-pest management programs.

V. SAMPLE SIZE ESTIMATION

Sampling plays an important role in the tactical or day-to-day management of cotton, and in basic research aimed at better understanding the population dynamics and interactions between organisms within a cotton agroecosystem. From a crop management perspective, the benefits of a sound monitoring program include an often substantial reduction in the misapplication of control measures, and improvement in the timing of necessary control measures. Many of the currently available crop monitoring procedures provide the ability to reliably estimate the control status of a range of pest species at a fraction of the cost of previously available methods. A less tangible benefit obtained by using a quantitatively based monitoring procedure is peace of mind for the scout, consultant, or grower in knowing that they did the best job possible with the available resources. From a research perspective, appropriate sampling procedures enable the abundance of different species and interactions between species to be quantified with a defined degree of accuracy and reliability. The use of appropriate estimation procedures can greatly clarify our understanding of ecosystem interactions. In contrast, research based on inappropriate estimation procedures represents, at best, a misdirected resource, and at worst, a clouding of our understanding of processes and relationships that are sufficiently complex in themselves.

A. DEFINING THE OBJECTIVES OF SAMPLING

When sampling, the objective may be to understand and predict the distribution, abundance, and possibly the interaction of a population(s) with the host crop. Or, the objective may be to apply the information gathered to aid in managing a crop. The reason for the distinction is that population sampling in the first case and decision sampling in the second have different objectives.[103] The emphasis on population sampling is on the reliability of the parameter being measured, i.e., the closeness of the sample mean to the true population mean.[70] Decision sampling emphasizes the reliability of the resulting estimate of the pest's status, i.e., whether the pest (or other group) is above or below the economic threshold. It may not be necessary to have a highly reliable estimate of population density when determining whether or not a pest requires control, if the density is far above or far below its economic threshold (Figure 10a and b).

The objective of population sampling may be to estimate the number of organisms on an absolute basis (i.e., number per unit area). Since the results of this type of sampling may be used to derive functional relationships, a greater number of sample units may have to be examined to achieve sufficient reliability in the estimate (Figure 11a and b). Population sampling and decision sampling may also require different sample units. An accurate sampling method may not be necessary for commercial monitoring, particularly when estimates of the efficiency of sampling procedures at capturing the target species are known, enabling absolute density estimation.

Failure to distinguish between the objectives of population and management-decision sampling may result in scouts spending much more time sampling than is necessary. The result is the expenditure of time, effort, and money for population sampling when only decision sampling is needed. Similarly, unless researchers define their hypotheses sufficiently well and understand the constraints associated with a

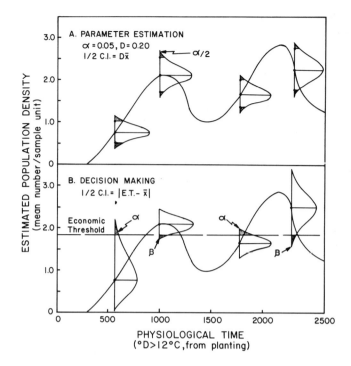

FIGURE 10. Comparison of confidence intervals for (A) population (parameter estimation) sampling and (B) commercial (decision) monitoring. (From Wilson et al.,[2] reprinted with permission of John Wiley & Sons, New York.)

particular sampling procedure, they may end up with data and results that are too variable and therefore of little use.

Although the objectives of population sampling and commercial monitoring differ, it is important to keep in mind that the development of an accurate, reliable, and easy-to-use commercial monitoring program is dependent on adequate knowledge of the target species, including distribution patterns and factors responsible for the observed patterns. Ecological principles and appropriate commercial monitoring procedures can, in turn, be developed from the more detailed spatial data.

B. POPULATION SAMPLING (PARAMETER ESTIMATION)

A considerable investment in time may be required to derive a parameter estimate, while a farmer or crop consultant is often interested in knowing only whether a pest is above or below some economic or management threshold. The greater the reliability necessary for the estimate, the larger must be the sample size. Too often the sample size is not related to the number required, and the reliability of the estimate is often too low or too high. This equates to poor data at one extreme and unnecessary costs at the other.

Karandinos[104] presented a series of equations for use in estimating sample sizes. Ruesink[105] and Wilson and Room[106] incorporated Taylor's equation into Karandino's equations. Wilson and Room,[106] however, use the t statistic instead of the z statistic, to account for the effect of sample size on the reliability of the sample estimate.

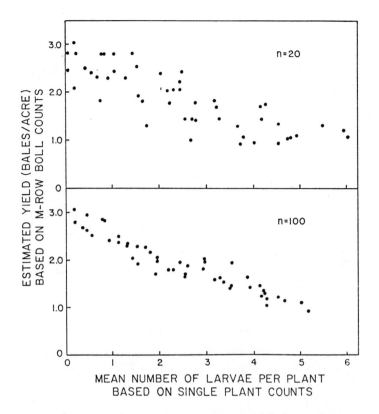

FIGURE 11. Effect of sample size on the reliability of a simulated yield response function. (From Wilson et al.,[2] reprinted with permission of John Wiley & Sons, New York.)

Equations 10a and 10b are generalized equations for estimating sample sizes for population sampling:[106]

$$n = t_{\alpha/2}^2 D_x^{-2} a \bar{x}^{b-2} \tag{10a}$$

$$n = t_{\alpha/2}^2 D_p^{-2} p^{-1} q \tag{10b}$$

where $t_{\alpha/2}$ = standard normal variate for a two-tailed confidence interval, D_x = a proportion defined as the ratio of half the desired confidence interval to the mean ($D_x = [C.I./2]/\bar{x}$ for enumerative sampling), D_p = proportion defined as the ratio of half the desired confidence interval to the proportion of infested sample units ($D_p = [C.I./2]/p$ for binomial or presence-absence sampling), and a and b = Taylor's coefficients. A general feature of Equation 10a is that species whose distribution patterns are more aggregated, reflected by higher values for Taylor's coefficients, require a greater sample size for a given level of reliability (Figure 12). This is the opposite of what is required when determining whether a population has an aggregated, random, or uniform spatial pattern. When categorizing spatial aggregation, variances are compared, while for research sampling or commercial monitoring the objective is to estimate a density or a density relationship, or to categorize the density as being above, below, or statistically indistinguishable from an economic threshold population density.

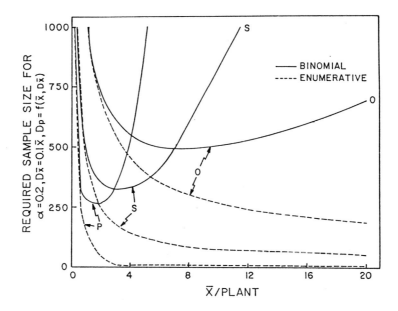

FIGURE 12. Impact of species distribution pattern on the number of sample units required to obtain a population estimate with a given level of reliability. P represents a Poisson-distributed variate, S represents cotton squares (flower buds), and O represents *Oxycarenus luctuosis* M. & S., a hemipteran found on cotton in Australia, having a highly aggregated spatial pattern. (From Wilson and Room.[21])

A smaller sample size is required at higher densities for a given level of reliability using the enumerative sample size equation compared with using the binomial equation.[45] This is due to the sampler using a lesser amount of spatial information with binomial than with enumerative sampling. The smaller the value of D_x and D_p, the greater the required sample size. Wilson[45] also indicated that for binomial sampling, unlike enumerative sampling, sample size first decreases then increases as p and corresponding density increase. This is due to a small confidence interval about p, corresponding to a very large confidence interval about x at p values approaching unity (Figure 13) (also see Wilson et al.[2]).

C. COMMERCIAL MONITORING (DECISION ESTIMATION)

When a field is monitored to determine a pest's economic status or to determine whether a natural enemy is at a level capable of suppressing a pest's population density, the farther $\bar{x}$ or p are from the economic threshold, the smaller the sample size required to estimate whether the pest is above or below the threshold. While the number of sample units required to estimate a population density with a given level of reliability is independent of an economic threshold, the threshold plays an integral part in determining the required sample size when a management decision is being made. Figure 14 illustrates the number of cotton terminals (upper 12 cm of the plant) that would need to be examined comparing population sampling and commercial monitoring for the cotton bollworm. The number of sample units that would be required for commercial monitoring increases exponentially as density approaches the economic threshold from either above or below. Unlike research sampling, the number of units that would have to be examined changes as the threshold changes (Figure 13). The number of sample units requiring examination for commercial

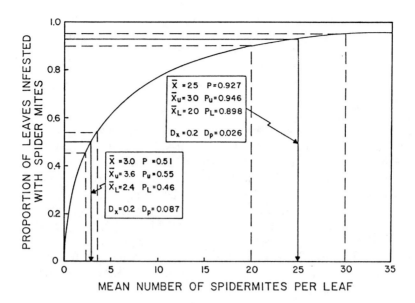

FIGURE 13. Relationship between D_x and the confidence interval about a mean and D_p, and the confidence about p. (From Wilson et al.,[2] reprinted with permission of John Wiley & Sons, New York.)

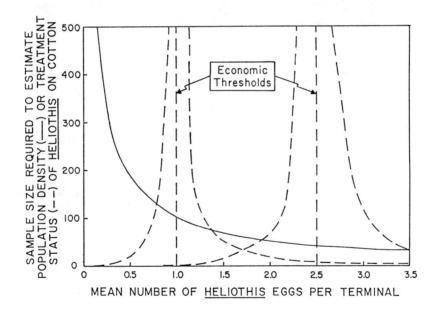

FIGURE 14. Number of cotton terminals that would need to be examined comparing population sampling and commercial monitoring, with the upper 12 cm of the plant as the sample unit. (From Wilson et al.,[2] reprinted with permission of John Wiley & Sons, New York.)

monitoring increases as the species density approaches an economic threshold, and changes as the threshold changes.

As with any estimation procedure, there are errors associated with making management decisions. Two types of errors in decision making are (1) the probability of concluding that the population level is above the threshold when it is not, and (2) the probability of concluding that the population level is below the threshold when it is not[107] (see Figure 10b). Wald[108] classified these as α and β errors, respectively. In theory, an acceptable level of error is determined for α by minimizing the total cost involved with monitoring and the costs involved with spraying when not required (labor, machinery, pesticides, resurgence, secondary outbreaks, etc.) and for β by minimizing the total cost involved with monitoring and the cost of pest damage due to not spraying when justified.[109]

In its simplest case, consider two cotton crops, one irrigated and the other dryland. In the case of a pest such as spider mites or root-knot nematode, the yield loss of both crops is linearly related (within bounds) to pest density above some threshold response level (see Figure 11a and b), with the absolute loss per pest unit being greatest on the irrigated cotton. Because the relationship between α and monitoring and pesticide disruption-related costs are about equal for both irrigated and nonirrigated cotton, Figure 15A indicates that both fields would justify equivalent alpha error rates, while the irrigated field would have a much lower β error rate due to the greater potential for yield loss per pest unit (Figure 15B and C). It also can be shown that as pesticide costs increase or crop value decreases, more intensive sampling or lower α and β errors are justified.

Equations 11a and 11b,[45] which incorporate the α and β error rates, were derived from the central limit theorem:

$$n = t^2_{\alpha \text{ or } \beta} |\bar{x} - T|^{-2} a\bar{x}^b \qquad (11a)$$

$$n = t^2_{\alpha \text{ or } \beta} |p - T|^{-2} pq \qquad (11b)$$

where T is the economic threshold expressed as a mean density per sample unit for enumerative sampling (Equation 11a), or as a proportion of infested units for presence-absence or binomial sampling (Equation 11b).

Equations 11a and 11b and the equations developed by Wald[108] fit into the specialized area of sampling referred to as sequential sampling. Many sequential sampling plans have been developed to aid in the management of cotton insect

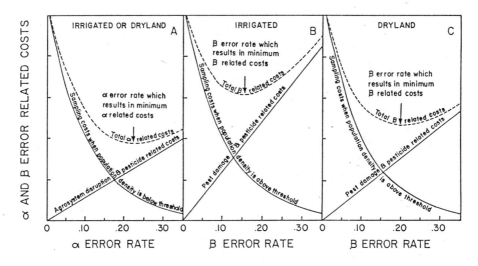

FIGURE 15A–C. Effect of pest damage on optimal α and β error rates for irrigated and dryland cotton. (From Wilson.[106])

pests.[18,107,110-115] Sequential sampling plans are available for most of the key pests and for both enumerative and binomial sampling methods, but they are not extensively used by crop scouts or consultants. Adoption of sequential sampling as an aid in pest management decision making has been shown to reduce sampling costs by 40 to 60%, compared with methods having similar average error rates. Only when a pest's population density approaches its economic threshold does the required sample size begin to increase rapidly. During the remainder of a season the required sample size for sequential sampling will be considerably less than that for other quantitative procedures.

From a statistical point of view, the use of sequential sampling can result in considerable saving in sampling costs compared to methods having comparable average error rates. Unfortunately, the use of one or more existing quantitative sampling methods as a baseline for evaluating the merits of sequential sampling means very little to a field scout. The problem is not necessarily due to scouts using inherently inferior sampling methods; the problem is that most scouts make treatment decisions based on sample sizes that are too small. The result is that scouts commonly have unacceptably high error rates in their decision making relative to the value of the fields that they manage. Too often this equates to an excessive use of less than optimally timed pesticides. This problem is not easily solved. On the one hand, growers and consultants rarely understand the economic consequences of inadequate sampling. As long as the direct farmer cost of pesticide misuse (spray costs and resistance to pesticide costs) remains relatively low, and the value of accurate and reliable management decisions is perceived as being relatively low, we can expect farm managers to err on the conservative side. On the other hand, slow adoption (but generally high level of adaptation) by growers and consultants is at least in part due to researchers not having provided convincing economic justification for adopting a new approach or new technology.

This problem is exacerbated by the fact that, although it is easy to document an economic advantage to a grower, a commercial consultant may find that adoption of sequential sampling may cause problems when allocating work between employees. During periods of high or low pest densities, relative to economic thresholds, fewer scouts are required to determine the treatment status of a pest species. In contrast, when pest population densities are near economic thresholds, a greater number of scouts will be required to sample all of the fields. Unfortunately, qualified individuals are generally unavailable on short notice.

A deserving criticism of sequential sampling is its use for assessing the relative abundance or treatment status of more than one insect species. The greater the number of species that are assessed at any time, the higher the probability that one or more of these will be sufficiently close to its economic threshold to warrant examining a predetermined upper limit number of sample units. A general rule of thumb is that sequential sampling will result in a savings in time when using a particular sampling procedure if the number of species or separate age classes being recorded is less than five.[45] Above that number, a scout can obtain comparable error rates using a fixed-sample-size approach.

A related but more important constraint that is often associated with the use of sequential sampling, but also a constraint associated with conventional sampling methods, is its general inability to categorize the collective economic impact of a complex of herbivorous arthropods. Dr. Winfield Sterling, a pioneer in the development of quantitative sampling principles for cotton arthropods, has noted that economic thresholds, as currently used, often fail to adequately capture the dynamics of crop-pest interactions. In particular, damage response functions developed for many of the pest species, and which are used to estimate economic thresholds, rarely integrate injury from a complex of pests. As a result, subthreshold densities of two or more herbivores will in some instances cause sufficient injury to warrant management intervention. This suggests that economic criteria, as currently used, both within and outside of the context of sequential sampling, provide inappropriate estimates of the need for management intervention when dealing with a complex of herbivorous arthropods. Use of an herbivore-equivalence index, where the injury by each herbivore is equated to a common unit, might overcome this problem with some crops. However, the dynamic manner in which cotton responds to injury of different types and changes with each stage of crop development suggests that the use of indices of herbivore equivalence in cotton management may be limited to systems or stages of crop growth during which injury is overwhelmingly caused by a single herbivore species or several species, causing the same general type of injury. A practical but future solution to this problem lies with the development of systems models which dynamically integrated interactions between the environment and the crop, herbivore, and entomophage populations. Only through the development and use of integrated systems models can we hope to realistically address this complex problem. However, until such models are readily available, pest management advisors must rely on available sampling methods and indices of damage.

VI. OPTIMIZING SAMPLING COST RELIABILITY

An optimal sampling method, compared to one or more alternative sampling methods, is one which provides the most reliable estimate of either a species density or management level for a given expenditure of time or effort. Reliability in this context refers to minimizing the time (cost) required to achieve a given confidence interval about a parameter estimate, in the case of research sampling, or minimizing α and β error rates, in the case of commercial monitoring.

Costs required to estimate a population parameter or make a management decision with a given level of reliability can be compared to determine which of two (or more) sampling procedures is better. The ratio of the costs of two sampling methods is referred to as relative cost reliability[35] and can be estimated comparing two sampling methods.

$$C_1/C_2 = n_1(\theta_1 + \Phi_1)/[n_2(\theta_2 + \Phi_1)] \qquad (12)$$

where C_i = cost per sample for a given level of reliability for the ith sampling method (not to be confused with c in Equations 3a and 3b); n_i = number of sample units required for an estimate with a given level of reliability (Equations 10a and 10b for research sampling and Equations 11a and 11b for commercial monitoring), using the ith sampling method; θ_i = time (cost) required to examine an individual sample unit using the ith sampling method; and Φ_i = time (cost) required to move from sample unit to sample unit for the ith procedure; assumes that the time taken to walk into a field to the first sample unit is equal to the time taken to walk between sampling units.

For species, or age class thereof, four major factors affect the cost reliability of a sampling method: (1) density of the sampled species, (2) sample unit size, (3) sampler efficiency and bias, and (4) the cost per sample unit. The first three affect the estimated variance-mean relationship which in turn affects the sample size needed to estimate a mean or management level with a given level of reliability. The cost per sample unit and the cost to move between sample units directly affect the cost of taking a sample.

Wilson et al.[35] provide estimates of the cost reliability of three sampling methods for estimating the abundance of *T. ni*. The three sample units were (1) whole plants, where every structure on the plant was examined, (2) all mainstem node leaves on each sampled plant, and (3) terminal bottom surface mainstem node leaves, where the bottom surface of each mainstem node leaf in the upper 50% of the nodes was examined. The smaller the sample unit size, the greater the number of sample units required for a given level of reliability, but conversely, the lower the cost to examine each sample unit. From their analyses, the authors concluded that sampling all mainstem node leaves was slightly more cost reliable than sampling only mainstem node leaves in the upper 50% of the nodes, while both of the methods were considerably more cost reliable than the whole plant method. Also, because the authors had quantified the within-plant distribution of *T. ni*, estimates of absolute whole plant and unit area density were readily obtained using the simpler but more cost effective mainstem node leaf sampling method.

Cost reliability analysis is considerably more complicated when simultaneously sampling more than one species and, as a result, has not been adequately addressed. When sampling multiple species, each with distinct within-plant distributions and between-plant spatial patterns, a sampling plan can either be optimized for one species, such as one having the greatest damage potential, or it can be a global optimization, which in general is not optimal for any single species. If a global optimization approach is chosen, the optimization should meet minimally accepted criteria in terms of estimates of reliability for each species sampled. Ideally, we should be able to quantify the relative value of the information for each arthropod and develop a sampling program that maximizes the economic benefits. However, at this stage in our understanding of agroecosystems, this remains an impossibility. Although sophisticated mechanistic crop-arthropod simulation models have been developed for cotton,[116-128] it may be several years before any are structured to allow appropriate economic analysis.

VII. BINOMIAL VS. ENUMERATIVE SAMPLING

Up to this point, I have focused largely on enumerative sampling, where each sample unit is examined and the number of individuals is counted with a given degree of efficiency. Ingram and Green[129] were the first to introduce the concept of binomial sampling to the cotton literature. Since then, several binomial or presence-absence sampling methods have been developed for cotton.[2,18,21,45,107,109,113,130-135] Unlike enumerative sampling, where all the individuals are recorded for a sample unit, binomial sampling requires that a sample unit be recorded as infested (> 0) or uninfested (0). While simple to use, binomial sampling has contributed more toward improving commercial crop monitoring than any other method.

An advantage of binomial sampling, except at very low densities, is that it is much quicker to categorize a sample unit as having either zero, or one or more individual, than it is to count the number of individuals on each unit. When the arthropod species are small and abundant, such as aphids, mites, thrips, and whiteflies, it may take several minutes or hours to count all individuals on a sample unit.[18] If baseline data have previously been gathered relating the proportion of infested units to mean density per unit,[45] and if the proportion of the population on the sample unit (i.e., on the part of the plant sampled) is known (see Section III), the proportion of infested sample units can be converted to a density per sample unit and a density per unit area (see Wilson,[45,108] Wilson et al.,[2] and Chapter 9 for details).

Wilson and Room[21] described the relationship between the proportion of infested sample units, P, and the variance and mean density per sample unit.

$$P = 1 - e^{-\bar{x} \log_e(a\bar{x}^{b-1})/(a\bar{x}^{b-1}-1)} \tag{13a}$$

This equation is unique among binomial equations in that it estimates the proportion of infested sample units as a function of both variance and mean, and, although derived from the negative binomial distribution, it is applicable for species having aggregated, random, and uniform spatial patterns. For the special case where $a\bar{x}^{b-1}$ approaches a value of 1.0, as a limit Equation 13a is equal to the zero term of the Poisson distribution:[21]

$$P = 1 - e^{-\bar{x}} \tag{13b}$$

Recently, there have been attempts to develop truncated binomial sampling methods, whereby a sampler records a sample unit as infested only if it has a minimum number of individuals[136-138] (also see Chapter 9). An advantage of this approach is that there exists one or more nonzero truncation values that have less variability about the estimated proportion of infested sample units than does a conventional binomial sampling method. It also has the advantage of enabling the use of binomial sampling to estimate an absolute density when conventional binomial sampling would show 100% of the sampled units occupied, an infestation level at which conventional binomial sampling is not able to discriminate between densities.[45] However, a more straightforward approach to estimating density while using conventional binomial sampling would be to sample an area of the plant that contains a lower proportion of the total population of the sampled species and, as a result, is less likely to reach 100% infested.

In contrast to its advantages, truncated binomial sampling has the disadvantage of greater cost (time) per sample unit, particularly at densities that are close to or greater than the truncation level.[139] For example, if the truncation level is six spider mites per leaf and the density of mites per leaf is about six per leaf or greater, the cost of taking a sample will offset the advantage of greater reliability per sample unit with the truncated binomial. In addition, while it is fairly easy to distinguish infested from uninfested sample units using conventional binomial sampling, there is greater sampler error when doing so using truncated binomial sampling.

VIII. INTEGRATING NATURAL ENEMIES INTO THE DECISION PROCESS

Within the U.S., only a limited area of the Cotton Belt continues to practice the heavy pesticide approach to management, and, although farmers in many areas have a long way to go before they can be considered practitioners of sound pest management, the pendulum has swung in the direction of low pesticide input management.[1,12] From 1971 to 1982, the volume of pesticides used to control cotton arthropod pests was reduced by about 75%, while the acreage that was treated was reduced by about 40%.[140] As pesticide use decreases, growers must increasingly rely on naturally occurring biological control agents and on the use of augmentative and inoculate releases of natural enemies.

The key to using biological control agents in pest management is identifying when natural enemies are not sufficiently abundant to prevent herbivores from inflicting economic loss. Plant and Wilson[141] using Bayesian forecasting, and Wilson et al.[45,142] using nonlinear regression, integrated the collective impact of the natural enemies complex upon the rate of spider mite population increase. The advantage of these approaches is that they implicitly consider the combined effect of all natural enemies on a single, easily measured variable, rather than quantifying the abundance of each natural enemy individually, which is costly. These approaches worked particularly well for spider mites, largely because of their short generation time which results in a population pattern that typically increases then decreases monotonically.

Tijerina-Chavez[143] addressed the question of the effectiveness of inoculative releases of phytoseiids for control of spider mites on cotton. Results from his experiments are in agreement with work by Wilson et al.[144] and Zalom et al.[145] on almonds which shows that a 1:10 predator:prey ratio is sufficient to suppress the spider mite density within a few weeks following release. Wilson et al.,[146] in summarizing research on several cropping systems, concluded that a ratio of between 1:10 and 1:20 was sufficient, with lower ratios being adequate on varieties or host crops that afforded a lower spider mite reproductive potential. These numerical ratios are approximately equal to a 1:1 to 1:2 binomial ratio and are about the same as Flaherty et al.[147,148] reported for the phytoseiid-spider mite system in grapes.

Nyrop,[149] studying spider mites in an apple orchard system, was the first to develop a sampling plan based on a predator:prey ratio. Wilson et al.[144] presented data showing that predatory mites on almonds greatly reduced the degree of spatial aggregation of spider mites, which in turn changes the binomial threshold for this pest. Biotic and physical factors that affect the spatial pattern of a herbivore are also likely to change the effective predator:prey ratio.

With the exception of spider mite predation, remarkably little research has been done on integrating the impact of natural enemies into monitoring programs for other cotton herbivores. This possibly can be attributed to the less predictable nature (compared with spider mites) in which the majority of herbivores build-up and inflict damage during the course of a season. The general lack of foresight of local, state, and federal agencies to provide long-term funding for integrated field research on predation and parasitism suggests that convenient and cost-effective monitoring programs which integrate the role of natural enemies will be slow in coming. Developing sampling plans which *a priori* attempt to evaluate the combined impact of each entomophage on the target herbivore is not the solution. As when attempting to use sequential sampling for systems which have several herbivores that simultaneously feed upon a crop, rarely can one develop a sampling plan which does more than estimate the abundance of each herbivore and entomophage. Instead, the solution might be to integrate basic field data into crop-herbivore-natural enemy models that estimate and forecast the interaction between these trophic levels. Building on the approach used by Gutierrez and co-workers[150-152] to quantify the relative role of each of several natural enemies in an alfalfa agroecosystem, one could greatly contribute to integrating natural enemies into the decision process.

IX. FUTURE DIRECTIONS

What will the future hold for cotton production and pest management, and how will these changes affect the day-to-day commercial management of cotton? Frisbie et al.[153] predict that by the year 2012 cotton production will appear to be very similar to what is practiced today. However, closer examination will reveal a major decrease in pesticide use, greater use of cultural and biological controls, a large-scale adoption of transgenic cultivars conferring resistance or tolerance to pest species, and the broad-scale use of sophisticated computer decision aids. A far-reaching prediction was the creation of an entirely new professional position that the authors refer to as a *Crop Practitioner Doctor*. A major responsibility of the *Crop Practitioner* will be to provide state-of-the-art assessments of the causes for production or management problems, and, similar to a medical doctor, prescribe suitable management intervention when appropriate. Although the trend will be toward development of highly trained individuals who have ready access to sophisticated management tools that enable greater integration of field data, these tools will not replace the need for accurate and reliable field data. Instead, they will help to fine-tune monitoring programs so as to achieve the greatest value per unit effort. Even as the management of cotton agroecosystems continues to evolve, and state-of-the-art technologies become commonplace, simple, reliable, and convenient to use, crop and pest monitoring procedures will remain one of the major foundation blocks for an ecologically sound and economically feasible pest management program.[2]

ACKNOWLEDGMENTS

I would like to thank R. E. Frisbie, R. Hanna, M. E. Makela, G. A. Rowell, P. J. Trichilo, and G. Wu for useful comments while preparing this manuscript. A special thanks to Paul Trichilo for his typically superb editorial comments, Guowei Wu for his thorough checking of many of my more detailed equations, and Rachid Hanna for his useful comments on clarifying the vertical distribution section. I would also like to thank John Wiley & Sons for allowing use of the *Sample Size Estimation* section of the chapter *Quantitative Sampling Principles in Cotton IPM*[2] from the book *Integrated Pest Management Systems and Cotton Production*©.

REFERENCES

1. Frisbie, R. E., El-Zik, K. M., and Wilson, L. T., Eds., *Integrated Pest Management in Cotton Production*, John Wiley & Sons, New York, 1989.
2. Wilson, L. T., Sterling, W. L., Rummel, D. R., and DeVay, J. E., Quantitative sampling principles in cotton IPM, in *Integrated Pest Management in Cotton Production*, Frisbie, R. E., El-Zik, K. M., and Wilson, L. T., Eds., John Wiley & Sons, New York, 1989, 85.
3. Southwood, T. R. E., *Ecological Methods*, Chapman & Hall, London, England, 1978.
4. Supplement for 1981 to Statistics on Cotton and Related Data, 1960–78, USDA Economic and Statistics Services Bulletin 617, USDA, Washington, DC, 1981.
5. Niles, G. A. and Feaster, C. V., Breeding, in *Cotton*, Kohel, R. J. and Lewis, C. F., Eds., American Society of Agronomy, Inc.; Crop Science Society of America, Inc.; Soil Science Society of America, Inc., Madison, WI, 1984, 201.
6. Lee, J. A., Cotton as a world crop, in *Cotton*, Kohel, R. J. and Lewis, C. F., Eds., American Society of Agronomy, Inc.; Crop Science Society of America, Inc.; Soil Science Society of America, Inc., Madison, WI, 1984, 1.
7. Cherry, J. P. and Leffler, H. R., Seed, in *Cotton*, Kohel, R. J. and Lewis, C. F., Eds., American Society of Agronomy, Inc.; Crop Science Society of America, Inc.; Soil Science Society of America, Inc., Madison, WI, 1984, 511.
8. Wilson, L. T., The compensatory response of cotton to fruit and leaf damage, 1986 Proceedings Beltwide Cotton Production Research Conferences, National Cotton Council of America, Memphis 1986, 149.
9. Rahman, M. A., Response of the Cotton Plant (*Gossypium hirsutum* cv. Resx S. L. Okra) to Defoliation and Disbudding, Ph.D. thesis, University of Queensland, Brisbane, Australia, 1977.
10. Wilson, L. T. and Bishop, A. L., Responses of Deltapine 16 cotton *Gossypium hirsutum* L. to simulated attacks by known populations of *Heliothis* larvae (Lepidoptera: Noctuidae) in a field experiment in Queensland, Australia, *Prot. Ecol.*, 4, 371, 1982.

11. Ellington, J., George, A. G., Kempen, H. M., Kerby, T. A., Moore, L., Taylor, B. B., and Wilson, L. T. (Technical Coordinators), *Integrated Pest Management for Cotton in the Western Region of the United States*, U.C. Division Agriculture and Natural Resource Publication 3305, 1984.
12. Sterling, W. L., El-Zik, K. M., and Wilson, L. T., Biological control of pest populations, in *Integrated Pest Management in Cotton Production*, Frisbie, R. E., El-Zik, K. M., and Wilson, L. T., Eds., John Wiley & Sons, New York, 1989, 155.
13. Whitcomb, W. H. and Bell, K., Predaceous insects, spiders, and mites of Arkansas cotton fields, *Arkansas Agric. Exp. Stn. Bull.* 690, 1964.
14. van den Bosch, R. and Hagen, K. S., Predaceous and Parasitic Arthropods in California Cotton Fields, *Calif. Agric. Exp. Stn. Bull.* 820, 1966.
15. Ma, C. and Wilson, L. T., Unpublished data, 1989.
16. Goodell, P. B. and Roberts, B., Implementation of a presence-absence sampling method for spider mites on cotton *1985 Proceedings Beltwide Cotton Production Research Conference*, National Cotton Council of America, Memphis, 1985, 170.
17. Wilson, L. T., Developing economic thresholds in cotton, in *Integrated Pest Management on Major Agricultural Systems*, MP-1616, Frisbie, R. E. and Adkisson, P. L., Eds., Texas Agricultural Experiment Station, College Station, 1986, 308.
18. Wilson, L. T., Gonzalez, D., Leigh, T. F., Maggi, V., Foristiere, C., and Goodell, P., Within-plant distribution of spider mites (Acari: Tetranychidae) on cotton: a developing implementable monitoring program, *Environ. Entomol.*, 12, 128, 1983.
19. Wilson, L. T., Trichilo, P. J., and Gonzalez, D., Spider mite (Acari: Tetranychidae) infestation rate and initiation: impact on cotton yield, *J. Econ. Entomol.*, 84, 593, 1991.
20. Taylor, L. R., Woiwod, I. P., and Perry, J. N., The density-dependence of spatial behavior and the rarity of randomness, *J. Anim. Ecol.*, 47, 383, 1978.
21. Wilson, L. T. and Room, P. M., Clumping patterns of fruit and arthropods in cotton, with implications for binomial sampling, *Environ. Entomol.*, 12, 50, 1983.
22. Reilly, J. J. and Sterling, W. L., Interspecific association between the red imported fire ant (Hymenoptera: Formicidae), aphids, and some predaceous insects in a cotton agroecosystem, *Environ. Entomol.*, 12, 541, 1983.
23. Schoenig, S. E. and Wilson, L. T., Patterns of spatial association between spider mites (Acari: Tetranychidae) and their natural enemies on cotton, *Environ. Entomol.*, 21, 471, 1992.
24. Fye, R. E., Kuehl, R. O., and Bonham, C. D., *Distribution of Insect Pests in Cotton Fields*, USDA ARS Misc. Publ. 1140, USDA/ARS, Washington, DC, 1969.
25. Wilson, L. T., Gutierrez, A. P., and Leigh, T. F., Within-plant distribution of the immatures of *Heliothis zea* (Boddie) on cotton, *Hilgardia*, 48, 12, 1980.
26. Hassan, S. T., Distribution of *Heliothis armiger* (Hübner) and *Heliothis punctiger* Wallengren (Lepidoptera: Noctuidae) Eggs and Larvae, and Insecticide Spray Droplets on Cotton Plants, Ph.D. dissertation, University of Queensland, St. Lucia, Australia, 1983.
27. Mitchell, R., An analysis of dispersal in mites, *Am. Nat.*, 104, 425, 1970.
28. Beeden, P., Bollworm oviposition on cotton in Malawi, *Cotton Grower Rev.*, 51, 52, 1974.
29. Bishop, A. L., The spatial dispersion of spiders in cotton ecosystem, *Aust. J. Zool.*, 29, 15, 1981.
30. Boyer, W. P., Warren, L. O., and Lincoln, C., Cotton insect scouting in Arkansas, Univ. Ark. Agric. Exp. Stn. Bull. 656, 1962.
31. Fye, R. E., Preliminary investigation of vertical distributions of fruiting forms and insects on cotton plants, *J. Econ. Entomol.*, 65, 1410, 1972.
32. Hillhouse, T. L. and Pitre, H. N., Oviposition by *Heliothis* on soybeans and cotton, *J. Econ. Entomol.*, 69, 144, 1976.
33. Matthews, G. A. and Tunstall, J. P., Scouting for pests and the timing of spray applications, *Cotton Grower Rev.*, 45, 115, 1968.
34. Wilson, L. T. and Gutierrez, A. P., Within-plant distribution of predators on cotton: comments on sampling and predator efficiencies, *Hilgardia*, 48, 3, 1980.
35. Wilson, L. T., Gutierrez, A. P., and Hogg, D. B., Within-plant distribution of cabbage looper, *Trichoplusia ni* (Hübner) on cotton: development of a sampling plan for eggs, *Environ. Entomol.*, 11, 251, 1982.
36. Wilson, L. T., Booth, D. R., and Morton, R., The behavioural activity and vertical distribution of the cotton harlequin bug *Tectocoris diophthalmus* (Thunberg) (Heteroptera: Scutelleridae) on cotton plants in a glasshouse, *J. Aust. Entomol. Soc.*, 22, 311, 1983.
37. Pickett, C. H., Wilson, L. T., and Gonzalez, D., Population dynamics and within-plant distribution of the western flower thrips (Thysanoptera: Thripidae), an early-season predator of spider mites infesting cotton, *Environ. Entomol.*, 17, 551, 1988.

38. Wilson, L. T., Leigh, T. F., Gonzalez, D., and Foristiere, C., Distribution of *Lygus hesperus* (Knight) (Miridae: Hemiptera) on cotton, *J. Econ. Entomol.*, 77, 1313, 1984.
39. Gonzalez, D. and Wilson, L. T., A food-web approach to economic thresholds: a sequence of pests/predaceous arthropods on California cotton, *Entomophaga*, 27, 31, 1982.
40. Gonzalez, D., Patterson, B. R., Leigh, T. F., and Wilson, L. T., Mites: a primary food source for two predators in San Joaquin Valley cotton, *Calif. Agric.*, 36(2,3), 18, 1982.
41. Trichilo, P. J. and Leigh, T. F., Predation of spider mite eggs by the western flower thrips, *Frankliniella occidentalis* (Thysanoptera: Thripidae), an opportunist in a cotton agroecosystem, *Environ. Entomol.*, 15, 821, 1986.
42. Jones, R. E. and Gutierrez, A. P., unpublished data, 1976.
43. Dumas, B. A., Boyer, W. P., and Whitcomb, W. H., Effect of time of day on surveys of predaceous insects in field crops, *Fla. Entomol.*, 45, 121, 1962.
44. Hutchison, W. D. and Pitre, H. N., Diurnal variation in sweepnet estimates of *Geocoris punctipes* (Say) (Hemiptera: Lygaeidae) density in cotton, *Fla. Entomol.*, 65, 578, 1982.
45. Wilson, L. T., Estimating the abundance and impact of arthropod natural enemies in IPM systems, in *Biological Control in Agricultural Integrated Pest Management Systems*, Hoy, M. A. and Herzog, D. C., Eds., Academic Press, New York, 1985, 303.
46. Steel, R. G. D. and Torrie, J. H., *Principles and Procedures of Statistics*, McGraw-Hill, New York.
47. Iwao, S., A new regression method for analyzing the aggregation pattern of animal populations, *Res. Popul. Ecol.*, 10, 1, 1968.
48. Iwao, S. and Kuno, E., Use of the regression of mean crowding on mean density for estimating sample size and the transformation of data for the analysis of variance, *Res. Popul. Ecol.*, 10, 210, 1968.
49. Iwao, S. and Kuno, E., An approach to the analysis of aggregation pattern in biological populations, in *Statistical Ecology*, Vol. 1, Patil, G. P., Pielou, E. C., and Waters, W. E., Eds., Pennsylvania State University Press, University Park, 1971, 461.
50. Lloyd, M., Mean crowding, *J. Anim. Ecol.*, 36, 1, 1967.
51. Morisita, M., I_σ-index, a measure of dispersion of individuals, *Res. Popul. Ecol.*, 4, 1, 1962.
52. Morisita, M., Application of I_σ-index to sampling techniques, *Res. Popul. Ecol.*, 6, 43, 1964.
53. Morisita, M., Composition of the I_σ-index, *Res. Popul. Ecol.*, 13, 1, 1971.
54. Fracker, S. B. and Brischle, H. A., Measuring the local distribution of *Ribes*, *Ecology*, 25, 283, 1944.
55. Hayman, B. I. and Lowe, A. D., The transformation of counts of the cabbage aphid (*Brevicoryne brassicae* (L.)), *N.Z. J. Sci.*, 4, 271, 1961.
56. Taylor, L. R., Aggregation, variance and the mean, *Nature*, 189, 732, 1961.
57. Taylor, L. R., Aggregation as a species characteristic, in *Statistical Ecology*, Vol. 1, Patil, G. P., Pielou, E. C., and Waters, W. E., Eds., Pennsylvania State University Press, University Park, 1971, 357.
58. Miller, W. E., Discussion section, in *Statistical Ecology*, Vol. 1, Patil, G. P., Pielou, E. C., and Waters, W. E., Eds., Pennsylvania State University Press, University Park, 1971, 372.
59. Bliss, C. I., The aggregation of species within spatial units, in *Statistical Ecology*, Vol. 1, Patil, G. P., Pielou, E. C., and Waters, W. E., Eds., Pennsylvania State University Press, University Park, 1971, 311.
60. Morisita, M., Measuring of the dispersion of individuals and analysis of the distributional patterns, *Mem. Fac. Sci., Kyuchu Univ. Ser. E (Biol.)*, 2, 215, 1959.
61. Wilson, L. T., unpublished data, 1992.
62. Smith, J. W., Stadelbacher, E. A., and Gant, C. W., A comparison of techniques for sampling beneficial arthropod populations in cotton, *Environ. Entomol.*, 5, 435, 1976.
63. Lincoln, C., Procedures for scouting and monitoring for cotton insects, *Ark. Agric. Exp. Stn. Bull.*, 829, 1978.
64. Banerjee, B., Variance to mean ratio and the spatial distribution of animals, *Experientia*, 32, 993, 1976.
65. Fisher, R., The negative binomial distribution, *Ann. Eugen.*, 11, 182, 1941.
66. Haldane, J. B. S., The fitting of binomial distributions, *Ann. Eugen.*, 11, 179, 1941.
67. Katti, S. K., Interrelations among generalized distributions and their components, *Biometrics*, 22, 44, 1966.
68. Gates, C. E. and Ethridge, F. G., A generalized set of discrete frequency distributions with FORTRAN program, *Math. Geol.*, 4, 1, 1972.
69. Wilson, L. T., Room, P. M., and Bourne A. S., Dispersion of arthropods, flower buds and fruit in cotton fields: effects of population density and season on the fit of probability distributions, *J. Aust. Entomol. Soc.*, 22, 129, 1983.
70. Sokal, R. R. and Rohlf, F. J., *Biometry: The Principles and Practices of Statistics in Biological Research*, W. H. Freeman, San Francisco, CA, 1969.
71. Zar, J. H., *Biostatistical Analysis*, Prentice-Hall, Englewood Cliffs, NJ, 1974.
72. Myers, J. H., Selecting a measure of dispersion, *Environ. Entomol.*, 7, 619, 1978.

73. Taylor, L. R., Assessing and interpreting the spatial distributions of insect populations, *Annu. Rev. Entomol.*, 29, 321, 1984.
74. Taylor, L. R., Woiwod, I. P., and Perry, J. N., The negative binomial as a dynamic ecological model for aggregation, and the density dependence of k, *J. Anim. Ecol.*, 48, 289, 1979.
75. Williams, C. B., Some experiences of a biologist with R. A. Fisher and statistics, *Biometrics*, 20, 301, 1964.
76. Anscombe, F. J., The statistical analysis of insect counts based on the negative binomial distribution, *Biometrics*, 5, 165, 1949.
77. Anscombe, F. J., Sampling theory of the negative binomial and logarithmic series distributions, *Biometrika*, 37, 358, 1950.
78. Axelsson, B., Falk, H., Gärdefors, D., Lohm, U., Persson, T., and Tenow, O., Confidence intervals of some animal populations with non-normal distributions, *Zoon*, 3, 115, 1975.
79. Bliss, C. E. and Fisher, R. A., Fitting the negative binomial distribution to biological data, *Biometrics*, 9, 176, 1953.
80. Bliss, C. I. and Owen, A. R. G., Negative binomial distribution with a common k, *Biometrika*, 45, 37, 1958.
81. Fisher, R. A., Corbet, A. S., and Williams, C. B., The relation between the number of species and the number of individuals in a random sample of an animal population, *J. Anim. Ecol.*, 12, 42, 1943.
82. Gerrard, D. J. and Cook, R. D., Inverse binomial sampling as a basis for estimating negative binomial densities, *Biometrics*, 28, 971, 1972.
83. McGuire, J. U., Brindley, T. A., and Bancroft, T. A., The distribution of European corn borer larvae *Pyrausta nubalalis* (Hbn.), in field corn, *Biometrics*, 13, 65, 1957.
84. Pieters, E. P. and Sterling, W. L., Aggregation indices of cotton arthropods in Texas, *Environ. Entomol.*, 3, 598, 1974.
85. Brooker, P., Kriging, *Eng. Min. J.*, 180(9), 148, 1979.
86. Cressie, N. and Hawkins, D. M., Robust estimation of the variograms. I, *Math. Geol.*, 12(2), 115, 1980.
87. Edwards, C. and Penney, D., *Calculus and Analytical Geometry*, Prentice-Hall, Englewood Cliffs, NJ, 1982.
88. Isaaks, E. H. and Srivastava, R. M., Spatial continuity measures for probabilistic deterministic geostatistics, *Math. Geol.*, 20(4), 313, 1988.
89. Isaaks, E. H. and Srivastava, R. M., *An Introduction to Applied Geostatistics*, Oxford University Press, New York, 1989.
90. Borth, P. W. and Huber, R. T., Modeling bollworm establishment and dispersal in cotton with the kriging technique, in *Proceedings Beltwide Production Research Conference*, National Cotton Council of America, Memphis, 1987, 267.
91. Okubo, A., *Diffusion and Ecological Problems: Mathematical Models*, Springer-Verlag, Berlin, 1980.
92. Culin, J., Brown, S., Rogers, J., Scarborough, D., Swift, A., Cotterill, B., and Kovach, J., A simulation model examining boll weevil dispersal: historical and current situations, *Environ. Entomol.*, 19, 195, 1990.
93. McKibben, G. H. and Smith, J. W., Weather factors affecting long-range dispersal of the boll weevil, in *1989 Proceedings Beltwide Cotton Production Research Conference*, National Cotton Council of America, Memphis, 1989, 250.
94. McKibben, G. H., Willers, J. L., Smith, J. W., and Wagner, T. L., Stochastic model for studying boll weevil dispersal, *Environ. Entomol.*, 20, 1327, 1991.
95. Corbett, A. and Plant, R. E., The role of movement in the response of natural enemies to agroecosystem diversification: a theoretical evaluation, *Environ. Entomol.*, 22, 519, 1993.
96. Bottrell, D. G., White, J. R., Moody, D. S., and Hardee, D. D., Overwintering habitats of the boll weevil in the Rolling Plains of Texas, *Environ. Entomol.*, 1, 633, 1972.
97. Rummel, D. R. and Adkisson, P. L., Distribution of boll weevil infested cotton fields in relation to overwintering habitats in the High and Rolling Plains of Texas, *J. Econ. Entomol.*, 63, p. 1906, 1970.
98. Wilson, L. T., Bozkurt, S., Tapadiya, P. K., Trichilo, P. J., Zaman, A. U., Haldenby, R. K., Rummel, D. R., Carroll, S. C., Fuchs, T. W., Slosser, J. E., and Frisbie, R. E., Development of a comprehensive forecasting program to complement boll weevil control and eradication efforts in the Rolling Plains and High Plains of Texas, in *1993 Proceedings Beltwide Cotton Production Research Conference*, National Cotton Council of America, Memphis, 947, 1993.
99. Trichilo, P. J., Wilson, L. T., Haldenby, R. K., Rummel, D. R., Carroll, S. C., Fuchs, T. W., Slosser, J. E., and Frisbie, R. E., Use of geographic information systems to assess risk of boll weevil infestations, in *1993 Proceedings Beltwide Cotton Production Research Conference*, National Cotton Council of America, Memphis, 944, 1993.

100. Aronoff, S., *Geographic Information Systems: A Management Perspective*, WDL Publications, Ottawa, Canada, 1989.
102. Star, J. and Estes, J., *Geographic Information Systems*, Prentice-Hall, Englewood Cliffs, NJ, 1990.
102. Phillips, J. R., Gutierrez, A. P. and Adkisson, P. L., General accomplishments toward better insect control in cotton, in *New Technology of Pest Control*, Huffaker, C. B., Ed., John Wiley & Sons, New York, 1980, 123.
103. Ruesink, W. G. and Kogan, M., the quantitative basis of pest management: sampling and measuring, in *Introduction to Insect Pest Management*, Metcalf, R. L. and Luckmann, W. H., Eds., John Wiley & Sons, New York, 1982, 315.
104. Karandinos, M. G., Optimum sample size and comments on some published formulae, *Entomol. Soc. Am. Bull.*, 22, 417, 1976.
105. Ruesink, W. G., Introduction to sampling theory, in *Sampling Methods in Soybean Entomology*, Kogan, M. and Herzog, D. C., Eds., Springer-Verlag, New York, 1980, 61.
106. Wilson, L. T. and Room, P. M., The relative efficiency and reliability of three methods for sampling arthropods in Australian cotton fields, *J. Aust. Entomol. Soc.*, 21, 175, 1982.
107. Sterling, W. L. and Pieters, E. P., Sequential decision sampling, in *Economic Thresholds and Sampling of Heliothis Species on Cotton, Corn, Soybeans, and Other Host Crops, Southern Cooperative Series Bulletin* 231, Sterling, W. L., Ed., 1979, 85.
108. Wald, A., *Sequential Analysis*, John Wiley & Sons, New York, 1947.
109. Wilson, L. T., Development of an optimal monitoring program in cotton: emphasis on spider mites and *Heliothis* spp., *Entomophaga*, 27, 45, 1982.
110. Allen, J., Gonzalez, D., and Gokhale, D. V., Sequential sampling plans for the bollworm, *Heliothis zea*, *Environ. Entomol.*, 1, 771, 1972.
111. Rothrock, M. A. and Sterling, W. L., A comparison of three sequential sampling plans for arthropods of cotton, *Southwest. Entomol.*, 7, 39, 1982.
112. Rothrock, M. A. and Sterling, W. L., Sequential sampling for arthropods of cotton: its advantages over point sampling, *Southwest. Entomol.*, 7, 70, 1982.
113. Sterling, W. L., Sequential sampling of cotton insect populations, in *1975 Proceedings Beltwide Cotton Production Research Conference*, National Cotton Council of America, Memphis, 1975, 133.
114. Sterling, W. L., Sequential decision plans for the management of cotton arthropods in south-east Queensland, *Aust. J. Ecol.*, 1, 265, 1976.
115. Sterling, W. L. and Frisbie, R., Sequential sampling, in *Cotton Pest Management Scouting Handbook*, Hamer, J., Ed., Miscellaneous Cooperative Extension Publication, 1981, 24.
116. Blood, P. R. B. and Wilson, L. T., Field validation of a crop/pest management descriptive model, *Proceedings of SIMSIG-78, Simulation Conference*, Australian National University, Canberra, 1978, 91.
117. Gutierrez, A. P. and Curry, G. L., Conceptual framework for studying crop-pest systems, in *Integrated Pest Management in Cotton Production*, Frisbie, R. E., El-Zik, K. M., and Wilson, L. T., Eds., John Wiley & Sons, New York, 1989, 267.
118. Gutierrez, A. P. and Daxl, R., Practical and evolutionary considerations: estimating economic thresholds for bollworm and boll weevil damage in Nicaraguan cotton, in *Pest and Pathogen Control: Strategic, Tactical, and Policy Models*, Conway, Ed., John Wiley & Sons, Chichester, England, 1984.
119. Gutierrez, A. P. and Wilson, L. T., Development and use of pest models, in *Integrated Pest Management in Cotton Production*, Frisbie, R. E., El-Zik, K. M., and Wilson, L. T., Eds., John Wiley & Sons, New York, 1989, 65.
120. Gutierrez, A. P., Falcon, L. A., Loew, W., Leipzig, P. A., and van den Bosch, R., An analysis of cotton production in California: a model for Acala cotton and the effects of defoliators on its yield, *Environ. Entomol.*, 4, 125, 1975.
121. Gutierrez, A. P., Leigh, T. F., Wang, Y., and Cave, R., An analysis of cotton production in California: *Lygus hesperus* (Heteroptera: Miridae) injury—an evaluation, *Can. Entomol.*, 109, 1375, 1977.
122. Gutierrez, A. P., Wang, Y., and Daxl, R., The interaction of cotton and boll weevil (Coleoptera: Curculionidae)—a study of co-adaptation, *Can. Entomol.*, 111, 357, 1979.
123. Gutierrez, A. P., Daxl, R., Leon Quant, G., and Falcon, L. A., Estimating economic thresholds for bollworm, *Heliothis zea* Boddie, and boll weevil, *Anthonomus grandis* Boh., damage in Nicaraguan cotton, *Gossypium hirsutum* L., *Environ. Entomol.*, 10, 872, 1981.
124. Gutierrez, A. P., Dos Santos, W. J., Pizzamiglio, M. A., Villacorta, A. M., Ellis, C. K., Fernandes, C. A. P., and Tutida, I., Modelling the interactions of cotton and the cotton boll weevil. II. Boll weevil (*Anthonomus grándis*) in Brasil, *J. Appl. Ecol.*, 28, 398, 1991.
125. Gutierrez, A. P., Dos Santos, W.J., Villacorta, A. M., Pizzamiglio, M. A., Ellis, C. K., Carvalho, L. H., and Stone, N. D., Modelling the interactions of cotton and the cotton boll weevil. I. A comparison of growth and development of cotton varieties, *J. Appl. Ecol.*, 28, 371, 1991.

126. Wilson, L. T., Changing perspectives on the use of simulation models in IPM, National IPM Symposium, Las Vegas, Nevada, April 1989; National IPM Coordinating Committee, New York State Agricultural Experiment Station, Cornell University, Ithaca, NY, 1989, 129.
127. Wilson, L. T., Plant, R.E., Kerby, T. A., Zelinski, L., and Goodell, P. B., Transition from a strategic to a tactical crop and pest management model: use as an economic decision aid, in *1987 Proceedings Beltwide Cotton Production Research Conferences*, National Cotton Council of America, Memphis, 1987, 207.
128. Wilson, L. T., Corbett, A., Trichilo, P. J., Kerby, T. A., Plant, R. E., and Goodell, P. B., Strategic and tactical modelling: cotton-spider mite agroecosystem management, *Exp. Appl. Acarol.*, 14, 357, 1992.
129. Ingram, W. W. and Green, S. M., Sequential sampling for bollworms on rain grown cotton in Botswana, *Cotton Grower Rev.*, 49, 265, 1972.
130. Hutchison, W. D., Henneberry, T. J., and Beasley, C. A., Rationale and potential applications for monitoring pink bollworm egg populations in cotton, in *1986 Proceedings Beltwide Cotton Production Research Conferences*, National Cotton Council of America, Memphis, 1986, 183.
131. Hutchison, W. D., Beasley, C. A., Henneberry, T. J., and Martin, J. M., Sampling pink bollworm (Lepidoptera: Gelechiidae) eggs: potential for improved timing and reduced use of insecticides, *J. Econ. Entomol.*, 81, 673, 1988.
132. Stroschein, D. L., Beasley, C. A., Hutchison, W. D., Martin, J. M., and Henneberry, T. J., Field evaluation of a sequential sampling plan for pink bollworm eggs, in *1988 Proceedings Beltwide Cotton Production Research Conferences*, National Cotton Council of America, Memphis, 1988, 315.
133. Pizzamiglio, M. A., Gutierrez, A. P., Dos Santos, W. J., Gallagher, K. D., De Oliveira, W. S., and Fujita, Z. H., Phenological patterns and sampling decision rules for arthropods in cotton fields in Parana, Brazil: before boll weevil, *Pesq. Agropec. Brasil*, 24, 337, 1989.
134. Sterling, W. L., Binomial sampling of cotton arthropods, *Folia Entomol. Mex.*, 39–40, 59, 1978.
135. Wilson, L. T., Leigh, T. F., and Maggi, V., Presence-absence sampling of spider mite densities on cotton, *Calif. Agric.*, 35, 10, 1981.
136. Binns, M. R., Robustness in binomial sampling for decision-making in pest incidence, in *Monitoring and Integrated Management of Arthropod Pests of Small Fruit Crops*, Bostanian, N. J., Wilson, L. T., and Dennehy, T. J., Eds., Intercept, Dorset, England, 1990, 63.
137. Binns, M. R. and Bostanian, N. J., Binomial and censored sampling in estimation and decision making for the negative binomial distribution, *Biometrics*, 44, 473, 1988.
138. Binns, M. R. and Bostanian, N. J., Robustness in empirically based binomial decision rules for integrated pest management, *Econ. Entomol.*, 83, 420, 1990.
139. Wilson, L. T., unpublished data.
140. Frisbie, R. E. and Adkisson, P. L., IPM: Definitions and current status in U.S. agriculture, in *Biological Control in Agricultural Integrated Pest Management Systems*, Hoy, M. A. and Herzog, D. C., Eds., Academic Press, New York, 1985, 41.
141. Plant, R. E. and Wilson, L. T., A Bayesian method for sequential sampling and forecasting in agricultural pest management, *Biometrics*, 41, 203, 1985.
142. Wilson, L. T., Gonzalez, D., and Plant, R. E., Predicting sampling frequency and economic status of spider mites on cotton, in *1985 Proceedings Beltwide Cotton Production Research Conferences*, National Cotton Council of America, Memphis, 1985, 168.
143. Tijerina-Chavez, A., Biological Control of Spider Mites (Acari: Tetranychidae) on Cotton through Inoculate Release of Predatory Mites *Metaseiulus occidentalis*, *Amblyseius californicus* (Acari: Phytoseiidae) in the San Joaquin Valley of California, Ph.D. dissertation, University of California, Davis, 1991.
144. Wilson, L. T., Hoy, M. A., Zalom, F. G., and Smilanick, J. M., Sampling mites in almonds. I. Within-tree distribution and clumping pattern of mites with comments on predator-prey interactions, *Hilgardia*, 52, 1, 1984.
145. Zalom, F. G., Hoy, M. A., Wilson, L. T., and Barnett, W. W., Sampling mites in almonds. II. Presence-absence sequential sampling for *Tetranychus* mite species, *Hilgardia*, 52, 14, 1984.
146. Wilson, L. T., Trichilo, P. J., Flaherty, D. L., Hanna, R., and Corbett, A., Natural enemy-spider mite interactions: comments on implications for population assessment, *Mod. Acarol.*, 1, 167, 1991.
147. Flaherty, D. L., Hoy, M. A., and Lynn, C. D., Spider mites, in *Ground Pest Management*, Flaherty, D. L., Jensen, F. L., Kasimatis, A. N., Kido, H., and Moller, W. J., Eds., University of California Agricultural Sciences Publication 4105, Berkeley, 1981, 111.
148. Flaherty, D. L., Wilson, L. T., Welter, S., Lynn, C. D., and Hanna, R. A., Spider mites, in *Grape Pest Management*, Flaherty, D. L., Christensen, P., Lannini, T., Marois, J., and Wilson, L. T., Eds., University of California Press, Berkeley, 1992, in press.

149. Nyrop, J. P., Sequential classification of prey/predator ratios with application to European red mite (Acari: Tetranychidae) and *Typhlodromus pyri* (Acari: Phytoseiidae) in New York apple orchards, *J. Econ. Entomol.*, 81, 14, 1988.
150. Gutierrez, A. P. and Baumgaertner, H. U., Age-specific energetics models-pea aphid *Acyrthrosiphon pisum* (Homoptera: Aphididae) as an example. I., *Can. Entomol.*, 116, 924, 1984.
151. Gutierrez, A. P. and Baumgaertner, H. U., A realistic model of plant-herbivore-parasitoid-predator interactions. II., *Can. Entomol.*, 116, 933, 1984.
152. Gutierrez, A. P., Baumgaertner, H. U., and Summers, C. G., A case study in an alfalfa ecosystem. III., *Can. Entomol.*, 116, 950, 1984.
153. Frisbie, R. E., Hardee, D. D., and Wilson, L. T., Biologically intensive integrated pest management: future choices for cotton, in *Food, Crop Pests, and the Environment*, Zalom, F. G. and Fry, W. E., Eds., APA Press, St. Paul, MN, 1992, 57.
154. Head, R. B., Cotton insect losses—1989, in *1990 Proceedings Beltwide Cotton Production Research Conference*, National Cotton Council of America, Memphis, 1990, 157.

Chapter 18

SAMPLING ARTHROPODS IN LIVESTOCK MANAGEMENT SYSTEMS

Timothy J. Lysyk and Roger D. Moon

TABLE OF CONTENTS

I. Livestock Production Systems..................................516
   A. Overview of Livestock Industries........................516
   B. Arthropod Pests of Livestock............................517
   C. Damage..517

II. Sampling Plans...518
   A. Direct Examination of Host..............................519
      1. Enumeration of Total Population on the Host........519
      2. Partial Enumeration of Parasite Populations.........524
      3. Presence or Absence Sampling........................526
   B. Sampling Off-Host Populations...........................530

III. Summary...534

References..534

I. LIVESTOCK PRODUCTION SYSTEMS

A. OVERVIEW OF LIVESTOCK INDUSTRIES

North American livestock industries are diverse and of high value, consisting of very different animals and industries: beef cattle are raised for calves, meat, and leather; dairy animals for milk and veal; swine for pork; poultry for eggs and meat; and sheep for meat and wool. In 1990 in the U.S.,[1] there were approximately 98 million beef cattle, 10 million dairy cattle, 55 million pigs, 11 million sheep and lambs, 350 million chickens, and 5.9 billion broiler chickens. Cattle were valued at over $60 billion and produced over $20 billion worth of milk. Hogs were valued at over $4.5 billion, sheep at $0.9 billion, and broilers at over $8 billion. In Canada in 1990,[1] there were 11.2 million beef cattle, 1.4 million dairy cattle, and 10.7 million hogs. Poultry mean and egg production in Canada were 3.7 and 8.3% of U.S. production, respectively.

Each species is affected by a variety of arthropod pests including mites, ticks, lice, and numerous species of Diptera. Drummond[2] estimated that losses to cattle alone amounted to over $1.5 billion annually in the U.S. Previous estimates[3] of annual losses were over $2.2 billion for cattle, $54 million for sheep, $230 million for swine, and $499 million for poultry. Haufe and Weintraub[4] estimated that $445 million dollars were lost annually in Canada due to livestock pests.

The pest complex associated with each host commodity is influenced by the producing species itself, as well as production practices and management systems. Many arthropod species show a high degree of host specificity and are pests of a particular commodity, whereas others have broad host ranges[5] and occur among several commodity systems. Host size, behavior, and accessibility are important constraints to be considered when sampling ectoparasites. Within each commodity, animals are housed in a variety of ways, ranging from intensive indoor systems to pastoral rangeland systems.

Axtell[6] classifies livestock production systems as (1) pasture or range, (2) outdoor confined, and (3) indoor confined. Pasture or range systems are typically used by beef and dairy cattle and sheep. Animal densities can vary greatly from one animal unit per 50 acres on poor native range to more than two animal units per acre of well-managed, irrigated pasture.[7] An animal unit is the equivalent of an animal that consumes 11.8 kg of forage per day. This unit could be a mature cow (450 kg) or five ewes. Tame or improved pastures are distinguished from rangeland in that desired forage species may be sown, irrigated, and fertilized, and selective grazing may be employed to increase the carrying capacity of the land. Pasture is the habitat for off-host segments of parasite populations, and animal feces in pastures are the developmental medium for several species of pest flies. Weather can be extremely variable, and pests have developed a variety of mechanisms for surviving the host-free winter. Studies are complicated by limited access to animals for sampling due to the large areas over which they roam.

Animals in outdoor confined systems are kept at high densities within pens and are fed conserved feeds. Pens may include feedlots for cattle, sheep, and swine.[6] There is little if any vegetation, and the high density of animals tends to compact the soil and trample manure except in low traffic areas and along fencelines. Producers may push the manure into a mound in the center of the lot, which may create additional habitat for filth flies. Buildings and fences provide some degree of shelter from the weather. Often, animals may be kept on pasture during the summer and moved to lots for feeding during the winter. Populations of parasitic lice and mites generally increase during the winter, a trend attributed to declining temperatures and increased host stress associated with winter confinement. Animals are more accessible due to the limited areas in which they can move.

Dairy, swine, and most poultry are generally kept in indoor confined systems.[6] Animals are housed continuously at extremely high densities in environmentally controlled buildings. Manure may accumulate beneath the flooring of cages and pens, or be scraped and flushed to outdoor settling basins. Accumulations of manure, if mismanaged, may favor production of filth breeding flies, and high animal densities favor growth and transmission of ectoparasites. Numerous surfaces are available for perching insects, and the modified environments may allow pests such as filth flies to survive throughout the winter. Animals can be examined easily in pens and cages.

B. ARTHROPOD PESTS OF LIVESTOCK

Detailed descriptions of the biology, taxonomy, host range, and management of livestock pests are available[8] and should be consulted before starting a sampling program. Arthropod pests vary according to their degree of association with the host animal. Some species, such as mites, lice, and keds, live as continuous, permanent ectoparasites and complete most if not all of their life cycle on the host. These parasites tend to be host specific. Other species, such as ticks, grubs, and some muscoid flies, have free-living and parasitic phases interspersed in their life cycle. The free-living and parasitic phases may correspond with particular stages. In the instance of the horn fly, *Haematobia irritans irritans* (L.), eggs, larvae, and pupae are free-living, whereas adults are continuously parasitic. Free-living and parasitic phases also may alternate within a stage. Adult stable flies, *Stomoxys calcitrans* (L.), visit their hosts daily to obtain a blood meal, which takes 5 to 15 min to complete. These flies spend the remainder of their time at rest in sheltered areas off their hosts. Finally, some insects that live in close association with livestock are not truly parasitic, but are nevertheless considered serious pests, e.g., the house fly *Musca domestica* (L.).

Not all pests originate within the production system. Blackflies, biting midges, tabanids, and mosquitoes may breed in nearby rivers, wetlands, and artificial water catchment areas. Adults of these pests can fly into confined production systems to feed on the animals, where severe outbreaks can occur.[9]

In summary, there is wide variation in production systems, in pest life histories, and in degrees of physical association with livestock. Samplers must consider the purposes of their sampling programs, the stages to be sampled, and the habitats to be included in their sample universes.

C. DAMAGE

Arthropods affect animal production in a variety of ways. Besides causing direct injury to host skin, arthropods can induce changes in host behavior and elicit immunological and physiological responses that ultimately reduce productivity.[10] Behavioral responses include twitching, bunching,[11] grooming, foot stomping, rubbing, and gadding. Physiological responses include elevated serum cortisone,[12,13] induction of analgesic opioids,[14,15] and other associated stress responses. Physiological responses may also include increased heart and respiration rates, elevated body temperatures, and decreased nitrogen retention.[12,13] These changes have been demonstrated with several species of biting flies. Immunochemical responses include analgesia, altered blood profiles expressed as anemia or altered cellular profiles,[10] and hypersensitivity.[16]

The ultimate effects of arthropod parasitism on animal production may be expressed as diminished feed intake, reduced feed conversion efficiency, decreased rate of growth and lactation, and prolonged times to reach slaughter weights. Effects may be passed on to nursing young; high numbers of horn flies on cows can reduce weaning weights of calves.[17] Other effects of parasites include paralysis and pathogen transmission. Punctures and feeding activity can damage carcasses and hides and devalue the product. Finally, Northern fowl mites, *Ornithonyssus sylviarum* Canestrini & Fanzango, and house flies cause serious annoyance and irritation of laborers or nearby residents.

Economic thresholds have been developed under field conditions for some pests and production systems. Effects of horn flies,[18] stable flies,[19,20] and ticks[21] on cattle have been demonstrated under experimental conditions or in screened feedlots. Effects of mites on swine[22] and poultry[23,24] also have been demonstrated experimentally. However, considerable information is still required to develop workable economic thresholds for the major production systems. Experimental manipulation is constrained by an inability to replicate adequately[25] and to control numbers of attacking parasites. Thresholds for many parasites in livestock production systems remain to be studied.

Haufe[26] proposed a general model for the effects of blood-feeding parasites on cattle and illustrated it using the horn fly as a model system. This model was reviewed by Nelson.[10] Haufe proposed that two mechanisms affect productivity, each with their respective threshold densities. The first mechanism involves a quantal response to low levels of fly infestation, with a threshold near 12 horn flies per animal. When fly densities exceed this threshold, productivity is reduced by 17 to 20% below levels set by interactions among breed, pasture productivity, and weather. Reduced production is believed to be caused by the collective effects of flies on neurosecretory responses, grazing, and social behavior, and utilization of forage. This low threshold of 12 flies per animal is likely to be undetectable in most field experiments; consequently, effects of horn flies measured to date may have been seriously underestimated.

The second mechanism, according to Haufe's model, elicits a graded response, but the mechanism operates at higher levels of infestations. Above 230 to 250 horn flies per animal, animals are incapable of adapting to the additional stress caused by fly feeding, and productivity is reduced as a linear function of fly density. The actual number of insects required to elicit the quantal and graded responses varies with the pest species, the degree to which it is capable of annoying the host, and the nature of the damage it causes to the host.

From a pest management standpoint, there are both theoretical and practical justifications for measuring parasite populations. Detection of a parasite, even at low densities, may be required by law to achieve or maintain local eradication, as is the case with cattle grubs in Alberta and psoroptic mange throughout the U.S. and Canada. Also, the high potential rate of increase of some parasitic species and their ability to spread rapidly to uninfested areas has led some workers to suggest that a practical action threshold is the presence of a single parasite.[27]

II. SAMPLING PLANS

Techniques for sampling arthropods in livestock systems have been previously summarized.[28] Ectoparasites are sampled by direct examination of the host. Mobile, free-living stages are generally sampled using visual indices of specimens per host, or by catch rates on traps near hosts. Less mobile, free-living stages are sampled by examining units of habitat. The techniques used depend to a large extent on the objectives of the sampling program, and on biological, ecological, and statistical details of the universe being sampled. Although the basic techniques for sampling many arthropods on livestock have been studied, development of these techniques into sampling programs which consider selection of sampling units, sub-sampling units, and respective sample sizes requires more attention. The nature of the sampling plan will vary according to its specific purpose.

From a pest management standpoint, a producer may be concerned merely with detecting the presence of a relatively immobile ectoparasite in a herd or flock to identify animals for spot treatment. In this instance, routine examination of all animals is warranted and sampling in a strict sense is not a consideration. The producer also may be concerned with determining pest intensity of a mobile ectoparasite that moves from animal to animal. In this instance, a treatment would be applied to the entire herd or farmstead when the pest population becomes excessive. Here, sampling animals within a herd is a useful consideration.

For research, detailed estimates of density on and off the host may be required. Furthermore, the researcher is usually concerned with monitoring populations on more than one herd or flock. In such instances, efficient sampling may involve selection of herds or flocks within a geographic area, selection of animals within herds, and perhaps even selection of sites within chosen animals to estimate density.

A. DIRECT EXAMINATION OF HOST

Direct examination involves either counting the number of arthropods on the host at a particular point in time, categorizing hosts according to level of infestation using a scoring or rating system, or determining the presence or absence of the parasite on the chosen host. Presence or absence sampling can be accomplished either by detecting the actual arthropod or examining the host for clinical signs of infestation. All techniques require access to the host animal. It may be possible to approach tame animals in the field and count large arthropods or view the animals from a distance using binoculars. More detailed or accurate examination may require rounding up animals and physically restraining them if the animals are not accustomed to being inspected. Smaller animals, such as poultry, can be handled directly by hand. Some types of surveys require examination of hides or carcasses.

1. Enumeration of Total Population on the Host

Adults of the common cattle grub (*Hypoderma lineatum* L.) and northern cattle grub (*H. bovis* de Vill.) are active in summer. Females attach eggs to hairs on the legs of cattle. Hatching larvae penetrate the skin and migrate through the tissues, eventually reaching their hosts' backs where they form warbles, pus-filled lesions, on the backs of the animals in January through April.[29] Mature larvae then exit the warbles and drop from the hosts from March through June, pupate in soil, and eventually emerge as adults to complete the cycle. The basic technique for sampling cattle grub larvae is to palpate the animals' backs in winter when animals are accessible in drylots. Palpation is simple; cattle are restrained, inspectors rub their hands over the backs, and warbles can be counted or removed.[30] The results for a herd can be summarized as the average number of grubs per animal. During the course of winter and spring, the number of warbles in an animal's back is a balance between the rate at which grubs arrive at the back and the rate at which they exit to the soil (Figure 1). As a result, peak grub abundance in the back occurs for only a very short period. If palpations are conducted too early or late, an infection in an animal may be missed, particularly if the number of parasites per host is low. If palpations are conducted more than once per animal, grub density for the individual animal can be expressed as the greatest number of grubs found over the sequence of observations. Exiting grubs can be collected in a girdle attached to the animal.[31] Detailed information on the appearance and disappearance of grubs from individual animals can be obtained by mapping locations of the animal's warbles.[31]

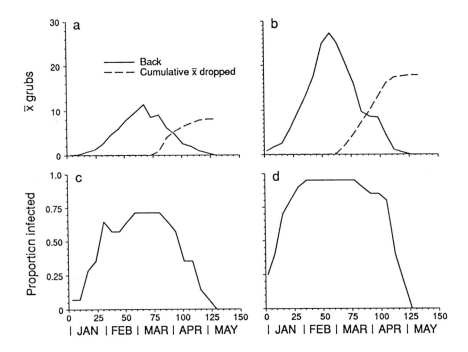

FIGURE 1. Mean number of cattle grubs in the back (solid line) and cumulative number of grubs dropped (broken line) from cattle during (a) 1991 and (b) 1992. Proportion of cattle positive for cattle grubs determined by manual palpation in (c) 1991 and (d) 1992. (Data supplied courtesy of Dr. D. D. Colwell, Agriculture Canada, Lethbridge, Alberta.)

Herds can be sampled to estimate the mean peak density of grubs per animal. Required sample size can be estimated using published formulas and the variance-mean relationship among maximum counts of grubs. The sample size necessary to estimate a mean with a half confidence interval of chosen length, expressed as a fixed proportion of the mean (D) is:[32]

$$n = \left(\frac{z_{\alpha/2}}{D}\right)^2 \frac{\sigma^2}{\mu^2} \quad (1)$$

where n = required sample size, $z_{\alpha/2}$ = the upper $\alpha/2$ point of the normal distribution, μ = the anticipated population mean, and σ^2 is the population variance. The variance usually increases with the mean[33] and is frequently modeled as:

$$\ln s^2 = a + b \ln \bar{x} \quad (2)$$

where s^2 and $\bar{x}$ are sample estimates of σ^2 and μ, respectively. By substituting Equation 2 into Equation 1, the formula for required sample size becomes:[34]

$$n = \left(\frac{z_{\alpha/2}}{D}\right)^2 e^a \bar{x}^{(b-2)} \quad (3)$$

which can be used to calculate sample size over a range of values of mean grub density, if the number of sample units were from an infinite population. When sampling cattle in finite herds,[35,36] a finite population correction:[37]

$$n' = \frac{n}{1 + (n/N)} \quad (4)$$

is used where N is the total number of cattle within a herd, n is defined as above, and n' is the adjusted sample size. When n or n' is < 30, values of Student t-distribution should be used in Equation 3 instead of z. Values of n' to estimate means from herds of 10 to 100 animals, with 90% confidence intervals ($\alpha = 0.1$) that are 10, 20, and 40% of the mean ($D = 0.05, 0.1,$ and 0.2) are shown in Table 1. Precise estimates ($D = 0.05$, 90% confidence interval = 10% of the mean) require that a relatively high fraction of the herd be examined, even when grub densities are high. Less precision ($D = 0.1$ or 0.2, 90% confidence interval = 20 and 40% of the mean, respectively) requires inspection of fewer animals (Table 1).

Palpation and visual examination also are used to estimate the numbers of ticks attached to cattle.[35,38-40] Palpations can provide repeatable estimates of the numbers of ticks on an animal, but searching the entire body surface is costly, tedious, and may be dangerous. Some species of ticks attach to certain host body parts;[41] as a result animals can be subsampled with preference for body regions where ticks prevail. Although the lone star tick, *Amblyomma americanum* (L.), attaches to all body regions, counting the ticks on selected regions can provide reliable estimates of extrapolated totals. Regression equations have been developed to convert densities on subsampled body regions to whole body estimates.[35]

Davis and Moon[36] developed a hide sampling plan for the itch mite, *Sarcoptes scabiei* (De Geer). This mite is an ectoparasite that causes mange in swine and a wide variety of other mammals.[42] Sampling itch mites is complicated by small mite size, mite burrowing habits, and mite distribution over their hosts. The mite lives within burrows in the hosts' skin and cannot be detected by simple visual examination. Two forms of mange are recognized based on symptoms.[43] Hypersensitive mange is characterized by pruritus or intense itching and raised red lesions. In this form, mites are distributed on the face, throat, dorsum, groin, ears, and to a minor extent on the belly. In some animals, hypersensitive mange develops into crusted mange (or hyperkeratotic mange)[42] and is recognized by the presence of dry, grayish-white, crusty lesions which form on the ears. Although less common, this form of mange has been reported to reach levels of 15% in a herd.[44] Mite populations in ear crusts can account for over 85% of the mites on an animal. Animals with crusty lesions may be a major reservoir within a herd.

Quantifying the number of mites on a hide requires KOH digestion of the hide and extraction of the mites.[36] Extraction efficiency was over 92% for adult females but was 49% for other stages. Eggs and possibly larvae are destroyed by the extraction procedure. Extraction is laborious, and sampling a single hide can take an entire day. However, sampling can be improved by using a double stratified method. Hide surface area is measured. If ears contain lesions, the ears should be entirely processed for mites. If lesions are not present, then 25-cm^2 skin sections should be removed from the dorsum and other regions, if affordable, and processed. The number of skin sections required for various levels of precision are given by Davis and Moon.[36] The resulting estimates per body region then can be adjusted for extraction efficiency and extrapolated to estimate the total number of mites on the animal using equations presented by Davis and Moon.[36] The number of hides required to estimate the average abundance of mites in a herd also was presented.

TABLE 1
Required Number of Animals per Herd of Various Sizes (N) to Estimate Mean Grub Abundance with 90% Confidence Intervals ($2D$) Equal to 10, 20, and 40% of the Mean

$\bar{x}$ Grubs per animal	Herd size					
	10	20	30	40	50	100
$D = 0.05$						
1.0	10	20	30	40	49	97
5.0	10	20	29	39	48	92
10.0	10	19	29	38	47	88
15.0	10	19	29	37	46	85
20.0	10	19	28	37	45	83
25.0	10	19	28	36	45	81
30.0	10	19	28	36	44	79
35.0	10	19	28	36	43	77
40.0	10	19	27	35	43	75
$D = 0.10$						
1.0	10	20	29	38	47	89
5.0	10	19	27	35	43	74
10.0	9	18	26	33	39	65
15.0	9	18	25	31	37	59
20.0	9	17	24	30	35	54
25.0	9	17	23	29	34	51
30.0	9	16	23	28	32	48
35.0	9	16	22	27	31	46
40.0	9	16	22	26	30	43
$D = 0.20$						
1.0	10	18	26	33	40	67
5.0	9	16	21	26	30	42
10.0	8	14	18	22	24	32
15.0	8	13	16	19	21	26
20.0	7	12	15	17	19	23
25.0	7	11	14	16	17	21
30.0	7	11	13	15	16	19
35.0	7	10	12	14	15	17
40.0	7	10	12	13	14	16

Sampling mobile ectoparasites presents its own problems. Adults of the horn fly spend their entire life on cattle, leaving only to oviposit on freshly deposited dung pats and then redistribute among animals. Estimates of fly density per animal can be made by approaching animals and counting the number of flies from a distance.[17,45-47] Horn flies exhibit marked preferences for certain animals within a herd. These animals are termed "fly-tolerant"[26] or "fly susceptible"[17] and may carry a significant portion of the fly population. Without prior knowledge about the identity of the fly-tolerant animals, estimating horn fly density with any degree of precision requires that a fairly large proportion of the herd be sampled to estimate the mean numbers of flies per animal (Table 2). This table was developed using Equations 1 through 4 with a and b in Equation 2 equal to 1.331 and 1.561, respectively. The parameters are based on unpublished analysis of variance-mean relationships of the total number of horn flies per animal at Lethbridge, 1989 to 1991.

Unlike the other ectoparasites discussed so far, adult horn flies are mobile and exhibit diurnal changes in their distribution on the host. A higher proportion of flies can be found on the bellies of cattle than on the sides during the hottest part of the

TABLE 2
Required Number of Animals per Herd of Various Sizes (N) to Estimate Mean Horn Fly Abundance with 90% Confidence Intervals ($2D$) Equal to 10, 20, and 40% of the Mean

$\bar{x}$ Flies per animal	Herd size					
	10	20	30	40	50	100
$D = 0.05$						
10	10	20	29	39	48	94
50	10	19	29	38	47	88
100	10	19	28	37	46	84
150	10	19	28	37	45	82
200	10	19	28	36	44	80
250	10	19	28	36	44	78
300	10	19	28	36	44	77
350	10	19	27	35	43	76
400	10	19	27	35	43	75
$D = 0.10$						
10	10	19	28	36	44	79
50	9	18	26	33	39	65
100	9	17	25	31	37	58
150	9	17	24	30	35	53
200	9	17	23	29	33	50
250	9	16	23	28	32	48
300	9	16	22	27	31	46
350	9	16	22	26	31	44
400	9	16	21	26	30	42
$D = 0.20$						
10	9	16	23	28	33	48
50	8	14	18	21	24	31
100	8	13	16	18	20	25
150	7	12	15	17	18	22
200	7	11	14	15	17	20
250	7	11	13	14	16	18
300	7	10	12	14	15	17
350	7	10	12	13	14	16
400	6	10	11	13	13	16

day and will not be visible if counts are made from a distance. In addition, the proportion of flies on the belly exhibits seasonal variation and also varies with fly abundance (Figure 2). At low densities, fewer than 10% of the flies will be on the belly. At higher densities, from 15 to 55% of the flies may be on the belly. If visual counts do not include the belly and both sides of the animal, horn fly numbers are greatly underestimated. For detailed population work, cattle may be loaded into a chute, and the number of flies on each side and belly can be counted using a mirror attached to a long handle.

2. Partial Enumeration of Parasite Populations

Many species of mobile ectoparasites visit the host for short periods of time and are indexed by counting the number of insects on a portion of the host. Because the numbers on the host only represent a small fraction of the entire population, these estimates are properly considered indices of relative abundance. Adult face flies, *Musca autumnalis* De Geer, visit pastured cattle and feed on eye and nasal secretions, but spend a considerable portion of their lives elsewhere at rest on vertical substrates, or ovipositing in dung. The number of flies per face is frequently taken as a measure of pest intensity[48-51] and is commonly determined by counting flies on the faces of 15 to 25 cattle per herd. Sticky traps scattered within pastures also have been used to index the abundance of the off-host population. Moon and Kaya[51] found that face counts increased with corrected trap catch rates but leveled off at catch rates above 50 females per trap per hour, suggesting there is either a limit to the number of flies that could occur on a host's face or to the number of flies an observer could count.

Adult stable flies are common in feedlots and confined situations and visit hosts to obtain blood meals. On cattle, fly activity can be evaluated by visual examination of a host's front legs. Front legs are used to avoid the disturbing effects of cattle tail flicks. The typical method is to count the flies from the shoulder to the hoof on the outside of one front leg and the inside of the other, when viewing animals from the side.[52] Cattle usually can be viewed easily from the outside of their pens, using the naked eye or binoculars. These counts represent feeding indices rather than measures of population density, because fly activity is influenced by both temperature and time of day.[53] Stable flies seem most active when temperatures are between 20 and 38°C.[54] Feeding indices have been shown to be correlated with other measures of abundance.[55,56] Tame animals face the observer when feeding and will often approach the feed apron when the observer is present; an alternate method is to count the number of flies on the front and sides of each leg when viewing the animal head on.[57]

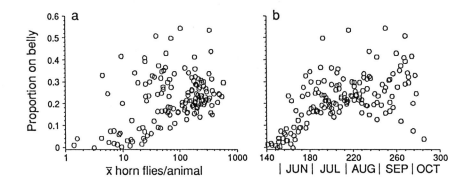

FIGURE 2. Proportion of horn fly population on cattle's bellies during 1989 to 1991.

This method provides indices that also are correlated with other sampling methods. It offers the additional advantage that counts on one leg of an animal are independent from those on the other, so two counts per animal can be obtained, effectively doubling sample size.

Many species of flightless ectoparasites complete their entire life cycle on the host. These parasites are too small, numerous, or dispersed throughout the host's body to be enumerated. Such populations may be indexed using relative measures of parasite number on specific portions of the host's body or, more commonly, by rating infestations according to parasite density or severity of resulting symptoms. These rating systems are typically used for ectoparasites that are thought to have negligible economic effects at low densities.[24,58] In these instances, practitioners only need to detect high levels of infestation.

Cattle are attacked by a variety of lice, the most important of which are the short-nosed cattle louse, *Haematopinus eurysternus* (Nitzsch); the long-nosed cattle louse, *Linognathus vituli* (L.); and the cattle biting louse, *Bovicola bovis* (L.).[59] These species complete their life cycle on the host in <1 month. Infestations are most dense in the winter, probably because high body temperatures during the summer inhibit population growth.

Cattle lice occur on particular parts of their host's body. The most commonly infested areas are the neck, back, dewlap, shoulder, and tail. However, as populations increase, lice spread over the entire body.[59] Frequently, animals are assigned a ranked classification according to parasite density, such as lightly, moderately, or heavily infested. Scarff[60] examined animals for lice and assigned scores of 1 = very light, 2 = light, 3 = moderate, 4 = heavy, and 5 = very heavy. Collins[61] examined an area on the neck of each animal and assigned subjective scores of 1 = louse free, 2 = lightly infested, 3 = moderately infested, and 4 = heavily infested. Gibney et al.[58] used a more quantitative method of scoring animals, where five 6.5-cm^2 areas of each animal, one each on the face, dewlap, neck, back, and tailhead were examined and scored using categories of Hoffman et al.,[62] viz., light = 0 to 3 lice, moderate = 3 to 10, and heavy = > 10 lice.

Shemanchuk et al.[63] counted *H. eurysternus* among sample units defined as areas 4 cm × 0.5 cm, delineated with a pencil or spatula to part the host's hair. Lice in each area were counted up to 70, and recorded as 70+ if higher. A total of 26 units per animal side were distributed among the ears, the flanks, the sides and crest of the neck, the brisket, the back line, and the tailhead. The numbers of lice per site were summed and provided a measure of louse abundance for the animal. Heath et al.[64,65] used a similar procedure to measure louse density on sheep. Geden et al.[66] inspected the neck, midback, hips, and tailhead of dairy cattle and expressed louse density as the average number of lice per unit area chosen from heavily infested body parts.

Direct enumeration of parasites is most difficult in chickens infested with northern fowl mite. Matthysse et al.[23] examined populations on hens and estimated that mite populations exceeded 20,000 per hen, within 30 d after initial infestation. Because mite populations can reach such high numbers, infestations are typically categorized according to the number of mites on the vent where they are most prevalent. Each bird is removed from its cage and held upside down to examine the vent feathers with the aid of a headlamp.[67]

Several systems have been used to rate extent of infestation, based on gross symptoms or on ranges in mite abundance. Foulk and Matthysse[68] rates birds according to visible symptoms rather than counts of mites themselves. Their categories were 0 = no mites; 1 = low numbers of mites; 2 = numerous mites, feathers beneath vent blackening; 3 = mites very numerous, vent feathers blackened and matted, some skin lesions; and 4 = several thousand mites present, vent feathers completely blackened, severe reddening and scabbing of skin. Other workers based rating systems on estimates of the numbers of mites present. DeVaney[24] used a logarithmic rating scale with 0 = no mites, 1 = 1 to 10 mites, 2 = 11 to 100 mites, 3 = 101 to 1000 mites, 4 = 1001 to 10,000 mites, and 5 = > 10,000 mites. Arthur and Axtell[67] and Lemke and Collison[69] used similar systems, and both groups of workers evaluated their index system by comparing the categorical density ratings from a set of birds with counts of mites obtained through extraction of mites from the same birds. Both groups found that their categorical systems grossly underestimated the number of mites actually present on the birds. Lemke and Collison[69] noted that ratings from different observers were generally consistent and that ratings were proportional to actual mite abundance.

Use of rating systems to compare infestations among groups of hosts should be approached with caution. Rating systems are fundamentally categorical rather than continuous, and their proper analysis would entail methods designed for binomial or multinomial responses. Least-squares procedures (ANOVA, least-squares regression) are derived from the assumption that responses are normally distributed and should be reserved for analysis of continuous response variables.

Northern fowl mite infestations can spread from a point source throughout a caged-layer house in a matter of weeks[70] and can spread readily to adjacent houses. Because treatment is expensive and time consuming, Axtell and Arends[27] recommend a given house should be treated if any ectoparasites are detected on birds. In caged-layer houses, mites should be monitored routinely by examining birds throughout the entire house. Emphasis should be placed on birds caged singly as populations are usually higher on these than on birds caged in groups.[67] Essentially this procedure amounts to a presence or absence sampling program.

Itch mites cause intense itching, also known as pruritus, in swine. Courtney et al.[71] ranked infested pigs according to the degree of pruritus exhibited, with 0 = no pruritus, 0.5 = questionable pruritus, and 1 to 5 reflecting increasing levels of pruritus. Martineau et al.[44] ranked infestations as 0 = no pruritus or active lesions, 1 = pruritus but no hyperkeratotic lesions, 2 = pruritus and a few gray lesions, 3 = intense pruritus and large lesions, and 4 = intense pruritus, large lesions, and heavily scabbed areas. The system of Martineau et al.[44] is less subjective than that of Courtney et al.,[71] and both studies indicate pruritus is a better indicator of infestations than direct recovery of mites. Davis and Moon[72] developed an objective method for evaluating pruritus. Pigs within pens were scanned[73] visually every 5 s, for an interval of 4 min, to count the number of pigs seen rubbing or scratching. Pruritus was then expressed as the proportion of possible pig sightings where rubbing or scratching was observed. In experimental herds that were not disturbed by observers, infested and uninfested pigs spent 0.4 to 2.2% of their time scratching and rubbing. However, pruritus increased four- to ten-fold if the pigs were disturbed with a cold water spray before scanning, and infested pigs showed significantly greater pruritus than uninfested pigs (about 5% more in early stages of infestations). This behavioral assay could be developed for general use for the purpose of making treatment decisions in herds with incipient mange.

3. Presence or Absence Sampling

Detection of a single parasite or early symptoms of their presence can justify treatment decisions in instances where the high reproductive potential of parasites makes an intolerable outbreak virtually certain or where local eradication is required by law. In these circumstances, presence or absence sampling is frequently employed.

Sampling is accomplished by inspecting animals to detect the parasite directly, to detect symptoms, or to detect immunological signs of parasitism.

The goal of swine mange management is to effect local eradication within a herd by treating breeding stock and isolating piglets to prevent infestation from untreated reservoirs.[43] The basic method for detecting itch mites is to collect scrapings from the skin of suspect animals. Skin of the ears and face are scraped with a scalpel or sharp instrument, and then the scrapings are mixed with mineral oil for microscopic examination. Scraping a fixed area of the skin, such as 1 or 3 cm^2, allows further expression of mite density in quantitative terms. Mites, however, are difficult to locate, and random scrapings may not be sensitive enough to detect incipient infestations.[44,71]

Immunological methods may be used in the future to detect infestations and infections of immunogenic parasites. In swine, crude extracts of itch mites, when injected into hypersensitive pigs, elicited edema and reddening at injection sites within 24 h.[74] Hypersensitivity in experimental herds developed within 2 to 4 weeks after initial infestation.

A great deal of effort is directed toward control of cattle grubs. Recently, a joint Canada-U.S. pilot study demonstrated that a combination of released sterile males and insecticide treatments could eradicate the cattle grub.[75] The Alberta Warble Control Program requires treatment of all cattle (except lactating dairy cattle) with a systemic insecticide in the fall and treatment of all infested cattle in the spring.[76] Animals also are inspected at public stockyards between February and June, and a subset of on-farm cattle are inspected each year. Inspection of live animals is conducted through palpation, which fails to detect infections if larvae have yet to reach their hosts' backs. Furthermore, palpations fail to detect populations in a specific geographic region, if inspection occurs after the herds have been treated with systemic insecticides.

Serological diagnosis of cattle grub infections with an ELISA test has been developed and used to detect infected animals in the U.K.[77] Early infections of *H. lineatum* and mixed *H. lineatum* and *H. bovis* can be detected.[78] This method requires collection of blood samples from cattle and extraction of the sera. Sera are exposed to cattle grub antigen, and the optical density of the reaction mixture assessed using a labeled antibody. Antibody activity greater than 2 × the standard deviation of negative sera are considered positive. This method is easier, faster, and more reliable than palpation for detecting grubs, and can be applied over large geographic areas where area-wide management or eradication are being considered. Because the method can accurately detect infected animals in the fall, it provides enough lead time for application of systemic insecticides before the December 1 cutoff date in Alberta, or for production of a population of sterile flies for release in an infested area.

Serodiagnosis can detect cattle grubs in a herd, but cannot provide a direct measure of actual levels of grub density, information that is important for predicting the number of sterile insects required for release. Several approaches have been used to estimate density from the proportion of uninfected animals in a herd. The first attempt was to use the negative binomial distribution to relate mean density to the estimated proportion of uninfected animals in the herd, p_0.[79] From the negative binomial distribution,

$$p_0 = \left(\frac{k}{m+k}\right)^k \qquad (5)$$

where p_0 is the probability an animal is not infected, m is the mean, and k is the dispersion parameter. The mean of the distribution can be estimated then by rearranging Equation 5 as

$$m = k\left(\frac{1 - p_0^{1/k}}{p_0^{1/k}}\right) \quad (6)$$

Although this equation has two unknowns, $1/k$ and p_0, Burillon and Messean[79] used an empirical relationship between $1/k$ and p_0 to eliminate one of the unknowns. Using the negative binomial distribution requires the assumption that k is constant for various levels of the mean. However, field studies of grub abundance in Alberta indicate k varies with the mean in an unpredictable fashion, so a common k cannot be estimated.[80]

As an alternative, an empirical model was developed to estimate mean grub density in a herd ($\bar{x}$) from the proportion of uninfested animals measured either with ELISA (p_{0e}) or with palpations (p_0) using the following steps:[80]

$$p_{0e} \to p_0 \to \ln(-\ln p_0) \to \ln \bar{x} \to \bar{x} \quad (7)$$

The arrows indicate calculations. Given estimates of p_{0e} or p_0, the user can calculate $\ln(-\ln p_0)$ directly and then estimate $\ln \bar{x}$ from the relationship:

$$\ln \bar{x} = 1.257 + 1.124 \ln(-\ln p_0) \quad (8)$$

Because this relationship was developed using palpations, it is necessary to use p_{0e} as an estimate of p_0. This is acceptable because $p_0 \approx p_{0e} + \epsilon$ where ϵ is an error term. The variance of a predicted value of $\ln \bar{x}$ assuming p_0 is known exactly can be calculated using:

$$V(\ln \bar{x} | p_0) = 0.7108\left(1 + \frac{1}{35} + \frac{(\ln[-\ln(p_0)] - 0.7086)^2}{32.3649}\right) \quad (9)$$

If p_{0e} is used, then variation arising from using p_{0e} as an estimate of p_0 must be accounted for, and the variance of a predicted value of $\ln \bar{x}$ is

$$V(\ln \bar{x} | p_{0e}) = V(\ln \bar{x} | p_0) + \left(\frac{0.2931}{p_{0e} \ln(p_{0e})}\right)^2 \quad (10)$$

Finally, the variance of the predicted value of $\bar{x}$ is simply:

$$V(\bar{x}) = \bar{x}^2 V(\ln \bar{x}) \quad (11)$$

The serodiagnosis sampling model could provide two major benefits for future work on eradication with sterile releases. The first is to determine the number of sterile insects required for release in a given area. The procedure[81] would be to (1) collect serum samples in the fall and use ELISA to determine p_{0e}, (2) use p_{0e} in Equation 8 to calculate $\ln \bar{x}$, and then (3) calculate $\bar{x}$ from $\ln \bar{x}$. Table 3 provides values of $\bar{x}$ for various levels of p_0. Since $\bar{x}$ is a measure of the mean maximum number of grubs per animal, the total number of grubs in a herd can be calculated and used to plan rearing requirements for sterile male releases. Because *in vitro*

TABLE 3
Predicted Number of Cattle Grubs per Animal for Various Levels of the Proportion of Uninfested Animals

p_0	$\bar{x}$	p_0	$\bar{x}$	p_0	$\bar{x}$	p_0	$\bar{x}$
0.05	12.1	0.30	4.3	0.55	2.0	0.80	0.7
0.10	9.0	0.35	3.7	0.60	1.7	0.85	0.5
0.15	7.2	0.40	3.2	0.65	1.4	0.90	0.3
0.20	6.0	0.45	2.7	0.70	1.1	0.95	0.1
0.25	5.1	0.50	2.3	0.75	0.9	1.00	—

rearing of the cattle grub is not feasible, serodiagnosis also can identify infested animals early enough in the season so they can be sequestered as a source for sterile insects.

Presence or absence treatment decision plans make use of the proportion of sample units that have a number of insects less than or equal to some critical value T[82] (see also Chapter 9). This proportion is expressed as P_T. Usually $T = 0$, but could be any integer. Choice of T can affect the robustness of the method for decision making. Lysyk and Schaalje[57] developed binomial sampling plans for the stable fly in beef feedlots, based on the relationship between the mean number of flies per leg and P_T, and compared results if $T = 0$, 1, or 2 flies per leg. They assumed a low treatment threshold of one stable fly per leg based on studies conducted by Campbell et al.[20] In this instance, using different levels of T had little impact on the probability of making a decision error; however, choice of T influenced the range for which the mean number of flies per leg could be predicted (Figure 3). If evaluating whether or not populations are above one fly per leg, the observer can examine 30 legs (or 15 animals) for the presence of flies. Infested legs are scored as 1, uninfested legs as 0. The mean of these scores represents the proportion of infested legs. If the proportion infested is above 0.53, the mean number of flies per leg will be about 1.0, and treatment should be considered. The advantage of this method is that infestations can be quickly classified as being above or below the treatment threshold without unnecessary counting.

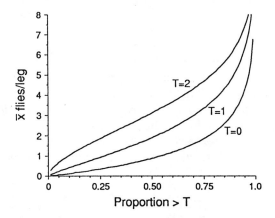

FIGURE 3. Relationship between the mean number of stable flies per leg and the proportion of legs above thresholds of $T = 0$, 1, and 2 flies per leg.

Sequential sampling can be used to determine whether or not a herd or flock is infested. This decision is testing the null hypothesis

$$\text{Ho: } P(\text{infested}) = 0$$

vs. the alternate hypothesis

$$\text{Ha: } P(\text{infested}) > 0$$

where $P(\text{infested})$ is the proportion of the herd or flock that is infested. Obviously, when the first infested animal is detected, sampling stops and the unit is declared infested. However, if an accumulating sample were yet to contain an infestation, then the sampler would need to decide whether to continue sampling or to terminate before inspecting every animal.

The number of animals, n, required to test Ho should be governed mainly by the power with which the observer wishes to discriminate against Ho. The power of a test is the chance of rejecting Ho when Ha is true, which equals $1 - P(0)$, or $1 -$ the true proportion of the herd or flock that is not infested. For a chosen power, required sample size is a function of the size of the herd or flock, N, and the number of infested animals in the unit, I.

When just one animal is infested ($I = 1$), then the power to detect that animal is equal to the sampling fraction, n/N, when every animal has an equal chance of being examined, and sampling is without replacement. For example, if five animals in a herd of 100 were inspected, then the power to detect a single infested animal would be 0.05 or 5%. If 50 of the 100 were inspected, the power would be 50%.

When more than one infested animal is present, calculation of power is more complex. Power to detect I infested animals is given by:

$$\text{Power} = 1 - P(0) = 1 - \left(\frac{(N-n)!}{(N-n-I)!}\right)\left(\frac{(N-I)!}{N!}\right) \qquad (12)$$

This equation can be solved iteratively to calculate n for selected N, I, and powers. Required sample sizes for infestation rates of $100*(I/N) = 1, 5, 10,$ and 25% are presented in Table 4. At low rates of infestation, comparatively large samples (or sampling fractions) are required to be relatively certain one is not falsely accepting Ho. However, at higher infestation rates, sampling fractions become insensitive to N as management units become large, such as might occur when sampling from large units of sheep or birds.

To use Table 4 for setting an upper bound on sample size, the sampler would choose an appropriate rate of infestation that is desired for detection and choose the desired power. Sampling then would terminate when the corresponding number of animals had been examined for the appropriate unit size (N). For example, if a 1% infestation rate in a herd of 100 animals is satisfactory, then a sample of 50 animals, none infested, would be required to correctly declare, with a 50% chance of being wrong, that no animals were infested. In contrast, a sample of 90 animals would be required before one could be 90% sure that no animals were infested. If a 5% infestation rate was desired, then corresponding sample sizes would be 13 and 37 animals, respectively.

B. SAMPLING OFF-HOST POPULATIONS

Various livestock pests have free-living stages or periods in their life cycle. Mobile insects such as stable flies, face flies, house flies, and unattached ticks can be sampled using a variety of traps. Less mobile stages, such as fly larvae, must be sampled by removing units of habitat.

TABLE 4
Sample Sizes (n) Required to Achieve Chosen Power, by Rate of Infestation ($100 * [I / N]$) and Size of Herd or Flock

No. animals in unit	Power				
	0.5	0.8	0.9	0.95	0.99
1% Infestation					
10	5	8	9	10	10
20	10	16	18	19	20
50	25	40	45	48	50
100	50	80	90	95	99
200	59	111	137	155	180
500	65	138	184	225	300
1,000	67	148	205	258	368
10,000	68	159	227	294	448
5% Infestation					
10	5	8	9	10	10
20	10	16	18	19	20
50	11	21	27	31	39
100	13	27	37	45	59
200	14	30	41	51	73
500	14	31	43	56	82
1,000	14	31	44	57	86
10,000	14	32	45	59	90
10% Infestation					
10	5	8	9	10	10
20	6	11	14	16	18
50	7	14	18	22	29
100	7	15	20	25	36
200	7	15	21	27	40
500	7	16	22	28	42
1,000	7	16	22	29	43
10,000	7	16	22	29	44

Trapping provides relative estimates of fly abundance that usually are more reliable than instantaneous visual estimates, because traps accumulate flies and thereby average out probable short-term variations in fly activity. However, catch rates can be influenced by dust, if trap capture depends on adhesives.[83] This problem can be circumvented by regularly servicing traps and leaving the adhesive only for 24-h periods. Similarly, traps with volatile baits can vary in attraction, depending on age and condition of the bait.

A serious limitation of traps arises when the sampler is interested in comparing fly densities among different livestock facilities. Trap productivity is influenced strongly by where a trap is positioned relative to animals in pens, to buildings, or to sheltered areas, and a method is not yet available for standardizing trap placement among facilities. As a result, traps are most useful for indexing relative changes in abundance at a specific farm.

Stable flies are monitored using traps constructed from Alsynite fiberglass. Two basic designs have been used. The Williams trap[84] consists of two 35 × 45 cm panels joined at the center to form four wings which are coated partially or completely with an adhesive. The cylindrical trap[85] consists of a 90 × 30 cm sheet of Alsynite wrapped in a cylinder and covered with adhesive. Both traps are mounted on stakes 1 m or so above the ground. The cylindrical trap is more convenient to use because it is smaller and less susceptible to wind damage. Trap surfaces can be covered with a thin sheet of plastic wrap before application of adhesive to facilitate cleaning. Both traps are equally efficient, catching equivalent numbers of flies per unit adhesive surface.

Face flies in pastures have been sampled using a variety of white sticky traps with shapes of pyramids, panels, or cylinders.[49,86-88] Trapping efficiency is known to vary with temperature, and crude catch rates can be standardized accordingly, using methods outlined by Moon and Kaya.[51] Further variation in catch rates may arise from trap position in pastures and from differences in activity among component fly-age classes.

Sticky traps are impractical in dusty environments such as confined poultry houses. As an alternative, baited jug-traps have been employed.[89] These are 3.8-l translucent white plastic jugs with access holes in their sides. The bait is a standardized, commercial preparation containing the insecticide methomyl and the pheromone analogue muscamone. As in outdoor environments, trap location is influential and locations of replicate traps must be standardized if densities among dates in the same house are to be compared. In controlled studies,[90] density of released house flies accounted for 60% of the variation in trap catch rates, whereas temperature was relatively uninfluential.[90,91] The relationship between sample variance and mean house fly catch rates was used to calculate sample sizes for specific levels of precision.[92] With a commonly recommended treatment threshold of 300 flies per trap per week, 10 baited jug-traps per poultry house are required to obtain estimates with the half-length of a 90% confidence interval equal to 33% of the threshold.

Also, house flies can be effectively monitored using "spot cards" attached to rafters within the poultry house.[93] Spot cards are white index cards measuring 7.5 × 12.5 cm, attached flush against the rafters of the house. Flies landing on the cards leave vomit and fecal spots. The number of spots which accumulate over 1 week is used as an index of fly abundance. Spot cards provide a measure of fly activity as indices are influenced by several other species of flies common in poultry houses.[92] The number of spots per card is mainly influenced by fly density.[90] Temperature has relatively little effect on spot card indices. Twelve spot cards per house should be used to obtain fly activity indices with the half-length of a 90% confidence interval equal to 33% of a threshold of 50 spots per card per week. Again, the cards should be uniformly dispersed throughout the house to minimize location effects.

The house fly is a nuisance pest in surrounding communities and not of direct concern in most poultry systems. If traps or spot cards are to be useful as a guide for fly management decisions, producers will need to routinely monitor populations on their premises and communicate concurrently with their neighbors to calibrate on-premise abundance with off-premise annoyance. This relationship is likely to vary from premise to premise, depending on fly emigration rates, proximity of neighbors, and tolerance of the neighbors to flies.

Steelman et al.[94-96] was able to relate weight gain reductions of beef cattle to mosquito attack by housing animals in sheds and counting the number of blood-engorged females on grids painted on the walls of the sheds. These grid counts were later correlated with two other indices of mosquito abundance commonly used in abatement programs. The indices were landing rates and catch rates in mechanical traps. Landing rates are defined as the number of mosquitos landing below the waist of a human host in 1 min observation intervals. Catch rates were derived from overnight catches of standard CDC and New Jersey light traps. Steelman et al. suggested that landing rates and catch rates could be used by producers to make treatment decisions, providing attack on the producer's herd could be related to the more convenient mosquito sampling methods.

Free-living stages of ticks can be sampled with dry ice-baited traps.[97-99] A trap consists of a cube of dry ice placed on a piece of cotton cloth. Traps are visited after a specified period of exposure, and a count of ticks on the trap is an index of abundance. Dry ice-baited traps are more effective for attracting lone star ticks than traps using other types of attractants, including live animals.[97] Koch and McNew[98] found that initial cube weights of 227 to 367 g were equally effective for capturing lone star ticks but that capture decreased with less dry ice. Speed of dry ice sublimation is proportional to temperature, and a sublimation rate of at least 80 g/h is required for effective trapping. Adults and nymphs seem to be recruited from a 3.1-m radius, but larvae are less mobile. Dry ice traps have been used in conjunction with mark-recapture studies to estimate absolute densities.[99]

Dry ice traps are impractical in extensive rangeland with steep slopes.[100,101] Under these conditions, sampling has been accomplished by dragging or flagging. Dragging is done by pulling a 0.9 × 1.4 m piece of white cotton cloth over the ground for a fixed distance.[102] Flagging[100] consists of waving a piece of 90 × 90 cm white cotton fastened to a pole over the surface of shrubby vegetation, a substrate less amenable to dragging. With both methods, ticks stimulated by the operator's breath adhere to the fabric, and the number of ticks per unit effort is assumed to be proportional to absolute density of the active stages. Both methods can be used within randomly chosen quadrats of a specified size.

Wilkinson and Gregson[100] examined several methods to measure relative abundance of ticks. These included enumeration on cattle and flagging in spatial units of their pasture arranged in fixed (systematic), random, and stratified random designs. Fixed units were 31-m line transects, chosen according to knowledge of where ticks were most likely to occur. Flags were swept from side to side in 3-m wide swaths along the length of the transect. Random quadrats were selected from a ground plan prepared using aerial photographs. The field was divided into 38 × 38 m plots which were numbered and selected for sampling using a random numbers table. The chosen plots were located in the field using marked fence posts. Each quadrat was flagged according to the same procedure used in the fixed quadrats. Finally, the range was stratified into areas with different amounts of shrubby vegetation, to distinguish among habitats with high, medium, and low likelihood of containing ticks. Quadrats for flagging were then allocated among strata according to cost and variation in tick density within strata. Indices obtained using fixed units may be biased by the selection of unrepresentative units; however, fixed units can cover a high proportion of a study area and be used by workers with minimal training. Use of 280-m^2 quadrats was considered optimal for randomized and stratified sampling. Vegetation and topographical features[103] can be used to identify strata.

Most muscoid flies spend the immature stages off the host, developing in either freshly deposited cattle dung or accumulated organic matter. Sampling immature stages is important for identifying breeding sites and to direct biological or cultural control measures. Horn flies and face flies oviposit in freshly deposited cattle dung in pastures. Pats quickly form a crust and within a few hours are no longer suitable for oviposition. Resulting larvae develop within the pads and then some, if not all, emigrate to surrounding soil to pupate.

Sampling immatures of these two species has required sampling of the population of inhabited pats. To control for pat age (or time of deposition), a cohort of freshly deposited pats is marked, and the sampler then returns within days to weeks to exhume the pats (or subsamples) and extract preimaginal stages or to install conical traps to catch adults as they emerge.[104-107] Resulting densities per pat are a measure of recruitment into the adult population from the sampled cohort of pats.

Immature stages of stable flies and house flies occur in manure mounds, silage mounds, animal pens, or anywhere else that decaying organic matter has accumulated.[108] Relative estimates of larval abundance have been made by determining the number of larvae in a fixed volume of breeding material. Schmidtmann[109] removed dry material above the collection site and collected material down to dry soil to make up a 1-l sample. He further refined his sampling method by placing a 30.5-cm^2 template over the site to be sampled and collected material from quadrants within the template to total 1 l. Skoda et al.[110] estimated densities of stable fly and house fly larvae in feedlots in eastern Nebraska. They defined five sampling strata in each lot: the feed apron, back fence, side fences, mound, and the remaining lot. Each stratum was divided into 0.1-m^2 sampling plots which were sampled randomly on chosen dates. The plots were sampled using a 10.5-cm diameter × 11-cm long coil corer, and subsequently larvae were extracted from the substrate. If the area of each stratum were known, then a feedlot manager would be able to calculate the contribution of each stratum to the lot's population of adult flies and direct sanitation efforts accordingly.

III. SUMMARY

A diversity of domestic animals and management systems constitute livestock industries in Canada and the U.S. As in other agricultural sectors, a variety of sampling approaches have been used to achieve management and research objectives. Choice of method depends on the sampler's goals and on the biological, ecological, and statistical details of the universes being sampled.

Host animals are routinely inspected to measure on-animal parasite densities, and, in some instances, subsampling of selected body regions is used to estimate totals for chosen animals. Treatment thresholds have been developed where measures of on-animal density have been related empirically to damage, but many combinations of host and parasite species remain to be studied. Similarly, relative abundance indices from traps also have been employed, but information from traps is more site-specific because catch rates can be influenced by trap location, insect activity, and weather. Binomial or presence-absence sampling based on observation of the pests themselves or the symptoms they elicit are being developed, and these may provide valuable tools for pest management in livestock systems.

REFERENCES

1. U.S. Department of Agriculture, *Agricultural Statistics 1991*, U.S. Department of Agriculture, Washington, D.C., 1992.
2. Drummond, R. O., Economic aspects of ectoparasites of cattle in North America, in *The Economic Impact of Parasitism in Cattle*, Leaning, W. H. D. and Guerrero, J., Eds., Proc. MSD AGVET Symposium, Montreal, 1987, 9.
3. Drummond, R. O., Lambert, G., Smalley, H. E., Jr., and Terril, C. E., Estimated losses of livestock to pests, in *CRC Handbook of Pest Management in Agriculture*, Vol. 1, Pimentel, D., Eds., CRC Press, Boca Raton, FL, 1981, 111.

4. Haufe, W. O. and Weintraub, J., Economics of veterinary-medical entomology, *Can. Entomol.*, 117, 901, 1985.
5. Haufe, W. O., Host-parasite interaction of blood-feeding dipterans in health and productivity of mammals, *Int. J. Parasitol.*, 17, 607, 1987.
6. Axtell, R. C., Status and potential of biological control agents in livestock and poultry pest management systems, *Entomol. Soc. Am. Misc. Publ.*, 61, 1, 1987.
7. Alberta Agriculture, *The Beef Cow-Calf Manual*, Alberta Agriculture, Edmonton, 1989.
8. Williams, R. E., Hall, R. D., Broce, A. B., and Scholl, P. J., Eds., *Livestock Entomology*, John Wiley & Sons, New York, 1985.
9. Fredeen, F. J. H., A review of the economic importance of black flies (Simuliidae) in Canada, *Quaest. Entomol.*, 13, 219, 1977.
10. Nelson, W. A., Metabolic responses of livestock to hemotophagous arthropod invasion, *Misc. Publ. Entomol. Soc. Am.*, 71, 15, 1989.
11. Wieman, G. A., Campbell, J. B., Deshazer, J. A., and Berry, I. L., Effects of stable flies (Diptera: Muscidae) and heat stress on weight gain and feed efficiency of feeder cattle, *J. Econ. Entomol.*, 85, 1835, 1992.
12. Schwinghammer, K. A., Knapp, F. W., Boling, J. A., and Schillo, K. K., Physiological and nutritional response of beef steers to infestations of the horn fly (Diptera: Muscidae), *J. Econ. Entomol.*, 79, 1010, 1986.
13. Schwinghammer, K. A., Knapp, F. W., Boling, J. A., and Schillo, K. K., Physiological and nutritional responses of beef steers to infestations of the stable fly (Diptera: Muscidae), *J. Econ. Entomol.*, 79, 1294, 1986.
14. Colwell, D. D. and Kavaliers, M., Exposure to mosquitoes, *Aedes togoi* (Theo.), induces and augments opioid-mediated analgesia in mice, *Physiol. Behav.*, 48, 397, 1990.
15. Colwell, D. D. and Kavaliers, M., Evidence for activation of endogenous opioid systems in mice following short exposure to stable flies, *Med. Veg. Entomol.*, 6, 159, 1992.
16. Akey, D. H., Luedke, A. J., and Osborn, B. I., Development of hypersensitivity in cattle to the biting midge (Diptera: Ceratopogonidae), *Misc. Publ. Entomol. Soc. Am.*, 71, 22, 1989.
17. Steelman, C. D., Brown, A. H., Gbur, E. E., and Tolley, G., Interactive response of the horn fly (Diptera: Muscidae) and selected breeds of beef cattle, *J. Econ. Entomol.*, 84, 1275, 1991.
18. Kinzer, H. G., Houghton, W. E., Reeves, J. M., Kunz, S. E., Wallace, J. W., and Urquhart, N. S., Influence of horn flies on weight loss in cattle, with notes on prevention of loss by insecticide treatment, *Southwest. Entomol.*, 9, 212, 1984.
19. Campbell, J. B., White, R. G., Wright, J. E., Crookshank, R., and Clanton, D. C., Effects of stable flies on weight gains and feed efficiency of calves on growing or finishing rations, *J. Econ. Entomol.*, 70, 592, 1977.
20. Campbell, J. B., Berry, I. L., Boxler, D. J., Davis, R. L., Clanton, D. C., and Deutscher, G. H., Effects of stable flies (Diptera: Muscidae) on weight gain and feed efficiency of feedlot cattle, *J. Econ. Entomol.*, 80, 117, 1987.
21. Barnard, D. R., Injury thresholds and production loss functions for the lone star tick, *Amblyomma americanum* (Acari: Ixodidae), on pastured, preweaner beef cattle, *Bos taurus*, *J. Econ. Entomol.*, 78, 852, 1985.
22. Cargill, C. R. and Dobson, K. J., Experimental *Sarcoptes scabei* infestations in pigs: (2) effects on production, *Vet. Rec.*, 104, 11, 1979.
23. Matthysse, J. G., Jones, C. J., and Purnasiri, A., Development of northern fowl mite populations on chickens, effects on the host, and immunology, *Search Agric.*, 4(9), 1, 1974.
24. De Vaney, J. A., The effects of the northern fowl mite, *Ornithonyssus sylviarum* on egg production and body weight of caged white leghorn hens, *Poultry Sci.*, 58, 191, 1979.
25. Palmer, W. A. and Bay, D. E., A review of the economic importance of the horn fly, *Haematobia irritans irritans* (L.), *Prot. Ecol.*, 3, 237, 1981.
26. Haufe, W. O., A modeling system for horn flies on cattle, in *Modeling and Simulation: Tools for Management of Veterinary Pests*, ARS-46, Miller, J. A., Ed., U.S. Department of Agriculture-ARS, Washington, D.C., 1986.
27. Axtell, R. C. and Arends, J. J., Ecology and management of arthropod pests of poultry, *Annu. Rev. Entomol.*, 35, 101, 1990.
28. Bram, R. A., *Surveillance and Collection of Arthropods of Veterinary Importance*, Agricultural Handbook No. 518, U.S. Department of Agriculture, Washington, D.C., 1978.
29. Weintraub, J., Rich, G. B., and Thomson, C. O. M., Timing the treatment of cattle with trolene for systemic control of the cattle grubs *Hypoderma lineatum* (De Vill.) and *H. bovis* (L.) in Alberta and British Columbia, *Can. J. Anim. Sci.*, 39, 50, 1959.

30. Weintraub, J. and Thomson, C. O. M., Comparison of Ronnel, Dowco 109, and Dowco 105 for systemic control of cattle grubs in Alberta, *J. Econ. Entomol.*, 54, 79, 1961.
31. Gregson, J. D. and Holland, G. P., Devices for charting and obtaining naturally emerged cattle warbles (Diptera:Oestridae), *Proc. Entomol. Soc. Br. Columb.*, 41, 31, 1944.
32. Karandinos, M. G., Optimum sample size and comments on some published formulae, *Bull. Entomol. Soc. Am.*, 22, 417, 1976.
33. Taylor, L. R., Aggregation, variance and the mean, *Nature*, 189, 732, 1961.
34. Regniere, J. and Sanders, C. J., Optimal sample size for the estimation of spruce budworm (Lepidoptera: Tortricidae) populations on balsam fir and white spruce, *Can. Entomol.*, 115, 1621, 1983.
35. Barnard, D. R. and Morrison, R. D., Density estimators for populations of the lone star tick, *Amblyomma americanum* (Acari: Ixodidae), on pastured beef cattle, *J. Med. Entomol.*, 22, 244, 1985.
36. Davis, D. P. and Moon, R. D., Density, location, and sampling of *Sarcoptes scabei* (Acari: Sarcoptidae) on experimentally infested pigs, *J. Med. Entomol.*, 27, 391, 1990.
37. Cochran, W. G., *Sampling Techniques*, 3rd ed., John Wiley & Sons, New York, 1977.
38. Ervin, R. T., Epplin, F. M., Byford, R. L., and Hair, J. A., Estimation and economic implications of lone star tick (Acari: Ixodidae) infestation on weight gain of cattle, *Bos taurus* and *Bos taurus* X *Bos indicus*, *J. Econ. Entomol.*, 80, 443, 1987.
39. Latif, A. A., Punya, D. K., Nokoe, S., and Capstick, P. B., Tick infestations on zebu cattle in western Kenya: individual host variation, *J. Med. Entomol.*, 28, 114, 1991.
40. Punya, D. K., Latif, A. A., Nokoe, S., and Capstick, P. B., Tick (Acari: Ixodidae) infestations on Zebu cattle in western Kenya: seasonal dynamics of four species of ticks on traditionally managed cattle, *J. Med. Entomol.*, 28, 630, 1991.
41. Barnard, D. R., Seasonal activity and preferred attachment sites of *Ixodes scapularis* (Acari: Ixodidae) on cattle in southeastern Oklahoma, *J. Kan. Entomol. Soc.*, 54, 547, 1981.
42. Arlian, L. G., Biology, host relation, and epidemiology of *Sarcoptes scabei*, *Annu. Rev. Entomol.*, 34, 139, 1989.
43. Davis, D. P. and Moon, R. D., Dynamics of swine mange: a critical review of the literature, *J. Med. Entomol.*, 27, 727, 1990.
44. Martineau, G. P., Vaillancourt, J., and Frechette, J. L., Control of *Sarcoptes scabei* infestation with Ivermectin in a large intensive breeding piggery, *Can. Vet. J.*, 25, 235, 1984.
45. Haufe, W. O., Growth of range cattle protected from horn flies (*Haematobia irritans*) by ear tags impregnated with fenvalerate, *Can. J. Anim. Sci.*, 62, 567, 1982.
46. Ernst, C. M. and Krafsur, E. S., Horn fly (Diptera: Muscidae): sampling considerations of host breed and color, *Environ. Entomol.*, 13, 892, 1984.
47. Schreiber, E. T. and Campbell, J. B., Horn fly (Diptera: Muscidae) distribution on cattle as influenced by host color and time of day, *Environ. Entomol.*, 15, 1307, 1986.
48. Ode, P. E. and Matthysse, J. G., Bionomics of the face fly, *Musca autumnalis* DeGeer, *Cornell University Agric. Exp. Stn. Mem.*, 402, 1967.
49. Easton, E. R., The value of the pyramid sticky trap for sampling face flies, *Musca autumnalis*, under field conditions in South Dakota, *Environ. Entomol.*, 8, 1161, 1979.
50. McGuire, J. U. and Sailer, R. I., *A Method of Estimating Face Fly Populations on Cattle*, ARS Rep. 33-80, U.S. Department of Agriculture, Washington, D.C., 1962.
51. Moon, R. D. and Kaya, H. K., A comparison of methods for assessing age structure and abundance in populations of nondiapausing female *Musca autumnalis* (Diptera: Muscidae), *J. Med. Entomol.*, 18, 289, 1981.
52. Campbell, J. B., Arthropod pests of confined beef, in *Livestock Entomology*, Williams, R. E., Hall, R. D., Broce, A. B., and Scholl, P. J., Eds., John Wiley & Sons, New York, 1985, 335.
53. Berry, I. L. and Campbell, J. B., Time and weather effects on daily feeding patterns of stable flies (Diptera: Muscidae), *Environ. Entomol.*, 14, 336, 1985.
54. Semakula, L. M., Taylor, R. A., and Pitts, C. W., Flight behavior of *Musca domestica* and *Stomoxys calcitrans* (Diptera: Muscidae) in a Kansas dairy barn, *J. Med. Entomol.*, 26, 501, 1989.
55. Mullens, B. A. and Meyer, J. A., Seasonal abundance of stable flies (Diptera: Muscidae) on California dairies, *J. Econ. Entomol.*, 80, 1039, 1987.
56. Thomas, G. D., Berry, I. L., Berkebile, D. R., and Skoda, S. R., Comparison of three sampling methods for estimating adult stable fly (Diptera: Muscidae) populations, *Environ. Entomol.*, 18, 513, 1989.
57. Lysyk, T. J. and Schaalje, G. B., Binomial sampling for pest management of stable flies (Diptera: Muscidae) that attack dairy cattle, *J. Econ. Entomol.*, 85, 130, 1992.
58. Gibney, V. J., Campbell, J. B., Boxler, D. J., Clanton, D. C., and Deutscher, G. H., Effects of various infestation levels of cattle lice (Mallophaga: Trichodectidae and Anoplura: Haematopinidae) on feed efficiency and weight gains of beef heifers, *J. Econ. Entomol.*, 78, 1304, 1985.
59. Matthysse, J. G., Cattle lice: their biology and control, *Cornell Univ. Agric. Exp. Stn. Bull.*, 832, 1946.
60. Scharff, D. K., An investigation of the cattle louse problem, *J. Econ. Entomol.*, 55, 685, 1962.

61. Collins, R. C. and Dewhirst, L. W., Some effects of the sucking louse, *Haematopinus eurysternus*, on cattle on unsupplemented range, *J. Am. Vet. Med. Assoc.*, 146, 129, 1965.
62. Hoffman, R. A., Drummond, R. O., and Graham, O. H., Survey methods for livestock insects, in *Survey Methods for Some Economic Insects*, ARS 81-31, U.S. Department of Agriculture, Washington, D.C., 1969, 87.
63. Shemanchuk, J. A., Haufe, W. O., and Thompson, C. O. M., Effects of some insecticides on infestations of the short-nosed cattle louse, *Can. J. Anim. Sci.*, 43, 56, 1963.
64. Heath, A. C. G., Nottingham, R. M., Bishop, D. M., and Cole, D. J. W., An evaluation of two cypermethrin-based pour-on formulations on sheep infested with the biting louse, *Bovicola ovis*, *N.Z. Vet. J.*, 42, 104, 1992.
65. Heath, A. C. G., Cole, D. J. W., and Bishop, D. M., Some currently available insecticides and their comparative efficacy on louse-infested, long-wooled sheep, *N.Z. Vet. J.*, 40, 101, 1992.
66. Geden, C. J., Rutz, D. A., and Bishop, D. R., Cattle lice (Anoplura, Mallophaga) in New York: seasonal population changes, effects of housing type on infestations of calves, and sampling efficiency, *J. Econ. Entomol.*, 83, 1435, 1990.
67. Arthur, F. H. and Axtell, R. C., Northern fowl mite population development on laying hens caged at three colony sizes, *Poultry Sci.*, 62, 424, 1983.
68. Foulk, J. D. and Matthysse, J. G., Experiments on control of the northern fowl mite, *J. Econ. Entomol.*, 56, 321, 1963.
69. Lemke, L. A. and Collison, C. H., Evaluation of a visual sampling method used to estimate northern fowl mite, *Ornithonyssus sylviarum* (Acari: Macronyssidae), populations on caged laying hens, *J. Econ. Entomol.*, 78, 1079, 1985.
70. De Vaney, J. A., Dispersal of the northern fowl mite, *Ornithonyssus sylviarum* (Canestrini and Fanzago), and the chicken body louse, *Menacanthus stramineus* (Nitzsch), among thirty strains of egg-type hens in a caged-layer house, *Poultry Sci.*, 59, 1745, 1979.
71. Courtney, C. H., Ingallis, W. L., and Stitzlein, S. L., Ivermectin for the control of swine scabies: relative values of prefarrowing treatment of sows and weaning treatment of pigs, *Am. J. Vet. Res.*, 44, 1220, 1983.
72. Davis, D. P. and Moon, R. D., Pruritus and behavior of pigs infested by itch mites *Sarcoptes scabei* (Acari: Sarcoptidae), *J. Econ. Entomol.*, 83, 1439, 1990.
73. Altmann, J., Observational study of behavior: sampling methods, *Behaviour*, 49, 227, 1974.
74. Davis, D. P. and Moon, R. D., Density of itch mite, *Sarcoptes scabei* (Acari: Sarcoptidae) and temporal development of cutaneous hypersensitivity in swine mange, *Vet. Parasitol.*, 36, 285, 1990.
75. Kunz, S. E., Scholl, P. J., Colwell, D. D., and Weintraub, J., Use of sterile insect releases in an IPM program for control of cattle grubs (*Hypoderma lineatum*, *H. bovis*: Diptera: Oestridae): a pilot test, *J. Med. Entomol.*, 27, 523, 1990.
76. Klein, K. K. and Jetter, F. P., Economic benefits from the Alberta Warble Control Program, *Can. J. Agric. Econ.*, 35, 289, 1987.
77. Tarry, D. W., The monitoring and diagnosis of hypodermosis, in *Warble Fly Control in Europe*, Boulard, C. and Thornberry, H., Eds., A. A. Balkema, Rotterdam, 1984, 125.
78. Colwell, D. D. and Baron, R. W., Early detection of cattle grub infestation (*Hypoderma lineatum* de Vill and *H. bovis* L.) (Diptera: Oestridae) using ELISA, *J. Med. Entomol.*, 4, 35, 1990.
79. Burillon, G. and Messean, A., Comparison of two methods of estimation of the warble fly infestation rate, in *Warble Fly Control in Europe*, Boulard, C. and Thornberry, H., Eds., A. A. Balkema, Rotterdam, 1984, 131.
80. Lysyk, T. J., Colwell, D. D., and Baron, R. W., A model for estimating abundance of cattle grub (Diptera: Oestridae) from the proportion of uninfested cattle as determined by serology, *Med. Vet. Entomol.*, 5, 253, 1991.
81. Lysyk, T. J., Colwell, D. D., and Baron, R. W., Using the ELISA blood test to determine cattle grub abundance, in *Research Highlights 1990*, Sears, L. J., Ed., Agriculture Canada, Lethbridge, Alberta, 1991, 29.
82. Binns, M. R. and Bostanian, N. J., Robustness in empirically based binomial decision rules for integrated pest management, *J. Econ. Entomol.*, 83, 420, 1990.
83. Rutz, D. A. and Axtell, R. C., House fly (*Musca domestica*) control in broiler-breeder poultry houses by pupal parasites (Hymenoptera: Pteromalidae): indigenous parasite species and releases of *Muscidifurax raptor*, *Environ. Entomol.*, 10, 343, 1981.
84. Williams, D. E., Sticky traps for sampling populations of *Stomoxys calcitrans*, *J. Econ. Entomol.*, 66, 1279, 1973.
85. Broce, A. B., An improved alsynite trap for stable flies, *Stomoxys calcitrans* (Diptera: Muscidae), *J. Med. Entomol.*, 25, 406, 1988.
86. Pickens, L. G., Miller, R. W., and Grasela, J. J., Sticky panels as traps for *Musca autumnalis*, *J. Econ. Entomol.*, 70, 549, 1977.

87. Moon, R. D., Schwinghammer, K., Knapp, F. W., Meyer, H. J., Schmidtmann, E. T., and Hall, R. O., Sampling bias and habitat variation in gonotrophic age of female *Musca autumnalis* (Diptera: Muscidae), *J. Med. Entomol.*, 23, 269, 1986.
88. Kaya, H. K. and Moon, R. D., The nematode *Heterotylenchus autumnalis* and face fly *Musca autumnalis*: a field study in northern California, *J. Nematol.*, 10, 333, 1978.
89. Burg, J. G. and Axtell, R. C., Monitoring house fly (*Musca domestica*) (Diptera: Muscidae) populations in caged-layer poultry houses using a baited jug-trap, *Environ. Entomol.*, 13, 1083, 1984.
90. Lysyk, T. J. and Axtell, R. C., Comparison of baited jug-trap and spot cards for sampling house fly, *Musca domestica* (Diptera: Muscidae), populations in poultry houses, *Environ. Entomol.*, 14, 815, 1985.
91. Stafford, K. C., III, Collison, C. H., and Burg, J. G., House fly (Diptera: Muscidae) monitoring method comparisons and seasonal trends in environmentally controlled high-rise, caged-layer poultry houses, *J. Econ. Entomol.*, 81, 1426, 1988.
92. Lysyk, T. J. and Axtell, R. C., Field evaluation of three methods for monitoring populations of house flies (*Musca domestica*) (Diptera: Muscidae) and other filth flies in three types of poultry housing systems, *J. Econ. Entomol.*, 79, 144, 1986.
93. Axtell, R. C., Integrated fly control program for caged-poultry houses, *J. Econ. Entomol.*, 63, 400, 1970.
94. Steelman, C. D., White, T. W., and Schilling, P. E., Effects of mosquitoes on the average daily gain of feedlot steers in southern Louisiana, *J. Econ. Entomol.*, 65, 462, 1972.
95. Steelman, C. D., White, T. W., and Schilling, P. E., Effects of mosquitoes on the average daily gain of Hereford and Brahman breed steers in southern Louisiana, *J. Econ. Entomol.*, 66, 1081, 1973.
96. Steelman, C. D., White, T. W., and Schilling, P. E., Efficacy of Brahman characters in reducing weight loss steers exposed to mosquito attack, *J. Econ. Entomol.*, 69, 499, 1976.
97. Koch, H. G. and McNew, R. W., Comparative catches of field populations of lone star ticks by CO_2 emitting dry-ice, dry-chemical, and animal-baited traps, *Ann. Entomol. Soc. Am.*, 74, 498, 1981.
98. Koch, H. G. and McNew, R. W., Sampling of lone star ticks (Acari: Ixodidae): dry ice quantity and capture success, *Ann. Entomol. Soc. Am.*, 75, 579, 1982.
99. Koch, H. G., Estimation of absolute numbers of adult lone star ticks (Acari: Ixodidae) by dry ice sampling, *Ann. Entomol. Soc. Am.*, 80, 624, 1987.
100. Wilkinson, P. R. and Gregson, J. D., Comparisons of sampling methods for recording the numbers of rocky mountain wood ticks (*Dermacentor andersoni*) on cattle and range vegetation during control experiments, *Acarologia*, 26, 131, 1985.
101. Kinzer, D. R., Presley, S. M., and Hair, J. A., Comparative efficiency of flagging and carbon dioxide-baited sticky traps for collecting the lone star tick, *Amblyomma americanum* (Acarina: Ixodidae), *J. Med. Entomol.*, 27, 750, 1990.
102. Barnard, D. R., *Ambylomma americanum*: comparison of populations of ticks free living on pasture and parasitic on cattle, *Ann. Entomol. Soc. Am.*, 74, 507, 1981.
103. Schaalje, G. B. and Wilkinson, P. R., Discriminant analysis of vegetational and topographical factors associated with the focal distribution of rocky mountain wood ticks, *Dermacentor andersoni* (Acari: Ixodidae), on cattle range, *J. Med. Entomol.*, 22, 315, 1985.
104. Kunz, S. E., Blume, R. R., Hogan, B. F., and Matter, J. J., Biological and ecological investigations of horn flies in central Texas: influence of time of manure deposition on oviposition, *J. Econ. Entomol.*, 63, 930, 1970.
105. Blume, R. R., Kunz, S. E., Hogan, B. F., and Matter, J. J., Biological and ecological investigations of horn flies in central Texas: influence of other insects in cattle manure, *J. Econ. Entomol.*, 63, 1121, 1970.
106. Kunz, S. E., Hogan, B. F., Blume, R. R., and Eschle, J. L., Some bionomical aspects of horn fly populations in central Texas, *Environ. Entomol.*, 1, 565, 1972.
107. MacQueen, A. and Beirne, B. P., Influence of other insects on production of horn fly, *Haematobia irritans* (Diptera: Muscidae), from cattle dung in south-central British Columbia, *Can. Entomol.*, 107, 1255, 1975.
108. Meyer, J. A. and Petersen, J. J., Characterization and seasonal distribution of breeding sites of stable flies and house flies (Diptera: Muscidae) on eastern Nebraska feedlots and dairies, *J. Econ. Entomol.*, 76, 103, 1983.
109. Schmidtmann, E. T., Exploitation of bedding in dairy outdoor calf hutches by immature house and stable flies (Diptera: Muscidae), *J. Med. Entomol.*, 25, 484, 1988.
110. Skoda, S. R., Thomas, G. D., and Campbell, J. B., Developmental sites and relative abundance of immature stages of the stable fly (Diptera: Muscidae) in beef cattle feedlot pens in eastern Nebraska, *J. Econ. Entomol.*, 84, 191, 1991.

Chapter 19

SAMPLING PROGRAMS FOR SOYBEAN ARTHROPODS

Michael R. Zeiss and Thomas H. Klubertanz

TABLE OF CONTENTS

I. Introduction 540

II. Soybean Agroecosystem as a Sampling Universe 541
    A. Soybean Production 541
        1. Soybean Production Practices 541
        2. U.S. Production Regions 542
    B. Soybean Development and Phenology 542
        1. Soybean Development Stages 542
        2. Dates of Flowering and Maturity 542
        3. Plant Size, Canopy Closure, and Lodging 543
    C. Principal Soybean Arthropods 543
        1. Feeding Guild Concept 543
        2. Principal Arthropod Species 544
            a. Major Pests 544
            b. Minor Pests 544
            c. Natural Enemies 545

III. Need for Sampling Efficiency 545
    A. Economics of Soybean Sampling 545
    B. Reality vs. Economics: Grower Acceptance of Soybean Sampling 554

IV. Methods for Improving Sampling Efficiency 554

V. When and Where to Sample: Patterns of Arthropod Density in Time and Space 556
    A. When to Sample: Arthropod Concentrations Over Time 557
        1. Density Differences among Years 557
        2. Density Differences within Each Year 559
            a. Soybean Growth Stage or Calendar Date 559
            b. Temperature or Degree-Day Accumulations 561
            c. Other Biological Considerations 561
        3. Density Differences within Each 24 h (Including Diel Variation) 562
    B. Where to Sample: Arthropod Concentrations Over Space 563
        1. Density Differences among Geographic Regions 563
        2. Density Differences within Each Region 564
            a. Relative Maturity of Crop 564
            b. Soybean Variety 569
            c. Other Plants in Soybean Field (Weeds or Intercropped Species) ... 569
            d. Previous Crop in Soybean Field 570
            e. Insecticide Use 570
            f. Irrigation 570

	g. Soil Type...570
	3. Density Patterns within Each Field 570
	a. Field Edges..571
	b. Weedy Areas within Soybean Field 571
	c. Within-Row and Within-Plant Patterns 571
VI.	How to Sample: Existing Sampling Techniques and Programs 571
	A. Sampling to Determine Arthropod Density 573
	1. Absolute Techniques.....................................573
	2. Relative Techniques....................................575
	3. Fixed-Precision Sampling Plans.........................577
	4. Population Indices....................................577
	B. Sampling to Determine Population Dispersion Pattern 580
	C. Sampling to Classify Arthropod Density 580
	1. Economic Injury Levels and Economic Thresholds 580
	2. Sequential Sampling Plans 581
	3. Time-Sequential Sampling Plans..........................581
	4. Predictive Computer Models and Kalman Filters 581
	D. Sampling to Determine Impact of Natural Enemies 584
VII.	Recommendations: Future Directions for Sampling Soybean Arthropods...585
	A. Developing Programs for Additional Arthropod Species 586
	1. Programs for Minor Pests................................586
	2. Programs to Incorporate Natural Enemy Impacts 586
	B. Improving Sampling Efficiency 586
	1. Sample Only High-Risk Fields............................586
	2. Substitute Easy-To-Sample "Proxies" for Arthropod Counting......587
	3. Develop Efficient Sampling Plans 587
	4. Develop Programs for Sampling Multiple Species 587
	C. Improving Delivery Systems to Simplify Program Implementation 587

Acknowledgments...588

References...588

I. INTRODUCTION

"Sampling has no intrinsic merit, but is only a tool which the entomologist should use to obtain certain information, provided there is no easier way to get the information."[1]
R. F. Morris (to whom this book is dedicated)

In this chapter, we review programs for sampling arthropods of economic importance in U.S. soybean. Specifically, we focus on ways to make soybean sampling programs more acceptable to growers and commercial scouts. Researchers have already developed many effective sampling plans and techniques for soybean arthropods.[2] Nonetheless, the lack of grower-acceptable sampling programs is one of the major roadblocks to soybean integrated pest management (IPM). We believe that the ultimate test of a sampling program is whether it is put into practice by its intended users. As Herzog[3] said in a paper about soybean sampling:

> There are still those who enter the field to start sampling without having given thought to design of the sampling program. But, I'm not convinced that is worse than spending all time and effort in developing sampling programs with no thought given to acceptability or to implementation of the sampling program by extension personnel, private consultants or producers. It is the responsibility of research to develop new information and methodologies. But, unless there is planning for and a mechanism of implementation, that information may be of use to no one other than those who developed it. How many sequential sampling [plans] for management have been developed? How many are actually implemented in the field?

There are several ways by which sampling programs can be made more acceptable to growers and scouts. One way is to make sampling as efficient as possible, so that sampling requires the minimum possible effort and expense. This is the approach that we emphasize in this chapter. Morris[1] correctly points out that collecting and counting arthropods has no intrinsic merit; sampling is merely a means to an end. Thus, sampling programs should be designed to minimize the number of dates and locations sampled. Specifically, sampling efficiency can be increased by developing or selecting efficient sampling plans (discussed below in Sections IV and VI) and by taking advantage of patterns of arthropod abundance (Section V). However, even an efficient sampling program will gather dust unless it meets the needs of the intended users. Sampling program designers should therefore consult growers and scouts about sampling needs before developing a sampling program. Once a preliminary program has been developed, it should be formally pretested with the intended users.[4] In short, the needs and perceptions of users (as well as efficiency) should be considered throughout the development of a sampling program.

One brief note about terminology in this chapter. When discussing arthropod feeding on plants, we distinguish between injury (the effect of arthropod activities on plant physiology that is usually deleterious) and damage (the measurable loss of plant utility, most often including yield quantity or quality).[5] For example, some arthropods injure soybean by chewing holes in leaves, but this injury does not always result in damage (reduced yield or quality). When discussing sampling, we use the terminology outlined in Chapter 1 (such as sampling program, sampling technique, etc.). In addition, however, we define "sampling plan" as a mathematical model that makes an estimate or a prediction based on the number of sampling units already examined and the results obtained from those sampling units (example: sequential sampling plans). A sampling plan is used as part of a sampling program, and makes sampling more efficient by reducing the average number of sampling units required.

II. SOYBEAN AGROECOSYSTEM AS A SAMPLING UNIVERSE

A. SOYBEAN PRODUCTION
1. Soybean Production Practices

As discussed throughout this chapter, soybean production practices can affect both arthropod density and sampling efficiency. When planning a sampling program, it is therefore useful to understand common production practices. Soybean varieties are discussed in Section II.B; insecticide use is discussed in Section V.B. Readers wishing a complete review of soybean production practices should consult Johnson[9] and references therein. In addition, recent references provide information about the use and effects of tillage,[10-12] plant density and row spacing,[13-15] intercropping,[16-21] and rotations and double cropping.[10,22-24]

2. U.S. Production Regions

The major soybean producing states can be grouped into four regions based on soil, climate, and cultural practices:[6,7] North Central (Illinois, Indiana, Iowa, Kansas, Kentucky, Michigan, Minnesota, central and northern Missouri, Nebraska, Ohio, Wisconsin); Southeast (Alabama, Florida, Georgia, southern North Carolina, South Carolina); Mississippi Valley (Arkansas, Louisiana, Mississippi, southern Missouri, Tennessee); and Mid-Atlantic (Delaware, Maryland, New Jersey, northern North Carolina, Pennsylvania, Virginia). The North Central region produces about 70% of U.S. soybean, followed by the Mississippi Valley, which produces about 10%.[8]

B. SOYBEAN DEVELOPMENT AND PHENOLOGY

Within a soybean field, microclimate and the availability of resources (leaves, flowers, etc.) change dramatically during the course of a growing season.[25,26] These changes can affect arthropod density (Section V.A) and sampling efficiency. For example, it is difficult to get representative samples with a sweepnet when plants are very small. Conversely, dense foliage late in the season may slow a sweepnet and allow arthropods to escape.[27] Further, soybean growth stages differ in susceptibility to arthropod injury (Section VI.C.1). Thus, economic thresholds (and therefore sampling programs) change with soybean development.

1. Soybean Development Stages

Soybean development has been divided into standardized stages[28] that are described and illustrated in an excellent publication.[29] Prior to flowering, vegetative (V) stages are named for the number of main-stem leaves (for example, V3 plants have three fully expanded trifoliate leaves). Flowering begins the reproductive (R) stages. These comprise R1 (beginning bloom), R2 (full bloom), R3 (beginning pod), R4 (full pod), R5 (beginning seed), R6 (full seed), R7 (beginning maturity), and R8 (full maturity). Detailed reviews of growth and development are available.[30,31]

2. Dates of Flowering and Maturity

If several soybean fields in the same region are in different developmental stages, they will often have different arthropod densities (Section V.B). It is therefore helpful to understand the factors that determine when soybean enters reproductive stages (begins to flower). In most soybean varieties, flowering is triggered by long nights. Soybean varieties are classified into 13 "maturity groups", 000 to X. Varieties in group 000 mature earliest, and therefore are adapted for Canada; group X varieties mature latest, and therefore are adapted to southern Florida and Louisiana.[32]

Even young plants can be induced to flower by long nights. For this reason, dates of flowering and maturity may be similar in early planted vs. late planted fields of the same variety. In other words, late planted fields tend to catch up with early planted fields.[33,34] However, several factors can affect the synchrony between fields planted on different dates, including temperature[35,36] and arthropod injury.[38,63] Any of these factors may prevent late planted fields from catching up. For example, fields planted on May 9 flowered in late June, while fields planted on June 15 flowered in mid-July.[79] In other experiments, each month delay in planting (from mid-April to mid-June) caused approximately a 2-week delay in flowering.[40,41]

3. Plant Size, Canopy Closure, and Lodging

Plant size affects the degree of canopy closure (i.e., the extent to which plants in adjacent rows overlap) and the degree of lodging. Canopy closure and lodging, in turn, affect arthropod density (Section V.B) and sampling. Many factors can affect soybean plant size. For example, plants often grow larger in early planted vs. late planted fields.[79] In addition, biological stressors such as weeds, nematodes, and arthropods can reduce plant size and delay canopy closure.[42] Further, high planting densities reduce biomass per plant,[43] but may nonetheless speed canopy closure.[32,44] Large plants (especially those grown at high densities under optimal fertility and high moisture) tend to lodge during mid to late season.[45] Sampling program designers should not underestimate the sampling difficulties caused by lodging. Foliage must be moved, thereby disturbing arthropods, to spread a ground cloth between lodged plants. When sampling with a sweepnet, samplers can trip over or crush lodged plants. Lodging can also affect plant susceptibility to injury,[46] perhaps requiring changes in economic thresholds (ETs).

C. PRINCIPAL SOYBEAN ARTHROPODS

Many excellent resources are available for anyone developing sampling programs for soybean arthropods. Detailed information on the biology of pest and beneficial soybean arthropods is presented in several thorough, clearly written reviews.[2,32,47-53] A bibliography of over 5000 references has been prepared and indexed by taxonomic group, common name, and subject.[54] Several superb compilations of photographs, drawings, keys, and distribution maps are available to aid in arthropod identification.[2,47,48,55-57] Rather than trying to duplicate existing resources, this section simply tabulates the soybean arthropods of economic importance.

1. Feeding Guild Concept

A feeding guild is a group of species that uses the same resource (such as soybean pods) in the same way (such as chewing holes in the pod wall).[58] Grouping soybean arthropods into feeding guilds is a useful first step toward multispecies economic thresholds,[59-61] which are required to develop programs for sampling several species at the same time. Such multispecies programs would be very efficient because information about several species could be obtained from a single sample. Because the feeding guild concept facilitates the development of multispecies ETs (and therefore of multispecies sampling programs), arthropods are grouped into feeding guilds throughout this chapter.

The feeding guild concept was reviewed by Pedigo et al.[5] and Hutchins and co-workers,[59,60] and was recently extended to include nonarthropod stresses.[61] The following feeding guilds contain important soybean arthropods.

DEFOLIATORS remove pieces of leaves or entire leaves. LEAF SURFACE GRAZERS injure patches of leaf surface without removing them from the leaf. PLANT FLUID REMOVERS use hypodermic needle-like mouthparts to suck sap from phloem tubes in leaves or stems. POD FEEDERS feed on pod surfaces or seeds. STEM FEEDERS either pierce stems from the outside or tunnel inside them.

UNDERGROUND PESTS include several feeding guilds. They are grouped together because all require specialized sampling techniques. Thus it would probably be easier to develop a multispecies sampling program for all underground pests than to include a specific underground pest in a sampling program designed for aboveground members of its feeding guild. For example, GERMINATING SEED FEEDERS remove pieces of the developing plant that are still below the soil surface.

UNDERGROUND STEM FEEDERS cause the same injury as above-ground stem feeders. ROOT FEEDERS eat small roots, root hairs, and nodules.

Like pests, natural enemies can be grouped into guilds. PREDATORS are beneficial arthropods that attack and eat more than one pest individual during their life cycle. They are often larger than their prey. PARASITOIDS are beneficial arthropods that attack and gradually eat a single pest individual during their life cycle. Like true parasites, they are smaller than the pests on which they feed. Unlike parasites, parasitoids eventually kill their host. PARASITES are beneficial arthropods that feed on pests without killing them. PATHOGENS are microorganisms (fungi, bacteria, or viruses) that can cause fatal diseases in arthropods.

2. Principal Arthropod Species

Nearly 500 species of plant-feeding arthropods have been collected from soybean in the U.S.[47,62,64,65,169] However, most current sampling programs focus on the major pests and their natural enemies.

a. Major Pests

Surveys of entomologists in 18 states indicated that, as of 1977, over 80% of arthropod damage to U.S. soybean was caused by eight species[66] in the defoliator and pod feeder guilds. Research published after the survey[71,72] indicated that a stem feeder species, the three-cornered alfalfa hopper, caused more yield loss than previously had been thought.[53] In addition, spider mites require careful management because of their ability to cause devastating losses in drought years.[73] Thus, ten species account for most of the arthropod damage to U.S. soybean.[7,32,52,53,67] These species are summarized in Table 1, and are referred to as "major pests" in this chapter.

b. Minor Pests

Many other soybean arthropods are occasional pests (causing economic damage only in sporadic locations and years) or are subeconomic pests (not causing economic damage by themselves, but contributing to overall damage).[74] Such arthropods are referred to as "minor pests" in this chapter, and are listed in Table 2. Minor pests should not be neglected when developing sampling programs. In the long run, sampling programs for minor pests (especially occasional pests) may prove vital for improving pest management. Major pests are best managed by strategies of reducing the environmental carrying capacity, or the damage per individual, for the pest species. Such strategies rely on preventive tactics (such as biological control or tolerant varieties) that are usually implemented without pest sampling.[74,75] In contrast, a curative management strategy based on pest sampling is appropriate for occasional and subeconomic pests. "A wait and see attitude is assumed, with reliance on early detection, prediction of impending outbreaks, and employment of [curative] tactics only when the economic threshold is reached."[74]

A second justification for investing effort to develop sampling programs for minor pests is that these species may cause increased damage in the future, i.e., may become major pests. For example, there is evidence that the tobacco budworm, *Helicoverpa virescens* (F.), is becoming an important soybean pest.[52] Alternatively, entomologists may be underestimating the current damage from these species. For example, recent research has shown that both the three-cornered alfalfa hopper[53] and the soybean nodule fly[52] cause more damage than previously had been thought. Further, very little is known about the economic impact of underground feeders.[51,76]

c. Natural Enemies

In addition to pests, natural enemies (beneficial arthropods and pathogens) may need to be sampled for effective pest management. First, the density of natural enemies may determine whether an occasional pest reaches economic density. Further, many future pest management systems probably will include deliberate augmentation of natural enemies,[77] requiring natural enemy sampling programs. For example, sampling programs might be required to quantify the degree to which environmental manipulations had increased natural enemy populations, or to determine when additional releases of natural enemies were needed. Major natural enemies in soybean are listed in Table 3.

III. NEED FOR SAMPLING EFFICIENCY

"The object of sampling in pest management is to collect either the maximum information affordable, or the minimum amount of information necessary."[98] Researchers usually sample to collect as much information as possible. Because budgets are always limited, efficiency is important even when sampling for research purposes. However, researchers may tolerate an inefficient sampling program if it is the only way to obtain necessary information. In contrast, pest managers need only the minimum information required to make a correct decision. With a low-value crop such as soybean, growers may decide not to sample rather than pay for an inefficient sampling program. This section explores the economic factors that must be considered when developing or choosing a sampling program for use by growers or commercial scouts.

A. ECONOMICS OF SOYBEAN SAMPLING

As discussed in Section V.B, pest impact (and thus perhaps the benefit of scouting) is generally higher in southern vs. North Central states. Perhaps for this reason, few data are available on how scouting soybean affects profit or risk in the North Central region. A 1982 unpublished report (reviewed by Rajotte et al.[99]) showed that the use of IPM (including scouting) decreased risk of low net returns for Illinois soybean producers.

Most cost/benefit studies of soybean sampling have been based on data from southern states. Early studies[100,101] showed large benefits from scouting, but this was probably due to inefficient insecticide use on unscouted farms. For example, over 87% of insecticide applications on unscouted Virginia fields were made before ETs were reached.[101] Since those studies, the CIPM project[102] (1979 to 1985) and other efforts have greatly improved soybean pest management.[53,102,103] As a result, unnecessary insecticide use presumably has declined even on unscouted soybean fields. Nevertheless, most recent studies show increased profit, reduced variability in profit, and reduced insecticide use on scouted vs. unscouted soybean in southern states.[103-107]

There are two notable exceptions. First, a computer simulation based on Maryland conditions showed that when only Mexican bean beetle damage was considered, calendar insecticide applications gave higher benefit:cost ratios than scouting.[108] However, inclusion of damage from other arthropods might have changed the outcome. Second, the Soybean Integrated Crop Management (SICM) computer simulation model[109] was used to predict profits based on simulated insect damage and costs (including scouting costs). Unscouted fields received two simulated insecticide applications (on August 15 and September 10), regardless of insect abundance. In contrast, scouted fields received simulated applications only if insect abundance exceeded an ET. The calendar application strategy gave 13 to 32% higher profits and

TABLE 1
Major Arthropod Pests of U.S. Soybean[a]

Guild and common name	Scientific name	Order	Family	Injurious stage	Nature of injury
DEFOLIATORS					
Velvetbean caterpillar	*Anticarsia gemmatalis* (Hubner)	Lepidoptera	Noctuidae	Larva	Chew large, irregular holes starting with upper canopy.
Soybean looper	*Pseudoplusia includens* (Walker)	Lepidoptera	Noctuidae	Larva	Chew large, irregular holes starting with middle canopy.
Green cloverworm	*Plathypena scabra* (F.)	Lepidoptera	Noctuidae	Larva	Chew large, irregular holes starting with upper canopy.
Mexican bean beetle	*Epilachna varivestis* (Mulsant)	Coleoptera	Coccinellidae	Larva, adult	Larvae scrape surface tissue, leaving lace-like strips of crushed tissue. Adults chew rounded holes in leaves.
LEAF SURFACE GRAZERS					
Spider mites	*Tetranychus* spp.	Acari	Tetranychidae	Nymph, adult	Suck out contents of cells of leaf under-surfaces, injured leaves appear sandblasted, may wilt and die.
POD FEEDERS					
Bean leaf beetle[b]	*Cerotoma trifurcata*	Coleoptera	Chrysomelidae	Adult	Chew small, rounded holes (Forster) in leaves; vector bean pod mosaic virus. Pods: rounded holes in surface, seldom deep enough to reach seeds.

Common name	Species	Order	Family	Stage[b]	Type of damage
Southern green stink bug	*Nezara viridula* (L.)	Hemiptera	Pentatomidae	Nymph, adult	Pierce pods with sucking mouthparts, causing distorted pods with seeds missing or discolored.
Green stink bug	*Acrosternum hilare* (Say)	Hemiptera	Pentatomidae	Nymph, adult	As for southern green stink bug.
Corn earworm (= bollworm, tomato fruitworm)	*Helicoverpa zea* (Boddie)	Lepidoptera	Noctuidae	Larva	Leaves: large, irregular holes Pods: irregular holes in surface, large larvae enter pods to feed on seeds.
STEM FEEDERS					
Three-cornered alfalfa hopper	*Spissistilus festinus* (Say)	Homoptera	Membracidae	Nymph, adult	Girdle stems and leaf petioles with sucking mouthparts. This diverts plant sugars, allows disease entry, and kills young plants.

[a] Sources: References 2, 32, 47, 51, 55, 57, 66, 67, 68.
[b] Adult bean leaf beetles also cause injury as defoliators, and larvae are underground pests, but the species probably causes most of its economic damage via pod feeding.[69,70]

TABLE 2
Minor Arthropod Pests of U.S. Soybean[a]

Guild and common name	Scientific name	Order	Family	Injurious stage	Nature of injury
DEFOLIATORS					
Blister beetles	*Epicauta* spp.	Coleoptera	Meloidae	Adult	Chew large, rounded holes in leaves, leaving the weblike network of veins intact.
Japanese beetle	*Popillia japonica* Newman	Coleoptera	Scarabeidae	Adult	As for blister beetles.
Beet armyworm	*Spodoptera exigua* (Hubner)	Lepidoptera	Noctuidae	Larva	Chew large, irregular holes in leaves.
Cabbage looper	*Trichoplusia ni* (Hübner)	Lepidoptera	Noctuidae	Larva	As for beet armyworm.
Grasshoppers	*Melanoplus* spp.	Orthoptera	Acrididae	Nymph, adult	Chew large, irregular identations in leaf margins. Chew large, irregular holes in pods, eat exposed seeds.
LEAF SURFACE GRAZERS					
Thrips	*Sericothrips variabilis* (Beach)	Thysanoptera	Thripidae	Nymph, adult	Rasp leaf surfaces with their mouthparts, then suck up the shredded tissue.
PLANT FLUID REMOVERS					
Aphids	*Aphis craccivora* Koch, and other species	Homoptera	Aphididae	Nymph, adult	Suck sap from phloem tubes in leaves and transmit viruses.
Potato leaf hopper	*Empoasca fabae* (Harris)	Homoptera	Cicadellidae	Nymph, adult	As for aphids, and inject toxins.
POD FEEDERS					
Brown stink bug	*Euschistus servus* (Say)	Hemiptera	Pentatomidae	Nymph, adult	Pierce pods with sucking mouthparts, causing distorted pods with seeds missing or discolored.
Tobacco budworm	*Helicoverpa virescens* (F.)	Lepidoptera	Noctuidae	Larva	Leaves: large, irregular holes Pods: chew irregular holes in surface, large larvae enter pods to feed on seeds.

TABLE 2 (continued)
Minor Arthropod Pests of U.S. Soybean[a]

Guild and common name	Scientific name	Order	Family	Injurious stage	Nature of injury
STEM FEEDERS					
Dectes stemborer[b]	*Dectes texanus* LeConte	Coleoptera	Cerambycidae	Larva	Larvae tunnel in leaf petioles and main stem, and often girdle stem.
UNDERGROUND PESTS					
GERMINATING SEED FEEDERS					
Seed corn maggot	*Hylemya platura* (Meigen)	Diptera	Anthomyiidae	Larva	Chew irregular holes in germinating seeds. Seedlings that survive often have two growing points ("Y"-plants).
UNDERGROUND STEM FEEDERS					
Cutworms	*Agrotis* spp.	Lepidoptera	Noctuidae	Larva	Chew into bases of seedling stems, usually clipping seedlings at soil level.
Lesser corn stalk borer	*Elasmopalpus lignosellus* (Zeller)	Lepidoptera	Pyralidae	Larva	Tunnel inside bases of seedling stems.
ROOT FEEDERS					
Bean leaf beetle	*Cerotoma trifurcata* (Forster)	Coleoptera	Chrysomelidae	Larva (adults are pod feeders)	Consume small roots and root hairs, and bore into root nodules.
Soybean nodule fly	*Rivellia quadrifasciata* (Macquart)	Diptera	Platystomatidae	Larva	Bore into root nodules.

[a] Sources: References 2, 32, 47, 51, 52, 55, 57, 66, and 67.
[b] The Entomological Society of America does not recognize any common name for *Dectes texanus* LeConte.

was less risky (had a higher probability of better profits under most conditions) than scouting.[110,111]

The simulation results are difficult to reconcile with the results of case studies from southern states.[100,101,103-107] Perhaps the researchers who set the calendar application dates[110,111] were better managers than the non-IPM growers in the case studies. Indeed, a second study[112,113] using the same SICM model illustrates the importance of understanding the timing of insect population growth. Under simulated Florida conditions, the SICM model showed that a calendar application strategy gave *lower* profits than any scouting schedule. However, calendar sprays were simply applied at fixed intervals (the most profitable being an application every 7 d).[112,113] No effort was

TABLE 3
Common Natural Enemies of U.S. Soybean Pests[a]

Guild and common name	Scientific name	Order	Family	Stage that feeds on pests	Pests[b,c] attacked
FOLIAGE-INHABITING PREDATORS					
Minute pirate bug	*Orius insidiosus* (Say)	Hemiptera	Anthocoridae	Nymph, adult	Eggs, small larvae of GCW, CEW, SBL,
Big-eyed bugs	*Geocoris* spp.	Hemiptera	Lygaeidae	Nymph, adult	VBC, MBB. Eggs of SGSB and GSB.
Damsel bugs	*Nabis* spp.	Hemiptera	Nabidae	Nymph, adult	
Spined soldier bug	*Podisus maculiventris* (Say)	Hemiptera	Pentatomidae	Nymph, adult	Larvae of GCW, CEW, SBL, VBC, MBB. Nymphs of SGSB and GSB.
Spiders	Many	Araneae	Many	Immature, adult	Primarily small larvae of GCW, CEW, SBL, and VBC. Also MBB larvae, nymphs of GSB and SGSB.
Ladybird beetles	*Coleomegilla maculata* (DeGeer), *Hippodamia convergens* Guérin-Méneville	Coleoptera	Coccinellidae	Larva, adult	Eggs and small larvae of GCW, CEW, VBC, MBB. Eggs and nymphs of SGSB and GSB.
Lacewings, aphid lions	*Chrysopa* spp.	Neuroptea	Chrysopidae	Larva, adult	Eggs and small caterpillars, aphids.
GROUND-INHABITING PREDATORS					
Ground beetles	*Calleida decora* (F.), *Calosoma sayi* Dejean, *Lebia* spp.	Coleoptera	Carabidae	Larva, adult (*Lebia* larvae are parasites)	Medium to large larvae of GCW, CEW, SBL, VBC, and MBB. Nymphs of GSB and SGSB.
Earwig	*Labidura riparia* (Pallas)	Dermaptera	Labiduridae	Nymph, adult	Larvae of CEW, GCW, SBL, VBC.
Ants	*Solenopsis* spp.	Hymenoptera	Formicidae	Adult	Eggs and small larvae of GCW, VBC, CEW, SBL, MBB.
Spiders	Many	Araneae	Many	Immature, adult	As for foliar spiders.
PARASITOIDS[d]					
	Cotesia (=*Apanteles*) spp.	Hymenoptera	Braconidae	Larva	Larvae of GCW, BA, CEW, SBL, and VBC.

TABLE 3 (continued)
Common Natural Enemies of U.S. Soybean Pests[a]

Guild and common name	Scientific name	Order	Family	Stage that feeds on pests	Pests[b,c] attacked
PARASITOIDS (cont.)	*Meteorus* spp.	Hymenoptera	Braconidae	Larva	
	Microplitis spp.	Hymenoptera	Braconidae	Larva	Larvae of SBL, GCW, CEW.
	Orgilus spp.	Hymenoptera	Braconidae	Larva	Larvae of LCSB.
	Protomicroplitis facetosa (Weed)	Hymenoptera	Braconidae	Larva	Larvae of GCW.
	Rogas spp.	Hymenoptera	Braconidae	Larva	Larvae GCW, SBL.
	Brachymeria spp.	Hymenoptera	Chalcididae	Larva	Larvae of SBL.
	Copidosoma truncatellum (Dalman)	Hymenoptera	Encyrtidae	Larva	Oviposit into eggs, but emerge from larvae of SBL.
	Euplectrus spp.	Hymenoptera	Eulophidae	Larva	Larvae GCW, VBC.
	Pediobius foveolatus (Crawford)	Hymenoptera	Eulophidae	Larva	Larvae of MBB.
	Campoletis spp.	Hymenoptera	Ichneumonidae	Larva	Larvae of SBL, GCW, CEW.
	Charops annulipes Ashmead	Hymenoptera	Ichneumonidae	Larva	Larve of GCW.
	Colpotrochia trifasciata (Cresson)	Hymenoptera	Ichneumonidae	Larva	Larvae of GCW.
	Sinophorus validus (Cresson)	Hymenoptera	Ichneumonidae	Larva	Larvae of GCW.
	Venturia nigriscapus (Ashmead)	Hymenoptera	Ichneumonidae	Larva	Larvae of GCW.
	Vulgichneumon brevicinctor (Say)	Hymenoptera	Ichneumonidae	Larva	Larvae of GCW.
	Telenomus podisi Ashmead	Hymenoptera	Scelionidae	Larva	Eggs of SGSB, GSB.
	Trissolcus (=*Telenomus*) *basalis* Wollaston	Hymenoptera	Scelionidae	Larva	Eggs of SGSB, GSB.
	Trichogramma exiguum Pinto & Platner	Hymenoptera	Trichogrammatidae	Larva	Eggs of CEW.

TABLE 3 (continued)
Common Natural Enemies of U.S. Soybean Pests[a]

Guild and common name	Scientific name	Order	Family	Stage that feeds on pests	Pests[b,c] attacked
PARASITOIDS (cont.)	Trichogramma minutum Riley	Hymenoptera	Trichogrammatidae	Larva	Eggs of GCW, CEW.
	Trichogramma pretiosum Riley	Hymenoptera	Trichogrammatidae	Larva	Eggs of CEW.
	Blondelia hyphantriae (Tothill)	Diptera	Tachinidae	Larva	Larvae of GCW.
	Campylochaeta (=Chaetophlepsis) spp.	Diptera	Tachinidae	Larva	Larve of GCW and SBL.
	Celatoria diabroticae (Shimer)	Diptera	Tachinidae	Larva	Adult of BLB.
	Chetogena (=Euphorocera) (=Stomatomyia) spp.	Diptera	Tachinidae	Larva	Larvae of GCW, LCSB.
	Compsilura concinnata (Meigen)	Diptera	Tachinidae	Larva	Larvae of GCW.
	Copecrypta ruficauda (Wulp)	Diptera	Tachinidae	Larva	Larvae of GCW.
	Lespesia spp.	Diptera	Tachinidae	Larva	Larvae of CEW, GCW, SBL.
	Medina sp.	Diptera	Tachinidae	Larva	Adult of BLB.
	Oswaldia assimilis (Townsend)	Diptera	Tachinidae	Larva	Larvae of GCW.
	Strongygaster (=Hyalomyodes) triangulifer (Loew)	Diptera	Tachinidae	Larva	Adult of BLB.
	Trichopoda pennipes (F.)	Diptera	Tachinidae	Larva	Nymphs, adults of GSB and SGSB.
	Winthemia spp.	Diptera	Tachinidae	Larva	Larvae of CEW, GCW, SBL.

TABLE 3 (continued)
Common Natural Enemies of U.S. Soybean Pests[a]

Guild and common name	Scientific name	Order	Family	Stage that feeds on pests	Pests[b,c] attacked
PARASITES					
	Coccipolipus epilachnae Smiley	Acari	Podapolipidae	Nymph, adult	Adult of MBB.
	Trombidium spp.	Acari	Trombidiidae	Nymph, adult	Adult of BLB.
	Lebia spp.	Coleoptera	Carabidae	Larva	Pupae of MBB.
	Membracixenos jordani Pierce	Strepsiptera	Halictophagidae	Larva	Nymphs, adults of TCAH.
PATHOGENS					
	Neozygites floridana Weiser and Muma	Entomophthorales	Entomophthoraceae	Fungal hyphae	Spider mites.
	Zoophthora spp.	Entomophthorales	Entomophthoraceae	Fungal hyphae	Larvae of CEW, SBL, GCW, VBC.
	Hirsutella thompsonii Fisher	Hypocreales	N.A.	Fungal hyphae	Spider mites.
	Beauveria bassiana (Balsamo) Vuillemin	Moniliales	N.A.	Fungal hyphae	Larvae of CBL and MBB.
	Metarhizium anisopliae (Metchnikoff) Sorokin	Moniliales	N.A.	Fungal hyphae	Larvae of CBL, GCW, MBB.
	Nomuraea rileyi (Farlow)	Moniliales	N.A.	Fungal hyphae	Larvae of BA, CEW, GCW, SBL, and VBC.
Granulosis virus			Baculoviridae	Virions	Larvae of BA, CEW, GCW.
Nuclear polyhedrosis viruses			Baculoviridae	Virions	Larvae of SBL, BA, VBC, CEW.
Cytoplasmic polyhedrosis virus			Reoviridae	Virions	Larvae of BA and CEW.

[a] Sources: References 32, 49, 78, 79, 80–94 and references therein.
[b] BA = beet armyworm. BLB = bean leaf beetle. CBL = cabbage looper. CEW = corn earworm. GCW = green cloverworm. GSB = green stink bug. LCSB = lesser corn stalk borer. MBB = Mexican bean beetle. SBL = soybean looper. SGSB = southern green stink bug. TCAH = three-cornered alfalfa hopper. VBC = velvetbean caterpillar.
[c] Predators and pathogens may attack additional species not listed, including certain beneficial arthropods.
[d] Parasitoid scientific names in accordance with Wood,[95] Krombein et al.,[96] and Mason.[97]

made to time the calendar sprays to maximize their effectiveness as in the Georgia simulation.[110,111]

B. REALITY VS. ECONOMICS: GROWER ACCEPTANCE OF SOYBEAN SAMPLING

If scouting increases profits, soybean growers should be clamoring to adopt programs for sampling soybean arthropods. Although hard data are scarce, there is evidence that soybean growers' enthusiasm for sampling is mild at best. Soybean entomologists estimated that as of 1985, approximately 75% of soybean hectares in the Mississippi Valley and North Central regions were managed using IPM strategies that included scouting.[102] However, other published surveys report that only 12 to 51% of soybean growers scout their fields for arthropods.[103,106,110,114,115] Further, our 1993 discussions with soybean entomologists in many major soybean-producing states indicated that less than half of all soybean fields are scouted for arthropods.

There are at least two possible reasons for the apparent difference between what growers do and what research suggests they should do. First, soybean scouting may indeed be profitable, but cash flow considerations may not permit growers to invest in commercial scouting services, and time constraints may prevent growers from scouting themselves. Clearly, sampling costs can be a substantial part of the production costs of a comparatively low hectare-value crop such as soybean.[116] Based on national average market values during the past decade,[8] estimates of average profit per hectare of soybean range from $49 to $138.[103,106,107,110,111,117,118] In contrast, most estimates of soybean scouting costs per hectare range from $3.70 to $8.65.[104,105,107,110,111]

Grower surveys provide direct evidence that money and labor constraints limit scouting. Despite relatively low arthropod pressure in the North Central region, a substantial portion of North Central soybean producers that responded to a survey said that additional research was needed to improve scouting techniques for arthropods.[119] In surveys of producers of multiple crops, the two main reasons given for not scouting were that scouting services were too expensive, and that survey respondents themselves did not have enough time to scout.[37,99]

A second possible reason for low scouting rates is that soybean scouting is *not* profitable. Soybean growers may have discovered what the Georgia simulation[110,111] seems to indicate: it may be more profitable to predict arthropods than to sample them.

Both of these possible explanations can be addressed by making soybean sampling more efficient, and therefore less expensive.

IV. METHODS FOR IMPROVING SAMPLING EFFICIENCY

Sampling efficiency is measured by the amount of labor required to take a sample and the amount of information (precision or reliability) that the sample provides. If two sampling programs each provide the same information, whichever program requires less sampling labor is more efficient. Naturally, a sampler would prefer to use efficient sampling programs. The problem is that developing efficient sampling programs requires effort. In other words, researchers must invest time, labor, and money to gather the data necessary to develop efficient sampling programs (Figure 1). For example, researchers must invest effort to develop conversion equations before absolute density can be determined from relative techniques such as sweepnet sampling (Chapter 6). Similarly, researchers must invest effort to determine the dispersion pattern of a species before a sequential sampling plan can be developed (Chapter 8). Note that the information obtained from such an investment may be useful for more than one species.[120]

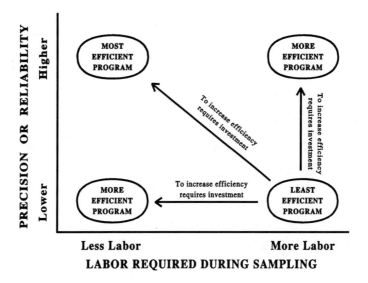

FIGURE 1. Relative efficiencies of hypothetical sampling programs.

When developing a sampling program, how can a researcher decide how much effort to invest in making the program efficient? There are at least three factors to consider. First, how much resources (time, labor, and money) does the researcher have? If resources are limited, it may be impossible to invest in the development of an efficient sampling program no matter how much effort the program would save if it was developed. Second, what is the projected life span of the sampling program? The longer the program will be in use, the more investment is justified to maximize its efficiency. Third, and most important, what are the characteristics of the final user of the program? Characteristics that should be considered include labor availability, degree of precision or reliability required, and willingness to use complicated sampling plans. If the user has limited labor available, he or she will not be able to use a labor-intensive program regardless of how much information it provides. If the user requires very precise estimates of arthropod density, it would be inappropriate to sacrifice precision in order to save labor. Finally, if the user is not comfortable with sampling plans that appear complicated (such as sequential sampling plans), he or she will not accept a complicated sampling program regardless of how efficient it is. If researchers already have developed several sampling programs, the same considerations apply for choosing which program would be most appropriate for a user.

Once a researcher has decided that investment of resources is justified, how can sampling efficiency be improved? Three general approaches can be used regardless of the sampling objective. First, instead of counting arthropods on soybean plants, sampling programs should (when possible) measure easy-to-sample phenomena that are highly correlated with arthropod density. Phenomena such as weather, plant injury, or trap catches may be highly correlated with eventual density of damaging arthropod stages on soybean. It is much less laborious to count (for example) the number of moths caught on a sticky trap than to quantify caterpillar densities in soybean. Labor requirements can be reduced (and sampling efficiency thereby increased) by measuring easy-to-sample "proxies" instead of arthropods. A second general approach for increasing sampling efficiency is to count groups of arthropods rather than individual arthropods. For example, it is much less laborious to determine the proportion of leaves that have at least one arthropod than it is to count the number of arthropods per leaf. This second approach, known as binomial sampling, is

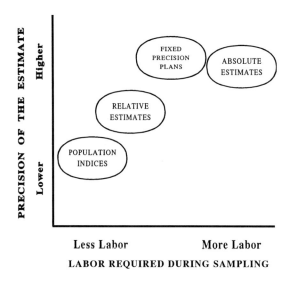

FIGURE 2. Relative efficiencies of programs for estimating arthropod densities.

discussed in detail in Chapter 9. Third, sampling efficiency can be increased by sampling only at times and locations that provide the most information (for example, provide the best prediction about whether a species will exceed its ET). Arthropod patterns in time and space are discussed in Section V.

Other approaches for improving sampling efficiency are applicable only for certain sampling objectives. As was discussed in Chapters 3 and 6, sampling is carried out to satisfy one of four objectives: (1) estimate arthropod density, (2) classify arthropod populations (for example, as being over or under an ET) without obtaining an estimate of density, (3) determine the population's dispersion pattern, or (4) determine the impact of natural enemies on pest species. Only researchers are likely to sample to meet objectives 3 or 4. In contrast, either pest managers or researchers may have reason to sample to satisfy objectives 1 or 2. Sampling objectives 1 and 2 can be achieved using several types of sampling programs, some of which are more efficient than others (Figures 2 and 3, respectively).

For many soybean arthropods and sampling objectives, efficient programs have been developed. For other arthropods and objectives, researchers have not yet invested the effort necessary to develop efficient programs. Progress toward developing efficient programs is discussed in Section VI. When designing or choosing a sampling program, remember that "the standard statistical criteria for sampling, such as precision and repeatability, must be weighed along with practical limitations of human resources, economics, and feasibility."[98] In particular, adoption of sampling programs by soybean farmers will depend on researchers' ability to make sampling both efficient and simple.[51]

V. WHEN AND WHERE TO SAMPLE: PATTERNS OF ARTHROPOD DENSITY IN TIME AND SPACE

As discussed in Section IV, sampling efficiency can be increased by restricting sampling to only those times and locations that provide the most information. Patterns of arthropod density in time and space are seldom consistent enough to eliminate completely the need for sampling. Nonetheless, knowledge about such

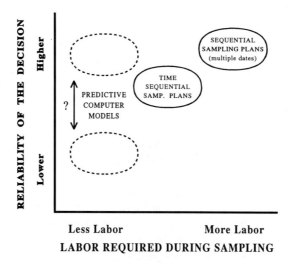

FIGURE 3. Relative efficiencies of programs for classifying arthropod populations as above or below a threshold.

patterns can help reduce unnecessary sampling and thereby increase sampling efficiency.

A. WHEN TO SAMPLE: ARTHROPOD CONCENTRATIONS OVER TIME

Many soybean arthropods are consistently more abundant at some times than at others. These patterns in time occur over a range of scales, from among years to within a single day. Knowledge about how arthropod density changes over time can greatly increase sampling efficiency. For example, when sampling pests for decision making, sampling can be restricted only to those dates that will provide the best predictions about whether pests will reach ETs. This greatly increases sampling efficiency by reducing the number of sampling dates.[52] As Wilson stated, "the number of unknowns or 'black boxes' in our understanding of pest management systems promoted calendar-type sampling in the same manner as it promoted a calendar or prophylactic spray strategy."[121] As discussed in Chapter 7, however, sampling efficiency must be balanced against risk. Decreasing the number of sampling dates may be more appropriate in North Central states than in southern regions where the risk of pests reaching economic levels is higher.[51]

In sampling to achieve other objectives (such as determining mortality rates from natural enemies), it may not be possible to reduce the number of sampling dates. Nonetheless, knowledge about patterns in time could be used (for example) to choose sampling dates most likely to represent a broad range of arthropod densities. Further, knowledge about how density changes over time is essential when comparing results of sampling conducted at different times.

1. Density Differences among Years

If it was possible to *predict* that a soybean pest would be below its economic threshold throughout a specific year, there would be no need to *sample* the pest during that year. Unfortunately, many factors can affect the average density of soybean arthropods from one year to the next. As discussed later, computer models will be probably the only practical way to combine all such factors to predict density.

However, the density of a few soybean pests can be predicted from a single factor—weather. For these species, it may be more efficient to "sample" the weather than to sample the pest.

One example is the green cloverworm. This species can overwinter only in southern states with relatively mild winters. Each year, moths migrate into North Central states from southern overwintering sites. The number of immigrating moths largely determines whether the subsequent generation of larvae will exceed the ET.[122] The number of immigrating moths can be predicted from synoptic weather patterns that are summarized on widely available daily weather maps. Specifically, moths migrate into Iowa during periods of at least two consecutive nights of wind flow from the south. These potential migration days occur when a low pressure system is located to the west of Iowa and a high pressure system is located to the east.[123]

The total number of potential migration days in a specific year was a good predictor of the total number of green cloverworm moths caught in light traps (correlation $R = 0.808$).[123] Thus, if the goal of sampling is to make insecticide treatment decisions, it is probably not necessary to sample North Central fields for green cloverworm in years with few potential migration days.

Michigan State University's "Pest Event Forecasting Group" uses synoptic weather patterns to predict spring migrations of the potato leafhopper from overwintering sites in Gulf Coast states into North Central states.[124] Synoptic weather patterns are probably also important for the velvetbean caterpillar and soybean looper, which emigrate from southern overwintering sites in early summer.[125,126] Similarly, thrips migrate into the North Central region from overwintering sites in southern states.[127] Analysis of weather patterns might increase efficiency of sampling for each of these pests.

Weather also can determine the density of nonmigratory arthropods (species that are present throughout the year). First, severe winters may kill a high enough proportion of overwintering pests to prevent the species from exceeding its ET the following summer. For example, spider mite densities are often higher in years that follow a mild winter.[57] Similarly, a very cold winter with little snow cover can drastically reduce the number of overwintering bean leaf beetles.[128]

In addition, nonmigratory pests can be greatly affected by weather during the growing season. In particular, mites usually reach high densities only in years with prolonged hot weather and drought, because cool, humid conditions ($< 27°C$ and $> 90\%$ RH) slow mite reproduction and favor growth of fungal pathogens.[129-135] Thus, when the goal of sampling is decision making, it probably is necessary to sample mites only in years when hot, dry conditions persist. In contrast, Mexican bean beetle usually reaches economic densities only in years when summers are *not* hot and dry.[105,136] Mexican bean beetle fecundity and larval survival are reduced severely by moderately high temperatures and low humidities.[137] Marcovitch and Stanley[138] developed a "climatic index" to rate summer weather patterns based on their favorability to Mexican bean beetle. High values of the climatic index are associated with hot, dry summers. In the 1980s, Mexican bean beetle reached economic densities in Atlantic Coast states only in years when the climatic index was below 2000.[136] Thus, for pest management purposes, it is probably necessary to sample Mexican bean beetle only in years when the climatic index is below 2000 (mild, rainy summers).

Whenever arthropod density can be predicted directly from weather patterns, it is often more efficient to sample the weather than the arthropod. However, weather can also affect pest density *indirectly*, by affecting the synchrony between arthropod and soybean phenologies. In such cases, the ultimate effect of weather on pest density depends on agronomic practices. For example, bean leaf beetles emerge from overwintering sites in response to warm spring temperatures.[139,140] If most growers in a region have planted soybean before beetles emerge, a high proportion of beetles will survive and reproduce, producing high densities later in the season. In contrast, a year in which beetle emergence was not synchronized with soybean planting is likely to have lower densities. For example, a warm spring (early emergence) together with wet conditions (delayed planting) can cause beetle populations to crash.[67,139]

The corn earworm is another arthropod whose density can be affected by a combination of weather and agronomic practices. Because corn earworm moths are most attracted to flowering soybean,[141-145] growers may select planting dates and varieties that cause soybean to flower before corn earworm moths emigrate from corn in most years. Hot, dry weather can cause corn to dry down early and can speed larval development, causing earlier than normal emigration of moths from corn.[13,146] Because soybean flowering is primarily governed by daylength, soybean phenology is less affected by hot weather than corn phenology. Moth flights in hot, dry years may therefore be better synchronized with soybean flowering, causing higher corn earworm oviposition in soybean.[13,146] Similarly, Herbert[147] found that years with earlier moth flights had higher soybean infestations.

If the indirect effects of weather are understood, they can be used to restrict sampling only to years in which pests are most likely to reach economic levels. However, indirect effects of weather on the synchrony between pests and soybean can be too complicated for most samplers to predict unless a simulation model is available (see Section VI.C.4).

2. Density Differences within Each Year

As was discussed in Section II.B, microclimate and the availability of resources within a soybean field change dramatically as soybean plants develop. Because arthropods may require specific microclimates or resources, the density of some arthropods can be principally determined by the stage of soybean development. In contrast, density of other species is determined more by degree-day accumulations, regardless of how mature soybean is at the time. Knowledge about how arthropod density (and especially arthropod injury to soybean) changes during a growing season is essential to optimize the timing of sampling and of control interventions.[5,121,148-151] For example, when the goal of sampling is economic decision making, sampling can be restricted only to those dates that will provide the best predictions about whether pests will reach ETs. As was discussed in Chapter 7, the optimal dates for monitoring population growth are usually much earlier than the dates on which a species reaches peak density.[5,121,148]

a. Soybean Growth Stage or Calendar Date

Figure 4 shows that many arthropods become most abundant in mid-season during soybean flowering and pod growth stages.[45,65,79] Indeed, Newsom[116] stated that, "There is no reason for starting an insect monitoring program on soybean in Louisiana before mid-July." Alternatively, scouting in Louisiana can be timed according to soybean growth stage, beginning at flowering and continuing until beans are mature.[145] As of 1974, scouting for economic decision making in Virginia could usually be limited to late August through late September.[101]

Changes in soybean production systems (rare weather patterns, changes in varieties or planting dates, etc.) could alter the seasonal patterns shown in Figure 4.[13,53] To adjust the timing of sampling to match new patterns, sampling program designers would need to understand the reasons for current seasonal patterns. For this reason, researchers have attempted to determine if arthropods become abundant in direct

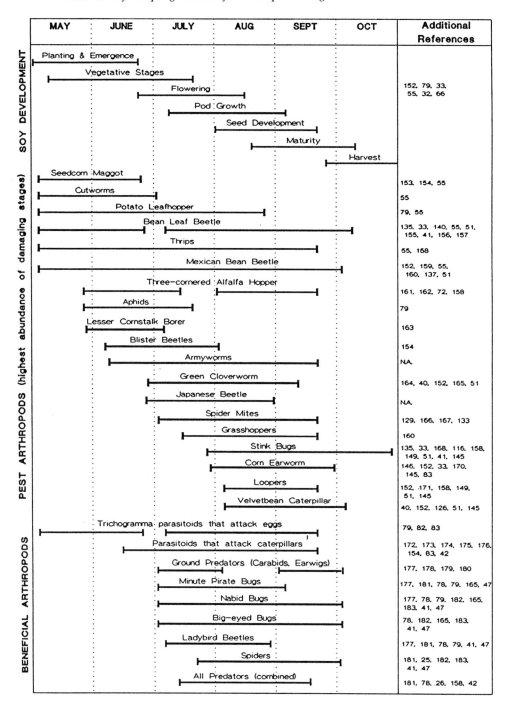

FIGURE 4. Phenology of soybean development and of peak arthropod densities in the U.S. References for most arthropods: 32, 66, 127, and 245.

response to soybean growth stage, or if instead they respond to other factors that merely coincide with a certain growth stage. Multiyear experiments, in which several soybean stages were present on each of several dates, have shown that neither soybean stage nor calendar date is always a reliable predictor of arthropod density.[33,34,40,41,79,160,182,184] For example, green cloverworm density peaked in August–

September regardless of growth stage in one experiment,[40] but peaked in growth stages R2 to R4 regardless of date in another experiment.[184] Similar differences between experiments were found for stink bugs.[33,34] Further, timing of velvetbean caterpillar peaks varied by 4 to 5 weeks from one year to the next; early peaks coincided with stages R3 to R4, while late peaks coincided with stages R5 to R7.[40] These examples indicate that additional factors besides date and soybean stage should be used to time sampling. The following two sections discuss other factors that can help time sampling.

b. Temperature or Degree-Day Accumulations

The use of degree-day accumulations for timing arthropod development was discussed in Chapter 7. Typically, the dates on which eggs were oviposited must be estimated, usually via monitoring. Then, daily temperature data are used to estimate when damaging stages will be present. Data on how temperature affects development rate are available for the velvetbean caterpillar,[185,186] green cloverworm,[187] corn earworm,[188-190] bean leaf beetle,[118,191] and seedcorn maggot.[153] Computerized predictive models can make degree-day models much easier to use. For example, temperature-driven development models for velvetbean caterpillar, corn earworm, and southern green stink bug have been incorporated into the SICM.[110,111] Similarly, the AUSIMM computer simulation includes temperature-driven models of velvetbean caterpillar, green cloverworm, soybean looper, bean leaf beetle, corn earworm, southern green stink bug, and green stink bug.[192] However, AUSIMM is easier to use for decision making than for timing sampling (see Section VI.C.1).

Other computer models use average (i.e., fixed) development rates to predict the timing of the southern green stink bug[149,151] and the Mexican bean beetle.[108,193] Alternatively, observational data on average development rates of field populations can be used to time sampling. For example, green cloverworm eggs were most abundant within 1 week of peak moth catches in light traps,[123] and larvae peaked 12 to 25 d later.[123,164] Similarly, density of corn earworm larvae peaked 8 to 14 d after peak oviposition.[44,147]

In addition to affecting development rate of immature stages, temperature may determine the timing of adult activity. In Illinois, a degree-day model gave better predictions than calendar date for timing of bean leaf beetle emergence from overwintering sites.[139] In North Carolina, overwintered bean leaf beetles began emerging when mean daily temperatures reached 26°C.[140]

c. Other Biological Considerations

i. Phenology of alternate food plants

The phenology of nonsoybean plants may determine when several important pests invade soybean. For example, both grasshoppers[127,135,145] and spider mites[129,132,135] feed in grassy or weedy noncrop areas early in the season. Similarly, the three-cornered alfalfa hopper invades soybean fields from nearby clover, vetch,[162] or weedy areas.[194] It may be possible to time sampling for these species according to when alternate food plants dry out or are cut. The same concept could apply to aphids, because soybean invasion by aphids is probably triggered by changes in food quality of their primary (woody) hosts.[195] Unfortunately, the primary hosts have not been identified for most aphid species.[195] Computer models have been developed to predict the phenology of alternate food plants and thereby predict the phenology of the corn earworm[188-190] and the southern green stink bug[149,151] (discussed in more detail in Section V.B).

ii. Phenology of arthropod prey in soybean

Predators may delay colonization of soybean until densities of their prey (plant-feeding arthropods) have begun to increase. For example, peak predator density coincided with or slightly followed the buildup of populations of small caterpillars.[41,183] Thus, any factors that directly affect the phenology of plant-feeding arthropods may indirectly affect predator phenology.

iii. Weather patterns

As discussed in Section V.A.1, certain synoptic weather patterns may promote arthropod migration from southern to northern latitudes. It may be possible to time sampling according to when these weather patterns occur.[124]

3. Density Differences within Each 24 h (Including Diel Variation)

As discussed in Chapter 4, daily shifts in location or behavior may make arthropods more accessible or susceptible to a sampling technique. For this reason, relative estimates of density may vary with time of day. If samples are taken at the time of day when counts tend to be highest, sample precision may be increased and the ability to detect low densities will be improved. However, practical considerations may limit options for timing sampling. For example, ground cloth sampling in a commercial program was restricted to daylight hours after the dew had dried and before sunset. At other times, dew on soybean plants made ground cloth sampling inefficient.[101]

Regular fluctuations of relative density over the course of a 24-h period are called "diel variation". Diel variation affects sampling most in species that are active only during either dark or light hours. For example, most noctuid moths are active only at night.[55,196,197] In contrast, grasshoppers are so inactive at night that they may remain motionless even when prodded with a finger. Nighttime sampling therefore provides precise absolute estimates of grasshopper density.[198] Timing of arthropod activity may vary with life stage. For example, pitfall catches of adult carabid beetles, *Calosoma sayi*, were highest at night, but larval catches were highest during the day.[199]

Some species exhibit regular fluctuations of relative density even during daylight hours. In Illinois, bean leaf beetle adults in sweepnet sampling were abundant at 0700, declined to a minimum at 1200, then increased sharply to a maximum at 1700.[200] In contrast, Smelser[156] caught significantly fewer bean leaf beetles via sweepnet sampling at 0830 and 1000 than at any other times, apparently because beetles did not enter the upper canopy until dew had evaporated. Counts peaked at 1130, and remained high during the afternoon.[156] For the three-cornered alfalfa hopper, sweep-net counts were higher in the afternoon than in the morning.[194] For minute pirate bugs and big-eyed bugs, both sweepnet samples and direct counts were significantly higher at 0700 than at 1300 or 1800.[181] In contrast, the foliage-inhabiting carabid beetle *Lebia analis* was less abundant at 1300 than at either 0700 or 1800.[81] For a given species, sweepnet and direct count techniques showed different within-day patterns of relative density.[181]

In addition to regular patterns of diel variation, density estimates may vary in response to less-regular fluctuations in sunlight intensity, relative humidity, wind speed, etc.[245] For example, although cutworm larvae usually feed above ground only at night, they may also emerge on cloudy days.[127] Similarly, sweepnet sample density of *Lebia analis* was significantly correlated with air relative humidity, but was not correlated with time of day per se.[181] Although it is difficult to time sampling to control for less-regular fluctuations, knowledge of such patterns may help in interpretation of sampling results.

B. WHERE TO SAMPLE: ARTHROPOD CONCENTRATIONS OVER SPACE

Because soybean is grown over a broad range of latitudes and ecological conditions within the U.S., it is not surprising that many soybean arthropods are consistently more abundant in some regions than in others.[66] In addition to differences *among* regions, many soybean arthropods show consistent patterns of density *within* regions and even within fields. By understanding spatial patterns at all scales (from a single field to the entire geographic range), sampling can be restricted only to locations likely to have the desired arthropod density. For example, in sampling pests for decision making, sampling typically would be concentrated in locations likely to have the *highest* pest densities. In contrast, sampling that was intended to achieve some other objective (such as determining dispersion pattern) might be concentrated where arthropod densities were lower.

1. Density Differences among Geographic Regions

Many soybean arthropods can overwinter only at southern latitudes. Further, warmer average temperatures and a longer growing season allow pests to develop more generations per year in southern states. For both these reasons, soybean arthropods are more likely to reach economic densities and, on average, cause greater losses in southern states (Figure 5). For example, in 1980 > 40% of the soybean

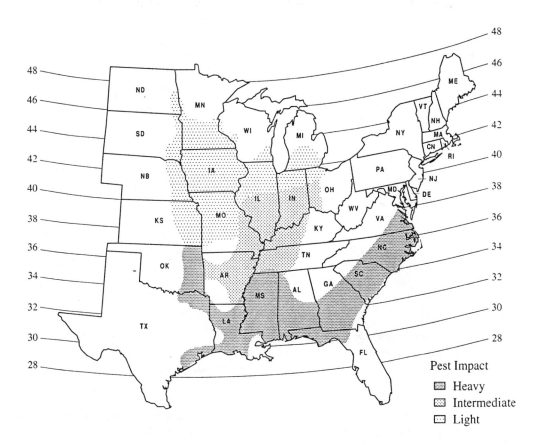

FIGURE 5. Zones of similar combined impact from soybean arthropod pests. Redrawn with permission from Kogan.[66]

hectarage in the Southeast region was treated for arthropods, while < 2% of the North Central hectarage was treated.[7] Similarly in 1982, over 3.2 million kg of insecticide active ingredient (AI) was applied to soybean in the Southeast region vs. approximately 76,200 kg AI in North Central states.[201]

In general, pest management interventions (including sampling) should be more profitable in regions where pest densities and damage are higher. When designing sampling programs, however, it is also necessary to consider how *predictable* is the damage. If arthropods cause economic damage every year, at most sampling will be needed to determine *when* economic densities will occur (see Chapter 7). In contrast, if arthropods cause economic damage only occasionally, sampling also must determine *whether* economic densities will occur in a given year. In other words, sampling program designers should consider both the mean and the variance of pest damage.

Pest density is probably more variable in northern than in southern soybean production regions. In Illinois, for example, conditions suitable for a pest outbreak may occur only in one year out of ten.[316] Occasional outbreaks nonetheless can be severe, such as the mite outbreak that affected over 5 million ha throughout the North Central region during the 1988 drought.[32]

Regions with intermediate pest impact may require the most sampling. In regions where at least one species causes economic damage nearly every year, it may be most profitable to time insecticide applications according to historical patterns of pest immigration instead of sampling.[110,111] At the other extreme, neither insecticide use nor sampling may be profitable in regions with extremely light pest impact. Indeed, Ignoffo et al. stated that, "...in most instances soybean growers in Missouri and probably throughout the Midwest do not need to do anything" about soybean arthropods.[154] Especially in regions with light pest impact, efforts should focus on predicting pest density so that only locations with the highest probability of pest outbreaks are sampled.[51]

2. Density Differences within Each Region

Regardless of geographic region, certain agroecosystem characteristics can be used to predict arthropod density and thereby reduce unnecessary sampling. "Through partitioning the crop by maturity group, planting dates, and row widths, the portion of the [hectareage] susceptible to a given pest at a specific time can be predicted and the monitoring efforts can be concentrated and restricted at these focal points in space."[52] The amount by which sampling can be reduced may vary from year to year, depending on the overall density of the pest. In years of heavy pest pressure, even fields that are only marginally favorable for the pest may suffer economic infestations.[13]

Some of the most useful factors for predicting density of soybean arthropods are summarized in Table 4. As discussed below, however, the effect of some factors can depend on weather or crop development.

a. Relative Maturity of Crop

The degree of maturity ("younger" or "older") of a soybean crop relative to other nearby soybean fields can be affected by variety maturity group, planting date, and other factors. Within the same region, fields planted early to early maturing varieties are likely to be more mature on a given date than late planted fields or late maturing varieties (see Section II.B.2). In other words, on a given date early planted fields are likely to be physiologically "old" while late planted fields are still "young". As shown in Table 4, differences in relative maturity among fields can affect arthropod density.

i. Mexican bean beetle

Early planted fields tend to have higher early season Mexican bean beetle populations, but the reason for this depends on geographic region. In southern latitudes, overwintered beetles invade and reproduce on snap and lima beans. First-generation beetles concentrate in early planted soybean fields because these fields have larger, more mature plants. Closed canopies reduce temperature and moisture stress experienced by larvae, and feeding on postbloom foliage increases beetle survival and egg production.[13,137] In contrast, in North Central states overwintered beetles invade soybean fields as soon as seedlings emerge. Overwintered beetles concentrate in early planted soybean fields because these fields are the first to have soybean seedlings available.[32,55,127,160] Late in the season, when early planted fields reach harvest maturity, beetles may migrate into younger (later planted) fields.[160]

ii. Bean leaf beetle

Because overwintered adults invade soybean fields as soon as seedlings emerge, early planted fields initially have higher beetle densities.[33,41,55,210] Populations may remain higher in early planted fields even during later beetle generations, resulting in greater total damage in early planted fields.[75] However, once pods begin to dry in early planted fields, beetles may migrate to late planted or late maturing fields that still have green pods.[41,157]

iii. Stink bugs

In general, stink bug densities are highest in older fields early in the season, but highest in younger fields later in the season. Stink bugs are highly mobile,[221] and local migrations to more preferred plant species and stages are common even when food is abundant.[33,34,149] Adults may invade and oviposit in early planted fields even while they are still in vegetative stages.[34,52] However, stink bugs are most attracted to soybean that has begun to set pods.[33,34,41,116,135,145] Early in the season, the oldest fields (early plantings of early maturing varieties) are most likely to have pods, and therefore likely to have the highest stink bug densities.[34,145]

The highest stink bug densities of the season usually occur in September.[33,41,51,52,168] By that time, pods in older fields have begun to turn brown, and stink bugs often migrate to nearby fields that still have young, green pods.[34,149] The highest stink bug densities of the season are therefore found in late planted fields[34] or late maturing varieties.[168] Indeed, early maturing varieties can escape most stink bug damage if their pods mature before peak stink bug populations occur.[222,223] However, if early maturing varieties are planted early, early season stink bug infestations can reduce yields.[168]

Computer simulation models are available for predicting the relative density of stink bugs among soybean fields that vary in maturity.[149,151]

iv. Corn earworm

As discussed in Section V.A.1, densities of corn earworm larvae are often highest in soybean fields that were flowering and that had open canopies (i.e., plants in adjacent rows not touching) during peak moth flight.[42,141-145] In the past, older fields (early plantings of early maturing varieties) usually flowered and produced closed canopies before peak moth flight. For this reason, corn earworm densities were usually highest in younger fields.[33,44,52,137,141]

TABLE 4
Some Within-Region Factors that Favor High Arthropod Abundance

Guild and common name	Relative maturity of crop[a]	Spacing between rows	Tillage	Plants in adjacent fields	Insecticide use[b]
DEFOLIATORS					
Velvetbean caterpillar	Young 22, 40, 52	Narrow 40		Noncrop areas 203	When VBC present 203
Soybean looper	Young 171, 23, 52	Narrow 23		Cotton 51, 116, 204, 173 Noncrop areas 203	Before loopers present 51 When loopers present 171, 203
Green cloverworm	Young 57, 40, 23 Old 22, 79, 356	Narrow 40, 43, 23, 22	Conventional 184 No effect 206	Noncrop areas 203	When GCW present 203
Mexican bean beetle	Old (early planting may have direct effect; see text) 13, 33, 41 but may migrate to young fields late in the season 160		Conventional 160 No effect 33	Garden beans 137	
Grasshoppers			Reduced 160, 206	Grass, weeds 135	
LEAF SURFACE GRAZERS					
Spider mites				Grass, weeds, or alfalfa 67, 127, 132, 207	Early season 207, 166
PLANT FLUID REMOVERS					
Aphids		Narrow 43			

Sampling Programs for Soybean Arthropods   **567**

Pest	Planting date	Row spacing	Tillage	Other cultural	Alternative hosts	Timing
Potato leafhopper			Conventional 23, 43		Alfalfa 209	
POD FEEDERS						
Bean leaf beetle	Early planted 33, 210, 40, 155, 55, 75, but may migrate to young fields at the end of the season 41, 157	Narrow 41 Variable results 33	Conventional 206, 208	Conventional 211, 212 Reduced 208	Woodlands 139, 213 Uncultivated fields 140 Alfalfa 156, 155	
Stink bugs (Southern green and Green)	Old early in the season, but young later in the season; see text 33, 34, 52, 145, 168				Dogwood or elderberry trees 55, 135, 168; Clover, corn cowpeas, small grain 33, 149	
Corn earworm (= bollworm, tomato fruitworm)	Depends on corn maturity and weather; see text	Conventional 145, 47, 44, 146, 42 Narrow 23; No effect 214, 40		Conventional 215		Before CEW present 216, 146
Stem feeders						
Three-cornered alfalfa hopper				Reduced 211	Clover, vetch 162 or weeds 194 Woodlands 213	Early season 217

TABLE 4 (continued)
Some Within-Region Factors that Favor High Arthropod Abundance

Guild and common name	Relative maturity of crop[a]	Spacing between rows	Tillage	Plants in adjacent fields	Insecticide use[b]
Dectes stem borer	Early planted 41		Reduced 52		
UNDERGROUND PESTS					
Seed corn maggot	Early planted 153, 32		Reduced 22		
			Disking of fresh residue or manure 20, 76, 127, 218		
			Variable results 219		
Lesser corn stalk borer			Reduced 22, 76		
FOLIAGE-INHABITING PREDATORS					
	Old 23, 22, 182, 214, 41	Narrow 43, 23, 141, 41, 78, 42	Conventional 220	Uncultivated areas 25	
	Young 41		Reduced 78		
			No effect 206		
GROUND-INHABITING PREDATORS					
			Reduced 215, 212, 178, 76	Uncultivated areas 25	Before earwigs present 179

[a] Definition: the degree of maturity (young or old) of the soybean crop in a field relative to other soybean fields in the region. Relative maturity can be affected by cultivar maturity group, planting date, and other factors. Within the same region, fields planted early to early maturing cultivars are likely to be more mature on a given date than late planted fields or late maturing cultivars. The effect of planting date per se on arthropod abundance is discussed in Section V.B.2.

[b] CEW = Corn earworm. GCW = Green cloverworm. VBC = Velvetbean caterpillar.

Recent changes in production systems may have shifted peak corn earworm density from younger to older soybean fields. In South Carolina during 1980 to 1983, older fields consistently had higher corn earworm populations than younger fields.[51] Indeed, Kogan and Turnipseed[53] stated that as of 1987, the only soybean at high risk from corn earworm in South Carolina are early plantings of early maturing varieties. Kogan and Turnipseed[53] suggested that the apparent shift in peak corn earworm density from young to old fields is due both to increased mid-July infestation and decreased mid-August survival. During the 1970s, southern growers switched from long season to early maturing corn varieties that are not attractive (have begun to mature) during mid-July moth flights. In the absence of long season corn, moths oviposit in old soybean fields, which usually flower in mid-July. The next peak moth flight (in early August) presumably oviposits in young soybean fields, which usually flower in early August. However, pathogens apparently kill a high proportion of corn earworm larvae in August. Thus in recent years, peak corn earworm densities have occurred in older soybean fields in July.[53]

Predictions about which soybean fields were likely to have highest corn earworm density could increase sampling efficiency. However, obviously it would be difficult for a sampler to keep track of the complex relationships among relative soybean maturity, relative corn maturity, weather, and corn earworm density. Fortunately, researchers have developed computer models to predict relative corn earworm densities.[188-190] One of these models, HELSIM-2, successfully predicted earlier moth flights in hot, dry years.[13] However, each model may need to be modified to account for the earlier maturity of newer corn varieties.[53]

v. Other species

Early planted fields attract and support the largest populations of soybean nodule fly.[51] Thrips densities were twice as high in early planted as in late planted fields.[79] Percent parasitization of the corn earworm by *Microplitis croceipes* (Cresson) was significantly higher in fields planted after June 10 than in earlier planted fields.[83]

b. Soybean Variety

As was discussed above, "one of the most important considerations in variety selection...is phenology of the plant and that of the insect population."[32] Another variety characteristic that would be expected to influence relative arthropod abundance is resistance. Specifically, varieties that were resistant (unattractive or toxic) to pests would be likely to have lower arthropod densities. Nearly all commercial soybean varieties are somewhat resistant to potato leafhopper due to trichomes (hairs) on their leaves. Glabrous (hairless) varieties are likely to have high potato leafhopper densities.[21,127] Although several soybean genotypes are resistant to additional pests,[32,171,224] 'Crockett' is currently the only soybean variety in commercial production with resistance to arthropods other than potato leafhopper.[32,225] Thus, aside from the impact of variety on soybean crop maturity, little is known about how commercial soybean varieties affect arthropod density.

c. Other Plants in Soybean Field (Weeds or Intercropped Species)

In general, predators tend to be more abundant and herbivores tend to be less abundant in weedy vs. weed-free soybean fields.[22,42,217,226] One exception is armyworms and cutworms, which tend to be more abundant in fields where grass weeds are present. Larvae first eat the grasses, then move to soybean.[22,127,145] Strip cropping soybean with alfalfa increased soybean populations of thrips, green cloverworm, and spiders relative to a soybean monoculture.[79] Relay intercropping soybean into small grain protected even a glabrous soybean variety from potato leafhopper until the small grain was harvested.[21]

d. Previous Crop in Soybean Field

Seed corn maggot,[219] Japanese beetle, and damsel bugs (predators)[206] were more abundant in soybean planted after soybean than in soybean where the previous crop had been corn. Soybean planted into sod or set-aside areas often has increased densities of white grubs or wireworms.[127] In an experiment that included row spacing, planting date, and tillage, the most important factor affecting arthropod density was whether soybean was or was not planted following barley. Some arthropods were more abundant, and others were less abundant, in soybean that followed barley.[33]

e. Insecticide Use

Densities of several pests can increase when insecticides are used early in the season (Table 4), presumably because the insecticides kill natural enemies such as predators[79,182] that suppress pest populations. Surprisingly, insecticide applications had little effect on an egg parasitoid,[227] a larval parasitoid,[90] or total number of parasitoids.[79]

f. Irrigation

Densities of bean leaf beetle, green cloverworm, soybean looper, three-cornered alfalfa hopper, and damsel bugs (predators) were usually higher in irrigated soybean.[228] In contrast, corn earworm and big-eyed bugs were equally or more abundant in dryland soybean.[228]

g. Soil Type

In Illinois, bean leaf beetle adult densities were generally higher in fields where soil had low clay content.[200] Other research suggests that organic matter content is more important than clay content per se. In North Carolina, bean leaf beetle is usually most abundant in fields with high moisture-holding capacity soils,[221] including those with high organic matter content.[47] In greenhouse experiments, bean leaf beetle oviposition rates and larval survival rates were highest in organic soils, lower in sandy clay loam, and lowest in loamy sand.[229,230] For the Japanese beetle, a computer model has been developed to predict relative density of eggs in fields of various soil types.[231]

3. Density Patterns within Each Field

Between-region and within-region patterns of density (discussed above) are most useful for deciding which soybean fields should be sampled. Once the decision has been made to sample a specific field, the next decision is which portions of the field should be sampled. It may be possible to increase sampling efficiency by concentrating sampling in those portions of the field where arthropods are likely to be most abundant. For example, it may be possible to determine whether a pest has exceeded its ET by sampling only the parts of the field where pest densities are highest. Such an approach would greatly increase sampling efficiency. In contrast, if the goal of sampling is to estimate the mean density throughout the field, the positions of sampling units should be stratified according to expected patterns of density (see Chapter 4). Alternatively, samplers can be instructed to avoid the areas of the field that are known to have arthropod densities that greatly exceed the field average.[44,127,221]

a. Field Edges

Some soybean arthropods are most abundant near field edges, especially those edges closest to plant species used by the arthropods early in the season.

i. Spider mites

Mites overwinter in alfalfa or clover fields or weedy areas.[132,135] As these areas mature or are cut, mites can move into adjacent soybean. Thus, mite infestations in soybean usually begin at field edges next to alfalfa, clover, or weedy areas.[55,57,129,132,133,135,207] If sampling fails to detect economic mite densities in field edges, it is probably unnecessary to sample the rest of the field. However, the entire field should be sampled once mites are discovered in field edges.[57,129,132,133,207]

ii. Three-cornered alfalfa hopper

This pest invades soybean fields from nearby clover, vetch,[162] or weedy areas.[194] Nonetheless, the type of adjacent vegetation did not affect the number of adults caught on sticky traps at different edges of a soybean field.[161] Further, egg sampling suggests that females disperse throughout a soybean field before ovipositing.[232] Thus, it is probably necessary to sample the entire field for this species.

iii. Other arthropods

Stink bugs first invade the margins of soybean fields. Stink bug densities were often highest in borders closest to wooded or weedy areas.[127,145,168] Early in the season, spider densities in the first 20 edge rows were three times those of the field interior.[25] Corn earworm densities likewise tend to be higher along field margins.[44,221] Grasshoppers tend to oviposit in grassy noncrop areas, and nymphs feed in these areas early in the season. As vegetation dries out or is mowed, nymphs begin invading soybean. Thus, grasshopper populations are usually concentrated (at least initially) at the edges of soybean fields.[127,135,145] An exception is no-till soybean planted into small-grain stubble. Grasshoppers may have oviposited throughout the small grain field, resulting in roughly uniform nymph densities throughout the soybean crop.[127,160]

b. Weedy Areas within Soybean Field

Cutworm densities are often highest in wet, weedy areas within soybean fields.[127] Mite infestations may begin in grassy areas within a soybean field.[129]

c. Within-Row and Within-Plant Patterns

Many arthropods are more abundant at certain positions on the soybean plant, or in the soil at certain horizontal distances from the plant. Sampling efficiency can be increased by concentrating sampling at these locations. For major pests, specific locations of each life stage are summarized in Table 5. Data on within-plant distribution are available for other soybean arthropods,[2] including potato leafhopper,[234] spiders,[25] and damsel and big-eyed bugs.[235,236]

VI. HOW TO SAMPLE: EXISTING SAMPLING TECHNIQUES AND PROGRAMS

The most important factor determining the design of a sampling program is the sampling objective. Each decision about the sampling program (technique, sampling unit, timing of samples, etc.) is affected more by the objective of the program than by the arthropod or crop for which the program will be used. In particular, the objective

TABLE 5
Within-Row and Within-Plant Locations of Major Soybean Pests

Guild and common name	Stage	Location	Ref.
DEFOLIATORS			
Velvetbean caterpillar	Egg	Laid singly on undersides of leaves and leaf petioles, close to leaf surface between plant trichomes.	237
	Larvae	Upper half to upper third of canopy; early instars on undersides of leaves.	237, 145, 51, 126
	Pupa	In soil. Most less than 2.5 cm below surface; most within 30 cm horizontally from soybean plant.	51, 237, 238
	Adult	Day: roosting in soybean canopy. Night: flying near soybean.	237
Soybean looper	Egg	Undersides of leaves; preference for canopy stratum unknown.	196
	Larvae	Lower half to two thirds of canopy on leaf undersides.	196, 171
	Pupa	In silken cocoon attached to soybean plant.	51, 196
	Adult	Day: roosting in soybean canopy. Night: flying to nearby nectar sources.	196
Green cloverworm	Egg	Laid singly on leaves; slight preference for upper surface; concentrated in middle third of canopy.	57, 239
	Larvae	Undersides of leaves. Young instars show no stratum preference, but 5th and 6th instars are concentrated in upper third of canopy.	240, 32, 55
	Pupa	In canopy (middle third), or at base of plant on soil surface or in leaf litter. Surrounded by cocoon.	240, 32, 51, 241
	Adult	Day: roosting on undersides of soybean leaves or on building walls. Beginning at dusk: flying near soybean.	197, 55, 242
Mexican bean beetle	Egg	Undersides of lower leaves in clumps of 40–60.	243, 244
	Larvae	Undersides of leaves (no data on stratum preference).	245
	Pupa	Undersides of lower leaves, or nearby weeds.	243, 32, 246
	Adult	Undersides of leaves.	As for larvae
LEAF SURFACE GRAZERS			
Spider mites	All	Undersurfaces of lower leaves preferred, but at high densities entire plant becomes infested. Colonies protected by thin layer of webbing.	247, 133, 132
POD FEEDERS			
Bean leaf beetle	Egg	In soil; less than 3.8 cm deep, and most < 2.5 cm horizontally from a soybean plant.	248
	Larvae	In soil; less than 7.6 cm deep, and most within 18 cm horizontally from a soybean plant.	249, 250
	Pupa	As for larvae.	249, 250
	Adult	Above ground (usually on plant); stratum preference apparently changes with time of day.	156, 200
Stink bugs	Egg	In clumps of 10–30. Most clumps on leaf undersides; some on upper leaf surface, stem, pods.	57, 55, 251
	Nymph	First 2 stages: clustered on or near egg mass. Last 3 stages: most on pods, also on the growing shoot.	251
	Adult	Most on pods, also on the growing shoot.	251
Corn earworm	Egg	Laid singly. Most on undersides of fully expanded trifoliate leaves; preference for upper 2/3 of canopy.	144, 143
	Larvae	Small larvae mostly on youngest (rolled) leaves, or flowers if available; medium and large larvae mostly on expanded leaves, also flowers and pods if available.	252

TABLE 5 (continued)
Within-Row and Within-Plant Locations of Major Soybean Pests

Guild and common name	Stage	Location	Ref.
	Pupa	In soil, within 15 cm of surface. Horizontal distance from soybean plant depends on row spacing.	13, 238
	Adult	Day: roosting in canopy. Active at night.	13
STEM FEEDERS			
Three-cornered alfalfa hopper	Egg	Clumps (up to 17) inserted in slits in stem just above leaf petioles; preferred height increases as plant grows.	232, 162
	Nymph	Main stem early in season; as plant develops, prefer leaf petioles and then stems of lateral branches.	253
	Adult	Feed in canopy. Fly at altitudes below 33 cm.	194, 161

of a sampling program determines the options available for increasing sampling efficiency. As discussed in Section IV, many methods for improving efficiency can be used only for a specific sampling objective. For this reason, the following section is organized by sampling objective rather than by arthropod species. Within each objective, sampling programs and plans are grouped according to efficiency. This structure duplicates the order of decisions that must be made when designing a sampling program. First, what is the objective of the program? Then, what categories of sampling plans or techniques are relevant to the objective? Next, has a relevant sampling plan or technique been developed for the specific arthropod? If not, does the need for efficiency justify investing time and labor to develop such a plan or technique (Section IV)?

Kogan and Herzog's 1980 reference, *Sampling Methods in Soybean Entomology*,[2] remains the preeminent guide to sampling techniques and plans for soybean arthropods. Anyone planning a soybean sampling program should consult it. However, both techniques and plans have improved considerably in the last decade, providing several new and exciting options to soybean researchers, scouts, and growers. These improvements are the result of a greater understanding of pest impact and biology, as well as advances in sampling theory.

A. SAMPLING TO DETERMINE ARTHROPOD DENSITY

Both researchers and pest managers frequently require programs for determining arthropod density. In pest management applications, growers and scouts primarily are concerned with pest population density and subsequent economic damage. Although the sampling efficiency of density-determining techniques is often low, these techniques are essential to the development of pest management programs. This is because all relatively advanced sampling plans, such as sequential count and sequential sampling, rely on a foundation of research data collected with density-determining techniques.

1. Absolute Techniques

Absolute density, defined as arthropods per ground area,[254] is estimated by collecting all individuals within a given area (see Chapter 1). Table 6 summarizes the utility of absolute techniques for selected soybean arthropod guilds. Sampling techniques that produce absolute estimates are available for most major pest species.

TABLE 6
Absolute Sampling Techniques Appropriate for Selected Guilds of Soybean Arthropods

Feeding guild	Direct counts	Suction nets	Fumigation cages	Plant removal	Flotation / sieving of soil cores	Emergence cages
DEFOLIATORS						
Lepidopteran (SL, VBC, GCW)	+		+	+		+
Coleopteran (BLB, MBB)	+		+	+		+
Orthopteran (grasshoppers)				At night[a]		
LEAF SURFACE GRAZERS (mites, thrips)	+			+		
PLANT FLUID REMOVERS (PLH, aphids)	+	+		+		
POD FEEDERS (BLB, SB, CEW)	+		+	+		+
STEM FEEDERS (Dectes, TCAH)	+		TCAH	+		Dectes
UNDERGROUND FEEDERS (BLB larvae, nodule fly larvae, white grubs, SCM larvae)					+	+
FOLIAR PREDATORS	+		+	+		
GROUND PREDATORS					+	
PARASITOIDS		+		+		

[a] Night sampling gives best absolute estimates for grasshoppers.[198]

Programs may include the use of direct observation,[197,255,256] emergence traps,[20,140,219,257,258] fumigation chambers or cages,[64,259,260] plant removal,[198,255,261-264] or soil sampling.[265-268]

Usually, absolute density measurements require bulky equipment and considerable effort, time, and money. In addition, several absolute devices do not result in immediately useful data because samples must be sorted in the laboratory. For these reasons, the sampling efficiency of absolute techniques usually is very low. This makes programs using absolute density techniques undesirable for most pest management applications. In contrast, the precision of these programs makes them indispensable for in-depth population research, including life table, population dynamics, sampling-program design, and plant-response studies.

Arthropod intensity (individuals per unit of available food supply) also can be considered an absolute measurement of density (see Chapter 1). The most common intensity measurements are arthropods per plant or arthropods per cm^2 of leaf area. Mayse et al.[264] used direct observation to determine population intensities of several soybean arthropods. Population intensity programs have been developed for sampling thrips,[269,270] three-cornered alfalfa hopper,[271] aphids,[272] and spider mites.[130,131] With adjustments for plant stand or leaf area, intensity data easily can be converted to absolute densities (arthropods per ground area). Absolute intensity programs are used more frequently in pest management than their density counterparts.

2. Relative Techniques

Relative techniques produce data expressed in terms of arthropod count per unit effort (e.g., bean leaf beetles per sweep) or count per trapping period. Such techniques provide information about pest densities with less time and effort than absolute techniques.[273] Data collected with relative sampling techniques can be compared across fields, across dates, or to an established threshold as long as sampler and device differences are taken into account. When relative estimates are sufficient for the objectives of the sampling program, using relative techniques can increase the efficiency of a sampling program considerably. Programs using relative techniques are available for most key soybean pests and usually provide sufficient information for management decisions.

In addition, when percent capture is known for a relative technique, relative data can be converted to estimates of absolute density.[182,198,274,275] This approach is useful for determining whether populations have exceeded a threshold expressed in terms of absolute density. Determining the mathematical relationship between absolute and relative data may require substantial research effort (see Chapter 1).

Table 7 summarizes the utility of various relative techniques for selected soybean arthropod guilds. The sweepnet is the most commonly used sampling device in row crop pest management.[273,276] Sweepnets are highly portable, simple and economical to operate, and capture a greater variety of arthropod species than do other sampling devices. Hillhouse and Pitre[278] found sweeps across one or two rows, depending upon the pest species sampled, to produce data with high relative net precision (high precision and low sampling costs). However, their study showed that for intensive sampling, the ground cloth technique was preferable on the basis of greater precision. In sampling green cloverworm larvae, Pedigo[259] determined that sweepnet samples had a high level of precision and closely followed absolute densities. Drawbacks of sampling with sweepnets have been noted by several authors, including variation due to climatic conditions, diurnal activity patterns, sampler effort, and plant architectural characteristics.[27,64,273,275,278] Sparks[72] also has shown that sweepnets ineffectively sample three-cornered alfalfa hopper nymphs.

Ground cloths also are used frequently to make relative-density estimates of arthropods in soybean. In comparison to sweepnets, ground cloths are easier to construct and are less susceptible to differences in sampler technique. Ground cloths are most effective for sampling less mobile arthropods such as lepidopterous larvae. For some species, ground cloths capture nearly all individuals in the canopy, with counts not significantly different from those of absolute techniques.[278] Hammond and Pedigo[279] reported up to 90% capture of green cloverworm larvae using ground cloths. For other species, notably the three-cornered alfalfa hopper, ground cloths may fail to capture some life stages.[194,271,277] Rudd and Jensen,[280] on the basis of an economic analysis, determined that ground cloths are less desirable than sweepnets in sampling foliar arthropods. Ground cloths also are considerably less effective in narrow-rowed or drilled soybean.[55,261,276,281] In addition, ground cloths may become less efficient in comparison to sweepnets under high pest densities.[282]

Many researchers have modified the basic ground cloth design for more rapid sampling of a wide variety of soybean arthropods. Herbert and Harper[283] developed a simple technique for research programs that rapidly collects samples, although laboratory analysis of samples is required. Felland and Pitre[284] developed a plant-shake technique that replaced the ground cloth with a polyethylene funnel. A recently developed alternative device for sampling narrow-rowed or drilled soybean is the rigid (vertical) beat sheet.[146,261,285] This device is held in a partially vertical position,

TABLE 7
Relative Sampling Techniques Appropriate for Selected Guilds of Soybean Arthropods

Feeding guild[a]	Ground cloth	Sweep-net	Sticky traps	Light traps	Pheromone traps	Food baits	Pitfall traps	Berlese soil extraction	Population indices[b]
DEFOLIATORS									
Lepidopteran: (SL, VBC, GCW)	Larvae	+	Adults	Adults	Adults				+
Coleopteran: (BLB, MBB)	+	+	Adults						+
Orthopteran (grasshoppers)		+							+
LEAF SURFACE GRAZERS									
(mites, thrips)			Thrips	Adult					+
PLANT FLUID REMOVERS									
(PLH, aphids)		Adults	Adults	Winged adults	Some				+
POD FEEDERS									
(BLB, SB, CEW)	+	+	Adults	Some adults					+
STEM FEEDERS									
(Dectes, TCAH)	Dectes Adults[c]	Adults[d]							+
UNDERGROUND FEEDERS									
(BLB larvae, nodule fly larvae, white grubs, SCM larvae)	Some adults	Adults	Some adults	Some adults		Some		Larvae	+
FOLIAR PREDATORS	+	+	Adults	Adults					Web-spinning spiders
GROUND PREDATORS							+	+	
PARASITOIDS		+				Use hosts as bait			Some

[a] TCAH = three-cornered alfalfa hopper. BLB = bean leaf beetle. SB = stink bugs. CEW = corn earworm. SL = soybean looper. VBC = velvetbean caterpillar. GCW = green cloverworm. MBB = Mexican bean beetle. SCM = seedcorn maggot.
[b] Population indices indirectly sample arthropod populations and are not true relative techniques. They are included here for completeness.
[c] Ground cloth not suitable for TCAH.[194,271,277]
[d] Sweepnet not suitable for TCAH nymphs.[72]

reducing the space requirements between rows during sampling (Figure 6). Although bulkier than the conventional ground cloth, with additional research and development the rigid beat sheet may have increased utility.[285]

Fewer relative sampling techniques are available for below-ground than above-ground arthropods. Anderson and Waldbauer,[249] however, have developed a relative sampling technique for bean leaf beetle larvae and determined calibration ratios for conversion to absolute values.

Additional relative techniques include pheromone traps,[163,256] sticky traps,[286] pan traps,[286,287] blacklight traps, and pitfall traps.[178] Pheromone and blacklight traps play key roles in monitoring regional infestation patterns of major pests in southern states, such as corn earworm[44,51,146,147] and velvetbean caterpillar.[126,288] Predictions of larval populations and damage potential are based on trap catches of adults and computerized models of pest population dynamics and soybean development/yield. Predictions from computer models are delivered to growers and scouts via conventional or computerized[51] cooperative extension bulletins and advise growers and scouts of potential pest problems at the local or county level. Information from a regional

FIGURE 6. Vertical beat sheet for sampling narrow-row soybean. Arthropods fall into the trough at the bottom of the beat sheet. Photographs courtesy of B. M. Drees and M. E. Rice.

network of surveillance traps allows concentration of sampling effort at times and locations with the greatest risk of economic pest populations.

3. Fixed-Precision Sampling Plans

Fixed-precision sampling plans, also known as sequential estimation, allow the sampler to collect data until the population density can be estimated with a desired level of precision.[289,290] After each sample, precision is calculated using a known relationship between density and sample variance. Therefore, collection of samples in excess of what is required for a satisfactory density estimate is avoided.[239,291]

Sampling programs based on sequential estimation of population density have been developed for several major pests (see Table 8). Sequential estimation programs can be adjusted for changes in plant phenology.[292] Although most sequential estimation plans have been developed for use in ecological research, some have pest management applications. Based on a sampling efficiency study by Funderburk and Mack,[293] sequential estimation of three-cornered alfalfa hopper at the 25% precision level potentially is useful for pest management decision making. Also, Bechinski et al.[291] have developed plans for sequential estimation of green cloverworm in research or pest management programs.

4. Population Indices

As soybean arthropods develop, feed, and reproduce, they leave behind evidence of their activity. These signs include plant injury, webbing, frass, and exuviae, and may be easier to sample than the actual arthropods. Sampling techniques that evaluate signs of arthropod activity are called population indices (or indirect counts) and are

TABLE 8
Progress Toward Sequential Sampling and Fixed Precision Plans

Guild and common name	Stage	Sampling technique	Distribution[a] on which sampling plans are based (Ref.)	Ref. for seq. sampling plan if developed	Ref. for fixed precision plan if developed
DEFOLIATORS					
Velvetbean caterpillar	Larva	Beat cloth	Neg. binomial (3, 305)	305	—
	Larva	Beat cloth	Poisson (306, 307)	307, 237	237
	Larva	Sweepnet	Poisson (237)	237	237
Soybean looper	Larva	Beat cloth	Poisson (306)	196	—
	Larva	Sweepnet	Poisson (196)	196	—
Green cloverworm	Egg	Direct count	Poisson (239)	—	239
	Larva	Beat cloth	Clumped TPL (291)	291	291
	Larva	Beat cloth	Poisson (279, 306)	279	—
	Pupa	Direct count	Clumped TPL (308)	—	308
	Adult	Flushing	Clumped TPL (197)	—	197
Mexican bean beetle	Larva	Beat cloth	Neyman's type A, Neg. binomial (306)	—	—
	Adult	Beat cloth	Neyman's type A, Neg. binomial (306)	—	—
	Defoliation (plant injury)	Visual estimate	Binomial (298)	298	—
LEAF SURFACE GRAZERS					
Spider mites	All	Direct count	Clumped TPL[b] (120)	—	120[b]
Thrips	Nymphs and adults	Direct count or sticky traps	Clumped TPL (309)	—	—
PLANT FLUID REMOVERS					
Potato leafhopper	Nymph	Plant removal	Neg. binomial (310)	—	—
	Adult	Vacuum net	Clumped TPL (310)	—	—
Aphids	Adult, winged	Sticky trap	Clumped TPL (195)	—	—
POD FEEDERS					
Bean leaf beetle	Egg	Soil sieving	Neg. binomial (311)	311	—
	Adult	Sweepnet	Neg. binomial and clumped TPL (292, 200, 312)	292	292, 312
	Adult	Beat cloth	Neg. binomial and clumped TPL (200)	311	—
	Adult	Direct count	Neg. binomial and clumped TPL (200)	311	—
	Injury to soybean pods	Plant removal	Poisson (312)	—	312
Stink bugs	Nymph or adult	Beat cloth	Neg. binomial (251)	251	251
	Nymph or adult	Sweepnet	Neg. binomial (251)	251	251
Corn earworm	Egg	Plant removal	Clumped TPL (313)	—	313
	Larva	Beat cloth	Poisson (306)	44	—

TABLE 8 (continued)
Progress Toward Sequential Sampling and Fixed Precision Plans

Guild and common name	Stage	Sampling technique	Distribution[a] on which sampling plans are based (Ref.)	Ref. for seq. sampling plan if developed	Ref. for fixed precision plan if developed
STEM FEEDERS					
Three-cornered alfalfa hopper	Egg	Plant removal	Neg. binomial (232)	—	—
	Nymph, both generations	Beat cloth	Neg. binomial (277)	277	—
	Nymph, both generations	Beat cloth	Clumped TPL (293)	—	—
	Nymph, 1st generation	Beat net	Neg. binomial (277)	277	—
	Nymph, 2nd generation	Beat net	Poisson (277)	277	—
	Adult	Sweepnet	Neg. binomial (277)	277	—
FOLIAGE-INHABITING PREDATORS					
Damsel bugs	Nymph or adult	Beat cloth	Poisson (314)	314	—
	Nymph	Beat cloth	Clumped TPL (315, 316)	—	315
	Adult	Beat cloth	Random TPL (315, 316)	—	315
Big-eyed bugs	Nymph or adult	Beat cloth	Poisson (314)	314	—
	Nymph	Beat cloth	Clumped TPL (317)	—	315
	Adult	Beat cloth	Random TPL (317)	—	—
Minute pirate bug	Nymph	Beat cloth	Clumped TPL (315)	—	315
	Adult	Beat cloth	Clumped TPL (315)	—	315

[a] TPL = Taylor's Power Law.
[b] Mite distribution may not apply to soybean (derived from several other plant species).

useful for a wide variety of feeding guilds (Table 6). Population indices easily sum the combined effects of multiple pest species with similar feeding habits. For example, although the density of each species in a feeding guild is subeconomic, the effect of the entire guild may cause economic damage.

Replacing arthropod counts with population indices can result in significantly increased sampling efficiency. For example, spider mites are extremely difficult to count in the field. Therefore, most sampling programs evaluate plant injury from mite feeding, with only a visual confirmation of active, feeding mites.[129,132,133,294-297] Population indices also are useful in evaluating defoliator populations. In Indiana, thresholds expressed in percent defoliation are recommended for Mexican bean beetle and bean leaf beetle injury.[128,246,298] Defoliation ratings also have been incorporated into a computerized management program developed in Alabama.[192,299]

Untrained samplers usually overestimate percent defoliation. With experience, defoliation estimates are rapid and repeatable, although defoliation intervals narrower than 10% become increasingly difficult to differentiate.[45,273] The ability to differentiate between defoliation levels may be improved by providing scouts with pictorial field guides to evaluating defoliation levels. Similarly, Witkowski[300] has developed a hand-held instrument marked with a grid-shaped pattern. With this system, the number of squares visible through chewed portions of the leaf are counted and used to estimate percent defoliation.

If yield/damage/injury-rating relationships are known with adequate precision, population-index data can predict yield potential accurately. However, it is important to note the difference between predicting yield potential (i.e., assuming no additional injury) and predicting potential yield loss (i.e., incorporating present and predictions of future injury). Population indices usually predict yield potential from a single point in time. For example, it is difficult to differentiate between 15% defoliation by second instars and 15% defoliation by mature larvae with a population index. In contrast, pest density data often can predict future as well as past yield losses. This is especially true for age and stage-specific sampling or with the calculation of injury equivalents.[150] This limitation must be kept in mind when developing pest management programs.

B. SAMPLING TO DETERMINE POPULATION DISPERSION PATTERN

The dispersion pattern of a population is the result of a combination of behavioral characteristics, population density, resource availability, and other factors.[254] The level of aggregation, or clumping, of individuals in a population must be determined when developing sampling plans such as sequential sampling. Herzog[3] correctly points out that for a given species, estimates of aggregation can vary depending on sampling unit size, sample location, etc. Reviews of how to collect and apply dispersion data have been published by Southwood,[254] Ruesink,[301] and Ruesink and Kogan.[273] Dispersion patterns for principal soybean arthropods are listed in Table 8. Dispersion patterns have also been determined for white grubs,[302] soybean agromyzid fly,[303] adult lesser cornstalk borer,[256] and general soil arthropods.[265] In some cases, dispersion data from one species may be applicable to a second, closely-related species.[120]

C. SAMPLING TO CLASSIFY ARTHROPOD DENSITY
1. Economic Injury Levels and Economic Thresholds

One of the keystones of pest management is the use of economic injury levels (EILs) and ETs to aid in decision making. Economic thresholds and EILs incorporate feeding rates, potential yield loss, crop value, and cost of control tactics to determine if control measures are economically justified.[5] Using 1991 data, Hammond[32] determined that approximately 3 to 6% loss of potential soybean yield is required to offset the costs of an insecticidal application. Such calculated thresholds are a significant refinement over subjective, nominal thresholds often based solely on the sampler's "impression" of plant yield status.[318] Table 9 summarizes current ETs for most major soybean pests. Thresholds are expressed as absolute density (arthropods per row meter), relative density (arthropods per sweep), or plant injury rating. Because thresholds are substantially affected by several regional and economic variables, we recommend consulting the local Cooperative Extension office for precise thresholds.

Economic thresholds (and sampling plans that are based on them) are usually calculated for a single arthropod species (Table 9). However, soybean yield and profit are affected by the combined injury from all phytophagous species.[61,323,324] The combined damage from all species may justify treatment even if each species is slightly below its ET.[51] Multispecies ETs would therefore improve soybean pest management.

The flexibility and precision of pest management programs can be increased considerably by using computer models to assist in pest management decisions. Computers easily incorporate multiple variables such as expected crop value, treatment costs, plant phenology, and environmental conditions to optimize management decisions on a field-by-field basis. Several computerized pest management decision models have been developed for use in soybean, including AUSIMM,[192] SICM,[110,111,319]

SMARTSOY,[320,321] and SOYBUG.[322] These models are appropriate for a variety of defoliating and pod-feeding pests, mostly in southern states.

2. Sequential Sampling Plans

Because the sampling objective in pest management is to determine if treatment is warranted, a precise quantification of pest population density usually is not necessary. Instead, it only is necessary to determine whether pest density is above or below the ET. Sequential sampling programs allow sampling to cease as soon as the economic status of the pest population can be determined with a desired level of reliability. This provides a substantial savings in total sampling time when pest populations are very low or very high relative to the ET.[327] Because of these reductions in labor costs, sequential sampling plans have very high sampling efficiency. Information on the development and use of sequential sampling plans can be found in Chapter 8.

Sequential sampling plans have been developed for several soybean pests (Table 8). Sequential sampling plans can be improved by developing decision lines that are responsive to changes in plant phenology or arthropod age/stage. Strayer et al.[307] developed sequential sampling plans for sampling velvetbean caterpillar in either pre or postbloom soybean. The sampling efficiency of sequential sampling plans can be increased by replacing arthropod density estimates with plant injury ratings. Such programs have been developed by Bellinger and Dively.[298,328]

3. Time-Sequential Sampling Plans

Recently, Pedigo and van Schaik[148] developed a new class of sampling plans called time-sequential sampling. In most applications of time-sequential sampling, field population data over time is compared to endemic and outbreak models of population growth. Time-sequential sampling is similar to standard sequential sampling in allowing sampling to terminate when the risk of an incorrect conclusion is reduced to an acceptable level. However, the temporal component is unique to time-sequential sampling.

Time-sequential sampling was used by Pedigo and van Schaik[148] to classify population growth curves of adult green cloverworm. As is the case for most types of programs, sampling the adult stage of a larval pest provides an early warning of potential yield losses. Conversely, if the endemic configuration of population growth is confirmed, then field scouting of subsequent larvae will not be required. In Chapter 12, Pedigo describes a similar program for predicting green cloverworm damage from larval counts early in the season. Although time-sequential sampling efficiently provides information about population status, much research is required to develop alternative (endemic and outbreak) models of population growth. In addition, population growth of many species cannot be categorized into distinct endemic and outbreak forms. These factors have limited the use of time-sequential sampling by growers and scouts despite its advantages.

4. Predictive Computer Models and Kalman Filters

Several computer models of pest population dynamics, especially AUSIMM[192] and SICM,[110,111,319] can be used to predict future pest densities. These predictions assist pest managers in classifying existing populations as potentially economic or subeconomic. Computer models rarely have the predictive ability to replace field sampling. Rather, these models and expert systems are best used to interpret and/or augment field sampling data. Specifically, predictive models can be coupled with field data via Kalman filters (Figure 7). This is accomplished by weighting the use of field and stored data, respectively, to produce a population prediction with maximum precision.

TABLE 9
Range[a] of Economic[b] Thresholds Published between 1988 and 1993[c]

Guild and common name	Soybean stage[d]	Arthropod density[e]	Soybean injury
DEFOLIATORS			
Velvetbean caterpillar or soybean looper	Vegetative	13–26/m row or 1.5–3/sweep	30–50% defoliation
	Flowering	13–26/m row or 0.8–3/sweep	15–35% defoliation
	Pod development	13–26/m row or 0.8–3/sweep	10–30% defoliation
Green cloverworm	Vegetative	26–66/m row (raise threshold further if larvae are diseased or parasitized)	30–50% defoliation
	Flowering	13–39/m row or 3/sweep	15–35% defoliation
	Pod development	13–39/m row or 3/sweep	10–30% defoliation
Grasshoppers	Vegetative	22/m² in adjacent noncrop areas; 17 nymphs or 9 adults/m² in soybean	30–50% defoliation
	Flowering		15–35% defoliation
	Pod development through seed maturation	13–20/m row	10–30% defoliation, or 0.5–1 injured pod/plant, or 5–10% injured pods
LEAF SURFACE GRAZERS			
Spider mites	All		Plants along field edges yellowed, and mites present on some plants in field interior. Decision must consider (1) plant moisture stress, (2) amount of field affected, (3) number of mites that remain in affected areas, and (4) weather forecast.
Thrips	Seedling through vegetative		Discoloration of 40–60% of leaf area or of 75% of leaflets, and plants under drought stress.

PLANT FLUID REMOVERS			
Potato leafhopper	Seedling	0.8–1.5/plant	
	Flowering	6/plant	
	Pod development	7–13 or more/plant	
	Seed maturation	14–28/plant	
POD FEEDERS			
Bean leaf beetle	Seedling	13–19/m row	20–30% of plants destroyed, or stand with gaps of 30 cm or more, or 30–65% defoliation, or 3–6 cotyledons destroyed/m row
	Vegetative	13–19/m row	30–50% defoliation
	Flowering	52/m row	15–35% defoliation
	Pod development	26–52/m row or 3–6/sweep	0.6–1.0 injured pod/plant, or 5–10% injured pods
	Seed maturation	26–32/m row	8–10% injured pods
Stink bugs	Flowering through seed maturation	1–3/m row or 0.2–0.4/sweep	
Corn earworm	Pod development through seed mutation	3–13/m row or 0.2–0.4/sweep	5–10% injured pods
STEM FEEDERS			
Three-cornered alfalfa hopper	Seedling		Stand counts 66% of optimal or 50% of seedlings girdled
	Flowering through pod development	8–10 nymphs/m row, or 1 adult/sweep	
UNDERGROUND PESTS			
All	Seedling		Stand counts 66% of optimal, or stand with gaps of 30 cm or more

[a] Thresholds change frequently, and vary with soybean variety and region. Consult the local Cooperative Extension office for precise thresholds.
[b] Calculations (if necessary) were based on market value of ca. $0.22–$0.31/kg ($5–$7/bushel), potential yield of ca. 1200–2200 kg/ha (20–40 bu/acre), and insecticide application costs of ca. $20–$27/ha ($8–$11/acre).
[c] Sources: References 55, 57, 127, 132, 133, 135, 145, 146, 245, 281, 300, 325, and 326.
[d] Seedling = stages VE–V3. Vegetative = stages V4–R1. Fowering = stages R1–R2. Pod development = stages R3–R5. Seed maturation = stages R6–R8. For stage descriptions, see Fehr et al.[97]
[e] /m row = number of arthropods detected by sampling per meter of row. For most pests, sampling technique is ground cloth. For grasshoppers and potato leafhopper, sampling technique is direct count.

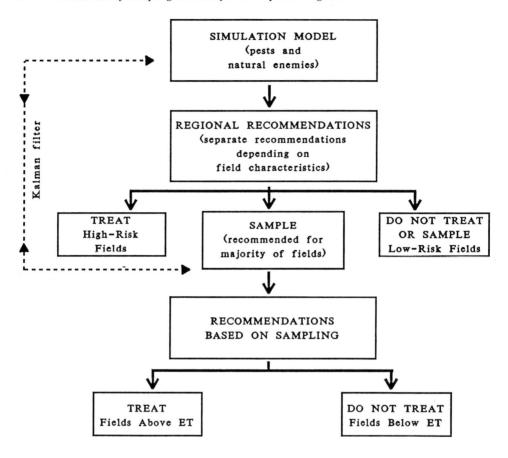

FIGURE 7. Conceptual model of an idealized pest management sampling program. A simulation model is used to determine which fields require sampling. A Kalman filter couples the simulation model with sampling data. Modified from Jackman.[98]

Empirically, predictions from Kalman filters have greater precision than predictions based on field data alone.[118] Kalman filtering requires fewer field samples to obtain the same level of precision, thereby decreasing sampling costs and increasing sampling efficiency. Despite the potential of this approach, Kalman filters have not been incorporated into existing soybean pest management systems.

D. SAMPLING TO DETERMINE IMPACT OF NATURAL ENEMIES

One of the most important influences on arthropod density is the action of natural enemies. Sampling programs designed to evaluate natural enemy impact differ sufficiently from general arthropod sampling in objectives and complexity to merit separate discussion. For example, some natural enemies, such as pathogens and nematodes, may require highly specialized sampling techniques. Although many sampling programs have been developed to estimate pest density, relatively few determine the impact of natural-enemy activity on pest populations.

Natural-enemy impact data may be collected for several reasons. First, the development of life tables and population dynamics models requires the study of biotic mortality factors.[329] Second, determining the impact of natural enemies improves our ability to conserve beneficial arthropods and manage pests.[183,316,329,330] Finally, with the ability to determine or predict the effect of natural enemies, mortality factors can be incorporated into management programs, reducing the risk of treating naturally controlled populations.[150]

Natural-enemy impacts can be measured directly by finding evidence of the activity of beneficials, such as with cadaver surveys and direct (visual) observation. The impact of beneficials also can be determined indirectly, if both population density and feeding/attack rates of the natural enemies are known. Reviews of sampling techniques for determining impacts of natural enemies in soybean have been made by several authors.[329-336] Sampling of natural enemies usually is conducted with techniques and devices similar to those used for pest species.[330] Sweepnets and ground cloths are effective in sampling a wide variety of foliar natural enemies.[23,83,93,94,173,174,177,183,184,304,315,316,330,336-343] Other techniques used to sample natural enemies include vacuum sampling, pitfall traps, and plant removal.[25,94,177,330,336,337,341,344-346] To evaluate parasitoid and pathogen activity, rearing of pest individuals may be required to complete physiological development of the natural enemy.[91,173,174,184,329,339] Direct observation of natural enemy activity in the field or in enclosures provides valuable information on prey/host selectivity and behavior and has been used in several studies.[159,343,345,347,348] Arthropod cadavers resulting from pathogen activity usually can be distinguished from remains left by other natural enemies. In some cases, infected individuals, such as two-spotted spider mites (*Tetranychus urticae* Koch) exposed to *Neozygites* sp. fungus, can be identified using microscopic techniques even before macroscopic symptoms are apparent.[130]

Although originally designed for determining pest densities, fixed-precision and sequential sampling plans also have been developed for natural enemies.[314,315] As in programs for pest species, the use of these types of sampling plans to evaluate natural enemies greatly increases sampling efficiency.

Two of the newest technologies for determining the causes of arthropod mortality are radiolabeled prey and serological (ELISA) techniques. Sampling programs utilizing these techniques precisely determine which natural enemies in a community prey upon a given pest species.[80,93,349-352] Pedigo et al.[329] noted that some recently developed serological techniques may produce quantitative data. However, the technical skills required to implement these techniques have limited their use to research programs, with few applications in field pest management.

Despite the potential advantages of incorporating natural enemy data into pest management programs, such programs have remained primarily in the developmental stages.[315] For many natural enemy impact programs, specimens must be transported to the laboratory for rearing or analysis, which delays data summary and analysis. Lab rearing of field-collected specimens to determine percent parasitism or percent infection must be planned carefully to avoid several forms of sampling bias.[330] Artificial diets are not always available, making rearing of some species difficult.[330] In addition, few programs adequately quantify actual feeding rates of natural enemies to allow prediction of pest population changes.[329] These factors have limited considerably the use of natural enemy impacts in pest management decision-making despite the potential benefits of such programs.

VII. RECOMMENDATIONS: FUTURE DIRECTIONS FOR SAMPLING SOYBEAN ARTHROPODS

Future developments in soybean sampling will be determined by future levels of environmental concern, economic incentives to reduce management costs, and availability of new technology and new biological knowledge. The difficulty of predicting the levels of these determining factors makes it equally difficult to recommend specific approaches for developing sampling programs. During the coming decade, public concern about environmental quality probably will create incentives for growers to minimize insecticide use.[353] This may result in increased pest scouting to ensure that insecticides are applied only when pest densities exceed ETs. Alternatively, pest management strategy may shift away from curative insecticide use and toward preventive deployment of nonchemical tactics (such as natural enemy augmentation) that require *less* pest scouting.[75] Regardless, three directions for soybean sampling should prove fruitful in the future: (1) developing programs for additional species, (2) improving sampling efficiency, and (3) improving delivery systems to make sampling programs simpler to implement.

A. DEVELOPING PROGRAMS FOR ADDITIONAL ARTHROPOD SPECIES

1. Programs for Minor Pests

As discussed in the Introduction, sampling has no intrinsic merit; it is only a means to an end. Growers and consultants should not be encouraged to scout for minor pests unless such scouting is economically justified. However, research is needed to determine ETs for many minor pests (Table 9), especially underground feeders.[51,76] The need for this research justifies investment of effort to develop sampling programs for minor pests. In particular, there is a need for efficient techniques for underground feeders (for example, research on correlating densities of injurious stages with adult densities or other easily sampled proxies).

2. Programs to Incorporate Natural Enemy Impacts

As discussed in Section II.C.2.c, future soybean pest management systems probably will require improved sampling programs for natural enemies. Good progress has been made on sampling foliar predators,[314,315] and similar programs are needed for other natural enemies. In addition, programs are needed for incorporating natural enemy impact into treatment decisions. Techniques exist for adjusting ET for *average* natural enemy impact.[150] Further, if natural enemy densities are high, extension publications recommend increasing the value of ET for soybean pests.[245] However, no sampling programs are available for making soybean treatment decisions based on both pest and natural enemy densities. In other crops, sequential sampling programs have been developed for classifying the ratio of natural enemies to pests (Chapter 11). This approach appears promising for soybean pest management.

B. IMPROVING SAMPLING EFFICIENCY

Regardless of future incentives for sampling, growers will be more likely to sample (and thus less likely to make unnecessary insecticide applications) if sampling can be made more efficient. As discussed in Section IV, the effort required to increase sampling efficiency will not be justified for every sampling program. Nonetheless, the following approaches seem promising for many soybean arthropods.

1. Sample Only High-Risk Fields

To the extent possible, pest management sampling should be organized according to the conceptual model shown in Figure 7. Patterns of arthropod abundance in time and space (Section V) should be used to rank soybean fields according to risk of economic pest densities. Computer models are useful but not necessary for evaluating risk. Sampling should then be concentrated at times and locations where risk is high. Both Virginia[146,147] and North Carolina[44] have used this approach to organize corn earworm sampling. Researchers and extension coordinators may be able to develop similar systems for other pests using research data reviewed in Section V. Note that a given arthropod density may be more damaging in soybean fields with small canopies (specifically, small leaf areas).[149,354] Projections of leaf area therefore should be included in any system for ranking field risk.

2. Substitute Easy-To-Sample "Proxies" for Arthropod Counting

As discussed in Section IV, it is often more efficient to sample "proxies" that are highly correlated with the ultimate density of damaging arthropod stages. One useful proxy, weather, was discussed in Section V.A.1. In addition, it may be possible to use remote sensing of soybean stress (Chapter 7) to replace early season arthropod counting. Further, population indices (Section V.I.A.4) have great promise as substitutes for costly arthropod counting. However, most existing population indices indicate only past arthropod density. Additional research is needed to correlate population indices with probable future arthropod densities. Trap catches also have great promise as a proxy for arthropod counting. Additional research is needed to correlate trap counts with ultimate densities of damaging arthropod stages on soybean.

3. Develop Efficient Sampling Plans

Many efficient sampling plans already have been developed (Table 8), but may need to be modified to reflect changes in ETs (Table 9). Further, binomial sampling (Chapter 9) and double sampling[355] approaches could increase the efficiency of some existing sampling plans. However, increases in efficiency would have to be balanced against possible decreases in ease of use. For many species, sampling to classify populations would be made more efficient if time-sequential sampling plans could be developed. It is hoped that the availability of a computer program for calculating sampling coefficients (Chapter 12) will catalyze development of time-sequential sampling plans for additional soybean arthropods. In addition, recent research[354] and computer modeling indicate what Rudd et al.[149] described as, "the dominant role leaf area plays in the growth of normal and damaged soybeans." For this reason, efficient plans are needed to sample the soybean plant itself (specifically, to quantify soybean leaf area).

4. Develop Programs for Sampling Multiple Species

For years, authors have emphasized the need for ETs that integrate the injury from multiple species.[51,355] Recent research has provided the tools necessary to develop such thresholds for the defoliator[59,60,354] and pod feeder[59] feeding guilds. However, implementation of multiple-species thresholds will require multiple-species sampling programs. This task is made easier by the fact that most members of each feeding guild can be sampled via sweepnet (Table 7). Nonetheless, details of an efficient sampling program (number and timing of samples, etc.) remain to be worked out. Additional research is needed to incorporate the effects of injury by other guilds.[61,205,323]

C. IMPROVING DELIVERY SYSTEMS TO SIMPLIFY PROGRAM IMPLEMENTATION

The success of all the approaches discussed above will depend on the ability of sampling program designers to make sampling programs easy to understand and carry out.[51] Simplicity and convenience should not be considered luxuries, to be added to a sampling program if time permits. If intended users reject a program because it is too complicated or inconvenient, the program should be judged a failure regardless of how elegant the analyses on which it is based. Delivery systems are discussed in detail in Chapter 23; design of computer interfaces is discussed by Hammond et al.[32] We will mention only two key considerations. First, sampling programs should whenever possible allow users to make their own choices about acceptable levels of risk and sampling effort (higher risk/lower effort vs. lower risk/higher effort). Second, and most important, sampling program designers should seek input from potential users at each design step.[4] Simply asking for user input is perhaps the most effective way to increase acceptance of soybean sampling.

ACKNOWLEDGMENTS

Our special thanks to Joseph Browde for assistance with the literature review.

We gratefully acknowledge the many soybean sampling experts who sent us literature and spoke with us on the telephone, including Jeffrey Andresen, Larry Bledsoe, David Boethel, Bastiaan Drees, Richard Edwards, Joseph Funderburk, Michael Gray, Ronald Hammond, Charles Helm, D. Ames Herbert, Armon Keaster, Arthur Mueller, Henry Pitre, Kevin Steffey, John Van Duyn, Michael Way, Hal Wilson, and Kenneth Yeargan. Your generous collegiality made our task much easier.

Thanks also to Leslie Lewis for assistance with pathogen taxonomy, and Leon Higley for discussions of soybean response to injury. Finally, we wish to express our gratitude to the editors, Larry Pedigo and David Buntin, for giving us the opportunity to contribute to this book.

REFERENCES

1. Morris, R. F., Sampling insect populations, *Annu. Rev. Entomol.*, 5, 243, 1960.
2. Kogan, M. and Herzog, D. C., *Sampling Methods in Soybean Entomology*, Springer-Verlag, New York, 1980, 587.
3. Herzog, D. C., Advances in sampling insects in soybeans, in *Proc. World Soybean Conf. III*, Shibles, R., Ed., Westview Press, Boulder, CO, 1985, 651.
4. Bertrand, J. T., Communications pretesting, *Media Monograph 6*, Community and Family Study Center, University of Chicago, Chicago, 1978, 144.
5. Pedigo, L. P., Hutchins, S. H., and Higley, L. G., Economic injury levels in theory and practice, *Annu. Rev. Entomol.*, 31, 341, 1986.
6. Jordan, T. N., Coble, H. D., and Wax, L. M., Weed control, in *Soybeans: Improvement, Production and Uses*, Wilcox, J. R., Ed., American Society of Agronomy, Madison, WI, 1987, 429.
7. Suguiyama, L. F. and Carlson, G. A., Field Crop Pests: Farmers Report the Severity and Intensity, Agric. Info. Bull., No. 487, U.S. Dept. of Agriculture, Economic Research Service, Washington, D.C., 1985, 54.
8. Anon., Soybeans: Area, yield, production, disposition, and value, United States, 1975-89, in *Agricultural Statistics 1990*, U.S. Dept. of Agriculture, Washington, D.C., 1990, 121.
9. Johnson, R. R., Crop management, in *Soybeans: Improvement, Production, and Uses*, 2nd ed., Wilcox, J. R., Ed., American Society of Agronomy, Madison, WI, 1987, 355.
10. Marra, M. C. and Carlson, G. A., Double-Cropping Wheat and Soybeans in the Southeast, USDA Agric. Econ. Rep., No. 552, U.S. Dept. of Agriculture, Washington, D.C., 1986.
11. Lessiter, F., No-till grows 24% in 2 years!, *No-Till Farmer*, Brookfield, WI, 1991, 4.
12. Anon., 1992-1993 Tillage survey: soybean, *No-Till Farmer*, Brookfield, WI, 1993.
13. Stinner, R. E., Bradley, J. R., Jr., and Van Duyn, J. W., Sampling *Heliothis* spp. on soybean, in *Sampling Methods in Soybean Entomology*, Kogan, M., Ed., Springer-Verlag, New York, 1980, 407.
14. Agricultural Statistics Board, Soybeans: row spacing-1992, *Crop Production*, U.S. Dept. of Agriculture, Washington, D.C., 1992, A31.
15. Lane, M. and Owen, G., Interest spreads for broadcast beans, *Soybean Dig.*, 52, 9, 1991.
16. Atwood, J. D., Johnson, S. R., Shogren, J. F., and Thompson, L. C., Analysis of 1990 Farm Bill Conservation Options, C.A.R.D. Staff Rep. 90-SR43, Center for Agricultural and Rural Development, Iowa State University, Ames, 1990.

17. Mangold, G., Farmers test stripcrops, *Soybean Dig.*, 52, 28, 1992.
18. West, T. D. and Griffith, D. R., Effect of strip-intercropping corn and soybean on yield and profit, *J. Prod. Agric.*, 5, 107, 1992.
19. Francis, C., Jones, A., Crookston, K., and Goodman, S., Strip cropping corn and grain legumes: a review, *Am. J. Altern. Agric.*, 1, 159, 1986.
20. Hammond, R. B. and Jeffers, D. L., Adult seedcorn maggots in soybeans relay intercropped into winter wheat, *Environ. Entomol.*, 12, 1487, 1983.
21. Hammond, R. B. and Jeffers, D. L., Potato leafhopper (Homoptera: Cicadellidae) populations on soybean relay intercropped into winter wheat, *Environ. Entomol.*, 19, 1810, 1990.
22. Pitre, H. N., Ecological effects of double-cropping on soybean insect populations, in *Proc. World Soybean Conf. III*, Shibles, R., Ed., Westview Press, Boulder, CO, 1985, 667.
23. Sprenkel, R. K., Brooks, W. M., Van Duyn, J. W., and Deitz, L. L., The effects of three cultural variables on the incidence of *Nomuraea rileyi*, phytophagous Lepidoptera, and their predators on soybeans, *Environ. Entomol.*, 8, 334, 1979.
24. Agriculture Statistics Board, Percent of soybean acreage planted following another crop, selected states, *Crop Production*, U.S. Dept. of Agriculture, Washington, D.C., 1987, A21.
25. LeSar, C. D. and Unzicker, J. D., Soybean spiders: species composition, population densities, and vertical distribution, *Ill. Nat. Hist. Surv. Biol. Notes*, 107, 3, 1978.
26. Price, P. W., Colonization of crops by arthropods: non-equilibrium communities in soybean fields, *Environ. Entomol.*, 5, 605, 1976.
27. De Long, D. M., Some problems encountered in the estimation of insect populations by the sweeping method, *Ann. Entomol. Soc. Am.*, 25, 13, 1932.
28. Fehr, W. R., Caviness, C. E., Burmwood, D. T., and Pennington, J. S., Stage of development descriptions for soybeans, *Glycine max* (L.) Merrill, *Crop Sci.*, 11, 929, 1971.
29. Ritchie, S. W., Hanway, J. L., Thompson, H. E., and Benson, G. O., How a Soybean Plant Develops, Ext. Serv. Spec. Rep. No. 53, Iowa Cooperative Extension Service, Ames, 1985, 20.
30. Lersten, N. R. and Carlson, J. B., Vegetative morphology, in *Soybeans: Improvement, Production, and Uses,* Wilcox, J. R., Ed., American Society of Agronomy, Madison, WI, 1987, 49.
31. Carlson, J. B. and Lersten, N. R., Reproductive morphology, in *Soybeans: Improvement, Production, and Uses,* 2nd ed., Wilcox, J. R., Ed., American Society of Agronomy, Madison, WI, 1987, 95.
32. Hammond, R. B., Higgins, R. A., Mack, T. P., Pedigo, L. P., and Bechinski, E. J., Soybean pest management, in *CRC Handbook of Pest Management in Agriculture*, Vol. 3, 2nd ed., Pimentel, D., Ed., CRC Press, Boca Raton, FL, 1991, 341.
33. Deighan, J., The Effect of Cropping Systems on the Pest Complex in Virginia Soybeans and Calibration of the Sweepnet and Ground Cloth Sampling Methods for Use in These Cropping Systems, M.S. thesis, Virginia Polytechnic Institute and State University, Blacksburg, 1983.
34. Schumann, F. W. and Todd, J. W., Population dynamics of the southern green stink bug (Heteroptera: Pentatomidae) in relation to soybean phenology, *J. Econ. Entomol.*, 75, 748, 1982.
35. Hadley, P., Roberts, E. H., Summerfield, R. J., and Martin, F. R., Effects of temperature and photoperiod on flowering in soya bean [*Glycine max* (L.) Merrill]: a quantitative model, *Ann. Bot.*, 53, 669, 1984.
36. Major, D. L., Johnson, D. R., Tanner, J. W., and Anderson, I. C., Effects of daylength and temperature on soybean development, *Crop Sci.*, 15, 174, 1975.
37. Wintersteen, W., Padgitt, S., and Stone, J., The impact of pesticide education on Iowa farmers, in *Practical Application in Health and Safety in Agriculture, Third Int. Symp. Issues in Health, Safety and Agric.*, Lewis Publishers, Boca Raton, FL, 1993 (in press).
38. Higley, L. G. and Pedigo, L. P., Soybean growth responses and intraspecific competition from simulated seedcorn maggot injury, *Agron. J.*, 82, 1057, 1990.
39. Livingston, J. M., McLeod, P. J., Yeargan, W. C., and Young, S. Y., Laboratory and field evaluation of a nuclear polyhedrosis virus of the soybean looper, *Pseudoplusia includens*, *J. Ga. Entomol. Soc.*, 15, 194, 1980.
40. Buschman, L. L., Pitre, H. N., and Hodges, H. F., Soybean cultural practices: effects on populations of green cloverworm, velvetbean caterpillar, loopers, and *Heliothis* complex, *Environ. Entomol.*, 10, 631, 1981.
41. Buschman, L. L., Pitre, H. N., and Hodges, H. F., Soybean cultural practices: effects on populations of *Geocoris*, nabids, and other soybean arthropods, *Environ. Entomol.*, 13, 305, 1984.
42. Alston, D. G., Bradley, J. R., Jr., Schmidt, D. P., and Coble, H. D., Relationship of *Heliothis zea* predators, parasitoids, and entomopathogens to canopy development in soybean as affected by *Heterodera glycines* and weeds, *Entomol. Exp. Appl.*, 58, 279, 1991.

43. Mayse, M. A., Effects of spacing between rows on soybean arthropod populations, *J. Appl. Ecol.*, 15, 439, 1978.
44. Bradley, J. R. and Van Duyn, J. W., Insect pest management in North Carolina soybeans, in *Proc. World Soybean Conf. II*, Corbin, F. T., Ed., Westview Press, Boulder, CO, 1980, 343.
45. Kogan, M. and Turnipseed, S. G., Soybean growth and assessment of damage by arthropods, in *Sampling Methods in Soybean Entomology*, Kogan, M. and Herzog, D. C., Eds., Springer-Verlag, New York, 1980, 3.
46. Johnston, T. J. and Pendleton, J. W., Contribution of leaves at different canopy levels to seed production of upright and lodged soybeans (*Glycine max* (L.) Merrill), *Crop Sci.*, 8, 291, 1968.
47. Deitz, L. L., Van Duyn, J. W., Bradley, J. R., Jr., Rabb, R. L., Brooks, W. M., and Stinner, R. E., A guide to the identification and biology of soybean arthropods in North Carolina, *N.C. Agric. Exp. Stn. Tech. Bull.* 238, 1976.
48. Higley, L. G. and Boethel, D., Eds., *Handbook of Soybean Insect Pests*, Entomological Society of America, College Park, MD, 1993, in press.
49. Harper, J. D., McPherson, R. M., and Shepard, M., Geographical and seasonal occurrence of parasites, predators and entomopathogens, in *Natural Enemies of Arthropod Pests in Soybean*, Pitre, H. N., Ed., Southern Coop. Serv. Bull. 285, 1983, 7.
50. Turnipseed, S. G. and Kogan, M., Soybean entomology, *Annu. Rev. Entomol.*, 21, 247, 1976.
51. Turnipseed, S. G. and Kogan, M., Integrated control of insect pests, in *Soybeans: Improvement, Production, and Uses,* 2nd ed., Wilcox, J. R. Ed., American Society of Agronomy, Madison, WI, 1987, 779.
52. Newsom, L. D., Kogan, M., Miner, F. D., Rabb, R. L., Turnipseed, S. G., and Whitcomb, W. H., General accomplishments toward better pest control in soybean, in *New Technology of Pest Control*, Huffaker, C. B., Ed., John Wiley & Sons, New York, 1980, 51.
53. Kogan, M. and Turnipseed, S. G., Ecology and management of soybean arthropods, *Annu. Rev. Entomol.*, 32, 507, 1987.
54. Kogan, J., Kogan, M., Brewer, E. F., and Helm, C. G., *World Bibliography of Soybean Entomology*, Vol. 1 and 2, Ill. Agric. Exp. Stn. Spec. Pub. 73, University of Illinois, Urbana-Champaign, 1988.
55. Kogan, M. and Kuhlman, D. E., Soybean insects: identification and management in Illinois, *Univ. Ill. Agr. Exp. Stn. Bull.*, 773, 1982.
56. Anon., Soybean insect identification and control, *Soybean Dig.*, 52, 49, 1992.
57. Higdon, M. L. and Keaster, A. J., Soybean insects in Missouri, *University of Missouri, Columbia Agric. Exp. Stn.*, SR358, 1987.
58. Root, R. B., The niche exploitation pattern of the blue-gray gnatcatcher, *Ecol. Monogr.*, 33, 317, 1967.
59. Hutchins, S. H. and Funderburk, J. E., Injury guilds: a practical approach for managing pest losses to soybean, *Agric. Zool. Rev.*, 4, 1, 1991.
60. Hutchins, S. H., Higley, L. G. and Pedigo, L. P., Injury equivalency as a basis for developing multiple-species economic injury levels, *J. Econ. Entomol.*, 81, 1, 1988.
61. Higley, L. G., Browde, J. A., and Higley, P. M., Moving towards new understandings of biotic stress and stress interactions, in *International Crop Science*, Crop Science Society of America, Madison, Wisconsin. Buxton, D. and Shibles, R., Eds., Ames, IA, 1993, in press.
62. Balduf, W. V., The insects of the soybean in Ohio, *Ohio Agr. Exp. Stn. Res. Bull.*, 366, 148, 1923.
63. Eckel, C. S., Bradley, J. R., Jr., and Van Duyn, J. W., Reductions in soybean yield and quality from corn earworm flower feeding, *Agron. J.*, 84, 402, 1992.
64. Kretzschmar, G. P., Soybean insects in Minnesota with special reference to sampling techniques, *J. Econ. Entomol.*, 41, 586, 1948.
65. Tugwell, P., Rouse, E. P., and Thompson, R. G., Insects in soybeans and a weed host (*Desmodium* sp.), *Ark. Agric. Exp. Stn. Rep. Ser.*, 214, 18, 1973.
66. Kogan, M., Insect problems of soybeans in the United States, in *World Soybean Research Conference II: Proceedings,* Corbin, F. T., Ed., Westview Press, Boulder, CO, 1980, 303.
67. Helm, C. G., Ecology of soybean pests, in *World Soybean Research Conference IV Proceedings*, Pascale, A. J., Ed., 1989, 1472.
68. Mack, T. P. and Backman, C. B., Management of soybean insects, in *Ala. Agricultural Experiment Station Soybeans*, Ala. Agric. Exp. Stn. Res. Rep. Ser. No. 4, 1986, 32.
69. Smelser, R. B. and Pedigo, L. P., Bean leaf beetle (Coleoptera: Chrysomelidae) herbivory on leaf, stem, and pod components of soybean, *J. Econ. Entomol.*, 85, 2408, 1992.
70. Smelser, R. B. and Pedigo, L. P., Soybean seed yield and quality reduction by bean leaf beetle (Coleoptera: Chrysomelidae) pod injury, *J. Econ. Entomol.*, 85, 2399, 1992.
71. Mitchell, E. R., Webb, J. C., Baumhover, A. H., Hines, R. W., Stanley, J. W., Endris, R. G., Lindquist,

D. A., and Masudo, S., Evaluation of cylindrical electric grids as pheromone traps for loopers and tobacco hornworms, *Environ. Entomol.*, 1, 365, 1972.
72. Sparks, A. N., Jr. and Boethel, D. J., Late-season damage to soybeans by three-cornered alfalfa hopper (Homoptera: Membracidae) adults and nymphs, *J. Econ. Entomol.*, 80, 471, 1987.
73. Higley, L. G. and Wintersteen, W., Consequences of the 1988 spider mite outbreak, *Crops, Soils, and Pests Newsletter*, Iowa Coop. Ext. Serv., IC-458, No. 5, Ames, 1989.
74. Pedigo, L. P., *Entomology and Pest Management*, Macmillan, New York, 1989, 646.
75. Pedigo, L. P., Integrating preventive and therapeutic tactics in soybean insect management, in *Pest Management in Soybean*, Copping, L. G., Green, M. B., and Rees, R. T., Eds., Elsevier, London, 1992, 10.
76. Hammond, R. B. and Funderburk, J. E., Influence of tillage practices on soil-insect population dynamics in soybean, in *Proc. World Soybean Conf. III*, Shibles, R., Ed., Westview Press, Boulder, CO, 1985, 659.
77. Parrella, M. P., Heinz, K. M., and Nunney, L., Biological control through augmentative releases of natural enemies: A strategy whose time has come, *Am. Entomol.*, 38, 172, 1992.
78. Ferguson, H. J., McPherson, R. M., and Allen, W. A., Effect of four soybean cropping systems on the abundance of foliage-inhabiting insect predators, *Environ. Entomol.*, 13, 1105, 1984.
79. Marston, N. L., Thomas, G. D., Ignoffo, C. M., Gebhardt, M. R., Hostetter, D. L., and Dickerson, W. A., Seasonal cycles of soybean arthropods in Missouri: effect of pesticidal and cultural practices, *Environ. Entomol.*, 8, 165, 1979.
80. Buschman, L. L., Whitcomb, W. H., Hemenway, R. C., Mays, D. L., Ru, N., Leppa, N. C., and Smittle, B. J., Predators of velvetbean caterpillar eggs in Florida soybeans, *Environ. Entomol.*, 6, 403, 1977.
81. Dumas, B. A., Boyer, W. P., and Whitcomb, W. H., Effect of the time of day on surveys of predaceous insects in field crops, *Fla. Entomol.*, 45, 121, 1962.
82. Thorpe, K. W., Seasonal distribution of *Trichogramma* (Hymenoptera: Trichogrammatidae) species associated with a Maryland soybean field, *Environ. Entomol.*, 13, 127, 1984.
83. Zehnder, G. W., Herbert, D. A., McPherson, R. M., Speese, J., III, and Moss, T., Incidence of *Heliothis zea* (Lepidoptera: Noctuidae) and associated parasitoids in Virginia soybeans, *Environ. Entomol.*, 19, 1135, 1990.
84. Turnipseed, S. and Kogan, M., Soybean pests and indigenous natural enemies, in *Natural Enemies of Arthropod Pests in Soybean*, Southern Coop. Ser. Bull., 285, Pitre, H. N., Ed., 1983, 1.
85. Loughran, J. C. and Ragsdale, D. W., *Medina* n. sp. (Diptera: Tachinidae): a new parasitoid of the bean leaf beetle (Coleoptera: Chrysomelidae), *J. Kan. Entomol. Soc.*, 59, 468, 1986.
86. Peterson, R. K. D., Smelser, R. B., Klubertanz, T. H., Pedigo, L. P., and Welbourn, W. C., Ectoparasitism of the bean leaf beetle (Coleoptera: Chrysomelidae) by *Trombidium hyperi* Vercammen-Grandjean, Van Driesche, and Gyrisco and *Trombidium newelli* Welbourn and Flessel (Acari: Trombidiidae), *J. Agric. Entomol.*, 9, 99, 1992.
87. Daigle, C. J., Boethel, D. J., and Fuxa, J. R., Parasitoids and pathogens of soybean looper and velvetbean caterpillar (Lepidoptera: Noctuidae) in soybeans in Louisiana, *Environ. Entomol.*, 19, 746, 1990.
88. Daigle, C. J., Boethel, D. J., and Fuxa, J. R., Parasitoids and pathogens of green cloverworm (Lepidoptera: Noctuidae) on an uncultivated spring host (vetch, *Vicia* spp.) and a cultivated summer host (soybean, *Glycine max*), *Environ. Entomol.*, 17, 90, 1988.
89. Hochmuth, R. C., Hellman, J. L., Dively, G., and Schroder, R. F. W., Effect of the ectoparasitic mite *Coccipolis epilachnae* (Acari: Podapolipidae) on feeding, fecundity, and longevity of soybean-fed adult Mexican bean beetles (Coleoptera: Chrysomelidae) at different temperatures, *J. Econ. Entomol.*, 80, 612, 1987.
90. McCutcheon, G. S., Turnipseed, S. G., and Sullivan, M. J., Parasitization of lepidopterans as affected by nematicide-insecticide use in soybean, *J. Econ. Entomol.*, 83, 1002, 1990.
91. Orr, D. B., Russin, J. S., Boethel, D. J., and Jones, W. A., Stink bug (Hemiptera: Pentatomidae) parasitism in Louisiana soybeans, *Environ. Entomol.*, 15, 1250, 1986.
92. Turnipseed, S. G., Herzog, D. C., and Chapin, J. W., Evaluation of biological agents for pest management in soybean, in *Integrated Pest Management on Major Agricultural Systems*, Frisbie, R. E. and Adkisson, P. L., Eds., Texas A & M University Press, College Station, 1986, 444.
93. Elvin, M. K., Stimac, J. L., and Whitcomb, W. H., Estimating rates of arthropod predation on velvetbean caterpillar larvae in soybeans, *Fla. Entomol.*, 66, 319, 1983.
94. Culin, J. D. and Rust, R. W., Comparison of the ground surface and foliage dwelling spider communities in a soybean habitat, *Environ. Entomol.*, 9, 577, 1980.
95. Wood, D. M., *Tachinidae*, in *Manual of Nearctic Diptera*, Vol. 2, Monograph No. 28, McAlpine, J. F., Ed., Agriculture Canada, Ottawa, 1987, 1193.

96. Krombein, K. V., Hurd, P. D., Smith, D. R., and Burks, B. D., *Catalog of Hymenoptera in America North of Mexico*, Vol. 1, *Symphyta and Apocrita (Parasitica)*, Smithsonian Institution Press, Washington, D.C., 1979, 2014.
97. Mason, W. R. M., The polyphyletic nature of *Apanteles foerster* (Hymenoptera: Braconidae): a phylogeny and reclassification of Microgastrinae, *Mem. Entomol. Soc. Can.*, 15, 147, 1981.
98. Jackman, J. A., Witz, J. A., Frisbie, R. E., and Skieth, R. W., The role of sampling in future pest management systems, in *Economic Thresholds and Sampling of Heliothis Species on Cotton, Corn, Soybeans, and Other Host Plants*, Southern Coop. Ser. Bull. 231, 1979, 152.
99. Rajotte, E. G., Kazmierczak, R. F., Jr., Norton, G. W., Lambur, M. T., and Allen, W. A., The National Evaluation of Extension's Integrated Pest Management (IPM) Programs, Publ. 491-010, Virginia Cooperative Extension Service, Blacksburg, Virginia, 1987, 123.
100. Carlson, G. A., IPM experience in major crops, *Tarheel Econ.*, September 1981.
101. Allen, W. A. and Roberts, J. E., Jr., Economic feasibility of scouting soybean insects in later summer in Virginia, *J. Econ. Entomol.*, 67, 644, 1974.
102. Adkisson, P. L., Frisbie, R. E., Thomas, J. G., and McWhorter, G. M., Impact of IPM on several crops of the United States, in *Integrated Pest Management on Major Agricultural Systems*, Publ. MP-1616, Frisbie, R. E. and Adkisson, P. L., Eds., Texas Agriculture Experimental Station, College Station, 1986, 663.
103. Hatcher, J. E., Wetzstein, M. E., and Douce, G. K., An economic evaluation of integrated pest management for cotton, peanuts, and soybeans in Georgia, *Univ. Ga. Coll. Agric. Exp. Stn. Res. Bull.*, 318, 1984.
104. Paxton, K. and Lavergne, D. R., Integrated pest management for soybeans, *La. Agric.*, 28, 7, 1984.
105. Greene, C. R., Rajotte, E. G., Norton, G. W., Kramer, R. A., and McPherson, R. M., Revenue and risk analysis of soybean pest management options in Virginia, *J. Econ. Entomol.*, 78, 10, 1985.
106. McPherson, R. M., Kazmierczak, R. F., Rajotte, E. G., and Allen, W. A., The Impacts of Integrated Pest Management (IPM) on Virginia Soybeans, Va. Coop. Exten. Serv. Publ. 500-051, Virginia Polytechnic Institute, Blacksburg, 1987, 56.
107. Boggess, W. G., Carlson, G. A., Zavaleta, L. R., and Paxton, K. W., Economics of improved soybean production systems, in *Integrated Pest Management on Major Agricultural Systems*, Frisbie, R. E. and Adkisson, P. L., Eds., Texas A & M University Press, College Station, 1986, 548.
108. Reichelderfer, K. H. and Bender, F. E., Application of a simulative approach to evaluating alternative methods for the control of agricultural pests, *Am. J. Agric. Econ.*, 61, 258, 1979.
109. Wilkerson, G. G., Mishoe, J. W., Jones, J. W., Stimac, J. L., Swaney, D. P., and Boggess, W. G., SICM Soybean Integrated Crop Management Model. Model Description and User's Guide, ver. 4.2, Univ. Fla. Dept. Agric. Eng. Rep. AGE 83-1, University of Florida, Gainesville, 1983, 216.
110. Szmedra, P. I., Wetztein, M. E., and McClemdon, R. W., Economic threshold under risk: a case study of soybean production, *J. Econ. Entomol.*, 83, 641, 1990.
111. Szmedra, P. I., McClendon, R. W., and Wetzstein, M. E., Risk efficiency of pest management strategies: a simulation case study, *Trans. Am. Soc. Agric. Eng.*, 31, 1642, 1988.
112. Boggess, W. G., Cardelli, D. J., and Barfield, C. S., A bioeconomic simulation approach to multi-species insect management, *South. J. Agric. Econ.*, 17, 43, 1985.
113. Barfield, C. S., Cardelli, D. J., and Boggess, W. G., Major problems with evaluating multiple stress factors in agriculture, *Trop. Pest Manage.*, 33, 109, 1987.
114. Pike, D. R., Glover, K. D., Knake, E. L., and Kuhlman, D. E., Pesticide Use in Illinois: Results of a 1990 Survey of Major Crops, Coop. Ext. Serv. Publ. DP-91-1, University of Illinois, Urbana-Champaign, 1991.
115. Herbert, D. A., *Virginia Soybean Board 1992 Project Report*, Virginia Polytechnic Institute and State University, Tidewater Agricultural Experiment Station, Suffolk, 1992.
116. Newsom, L. D., Jensen, R. L., Herzog, D. C., and Thomas, J. W., A pest management system for soybeans, *La. Agric.*, 18, 10, 1975.
117. Duffy, M. and Judd, D., Estimated Costs of Crop Production in Iowa, Iowa Coop. Ext. Serv. Publ. FM-1712, Iowa Cooperative Extension Service, Ames, 1992, 14.
118. Zavaleta, L. R. and Dixon, B. L., Economic benefits of Kalman filtering for insect pest management, *J. Econ. Entomol.*, 75, 982, 1982.
119. Edwards, C. R., Soybean pest research needs identified by user groups in the Midwest, *Am. Entomol.*, 38, 136, 1992.
120. Jones, V. P., Developing sampling plans for spider mites (Acari: Tetranychidae): those who don't remember the past may have to repeat it, *J. Econ. Entomol.*, 83, 1656, 1990.
121. Wilson, L. T., Estimating the abundance and impact of arthropod natural enemies in IPM systems, in *Biological Control in Agricultural IPM Systems*, Hoy, M. H. and Herzog, D. C., Eds., Academic Press, New York, 1985, 303.

122. Pedigo, L. P., Bechinski, E. J., and Higgins, R. A., Partial life tables of the green cloverworm (Lepidoptera: Noctuidae) in soybean and a hypothesis of population dynamics in Iowa, *Environ. Entomol.*, 12, 186, 1983.
123. Wolf, R. A., Pedigo, L. P., Shaw, R. H., and Newsom, L. D., Migration/transport of the green cloverworm, *Plathypena scabra* (F.), (Lepidoptera: Noctuidae), into Iowa as determined by synoptic-scale weather patterns, *Environ. Entomol.*, 16, 1169, 1987.
124. Carlson, J. D., Whalon, M. W., Landis, D. A., and Gage, S. H., Springtime weather patterns coincident with long-term migration of potato leafhopper into Michigan, *Agric. For. Meteorol.*, 59, 183, 1992.
125. Mitchell, E. R., Chalfant, R. B., Green, G. L., and Creighton, C. S., Soybean looper: populations in Florida, Georgia, and South Carolina as determined with pheromone-baited BL traps, *J. Econ. Entomol.*, 68, 747, 1975.
126. Greene, G. L., Pest management of the velvetbean caterpillar in a soybean ecosystem, in *Proc. World Soybean Res. Conf. 1976*, Hill, L. D., Ed., Interstate Printing, Danville, IL, 1976, 602.
127. Edwards, C. R., Obermeyer, J. L., and Jordan, T. N., Field Crops Pest Management Manual, Purdue Univ. Coop. Ext. Serv. IPM-1, Indiana Cooperative Extension Service, Lafayette, 1991.
128. Edwards, C. R., Obermeyer, J. L., and Bledsoe, L. W., Bean Leaf Beetle on Soybeans, Purdue Univ. Coop. Ext. Serv. E-51, Indiana Cooperative Extension Service, Lafayette, 1991.
129. Edwards, C. R., Obermeyer, J. L., and Bledsoe, L. W., Two-Spotted Spider Mites on Field Crops, Purdue Univ. Coop. Ext. Serv. E-39, Indiana Cooperative Extension Service, Lafayette, 1991.
130. Klubertanz, T. H., Pedigo, L. P., and Carlson, R. E., Impact of fungal epizootics on the biology and management of the two-spotted spider mite (Acari: Tetranychidae) in soybean, *Environ. Entomol.*, 20, 731, 1991.
131. Klubertanz, T. H., Pedigo, L. P., and Carlson, R. E., Effects of plant moisture stress and rainfall on population dynamics of the two-spotted spider mite (Acari: Tetranychidae), *Environ. Entomol.*, 19, 1773, 1990.
132. Higley, L., Wintersteen, W., Klubertanz, T., and Rice, M., Two-Spotted Spider Mites on Soybeans and Corn, Iowa Coop. Ext. Serv. PM-1363, Iowa Cooperative Extension Service, Ames, 1989.
133. Ostlie, K. R., Controlling Spider Mites in Soybean, University of Minnesota Extension Service, St. Paul, 1988.
134. Simpson, K. W. and Connell, W. A., Mites on soybeans: moisture and temperature relations, *Environ. Entomol.*, 2, 319, 1973.
135. Steffey, K., Gray, M., Royer, R., and Weinzierl, R., Insect pest management for field and forage crops, in *1992 Illinois Pest Control Handbook*, Illinois Cooperative Extension Service, Urbana, 1992.
136. Dively, G. P., Effect of weather on Mexican bean beetle populations, in *Mexican Bean Beetle Biocontrol Demonstration Project: Final Evaluation and Recommendations*, University of Maryland, College Park, 1985, 36.
137. Stinner, R. E., Regniere, J., and Wilson, K., Differential effects of agroecosystem structure on dynamics of three soybean herbivores, *Environ. Entomol.*, 11, 538, 1982.
138. Marcovitch, S. and Stanley, W. W., The climatic limitations of the Mexican bean beetle, *Ann. Entomol. Soc. Am.*, 23, 666, 1930.
139. Jeffords, M. R., Helm, C. G., and Kogan, M., Overwintering behavior and spring colonization of soybean by the bean leaf beetle (Coleoptera: Chrysomelidae) in Illinois, *Environ. Entomol.*, 12, 1459, 1983.
140. Boiteau, G., Bradley, J. R., Jr., and Van Duyn, J. W., Bean leaf beetle: emergence patterns of adults from overwintering sites, *Environ. Entomol.*, 8, 427, 1979.
141. Boyer, W. P., The effect of type of plant growth on bollworm (*Heliothis zea*) infestations in soybeans in Arkansas, *Ark. Coop. Econ. Insect Rep.*, 17, 83, 1967.
142. Johnson, M. W., Stinner, R. E., and Rabb, R. L., Ovipositional response of *Heliothis zea* (Boddie) to its major hosts in North Carolina, *Environ. Entomol.*, 4, 291, 1975.
143. Hillhouse, T. L. and Pitre, H. N., Oviposition by *Heliothis* on soybeans and cotton, *J. Econ. Entomol.*, 69, 144, 1976.
144. Terry, I., Bradley, J. R., and Van Duyn, J. W., Within-plant distribution of *Heliothis zea* (Boddie) (Lepidoptera: Noctuidae) eggs on soybeans, *Environ. Entomol.*, 16, 625, 1987.
145. Tynes, J. J. and Boethel, D. J., Control Soybean Insects 1992, La. Coop. Ext. Serv. Publ. 2211 (revised), Louisiana Cooperative Extension Service, Baton Rouge, 1992.
146. Herbert, A., Hull, C., and Day, E., Corn Earworm: Biology and Management in Soybeans, Virginia Coop. Ext. Serv. Publ. 444-770, Virginia Cooperative Extension Service, Blacksburg, 1992.
147. Herbert, D. A., Jr., Zehnder, G. W., and Day, E. R., Evaluation of a pest advisory for corn earworm (Lepidoptera: Noctuidae) infestations in soybean, *J. Econ. Entomol.*, 84, 515, 1991.

148. Pedigo, L. P. and van Schaik, J. W., Time-sequential sampling: a new use of the sequential probability ratio test for pest management decisions, *Bull. Entomol. Soc. Am.*, 30, 32, 1984.
149. Rudd, W. G., Ruesink, W. G., Newsom, L. D., Herzog, D. C., Jensen, R. L. and Marsolan, N. F., The systems approach to research and decision making for soybean pest control, in *New Technology of Pest Control*, Huffaker, C. B., Ed., John Wiley & Sons, New York, 1980, 99.
150. Ostlie, K. R. and Pedigo, L. P., Incorporating pest survivorship into economic thresholds, *Bull. Entomol. Soc. Am.*, 33, 98, 1987.
151. Marsolan, N. F. and Rudd, W. G., Modeling and optimal control of insect pest populations, *Math. Biosci.*, 30, 231, 1976.
152. Carner, G. R., Shepard, M., and Turnipseed, S. G., Seasonal abundance of insect pests of soybeans, *J. Econ. Entomol.*, 67, 487, 1974.
153. Funderburk, J. E., Higley, L. G., and Pedigo, L. P., Seedcorn maggot (Diptera: Anthomyiidae) phenology in central Iowa and examination of a thermal-unit system to predict development under field conditions, *Environ. Entomol.*, 13, 105, 1984.
154. Ignoffo, C. M., Marston, N. L., Putler, B., Hostetter, D. R., Thomas, G. D., Biever, K. D., and Dickerson, W. A., Natural biotic agents controlling insect pests of Missouri soybeans, in *World Soybean Research*, Hill, L. D., Ed., Interstate Printers, Danville, IL, 1976, 561.
155. Waldbauer, G. P. and Kogan, M., Bean leaf beetles: phenological relationship with soybean in Illinois, *Environ. Entomol.*, 5, 35, 1976.
156. Smelser, R. B. and Pedigo, L. P., Phenology of *Cerotoma trifurcata* on soybean and alfalfa in central Iowa, *Environ. Entomol.*, 20, 514, 1991.
157. Boiteau, G., Bradley, J. R., Jr., and Van Duyn, J. W., Bean leaf beetle: temporal and macro-spatial distribution in North Carolina, *J. Ga. Entomol. Soc.*, 15, 151, 1980.
158. Pacheco, F., Seasonal and daily fluctuation of soybean insect populations in the Yaqui Valley, Sonora, Mexico, in *World Soybean Research*, Hill, L. D., Ed., Interstate Printers, Danville, IL, 1976, 584.
159. Stevens, L. M., Steinhauer, A. L., and Coulson, J. R., Suppression of Mexican bean beetle on soybeans with annual inoculative releases of *Pediobius foveolatus*, *Environ. Entomol.*, 4, 947, 1975.
160. Sloderbeck, P. E. and Edwards, C. R., Effects of soybean cropping practices on Mexican bean beetle and redlegged grasshopper populations, *J. Econ. Entomol.*, 72, 850, 1979.
161. Johnson, M. P. and Mueller, A. J., Flight activity of the three cornered alfalfa hopper (Homoptera: Membracidae) in soybean, *J. Econ. Entomol.*, 81, 1101, 1989.
162. Mitchell, P. L. and Newsom, L. D., Seasonal history of the three cornered alfalfa hopper (Homoptera: Membracidae) in Louisiana, *J. Econ. Entomol.*, 77, 906, 1984.
163. Funderburk, J. E., Herzog, D. C., and Lynch, R. E., Seasonal abundance of lesser cornstalk borer (Lepidoptera: Pyralidae) adults in soybean, peanut, corn, sorghum, and wheat in northern Florida, *J. Entomol. Sci.*, 22, 159, 1986.
164. Myers, T. V. and Pedigo, L. P., Forecasting green cloverworm larval population peaks, *Iowa State J. Res.*, 51, 363, 1977.
165. Raney, H. G. and Yeargan, K. V., Seasonal abundance of common phytophagous and predaceous insects in Kentucky soybeans, *Trans. Ky. Acad. Sci.*, 38, 83, 1977.
166. Douglass, J. R., Portman, R. W., and Manis, H. C., Spider mites of beans and corn and their control, *Bull. Univ. Idaho Coll. Agric.*, 450, 1965.
167. Baker, J. E. and Connel, W. A., Mites on soybeans in Delaware, *J. Econ. Entomol.*, 54, 1024, 1961.
168. Miner, F. D., Biology and control of stink bugs on soybeans, *Ark. Agric. Exp. Stn. Bull.* 708, 1966.
169. Bickenstaff, C. C. and Huggans, J. L., Soybean insects and related arthropods in Missouri, *Mo. Agric. Exp. Stn. Res. Bull.*, 803, 51, 1962.
170. Miller, R. G., Soybean insects: their damage and practical control, *Soybean Dig.*, 32, 8, 1972.
171. Hamer, J., Soybean looper: biology and approaches for improved management, *Miss. Coop. Ext. Serv. Info. Sheet* 1400, 1991.
172. Lentz, G. L. and Pedigo, L. P., Population ecology of parasites of the green cloverworm in Iowa, *J. Econ. Entomol.*, 68, 301, 1975.
173. Burleigh, J. G., Population dynamics and biotic controls of the soybean looper in Louisiana, *Environ. Entomol.*, 1, 290, 1972.
174. Barry, R. M., Insect parasites of the green cloverworm in Missouri, *J. Econ. Entomol.*, 63, 1963, 1970.
175. McCutcheon, G. S. and Turnipseed, S. G., Parasites of lepidopterous larvae in insect resistant and susceptible soybeans in South Carolina, *Environ. Entomol.*, 10, 67, 1981.
176. Whiteside, R. C., Burbutis, P. P., and Kelsey, L. P., Insect parasites of the green cloverworm in Delaware, *J. Econ. Entomol.*, 60, 326, 1967.
177. Bechinski, E. J. and Pedigo, L. P., Ecology of predaceous arthropods in Iowa soybean agroecosystems, *Environ. Entomol.*, 10, 771, 1981.

178. House, G. J. and All, S. N., Carabid beetles in soybean agroecosystems, *Environ. Entomol.*, 10, 194, 1981.
179. Price, J. F. and Shepard, M., Striped earwig, *Labidura riparia*, colonization of soybean fields and response to insecticides, *Environ. Entomol.*, 6, 679, 1977.
180. Price, J. F. and Shepard, M., *Calosoma sayi*: seasonal history and response to insecticides in soybeans, *Environ. Entomol.*, 7, 359, 1978.
181. Dumas, B. A., Boyer, W. P., and Whitcomb, W. H., Effect of various factors on surveys of predaceous insects in soybeans, *J. Kan. Entomol. Soc.*, 37, 192, 1964.
182. McPherson, R. M., Smith, J. C., and Allen, W. A., Incidence of arthropod predators in different soybean cropping systems, *Environ. Entomol.*, 11, 685, 1982.
183. Shepard, M., Carner, G. R., and Turnipseed, S. G., Seasonal abundance of predaceous arthropods in soybeans, *Environ. Entomol.*, 3, 985, 1974.
184. Sloderbeck, P. E. and Yeargan, K. V., Green cloverworm (Lepidoptera: Noctuidae) populations in conventional and double-crop, no-till soybeans, *J. Econ. Entomol.*, 76, 785, 1983.
185. Johnson, D. W., Temperature-Dependent Developmental Models for the Velvetbean Caterpillar and an Associated Nucleopolyhedrosis Virus, Ph.D. dissertation, University of Florida, Gainesville, 1980, 110.
186. Menke, W. W. and Greene, G. L., Experimental validation of a pest management model, *Fla. Entomol.*, 59, 135, 1976.
187. Hammond, R. B., Poston, F. L., and Pedigo, L. P., Growth of the green cloverworm and a thermal-unit model for development, *Environ. Entomol.*, 8, 639, 1979.
188. Stinner, R. E., Rabb, R. L., and Bradley, J. R., Jr., Natural factors operating in the population dynamics of *Heliothis zea* in North Carolina, *Proc. Int. Congr. Entomol.*, 15, 622, 1977.
189. Harstack, A. W., White, J. A., Hollingsworth, J. P., Ridgway, R. L., and Lopez, J. D., MOTHZV-2: A Computer Simulation of *Heliothis zea* and *Heliothis virescens* Population Dynamics, User's Manual, USDA-ARS S-127, U.S. Govt. Printing Office, Washington, D.C., 1976.
190. Yu, Y., Gold, H. J., Stinner, R. E., and Wilkerson, G. G., Leslie model for the population dynamics of corn earworm in soybean, *Environ. Entomol.*, 21, 253, 1992.
191. Nordh, M. B., Zavaleta, L. R., and Ruesink, W. G., Estimating multidimensional economic injury levels with simulation models, *Agric. Syst.*, 26, 19, 1988.
192. Backman, P. A., Mack, T. P., Rodriguez-Kabana, R., and Herbert, D. A., A computerized integrated pest management model (AUSIMM) for soybeans grown in the southeastern United States, in *Proc. World Soybean Conference IV*, Pascale, A. J., Ed., Orientacion Grafica Editora, Buenos Aires, 1989, 1494.
193. Waddill, V., Shepard, M., Lambert, J. R., Carner, G. R., and Baker, D. N., A computer simulation model for populations of Mexican bean beetles on soybeans, *S.C. Exp. Stn. Tech. Bull.*, 590, 1975.
194. Mueller, A. J., Sampling three-cornered alfalfa hopper on soybean, in *Sampling Methods in Soybean Entomology*, Kogan, M. and Herzog, D. C., Eds., Springer-Verlag, New York, 1980, 382.
195. Irwin, M. E., Sampling aphids in soybean fields, in *Sampling Methods in Soybean Entomology*, Kogan, M. and Herzog, D. C., Eds., Springer-Verlag, New York, 1980, 239.
196. Herzog, D. C., Sampling soybean looper on soybean, in *Sampling Methods in Soybean Entomology*, Kogan, M. and Herzog, D. C., Eds., Springer-Verlag, New York, 1980, 141.
197. Pedigo, L. P., Buntin, G. D., and Bechinski, E. J., Flushing technique and sequential-count plan for green cloverworm (Lepidoptera: Noctuidae) moths in soybean, *Environ. Entomol.*, 11, 1223, 1982.
198. Browde, J. A., Pedigo, L. P., DeGooyer, T. A., Higley, L. G., Wintersteen, W. K., and Zeiss, M. R., Comparison of sampling techniques for grasshoppers (Orthoptera: Acrididae) in soybean, *J. Econ. Entomol.*, 85, 2270, 1992.
199. Price, J. F. and Shepard, M., *Calosoma sayi* and *Labidura riparia* predation on noctuid prey in soybeans and locomotor activity, *Environ. Entomol.*, 7, 653, 1978.
200. Kogan, M., Ruesink, W. G., and McDowell, K., Spatial and temporal distribution patterns of the bean leaf beetle, *Cerotoma trifurcata* (Forster), on soybeans in Illinois, *Environ. Entomol.*, 3, 607, 1974.
201. Osteen, C. D. and Szmedra, P. I., Agricultural Pesticide Use Trends and Policy Issues, Agric. Econ. Rep. 622, U.S. Dept. of Agriculture Economic Research Service, Washington, D.C., 1989, 75.
202. Moffitt, L. J., Farnsworth, R. L., Zavaleta, L., and Kogan, M., Economic impact of public pest information: soybean insect forecasts in Illinois, *Am. J. Agric. Econ.*, 68, 274, 1986.
203. Shepard, M., Carner, G. R., and Turnipseed, S. G., Colonization and resurgence of insect pests of soybean in response to insecticides and field isolation, *Environ. Entomol.*, 6, 501, 1977.
204. Beach, R. M. and Todd, J. W., Comparison of soybean looper (Lepidoptera: Noctuidae) populations in soybean and cotton/soybean agroecosystems, *J. Econ. Entomol.*, 21, 21, 1986.

205. Higgins, R. A., Approaches to studying interactive stresses caused by insects and weeds, in *Proc. World Soybean Conf. III*, Shibles, R., Ed., Westview Press, Boulder, CO, 1985, 641.
206. Hammond, R. B. and Stinner, B. R., Soybean foliage insects in conservation tillage systems: effects of tillage, previous cropping history, and soil insecticide application, *Environ. Entomol.*, 16, 524, 1987.
207. Smith, J. C. and McPherson, R. M., Spider mites, section 6, *Soybean Insect Management Guidelines*, Publ. 444-046, Virginia Cooperative Extension Service, Blacksburg, 1984, 2.
208. Smith, A. W., Hammond, R. B., and Stinner, B. R., Influence of rye-cover crop management on soybean foliage arthropods, *Environ. Entomol.*, 17, 109, 1988.
209. Poston, F. L. and Pedigo, L. P., Migration of plant bugs and the potato leafhopper in a soybean-alfalfa complex, *Environ. Entomol.*, 4, 8, 1975.
210. Newsom, L. and Herzog, D. C., Trap crops for control of soybean pests, *La. Agric.*, 20, 14, 1977.
211. Troxclair, N. N., Jr. and Boethel, D. J., Influence of tillage practices and row spacing on soybean insect populations in Louisiana, *J. Econ. Entomol.*, 77, 1571, 1984.
212. House, G. J. and Stinner, B. R., Arthropods in no-tillage soybean agroecosystems: community composition and ecosystem interactions, *Environ. Manage.*, 7, 23, 1983.
213. Wuensche, A. L., Relative Abundance of Seven Pest Species and Three Predaceous Genera in Three Soybean Ecosystems, M.S. thesis, Louisiana State University, Baton Rouge, 1976.
214. Terry, I., Bradley, J. R., and Van Duyn, J. W., Population dynamics of *Heliothis zea* (Boddie) (Lepidoptera: Noctuidae) as influenced by selected soybean cultural practices, *Environ. Entomol.*, 16, 237, 1987.
215. Landis, D. W., Effects of No-Tillage Corn and Soybean Production on the Behavior, Development and Survival of *Heliothis zea* (Boddie) Prepupae, M.S. thesis, North Carolina State University, Raleigh, 1983.
216. Morrison, D. E., Bradley, J. R., Jr., and Van Duyn, J. W., Populations of corn earworm and associated predators after applications of certain soil-applied pesticides to soybeans, *J. Econ. Entomol.*, 72, 97, 1979.
217. Isenhour, D. J., Todd, J. W., and Harper, E. W., The impact of toxaphene applied as a post-emergence herbicide for control of sicklepod, *Cassia obtusifolia* L., on arthropods associated with soybean, *Crop Prot.*, 4, 434, 1985.
218. Funderburk, J. E., Pedigo, L. P., and Berry, E. C., Seedcorn maggot (Diptera: Anthomyiidae) emergence in conventional and reduced-tillage systems in Iowa, *J. Econ. Entomol.*, 76, 131, 1983.
219. Hammond, R. B. and Stinner, R. D., Seedcorn maggots (Diptera: Anthomyiidae) and slugs in conservation tillage systems in Ohio, *J. Econ. Entomol.*, 80, 680, 1987.
220. Funderburk, J. E., Wright, D. L., and Teare, I. K., Preplant tillage effects on population dynamics of soybean insect predators, *Crop Sci.*, 28, 973, 1988.
221. Linker, H. M., Coble, H. D., Van Duyn, J. W., Dunphy, E. J., Bacheler, J. S., and Schmitt, D. P., Scouting Soybeans in North Carolina, N.C. Ext. Serv. Publ. AG-385, North Carolina State University, Raleigh, 1988, 14.
222. Jones, W. A. and Sullivan, M. J., Susceptibility of certain soybean cultivars to damage by stink bugs, *J. Econ. Entomol.*, 71, 534, 1978.
223. Daugherty, D. M., Newstadt, M. H., Gehrke, C. W., Cavanah, L. E., Williams, L. F., and Green, D. E., An evaluation of damage to soybeans by brown and green stink bugs, *J. Econ. Entomol.*, 57, 719, 1964.
224. Lambert, L. and Snodgrass, G. L., Tarnished plant bug (Heteroptera: Miridae) populations on a susceptible and a resistant soybean, *J. Entomol. Sci.*, 24, 378, 1989.
225. Kogan, M., Plant resistance in soybean insect control, in *World Soybean Research Conference IV Proceedings*, Pascale, A. J., Ed., Orientacion Grafica Editora, Buenos Aires, 1989, 1519.
226. Shelton, M. D. and Edwards, C. R., Effects of weeds on the diversity and abundance of insects in soybeans, *Environ. Entomol.*, 12, 296, 1983.
227. Orr, D. B., Boethel, D. J., and Layton, M. B., Effect of insecticide applications in soybeans on *Trissolcus basalis* (Hymenoptera: Scelionidae), *J. Econ. Entomol.*, 82, 1078, 1989.
228. Felland, C. M. and Pitre, H. N., Diversity and density of foliage-inhabiting arthropods in irrigated and dryland soybean in Mississippi, *Environ. Entomol.*, 20, 498, 1991.
229. Marrone, P. G. and Stinner, R. E., Effects of soil moisture and texture on oviposition preference of the bean leaf beetle, *Cerotoma trifurcata* (Forster) (Coleoptera: Chrysomelidae), *Environ. Entomol.*, 12, 426, 1983.
230. Marrone, P. G. and Stinner, R. E., Influence of soil physical factors on survival and development of the larvae and pupae of the bean leaf beetle (Coleoptera: Chrysomelidae), *Can. Entomol.*, 116, 1015, 1984.
231. Regniere, J., Rabb, R. L., and Stinner, R. E., *Popillia japonica* (Coleoptera: Scarabeidae): a mathematical model of oviposition in heterogeneous agroecosystems, *Can. Entomol.*, 111, 1271, 1979.

232. Daigle, C. J., Sparks, A. N., Jr., Boethel, D. J., and Mitchell, P. L., Distribution of three-cornered alfalfa hopper (Homoptera: Membracidae) eggs in vegetative stage soybean, *J. Econ. Entomol.*, 81, 1057, 1988.
233. Van Duyn, J. W., Scouting for Soybean Insects, Soybean Insect Pest Management Series, North Carolina Agric. Ext. Serv., North Carolina State University, Raleigh, 1987.
234. Simmons, A. M., Godfrey, L. D., and Yeargan, K. V., Ovipositional sites of the potato leafhopper (Homoptera: Cicadellidae) on vegetative stage soybean plants, *Environ. Entomol.*, 14, 165, 1985.
235. Braman, S. K. and Yeargan, K. V., Intraplant distribution of three Nabis species (Hemiptera: Nabidae), and impact of *N. roseipennis* on green cloverworm populations in soybean, *Environ. Entomol.*, 18, 240, 1989.
236. Isenhour, D. J. and Yeargan, K. V., Oviposition sites of *Orius insidiosus* (Say) and *Nabis* spp. in soybean (Hemiptera: Anthocoriidae and Nabidae), *J. Kan. Entomol. Soc.*, 55, 65, 1982.
237. Herzog, D. C. and Todd, J. W., Sampling velvetbean caterpillar on soybean, in *Sampling Methods in Soybean Entomology*, Kogan, M. and Herzog, D. C., Eds., Springer-Verlag, New York, 1980, 107.
238. Lee, J. H. and Johnson, S. J., Microhabitat distribution of velvetbean caterpillar (Lepidoptera: Noctuidae) pupae in soybean fields in Louisiana, *Environ. Entomol.*, 19, 740, 1990.
239. Buntin, G. D. and Pedigo, L. P., Dispersion and sequential sampling of green cloverworm eggs in soybeans, *Environ. Entomol.*, 10, 980, 1981.
240. Pedigo, L. P., Sampling green cloverworm on soybean, in *Sampling Methods in Soybean Entomology*, Kogan, M. and Herzog, D. C., Eds., Springer-Verlag, New York, 1980, 169.
241. Bechinski, E. J. and Pedigo, L. P., Microspatial distribution of pupal green cloverworms, *Plathypena scabra* (F.), in Iowa soybean fields, *Environ. Entomol.*, 12, 273, 1983.
242. Edwards, C. R., Bledsoe, L. W., and Turpin, F. T., Green Cloverworm on Soybeans, Purdue Univ. Coop. Ext. Serv. E-78, Indiana Cooperative Extension Service, Lafayette, 1991.
243. Turnipseed, S. G. and Shepard, M., Sampling Mexican bean beetle on soybean, in *Sampling Methods in Soybean Entomology*, Kogan, M. and Herzog, D. C., Eds., Springer-Verlag, New York, 1980, 189.
244. Hammond, R. B., Insect control, in *The Soybean in Ohio*, Beuerlein, J., Ed., Ohio State Univ. Bull. 741, Ohio Cooperative Extension Service, Columbus, 1987, 84.
245. Curran, W. S., Gray, M. E., and Shurtleff, M. C., Eds., *Field Scouting Manual*, Illinois Nat. Hist. Surv. X880a, Illinois Natural History Survey, Urbana, 1991.
246. Edwards, C. R., Bledsoe, L. W., and Turpin, F. T., Mexican Bean Beetle on Soybeans, Purdue Univ. Coop. Ext. Serv. E-76, Indiana Cooperative Extension Service, Lafayette, 1991.
247. Poe, S. L., Sampling mites on soybean, in *Sampling Methods in Soybean Entomology*, Kogan, M. and Herzog, D. C., Eds., Springer-Verlag, New York, 1980, 312.
248. Waldbauer, G. P. and Kogan, M., Position of bean leaf beetle eggs in soil near soybeans determined by a refined sampling procedure, *Environ. Entomol.*, 4, 375, 1975.
249. Anderson, T. E. and Waldbauer, G. P., Development and field testing of a quantitative technique for extracting bean leaf beetle larvae and pupae from soil, *Environ. Entomol.*, 6, 633, 1977.
250. Levinson, G. A., Waldbauer, G. P., and Kogan, M., Distribution of bean leaf beetle eggs, larvae, and pupae in relation to soybean plants: determination by emergence cages and soil sampling techniques, *Environ. Entomol.*, 8, 1055, 1979.
251. Todd, J. W. and Herzog, D. C., Sampling phytophagous pentatomidae on soybean, in *Sampling Methods in Soybean Entomology*, Kogan, M. and Herzog, D. C., Eds., Springer-Verlag, New York, 1980, 438.
252. Eckel, C. S., Terry, L. I., Bradley, J. R., Jr., and Van Duyn, J. W., Changes in within-plant distribution of *Helicoverpa zea* (Boddie) (Lepidoptera: Noctuidae) on soybeans, *Environ. Entomol.*, 21, 287, 1992.
253. Spurgeon, D. W. and Mueller, A. J., Girdle and plant-part association of three-cornered alfalfa hopper nymphs (Homoptera: Membracidae) on soybean, *Environ. Entomol.*, 21, 345, 1992.
254. Southwood, T. R. E., *Ecological Methods*, John Wiley & Sons, New York, 1978.
255. Mayse, M. A., Kogan, M., and Price, P. W., Sampling methods for arthropod colonization studies in soybean, *Can. Entomol.*, 110, 265, 1978.
256. Funderburk, J. E., Herzog, D. C., Mack, T. P., and Lynch, R. E., Sampling lesser cornstalk borer (Lepidoptera: Pyralidae) adults in several crops with reference to adult dispersion patterns, *Environ. Entomol.*, 14, 452, 1985.
257. Funderburk, J. E. and Pedigo, L. P., Sampling seedcorn maggots, *Hylemya platura* Meigen (Diptera: Anthomyiidae), in soybeans, *J. Kan. Entomol. Sci.*, 53, 625, 1980.
258. Higley, L. G. and Pedigo, L. P., Examination of some adult sampling techniques for the seedcorn maggot, *J. Agric. Entomol.*, 2, 52, 1985.
259. Pedigo, L. P., Lentz, G. L., Stone, J. D., and Cox, D. F., Green cloverworm populations in Iowa soybean with special reference to sampling procedure, *J. Econ. Entomol.*, 65, 414, 1972.

260. Marston, N. L., Morgan, C. E., Thomas, G. D., and Ignoffo, C. M., Evaluation of four techniques for sampling soybean insects, *J. Kan. Entomol. Soc.*, 49, 389, 1976.
261. Van Duyn, J. W., Sampling Narrow Row Soybeans for Corn Earworm, Soybean Insect Pest Management Series, North Carolina Agric. Ext. Serv., North Carolina State University, Raleigh, 1984.
262. Barstow, B. B. and Edwards, C. R., A new design of clam trap for sampling above-ground arthropods in row crops, *J. Kan. Entomol. Soc.*, 56, 229, 1983.
263. Luna, J. M., Cinker, H. M., Stimas, J. L., and Rutherford, S. L., Estimation of absolute larval densities and calibration of relative sampling methods for velvetbean caterpillar, *Anticarsia gemmatalis* (Hubner) in soybean, *Environ. Entomol.*, 11, 497, 1982.
264. Mayse, M. A., Kogan, M., and Price, P. W., Sampling abundances of soybean arthropods: comparison of methods, *J. Econ. Entomol.*, 71, 135, 1978.
265. Farrar, F. P., Jr. and Crossley, D. A., Jr., Detection of soil microarthropod aggregations in soybean fields using a modified Tullgren extractor, *Environ. Entomol.*, 12, 1303, 1983.
266. Mayse, M. A. and Tugwell, N. P., Sampling Methods for Grape Colaspsis Larvae in Arkansas Soybean/Rice Fields, Ark. Farm Res., No. 29, Univ. Ark. Agric. Exp. Stn., Fayetteville, 1980, 10.
267. Mayse, M. A., Waldbauer, G. P., and Maddox, J. V., Southern corn rootworm eggs in soybean fields, *Fla. Entomol.*, 58, 296, 1975.
268. Waldbauer, G. P. and Kogan, M., Sampling for bean leaf beetle eggs: extraction from the soil and location in relation to soybean plants, *Environ. Entomol.*, 2, 441, 1973.
269. Chang, N. T., Population trends of *Megalurothrips usitatus* (Bagnall) (Thysanoptera: Thripidae) on adzuki bean and soybean examined by four sampling methods, *Plant Prot. Bull. Taiwan*, 30, 289, 1988.
270. Chang, N. T., Seasonal abundance and developmental biology of thrips, *Megalurothrips usitatus*, on soybean at southern area of Taiwan, *Plant Prot. Bull. Taiwan*, 29, 165, 1987.
271. Spurgeon, D. W. and Mueller, A. J., Sampling methods and spatial distribution patterns for three-cornered alfalfa hopper nymphs (Homoptera: Membracidae) on soybean, *J. Econ. Entomol.*, 84, 1108, 1991.
272. Erwin, M. E., Sampling aphids in soybean fields, in *Sampling Methods in Soybean Entomology*, Kogan, M. and Herzog, D. C., Eds., Springer-Verlag, New York, 1980, 239.
273. Ruesink, W. G. and Kogan, M., The quantitative basis of pest management: sampling and measuring, in *Introduction to Insect Pest Management*, 2nd ed., Metcalf, R. L. and Luckmann, W. H., Eds., John Wiley & Sons, New York, 1982, 315.
274. Marston, N., Davis, D. G., and Gebhardt, M., Ratios for predicting field populations of soybean insects and spiders from sweepnet samples, *J. Econ. Entomol.*, 75, 976, 1982.
275. Marston, N. L., Dickerson, W. A., Ponder, W. W., and Booth, G. D., Calibration ratios for sampling soybean Lepidoptera *Heliothis zea*, *Pseudoplusia includens*, and *Anticarsia gemmatalis*: effect of larval species, larval size, plant growth stage, and individual sampler, *J. Econ. Entomol.*, 72, 110, 1979.
276. Kogan, M. and Pitre, H. N., Jr., General sampling methods for above-ground populations of soybean arthropods, in *Sampling Methods in Soybean Entomology*, Kogan, M. and Herzog, D. C., Eds., Springer-Verlag, New York, 1980, 30.
277. Sparks, A. N., Jr. and Boethel, D. J., Evaluation of sampling techniques and development of sequential sampling plans for three-cornered alfalfa hoppers (Homoptera: Membracidae) on soybeans, *J. Econ. Entomol.*, 80, 369, 1987.
278. Hillhouse, T. L. and Pitre, H. N., Comparison of sampling techniques to obtain measurement of insect populations on soybeans, *J. Econ. Entomol.*, 67, 411, 1974.
279. Hammond, R. B. and Pedigo, L. P., Sequential sampling plans for the green cloverworm in Iowa soybeans, *J. Econ. Entomol.*, 69, 181, 1976.
280. Rudd, W. G. and Jensen, R. L., Sweepnet and ground cloth sampling for insects in soybeans, *J. Econ. Entomol.*, 70, 301, 1977.
281. Hamer, J., Soybean insect control, Miss. Coop. Ext. Serv. Publ. 883, 1991.
282. Deighan, J., McPherson, R. M., and Ravlin, F. W., Comparison of sweepnet and ground cloth sampling methods for estimating arthropod densities in different soybean cropping systems, *J. Econ. Entomol.*, 78, 208, 1985.
283. Herbert, D. A. and Harper, J. D., Modification of the shake cloth sampling technique for soybean insect research, *J. Econ. Entomol.*, 96, 667, 1983.
284. Felland, C. M. and Pitre, H. N., Method for collecting arthropods dislodged from soybean plants, *Fla. Entomol.*, 72, 720, 1989.
285. Drees, B. M. and Rice, M. E., The vertical beat sheet: a new device for sampling soybean insects, *J. Econ. Entomol.*, 78, 1507, 1985.
286. Zettler, F. W., Louie, R., and Olson, A. M., Collections of winged aphids from black sticky traps compared with collections from bean leaves and water-pan traps, *J. Econ. Entomol.*, 60, 242, 1967.

287. Halbert, S. E., Zhang, G. X., and Pu, Z. Q., Comparison of sampling methods for alate aphids and observations on epidemiology of soybean mosaic virus in Nanjing, China, *Ann. Appl. Biol. Warwick*, 109, 473, 1986.
288. Menke, W. W., A computer simulation model: the velvetbean caterpillar in the soybean agroecosystem, *Fla. Entomol.*, 66, 665, 1973.
289. Rudd, W. G., Sequential estimation of soybean arthropod population densities, in *Sampling Methods in Soybean Entomology*, Kogan, M. and Herzog, D. C., Eds., Springer-Verlag, New York, 1980, 94.
290. Kuno, E., A new method of sequential sampling to obtain the population estimates with a fixed level of precision, *Res. Popul. Ecol.*, 11, 127, 1969.
291. Bechinski, E. J., Buntin, G. D., Pedigo, L. P., and Thorvilson, H. G., Sequential count and decision plans for sampling green cloverworm (Lepidoptera: Noctuidae) larvae in soybean, *J. Econ. Entomol.*, 76, 806, 1983.
292. Boiteau, G., Bradley, J. R., Jr., Van Duyn, J. W., and Stinner, R. E., Bean leaf beetle [*Cerotoma trifurcata*]: microspatial patterns and sequential sampling of field populations, *Environ. Entomol.*, 8, 1139, 1979.
293. Funderburk, J. E. and Mack, T. P., Population dynamics and dispersion patterns of nymphal three-cornered alfalfa hoppers (Homoptera: Membracidae), *Fla. Entomol.*, 72, 344, 1989.
294. Arthur, T., Defining thresholds for spider mites, *Agric. Consul.*, 45, 13, 1989.
295. Klubertanz, T. H., Environmental and Host Plant Effects on the Two-Spotted Spider Mite in Soybean, M.S. thesis, Iowa State University, Ames, 1989, 78.
296. Higley, L., Klubertanz, T., and Pedigo, L., Spider Mites, Iowa Coop. Ext. Serv. IC-456(17), Iowa Cooperative Extension Service, Ames, 1988.
297. Gray, M., Pepper, G., and Fredericks, J., The spider mite outbreak of 1988: did we learn enough to improve our decision-making capabilities?, in *Proc. Ill. Agric. Pest. Conf. 1989*, Steffy, K., Ed., Illinois Cooperative Extension Service, Urbana, 1989, 31.
298. Bellinger, R. G., Dively, G. P., II, and Douglas, L. W., Spatial distribution and sequential sampling of Mexican bean beetle defoliation on soybean, *Environ. Entomol.*, 10, 835, 1981.
299. Anon., AUSIMM, the Auburn University Soybean Integrated Management Model. Introduction and Program User's Guide, Alabama Cooperative Extension Service, Auburn, 1988, 18.
300. Witkowski, J. F., Wright, R., and Jarvi, K., The Bean Leaf Beetle in Soybeans, Nebguide G90-974, Nebraska Cooperative Extension Institute of Agriculture and Natural Resources, University of Nebraska, Lincoln, 1990, 4.
301. Ruesink, W. G., Introduction to sampling theory, in *Sampling Methods in Soybean Entomology*, Kogan, M. and Herzog, D. C., Eds., Springer-Verlag, New York, 1980, 61.
302. Wang, Z. R., Chu, Z. O., and Zhang, D. S., Underground distribution pattern of white grubs and sampling method in peanut and soybean fields, *Acta Entomol. Sin.*, 29, 395, 1986.
303. Xia, J., Wang, Z., and Li, Z., Preliminary studies on the distribution pattern and methods of sampling of the soybean agromyzid fly (*Melanagromyza sojae* Zehnte), *Acta Phytophylagica Sin. Peking*, 8, 227, 1981.
304. Sprenkel, R. K. and Brooks, W. M., Artificial dissemination and epizootic initiation of *Nomuraea rileyi*, an entomogenous fungus of lepidopterous pests of soybean, *J. Econ. Entomol.*, 68, 847, 1975.
305. Estefanel, V. and Barbin, D., Sequential sampling based on the sequential probability ratio test (SQRT) and its use in the determination of the time of control of the soybean caterpillar (*Anticarsiia gemmatalis* and *Plusia* spp.) in the state of Rio Grande de Sol, *Rev. Cent. Cienc. Rurals. Santa Maria. Univ. Fed. Santa Maria*, 9, 29, 1979.
306. Shepard, M. and Carner, G. R., Distribution of insects in soybean fields, *Can. Entomol.*, 108, 767, 1976.
307. Strayer, J., Shepard, M., and Turnipseed, S. G., Sequential sampling for management decisions on the velvetbean caterpillar, *Anticarsia gemmatalis* (Hubner), on soybeans, *J. Ga. Entomol. Soc.*, 12, 220, 1977.
308. Bechinski, E. J. and Pedigo, L. P., Development of a sampling program for estimation of pupal densities of green cloverworm (Lepidoptera: Noctuidae) in soybeans and evaluation of alternative sampling procedures, *Environ. Entomol.*, 12, 96, 1983.
309. Irwin, M. E. and Yeargan, K. V., Sampling phytophagous thrips on soybean, in *Sampling Methods in Soybean Entomology*, Kogan, M. and Herzog, D. C., Eds., Springer-Verlag, New York, 1980, 283.
310. Helm, C. G., Kogan, M., and Hill, B. G., Sampling leafhoppers on soybean, in *Sampling Methods in Soybean Entomology*, Kogan, M. and Herzog, D. C., Eds., Springer-Verlag, New York, 1980, 260.
311. Kogan, M., Waldbauer, G. P., Boiteau, G., and Eastman, C. E., Sampling bean leaf beetles on soybean, in *Sampling Methods in Soybean Entomology*, Kogan, M. and Herzog, D. C., Eds., Springer-Verlag, New York, 1980, 201.

312. Smelser, R. and Pedigo, L. P., Population dispersion and sequential count plans for bean leaf beetle (Coleoptera: Chrysomelidae) on soybean during late season, *J. Econ. Entomol.*, 85, 2404, 1992.
313. Terry, I., Bradley, J. R., Jr., and Van Duyn, J. W., *Heliothis zea* (Lepidoptera: Noctuidae) eggs in soybeans: within-field distribution and precision level sequential count plans, *Environ. Entomol.*, 18, 908, 1989.
314. Waddill, V. H., Shepard, B. M., Turnipseed, S. B., and Carner, G. R., Sequential sampling plans for *Nabis* spp. and *Geocoris* spp. on soybeans, *Environ. Entomol.*, 3, 415, 1974.
315. Beckinski, E. J. and Pedigo, L. P., Population dispersion and development of sampling plans for *Orius insidiosus* and *Nabis* spp. in soybean, *Environ. Entomol.*, 10, 956, 1981.
316. Funderburk, J. E. and Mack, T. P., Seasonal abundance and dispersion patterns of damsel bugs (Hemiptera: Nabidae) in Alabama and Florida soybean fields, *J. Entomol. Sci.*, 24, 9, 1989.
317. Funderburk, J. E. and Mack, T. P., Abundance and dispersion of *Geocoris* spp. (Hemiptera: Lygaeidae) in Alabama and Florida soybean fields, *Fla. Entomol.*, 70, 432, 1987.
318. Poston, F. L., Pedigo, L. P., and Welch, S. M., Economic injury levels: reality and practicality, *Bull. Entomol. Soc. Am.*, 29, 49, 1983.
319. Gold, H., Wilkerson, G. G., Yu, Y., and Stinner, R. E., Decision analysis as a tool for integrating simulation with expert systems when risk and uncertainty are important, *Comput. Electron. Agric.*, 4, 343, 1990.
320. Batchelor, W. D., McClendon, R. W., Jones, J. W., and Adams, D. B., An expert simulation system for soybean insect pest management, *Trans. Am. Soc. Agric. Eng.*, 32, 335, 1989.
321. Batchelor, W. D., McClendon, R. W., Adams, D. B., and Jones, J. W., Evaluation of SMARTSOY: an expert simulation system for insect pest management, *Agric. Syst.*, 31, 67, 1989.
322. Beck, H. W., Jones, P., and Jones, J. W., SOYBUG: an expert system for soybean insect pest management, *Agric. Syst.*, 30, 269, 1989.
323. Newsom, L. D. and Boethel, D. J., Interpreting multiple pest interactions in soybean, in *Integrated Pest Management on Major Agricultural Systems*, Frisbie, R. E. and Adkisson, P. L., Eds., Texas A & M University Press, College Station, 1986, 232.
324. Hutchins, S. H., Higley, L. G., Pedigo, L. P., and Calkins, P. H., Linear programming model to optimize management decisions with multiple pests: an integrated soybean pest management example, *Bull. Entomol. Soc. Am.*, 32, 96, 1986.
325. Johnson, D. R., Jones, B. F., Kimbrough, J. J., and Wall, M. L., Control Insects on Soybean, Arkansas Agric. Exp. Stn. Publ. FSA2067, Fayetteville, 1991.
326. Ogunlana, M. O. and Pedigo, L. P., Economic injury level of the potato leafhopper on soybeans in Iowa, *J. Econ. Entomol.*, 67, 29, 1974.
327. Shepard, M., Sequential sampling plans for soybean arthropods, in *Sampling Methods in Soybean Entomology*, Kogan, M. and Herzog, D. C., Eds., Springer-Verlag, New York, 1980, 79.
328. Bellinger, R. G. and Dively, G. P., Development of sequential sampling plans for insect defoliation on soybeans, *J. N.Y. Entomol. Soc.*, 86, 278, 1978.
329. Pedigo, L. P., Pitre, H. N., Whitcomb, W. H., and Young, S. Y., Assessment of the role of natural enemies in regulation of soybean pest populations, in *Natural Enemies of Arthropod Pests in Soybean*, Pitre, H. N., Ed., Soybean Coop. Ser. Bull. 285, Mississippi State, 1983, 39.
330. Shepard, M., Marston, N., and Carner, G., Sampling predators, parasites, and entomopathogens, in *Natural Enemies of Arthropod Pests in Soybean*, Pitre, H. N., Ed., Soybean Coop. Ser. Bull. 285, Mississippi State, 1983, 20.
331. Marston, N. L., Sampling parasitoids of soybean insect pests, in *Sampling Methods in Soybean Entomology*, Kogan, M. and Herzog, D. C., Eds., Springer-Verlag, New York, 1980, 481.
332. Irwin, M. E. and Shepard, M., Sampling predaceous hemiptera on soybean, in *Sampling Methods in Soybean Entomology*, Kogan, M. and Herzog, D. C., Eds., Springer-Verlag, New York, 1980, 505.
333. Price, J. F. and Shepard, M., Sampling ground predators in soybean fields, in *Sampling Methods in Soybean Entomology*, Kogan, M. and Herzog, D. C., Eds., Springer-Verlag, New York, 1980, 532.
334. Whitcomb, W. H., Sampling spiders in soybean fields, in *Sampling Methods in Soybean Entomology*, Kogan, M. and Herzog, D. C., Eds., Springer-Verlag, New York, 1980, 544.
335. Carner, G. R., Sampling pathogens of soybean insect pests, in *Sampling Methods in Soybean Entomology*, Kogan, M. and Herzog, D. C., Eds., Springer-Verlag, New York, 1980, 559.
336. Beckinski, E. J. and Pedigo, L. P., Evaluation of method for sampling predatory arthropods in soybeans, *Environ. Entomol.*, 11, 756, 1982.
337. Culin, J. D. and Yeargan, K. V., Comparative study of spider communities in alfalfa and soybean ecosystems: foliage-dwelling spiders, *Ann. Entomol. Soc. Am.*, 76, 825, 1983.
338. Thorvilson, H. G., Pedigo, L. P., and Lewis, L. C., *Plathypena scabra* (F.) (Lepidoptera: Noctuidae) populations and the incidence of natural enemies in four soybean tillage systems, *J. Econ. Entomol.*, 78, 213, 1985.

339. Ignoffo, C. M., Puttler, B., Marston, N. L., Hostetter, D. L., and Dickerson, W. A., Seasonal incidence of the entomopathogenic fungus *Spicaria rileyi* associated with noctuid pests of soybeans, *J. Invert. Pathol.*, 25, 135, 1975.
340. Newman, G. G. and Carner, G. R., Disease incidence in soybean loopers collected by two sampling methods, *Environ. Entomol.*, 41, 231, 1975.
341. Pitre, H. N., Hillhouse, T. L., Donahoe, M. C., and Kinard, H. C., Beneficial Arthropods on Soybean and Cotton in Different Ecosystems in Mississippi, Miss. Agric. For. Exp. Stn. Tech. Bull., Mississippi State, 1978, 90.
342. Roberts, S. J., Mellors, W. K., and Armbrust, E. J., Parasites of lepidopterous larvae in alfalfa and soybeans in central Illinois, *Great Lakes Entomol.*, 10, 87, 1977.
343. Kish, L. P., Greene, G. L., and Allen, G. E., A method of determining the number of potential conidia-forming cadavers of *Anticarsia gemmatalis* infected with *Nomuraea rileyi* in a soybean field, *Fla. Entomol.*, 59, 103, 1976.
344. Culin, J. D. and Yeargan, K. V., Comparative study of spider communities in alfalfa and soybean ecosystems: ground-surface spiders, *Ann. Entomol. Soc. Am.*, 76, 832, 1983.
345. Barry, R. M., Hatchett, J. H., and Jackson, R. D., Cage studies with predators of the cabbage looper, *Trichoplusia ni*, and corn earworm, *Heliothis zea*, in soybeans, *J. Ga. Entomol. Soc.*, 9, 71, 1974.
346. Barry, R. M., A note on the species composition of predators in Missouri soybeans, *J. Ga. Entomol. Soc.*, 8, 284, 1973.
347. Gregory, B. M., Jr., Banfield, C. S., and Edwards, G. B., Spider predation on velvetbean caterpillar moths (Lepidoptera: Noctuidae) in a soybean field, *J. Arachnol.*, 17, 120, 1989.
348. Richman, D. B., Hemenway, R. C., Jr., and Whitcomb, W. H., Field cage evaluation of predators of the soybean looper, *Pseudaletia includens* (Lepidoptera: Noctuidae), *Environ. Entomol.*, 9, 315, 1980.
349. Hagler, J. R., Cohen, A. C., Bradley-Dunlop, D., and Enriquez, F. J., New approach to mark insects for feeding and dispersal studies, *Environ. Entomol.*, 21, 20, 1992.
350. Allen, W. R., Trimble, R. M., and Vickers, P. M., ELISA use without host trituration to detect larvae of *Phyllonorycter blancardella* (Lepidoptera: Gracillariidae) parasitized by *Pholetesor ornigis* (Hymenoptera: Braconidae), *Environ. Entomol.*, 21, 50, 1992.
351. McCarty, M. T., Shepard, M., and Turnipseed, S. G., Identification of predaceous arthropods in soybeans by using autoradiography, *Environ. Entomol.*, 9, 199, 1980.
352. Ragsdale, D. W., Larson, A. D., and Newsom, L. D., Quantitative assessment of the predators of *Nezara viridula* eggs and nymphs within a soybean agroecosystem using ELISA, *Environ. Entomol.*, 10, 402, 1981.
353. Higley, L. G., Zeiss, M. R., Wintersteen, W. K., and Pedigo, L. P., National pesticide policy: a call for action, *Am. Entomol.*, 38, 139, 1992.
354. Higley, L. G., New understandings of soybean defoliation and their implication for pest management, in *Pest Management in Soybean*, Copping, L. G., Green, M. B., and Rees, R. T., Eds., Elsevier Applied Science, London, 1992, 56.
355. Binns, M. R. and Nyrop, J. P., Sampling insect populations for the purpose of IPM decision making, *Annu. Rev. Entomol.*, 37, 427, 1992.
356. McCutcheon, G. S. and Turnipseed, S. G., Effect of planting date on parasitization of soybean lepidopterans by *Cotesia marginiventris* (Hymenoptera: Braconidae), *J. Agric. Entomol.*, 6, 127, 1989.

Chapter 20

SAMPLING ARTHROPOD PESTS IN VEGETABLES

John T. Trumble

TABLE OF CONTENTS

I. Special Considerations in Sampling Vegetable Crops 604

II. Key Pests and Major Techniques Used in Sampling Vegetable Crops 605
    A. Root-Feeding Arthropods 605
    B. Stem-Feeding Arthropods 607
    C. Foliage- and Fruit-Feeding Arthropods 609

III. Representative Programs 611
    A. Tomatoes: Comparing Sampling Programs in Florida, California, and Sinaloa 611
       1. Economic, Horticultural, and Environmental Information 611
       2. Key Arthropod Pests of Tomatoes 612
          a. *Helicoverpa*, *Spodoptera*, and *Trichoplusia* Species 612
          b. *Keiferia lycopersicella* (Walsingham) 614
          c. *Liriomyza* Species of Leafminers 615
          d. Other Arthropod Pests 617
    B. Strawberries: Comparing Sampling Programs in Canada and California 618
       1. Economic, Horticultural, and Environmental Information 618
       2. *Tetranychus urticae* (Koch) 618

IV. Conclusions 620

Acknowledgments 620

References 620

ISBN 0-8493-2923-X/94/$0.00 + .50
© 1994 by CRC Press, Inc.

I. SPECIAL CONSIDERATIONS IN SAMPLING VEGETABLE CROPS

The word "vegetable" is not very descriptive. *Webster's Third Collegiate Dictionary* defines a vegetable as "a usually herbaceous plant grown for an edible part which is usually eaten with the principal part of a meal." Other dictionaries broaden this definition to include any leaf, pod, tuber, or small fruit from virtually any plant which is not grown as a tree. For the purposes of this chapter, "vegetable" is used in the broadest sense, but is limited to those plants which generally are grown in annual, biennial, or 3-year plantings.

This broad definition includes plant species of exceptional diversity, providing substantial challenges for pest population assessment to both researchers and pest control advisors (PCAs). Differences in plant morphology can be dramatic, with some crops such as strawberries growing in dense stands close to the ground, and others such as sweet corn growing over 1.5 m in height. Many plants exhibit substantial changes in morphology with age. The harvested portions of vegetable crops vary from underground tubers such as potatoes, to leaves of spinach, buds and flowers of broccoli and cauliflower, whole heads of lettuce and cabbage, petioles of celery, or above-ground fruit such as tomatoes and beans. Variability is further increased by temporal and geographic disparity in cropping cycles between production areas, use of different cultivars, and a wide disparity in horticultural practices among growers. Finally, the key insect pests of these crops have representatives from virtually all major insect taxa. Not surprisingly, their behavior and damage potential are remarkably diverse. Thus, covering all of the sampling programs for insect pests in vegetable cropping systems is impossible, and this chapter focuses on some representative systems which provide insight into how sampling programs have evolved.

The importance of sampling in vegetable production is increasing. Many vegetables are grown in short-term plantings of considerable value. Although the relatively low thresholds for insect damage or contamination in vegetables would appear to justify a greater effort and cost of sampling to detect low pest population densities, most producers demonstrated a risk-adverse behavior in the 1950s, 1960s, 1970s, and the early 1980s; they sprayed pesticides on a calendar schedule. In spite of the fact that vegetable crops are categorized as "minor use" for pesticides due to low acreage, historically this low priority for registration was not a significant problem since so many new pesticides were developed each year that an effective compound was nearly always available. However, since the late 1970s a substantial reduction in the introduction of new pesticides and increased costs for registration have greatly reduced insecticide availability.[1] This problem has been compounded by the development of pesticide resistance in many key pest species, increasing costs of pesticides to the grower, and a growing aversion of the public to pesticide use on food crops. In addition, until recently vegetable crops have been considered generally unsuitable for biological control efforts because of the high value, expectation of cosmetically perfect produce, and short-term nature of the plantings.[2] Thus, despite the high potential value, the reduced availability of pesticides and a relative lack of proven tactics using natural agents currently limit the available control strategies.

Limitations on the availability of control strategies have important implications for sampling and economic thresholds (ETs), which, in turn, will influence sampling strategy. For example, if all available controls are poor, then the ET will be low (e.g., locating the insect pests may be enough to trigger treatment), and a simple survey for presence/absence may be adequate. If the control strategies are good, and the ET is well defined (even if it is low), then a sampling program with a high level of precision would be desirable. Similarly, if control costs are high, or if pesticide resistance potential is substantial, a highly precise sampling plan is needed. The possible combinations of sampling plans and ETs for vegetables are astronomical. One of the few certainties in sampling in vegetable crops is that the economic impact of sampling errors which underestimate insect populations or damage will be intolerable to producers. For this reason, most PCAs tend to be conservative when selecting sampling plans.

Basic knowledge of ETs for vegetable production systems is, unfortunately, quite limited. Interactions between plant compensatory/tolerance mechanisms and arthropod damage are only poorly understood.[112] Recognizing key types of damage which affect yield is critically important, but has been defined for very few vegetable crop systems. The influence of horticultural practices, geographic location, and environmental variation all interact to make predictions of plant compensation and insect distribution difficult. Nonetheless, some generalizations are possible and are discussed in this chapter where applicable. Anyone interested in assessing arthropod populations must consider the final goal of growers; producing an economic return on their investment. Further, questions on the specific program must be addressed; most importantly, is the information to be used for research into ETs, for efficacy of control strategies in small plot trials, as a practical procedure for determining pest infestation density on commercial plantings, or some combination of the above. For use in commercial operations, the programs must be validated in typical commercial situations. In general, sampling programs tend to evolve from intensive research-oriented procedures to less time- and labor-intensive methods suitable for commercial use. As examples of this process, some representative pests, their associated sampling programs, and related ETs are described in the next section.

II. KEY PESTS AND MAJOR TECHNIQUES USED IN SAMPLING VEGETABLE CROPS

The following insects are representative examples within selected guilds. Space does not permit the listing of all pest-plant systems. For life history and control information for other pest species, the reader is referred to McKinlay,[3] and Metcalf et al.[4] and the references therein. Rather than simply listing all sampling techniques, an attempt has been made to describe the rationale by which procedures were tested and, whenever applicable, adopted.

A. ROOT-FEEDING ARTHROPODS

Among the most difficult arthropods to quantify are those which feed below the soil surface. Extensive destructive sampling, designed to quantify population levels, is both tedious and economically damaging. Few growers are willing to tolerate removal of substantial numbers of plants in the absence of any evidence of plant stress. In some circumstances, destructive sampling may be used to detect the presence of the pests, particularly if some plants obviously have been stunted or are stressed by root loss. However, most current sampling programs focus on the presence/absence or movement of the above-ground stages of the insects.

An exception is the use of attractant bait traps for some arthropods (see Chapter 16 for examples in corn). This technique has proven particularly effective for wireworms (Coleoptera: Elateridae or Tenebrionidae). Generally, food baits such as corn, potatoes, or oats are placed in $15 \times 15 \times 10$ cm holes and covered with soil and a layer of plastic. The plastic serves to retain moisture and elevate soil temperatures, both of which have been shown to improve collection efficiency.[5] However, in areas where soil temperatures and moisture levels are normally high, the plastic covering may not be necessary.[6] Because the traps draw the insects from a variable area depending on climatic conditions and attractiveness of the bait, the resulting data provide an indication of presence or absence of the pests and a relative estimate of population intensity.

Arthropods feeding beneath the soil surface often can be separated into two categories; those feeding on the marketable portion of the plant and those causing economic losses through root damage. Sampling programs or representatives of each of these categories are discussed here, beginning with the potato tuberworm, *Phthorimaea operculella* (Zeller), which feeds on the potato tuber. A second pest, the lettuce root aphid, *Pemphigus bursarius* (L.), feeds on the roots of vegetable crops such as lettuce and celery, for which only the above-ground portions of the plants are marketed.

The potato tuberworm is a significant pest of potatoes throughout the world.[7] Although foliar damage can cause yield losses if stems are mined,[8] most economic losses occur when larvae infest the tubers.[9] Historically, pesticides have been used on a weekly or calendar spray schedule following visual detection of adults or larvae in the foliage. The development of more precise economic injury levels (EILs) and sampling plans have been hampered because visual counts of larval densities in the foliage are tedious and time consuming, and attempts to correlate bait traps and light traps with foliar infestations have been unsuccessful.[10] More recently, the documentation of an effective pheromone, and the development of a suitable pheromone trap, allowed PCAs to relate pheromone trap catches to foliage counts and tuber damage.[11,12] However, several drawbacks to pheromone trapping have been noted, including the lack of correlation between trap counts and tuber infestation which can occur if horticultural practices are altered. For example, increasing hill sizes and use of drip irrigation can reduce tuber exposure, eliminating statistical relationships between tuber infestation and trap catches. In addition, the use of pesticides may affect sampling efficacy; Bacon[13] reported that high levels of larval suppression in the foliage did not assure that tuber infestations would be proportionately reduced. Thus, in spite of the worldwide importance of potatoes as a staple food source, the variable impacts of diverse environmental and horticultural factors on potato tuberworm populations have severely limited the development of EILs and precise sampling plans.

An example of a more effective sampling program for a root-feeding insect can be provided for the cabbage maggot, *Hylema brassica* (Weidemann). This dipteran is an important pest of *Brassica* crops in much of the northern hemisphere, including Europe, Canada, the U.S., and the former USSR.[14] The cabbage maggot overwinters in the pupal stage, emerging in the spring to start a seasonal cycle of 2 to 4 generations, depending on location. The highly mobile adult females oviposit in cracks in the soil near the stems of young plants, and the resulting larvae burrow down to feed on the roots. A single plant may support hundreds of larvae, whose feeding causes stunting, wilting, and entryways for plant pathogens.

Predicting spring emergence of the adults is critical both for timing of adulticide treatments as well as effective implementation of additional sampling strategies. Several researchers have reported success using thermal summation techniques (e.g., accumulated degree-days).[15-17] The use of this technique for predicting emergence of subsequent generations is complicated by the occurrence of a summer aestivation of the pupal stage when soil temperatures exceed 21 to 22°C,[18] but this problem can be solved mathematically.[19] Procedures for developing thermal summation models are readily available.[20,21]

An alternative approach to calculating degree-day accumulation is to relate emergence to the phenology of common plant species in the region. This relatively simple method relies on the plant to integrate key environmental parameters such as temperature, daylength, and soil moisture, and has proven effective for *H. brassicae* in commercial situations.[22] In New York, the initial flowering of yellow rocket (*Barbarea vulgaris*) has been demonstrated to reliably coincide with spring emergence of the pests.[23]

Sampling plans are currently available for the eggs, pupae, and adults of the cabbage maggot. Harcourt,[24] using visual counts of the eggs on the soil surface near the base of cabbage plants, developed a sequential sampling plan based on the common 'k' of the negative binomial distribution (see Chapter 8 as well as Shelton and Trumble[25] for an explanation of sequential sampling designs). An ET of 20 or 30 eggs per plant was proposed, but not verified from field trials. The beauty of Harcourt's sequential technique is that decision making potential for sampling efforts is maximized. Unfortunately, when Finch et al.[26] conducted a more detailed evaluation of the negative binomial approach, they found the common 'k' could not be consistently fit to the data. Ultimately they created a sampling plan based on Taylor's Power Law (TPL), which was designed to determine the number of samples necessary to estimate population density with a fixed-precision level (see Chapter 10). However, because the sampling program had been changed from a visual estimate to a 5 × 2 cm core of soil washed through a screen to collect eggs, the ET used by Harcourt was no longer appropriate, but a new ET was not suggested. This situation is typical of many vegetable cropping systems; sampling plans have been developed but ETs have not been determined. Similarly, many proposed ETs have not been validated in commercial operations.

In a related study, Finch et al.[27] developed a fixed-precision level sampling scheme for *H. brassicae* pupae. This scheme required removing a 15-cm soil core from around plants, stirring each sample into a 9-l bucket of water, and rinsing the soil through a screen sized to collect the pupae. Interestingly, the distribution of the overwintering population was different from the summer populations, and therefore mandated a separate sampling program. In continental Europe, a threshold of ten pupae per plant has been proposed as an ET which will prevent more than 5% crop damage.[28] Unfortunately, like the egg sampling technique, the pupal method has proven impractical for large-scale field sampling. The time, materials, and effort in collecting and washing up to eight soil samples per hectare becomes prohibitive when a PCA has only 2 to 3 h to spend in a 150-ha field.

Adult trapping has proven to be one of the most effective monitoring tools for cabbage maggots. A variety of sticky traps, yellow, white, and blue water traps, and cone traps have been tested for adult capture.[29,30] The most effective design seems to be a "marigold" yellow water trap baited with the synthetic attractant, allylisothiocyanate.[31] The color was selected over the standard fluorescent yellow because *H. brassicae* counts were statistically equivalent and fewer nontarget arthropods were collected, making sorting of catches more rapid. When used in conjunction with the thermal summation model, the traps need be deployed only for a short period, coinciding with the predicted emergence dates, to determine if pesticide applications are warranted.

B. STEM-FEEDING ARTHROPODS

Cutworms are among the most difficult stem-feeding insects to sample. The larvae usually sever the stems of young plants at night and then hide beneath the soil surface during the day. Tomatoes, beans, celery, cole crops, and a host of other vegetable species can be attacked. Two of the most common cutworms are the variegated cutworm, *Peridroma saucia* (Hübner), and the black or greasy cutworm, *Agrotis ipsilon* (Hufn.). These pests may appear in fields in several ways: (1) adults migrate in and oviposit throughout the field, (2) larvae may be present from a previous planting, or (3) larvae may migrate in from weed hosts along field margins.

Knowledge of the field entry mechanism is important, as this will affect the distribution of the pests and therefore the sampling program chosen. For immigrant moths or larvae remaining from the previous crop, the population will be dispersed throughout the field. For migrating larvae, the border rows will contain the bulk of the population, and sampling can therefore be concentrated on the edges of the field. Visual searches often can detect the population distribution; careful survey of the field for lines of cut or wilting plants is a necessary prerequisite to subsequent, more detailed searches. The University of California IPM Manual[32] for tomatoes suggests an ET for a visual search at dawn of one larvae per minute.

A baiting strategy can be useful for determining the presence of larvae remaining after a previous crop prior to planting of the next crop. A typical bait includes wheat bran plus molasses, and sometimes a pesticide.[33] The baits are placed in a pitfall trap such as the Missouri cutworm trap, a modified pitfall design developed by Story and Keaster.[34] This trap has since been altered to include a vertical screen for visual attraction, which has improved collections.[35] This technique represents a considerable saving in time and effort over the soil sieving approach used previously.[36] For some cutworm larvae, the sampling technique with the best relative net precision[37] is a piece of sacking (burlap) or black plastic which acts as refuge during the daylight hours.[38] Such materials are relatively inexpensive, and within-field distribution can be rapidly estimated with minimal effort.

Adult trapping has proven useful for documenting migrations and generation peaks. Because many vegetable crops are susceptible to stand loss by cutworms for only a relatively short period after germination or transplanting, detecting the arrival of immigrants just before or during the susceptible plant stage is of critical importance. Although blacklights have been used successfully for this purpose,[39] a variety of factors, including weather effects and lunar periodicity, can dramatically affect trap catches.[40] In addition, because of a notable lack of specificity, light trap collections often require substantial commitments in time for sorting and identifying the catches.

If the cutworm species is known, pheromone trapping can provide a more species-specific technique. A variety of pheromone blends and trap designs have been studied for many of the economically important cutworm species.[41-43] However, choice of trap color can affect both specificity and total catch. Hendrix and Showers[44] documented lower catches for *A. ipsilon* using green traps, and suggested that white traps would be more desirable than yellow traps in some locations due to lower attractiveness to nontarget insects such as bumblebees.

When selecting the trap design for a particular species, the sampler should consider the duration of sampling and the possible numbers of moths which may be collected. Some traps are capable of collecting large numbers of adults (bucket traps, some water traps), while other designs using adhesive surfaces may become "saturated" with moths and, although additional moths may be attracted, they may not be collected. Many commercially available traps are made from plastics or metal and will last for many years, while others have been designed for a single use or a single season. Selection of the appropriate trap therefore depends on the specific needs in each crop (efficiency, cost, convenience, etc.). Unfortunately, few ETs have been developed for insect pests of vegetable crops using pheromone traps; correlative relationships between adult catches and larval counts have been difficult to document.

C. FOLIAGE- AND FRUIT-FEEDING ARTHROPODS

Foliage- and fruit-feeding arthropods are among the most diverse and destructive pests occurring on vegetables. Virtually every order of insects, and most of the economically important groups of spider mites are represented. Of these pests, one of the most damaging and most studied insects is the diamondback moth, *Plutella xylostella* (L.) (Lepidoptera: Plutellidae). The diamondback moth is a pest of cruciferous crops throughout the world, causing millions of dollars in damage annually. Losses result when larval feeding destroys plants, causes imperfect heading or unacceptable produce for the consumer, or when the larvae or pupae become contaminants. The contamination problem is most acute for the frozen food processing industry, which is subject to strict limitations on insects or insect parts. The pupae, which are attached to the plant surface with silk, are a particularly important problem.

Some of the earliest sampling studies of *P. xylostella* were conducted by Harcourt.[45] He determined that the negative binomial distribution provided a better fit to larval count data than the Poisson distribution. These data were then used to develop a sequential sampling plan on cabbage.[46] Subsequently, Theunissen and den Ouden[47] compared a series of different sampling strategies, including systematic sampling (200 plants per hectare), simple random sampling (100 plants per hectare), cluster sampling (20 clusters of 5 plants per hectare) and pilot sampling (random sampling of 10 plants per hectare). They concluded that while systematic sampling produced the best estimate (tightest confidence intervals), "presampling" via pilot sampling to determine that the pest population is approaching a threshold level may be the most cost effective approach. Subsequently, they used this background information to generate sequential sampling programs for the major pest species in cabbage.[48] The sequential sampling technique integrates the efficiency of the rapid pilot sampling approach with the reduced confidence intervals provided by systematic sampling.

More recently, several significant efforts to improve sampling efficiency have been made. In numerous vegetable crops, studies on the within-plant distribution of insects have allowed efficient estimates of whole plant counts based on sampling a relatively small portion of the plant.[49-51] Sears et al.[52] tried a modified approach; instead of determining the distribution of *P. xylostella* within the plant, they based their sampling program on the portion of the plant to be harvested (e.g., the head + ten wrapper leaves as a buffer). Unfortunately, the correlation between this artificial, economically important sample and whole plant counts was not significant. Thus, experiences to date indicate that partial plant samples should be based on detailed within-plant studies using available information on the biology and behavior of the pests, rather than on the portion of the plant to be harvested.

Hoy et al.[53] noted that even sequential sampling programs using a plant subsample lose considerable practical application because such plans do not always allow an entire field to be surveyed. As a result, other arthropod, disease, or weed problems may be missed by the PCA. They developed a modified technique called variable-intensity sampling that uses an algorithm to select the most efficient number of plants to be sampled along a V-shaped transect through a field. The orientation of the transect is changed each time to ensure the entire field is seen at least every two samples. The precision of sampling is adjusted according to the mean and its importance to decision making. Like standard sequential sampling plans, decisions can easily be reached if populations are exceptionally high or exceptionally low. However, in the critical range where the standard approach requires additional sampling (often *ad infinitum*), the variable-intensity technique rapidly reaches a decision. Originally the technique was designed to be used with larval count data, but since has been found to be useful with binomial (presence-absence) data.[54] Thus, a researcher or PCA can now walk the field in a preselected pattern and simply count the presence or absence of larvae on a minimum number of plants and reach a decision.

Unfortunately, as every PCA rapidly discovers, the diamondback moth is not the only pest in cabbage. The plants are usually infested with a complex of lepidopterous insects including the cabbage looper, *Trichoplusia ni* (Hübner), the imported cabbageworm, *Pieris rapae* (L.), and the beet armyworm, *Spodoptera exigua* (Hübner). Using separate sampling programs for each of these pests would be prohibitively complicated in most field situations. Therefore, many researchers have attempted to develop composite sampling plans for all caterpillars. Some of these plans require that all larvae be counted regardless of species.[55-58] Others minimized the sampling effort by developing presence-absence programs where plants with any lepidopterous larvae are counted as infested.[59] Because larvae are often cryptic, and detailed searches of large plants is tedious and time consuming, sampling plans were developed based on the percentage of plants with presence of new damage.[60,62] Although this methodology requires that the sampler learn to distinguish new from old damage, the use of a presence-absence sequential sampling plan permits decisions to be reached rapidly. Not surprisingly, presence-absence sampling plans seem to be strongly preferred by most PCAs.

On occasion, it may be necessary to sample *P. xylostella* larval populations for the development of pesticide resistance, which has been documented to occur at levels causing field failures.[63] Several techniques are available, including standard topical application, uniform droplet exposures, leaf dips, and residual assays in petri dishes or vials. Each of these approaches has advantages and disadvantages. For example, the topical application approach requires the purchase of an expensive microapplicator and is fairly labor intensive, but provides a high level of reliability. However, this technique is not suitable for those pesticides which must be ingested. Resistance evaluation with uniformly sized droplets sprayed onto the leaf substrate also requires the purchase of expensive equipment, a modified on-demand uniform droplet generator, but allows both residual (contact or ingestion) and topical assays.[64] This technique has the advantage of allowing larvae to encounter a dose distribution similar to that occurring in the field. Such distributions have been shown to affect larval behavior and development, two factors capable of influencing control efficacy and resistance development.[65,66] The leaf dip technique, where leaf material is dipped in solutions of pesticides and allowed to dry before the insect is given access, is relatively inexpensive. Like the uniform droplet technique, the residue permits contact and/or ingestion toxicity to be measured. However, tests which last beyond 24 h become increasingly labor intensive because the leaf material must be replaced.

Magaro and Edelson[67] used the leaf dip technique to determine dose response curves for *P. xylostella* larvae. They further simplified the assessment of resistance by modifying a residual vial technique[68] such that disposable cups were coated with a discriminating lethal dose for 90% of a susceptible population. Field-collected third instars are placed in these inexpensive cups and held for 4 h prior to calculating percent mortality. This provides a relatively simple, inexpensive, and extremely rapid resistance assessment which can be used by most PCAs. The drawbacks to this technique include the requirement for collection of larvae of a specific stage and the inability to measure larval resistance against those pesticides which must be ingested.

Although not yet available for the diamondback moth, attractant trap assays for monitoring both resistance development and population dynamics have been constructed for other vegetable crop pests. These traps are similar in that they have pesticides impregnated in the polybutene sticker used to hold the adults captured. In addition, these techniques all require a substantial amount of developmental effort; effects of insect age, time of capture, and bioassay for mortality, etc. need to be determined. In some cases, if the adult is not the target stage, relationships between adult resistance and target stage resistance must be established. However, once available, they can provide a rapid and accurate assessment of resistance to contact pesticides. Like the residual vial technique, the attractant traps may be most efficient when used with a discriminating dose. Sanderson et al. developed such a technique for dipterous leafminers, *Liriomyza trifolii* (Burgess), using yellow sticky cards.[69] Pheromone-baited traps have been used successfully in vegetables to monitor insecticide resistance in *S. exigua*[70] and the tomato pinworm, *Keiferia lycopersicella* (Walsingham).[71] At least in the case of *S. exigua*, the attractant trap technique has been built into a sequential sampling program which maximizes the input effort by minimizing the numbers of traps and adult moths required to reach a decision on the potential effectiveness of a pesticide.[72]

III. REPRESENTATIVE PROGRAMS

A. TOMATOES: COMPARING SAMPLING PROGRAMS IN FLORIDA, CALIFORNIA, AND SINALOA

1. Economic, Horticultural, and Environmental Information

Tomatoes destined for the North American market are grown primarily in California, Florida, and in the state of Sinaloa on the west coast of Mexico. The value of the tomato crops in each of these areas exceeds $100 million annually; in Sinaloa alone the tomatoes grown on 50,000 ha are valued at nearly $1 billion each year.[73] Unlike Florida, where primarily fresh market tomatoes are produced, California and Sinaloa have extensive plantings of processing tomatoes. Although many arthropod pests feed on both processing and fresh market tomatoes, the impact of each specific pest can vary between the two types of tomatoes, and the following discussion focuses primarily on sampling programs in fresh market tomatoes.

Horticultural conditions which affect sampling strategies vary between locations. In Sinaloa and Florida, fresh market tomatoes are grown in a "staked" or "trellised" fashion. In addition, the tomatoes grown are indeterminate varieties, producing fruit continually over several months. While a few hundred acres are grown in a similar fashion in southern California, most tomatoes are bush-type (nonstaked) determinate cultivars, which are harvested only once. Thus, sampling strategies in Florida and Mexico need to take into account that all stages of fruit may be present at the same time. Most of the tomatoes in both California and Florida are produced during a single crop each year (summer in California, winter-spring in Florida; spring and fall tomatoes may still be planted, but on much less acreage), while most growers in Sinaloa plant three overlapping crops in the fall, winter, and spring. This pattern of cropping in Sinaloa allows exceptionally large pest populations to build up by the spring planting. Further, irrigation practices differ, with increasing acreage in California and Florida using drip systems, and Sinaloa growers relying on furrow irrigation. Environmental conditions also are variable, with Florida and Sinaloa experiencing occasional rainfall and high relative humidity, while California's tomatoes are grown during the dry season when rain is uncommon and humidity is low.

Selection of tomato cultivars is inconsistent within each area, and often does not overlap between locations. Several studies have documented variable susceptibility of selected cultivars to key arthropod pests.[74,75] Any factor which changes the acceptance or suitability of a crop has the potential to alter how the pests are distributed in the field. This phenomenon, which has been described for environmental and cultivar effects in other vegetable crop systems, causes substantial changes in fixed-precision level sampling programs.[76]

In addition to the horticultural and environmental differences between tomato production areas, differing political and societal influences have substantial effects on sampling programs. In the U.S., considerable effort is made to increase the efficiency of each sampling procedure in an effort to cut sampling time (cost) to the bare minimum. In Sinaloa, the situation is quite different. Many of the tomato growers are subsidized by the government because they provide not only an important "balance of trade" item with the U.S. and Canada, but also because they hire a substantial proportion of the population in the region.[77] In fact, much of the labor force lives on site in villages established by the grower. In comparison to the U.S., labor costs are dramatically lower in Mexico. As a result, labor in Sinaloa for sampling is readily available. This availability of labor allows the employment of sampling procedures which would be economically prohibitive in the U.S.

2. Key Arthropod Pests of Tomatoes
a. Helicoverpa, Spodoptera, *and* Trichoplusia *Species*

Some of the key arthropod pests of tomatoes and their relative importance in each tomato production area are listed in Table 1. Interestingly, at some locations the armyworms, fruitworms, and loopers are often sampled as a group, probably in an effort to minimize the training and sampling time required for identification.[78] This occurs despite the different feeding behaviors exhibited by each of the species. Even though all of these species can feed and successfully mature on foliage alone, they have divergent feeding strategies on fruit. For example, the tomato fruitworm, *Helicoverpa zea* (Boddie) (Lepidoptera: Noctuidae), may penetrate a fruit as a first instar, completely hollow out the fruit, and later emerge as an adult. Most *H. zea* larvae probably feed in two or three fruit during development. In contrast, the beet armyworm feed on ten or more fruit, penetrating half a body length into each fruit one or more times. Once the epidermis of a fresh market fruit is injured, the fruit is essentially unmarketable. Thus, for fresh market tomatoes, *S. exigua* would seem more damaging. However, mortality rates for first instar *S. exigua* often exceed 95%, possibly due to their positively phototactic behavior and susceptibility to UV radiation.[79] For processing tomatoes, the internally feeding fruitworms become economically significant contaminants while injury caused by the beet armyworm frequently heals and allows normal harvest and processing.

Sampling programs for the immature stages of armyworms and fruitworms are quite variable between the three production areas. In Florida the fields are partitioned into 1-ha sections and six contiguous plants from each section are sampled twice weekly. The suggested action threshold (essentially equivalent to an ET) is one larva per six plants in the prebloom stage.[80] After bloom, detection of any eggs or larvae is enough to trigger treatment.[81] In California, fields are divided in 4-ha sections and 25 plants are arbitrarily selected and sampled weekly in each section.[82] The suggested ET for fruitworms is 5% of the plants with first to third instar present. For armyworms, the ET was set at 20% of the plants with first to third instar present. In all locations, most of the sampling effort is expended in finding the small larvae because (1) pesticidal control is more effective than for larger larvae, (2) damage potential increases with increasing size, and (3) large larvae of the beet armyworm become negatively phototactic, hiding below ground or in the densest foliage in the canopy.[83]

In Sinaloa, *S. exigua* eggs and larvae were sampled in large experimental plantings by examining 60 to 120 plants per hectare on a diagonal transect.[84] In large-scale

TABLE 1
Relative Importance of Insect Pests of Tomatoes in Florida and California, U.S., and Sinaloa, Mexico[a]

Pest species	Florida	California	Sinaloa
Armyworms and fruitworms			
Beet armyworm	XXX	XXXX	XXXX
Spodoptea exigua (Hübner)			
Tomato fruitworm	XXXX	XXX	XXXX
Heliocoverpa zea (Boddie)			
Southern armyworm	XXXX	X	X
Spodoptera eridania (Cramer)			
Cabbage loopers			
Trichoplusia ni (Hübner)	XXX	XX	XX
Tomato pinworm			
Keiferia lycopersicella (Walsing.)	XXXX	XXXX	XXXX
Leafminers			
Liriomyza trifolii (Burgess)	XXXX	XXXX	XXXX
Liriomyza sativae Blanchard	XX	XX	XX
Aphids			
Potato aphid	XX	XXX	XXX
M. euphorbiae (Thomas)			
Green peach aphid	XX	XX	XXX
Myzus persicae (Sulzer)			
Thrips			
Western flower thrips	X	XX	XX
Frankliniella occidentalis Pergande			
Tobacco thrips	XX	XX	XX
F. fusca Hind.			
Stink bugs			
Southern green stink bug	XX	XX	X
Nezara viridula L.			
Whiteflys			
Sweetpotato white fly	XXXX	XXX	XXX
Bemisa tabaci (Gennadius)			

[a] XXXX = highly significant; X = nonexistent or unimportant; see text for documenting references.

commercial operations, current recommendations suggest monitoring 10 m of plants in rows 1, 5, and 10 on each side of the field, and a minimum of 100 plants from the central portion of the field.[85] An ET of one larvae per four plants is used. For *H. zea*, finding four or more viable eggs (not parasitized) in an arbitrary sample of 30 leaves from each station is enough to require implementation of pesticidal controls. Numbers of parasitized eggs from each station are also recorded as a measure of the effectiveness of *Trichogramma pretiosum* (Riley) releases. The sampling procedure in Sinaloa is the only fresh market program that routinely calls for assessment of egg parasitism.

Although the larval stages of *H. zea*, *S. exigua*, and *T. ni* are most commonly sampled, information on adult migrations and generational peaks also can be useful.[86,87] Regional trapping strategies provide information on the beginning and end points of emergence and flight activity.[88] Such data can be helpful in determining the probable onset of oviposition by *H. zea*[89] or *S. exigua*[90] in tomatoes. Often, the development and maintenance of regional trapping systems are not feasible given the resources of a single PCA. However, regional information may be available from state or county offices, loose networks of cooperating PCAs, or from the larger grower operations. Even the use of individual traps can be beneficial, providing some information on adult appearance and movements.[91] However, many of the same caveats mentioned earlier (sorting time, multiple species catches, environmental effects, etc.) for adult sampling apply here.

Despite the availability of pheromone trapping technology for more than 10 years, relatively few statistically accurate sampling programs have been based on this technique. An exception is the program developed by Brewer and colleagues[92,93] to monitor the occurrence of pyrethroid resistance in *S. exigua*. The program required the development of a considerable amount of background information, including a discriminating dose of fenvalerate in the sticker of pheromone traps (that is, a dose which will kill nearly all susceptible adults but few of the resistant moths) and information of the relationship between adult and larval resistance. Using this information, a sequential sampling plan based on the sequential probability ratio test[94] was constructed. The technique allows populations to be rapidly categorized as resistant or susceptible with a high degree of reliability. Several significant advantages were realized with this approach: (1) determination of probable field failure of the pesticide could be made prior to application, (2) sample sizes, and thus cost and effort, could be minimized, and (3) the rapidity of the test provided more timely information on resistance levels than the use of a topical application method. The use of such techniques is likely to become increasingly important as the availability and use of pesticides declines. However, this approach has some substantial drawbacks that include the need for a large amount of developmental effort and the probability that a practical relationship between insecticide resistance levels in the adult stage and larval stage may not be available for many pest species.

b. Keiferia lycopersicella *(Walsingham)*

The tomato pinworm is frequently the single most damaging pest in tomatoes. High populations of this cosmopolitan gelichiid moth can result in the abandonment of entire fields prior to harvest. Although larval mining and folding of leaves cause substantial reductions in photosynthetic activity,[95] most of the injury is caused when the larvae penetrate the tomato fruit, often just beneath the calyx.

To date, attempts to model the population dynamics of *K. lycopersicella* in fresh market tomatoes have not been successful. Such models have been used effectively for many years in other crops to predict the occurrence and development of key pest species.[96] Despite considerable background information on the growth and development of *K. lycopersicella* at different temperatures and on various stages of plant growth,[97,98] and knowledge of temperature relationships between local weather stations and within the tomato canopy,[99] no models have been published. In southern California, variation in larval development and survival due to local environmental conditions has resulted in models too complicated for practical application (Trumble and Wiesenborn, unpublished). For example, air pollutants such as ozone dramatically affect plant chemistry, which will allow an increase in *K. lycopersicella* developmental rate by up to 10%; survival can more than double.[100] Interfield migration and large irregular immigrations also tend to confound efforts to model local populations.

Like the lepidopterous pests previously discussed, most sampling plans for *K. lycopersicella* rely on assessment of the larval stage. In Florida the fields are divided into approximately 1-ha sections and six plants are counted in each section. Initially, the number of larvae and mines are counted on all foliage when plants are small (with zero to three true leaves present), or the top three or four leaves (three or more true leaves present) as plants increase in size.[101] More recent recommendations have been modified to include whole-plant counts when zero to seven true leaves are present, and leaf samples from the lower portion of the tomato canopy. The latter change was made in response to some within-plant distribution analyses conducted by Pena,[102] who determined that the most statistically reliable estimates of populations could be achieved from sampling the lower half of the canopy. The ETs currently proposed include 0.7 larvae per plant for whole plant samples (< 7 leaves), and 0.7 larvae per leaf for plants with more than seven true leaves.[103]

Recommended ETs are much lower in California and Sinaloa than in Florida. In California, several recommendations have been tested. Using the previously mentioned sampling procedures for California, Toscano et al.[104] field tested a threshold based on percent infestation: if 5% of the plants had live, unparasitized larvae, treatments were considered justified. The resulting savings over a weekly spray schedule were substantial. This provides an example of one of a small group of successful sampling plans that were designed for convenience and practicality, rather than on extensive previous research showing statistical relationships between insect counts and population density or damage. However, even this approach requires fairly extensive experience with the pest-crop relationship. The drawback of this method is that although the savings were substantial with the 5% infestation EIL, potential additional savings generated by higher EILs may have been lost. More traditional approaches also have proven successful in California. Wiesenborn et al.[105] determined that a threshold of 0.5 larvae per plant provided a superior economic return over weekly or biweekly treatments, or the use of higher or lower larval ETs. In Sinaloa, a modified treatment level of 0.25 larvae per plant along 10-m sections of row provided a significant reduction in *K. lycopersicella* damage over grower practices, leading to a substantial economic benefit.[106]

Unlike Florida and California, labor costs in Sinaloa allow sampling for *K. lycopersicella* eggs. Because the eggs are small, ~1 mm, and placed between trichomes on the abaxial side of the leaf, locating and counting them requires both patience and time. Random removal and examination of 30 leaves from rows 1, 5, and 10 for each side of a planting, as well as an additional sample from a central location, can provide useful information on the oviposition and parasitization rates occurring in the field. This type of datum can be particularly useful if a pheromone confusion control technique is being employed. When pheromones are applied over large areas, collections in pheromone-containing traps are often reduced, and insect pest populations may still be high in the foliage. This egg sampling strategy has proven effective in Sinaloa for determining the degree of success offered by a pheromone confusion program for *K. lycopersicella*.[107]

Pheromone traps as sampling tools for *K. lycopersicella* have been used with mixed success. Van Steenwyk et al. developed an ET of 20 moths per trap per night in California which worked in experimental plantings. A statistically significant relationship was presented between adult numbers in traps and larval numbers per plant. Unfortunately, when subsequently tested on large-scale plantings, no relationship could be established.[108] In Sinaloa, the potential use of an ET based on pheromone trap catches has not been considered practical because populations often exceed 20 moths per trap per night prior to transplanting.[109] Nonetheless, pheromone traps can be used to determine the onset and termination of major migrations and generational peaks. In addition, pheromone traps with insecticide-impregnated inserts have proven effective in documenting pesticide resistance levels in California and Sinaloa.[110]

c. Liriomyza *Species of Leafminers*

Few insects in tomatoes have been as intensively studied as the leafminers *Liriomyza trifolii* (Burgess) and *L. sativae* Blanchard. These insects damage tomatoes by mining in the palisade mesophyll tissue of the leaf. A single mine can cause a reduction in photosynthetic output of a leaflet of more than 60%.[111] However, documenting an EIL for leafminers has proven difficult, partly due to an excess in photosynthate production by the plant (see Trumble et al.[112] and references therein). Regardless, high population densities can cause premature leaf abcission, resulting in excessive leaf loss, potential reduction in fruit size, and "sunburn" of the fruit.

Although these insects are frequently held below economic densities by a complex of parasitic Hymenoptera, the use of many agrichemicals for control of lepidopterous pests can selectively remove the natural control agents, allowing the *Liriomyza* species to increase dramatically.[113] Thus, the most effective sampling plans survey for both the leafminers and their associated parasites. The potential value of the parasites is such that a rapid resistance monitoring technique based on pesticide residues in vials has been developed for some key parasite species.[114]

Sampling programs based on the larval, pupal, and adult stages of leafminers are available. In Florida and California, the larval populations have been quantified using foliar searches. Initial attempts to use total numbers of mines per plant[115] have proven less practical than counting numbers of live larvae on just a portion of the plant.[116,117] Not only are whole-plant counts of all mines on large tomato plants time consuming, but first instar mines are difficult to detect. In addition, not all leaves with leafminer damage are abcissed; as a result, the irregular accumulation of empty leafmines or even new leafmines contained parasitized larvae can trigger an unnecessary treatment. In Florida, most scouts seemingly have adopted a sampling strategy whereby live larvae, or live and dead larvae, are counted on (1) all shoots if plants have zero to two true leaves, (2) the terminal trifoliate of the third fully expanded leaf from the top of the plant (prebloom), or (3) the terminal trifoliate of the fourth fully expanded leaf of each plant in a six-plant sample (postbloom).[118,119] The practicality of the latter program is increased by using these same leaf samples for counts of aphids and eggs of the lepidopterous pests.

The larval counting technique generally requires examining each larva with a hand lens or microscope, and can be time consuming until the observer becomes proficient in discriminating between live, dead, and parasitized larvae. However, Schuster and Beck[120] recently developed a presence-absence procedure which minimizes sampling time. A significant linear relationship ($R = 0.99$) between larval numbers and the proportion of infested leaf samples allows a rapid estimation of leafminer larval population density. The procedure is most effective when populations are < 1.6 larvae per the terminal three leaflets. Above this density, the regression is no longer linear.

In California and Sinaloa, a trapping system has been designed to take advantage of the leafminer larval behavior of dropping from the leaves to the ground just prior to pupation. Johnson et al.[121,122] determined that inexpensive styrofoam trays (12" × 8", ~ 1¢ each) placed beneath tomato plants would capture larvae in proportion to the numbers of larvae in the foliage. Because the key parasite species in California kill the leafminer larvae before they exit the leaf,[123] the tray technique also integrates information on leafminer parasite activity. If new mines are developing in the foliage, but few leafminer larvae or pupae are found in the trays, then the parasites are effectively suppressing the leafminer populations.[124]

The efficiency of the tray technique has been refined by determining the spatial dispersion of the leafminers using Iwao's[125] and Taylor's[127] regression techniques.[127] Both regressions provided a good fit to the data, with no significant variation in slope coefficients between years. Using the coefficients from Taylor's regression, constant precision level sampling schemes were generated that can be used to obtain rapid and accurate estimates of larval density with minimal effort.

The tray technique is not without problems. In Florida, the parasite populations include some species that emerge from *Liriomyza* pupae.[128] Thus, overestimates of leafminer populations based on numbers of larvae and pupae in trays could result in an unjustified treatment. In California and Sinaloa, occasional high winds and/or rain may artificially reduce collections. The losses due to the overflow of rainwater can be mitigated by using a fine screen to allow water, but not larvae, to pass through the trap. However, a design suitable for the extensive rainfall in Hawaii has been developed (M. W. Johnson, personal communication) that will largely alleviate this problem.

Adult trapping with colored sticky cards has evolved substantially since the technique was first reported as suitable for rapid detection of leafminers by Musgrave et al.[129] in 1975. Several researchers established that yellow cards collected adults more efficiently than other colors.[130,131] The height of trap placement in the canopy and time of day were determined to influence trapping efficiency; *L. sativae* was trapped most often in the centrally located traps while *L. trifolii* was found more commonly on the lowest traps.[132] An unintentional selection of species by trap placement could be of considerable significance, as the pesticide resistance profiles of these two pests differ substantially.[133] In California, a relationship between larval/pupal numbers on trays and subsequent adult collections (2 weeks later) on sticky cards has proven suitable for predictive purposes ($R^2 = 0.77$ to 0.97).[134] More recent studies in Florida also have found a strong correlation between adult counts in the foliage and sticky trap collections ($R^2 = 0.87$ to 0.93).[135] However, the latter two studies were conducted on research plantings, and need to be validated for use in large-scale tomato production.

The usefulness of yellow sticky traps may be extended by employing the technique for pesticide resistance monitoring. This procedure, in which insecticides are incorporated in the polybutene sticker, has proven effective for several contact insecticides, including chlorpyriphos and permethrin.[136,137] However, development of this method is tedious, requiring information on effects of duration of exposure, amount of sticker needed, insect age, size and sex, and relationship to standard topical bioassay procedures. In addition, this approach still needs to be validated in large-scale tomato plantings.

Like other trapping systems utilizing polybutene stickers, yellow sticky traps are not without problems. The sticky material can lose effectiveness if large numbers of flies are caught, or if dust or debris collects on the trap. Many other insects, including aphids and thrips, are attracted to the color yellow, and may make counting difficult. Finally, the technique may meet resistance from some PCAs who dislike working with stickers or paying a premium for commercially available cards.

d. Other Arthropod Pests

A wide variety of additional pests attack tomatoes. Most of these use similar types of sampling programs, including visual searches of foliage, pheromone trapping, or colored sticky cards. Some insects, such as *Bemesia tabaci* (Gennadius), recently have become major pests, and have not had statistically valid sampling plans verified for commercial fields. Most of the effort to date has focused on providing control, rather than determining thresholds or population distributions. One sporadic pest, the tomato russet mite, *Aculops lycopersici* (Massee), can develop populations capable of killing tomato plants in a very short period of time. Because the pest is quite small, detection is usually made when mite populations have begun to cause noticeable plant stress. Other pests also occur intermittently, including thrips, stink bugs, flea beetles, etc. Because many of these pests can occur in localized areas in the field, most vegetable crop PCAs vary the pattern of sample collection such that the entire field is surveyed a minimum of once every 2 weeks.

B. STRAWBERRIES: COMPARING SAMPLING PROGRAMS IN CANADA AND CALIFORNIA

1. Economic, Horticultural, and Environmental Information

Strawberries are grown throughout North America, including: Ontario and British Columbia, Canada; California, Florida, Louisiana, Michigan, Oregon, and Washington in the U.S.; and Baja, Mexico. California is the largest producer, with over 68% of the total market share.[138] With the exception of California, most strawberries are planted as multiple-year crops, often held for 3 to 6 years. Usually there is little production during the first year. Nonetheless, the plants must be protected from arthropod damage. In northern California, most plantings are maintained for just 1 or 2 years, with harvests during the late winter and spring. In southern California, plantings are annual, with most crops planted in October, harvested from December through May or June, and then removed. A second summer crop may be planted in June and harvested from August until early October, when they will be removed in preparation for the winter planting. Thus, the horticultural and environmental conditions are more diverse than for most other crops.

2. *Tetranychus urticae* (Koch)

The primary arthropod pest of strawberries throughout the world is the two-spotted spider mite, *Tetranychus urticae* (Koch) (Acari: Tetranychidae).[139-141] This pest, which feeds on the foliage and not the fruit, causes significant yield decline through a reduction in photosynthate production.[142] Other pests include a complex of aphid species, which can affect yields through plant stress, but primarily reduce yields by contaminating the fruit with honeydew.[143] Thrips, earwigs, tomato fruitworms, lygus bugs, and several other arthropods are occasionally of regional importance. Sequential sampling plans have been developed for a few of these pests.[144,145]

The potential profits from strawberry production justify a substantial research effort. However, most initial research focused on chemical control strategies. Subsequent studies were intended to document within-plant distributions for the purpose of building practical sampling strategies for commercial fields. As a result, the earliest sampling plans were designed for collecting foliar counts of *T. urticae* from small plots where intensive sampling could be used to determine population densities with a high degree of precision. Because foliar counts of all life stages generally require the use of microscopes, leaves are usually harvested, placed in bags either singly or in groups, and transported to the laboratory where they usually were held at 4°C.[146,147] Unfortunately, leaves in cold storage undergo a combination of dehydration and discoloration due to oxidation and hydrolysis of starches to sugars that can make counting difficult if leaves are held beyond 24 h.[148] In addition, movement of mites between leaves collected in groups either during transportation or after removal from cold storage, could confound results. Problems with spider mite movement and leaf deterioration were solved, in part, by dipping the leaves in inexpensive floor wax; the wax "fixed" the mites to the leaf, stopped development, and preserved the leaves for examination for at least 2 weeks.[149] However, the key complications of labor costs and slow data collection were not eliminated.

Attempts to reduce sampling effort by using mite brushing machines[150,151] or an imprinting technique[152,153] were not completely successful. The mite brushing machines, which use opposed, rotating bristles to brush the mites from detached leaves onto a collection plate, also accumulate plant hairs, dust, spider mite webbing, and associated trash. Discriminating and counting all the mite stages can be difficult. In addition, the technique still requires separate samples for each location to prevent spider mite movement during transport, and examination under a microscope. Nonetheless, in many circumstances, the mite brushing machines provides a significant improvement in sampling efficiency over foliar counts.

Although the time-consuming use of a microscope could be eliminated with the imprinting technique, discrimination between life stages of the spider mites is not possible. For the imprinting procedure, leaves are pressed into a filterpaper disk impregnated with a protein sensitive dye (bromo phenol blue) and the protein from the crushed mites produces a green response on the yellow paper. This procedure, which is extremely rapid if flat plates capable of holding the filterpaper disks are mounted on pliers, has proven most effective for determining the presence or absence of pest populations. High numbers of *T. urticae* cause overlapping stains, making numerical determinations impossible.[154]

Not surprisingly, all of the techniques requiring spider mite counts on foliage are difficult to employ effectively on large-scale commercial plantings. However, several researchers have used these techniques to determine the distribution of *T. urticae* in strawberries. From this information, several binomial (presence-absence) sampling plans have been developed that seemingly have gained acceptance in commercial production. Raworth[155] initially used foliar counts in untreated experimental plots to determine the coefficients of Taylor's power law,[156] and then developed a binomial sampling approach using Nachman's[157] statistical procedures. A predictive relationship was found between spider mite density and the proportion of uninfested leaflets sampled. Because the PCA can simply pull leaflets and examine them for presence of the pests, and the time-consuming laboratory pest counts are eliminated, this technique allows rapid and timely estimates of population density.

Binomial sampling programs in strawberries have been refined further by accounting for within-plant variation in population distribution. Reports from New Zealand and California have agreed that *T. urticae* populations prefer mature leaflets on plants.[158,159] Selection of mature leaflets from a predetermined location (basal, middle, or upper strata of the canopy) provides a sampling unit that has a relatively constant proportion of the population to be sampled.[160] Therefore, variability (and thus sampling time and costs) can be reduced by selective choice of leaflet location. Both Butcher et al.[161] and Trumble[162] used this information to develop presence-absence sequential sampling plans based on thresholds of five to seven *T. urticae* per leaflet and the statistical procedures of Wilson and Room.[163]

The sampling program in California has been complicated by the determination that *T. urticae* populations change distribution over time and following pesticide application.[164] Initial immigrant females are dispersed randomly, but with the production of offspring, spider mite populations become highly aggregated; a few plants have substantial numbers of mites while most have none. The population distribution subsequently becomes increasingly random as populations continue to increase and the mites disperse, infesting more of the available plants. Following pesticide application, small numbers of mites typically survive on many plants, allowing population regeneration from a relatively randomly distribution. Because the dispersion changes following pesticide use, a second sequential sampling plan was devised for post-treatment assessment.

This latter example points out a common flaw in many sampling programs for vegetables. If sampling plans are developed from fields that are not treated using commercial practices, the assumptions (distribution coefficients) used to create the program may be in error. In the case of strawberries, the use of the sampling program designed for aggregated, pretreatment populations would overestimate post-treatment spider mite density. As a result, additional and unnecessary pesticide applications would be recommended. In addition, many pesticides can dramatically affect pest behavior and development,[165-167] which in turn will impact distributions. Development of potentially inaccurate sampling programs is probably a common occurrence because many researchers do not have access to commercial application equipment or to the protective supplies needed to safely work in pesticide-treated fields. In terms of overuse of pesticides, environmental contamination, and human health concerns, the costs of overestimation of pest populations are probably quite significant.

The sampling programs that are employed in strawberries and many other crops probably need revision on a frequent basis. At every production location, different cultivars are used for a few years and then fall out of favor. This shift in cultivar preference by growers is of considerable importance because cultivars can vary substantially in their resistance to *T. urticae*.[168,169] While little is known of the impact of cultivar variability on pest dispersion, it seems reasonable that any factor that changes attractiveness, within-plant location, or suitability of a crop for a given pest has the potential to alter within-field distribution, thereby effecting sampling programs. This may explain in part why sampling plans developed at one location may not be transferable to another geographic area.[170,171]

IV. CONCLUSIONS

Universal sampling programs suitable for all vegetable cropping systems do not exist. Variation in environmental and horticultural conditions, the cultivars planted, pesticidal effects on arthropod distribution, and pesticide-induced behavioral changes, etc. can influence how pests are sampled. However, there is a clear trend toward development of binomial sampling procedures; these are the most readily accepted by crop protection consultants. Unfortunately, suitable relationships between infested and uninfested plants or plant parts cannot be, or have not been, documented for all pests. Therefore, many of the other methods described in this chapter are still in common use. Also, regardless of how sophisticated the sampling technique for the key pest or pests, the presence of a wide variety of other arthropods on most vegetable crops insures that failure to periodically examine the entire field will result in reduced productivity. Finally, the available literature is filled with papers describing sampling plans which will never be used commercially; failure to validate a proposed sampling plan in large-scale commercial operations is probably the single most important reason that such programs are not adopted.

ACKNOWLEDGMENTS

The reviews of Drs. S. Eigenbrode and M. Diawara improved this chapter and are greatly appreciated.

REFERENCES

1. Georghiou, G. P. and Lagunes-Tejeda, A., *The Occurrence of Resistance to Pesticides in Arthropods*, Food and Agriculture Organization of the United Nations, Rome, 1991.
2. Trumble, J. T., Vegetable insect control with minimal use of insecticides, *HortSci.*, 25, 159, 1990.
3. McKinlay, R. G., Ed., *Vegetable Crop Pests*, CRC Press, Boca Raton, FL, 1992.

4. Metcalf, C. L., Flint, W. P., and Metcalf, R. L., *Destructive and Useful Insects*, 4th ed., McGraw-Hill, New York, 1962.
5. Bynum, E. D. J. and Archer, T. L., Wireworm (Coleoptera: Elateridae) sampling for semiarid cropping systems, *J. Econ. Entomol.*, 80, 164, 1977.
6. Jansson, R. K. and Lecrone, S. H., Evaluation of food baits for pre-plant sampling of wireworms (Coleoptera: Elateridae) in potato fields in southern Florida, *Fla. Entomol.*, 73, 503, 1989.
7. Horne, P. A., The influence of introduced parasitoids on the potato moth, *Phthorimaea operculella* (Lepidoptera: Gelichiidae) in Victoria, Australia, *Bull. Entomol. Res.*, 80, 159, 1990.
8. Bald, J. G. and Helson, G. A., Estimation of damage to potato foliage by potato moth *Gnorimoschema opeculella* (Zell.), *J. Counc. Sci. Ind. Res.*, 17, 31, 1944.
9. Shelton, A. M. and Wyman, J. A., Potato tuberworm damage to potatoes under different irrigation and cultural practices, *J. Econ. Entomol.*, 72, 261, 1979.
10. Poos, F. W. and Peters, H. S., The potato tuber worm, *Bull. Va. Truck Exp. Stn.*, 61, 597, 1927.
11. Lal, L., Relationships between pheromone catches of adult moths, foliar larval populations and plant infestations by potato tuberworm in the field, *Trop. Pest Manage.*, 35, 157, 1989.
12. Shelton, A. M. and Wyman, J. A., Time of tuber infestation and relationships between pheromone catches of adult moths, foliar larval populations, and tuber damage by the potato tuberworm, *J. Econ. Entomol.*, 72, 599, 1979.
13. Bacon, O. G., Control of the potato tuberworm in potatoes, *J. Econ. Entomol.*, 53, 868, 1960.
14. Bonnemaison, L., Insect pests of crucifers and their control, *Annu. Rev. Entomol.*, 10, 233, 1965.
15. Eckenrode, C. J. and Chapman, R. K., Seasonal adult cabbage maggot populations in the field in relation to thermal unit accumulations, *Ann. Entomol. Soc. Am.*, 65, 151, 1972.
16. Finch, S., Monitoring insect pests of cruciferous crops, in *Brit. Crop. Prot. Conference—Pests and Diseases*, Boots Co. Ltd., Nottingham, 1977, 219.
17. Wyman, J. A., Libby, J. L., and Chapman, R. K., Cabbage maggot management aided by predictions of adult emergence, *J. Econ. Entomol.*, 70, 327, 1977.
18. (See Reference 14).
19. (See Reference 16).
20. Baskerville, G. L. and Emin, P., Rapid estimation of heat accumulation from maximum and minimum temperatures, *Ecology*, 50, 515, 1969.
21. Wilson, L. T. and Barnett, W. W., Degree-days: an aid in crop and pest management, *Calif. Agric.*, 37, 4, 1983.
22. Croaker, T. H. and Wright, D. W., The influence of temperature on the emergence of the cabbage root fly *Erioischia brassicae* (Bouche) from overwintering pupae, *Ann. Appl. Biol.*, 52, 337, 1963.
23. Pedersen, L. H. and Eckenrode, C. J., Predicting cabbage maggot flights in New York using common wild plants, in *A Grower's Guide to Cabbage Pest Management in New York*, Cornell University, Ithaca, NY, 1987.
24. Harcourt, D. G., Spatial arrangement of the eggs of *Hylema brassicae* (Bouche), and a sequential sampling plan for use in control of the species, *Can. J. Plant Sci.*, 47, 461, 1967.
25. Shelton, A. M. and Trumble, J. T., Monitoring insect populations, in *Handbook of Pest Management in Agriculture*, 2nd ed., Pimentel, D., Ed., CRC Press, Boca Raton, FL, 1990, 45.
26. Finch, S., Skinner, G., and Freeman, G. H., The distribution and analysis of cabbage root fly egg populations, *Ann. Appl. Biol.*, 79, 1, 1975.
27. Finch, S., Skinner, G., and Freeman, G. H., Distribution and analysis of cabbage root fly pupal populations, *Ann. Appl. Biol.*, 88, 351, 1978.
28. Crüger, G. and Maack, G., Using an economic threshold to reduce the amounts of insecticide applied to control *H. brassicae*, in *Integrated Control in Brassica Crops*, Pelerents, C., Ed., IOBC, SROP/WPRS Bulletin, Gand, 1980, 27.
29. Finch, S., The effectiveness of traps used currently for monitoring populations of the cabbage root fly, *Ann. Appl. Biol.*, 116, 447, 1990.
30. Finch, S., Improving the selectivity of water traps for monitoring populations of the cabbage root fly, *Ann. Appl. Biol.*, 120, 1, 1992.
31. (See Reference 30).
32. Flint, M. L., Ed., *Integrated Pest Management for Tomatoes*, University of California Press, Berkeley, 1982.
33. Salama, H. S., Moawed, S., Saleh, R., and Ragaei, M., Field tests on the efficacy of baits based on *Bacillus thuringiensis* and chemical insecticides against the greasy cutworm *Agrotis ypsilon* Rdf. in Egypt, *Anz. Schädlingskde., Pflanzenschutz, Umwelt.*, 63, 33, 1990.
34. Story, R. N. and Keaster, A. J., Modified larval bait trap for sampling black cutworm (Lepidoptera: Noctuidae) populations in field corn, *J. Econ. Entomol.*, 76, 662, 1983.

35. Whitford, F. and Showers, W. B., Olfactory and visual response by black cutworm larvae (Lepidoptera: Noctuidae) in locating bait trap, *Environ. Entomol.*, 13, 1269, 1984.
36. Danielson, S. D. and Berry, R. E., Redbacked cutworm (Lepidoptera: Noctuidae) sequential sampling plants in peppermint, *J. Econ. Entomol.*, 71, 323, 1978.
37. Southwood, T. R. E., *Ecological Methods*, Chapman & Hall, London, 1966.
38. Duffus, S. R., Busacca, J. D., and Carlson, R. B., Evaluation of sampling methods for dingy cutworm larvae (Lepidoptera: Noctuidae), *J. Econ. Entomol.*, 76, 1260, 1983.
39. Novák, I., An efficient light trap for collecting insects, *Acta Ent. Bohemoslov.*, 80, 29, 1983.
40. Nag, A. and Nath, P., Effect of moon light and lunar periodicity on the light trap catches of cutworm *Agrotis ipsilon* (Hufn.) moths, *J. Appl. Entomol.*, 111, 359, 1991.
41. Gray, T. G., Shepard, R. F., Struble, D. L., Byers, J. B., and Maher, T. F., Selection of pheromone trap and attractant dispenser load to monitor black army cutworm, *Actebia fennica*, *J. Chem. Ecol.*, 17, 309, 1991.
42. Hendrix, W. H. I. and Showers, W. B., Evaluation of differently colored bucket traps for black cutworm and armyworm (Lepidoptera: Noctuidae), *J. Econ. Entomol.*, 83, 596, 1990.
43. Struble, D. L. and Byers, J. R., Identification of sex pheromone components of darksided cutworm, *Euxoa messoria*, and modification of sex attractant blend for adult males, *J. Chem. Ecol.*, 13, 1187, 1987.
44. (See Reference 42).
45. Harcourt, D. G., Distribution of the immature stages of the diamondback moth, *Plutella maculipennis* (Curt.) (Lepidoptera: Plutellidae) on cabbage, *Can. Entomol.*, 92, 515, 1960.
46. Harcourt, D. G., Design of a sampling plan for studies on the population dynamics of the diamondback moth, *Plutella maculipennis* (Curt.) (Lepidoptera: Plutellidae), *Can. Entomol.*, 93, 820, 1961.
47. Theunissen, J. and den Ouden, H., Comparison of sampling methods for insect pests in brussels sprouts, *Med. Fac. Landbouww. Rijksuniv.*, 48, 281, 1983.
48. Theunissen, J. and den Ouden, H., Tolerance levels and sequential sampling tables for supervised control in cabbage crops, *Bull. Soc. Entomol. Suisse*, 60, 243, 1987.
49. Hollingsworth, C. S. and Gastonis, C. A., Sequential sampling plans for green peach aphid (Homoptera: Aphididae) on potato, *J. Econ. Entomol.*, 83, 1365, 1990.
50. Trumble, J. T., Within-plant distribution and sampling of aphids (Homoptera: Aphididae) on broccoli in southern California, *J. Econ. Entomol.*, 75, 587, 1982.
51. Walker, G. P., Madden, L. V., and Simonet, D. E., Spatial dispersion and sequential sampling of the potato aphid, *Macrosiphum euphorbiae* (Homoptera: Aphididae), on processing-tomatoes in Ohio, *Can. Entomol.*, 116, 1069, 1984.
52. Sears, M. K., Shelton, A. M., Quick, T. C., Wyman, J. A., and Webb, S. E., Evaluation of partial plant sampling procedures and corresponding action thresholds for management of lepidoptera on cabbage, *J. Econ. Entomol.*, 78, 913, 1985.
53. Hoy, C. W., Jennison, C. M. S. A., and Andaloro, J. T., Variable-intensity sampling: a new technique for decision making in cabbage pest management, *J. Econ. Entomol.*, 76, 139, 1983.
54. Hoy, C. W., Variable-intensity sampling or proportion of plants infested with pests, *J. Econ. Entomol.*, 84, 148, 1991.
55. Cartwright, B., Edelson, J. V., and Chambers, C., Composite action thresholds for the control of lepidopterous pests on fresh-market cabbage in the lower Rio Grande Valley of Texas, *J. Econ. Entomol.*, 80, 175, 1987.
56. Kirby, R. D. and Slosser, J. E., Composite economic threshold for three lepidopterous pests of cabbage, *J. Econ. Entomol.*, 77, 725, 1984.
57. Simonet, D. E. and Morisak, D. J., Utilizing action thresholds in small-plot insecticide evaluations against cabbage-feeding, lepidopterous larvae, *J. Econ. Entomol.*, 75, 43, 1982.
58. (See Reference 48).
59. Morisak, D. J., Simonet, D. E., and Lindquist, R. K., Use of action thresholds for management of lepidopterous larval pests of fresh-market cabbage, *J. Econ. Entomol.*, 77, 476, 1984.
60. Chalfant, R. B., Denton, W. H., Schuster, D. J., and Workman, R. B., Management of cabbage caterpillars in Florida and Georgia by using visual damage thresholds, *J. Econ. Entomol.*, 72, 411, 1979.
61. Leibee, G. L., Chalfant, R. B., Schuster, D. J., and Workman, R. B., Evaluation of visual damage thresholds for management of cabbage caterpillars in Florida and Georgia, *J. Econ. Entomol.*, 77, 1008, 1984.
62. Sheldon, A. M., Sears, M. K., Wyman, J. A., and Quick, T. C., Comparison of action thresholds for lepidopterous larvae on fresh-market cabbage, *J. Econ. Entomol.*, 76, 196, 1983.

63. Tabashnik, B. E., Cushing, N. L., and Johnson, M. W., Diamondback moth (Lepidoptera: Plutellidae) resistance to insecticides in Hawaii: intra-island variation and cross resistance, *J. Econ. Entomol.*, 80, 1091, 1987.
64. Hall, F. R. and Adams, A. J., Microdroplet application for determination of comparative topical and residual efficacy of formulated permethrin to two populations of diamondback moth (*Plutella xylostella* L.), *Pest. Sci.*, 28, 337, 1990.
65. Adams, A. J., Hall, F. R., and Hoy, C. W., Evaluating resistance to permethrin in *Plutella xylostella* (Lepidoptera: Plutellidae) populations using uniformly sized droplets, *J. Econ. Entomol.*, 83, 1211, 1990.
66. Hoy, C. W., Adams, A. J., and Hall, F. R., Behavioral response of *Plutella xylostella* (Lepidoptera: Plutellidae) populations to permethrin deposits, *J. Econ. Entomol.*, 83, 1216, 1990.
67. Magaro, J. J. and Edelson, J. V., Diamondback moth (Lepidoptera: Plutellidae) in south Texas: a technique for resistance monitoring in the field, *J. Econ. Entomol.*, 83, 1201, 1990.
68. McCutchen, B. F. and Plapp, F. W., Monitoring procedure for resistance to synthetic pyrethroids in tobacco budworm larvae, in *Proc. Beltwide Cotton Prod. Res. Conf.*, J. M. Brown and D. A. Richter Ed., National Cotton Council, Memphis, 1988, 356.
69. Sanderson, J. P., Parrella, M. P., and Trumble, J. T., Monitoring insecticide resistance in *Liriomyza trifolii* (Diptera: Agromyzidae) with yellow sticky cards, *J. Econ. Entomol.*, 82, 1011, 1989.
70. Brewer, M. J., Schuster, D. J., Trumble, J. T., and Alvarado-Rodriquez, B., Field monitoring for insecticide resistance in the beet armyworm, *Spodoptera exigua* (Hübner), *J. Econ. Entomol.*, 82, 1520, 1989.
71. Brewer, M. J. and Trumble, J. T., Tomato pinworm (Lepidoptera: Gelichiidae) resistance to fenvalerate from localities in Sinaloa, Mexico and California, U.S.A., *Trop. Agric.*, 70, 179, 1993.
72. Brewer, M. J. and Trumble, J. T., Classifying resistance severity in field populations: sampling inspection plans for an insecticide resistance monitoring program, *J. Econ. Entomol.*, 84, 379, 1991.
73. Trumble, J. T. and Alvarado-Rodriquez, B., Development and economic evaluation of an IPM program for fresh market tomato production in Mexico, *Agric. Ecosyst. Environ.*, 43, 267, 1993.
74. Farrar, R. A. and Kennedy, G. G., Sources of insect and mite resistance in tomato in *Lycopersicon* spp., in *Monographs on Theoretical and Applied Genetics. 14. Genetic Improvement of Tomato*, Kalloo, G., Ed., Springer-Verlag, New York, 1992, 121.
75. Sinha, N. K. and McLaren, D. G., Screening for resistance to tomato fruitworm and cabbage looper among tomato accessions, *Crop Sci.*, 29, 861, 1989.
76. Trumble, J. T., Brewer, M. J., Shelton, A. M., and Nyrop, N. P., Transportability of fixed-precision level sampling plans, *Res. Popul. Ecol.*, 31, 325, 1989.
77. Ramirez Díaz, J. M., Investigacion sobre plagas de tomate en Silaloa, in *Taller Sobre el Manejo Integrado de Plagas en Tomate*, Ed., CAADES, Culiacan, Sinaloa, Mexico, 1988, 178.
78. Welter, S. C., Johnson, M. W., Toscano, N. C., Perring, T. M., and Varela, L., Herbivore effects on fresh and processing tomato productivity before harvest, *J. Econ. Entomol.*, 82, 935, 1989.
79. Trumble, J. T., Moar, W. J., Brewer, M. J., and Carson, W. C., The impact of UV radiation on activity of linear furanocoumarins and *Bacillus thuringiensis* var *kurstaki* against *Spodoptera exigua*: implications for tritrophic interactions, *J. Chem. Ecol.*, 17, 973, 1991.
80. Schuster, D. J. and Pohronezny, K., Practical application of pest management on tomatoes in Florida, in *Biological Control and IPM: The Florida Experience,* Rosen, D., Bennett, F., and Capinera, J., Eds., Intercept, Ltd., Andover, MA, in press.
81. Pohronezny, K., Waddill, V. H., Schuster, D. J., and Sonoda, R. M., Integrated pest management for Florida tomatoes, *Plant Dis.*, 70, 96, 1986.
82. Toscano, N. C., Youngman, R. R., Oatman, E. R., Phillips, P. A., Jiminez, M., and Munoz, F., Implementation of an integrated pest management program for fresh market tomatoes, *Appl. Agric. Res.*, 1, 315, 1987.
83. Griswold, M. J. and Trumble, J. T., Responses of *Spodoptera exigua* (Lepidoptera: Noctuidae) larvae to light, *Environ. Entomol.*, 14, 650, 1985.
84. (See Reference 73).
85. Alvarado-Rodriquez, B. and Trumble, J. T., *El Manejo Integrado de las Plagas en el Cultivo de Tomate en Sinaloa*, Confederacion de Asociaciones Agricolas del Estado de Sinaloa, Culiacan, Mexico, 1989.
86. Fritt, G. P., The ecology of *Heliothis* species in relation to agroecosystems, *Annu. Rev. Entomol.*, 34, 17, 1989.
87. Wiesenborn, W. D. and Trumble, J. T., Optimal oviposition by the corn earworm (Lepidoptera: Noctuidae) on whorl-stage sweet corn, *Environ. Entomol.*, 17, 722, 1988.
88. Riedhl, H., Croft, R. A., and Howitt, A. J., Forecasting codling moth phenology based on pheromone trap catches and physiological time models, *Can. Entomol.*, 108, 449, 1976.

89. Hoffmann, M. P., Wilson, L. T., and Zalom, F. G., Area-wide pheromone trapping of *Helicoverpa zea* and *Heliothis phloxiphaga* (Lepidoptera: Noctuidae) in the Sacramento and San Joaquin Valleys in California, *J. Econ. Entomol.*, 84, 902, 1991.
90. Trumble, J. T. and Baker, T. C., Flight phenology and pheromone trapping of *Spodoptera exigua* (Hübner) (Lepidoptera: Noctuidae) in southern coastal California, *Environ. Entomol.*, 13, 1278, 1984.
91. Lange, W. H. and Bronson, L., Insect pests of tomatoes, *Annu. Rev. Entomol.*, 26, 345, 1981.
92. (See Reference 70).
93. (See Reference 72).
94. Wald, A., *Sequential Analysis*, John Wiley & Sons, New York, 1947.
95. Johnson, M. W., Welter, S. C., Toscano, N. C., Ting, I. P., and Trumble, J. T., Reduction of tomato leaflet photosynthesis rates by mining activity of *Liriomyza sativae* (Diptera: Agromyzidae), *J. Econ. Entomol.*, 76, 1061, 1983.
96. Whalon, M. E. and Smilowitz, Z., GPA-CAST, a computer forecasting system for predicting populations and implementing control of the green peach aphid on potatoes, *Environ. Entomol.*, 8, 908, 1979.
97. Lin, S. Y. and Trumble, J. T., Bibliography of the tomato pinworm, *Keiferia lycopersicella* (Walsingham) (Lepidoptera: Gelichiidae), *Bibliogr. Entomol. Soc. Am.*, 1, 65, 1983.
98. Lin, S. Y. and Trumble, J. T., Influence of temperature and tomato maturity on development and survival of *Keiferia lycopersicella* (Lepidoptera: Gelichiidae), *Environ. Entomol.*, 14, 855, 1985.
99. Davidson, N. A., Wilson, L. T., Hoffmann, M. P., and Zalom, F. G., Comparisons of temperature measurements from local weather stations and the tomato plant canopy: implications for crop and pest forecasting, *J. Am. Soc. Hort. Sci.*, 115, 861, 1990.
100. Trumble, J. T., Hare, J. D., Musselman, R. C., and McCool, P. M., Ozone-induced changes in host plant suitability: interactions of *Keiferia lycopersicella* and *Lycopersicon esculentum*, *J. Chem. Ecol.*, 13, 203, 1987.
101. Schuster, D. J., Montgomery, R. T., Gibbs, D. L., Marlowe, G. A., Jones, J. P. et al., The tomato pest management program in Manatee and Hillsborough Counties, 1978–1980, *Proc. Fla. State Hort. Soc.*, 93, 235, 1980.
102. Pena, J. E., Tomato pinworm, *Keiferia lycopersicella* (Walsingham): Population Dynamics and Assessment of Plant Injury in Southern Florida, Ph.D. dissertation, University of Florida, Gainesville, 1983.
103. (See Reference 80).
104. (See Reference 82).
105. Wiesenborn, W. D., Trumble, J. T., and Oatman, E. R., Economic comparison of insecticide treatment programs for managing the tomato pinworm (Lepidoptera: Gelichiidae) on fall tomatoes, *J. Econ. Entomol.*, 83, 212, 1990.
106. (See Reference 73).
107. (See Reference 73).
108. (See Reference 82).
109. (See Reference 73).
110. (See Reference 71).
111. (See Reference 95).
112. Trumble, J. T., Kolodny-Hirsch, D. M., and Ting, I. P., Plant compensation for arthropod herbivory, *Annu. Rev. Entomol.*, 38, 93, 1993.
113. Oatman, E. R. and Kennedy, G. G., Methomyl induced outbreak of *Liriomyza sativae* on tomato, *J. Econ. Entomol.*, 69, 667, 1976.
114. Rathman, R. J., Johnson, M. W., Rosenheim, J. A., and Tabashnik, B. E., Carbamate and pyrethroid resistance in the leafminer parasitoid *Diglyphus begini* (Hymenoptera: Eulophidae), *J. Econ. Entomol.*, 83, 2153, 1990.
115. Pohronezny, K. and Waddill, V., Integrated Pest Management-Development of an Alternative Approach to Control of Tomato Pests in Florida, IFAS, Ext. Plant Path. Rep. 22, University of Florida, Gainesville, 1978.
116. Johnson, M. W., Oatman, E. R., and Toscano, N. C., Potential sampling plan for *Liriomyza sativae* on pole tomatoes, in *Proc 3rd Annu. Ind. Conf. on the Leafminer*, Poe, S. L., Ed., Soc. Amer. Floriculturists, Alexandria, VA, 1982, 50.
117. (See Reference 81).
118. (See Reference 81).
119. (See Reference 80).
120. Schuster, D. J. and Beck, H. W., Presence-absence sampling for assessing densities of larval leafminers in field-grown tomatoes, *Trop. Pest Manage.*, 38, 254, 1992.
121. Johnson, M. W., Oatman, E. R., and Wyman, J. A., Effects of insecticides on populations of the vegetable leafminer and associated parasites on fall pole tomatoes, *J. Econ. Entomol.*, 73, 67, 1980.

122. Johnson, M. W., Oatman, E. R., and Wyman, J. A., Effects of insecticides on populations of the vegetable leafminer and associated parasites on summer pole tomatoes, *J. Econ. Entomol.*, 73, 61, 1980.
123. Zehnder, G. W. and Trumble, J. T., Host selection of *Liriomyza* species (Diptera: Agromyzidae) and associated parasites in adjacent plantings of tomatoes and celery, *Environ. Entomol.*, 13, 492, 1984.
124. (See Reference 2).
125. Iwao, S., A new regression method for analyzing the aggregation pattern of animal populations, *Res. Popul. Ecol.*, 12, 249, 1968.
126. Taylor, L. R., Assessing and interpreting the spatial distribution of insect populations, *Annu. Rev. Entomol.*, 29, 321, 1984.
127. Zehnder, G. W. and Trumble, J. T., Sequential sampling plans with fixed levels of precision for *Liriomyza* species (Diptera: Agromyzidae) in fresh market tomatoes, *J. Econ. Entomol.*, 78, 138, 1985.
128. Lema, K. and Poe, S. L., Age specific mortality of *Liriomyza sativae* due to *Chrysonotomyia formosa* and parasitization by *Opius dimidiatus* and *Chrysonotomyia formosa*, *Environ. Entomol.*, 8, 935, 1979.
129. Musgrave, C. A., Poe, S. L., and Bennett, D. R., Leafminer population estimation in polycultured vegetables, *Proc. Fla. State Hort. Soc.*, 88, 156, 1975.
130. Affeldt, H. A., Thimijan, R. W., Smith, F. F., and Webb, R. E., Response of the greenhouse whitefly (Homoptera: Aleyrodidae) and the vegetable leafminer (Diptera: Agromyzidae) to photospectra, *J. Econ. Entomol.*, 76, 1405, 1983.
131. Tryon, E. H., Poe, S. L., and Cromroy, H. L., Dispersal of vegetable leafminer onto a transplant production range, *Fla. Entomol.*, 63, 292, 1980.
132. Zehnder, G. W. and Trumble, J. T., Spatial and diel activity of *Liriomyza* species (Diptera: Agromyzidae) in fresh market tomatoes, *Environ. Entomol.*, 13, 1411, 1984.
133. (See Reference 123).
134. (See Reference 123).
135. Zoebisch, T. G., Stimac, J. L., and Schuster, D. J., Methods for estimating adult densities of *Liriomyza trifolii* (Burgess) (Diptera: Agromyzidae) in staked tomato fields, *J. Econ. Entomol.*, 86, 523, 1993.
136. Haynes, K. F., Parrella, M. P., Trumble, J. T., and Miller, T. A., Monitoring insecticide resistance with yellow sticky cards, *Calif. Agric.*, 40, 1, 1986.
137. (See Reference 69).
138. Anon., Strawberry Acreage and Production, Processing Strawberry Advisory Board, 1986, Watsonville, CA.
139. Butcher, M. R., Penman, D. R., and Scott, R. R., Population dynamics of the twospotted spider mites in multiple year strawberry plantings in Canterbury, *N.Z. J. Zool.*, 14, 509, 1987.
140. Oatman, E. R., Sances, V. F., LePre, L. F., Toscano, N. C., and Voth, V., Effects of different infestation levels of the twospotted spider mite on strawberry yield in winter plantings in southern California, *J. Econ. Entomol.*, 75, 94, 1982.
141. Raworth, D. A., Sampling statistics and a sampling scheme for the twospotted spider mite, *Tetranychus urticae* (Acari: Tetranychidae), on strawberries, *Can. Entomol.*, 118, 807, 1986.
142. Sances, V. F., Wyman, J. A., and Ting, I. P., Physiological responses to spider mite infestation of strawberries, *Environ. Entomol.*, 8, 711, 1979.
143. Trumble, J. T., Oatman, E. R., and Voth, V., Thresholds and sampling or aphids on strawberries, *Calif. Agric.*, 37, 10, 1983.
144. Mailloux, G. and Bostanian, N. J., Economic injury level model for tarnished plant bug, *Lygus lineolaris* (Palisot de Beauvois) (Hemiptera: Miridae), in strawberry fields, *Environ. Entomol.*, 17, 581, 1988.
145. (See Reference 143).
146. (See Reference 141).
147. Wyman, J. A., Oatman, E. R., and Voth, V., Effects of varying twospotted spider mite infestation levels on strawberry yield, *J. Econ. Entomol.*, 72, 747, 1979.
148. Trumble, J. T., Nakakihara, H., and Voth, V., Development and evaluation of a wax immersion technique designed for studies of spider mite (Acari: Tetranychidae) populations on strawberries, *J. Econ. Entomol.*, 77, 262, 1984.
149. (See Reference 148).
150. Chant, D. A. and Muir, R. C., A Comparison of the Imprint and Brushing Machine Methods for Estimating the Numbers of Tree Fruit Red Spider Mite, *Metatetranychus ulmi* (Koch), on Apple Leaves, Annu. Rep. East Malling Res. Stn. Kent, 1954, 141.
151. Morgan, C. V. G., Chant, D. A., Anderson, N. H., and Ayre, G. L., Methods of estimating orchard mite populations, especially with the mite brushing machine, *Can. Entomol.*, 87, 189, 1955.

152. (See Reference 150).
153. (See Reference 148).
154. (See Reference 148).
155. (See Reference 141).
156. (See Reference 126).
157. Nachman, G., Estimates of mean population density and spatial distribution of *Tetranychus urticae* (Acarina: Tetranychidae) and *Phytoselius persimilus* (Acarina: Phytoseiidae) based on the proportion of empty sampling units, *J. Appl. Ecol.*, 21, 903, 1984.
158. (See Reference 139).
159. Trumble, J. T., Implications of changes in arthropod distribution following chemical application, *Res. Popul. Ecol.*, 277, 1985.
160. (See Reference 159).
161. Butcher, M. R., Penman, D. R., and Scott, R. R., A binomial sequential decision plan for control of twospotted spider mite on strawberries in Canterbury, *N.Z. J. Exp. Agric.*, 15, 371, 1987.
162. (See Reference 159).
163. Wilson, L. T. and Room, P. M., Clumping patterns of fruit and arthropods in cotton with implications for binomial sampling, *Environ. Entomol.*, 12, 50, 1983.
164. (See Reference 159).
165. Dittrich, V., Streibert, P., and Bathe, P. A., An old case reopened: mite stimulation by insecticide residues, *Environ. Entomol.*, 3, 534, 1974.
166. Franklin, E. J. and Knowles, C. O., Influence of formamidines on twospotted spider mite (Acari: Tetranychidae) dispersal behavior, *J. Econ. Entomol.*, 77, 318, 1984.
167. Schiffhauer, D. E. and Mizell, R. F., III, Behavioral response and mortality of nursery populations of twospotted spider mite (Acari: Tetranychidae) to residues of six acaricides, *J. Econ. Entomol.*, 81, 1155, 1988.
168. Schuster, D. J., Price, J. F., Martin, F. G., Howard, C. M., and Albregts, E. E., Tolerance of strawberry cultivars to twospotted spider mites in Florida, *J. Econ. Entomol.*, 73, 52, 1980.
169. Shanks, C. H., Jr. and Barritt, B. H., Twospotted spider mite resistance of Washington Strawberries, *J. Econ. Entomol.*, 73, 419, 1980.
170. (See Reference 76).
171. Trumble, J. T., Edelson, J. V., and Story, R. N., Conformity and incongruity of selected dispersion indices in describing the spatial distribution of *Trichoplusia ni* (Hübner) in geographically separate cabbage plantings, *Res. Pop. Ecol.*, 29, 155, 1987.

Chapter 21
SAMPLING ARTHROPOD PESTS OF WHEAT AND RICE

Norman C. Elliott, Gary L. Hein, and B. Merle Shepard

TABLE OF CONTENTS

I. Introduction 628

II. Arthropod Pests of Wheat and Rice 630
    A. Wheat Pests 632
    B. Rice Pests 632

III. Sampling Methods for Arthropod Pests of Wheat and Rice 633
    A. Sampling Plans Based on Direct Count Methods 637
        1. Sampling Armyworms in Wheat and Barley 637
        2. Sampling Sunn Pests in Wheat 637
        3. Sampling Rice Black Bugs in the Philippines 638
        4. Sampling Rice Leaffolders in the Philippines 638
    B. Sampling Plans Based on Incidence Counts 638
        1. Sampling Cereal Aphids in Western Europe 639
        2. Sampling Cereal Aphids in the Northern Great Plains 640
        3. Sampling Russian Wheat Aphids in the Great Plains 644
        4. Sampling Brown- and Whitebacked Planthoppers in Rice in the Philippines 645
    C. Sampling Plans Based on Population Indices 645
        1. Sampling Rice Stem Borer Damage in Thailand and Japan 646
        2. Sampling Rice Water Weevil Adult Feeding Scars in Arkansas 647
        3. Sampling Tadpole Shrimp Injury to Rice in California 649
        4. Sampling Armyworm Injury to Rice in California 649
        5. Sampling Wheat Bulb Fly Injury 650
    D. Other Methods for Sampling and Monitoring Arthropod Pests of Wheat and Rice 650
        1. Sampling Arthropods Using Vacuum Samplers 650
        2. Sampling Arthropods Using a Sweepnet 650
            a. Sweepnet Sampling for Rice Stinkbags in Texas 651
        3. Sampling and Monitoring Arthropods Using Impaction Traps 651
        4. Sampling Soil-Dwelling Arthropod Pests of Wheat and Rice 652
            a. Sampling Wheat Bulb Flies in Great Britain 652
            b. Sampling Wireworms in Wheat in the U.S. 652
            c. Sampling Rice Water Weevils in the U.S. 653

IV. Monitoring Migrant BYDV Vectors Using Aerial Suction Traps 653

V. Sampling Natural Enemies in Wheat and Rice 655

VI. Recommendations for Further Research 656

Acknowledgments 659

References 659

ISBN 0-8493-2923-X/94/$0.00 + .50
© 1994 by CRC Press, Inc.

I. INTRODUCTION

Wheat, barley, oats, rye, rice, and sorghum are among the major small grains. Worldwide, rice, wheat, corn, and potatoes are the leading food staples in rank order of importance, but sorghum, oats, and barley are also important as food for both humans and domestic animals.[1] In the U.S., wheat is the most important small grain in terms of total production, at approximately 2.2 billion bu/year, followed in rank order by sorghum (0.59 billion bu), barley (0.37 billion bu), oats (0.35 billion bu), rice (0.34 billion bu), and rye (0.015 billion bu).[2] The majority of research dealing with methods for sampling small grain arthropods is being conducted on wheat and rice because of their importance as food staples. Therefore, our discussion is focused primarily on sampling methods for arthropod pests in these crops.

Wheat is grown in a variety of climates and soils using a wide range of agronomic practices. Tillage varies from no-till to moldboard plowing. Wheat is grown from temperate regions to the tropics, and in arid (under irrigation), semiarid, subhumid, and humid climates.[1] Cropping systems vary from intercropping with other crops[3] to wheat-fallow rotation where a crop is harvested only every other year.[4] Rice also is grown in a wide variety of climates and soils using various agronomic practices and cropping systems.[5]

In regions with low yield potential, small grains are usually grown using low levels of external inputs, whereas in regions with high potential yields they are often intensively managed.[4,6] As a consequence, average grain yields vary greatly among geographic regions. The profitability of small grain production also varies widely among regions due to differences in attainable yields, agronomic practices, commodity prices, and government price supports.[7] For example, high-yield technology is used in wheat production in several western European countries because of high attainable yields in the region and production price supports that maintain wheat prices at well above world averages.[7] In general, however, profit margins for small grain production are relatively low when compared with those for many crops, especially vegetable and orchard crops.

The diversity of management practices, environments, and economic situations under which small grains are grown create some unique problems for those developing integrated pest management (IPM) programs. First, the generally low profit margins associated with small grain production require that IPM tactics be relatively inexpensive if they are to be of practical value to producers. Consequently, methods for monitoring pest arthropod populations must be inexpensive to perform. Second, because of the worldwide distribution of cereal crops, IPM practices must address regional variation in major pest complexes and the economics of production. In particular, economic thresholds vary regionally due to differences in plant growth, cultivars, management practices, insect pests and diseases, and economics.

Sampling methods for arthropod pests of rice have been developed primarily in eastern Asia and the U.S. Likewise, methods for sampling arthropods in wheat have been reported mainly from Europe and the U.S. Major or potentially important wheat and rice pests are listed for each continent to illustrate the extent of geographical variation (Tables 1 and 2). While numerous arthropod species have been recorded as pests of wheat and rice (see for example References 5, 6, and 8 to 13), a thorough discussion of sampling methods for all known pests of these crops worldwide is impractical in a single chapter. We thus restrict our coverage to important arthropod pests of wheat and rice in the above mentioned regions. Although coverage is limited to wheat and rice, techniques reported here sometimes apply to sampling pest arthropods in other small grains such as barley, oats, and rye.

TABLE 1
Important Arthropod Pests of Wheat in Europe and the U.S., the Portion of the Geographic Region in Which Each Species Attains Significant Pest Status, Its Feeding Habit and Other Injury Caused, and References to Publications on Population Spatial Distribution and Sampling

Region / status / species (order)	Geographic distribution	Feeding habit / other damage	Ref.
Europe			
Major pests			
Rhopalosiphum padi	Cosmopolitan	Sap feeder/	13, 62, 63, 73–75,
Bird cherry-oat aphid (Homoptera)		virus vector	113, 159–164
Sitobion avenae	Western Europe	Sap feeder/	13, 62, 73, 74, 77,
English grain aphid (Homoptera)		virus vector	111–113, 159–162, 165–168
Delia coarctata	Great Britain	Stem borer	120, 131, 132
Wheat bulb fly (Diptera)			
Oulema melanopus	Eastern Europe	Leaf feeder	31, 169
Cereal leaf beetle (Coleoptera)			
Zabrus spp.	Southeastern Europe	Root feeders	
Carabids (Coleoptera)			
Ctenicera and *Agriotes* spp.	Southeastern Europe	Seed and root feeders	121, 122, 126–129
Wireworms (Coleoptera)			
Eurygaster integriceps	Southeastern Europe	Sap feeder	43, 115
Sunn pest (Heteroptera)			
Potential pests			
Schizaphis graminum	Southern Europe	Sap feeder/	13, 38, 68, 73, 159,
Greenbug (Homoptera)		virus vector	160, 163, 168
Metopolophium dirhodum	Cosmopolitan	Sap feeder/	13, 62, 74, 75,
Rose-grass aphid (Homoptera)		virus vector	112, 113, 168
Diuraphis noxia	Southern Europe	Sap feeder/	13, 37, 38, 78, 80,
Russian wheat aphid (Homoptera)		leaf necrosis	81, 85, 86, 156, 168, 175–178
Mayetiola destructor	Cosmopolitan	Sap feeder	
Hessian fly (Diptera)			
Cephus pygmaeus	Cosmopolitan	Stem borer	170
European wheat stem sawfly (Hymenoptera)			
Chlorops pumilionis	Western Europe	Stem feeder	
Gout fly (Diptera)			
Opomyza florum	Western Europe	Stem feeder	
Yellow cereal fly (Diptera)			
Oscinella frit	Western Europe	Sap feeder	
Frit fly (Diptera)			
Haplodiplosis equestris	Western Europe	Stem feeder	171
Saddle gall midge (Diptera)			
Sitodiplosis mosellana	Western Europe	Flower/ grain feeder	172
Orange wheat blossom midge (Diptera)			
Contarinia tritici	Western Europe	Flower/ grain feeder	
Lemon wheat blossom midge (Diptera)			
Limothrips denticornis	Cosmopolitan	Leaf feeder	173, 174
Grain thrip (Thysanoptera)			

TABLE 1(*continued*)
Important Arthropod Pests of Wheat in Europe and the U.S., the Portion of the Geographic Region in Which Each Species Attains Significant Pest Status, Its Feeding Habit and Other Injury Caused, and References to Publications on Population Spatial Distribution and Sampling

Region / status / species (order)	Geographic distribution	Feeding habit / other damage	Ref.
U.S.			
Major pests			
Mayetiola destructor Hessian fly (Diptera)	Cosmopolitan	Sap feeder	
Diuraphis noxia Russian wheat aphid (Homoptera)	Great Plains and northwest	Sap feeder / leaf necrosis	See above
Schizaphis graminum Greenbug (Homoptera)	Great Plains and northwest	Sap feeder / leaf necrosis	See above
Oulema melanopus Cereal leaf beetle (Coleoptera)	midwest	Leaf feeder	See above
Rhopalosiphum padi Bird cherry-oat aphid (Homoptera)	Cosmopolitan	Sap feeder / virus vector	See above
Eriophyes tulipae Wheat curl mite (Acari)	Great Plains	Leaf feeder / virus vector	185, 186
Cephus cinctus Wheat stem sawfly (Hymenoptera)	Northern Great Plains	Stem borer	179, 180
Potential pests			
Cephus pygmaeus European wheat stem sawfly (Hymenoptera)	Northeast	Stem borer	See above
Pseudaletia unipuncta Armyworm (Lepidoptera)	Cosmopolitan	Stem and leaf feeder	39
Spodoptera frugiperda Fall armyworm (Lepidoptera)	South	Stem and leaf feeder	
Euxoa auxiliaris Army cutworm (Lepidoptera)	Central Great Plains and northwest	Stem and root feeder	181
Agrotis orthogonia Pale western cutworm (Lepidoptera)	Central Great Plains	Stem and root feeder	
Ctenicera and *Agriotes* spp. Wireworms (Coleoptera)	Cosmopolitan	Seed and root feeder	See above
Blissus leucopterus Chinch bug (Heteroptera)	Central Great Plains	Sap feeder	
Limothrips denticornis Grain thrip (Thysanoptera)	Northern Great Plains	Leaf feeder	See above
grasshoppers Several species (Orthoptera)	Great Plains	Leaf, stem and grain feeder	182
Petrobia latens Brown wheat mite (Acari)	Central Great Plains	Leaf feeder	183, 184
Meromyza americana Wheat stem maggot (Diptera)	Great Plains	Stem borer	

II. ARTHROPOD PESTS OF WHEAT AND RICE

We classify arthropod pests of wheat and rice as major or potentially important pests (Tables 1 and 2). In our classification, major pests consistently reduce yields over a broad geographic area. Potentially important pests only occasionally cause economic damage or are restricted to a small portion of the geographic region under considera-

TABLE 2
Important Arthropod Pests of Rice in Asia and the U.S., the Portion of the Specified Geographic Region Over Which Each Species Attains Significant Pest Status, Its Feeding Habit and Other Injury Caused, and References to Publications on Population Spatial Distribution and Sampling

Region / status / species (order)	Geographic distribution	Feeding habit / other damage	Ref.
Asia			
Major pests			
Nilaparvata lugens Brown planthopper (Homoptera)	Region-wide	Sap feeder / virus vector	47, 87–96, 99
Scirpophaga spp., *Chilo* spp. Stem borers (Lepidoptera)	Region-wide	Stem borer	36, 99, 101, 102, 187
Sogatella furcifera Whitebacked planthopper (Homoptera)	Region-wide	Sap feeder	47, 87, 89–95
Potential pests			
Nephotettix spp. Green leafhoppers (Homoptera)	Region-wide	Sap feeder / virus vector	47, 94, 95, 114, 188
Cnaphalocrocis spp. Rice leaffolders (Lepidoptera)	Region-wide	Leaf feeder	47, 92, 99, 189
Orseolia oryzae Rice gall midge (Diptera)	Tropical Asia	Stem feeders / forms gall	99
Laodelphax striatellus Small brown planthopper (Homoptera)	Eastern Asia	Sap feeder	
Scotinophara coarctata Rice black bug (Heteroptera)	Tropical Asia	Sap feeder	46, 47
Lissorhoptrus oryzophilus Rice water weevil (Coleoptera)	Japan, Korea	Root feeder	See below
U.S.			
Major pests			
Lissorhoptrus oryzophilus Rice water weevil (Coleoptera)	Region-wide	Root feeder	35, 103, 105, 107–109, 123, 136, 137
Oebalus pugnax Rice stinkbug (Heteroptera)	Southern U.S.	Developing grain feeder	35
Potential pests			
Triops longicaudatus Tadpole shrimp (Notostraca)	California	Root feeder	108, 109
Pseudaletia unipuncta Armyworm (Lepidoptera)	California	Leaf and stem feeder	108, 109
Spodoptera frugiperda Fall armyworm (Lepidoptera)	Southern U.S.	Stem feeder	35
Hydrellia griseola Rice leafminer (Diptera)	California	Leaf miner	108, 109
Macrosteles fascifrons Aster leafhopper (Homoptera)	California	Sap feeder	108, 109
Chironomus spp., *Tanytarsus* spp., *Cricotopus* spp. Rice seed midges (Diptera)	California	Developing grain feeder	108, 109

tion. Literature relating to population distribution or sampling methods is also cited for these pests on wheat, rice, or related crops.

The biology, ecology, and management of important rice and wheat arthropod pests has been reviewed.[5,6,8-19] Therefore, we present only a brief account of information on arthropod pests, focusing primarily on regional differences in species occurrence and pest status.

A. WHEAT PESTS

Major arthropod pests of wheat in western Europe are mostly indigenous (Table 1). The natural climax vegetation of the region is forest, and several pests, e.g., aphids and the wheat bulb fly, are colonizers of ephemeral grasses that establish on recently exposed bare soil.[10] As a consequence, they are well adapted for exploiting wheat, especially during its early growth stages. The bird cherry-oat aphid, *Rhopalosiphum padi*, and the English grain aphid, *Sitobion avenae*, are the most important economic pests of wheat in most of western Europe; stem-feeding Diptera such as the wheat bulb fly, *Delia coarctata*, are also of considerable economic importance.[10,19]

The species composition of the arthropod pest complex of wheat in southeastern Europe more resembles that of the Middle East (the region where wheat is believed to have evolved) than it does the pest complex of western Europe (Table 1).[10] This is a reflection of the geographic proximity and climatic similarity of the Middle East and southeastern Europe. Economic pests of southeastern Europe include plant-feeding bugs, of which the sunn pest, *Eurygaster integriceps*, is the most important, and root-feeding carabids, most notably *Zabrus tenebriodes*.[10,20] The cereal leaf beetle, *Oulema melanopus*, a species indigenous to the region but considered a minor pest, has recently achieved major pest status in eastern and southeastern Europe.[20] This change may be related to the development and widespread use of new wheat varieties or the heavy use of chemical pesticides.

Wheat in the U.S. is grown in a wide range of climates. Some of those (midwest and eastern U.S.) resemble those of western Europe, while others (southern Great Plains) are similar to the climate in the Middle East. As a consequence, the arthropod pest complex of wheat in the U.S. varies among wheat-growing regions. For example, the greenbug, *Schizaphis graminum*, and Russian wheat aphid, *Diuraphis noxia*, are probably the most important economic pests of wheat in the Great Plains region, whereas the Hessian fly, *Mayetiola destructor*, is the most important pest in the central and southeastern U.S. Species native to Europe as well as those indigenous to North America are present (Table 1).

Wheat yield losses resulting from feeding by the bird cherry-oat aphid can be of economic importance in the U.S., especially if large populations occur during the early growth stages.[21-24] However, the economic importance of this cosmopolitan species is primarily due to a particularly damaging strain of barley yellow dwarf virus (BYDV) that it transmits.[25,26] Monitoring populations of disease vectors for pest management poses some difficult problems because economic damage can sometimes result from very low pest population densities, depending on the level of infectivity in the vector population.

B. RICE PESTS

A survey of 50 rice entomologists from 11 Asian countries estimated average yield losses of rice production in Asia caused by arthropod pests at 18.5%, but higher estimates (31.5%) have been reported.[12] In Table 2 are listed the most important rice arthropods causing economic damage and the approximate region in which each species causes significant damage. Also included in the table are the general feeding habit of each species and literature citations on sampling methods and spatial distribution patterns of pest populations in rice fields.

During the last 20 years, rice cropping practices in much of Asia have changed from wet-season production to multiple cropping during both wet and dry seasons. Concomitantly, the area planted to rice has increased at the expense of other crops and the use of agricultural chemicals has increased. These changes have resulted in a gradual shift in the rice arthropod pest complex from mostly polyphagous species to species with monophagous and oligophagous habits.[11,12] For example, before the release of short-statured, high-yielding rice varieties, stem borers were the most serious pests of rice. However, their importance has decreased relative to planthoppers and leafhoppers over the past two decades.[11,12] In spite of this decline, rice stem borers remain a problem and in India and Pakistan these pests are considered to be most important.[5,12] The brown planthopper, *Nilaparvata lugens*, has become a major pest of rice throughout tropical Asia (Table 2), but its pest status could be virtually eliminated through reduction in the use of broad-spectrum insecticides.[12]

Numerous arthropod species occasionally cause economic damage to rice in Asia or are consistent pests in localized regions (Table 2). For example, the rice black bug is an important pest of rice only on the island of Palawanin in the Philippines and in certain parts of Malaysia.[12] Several species of green leafhoppers, of which *Nephotettix virescens* is the most widespread economic pest, are considered important pests of rice in parts of tropical Asia.[5,6,11,12,27] Their damage to rice by direct feeding is seldom of economic importance, but most important is their ability to transmit several nonpersistent rice viral diseases.[27]

The complex of arthropod pests of rice in the U.S. differs markedly from that attacking rice in Asia (Table 2). The most important arthropod pest of rice in the U.S. is the rice water weevil, *Lissorhoptrus oryzophilus*, which is distributed throughout the rice-growing states (Arkansas, California, Florida, Louisiana, Mississippi, Missouri, and Texas).[5,14] Both adult and larval rice water weevils feed on rice, but only larval feeding damage is of economic significance. The rice stinkbug, *Oebalus pugnax*, is a major pest of rice in the southern U.S. This insect feeds on the contents of the developing rice kernels. Feeding by the rice stinkbug reduces yield and causes injury commonly called "peck".[28] Pecky rice is of reduced quality because of discoloration and kernels that tend to break more easily during milling.[14] Potential pests of rice vary among rice growing regions of the U.S. (Table 2), and many are polyphagous species that damage a wide range of agricultural crops.

III. SAMPLING METHODS FOR ARTHROPOD PESTS OF WHEAT AND RICE

A wide variety of methods have been used to sample arthropod populations in wheat and rice. Some sampling methods require no special equipment, but other methods require equipment developed especially for sampling small grain arthropods or for sampling arthropods in other ecosystems. This section provides an overview of methods commonly used for sampling pest and beneficial arthropods in wheat and rice and outlines representative sampling programs.

Population estimates derived from sampling can be divided into three major types: absolute, relative, and population indices (see Chapter 1). Absolute population densities are estimated by direct counts of all the arthropods per unit area in the field. Other sampling techniques also may provide estimates of absolute population density. For example, the aerial suction trap can be used to estimate the number of flying insects per unit volume of air.[29] Aerial suction traps have assumed an important role in monitoring migrant populations of cereal aphids that transmit BYDV. Vacuum sampling machines such as the D-Vac can sometimes be used to obtain estimates of absolute population density.[30] But the efficiency with which these machines extract arthropods can vary with the air speed at the sampling nozzle, the arthropod species being sampled, the habitat from which it is sampled, and abiotic and biotic variables, such as temperature, time of day, and plant phenology.[29] Thus, in population studies of arthropods, the efficiency of vacuum samplers should be determined over a range of conditions to insure that estimates are accurate or so they can be corrected for these variations.

Population sampling methods that remove only a proportion of the individuals (often an unknown proportion) from a unit area in a field provide estimates of relative population density. Most frequently, estimates of relative population density involve counting the number of individuals on a portion of a plant (e.g., tiller, leaf, or seed head), on an entire plant or a natural cluster of stems (e.g., a rice hill), or in a prescribed length of row. Most traps provide estimates of relative population because their efficiency is dependent on several biotic and abiotic variables.[29] Sweepnet sampling also provides estimates of relative population density.[29]

Estimates made by counting arthropods on such sample units such as tillers or leaves can be converted to estimates of absolute population density provided that (1) the arthropod occurs in the field only on the type of sample unit selected for inspection, (2) the sampling efficiency is known or can be determined, and (3) an estimate of the number of sample units per unit area is available. It is sometimes possible to develop mathematical models for converting estimates of relative population density obtained from sweepnet sampling or trapping of rice or wheat arthropods to estimates of absolute population density.[31,32]

Population indices are obtained by sampling damaged plants or parts of plants (e.g., number of white heads or percent defoliation).[29] The choice of sample unit and the method used for sampling are usually based on convenience and the objectives of the sampling program. Population indices usually require at least a crude estimate of the pests' population density because it is unwise to recommend insecticide treatment when a specified level of injury is reached unless live arthropods are also present.

Sampling based on incidence counts (presence or absence) of a specified number of individuals (usually a single individual) is an important method for sampling arthropod pests in wheat and rice. It is often possible to obtain an estimate of the mean number of arthropods per sample unit from an estimate of the proportion of infested sample units.[29,33,34] Incidence counts have been used extensively for sampling aphids and leafhoppers because their small size and large populations make direct counting extremely time consuming.

Published reports on economic injury levels and sampling schemes are available for several arthropod pests of wheat in various western European countries and regions of the U.S., though there are several pests for which published economic injury levels and sampling schemes do not exist. Published information on sampling and economic thresholds for rice pests is in a similar state. Our coverage of information on sampling and economic thresholds from unpublished reports is incomplete.

Schemes for sampling several of the important arthropod pests of wheat and rice are summarized in Tables 3 and 4 including parameters for several sequential sampling schemes. Many of the schemes have been field tested and shown to be reliable for IPM purposes; others have not. Whether field tested or not, most sampling schemes for pest management decision-making, and the economic thresholds on which they are based, were developed in a limited portion of the region over which a particular pest injures the crop, and none have been field tested over the entire region. As a result, most sampling schemes listed in Tables 3 and 4 are used for IPM in only a portion of the region under consideration, and other schemes may have been developed in other portions of the region. We have made no attempt to describe all methods available for sampling each pest, but the list of published references on population spatial distribution and sampling schemes (see Tables 1 and 2) includes several of them.

TABLE 3
Outline of Sampling Programs for Several Important Arthropod Pests of Wheat in Europe and the U.S.

Region / species	Sampling method	Sample unit	Sampling pattern	Economic threshold	Error probs. (α / β)	Sample size (n) or stop-lines[a]
Europe						
Cereal aphids						
Great Britain	Incidence counts (≥ 1 aphid)	One tiller	Diagonal transect	Dynamic	Not specified	$n = 50$
Netherlands	Incidence counts (≥ 1 aphid)	One tiller	Diagonal transect	Dynamic	Not specified	$n = 100$
Sweden	Incidence counts (≥ 1 aphid)	One tiller	Diagonal transect	5 to 10 aphids/tiller	Not specified	$n = 100$–200
Wheat bulb fly	Direct counts	10.2-cm diameter ×7.5-cm-deep soil core	Not specified	2.5×10^6 eggs/ha	Not specified	$n = 20$
Sunn pest						
Spring wheat	Direct counts	0.5×0.5 m quadrat	Diagonal transect	0.25 bugs/0.25 m²	0.10/0.10	$y_u = 3.17 + 0.180n$ $y_l = -3.17 + 0.180n$
Winter wheat	Direct counts	0.5×0.5 m quadrat	Diagonal transect	0.5 bugs/0.25 m²	0.10/0.10	$y_u = 6.09 + 0.361n$ $y_l = -6.09 + 0.361n$
U.S.						
Cereal aphids (mixed species)	Incidence counts (≥ 1 aphid)	One tiller	Zig-zag pattern	12.5 aphids/tiller	Derived by simulation	See Table 5
Russian wheat Aphid	Incidence counts (≥ 1 aphid)	One tiller	Zig-zag pattern	Variable: 5–50% infested tillers	Derived by simulation	See Reference 80 for tables
Wireworms	Direct counts	Soil core (or) 50–200 ml bait	Not specified	Not available	Not specified	See References 60, 108 for guidelines
Armyworm	Direct counts	0.9 m between two rows	Not specified	3 larvae/0.9 m	0.10/—	$y_u = 11.34 + 3.46n$ $y_l = -11.34 + 2.54n$

[a] Tables for use in sequential sampling for a particular pest species can be derived from the stop-line equations by substituting various integer values for n in both equations and calculating y_l and y_u, which are then rounded to integer values.

TABLE 4
Outline of Sampling Programs for Several Important Arthropod Pests of Rice in Eastern Asia and the U.S.

Region / species	Sampling method	Sample unit	Sampling pattern	Economic threshold	Error probs. (α / β)	Sample size or stop-lines[a]
Asia						
Planthoppers	Incidence counts (≥ 10 hoppers)	One hill	Zig-zag transect	40% infested	0.20/0.20	$y_l = -1.64 + 0.40n$ $y_u = 1.64 + 0.40n$
Planthoppers (with predators)	Incidence counts (≥ 10 hoppers)	One hill	Zig-zag transect	50% infested	0.20/0.20	$y_l = -1.71 + 0.50n$ $y_u = 1.71 + 0.50n$
Rice black bug	Direct counts of adults	One hill	w-Shaped transect	3 bugs	0.20/0.20	$y_l = -13.16 + 2.79n$ $y_u = 13.16 + 2.79n$
Rice leaffolders	Direct counts of larvae	One hill	w-Shaped transect	1.5 larvae	0.20/0.20	$y_l = -4.56 + 1.41n$ $y_u = 4.56 + 1.41n$
Rice stemborers						
Japan	Incidence counts (damaged tiller)	Random groups of 40 tillers/hill	Linear transect	30% hills with damage	0.05/0.05	$y_l = -4.25 + 0.289n$ $y_u = 4.25 + 0.289n$
Thailand	Incidence counts (damaged tiller)	One hill	Not specified	68% hills with damage	0.20/0.10	$y_l = -2.76 + 0.592n$ $y_u = 2.00 + 0.592n$
U.S.						
Rice water weevil	Incidence counts of adult feeding scars	Newest leaf of 40 plants	Random locations in the bay	60% leaves scarred	0.10/0.10	$y_l = -22.2 + 16.7n$ $y_u = 22.2 + 16.7n$
Rice stinkbug	38-cm-diameter sweepnet	10 sweeps	Random locations	Variable	Not specified	$n = 10$
Tadpole shrimp	Complete count of seedlings	Ring enclosing 0.09 m²	Diagonal transect	< 25 healthy seedlings and shrimp present	Not specified	$n = 10$
Armyworms						
Stem elongation	Visual inspection for damage	One plant	Diagonal transect through infested portion of field	5 plants with > 25% defoliation and larvae present	Not specified	$n = 10$
Grain filling	Visual inspection for damage	Ring enclosing 0.09 m²	Diagonal transect through infested portion of field	> 10% white heads and larvae present	Not specified	$n = 10$

[a] Tables for use in sequential sampling or a particular pest species can be derived from the stop-line equations by substituting various integer values for n in both equations and calculating y_l and y_u, which are then rounded to integer values.

A. SAMPLING PLANS BASED ON DIRECT COUNT METHODS

In sampling to survey populations over a wide geographic region, estimates of absolute population density, or a related measure, are desirable so that changes in arthropod numbers per sample unit over time or space maintain a consistent relationship to changes in population density. For IPM purposes, it is not usually necessary to obtain estimates of absolute population density because plant damage relationships are often developed by expressing plant damage (yield loss) in relation to the number of pest arthropods in a convenient sample unit.

Most sampling schemes based on direct counting involve enumerating insects on plants, parts of plants, or in a quadrat. However, sampling techniques have been developed for making direct counts of arthropods in fields that are difficult to access, such as flooded rice fields. For example, rice stinkbugs can be counted from outside the field using binoculars,[35] while tillers damaged by rice stem borers can be counted using a telescope.[36] Four sequential sampling schemes for wheat and rice arthropods based on direct counting are described in the remainder of this section.

1. Sampling Armyworms in Wheat and Barley

Many states provide pest management guidelines for armyworms in wheat and other small grains. However, guidelines for sampling are often vague. For example, pest management guidelines for Texas and Oklahoma discuss damage thresholds for armyworms based on the number of larvae per 0.09 m^2 (1 ft^2), but information on appropriate sample size and spatial pattern with which to sample fields is not provided.[37,38]

Coggin and Dively[39] developed a sequential sampling scheme for armyworms infesting wheat and barley in Maryland. The sample unit found most efficient from among those tested was a 0.9-m (3-ft) length of soil surface bordered by two adjacent rows. Larvae on the soil between the two rows are counted. The sampling pattern was not described, but any unbiased method that gives good coverage of the field should suffice. Iwao's[40] confidence interval method was used to construct sequential sampling stop-lines based on the variance-mean relationship obtained from the regression of Lloyd's mean crowding index[41] on mean density[42] (Table 3).

Five sample units are taken before consulting a table constructed from the stop-lines (Figure 1). If the stop-lines have not been crossed after 25 samples, sampling is terminated, the mean number of larvae per 0.9 m is calculated, and a control decision is made. The sampling scheme was validated in 23 fields from which samples of size 25 to 50 0.9-m lengths of row were taken. Using these validation data to simulate sequential sampling, 88% agreement was achieved between decisions based on sequential sampling and those based on the fixed sample of 25 to 50 sample units; in simulated sequential sampling, sampling continued after the limit of 25 was reached in 4% of fields, while in another 8% of fields decisions based on sequential sampling did not agree with those based on the larger fixed sample.

2. Sampling Sunn Pests in Wheat

Viktorov[43] developed a sequential sampling scheme for the overwintering generation of the sunn pest (Table 3). The sample unit selected was a 0.5 × 0.5-m (1.64 × 1.64-ft) quadrat within which all sunn pests are counted. Overwintering sunn pests are approximately randomly distributed within fields, and the Poisson distribution provided an adequate probability distribution from which to develop a sequential sampling plan using the sequential probability ratio test method.[44] The economic threshold for sunn pests in winter wheat is 0.5 bugs/0.25 m^2 (2 bugs/m^2), and control is not necessary below 0.25 bugs/0.25 m^2. In spring wheat, corresponding thresholds are

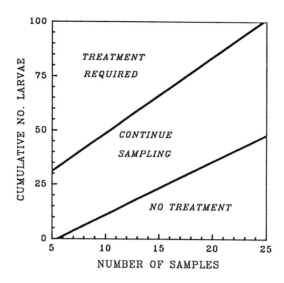

FIGURE 1. Sequential sampling stop-lines for armyworm sampling in small grain fields. (Redrawn from Coggin and Dively.[39])

0.25 and 0.125 bugs/0.25 m².[45] Samples are taken at regular intervals along a diagonal transect across the field. Sequential sampling stop-lines are consulted after the sixth sample and are then consulted after each successive sample.

3. Sampling Rice Black Bugs in the Philippines

A sequential sampling scheme was developed for the rice black bug based on the negative binomial distribution[46] (Table 4). To develop the scheme, densities estimated by visual counts of black bugs were correlated with estimates of absolute population; the two methods were highly correlated. Thus, sampling for black bugs was conducted by counting the number of black bugs seen per rice hill (i.e., a cluster of transplanted rice plants). The sequential sampling scheme was tested in rice fields at three locations in the Philippines by comparing it with a standard method using a fixed sample size of 20 hills per field. The sequential sampling scheme resulted in a 44 to 56% savings in number of samples and greater than 90% agreement in decisions made, compared with the standard sampling method.

4. Sampling Rice Leaffolders in the Philippines

A sequential sampling scheme for rice leaffolders was developed and tested in the Philippines[47] (Table 4). The method involves counting all the leaffolders per hill, which fitted the negative binomial distribution most closely from among those distributions tested. A common k of 1.1 and a damage threshold of 1.5 leaffolders per hill[48] were used in developing decision lines. Comparing the sequential sampling scheme in field tests with a standard sampling method based on a fixed sample size of 20 hills per field resulted in a 79% savings in time and greater than 90% agreement in decisions made compared with the fixed sample size method.

B. SAMPLING PLANS BASED ON INCIDENCE COUNTS

Direct counting is frequently used for experimental purposes and is useful in IPM programs for some arthropod pests, but for others it is too time consuming. For many pests it is desirable to explore the potential offered by more time-efficient sampling methods. Incidence count sampling is a time-efficient method that is valuable for pest management of several arthropod pests of wheat and rice.

1. Sampling Cereal Aphids in Western Europe

The English grain aphid, rose-grass aphid (*Metopolophium dirhodum*), and bird cherry-oat aphid are important pests of wheat in western Europe because of their direct feeding damage and by the secondary effects of their honeydew which covers leaf surfaces; promoting leaf senescence and reducing carbon-dioxide diffusion.[49-51] In most of western Europe, the English grain aphid often builds large populations on wheat heads and causes extensive yield loss. The rose-grass aphid occasionally causes significant damage.[19] The bird cherry-oat aphid is a minor pest in most of western Europe, but is the most important species in Scandinavia.[19] Economic thresholds for these aphids vary among countries (see References 52 to 57) due to differences in factors such as fixed and variable costs of production, the timing of arrival of aphids in the crop, and climatic conditions. Most countries use fixed thresholds, for which aphid control is recommended if populations exceed a specified number per tiller at a given stage of plant growth and the aphid population is expected to increase in the future.[19]

Computer-based IPM advisory systems have been developed for predicting the need for chemical control of crop pests in western Europe. Two such systems, EPIPRE and Farmlink, incorporate packages for cereal aphid IPM and are used by pest management specialists and farmers.[58-61] Use of these systems for a particular field requires the entry of aphid sampling data obtained from the field in addition to information on insecticide and application costs, expected yield, expected value of the crop, and crop growth stage. Simulation models are used to forecast aphid population growth and yield loss which then serve as a basis for decisions regarding the economic benefits of spraying insecticides to control aphids. The use of these systems for decision-making in cereal aphid IPM is described in detail by Mann et al.[58,59] and Rabbinge and Carter.[61] Burn[19] discusses the utility of these and other forecasting systems for cereal aphid IPM and directions for future research and improvements in aphid forecasting and advisory systems. Aphid population monitoring in both systems is based on incidence counts of aphids on individual tillers (Table 3). Farmlink requires that 50 tillers be taken at random locations along a diagonal transect through the field, whereas EPIPRE requires 100 tillers also taken along a diagonal transect. Each tiller is inspected for the presence of aphids, and the number of infested tillers is recorded. A single empirical relationship between the mean number of aphids per tiller and the proportion of infested tillers[62] was acceptable for sampling the English grain aphid, rose-grass aphid, bird cherry-oat aphid, and for mixed-species infestations:

$$\text{probit } p = 4.63 + 1.51 \log_{10} m \qquad (1)$$

where p is the proportion of tillers infested and m is the mean number of aphids per tiller. Equation 1 is used in both Farmlink and EPIPRE to estimate the average number of aphids per tiller.

Incidence count sampling for the bird cherry-oat aphid in spring cereals in Sweden is described by Ekbom[63] (Table 3). The suggested method for sampling fields is to walk the field along a diagonal transect and inspect 5 to 10 tillers at each of 20 sites for aphid presence. The mean number of aphids per tiller is estimated using the following equation:

$$\log_{10} m = 2.28 + 0.94 \log_{10}\{-\log_e(1-p)\} \qquad (2)$$

where m is the mean number of aphids per tiller and p is the proportion of infested tillers. The estimate is compared with an economic threshold. Equation 2 is one of three developed by Ekbom and is suggested for use in sampling for IPM purposes.

2. Sampling Cereal Aphids in the Northern Great Plains

Most wheat in the Great Plains region of the U.S. is grown in semi-arid climates under dryland conditions. The greenbug and Russian wheat aphid are the most economically important aphid pests of wheat in the region, but other species, the English grain aphid and bird cherry-oat aphid, typically infest wheat and can cause economic damage.[21-23]

In the northern Great Plains region, bird cherry-oat aphid populations occasionally develop in winter wheat shortly after emergence and persist throughout autumn.[64] Greenbugs, bird cherry-oat aphids, and English grain aphids typically invade small grains in the northern Great Plains in spring, and remain throughout the growing season in mixed species infestations.[65,66] The corn leaf aphid, *Rhopalosiphum maidis*, is uncommon in wheat, but often occurs with these species in barley.[66] Each species can cause economic injury if large populations develop prior to heading.[22-24,67]

Extension publications from most Great Plains states provide guidelines for sampling and controlling greenbugs and Russian wheat aphids, but usually do not consider less economically important species.[37,38,68,69] Counting the number of greenbugs per linear foot of row is commonly recommended (e.g., Texas, Oklahoma, and Nebraska);[37,38,68] other states recommend counting the number of greenbugs on individual tillers.[13,69] Sampling plans to use in the field to make control decisions usually are not specified, nor are economic thresholds precisely defined.

Relationships between cereal aphid density (number per tiller) and the injury they cause to plants has been studied in North America. The extent of damage caused by cereal aphids varies depending on the species and its population density, the growth stage of the crop, the time interval over which the aphids feed, plant variety, and the general health of the crop.[21-24,67,70-72] As a result, accurate economic thresholds are difficult to establish. Cereal aphids usually occur in mixed species infestations, thus complicating the aphid number-yield loss relationship.[22,23,67,70-72]

Elliott et al.[73] adopted 12.5 aphids per tiller as an economic threshold for cereal aphids in the northern Great Plans because it represented a compromise among published estimates. This threshold is acceptable for cereal aphids in small grains in both the autumn and spring growing seasons (up to boot stage). Economic thresholds for later stages of plant development would be larger than the threshold adopted.[22,23,67,72] However, infestations in later growth stages were not considered because populations that develop after boot stage usually do not cause economic injury to small grains in the northern Great Plains.[22,23,67,72]

Elliott et al.[73] developed a truncated-incidence-count sequential sampling scheme for sampling mixed species aphid populations in small grains in the northern Great Plains (Table 3). Using the scheme, a scout traverses the field in a zig-zag pattern stopping at approximately equally spaced intervals to inspect a single tiller for cereal aphids. The distance to walk between inspected tillers should vary with field size as it should be great enough to cover one third of the field by the time 25 tillers are inspected. This is important because cereal aphid populations are often localized in "pockets" within the field.[74,75] Twenty-five tillers are inspected for the presence or absence of cereal aphids prior to consulting decision tables. Very little time is required to inspect each tiller, and inspecting a minimum of 25 tillers from one-third of the field helps insure adequate coverage prior to making a pest management decision. To avoid biasing the sample, each tiller should be selected without looking at it. After 25 tillers have been inspected the scout compares the number of aphid-infested tillers with the upper and lower stop limits (Table 5). If the number of infested tillers is greater than the larger stop limit, the scout quits sampling and classifies the aphid population in the field as greater than the economic threshold. If the number of infested tillers is less than the lower stop limit, the scout quits sampling and classifies the population as below the threshold. If the number of infested tillers falls between the upper and lower stop limits or is equal to one of the limits, another five tillers are inspected. After inspecting five additional tillers, the table is consulted again, this time comparing the number of infested tillers out of a total of 30 tillers. This process is repeated until a decision is reached or a total of 100 tillers is inspected. If after 100 tillers are inspected, the number of infested tillers still exceeds the smaller of the two numbers corresponding to a sample size of 100 tillers, the scout classifies the population as above the threshold. The usual recommendation for fields classified above the economic threshold is to apply an insecticidal treatment to suppress the aphid population. However, sometimes it might be wise to sample the field again at a later date especially if the crop has progressed to jointing and 100 tillers were inspected prior to making a decision.

Fields larger than 40 ha (100 acres) should be divided into two or more approximately equal-sized sections, and each section should be sampled independently to arrive at a decision. Discretion is required with very large fields because the cost of sampling could be excessive. For such fields, sampling a few sections chosen as representative and basing a decision for the entire field on the results of sampling in those sections is probably appropriate. If the field has a known history of infestation (e.g., if aphid populations are always greater in certain sections of the field), the pattern of sampling should be modified to take advantage of that knowledge.

The frequency of errors in classifying populations by this method with respect to the economic threshold was estimated.[73] Error rates were less than 0.05 for popula-

TABLE 5
Decision Table for Incidence Count (Presence / Absence Sampling) for Cereal Aphids in Small Grains in the Northern Great Plains

Number of tillers inspected (n)	Lower stop limit	Upper stop limit
25	19	24
30	23	29
35	28	34
40	32	39
45	36	43
50	41	48
55	45	53
60	49	58
65	54	62
70	58	67
75	62	72
80	67	77
85	71	81
90	76	86
95	80	91
100	If < 84 tillers infested do not treat	Otherwise treat with insecticide

tions of less than 6 or greater than 14 aphids/tiller, but increased rapidly for populations greater than 8 aphids/tiller, and peaked near 0.9 at the economic threshold (12.5 aphids/tiller). Error rates climb to such high levels because of the decision rule used when sampling is truncated at 100 tillers, i.e., assuming that the population is above the economic threshold whenever 100 tillers are examined prior to making a decision. As populations approach the economic threshold the frequency of such decisions increases, resulting in a rapid increase in the frequency of erroneous decisions. However, considering the present uncertainty in economic thresholds and the potential for explosive growth by aphid populations, the decision rule used in developing the sampling plan provides a conservative solution to the problem of balancing the cost of sampling with that of obtaining reliable information upon which to base management decisions.[73,76]

The sequential sampling scheme was developed using an empirical relationship between the number of aphids per tiller and the proportion of infested tillers[33] (Figure 2):

$$\log_{10} m = 0.509 + 1.28 \log_{10}\{-\log_e(1-p)\} \tag{3}$$

where m is the average number of aphids per tiller and p is the proportion of infested tillers. Iwao's[40] confidence interval method was used to construct sequential sampling stop-lines (Table 3) from the approximate standard error of $\log_{10} m$ derived by Ward.[77] Ward's equation for the variance of $\log_{10} m$ is equivalent to one derived by Kuno[34] for $\ln m$ when natural logarithms are used in Equation 3 in place of base 10 logarithms. Their equations for variance account for variation arising from sampling populations using incidence counts and from using Equation 3 to estimate $\log_{10} m$, but lack terms to account for variance resulting from having estimated the relationship between $\log m$ and $\log\{-\log(1-p)\}$ by linear least-squares regression using sampling data from numerous fields.[33,78,79] The effect of omitting this source of variation in calculating sequential sampling stop-lines was not considered by Elliott et al.[73] To investigate this effect we conducted Monte Carlo simulations to assess error

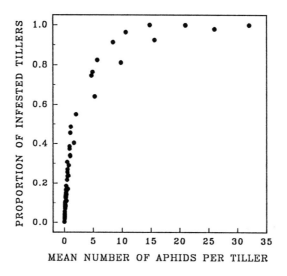

FIGURE 2. Proportion of cereal aphid infested tillers versus the mean number of aphids per tiller.

rates in sequential sampling using stop-lines constructed by Iwao's method with variances of $\log m$ derived by Kuno,[34] Binns and Bostonian,[79] and Schaalje et al.[78] Error rates arising from misclassifying populations with respect to the economic threshold of 12.5 aphids per tiller were similar for the three sets of stop-lines (Figure 3). Error rates for stop-lines constructed using the variance presented by Binns and Bostonian were generally similar to those arising using Kuno's variance, except for populations slightly below the economic threshold where they are as much as 12% greater (Figure 3). Error rates for stop-lines constructed using the variance from Schaalje et al. were also similar to those constructed from Kuno's variance, but deviated for populations near the economic threshold by as much as 4%, and were 2 to 4% larger for populations ranging from about 12.7 to 15 aphids per tiller. However, the increase in error rates resulting from decisions made during sequential sampling is counteracted by lower terminal error rates arising from misclassifying populations near but below the threshold as above it (Figure 3). For populations near the

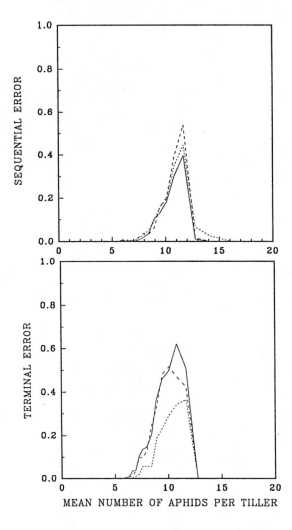

FIGURE 3. Probability of error and average sample size versus number of aphids per tiller for sequential sampling plan for cereal aphids using three published derivations for the variance of $\log_e m$: solid line variance from Kuno; long-dashed line variance from Binns and Bostonian; short-dashed line variance from Schaalje et al.

economic threshold, a large proportion of samples proceed to the maximum of 100 tillers.[73] Combined error rates (sequential plus terminal error) for sequential sampling differ very little. Until uncertainties are removed regarding the appropriate form of the variance for log m and appropriate economic thresholds for cereal aphids, the sequential sampling scheme[73] is adequate for cereal aphid pest management decision making in the Great Plains.

The relatively small difference in error rates using different variance formulae is due to the small contribution of prediction error from estimating Equation 3, which in turn is a consequence of three factors: the relatively large number of samples ($n = 58$) used in developing the regression relationship, the broad range of values of the proportion of infested tillers encountered in sampled populations, and the relatively small variance of points around the regression line, reflected in the large coefficient of determination ($r^2 = 0.98$). The last factor may have resulted from sampling a large number of tillers (200 to 400) from each field in obtaining data for estimating the regression relationship, thus obtaining relatively precise estimates of the proportion of infested tillers and corresponding estimates of mean number of aphids per tiller.

3. Sampling Russian Wheat Aphids in the Great Plains

A sequential sampling scheme based on incidence counts was developed for the Russian wheat aphid in wheat in the Great Plains.[80,81] Russian wheat aphid damage to small grains is generally more severe in early growth stages than in later ones;[82-84] therefore, a range of damage thresholds is provided. Thus, economic thresholds for Russian wheat aphid depend on wheat plant growth stage, management costs, and expected yield. The economic threshold for preflowering wheat is calculated by the following equation:

$$ET = 200 \times CC/EY \times MV \qquad (4)$$

where ET is the economic threshold in percentage infested tillers, CC is the control cost (insecticide plus application), EY is the expected yield, and MV is market value of the crop.[80] The quantities, CC, EY, and MV need to be calculated based on the same unit area (e.g., acre, hectare, etc.). After flowering, 500 is substituted for 200 in Equation 4. Upper and lower values for terminating sampling are available for infestations ranging from 5 to 50% in 5% increments.[80] In practice, the user consults the table with the percentage closest to that calculated from Equation 4. The maximum area in a field that should be sampled is 32 ha (80 acres). Fields larger than 32 ha should be divided into uniform areas of 32 ha or smaller, and each area should be sampled independently. If a sequential sample proceeds to 100 tillers before a decision is made, sampling is terminated. In this case, the infestation does not differ significantly from the economic threshold and growers are advised to treat with insecticide if warm, dry weather conducive to Russian wheat aphid population growth is expected. If weather conditions less favorable for population growth are expected, growers are advised to return to the field in a few days to resample.[80]

A novel procedure termed the "sequential interval procedure" was used to construct nonlinear sequential sampling stop-curves in the Russian wheat aphid sequential sampling scheme.[81,85,86] Using this procedure, stop-curves are developed from the confidence interval method, but with a variable error rate that is predetermined so that a specified maximum allowable error rate occurs for population densities equal to the economic threshold. Stop-curves are then constructed through an iterative process of simulation and curve fitting.

4. Sampling Brown- and Whitebacked Planthoppers in Rice in the Philippines

Several sampling techniques and methods have been used to sample brown- and whitebacked planthoppers in transplanted rice, including a variety of vacuum sampling machines, a carbon dioxide cone sampling device, pans containing water to catch insects dislodged from plants, sticky traps, and direct visual counting methods.[47,87-96] Most sampling methods for planthoppers use a single hill (cluster of transplanted plants) as the sample unit. Sequential sampling schemes have been developed for planthoppers based on direct counts and incidence counts.[47,88,90-93]

A sequential sampling scheme based on incidence counts was developed and is currently used in IPM programs for both species of planthoppers in flooded rice in the Philippines[47,91] (Table 4). The scheme is based on dislodging planthoppers into a pan of water held under a plant to catch arthropods and counting third instars to adults. Nearly all rice plants are infested with third instar to adult planthoppers when the population density approaches the economic threshold (13.5 third instar to adult planthoppers per plant). Therefore, the percentage of sample units infested by one or more planthoppers does not provide a useful model from which to develop a sequential sampling decision rule based on the binomial distribution. However, only about 45% of hills are infested with ten or more planthoppers when the population density approaches the economic threshold. Therefore, the decision rule was developed based on the proportion of plants infested with ten or more third instar to adult planthoppers.

Because polyphagous arthropod predators sometimes play an important role in regulating planthopper populations,[97,98] the sequential sampling scheme includes two decision rules: one for rice fields with high predator densities, and another for fields where predators are sparse (Table 4). Incorporating predators into the decision rule was accomplished by increasing the damage threshold to account for future reductions in the planthopper population due to predation. Adjustments to decision lines to account for predation were conservative. To sample a hill, a scout holds the pan of water under a plant and dislodges arthropods from the hill into the pan. The scout determines whether or not ten or more planthoppers are present and counts the number of predators. After sampling five hills, a decision is made whether to use the "predators present" or "predators absent" decision rules (Figure 4). If a total of five or more predators are found in the first five samples, the "predators present" rule is used. From the sixth sample on, only the presence or absence of ten or more planthoppers is recorded. The stop-lines for "predators present" are closer together and slope more rapidly upwards. This has the net effect of increasing the number of leafhoppers required in a sample in order to contact the upper stop-line and decreasing the number required to contact the lower stop-line compared to the "predators absent" sampling scheme.

The sequential sampling scheme was adapted in the form of a sampling pegboard, so that field sampling could be easily undertaken by relatively untrained persons.[92] Comparisons of the incidence count sequential sampling scheme with a more intensive direct counting method based on a fixed sample size of 20 hills per field resulted in an average saving of 70% in the number of samples taken, and a 96 to 100% agreement in decisions made.[91]

C. SAMPLING PLANS BASED ON POPULATION INDICES

When an easily measurable amount of injury can be sustained before the economic injury level is exceeded, sampling numbers of injured plants or parts of plants is a useful approach for IPM sampling programs. This is especially true for damage caused by cryptic arthropods that are difficult to count. Limitations to this approach

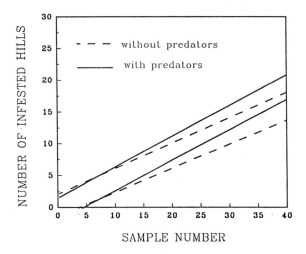

FIGURE 4. Sequential sampling stop-lines for rice planthopper sampling in fields with (solid lines) and without (dashed lines) predators. (Redrawn from Shepard and Ferrer.[47])

are that some yield may be lost before a decision to apply insecticide is made, and there is often little time between sampling and the time insecticide must be applied to the crop to avoid economic damage.

1. Sampling Rice Stem Borer Damage in Thailand and Japan

Four species of stem borers commonly occur on transplanted rice in Thailand, the most important of which are *S. incertulas* and *C. suppressalis*.[99] Larvae bore into tillers and remain there to feed. When this feeding occurs during the vegetative stages of plant growth, the newest leaves die, while the lower leaves remain healthy. This damage is easily seen upon visual inspection of plants and is known as "deadheart." The recommended economic threshold for stem borers is 1.5 deadhearts per hill,[99] although more recent experiments have shown that this level of damage can be compensated for by the rice plant.[100] Stem borers are sampled without regard for species and the number of "deadhearts" per hill is the usual sample unit employed. A sequential sampling scheme for rice stem borers was constructed based on a model for predicting the average number of deadhearts per hill from the proportion of hills with one or more deadhearts.[99] The economic threshold of 1.5 deadhearts per hill corresponds to 68% (0.68) of hills with one or more deadhearts. The lower level for which damage was considered tolerable was set at 50% (0.50) of hills infested. The probability of treating when the proportion infested hills is below the economic threshold, α, was set at 0.20, and the probability of not spraying when the proportion of infested hills is above the economic threshold, β, was set at 0.10. Sequential sampling stop-lines are listed in Table 4.

The sampling scheme was field tested in a total of 69 fields by comparing results of sequential sampling with those obtained from fixed sized samples of 20 hills per field. In 93% of fields (64 of 69), use of the sequential sampling scheme led to the same decision reached by sampling 20 hills. In three fields treatment was not recommended using sequential sampling when the results of the fixed sample indicated it was necessary. Using sequential sampling the average time required to sample a field was reduced to about one third that required to sample 20 hills.[99]

In Japan, sequential sampling schemes for rice infested with stem borer larvae were developed assuming uniform and random spatial distributions of stem borer-infested hills in paddy fields.[36,101,102] Assuming a random spatial distribution of borer-infested hills, a sequential sampling scheme based on the Poisson distribution (Table 4) involves visual inspection of rice hills to determine whether borer infested tillers are present. Sampling a field involves walking a diagonal transect through the field and stopping at randomly chosen rows along the route. Ten rice hills in each selected row are inspected for the presence of borer infested tillers. The cumulative number of borer-infested hills is recorded, and decision lines are consulted after inspecting each hill. If the cumulative number of infested hills exceeds the upper decision line, corresponding to a damage threshold of 40% of infested hills, insecticide treatment is recommended; correspondingly, if the cumulative number of infested hills falls below the lower decision line (corresponding to 20% infested hills), treatment is not recommended. To the best of our knowledge, the sampling method has not been field tested to determine its time efficiency or accuracy for decision making.

2. Sampling Rice Water Weevil Adult Feeding Scars in Arkansas

Larvae of the rice water weevil feed on and injure the roots of flooded rice. Nearly all the roots of rice plants can be pruned when dense populations occur. Gravid adult rice water weevils enter rice fields from overwintering habitats while fields are being flooded; they feed on rice leaves and begin depositing their eggs under water soon after flooding. Adult populations generally peak within a week of flooding and decrease to low levels a few weeks thereafter.[103] Adult feeding on leaves, which is usually of no economic consequence, is easily identifiable as elongated scars on leaves. The density of feeding scars on the newest leaves of rice plants is related to subsequent larval density (Figure 5B).[104] However, slightly less than 50% of the variance in larval density is accounted for by the density of feeding scars, suggesting that the method may lack precision for identifying damaging populations. Although leaf scar sampling lacks precision compared with sampling larvae directly, it has the advantages of providing a decision soon after flooding, before larval populations develop and begin to injure the crop, and it requires much less effort.[103]

Two sequential sampling schemes for scarred leaves were developed; one for sampling rice within about a 3-m (9.8-ft) distance from levees, and another for sampling rice in the interior portion of fields between levees, referred to as bays. Two schemes were developed because adult rice water weevils are more dense near levees than in bay areas;[105,106] however, a single sampling scheme could have been developed based on random sampling from rice fields. Both schemes are based on the negative binomial distribution. The scheme for sampling bay areas is summarized in Table 4. Counts from bays of the number of plants with scarred new leaves were found to be described well by a negative binomial distribution with $k = 5.19$. A relationship between the loss in yield of rice and the number of rice water weevil larvae per 9.2-cm soil core was established empirically (Figure 5A). Using knowledge of the cost of insecticide treatment and value of rice, the economic threshold was established at 10 larvae per core.[103] Based on the relationship between scarred leaves and larval density (Figure 5B), an action threshold of 24 scarred leaves in a sample unit consisting of the newest leaf of each of 40 plants, i.e., 60% of leaves scarred, was established.[103] To avoid having to count both plants and scarred leaves at the same time, it is suggested that samplers use a square frame of appropriate size to enclose 40 plants.[103] At typical stand densities, a 0.3-m (1-ft) square frame will enclose 13 to 14 plants, so that 40 plants can be sampled by placing the frame at three proximate locations. Fields should be sampled for rice water weevil feeding scars within 2 weeks of establishment of permanent flood, or within 2 weeks following emergence of leaves above the water in water-seeded fields.[35,103]

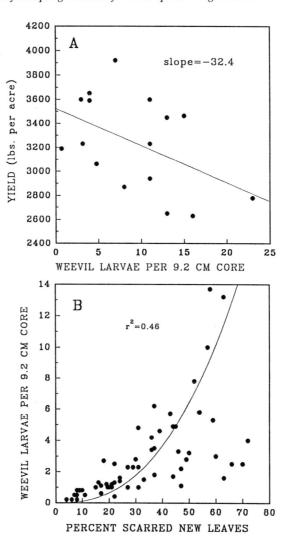

FIGURE 5. (A) Relationship between the number of rice water weevil larvae per soil core and yield of rice; (B) Relationship between the percent of new leaves scarred by adult rice water weevil feeding and the subsequent number of rice water weevil larvae per soil core. (Redrawn from Tugwell and Stephen.[103])

The average-sample-number curve for leaf scar sequential sampling reaches a maximum of approximately three 40-plant samples, so that relatively few samples are usually required to reach a decision.[103] In a 2-year validation study of the leaf scar sequential sampling scheme, control decisions based on an economic injury level of 10 larvae per soil core were made correctly for 95% of plots evaluated.[107] However, in the same study it was found that rice yield was not significantly reduced by larval populations above ten per core for the cultivars tested, raising doubt about the use of a constant economic threshold for all rice cultivars.

Because of the lead time for control actions obtained by leaf scar sampling, it is possible to verify decisions made using the leaf scar sampling scheme by sampling the damaging larvae from soil at a later date. This practice is not currently used in rice water weevil IPM, but might be a viable strategy as more research is conducted to reduce the time required for sampling larvae from the soil. We briefly review soil sampling methods for rice water weevils in a later section.

3. Sampling Tadpole Shrimp Injury to Rice in California

Tadpole shrimps are small and mobile and therefore difficult to count. As a result, sampling methods and economic thresholds for tadpole shrimps in IPM programs for rice in California are based on inspecting plants for damage characteristic of their feeding and digging activities, in addition to looking for evidence of the presence of the shrimps.[108,109] Tadpole shrimps begin to hatch within a few days after rice fields are flooded and develop to sexual maturity in about 9 d.[108] The shrimps damage seedling rice plants by feeding on emerging shoots and roots, and by uprooting seedling plants with their digging activities. The need for control measures is determined by counting numbers of seedlings at two stages of development, usually at 5 to 7 d after flooding and again 3 to 7 d later. Sampling is conducted using a metal or plastic cylinder enclosing 0.09 m^2, with length greater than the maximum depth of the water in the field. At the first sampling period, counts are made by placing the cylinder in the water all the way to the soil surface and counting the number of germinating seeds and seedlings contained in the cylinder. Ungerminated seeds and floating seeds should be examined for shrimp feeding damage, and the presence of shrimps or their cast skins should be noted. Ten such samples are made at points along a diagonal transect across the field. In large fields, two diagonal transects should be sampled. If there are less than an average of 30 healthy seedlings per 0.09 m^2, and tadpole shrimp are found and are damaging plants or seeds, treatment is recommended. At the second sampling date, ten stops are made along a diagonal transect to count the number of healthy and damaged or uprooted seedlings, and to inspect for evidence of shrimps. An average of 25 healthy seedlings per 0.09 m^2 at this time should provide a good stand. If the average number of healthy seedlings drops below 25 per 0.09 m^2 and there is evidence of damage and shrimps, the field should be treated.

4. Sampling Armyworm Injury to Rice in California

Two armyworm species, the true armyworm, *Pseudaletia unipuncta*, and the western yellowstriped armyworm, *Spodoptera praefica*, occasionally cause economic damage to rice in California.[108,109] These species have similar seasonal phenologies in rice, invading fields in mid- to late-summer, and cause similar damage. Armyworm infestations are typically limited to isolated portions of fields, so that spot treatment with insecticides is usually more desirable than treating the entire field. Serious economic damage by armyworms is most likely to occur at two periods during development of rice plants: first during the stem elongation stage, and later during grain filling. Armyworms remain hidden in the foliage during daytime and are difficult to count accurately. Therefore, inspecting plants for armyworm damage is the preferred sampling method for IPM purposes.

Formal sampling for foliar injury is conducted during the stem elongation stage if armyworm feeding is evident on casual inspection. To sample, choose a part of the field where injury is present, based on the initial inspection, and select ten plants at random from within that portion of the field. Each plant is inspected for the presence or absence of 25% or more defoliation. It is recommended that sampling be conducted in each area of the field that appears to have visible damage. Spot treatment with insecticide is recommended in any areas in which five or more of the ten plants sampled have 25% or more defoliation, and in which armyworm larvae are present. If plants are injured but armyworms are not present, treatment is not recommended because larvae have probably completed development.

Sampling during grain filling is conducted in parts of the field where white heads, characteristic of armyworm feeding on developing kernels, are present. Sampling is conducted using a ring that encloses an area of 0.09 m². The ring is placed at a random location within the infested site. The numbers of healthy and damaged heads in the ring are recorded, and the presence or absence of armyworms is recorded. This procedure is repeated every 2 to 3 m until a total of 10 sample units have been examined. The procedure should be repeated at other sites showing injury. Treatment of an infested site in the field is recommended if 10% or more of heads are injured and armyworm larvae are present.

5. Sampling Wheat Bulb Fly Injury

Pest management decision making for the wheat bulb fly in Great Britain is usually based on sampling eggs laid in the soil at the end of the oviposition period in autumn. An alternative to basing pest management decisions on egg sampling is to sample plant injury after the initiation of egg hatch, and treat fields with a systemic insecticide.[110] Sampling injured plants requires rather precise knowledge of the timing of egg hatch because injury can occur rapidly; it is probably not the preferred sampling method because of the greater cost of treating the field with a systemic insecticide compared with a seed treatment and the potential for yield loss before treatment can be applied.

D. OTHER METHODS FOR SAMPLING AND MONITORING ARTHROPOD PESTS OF WHEAT AND RICE

1. Sampling Arthropods Using Vacuum Samplers

Thornhill[111] compared the efficiency of the D-Vac®[30] with a modified vacuum sampler for sampling insects in grasslands and found similar efficiencies with both techniques. Dewar and Dean[112] compared the D-Vac to direct counting and a plant washing technique for sampling aphids in wheat and found that the efficiency of vacuum sampling decreased as aphid density increased. However, they indicated that the D-Vac may have utility for sampling low density populations that cannot be sampled precisely using other methods. The efficiency of D-Vac sampling was also shown to vary among aphid species in wheat and among life stages of a given species.[112,113] Hand[113] proposed that despite its drawbacks, the D-Vac could be used for estimating the life-stage structure of aphid populations in wheat, and for detecting the presence of immigrating alatae in wheat fields.

Perfect et al.[94] compared the efficiency of two vacuum samplers, the D-Vac and FARMCOP, for sampling cicadellids and delphacids and their arthropod predators in rice. D-Vac sampling from within an enclosure placed over a rice hill prior to vacuum sampling provided accurate estimates of population density for most arthropods. Its efficiency was not affected by crop age for most taxa, except for the spider *Microvelia douglasi atrolineata* and for brachypterous female brown and whitebacked planthoppers. Vacuum samplers are generally inappropriate for use in IPM programs because of their expense and the excessive time required for sampling. However, they are useful for ecological research and can be used to calibrate methods more appropriate for use in IPM.

2. Sampling Arthropods Using a Sweepnet

Sweepnet sampling is a technique that provides relative population estimates while providing a relatively large amount of information with minimal effort; its sampling efficiency for arthropods is nonetheless influenced by a variety of abiotic and biotic variables. For example, the efficiency of sweepnet sampling for green leafhoppers in rice is affected by the time of day sampling is conducted, with population estimates made in the morning or evening being greater than estimates made in the afternoon.[114]

Sampling cereal leaf beetles by sweepnet is influenced by wind speed, temperature, solar radiation intensity, and crop height.[31]

Sweepnet catch of some pest and beneficial arthropods inhabiting wheat has been calibrated to obtain estimates of absolute population density by modeling the effect of various factors on sampling efficiency.[31,32] Banks and Brown[115] compared the efficiency of quadrat, sweepnet, and mark-recapture methods for estimating sunn pest populations. Both sweepnet and quadrat sampling gave relatively precise estimates with a reasonable level of sampling effort. Population estimates obtained by sweepnet sampling adequately depicted population trends observed using the other methods.

a. Sweepnet Sampling for Rice Stinkbugs in Texas

Immature and adult rice stinkbugs enter rice fields from alternate habitats when rice plants are heading. They feed on the developing grain and are usually more abundant along field margins than in the interior of fields.[116] The recommended method for sampling in Texas is by making 180° sweeps using a 38-cm (15-in.) diameter sweepnet (Table 4).[35] Because most bugs congregate on rice heads, the net is held so that only the lower half of the opening is drawn through the foliage. Mid-day samples underestimate rice stink bug densities. Ten sets of ten sweeps are taken at random locations in the interior of the field, avoiding field margins. The average number of adult stinkbugs per ten sweeps is calculated and compared with a dynamic economic threshold that accounts for the effects of plant developmental stage, expected yield, market price, and cost of insecticide application on control decisions.[117] Tables required to estimate the economic threshold for particular circumstances are presented in Drees and Way.[35]

Recommended sampling methods for rice stinkbugs differ among states. For example, in Arkansas the recommended method is to count the number of rice stinkbugs on 100 rice heads from outside the field with the aid of binoculars.[35]

3. Sampling and Monitoring Arthropods Using Impaction Traps

Impaction traps such as sticky traps, water traps, Malaise traps, and light traps rely on active or passive movement of arthropods to the trap where they are caught.[118] Trap efficiency is affected by wind speed, trap color, size, and shape, and by other factors.[29] Sticky traps and water traps have been used extensively to capture alate aphids in agricultural crops,[119] although not for monitoring cereal aphids in IPM programs. Robert[119] suggests that because these traps are likely to catch aphids that are in their landing phase they may complement aerial suction traps in providing information on aphid migration.

Malaise traps, light traps, and sticky traps have been used to sample adult stem borers in rice for population dynamics studies,[36] but these techniques were not considered useful for IPM purposes. Sticky traps have been used for estimating brown planthopper populations in rice fields in Japan,[96] and as part of a region-wide surveillance system in Malaysia.[89] While sticky traps are useful for estimating arthropod populations, they are difficult to handle and relatively expensive. The primary utility of impaction traps in IPM in rice is for monitoring immigration of arthropods into fields, especially arthropods that transmit plant diseases. Such sampling techniques are valuable in research programs on arthropod pests, particularly for characterizing population trends over time or space.[29,36]

Impaction traps have sometimes been useful for population monitoring in wheat IPM programs. For example, catches of female wheat bulb flies in light traps correlate with the numbers of eggs laid in nearby wheat fields. Catches in light traps may be a basis for advising farmers to purchase insecticide-treated winter wheat seed for fields subject to infestation by the pest.[120] This method will require more research to verify that it has the necessary reliability for use in IPM programs. Sampling eggs in soil is currently recommended in IPM programs for the wheat bulb fly.[19]

4. Sampling Soil-Dwelling Arthropod Pests of Wheat and Rice

Several arthropod pests of wheat and rice live in the soil during their damaging life stages. Sampling these pests, like soil pests of most other crops, has received limited attention because of the unreasonable effort usually required. Soil corers or shovels are commonly used to collect a standard volume of soil; arthropods are then extracted from the soil for counting by sieving or by washing them from the soil and floating them in a salt solution from which they can be extracted.[121-125] Other methods for sampling soil insects involve burying baits to which the arthropods are attracted.[126-129] After a period of time, the bait is collected and the insects aggregated in it are counted. The efficiency of baiting is improved in some cases by covering the soil above the bait with transparent polyethylene or charcoal to collect heat.[126,127] Baiting techniques can concentrate arthropod numbers several times when compared with their density in the surrounding soil.[126-129]

a. Sampling Wheat Bulb Flies in Great Britain

Eggs of the wheat bulb fly are laid in bare soil in summer and hatch in autumn to feed on newly planted winter wheat.[19] The economic threshold in current use is 2.5×10^6 eggs/ha.[130] Earlier planted wheat suffers less injury than later planted wheat because of the more advanced growth stage by the time feeding begins. Insecticide-treated seed is the most economical method to control the pest.

Wheat bulb fly eggs are sampled in IPM programs by taking 20 10.2-cm (4-in.) diameter cores to a depth of 7.5 cm (3 in.) in early autumn, after oviposition is complete.[19,131,132] Eggs are separated from soil by wet sieving and magnesium sulfate flotation[124] and counted. Sampling and processing, however, are sometimes completed too late to advise farmers to order insecticide-treated seed.[19]

b. Sampling Wireworms in Wheat in the U.S.

Wireworms are important pests of several crops, and occasionally can severely damage small grains. Egg and larval wireworm populations have been estimated by sampling soil[133] and by using baits.[126-129] Adults have been monitored using light traps,[134] and emergence traps have been used to determine the seasonal production of adults.[121,135] Methods for sampling the soil-dwelling immature wireworms in wheat are time consuming. Time-efficient sampling methods have not been developed for wireworms, and these pests are not routinely sampled in pest management programs, except to determine their presence in fields with histories of wireworm infestations.

Doane[121] studied the spatial distribution patterns of two wireworm species in wheat fields in Saskatchewan, Canada. Eggs of both species were highly aggregated among sample units (5-cm [2-in.] diameter by 2-cm [0.8-in.] deep cores) taken from under dirt clods. Larvae were sampled by taking 10.2-cm (4-in.)-diameter by 30.5-cm (12-in.)-deep soil cores. The spatial distribution of larvae of the two species differed with age. *Ctenicera destructor* larvae started out highly aggregated, but became randomly distributed by the time they became large larvae. Conversely, the large *Hypolithus bicolor* larvae were more highly aggregated than small larvae. Doane[121] also calculated sample sizes required to achieve constant average levels of precision for soil core sampling. For sampling a population of wireworms in wheat with an average density of 0.5 larvae per square meter, 40 cores are required to achieve a standard error of 30% of the mean. Thus, sampling low-density populations with adequate precision for IPM decision making would be cost prohibitive. Landis and Onsager[133] proposed sampling 20 cores per acre and sieving the soil through a wire screen for counting to determine the potential for wireworm problems, still a time-consuming task.

Improved "baiting" methods for wireworm sampling in wheat have been developed.[126,129] The methods involve establishing bait stations by burying 0.2-l ($\approx$ 0.2-qt) of bait (wheat or a wheat/corn mixture) at 10-cm (3.9-in.) depth. Soil is mounded over the bait and covered with transparent polyethylene to collect heat. When attempting to determine damage potential from wireworms in spring-planted wheat these solar baits are most effective when placed in the autumn and sampled the following spring. The solar baiting technique can increase wireworm numbers in samples up to 25 times when compared with their density in the surrounding soil.[126] No estimates of the variability or required sample sizes for specified levels of precision have been reported for this technique, but increases in sample means relative to variances may result from greater larval presence in samples, thus reducing the number of samples required to estimate wireworm populations with acceptable precision for pest management purposes.

c. Sampling Rice Water Weevil Larvae in the U.S.

Rice water weevil larval sampling from soil is typically accomplished using a 9.3-cm (3.6-in.)-diameter by 7.6-cm (3-in.)-deep core sampler that is centered over a row. Larvae are washed from the soil core through a 40-mesh soil sieve, floated in a saturated salt solution, and counted.[123] Robinson et al.[123,136] found the 9.2-cm-diameter core sample a more efficient sample unit in most respects (accuracy and precision of estimates and time required for sampling) than a 15.2-cm (6-in.) or 30.5-cm (12-in.) soil cube.

While there is an approximately linear relationship between the number of rice water weevil larvae per core and yield loss (Figure 5A),[103] the economic threshold in current use varies among states. Ten larvae per 9.2-cm core is the recommended threshold in Arkansas,[103] whereas the Texas threshold is five larvae per 10-cm (3.9-in.) core.[35] Larval sampling is time consuming, requiring 2.5 to 3 h to estimate moderately dense populations with an average precision (mean/standard error) of 0.10.[137] Sampling rice leaves damaged by adult feeding, while less precise, is less time consuming and is currently used in IPM programs in several states. Soil sampling could be used to augment leaf scar sampling in fields with marginally economically damaging populations. Considering the current uncertainty of economic thresholds for the rice water weevil,[107] this approach may help reduce the number of fields unnecessarily treated with insecticides.

IV. MONITORING MIGRANT BYDV VECTORS USING AERIAL SUCTION TRAPS

The Rothamsted suction trap described by Taylor and Palmer,[118] or some modification of it has become a standard tool for sampling aerial populations of cereal aphids in Europe[119] and the U.S.[138] This device uses a fan located at the bottom of a tube which draws air from the top of the tube through a fine mesh screen funnel into a collecting jar. Suction traps have been used since the 1950s in the U.K. to monitor migrating aphids that transmit plant diseases. In 1964 the 12.2-m (40-ft)-tall Rothamsted trap was adopted as a standard and deployed in a nationwide network. Similar networks have been established in other European countries. Suction trap (Allison-Pike traps[138]) networks were established in the Western U.S. in order to monitor aphids that transmit BYDV. Suction trap networks were also established in several Western states to monitor the influx and spread of the Russian wheat aphid which was introduced in the U.S. in 1986.

The advantage of suction traps is that they continuously monitor migrating arthropods in the air. The volume of air sampled is relatively constant, but is influenced somewhat by ambient windspeed. The Rothamsted trap samples about five times more air per unit time than the Allison-Pike trap. The consistency of air flow through a trap is dependent on the motor type used for the fan and the cleanliness of the screen funnel. Air flow can be restricted by dirt and other debris that gets stuck in the screen. The primary disadvantages of suction traps are their expense, their cumbersome nature, the requirement of a power source to operate, and the expense of sorting samples.

Cereal aphids are trapped to determine the seasonal occurrence and magnitude of flights of vector populations. Several BYDV strains are persistently transmitted by different cereal aphids, and they differ in pathogenicity.[119] *Rhopalosiphum padi*-specific strains are among the most damaging.[119]

A network of Allison-Pike suction traps was established in eastern Washington state in 1983 to monitor the seasonal distribution and abundance of crop-damaging aphids.[138] In eastern Washington, the traps are established on farms or agricultural research stations and monitored weekly from April to November. Weekly samples are sent to Washington State University where aphids are counted and identified to species. A weekly report is prepared and distributed that includes suction trap counts of economically important species. Data on the seasonal pattern of abundance of migrating *R. padi* in autumn are used by farmers and consultants to predict appropriate dates for planting winter grains to avoid infestations by *R. padi*. In Washington, the seasonal distribution and abundance of *R. padi* are sufficient for IPM purposes because manipulating planting dates or using soil-applied systemic insecticides are viable approaches for controlling BYDV.

A network of Rothamsted suction traps was established in the U. K. in 1964 and has increased in size over time to include 23 traps operating at locations throughout the country.[139,140] Samples are collected daily from traps during the main aphid flight period (April to November). Data are processed and disseminated in several forms. The *Aphid Bulletin* is a weekly report to the agricultural industry that lists the total catch of 33 pest species or species groups recorded at each trapping station.[140] A more detailed set of summary tables is provided in a second written report, the *Aphid Commentary*.[140] In addition to information on aphid pests of other crops, the *Aphid Commentary* provides an infectivity index which measures the risk of BYDV infection in autumn-sown barley and wheat.[141] The BYDV forecasts, as well as other aphid-monitoring data in cereal crops, are used extensively by agricultural advisors and farmers.[142] The infectivity index assesses BYDV risk by integrating, vector aphid abundance, the proportion of migrant aphids infected with the disease, and the period of time during which the crop is exposed to infestation.[141] Earlier sown crops are exposed for a greater length of time and, hence, are at greater risk of infection. For each week, the index is calculated by multiplying the number of each vector species (*R. padi* and *S. avenae*) captured in Rothamsted insect survey suction traps with the proportion of each species found to be infective when trapped alive in suction traps operated at 1.5 m (4.9 ft).[143] The proportion of infective aphids of each species is determined by allowing individual aphids caught in 1.5-m traps to feed on test plants (cv. Blenda oats) for 2 to 3 d. BYDV symptoms appear on infected plants within 2 to 4 weeks, at which time the proportion of infective aphids is calculated. For a particular field, the index is accumulated from planting date (emergence date would be more accurate but is difficult to estimate) until the end of the vector flight period (usually early November) from weekly estimates based on the nearest Rothamsted suction trap. If the cumulative infectivity index exceeds a threshold value of 50, insecticide treatment is recommended.[143] The time lag in assessing the proportion of infective aphids presents some limitations for predicting disease incidence, but the procedure works reasonably well in practice because the density of infective aphids is usually relatively low, and there is usually a limited amount of secondary spread of disease within crops until after the end of aphid flights in November.[144] The threshold has been validated for the area represented by the Rothamsted trapping station,[144] but may not be valid for other regions.[145,146]

Improving the BYDV monitoring system in Great Britain might include adjusting counts of alate *R. padi* in Rothamsted suction trap samples for the proportion of males and gynoparae (sexually reproducing females), both of which seem to migrate to their alternate host (*Prunus* sp.) and not to cereals,[146,147] and by obtaining a better understanding of factors controlling within-field vector activity (e.g., weather).[145,146] The use of such information to develop refined indices may improve regional predictions.[146,147] Finally, the use of ELISA and ISEM techniques to determine whether test plants are infected with BYDV would decrease the time required to provide forecasts and thus increase their value to farmers.[119]

V. SAMPLING NATURAL ENEMIES IN WHEAT AND RICE

The role of natural enemies in IPM in wheat, rice, and other small grains is usually not considered either because quantitative information on the role of natural enemies in controlling pest arthropods is lacking, information on the impact of natural enemies on pest populations is unavailable to pest managers, or for both reasons. Although, this is the general state of affairs, there are some pests for which the impact of natural enemies is included in IPM decision making.

The use of sampling data on natural enemies of rice leafhoppers was described in this chapter in the section on incidence count sampling. A sequential sampling scheme for rice leafhoppers based on direct counts also incorporates sampling data on predators.[90] Using the sequential sampling scheme, the number of rice leafhoppers counted is reduced by five for each predator in the sample to account for the fact that an average predator can consume at least that many leafhoppers.[90] The net result is that it takes more leafhoppers in a sample in order to contact the upper stop-line and fewer to contact the lower stop-line when predators are present.

Attempts to incorporate natural enemies into sampling plans for pests of wheat have also been described. For example, Texas extension guidelines for sampling greenbugs in wheat advise farmers to delay control measures in fields with greenbug populations near economic thresholds until it can be determined whether greenbug populations will be controlled by natural enemies. This approach is advised when there are one or more coccinellids per 0.3 m (1 ft) of row or when 15% or more of greenbugs have been mummified by the parasitoid *Lysiphlebus testaceipes*.[37] The guidelines are relatively crude because methods are not provided for assessing the reliability of estimates of natural enemy density, and to our knowledge, there are no "hard" data from which to quantitatively relate natural enemy density to greenbug population suppression.

Aphidophagous coccinellids are a ubiquitous group of predators in the Great Plains region of the U.S., the importance of which has been established in the biological control of aphid pests in small grains in the region.[148,149] Perhaps due to their mobility in the constantly changing agricultural landscape, coccinellid beetle populations in small grain fields vary widely and unpredictably in time and space.[150] Because of this unpredictability, precise yet time-efficient sampling methods for coccinellids are needed for use in IPM programs. Recent studies describe sampling methods for coccinellids that provide satisfactory estimates of population densities.

Lapchin et al.[151] compared removal sampling and visual count sampling for estimating coccinellid population densities in wheat fields in Europe. They found that De Lury's method[152] could be used to calculate accurate estimates of population density of three coccinellid species using data obtained from removal sampling. Sampling efficiency remained constant among successive samples from a plot and a large proportion of the coccinellids contained within a 25-m^2 (269-ft^2) plot was removed in collections made during two 20- to 25-min inspections of the plot. Visual counts of three coccinellid species made while walking through plots for a constant amount of time were linearly related to population density estimated by De Lury's method.

A relatively simple method was developed for sampling adult coccinellids in wheat fields in the Great Plains.[32,76] Population density estimates obtained from De Lury's method were used to convert estimates of relative population density obtained from 2-min visual counts taken while walking at a constant velocity of 10 m/min (33 ft/min) through the field to absolute population density. The number of beetles observed in visual counts was influenced by temperature, and this variable was incorporated in a regression model to convert visual counts to population density:

$$\text{Density} = (0.18 + 0.0000094 \text{ Temperature}^2) \text{Visual Count} \quad (5)$$

A precise relationship between the mean and variance of the number of adult coccinellids per 2-min count was obtained using Taylor's power law:[153]

$$\log(s^2) = 0.021 + 1.34 \log m \quad (6)$$

where s^2 and m are the sample variance and sample mean of 2-min counts, respectively. Equation 6 was used to develop a sequential sampling scheme for estimating the mean number of adult coccinellids per 2-min count with known average precision (standard error/mean) using Green's method.[154] Using Equation 5, mean numbers of beetles per 2-min count obtained by sequential sampling were converted to estimates of absolute population density. The sampling scheme was modified to compensate for error in estimates of power law parameters and for practical limits on the number of samples taken. Using the modified scheme, a maximum of 18 2-min counts are made. Average precision varies depending on the actual mean number of beetles per 2-min count. For populations of less than 1.5 beetles per 2-min count, estimation is imprecise because sample sizes larger than 18 are required to achieve moderate precision. However, average precision is less than or equal to 0.25 whenever the actual mean number of beetles per 2-min count is greater than 1.5. When beetle populations are below 2 per 2-min count, coccinellids are unlikely to exert effective biological control, so imprecise estimates would be of little consequence in an IPM decision making context. Iperti et al.[155] had previously developed a sequential sampling scheme for the coccinellid, *Coccinella septempunctata*, in wheat using a similar approach.

VI. RECOMMENDATIONS FOR FURTHER RESEARCH

A significant amount of sampling research has been done on various small grain pests. The use of incidence counts, population indices, and other relative measures of arthropod populations allow development of inexpensive yet reliable sampling schemes. These approaches circumvent the often time-consuming task of estimating population density by direct counting. Further developments are needed in several areas to improve sampling methods for small grain pests.

First, there is a need for improving current sampling plans and developing new plans that are cost effective and practical. A major impediment to the acceptance of a sampling plan in relatively low-value small grain crops results from plans that are too complex to be carried out with reasonable time and effort. Sequential sampling has addressed this problem, and this approach should be continued and expanded. To aid in acceptance of such plans, sampling time and cost analyses, including analyses for optimal allocation of sampling resources, should be conducted so that prospective users can evaluate the expected costs of sampling relative to management based on other approaches. Researchers should work closely with end-users to insure that sampling methods are developed that satisfy constraints on time and other resources.

One way that sampling can be made more effective is to develop sampling plans for entire pest complexes of a crop. A step toward this approach is to develop sampling plans for complexes of closely related species in crops. This approach was taken for sampling several species of cereal aphids in small grains in the northern Great Plains of the U.S. and Europe.[62,73] In both cases, parameters of equations used in developing sampling plans were very similar (not statistically different) for all species, and although economic thresholds differed among species, these differences were often overshadowed by the influence of biotic and abiotic factors on damage relationships for a given species.

It may be possible to develop a 'general' incidence count sampling method for all economically important aphid species in small grains. Hein et al.[156] compiled published estimates of parameters of an emperical relationship between the average number of aphids per small grain tiller and the proportion of infested tillers:

$$\log_e m = a + b \log_e\{-\log_e(1 - p)\} \tag{7}$$

Estimates of parameters of Equation 7 were available for six aphid species (*R. maidis*, *R. padi*, *S. avenae*, *S. graminum*, *M. dirhodum*, and *D. noxia*) on three crops (spring wheat, winter wheat, and spring barley). Estimates were similar regardless of the aphid species sampled or small grain crop species and were averaged to obtain a single version of Equation 7 for predicting cereal aphid population intensity:

$$\log_e m = 1.62 + 1.175 \log_e\{-\log_e(1 - p)\} \tag{8}$$

Equation 8 was used to predict $\log_e m$ for each of four aphid species sampled from wheat fields in South Dakota, Nebraska, Oklahoma, and Texas. Equation 8 adequately predicted $\log_e m$ for each of the four species (Figure 6); coefficients of determination of linear regressions of predicted on observed $\log_e m$ were large, ranging from 0.74 to 0.98; and neither the slopes nor the intercepts of the regressions differed significantly from one or zero, respectively.

Sampling plans based on general parameters have proven useful for sampling mites in a variety of crops,[157] and for sampling white grubs in grain sorghum.[158] In practice, small differences that exist among species in the parameters of models used to construct sampling plans may be overshadowed by differences that occur among populations of a single species due to temporal or spatial variation in biotic or abiotic factors. The small differences that occur among species may not be great enough to warrant the time and resources required to develop separate plans for each species,

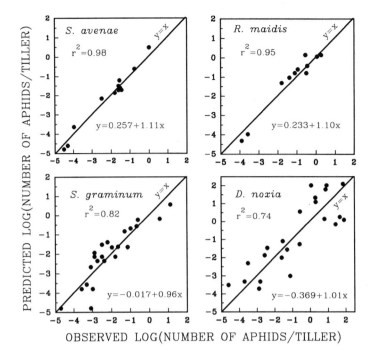

FIGURE 6. Predicted versus observed natural logarithms of the mean number of aphids per tiller for cereal aphids in spring and winter wheat and regression lines.

nor would they warrant the time required to use multiple sampling plans in IPM programs.

Another useful approach might be to develop very general survey sampling plans for detecting the presence of a pest, which could then be followed by the use of more time-consuming sampling plans for pests that pose a potential threat. Such sampling plans could change through the season as the composition of the pest and natural enemy complex changes. For example, survey sampling plans in wheat for most pest arthropods could be based on sweepnet sampling, provided threshold levels for catch were established that would trigger more intense sampling for decision making using appropriate methods. Threshold levels might need to vary over the growing season to account for differences in sampling efficiency for arthropods as the crop matures.

More thorough testing and validation of sampling plans should be undertaken as they are developed. Field validation should be an integral component of any program for developing a sampling plan. Once a sampling plan is proposed and field tested, it is much more likely to be adopted and utilized in an IPM program. In addition, field testing may uncover a flaw that, if left uncorrected, could doom the eventual acceptance of the plan.

Sampling research for a number of small grain pests has been very thorough. However, sampling plans are unavailable for some of the major and most of the potential pests of small grains. In order for IPM to reach its potential in small grains, sampling plans need to be developed or adapted for sampling many of these pests to facilitate informed decision making. An important shortcoming in developing sampling plans is lack of adequate information on pest-yield loss relationships. In many instances thresholds are merely best guesses as to the relationship between pest abundance and yield loss.

Increased emphasis should be given to developing sampling plans for important natural enemies, or complexes of natural enemies, of wheat and rice pests. A fundamental principle of IPM involves reliance as much as possible on natural pest controls, of which natural enemies are an important component. In order to incorporate the impact of natural enemies in the IPM decision making process, it is necessary to estimate natural enemy populations; and it may be necessary to determine the impact of natural enemies on pest populations and relate decreased pest numbers to economic thresholds. However, in many cases the impact of natural enemies on a pest may be consistent enough over time and space to facilitate development of robust sampling programs without precise knowledge of the intricacies of the natural enemy-pest interaction.

Finally, consultants and growers must be made aware of sampling plans and the benefits they will incur from their use. Extension personnel should rapidly adopt new sampling methods once they have been validated, and should transfer the new technology to the public as soon as possible. Researchers must also be involved in this process by providing information to extension personnel in a timely manner, and perhaps by involving them in the development or validation process. The direct involvement of IPM practitioners in the development of sampling plans will help insure their adoption as part of the pest management arsenal.

ACKNOWLEDGMENTS

Ruth Treat applied her editorial skills in making this chapter more readable than it otherwise would have been. Frank Peairs and Bob Kieckhefer provided several extremely helpful comments on the initial draft of the manuscript. Larry Godfrey, Mo Way, Phil Tugwell, John Robinson, and Al Knutson supplied information and several important published and unpublished reports.

REFERENCES

1. Briggle, L. W. and Curtis, B. C., Wheat worldwide, in *Wheat and Wheat Improvement*, 2nd ed., Heyne, E. G., Ed., American Society of Agronomy, Madison, Wis., 1987, 1.
2. United States Department of Agriculture, *Agricultural Statistics 1991*, United States Government Printing Office, Washington, D.C., 1991, 370.
3. Ofori, F. and Stern, W. R., Cereal-legume intercropping systems, *Adv. Agron.*, 41, 1986.
4. Cook, J. R. and Veseth, R. J., *Wheat Health Management*, APS Press, St. Paul, MN, 1991.
5. Way, M. O. and Bowling, C. C., Insect pests of rice, in *Rice Production*, Vol. 1, 2nd ed., Luh, B. S., Ed., Van Nostrand Reinhold, New York, 1991, 237.
6. Pathak, M. D., Ecology of common insect pests of rice, *Annu. Rev. Entomol.*, 13, 257, 1968.
7. Murphy, M., An economic perspective on cereal farming, food consumption and the environment, in *The Ecology of Temperate Cereal Fields*, Firbank, L. G., Carter, N., Darbyshire, J. F., and Potts, G. R., Eds., Blackwell Scientific, London, 1991, 69.
8. Bonnemaison, L., Principal animal pests, in *Wheat Documenta Ciba-Geigy*, Hafliger, E., Ed., Ciba-Geigy Ltd., Basel, 1980, 59.
9. Gallun, R. L., Breeding for resistance to insects in wheat, in *Biology and Breeding or Resistance to Arthropods and Pathogens in Agricultural Plants*, Harris, M. K., Ed., Texas A & M University Press, College Station, 1980, 245.
10. Way, M. J., Entomology of wheat, in *The Entomology of Indigenous and Naturalized Systems in Agriculture*, Harris, M. K. and Rogers, C. E., Eds., Westview Press, Boulder, CO, 1988, 183.
11. Loevinsohn, M. E., Litsinger, J. A., and Heinrichs, E. A., Rice insect pests and agricultural change, in *The Entomology of Indigenous and Naturalized Systems in Agriculture*, Harris, M. K. and Rogers, C. E., Eds., Westview Press, Boulder, CO, 1988, 161.

12. Shepard, B. M., Khan, Z. R., Pathak, M. D., and Heinrichs, E. A., Management of insect pests of rice in Asia, in *CRC Handbook of Pest Management in Agriculture*, Vol. 3, 2nd ed., Pimentel, D., Ed., CRC Press, Boca Raton, FL, 1991, 255.
13. Bacon, O. G., Burton, V. E., Chaney, W. E., Natwick, E. T., Stern, V. M., and Summers, C. G., Insects and mites, in *Integrated Pest Management for Small Grains*, Statewide Integrated Pest Management Project, University of California Division of Agriculture and Natural Resources Publ. 3333, 1990, 88.
14. Way, M. O., Insect pest management in rice in the United States, in *Pest Management in Rice*, Grayson, B. T., Green, M. B., and Copping, L. G., Eds., Elsevier, New York, 1990, 181.
15. Hatchett, J. H., Starks, K. J., and Webster, J. A., Insect and mite pests of wheat, in *Wheat and Wheat Improvement*, 2nd ed., Heyne, E. G., Ed., American Society of Agronomy, Madison, WI, 1987, 625.
16. Gair, R., Jenkins, J. E. E., and Lester, E., *Cereal Pests and Diseases*, 4th ed., Farming Press Ltd., Norwich, U.K., 1987, 32.
17. Wilde, G., Wheat arthropod-pest management, in *CRC Handbook of Pest Management in Agriculture*, Vol. 3, Pimentel, D., Ed., CRC Press, Boca Raton, FL, 1981, 317.
18. Grigarick, A. A., General problems with rice invertebrate pests and their control in the United States, *Prot. Ecol.*, 7, 105, 1984.
19. Burn, A. J., Cereal crops, in *Integrated Pest Management*, Burn, A.J., Coaker, T. H., and Jepson, P. C., Eds., Academic Press, London, 1987, 209.
20. Benedek, P., Igrc, H., Honek, A., Baicu, T., Pokacka, Z., and Adamczewski, K., A proposal for cooperative United States and Central and Eastern European integrated pest management demonstration program: maximizing biocontrol in small grain integrated pest management, 1991.
21. Pike, K. S. and Schaffner, R. L., Development of autumn populations of cereal aphids, *Rhopalosiphum padi* (L.) and *Schizaphis graminum* (Rondani) (Homoptera: Aphididae) and their effects on winter wheat in Washington State, *J. Econ. Entomol.*, 78, 676, 1985.
22. Kieckhefer, R. W. and Kantack, B. H., Yield losses in winter grains caused by cereal aphids (Homoptera: Aphididae) in south Dakota, *J. Econ. Entomol.*, 81, 317, 1988.
23. Kieckhefer, R. W. and Kantack, B. H., Losses in yield in spring wheat in South Dakota caused by cereal aphids, *J. Econ. Entomol.*, 73, 582, 1980.
24. Kieckhefer, R. W. and Gellner, J. L., Yield losses in winter wheat caused by low-density cereal aphid populations, *Agron. J.*, 84, 180, 1992.
25. Kieckhefer, R. W. and Stoner, W. N., Field infectivities of some aphid vectors of barley yellow dwarf virus, *Plant Dis. Rep.*, 51, 981, 1967.
26. Wyatt, S. D., Seybert, J. L., and Mink, G., Status of the barley yellow dwarf virus problem of winter wheat in eastern Washington, *Plant Dis.*, 72, 110, 1988.
27. Saxena, R. C. and Khan, Z. R., Factors affecting resistance of rice varieties to planthopper and leafhopper pests, *Agric. Zool. Rev.*, 3, 97, 1989.
28. Swanson, M. C. and Newsome, L. D., Effect of infestation by the rice stinkbug, *Oebalus pugnax*, on yield and quality of rice, *J. Econ. Entomol.*, 55, 877, 1962.
29. Southwood, T. R. E., *Ecological Methods*, Chapman and Hall, London, 1978, 2.
30. Dietrick, E. J., An improved backpack motor fan for suction sampling of insect populations, *J. Econ. Entomol.*, 54, 394, 1960.
31. Ruesink, W. G. and Haynes, D. L., Sweepnet sampling for the cereal leaf beetle, *Oulema melanopus*, *Environ. Entomol.*, 2, 161, 1973.
32. Elliott, N. C., Kieckhefer, R. W., and Kauffman, W. C., Estimating adult coccinellid populations in wheat fields by removal, sweepnet, and visual count sampling, *Can. Entomol.*, 123, 13, 1990.
33. Gerrard, D. J. and Chiang, H. C., Density estimation of corn root-worm egg populations based upon frequency of occurrence, *Ecology*, 51, 237, 1970.
34. Kuno, E., Evaluation for statistical precision and design of efficient sampling for the population estimation based on frequency of occurrence, *Res. Popul. Ecol.*, 28, 305, 1986.
35. Drees, B. M. and Way, M. O., in *1991 Rice Production Guidelines*, Texas Agricultural Extension Service Publ. D-1253, College Station, 1991, 32.
36. Nishida, T. and Torii, T., *A Handbook of Field Methods for Research on Rice Stem Borers and Their Natural Enemies*, IBP Handb. 14, Blackwell Scientific, Oxford, 1970, 85.
37. Patrick, C. D. and Boring, E. P., III, Managing insect and mite pests of Texas small grains, Texas Agricultural Extension Service Publ. B-1251, 1990.
38. Coppock, S. and Burton, R., Insects on small grains and their control, Oklahoma State University Cooperative Extension Service Publ. 7176, Stillwater, 1990.
39. Coggin, D. L. and Dively, G. P., Sequential sampling plan for the armyworm in Maryland small grains, *Environ. Entomol.*, 11, 169, 1982.
40. Iwao, S., A new method of sequential sampling to classify populations relative to a critical density, *Res. Popul. Ecol.*, 16, 281, 1975.

41. Lloyd, M., Mean crowding, *J. Anim. Ecol.*, 36, 1, 1967.
42. Iwao, S., A new regression method for analyzing the aggregation pattern of animal populations, *Res. Popul. Ecol.*, 10, 1, 1968.
43. Viktorov, G. A., Method of sequential counting of the number of hibernating bugs of *Eurygaster integriceps*, *Soviet J. Ecol.*, 6, 278, 1975.
44. Wald, A., *Sequential Analysis*, John Wiley & Sons, New York, 1947.
45. Grivanov, K. A. and Antononko, O. P., Biological basis for an integrated fight against the pest *Eurygaster integriceps* in the Saratov region, *Zool. Zh.*, 10, 50, 1971.
46. Ferrer, E. R. and Shepard, B. M., Sampling Malayan black bugs (Heteroptera: Pentatomidae) in rice, *Environ. Entomol.*, 16, 259, 1987.
47. Shepard, B. M. and Ferrer, E. R., Sampling insects and diseases in rice, in *Proc. Int. Workshop on Crop Loss Assessment to Improve Pest Management in Rice and Rice-Based Farming Systems in South and Southeast Asia*, Teng, P. S., Ed., The International Rice Institute, Los Banos, Philippines, 1992, 107.
48. Bautista, R. C., Heinrichs, E. A., and Rejesus, R. S., Economic injury levels for the rice leaffolder *Cnaphalocrocis medinalis* (Lepidoptera: Pyralidae): insect infestation and artificial leaf removal, *Environ. Entomol.*, 13, 439, 1984.
49. Rabbinge, R. and Vereijken, P. H., 1979, The effect of diseases upon the host, *Z. Pflanzenkrankh.*, 87, 409, 1978.
50. Carter, N. and Dewar, A., The development of forecasting systems of cereal aphid outbreaks in Europe, in *Proc. 9th Int. Congr. Plant Protection*, Washington DC, 1979.
51. Vereijken, P. H., Feeding and multiplication of three cereal aphid species and their effect on yield of winter wheat, Agric. Res. Rep. 888, Pudoc, Wageningen, The Netherlands, 1979.
52. George, K. S. and Gair, R., Crop loss assessment on winter wheat attacked by the grain aphid, *Sitobion avenae* (F.) 1974–77, *Plant Pathol.*, 28, 143, 1979.
53. Anderson, K., Situation of the integrated control of cereal aphids in Sweden, *Bull. IOBC/WPRS*, 8, 54, 1985.
54. Lindquist, B., Some information about cereal aphids in Finland, *Bull. IOBC/WPRS*, 8, 48, 1985.
55. Basedow, T., Bauers, C., and Lauenstein, G., The preliminary control threshold for cereal aphids in winter wheat in Western Germany, *Bull. IOBC/WPRS*, 8, 36, 1985.
56. Reitzel, J. and Jakobsen, J., The occurrence of and damage caused by aphids in cereal crops in Denmark, *Bull. IOBC/WPRS*, 3, 107, 1980.
57. Lescar, L., Current practice in integrated pest and disease control in North-Western Europe (excluding Great Britain), Proc. 1977 British Crop Protect. Conf.—Pests and Diseases, 1977, 763.
58. Mann, B. P., Wratten, S. D., and Watt, A. D., A computer-based advisory system for cereal aphid control, *Comput. Electron. Agric.*, 1, 263, 1986.
59. Mann, B. P. and Wratten, S. D., A computer-based advisory system for the control of *Sitobion avenae* and *Metopolophium dirhodum*, *Bull. IOBC/WPRS*, 10, 1, 1987.
60. Rabbinge, R. and Rijsdijk, F. H., EPIPRE, a disease and pest management system for winter wheat taking into account micrometeorological factors, *EPPO Bull.*, 13, 297, 1983.
61. Rabbinge, R. and Carter, N., Monitoring and forecasting of cereal aphids in the Netherlands: a subsystem of EPIPRE, in *Pest and Pathogen Control: Strategic, Tactical and Policy Models*, Conway, G. R., Ed., John Wiley & Sons, London, 1984, 242.
62. Rabbinge, R. and Mantel, W. P., Monitoring for cereal aphids in winter wheat, *Neth. J. Plant Pathol.*, 87, 25, 1981.
63. Ekbom, B. S., Incidence counts for estimating densities of *Rhopalosiphum padi* (Homoptera: Aphididae), *J. Econ. Entomol.*, 80, 933, 1987.
64. Kieckhefer, R. W. and Gustin, R. D., Cereal aphids in South Dakota. I. Observations of autumnal bionomics, *Ann. Entomol. Soc. Am.*, 60, 514, 1967.
65. Kieckhefer, R. W., Lytle, W. F., and Spuhler, W., Spring movement of cereal aphids into South Dakota, *Environ. Entomol.*, 3, 347, 1974.
66. Kieckhefer, R. W., Field populations of cereal aphids in South Dakota spring grains, *J. Econ. Entomol.*, 68, 161, 1975.
67. Kieckhefer, R. W. and Kantack, B. H., Yield losses in spring barley caused by cereal aphids (Homoptera: Aphididae) in South Dakota, *J. Econ. Entomol.*, 79, 749, 1986.
68. Kieth, D. L., Hagen, A. F., and Kalisch, J. A., Sphids in wheat, University of Nebraska Cooperative Extension Service Publ. G73-49, Lincoln, Rev. 1987.
69. Kantack, B. H., Kieckhefer, R. W., and Berndt, W. L. Controlling aphis on small grain, South Dakota State University Cooperative Extension Service Rep. FS-588, Brookings, 1981.
70. Johnston, R. L. and Bishop, G. W., Economic injury levels and economic thresholds for cereal aphids (Homoptera: Aphididae) on spring-planted wheat, *J. Econ. Entomol.*, 80, 478, 1987.

71. Ba-Angood, S. A. and Stewart, R. K., Economic thresholds and economic injury levels of cereal aphids on barley in Southwestern Quebec, *Can. Entomol.*, 112, 759, 1980.
72. Kieckhefer, R. W., Elliott, N. C., Riedell, W. E., and Fuller, B. W., A yield-loss model for spring wheat infested with greenbugs (Homoptera: Aphididae), *Can. Entomol.*, in press.
73. Elliott, N. C., Kieckhefer, R. W., and Walgenbach, D. D., Binomial sequential sampling methods for cereal aphids in small grains, *J. Econ. Entomol.*, 83, 1381, 1990.
74. Dean, G. J. and Luuring, B. B., Distribution of aphids in cereal crops, *Ann. Appl. Biol.*, 66, 485, 1970.
75. Dean, G. J. W., Distribution of aphids in spring cereals, *J. Appl. Ecol.*, 10, 447, 1973.
76. Elliott, N. C. and Kieckhefer, R. W., Sampling aphids and natural enemies, Proc. 1st Greenbug Workshop, February 4–5, Garden City, KS, 1992.
77. Ward, S. A., Rabbinge, R., and Mentel, W. P., The use of incidence counts for estimation of aphid populations. 1. Minimum sample size for required accuracy, *Neth. J. Plant Pathol.*, 91, 93, 1985.
78. Schaalje, G. B., Butts, R. A., and Lysyk, T. J., Simulation studies of binomial sampling: a new variance estimator and density predictor, with special reference to the Russian wheat aphid (Homoptera: Aphididae), *J. Econ. Entomol.*, 84, 140, 1991.
79. Binns, M. R. and Bostonian, N. J., Robustness in empirically based binomial decision rules for integrated pest management, *J. Econ. Entomol.*, 83, 420, 1990.
80. Legg, D. E., Hein, G. L., and Peairs, F. B., Sampling Russian wheat aphid in the western Great Plains, Russian Wheat Aphid Task Force, Great Plains Agric. Counc. Rep. GPAC-138, Fort Collins, CO, 1991.
81. Legg, D. E., Nowierski, R. M., Feng, M. G., Peairs, F B., Hein, G. L., Elberson, L. R., and Johnson, J. B., Binomial sequential sampling plans for the Russian wheat aphid (Homoptera: Aphididae) infesting small grains, *J. Econ. Entomol.*, in press.
82. Spackman, E., Russian wheat aphid and its control, University of Wyoming Cooperative Extension Bull. B-903.
83. Peairs, F. B., Beck, K. G., Brown, W. M., Jr., Schwartz, H. F., and Westra, P., Colorado Pesticide Guide, Colorado State University Cooperative Extension Publ. XCM-45, 1991.
84. Archer, T. L. and Bynum, E. D., Jr., Russian wheat aphid economic threshold, Proc. 2nd Russian Wheat Aphid Conference, Denver, CO, 1988, 127.
85. Legg, D. E., Kroening, M. E., and Peairs, F. B., A new procedure for developing binomial sequential sampling models for the Russian wheat aphid, Proc. 5th Russian Wheat Aphid Conference, Fort Worth, TX, 1992, 254.
86. Brewer, M. J., Legg, D. E., and Kaltenbach, J. E., Presence/absence sampling for Russian wheat aphid: a comparison of the sequential probability ratio test, sequential interval procedure, and fixed-sample inspection plan, Proc. 5th Russian Wheat Aphid Conference, Fort Worth, TX, 1992, 260.
87. Ferrer, E. R. and Shepard, B. M., Sampling methods for estimating population densities of planthoppers and predators in direct-seeded and transplanted rice, *J. Agric. Entomol.*, 5, 199, 1988.
88. Kuno, E., Distribution pattern of the rice brown planthopper and field sampling techniques, in *The Rice Brown Planthopper*, Food and Fertilizer Technology Center for the Asian and Pacific Region, Taipei, Taiwan, 1977, 135.
89. Ooi, P. A. C., A surveillance system for rice planthoppers in Malaysia, *Proc. Int. Conf. Plant Prot. Trop.*, 1982, 551.
90. Shepard, B. M., Ferrer, E. R., and Kenmore, P. E., Sequential sampling of planthoppers and predators in rice, *J. Plant Prot. Trop.*, 5, 39, 1988.
91. Shepard, B. M., Ferrer, E. R., Soriano, J., and Kenmore, P. E., Presence/absence sampling of planthoppers and major predators of rice, *J. Plant Prot. Trop.*, 6, 113, 1989.
92. Shepard, B. M., Minnick, D. R., Soriano, J. S., Ferrer, E. R., and Magistrado, O. N., A simplified method for sampling leaffolders (LFs) and planthoppers, *Int. Rice Res. Newsl.*, 13(6), 40, 1988.
93. Shepard, M., Ferrer, E. R., Kenmore, P. E., and Sumangil, J. P., Sequential sampling: planthoppers in rice, *Crop Prot.*, 5, 319, 1986.
94. Perfect, T. J., Cook, A. G., and Ferrer, E. R., Population sampling for planthoppers, leafhoppers (Hemiptera: Delphacidae & Cicadellidae) and their predators in flooded rice, *Bull. Entomol. Res.*, 73, 345, 1983.
95. Shepard, M., Aquino, G., Ferrer, E. R., and Heinrichs, E. A., Comparison of vacuum and carbon dioxide-cone sampling devices for arthropods in flooded rice, *J. Agric. Entomol.*, 2, 364, 1985.
96. Nagata, T. and Masuda, T., Efficiency of sticky boards for population estimation of the brown planthopper, *Nilaparvata lugens* Stal (Hemiptera: Delphacidae) on rice hills, *Appl. Entomol. Zool.*, 13, 55, 1978.
97. Dyck, V. A. and Orlido, G. C., Control of the brown planthopper (*Nilaparvata lugens*) by natural enemies and timely application of narrow-spectrum insecticides, in *The Rice Brown Planthopper*, Food and Technology Center for the Asian and Pacific Region, Taipei, Taiwan, 1977, 58.

98. Otake, A., Natural enemies of the brown planthopper, in *The Rice Brown Planthopper*, Food and Technology Center for the Asian and Pacific Region, Taipei, Taiwan, 1977, 42.
99. Heuel-Rolf, B. and Vungsilabutr, P., Development of simple sampling methods for pest surveillance in rice, *Proc. Southeast Asia Pesticide Management and Integrated Pest Management Workshop*, Pataya, Thailand, 1987, 113.
100. Rubia, E. G., Shepard, B. M., Yambao, E. B., Ingram, K. T., Arida, G. S., and Penning de Vries, F., Stem borer damage and grain yield of flooded rice, *J. Plant Prot. Trop.*, 6, 205, 1989.
101. Torii, T., Quantitative prediction of economic degree of infestation by the rice stem borer, *Chelo suppressalis* Walker, by the sequential method, *Kyusher Assoc. Plant Prot. Proc.*, 16, 27, 1970.
102. Torii, T., A sequential sampling method for predicting levels of infestations of rice by the stem borer *Chelo suppressalis* (Lepidoptera), *Pac. Sci. Congr. Proc.*, 1, 182, 1971.
103. Tugwell, N. P. and Stephen, F. M., Rice water weevil seasonal abundance, economic levels, and sequential sampling plans, University of Arkansas Agricultural Experiment Station Bull. 849, Fayetteville, 1981, 1.
104. Sooksai, S., Rice Water Weevil (*Lissofhoptrus oryzophilus* Kuschel) Feeding Scars—Larval Density Relationship, M.S. thesis, University of Arkansas, Fayetteville, 1976.
105. Soohsai, S. and Tugwell, P., Adult rice water weevil feeding symptoms: number samples required, spatial and seasonal distribution in a rice field, *J. Econ. Entomol.*, 71, 145, 1978.
106. Grigarick, A. A., The rice water weevil in California, in *Proc. 10th Rice Tech. Working Group*, 1965, 40.
107. Morgan, D. R., Tugwell, N. P., and Bernhardt, J. L., Early rice drainage for control of rice water weevil (Coleoptera: Curculionidae) and evaluation of an action threshold based upon leaf-feeding scars of adults, *J. Econ. Entomol.*, 82, 1757, 1989.
108. Grigarick, A. A. and Washino, R. K., Invertebrates, in *Integrated Pest Management for Rice*, Statewise Integrated Pest Management Project, Flint, M. L., Tech. Ed., University of California Division of Agriculture and Natural Resources Publ. 3280, Davis, 1983, 49.
109. Grigarick, A., Invertebrates, in *Rice Pest Management Guidelines*, Flint, M. L., Tech. Ed., University of California Division of Agriculture and Natural Resources, Pest Management Guidelines Publ. 23, Davis, 1991, 2.
110. Short, M., Decision making in cereal pest control, *Proc. 1982 British Crop Protection Council Symposium: Decision Making in the Practice of Crop Protection*, 121.
111. Thornhill, E. W., A motorized insect sampler, *Pest Artic. News Summ.*, 24, 205, 1976.
112. Dewar, A. M. and Dean, G. J., Assessment of methods for estimating the numbers of aphids (Hemiptera: Aphididae) in cereals, *Bull. Entomol. Res.*, 72, 675, 1982.
113. Hand, S. C., The capture efficiency of the Dietrick vacuum insect net for aphids on grasses and cereals, *Ann. Appl. Biol.*, 108, 233, 1986.
114. Estano, D. B. and Shepard, B. M., Influence of time of day and sweeping pattern on catches of green leafhoppers (GLH), *Int. Rice Res. Newsl.*, 13(2), 22, 1988.
115. Banks, C. J. and Brown, E. S., A comparison of methods of estimating population density of adult sunn pest, *Eurygaster integriceps* Put. (Hemiptera, Scutelleridae) in wheat fields, *Entomol. Exp. Appl.*, 5, 255, 1962.
116. Swanson, M. C. and Newsome, L. D., Effect of infestation by the rice stinkbug, *Oebalus pugnax*, on yield and quality of rice, *J. Econ. Entomol.*, 55, 877, 1962.
117. Harper, J. K., Developing Economic Thresholds for Rice Stinkbug Management in Texas Using Dynamic Programming, Ph.D. dissertation, Texas A & M University, College Station, 1988.
118. Taylor, L. R. and Palmer, J. M. P., Aerial sampling, in *Aphid Technology*, Van Emden, H. F., Ed., Academic Press, New York, 1972, 189.
119. Robert, Y., Aphid vector monitoring in Europe, in *Current Topics in Vector Research*, Vol. 3, Harris, K. F., Ed., Springer-Verlag, New York, 1987, 81.
120. Bowden, J. and Jones, M. G., Monitoring wheat bulb fly, *Delia coarctata* (Fallen) (Diptera: Anthomyiidae), with light traps, *Bull. Entomol. Res.*, 69, 129, 1979.
121. Doane, J. F., Spatial pattern and density of *Ctenicera destructor* and *Hypolithus bicolor* (Coleoptera: Elateridae) in soil in spring wheat, *Can. Entomol.*, 109, 807, 1977.
122. Doane, J. F., A method for separating the eggs of the prairie grain wireworm, *Ctenicera destructor*, from the soil, *Can. Entomol.*, 101, 1002, 1969.
123. Robinson, J. F., Smith, C. M., and Trahan, G. B., Sampling rice water weevil infestations, 70th Annual Progress Rep., Rice Experiment Station, Louisiana State University Agricultural Experiment Station, Crowley, 1978, 168.
124. Gough, H. C., Studies on wheat bulb fly (*Leptohylemyia coarctata* Fall.) II. Numbers in relation to crop damage, *Bull. Entomol. Res.*, 36, 439, 1947.
125. Kempton, R. A., Bardner, R., Fletcher, K. E., Jones, M. G., and Maskell, F. E., Fluctuations in wheat bulb fly egg populations in Eastern England, *Ann. Appl. Biol.*, 77, 102, 1974.

126. Morrill, W. L., Wireworms: control, sampling methodology, and effect on wheat yield in Montana, *J. Ga. Entomol. Soc.*, 19, 61, 1984.
127. Bynum, E. D., Jr. and Archer, T. L., Wireworm (Coleoptera: Elateridae) sampling for semiarid cropping systems, *J. Econ. Entomol.*, 80, 164, 1987.
128. Kirfman, G. W., Keaster, A. J., and Story, R. N., An improved wireworm (Coleoptera: Elateridae) sampling technique for midwest cornfields, *J. Kan. Entomol. Soc.*, 59, 37, 1986.
129. Toba, H. H. and Turner, J. E., Evaluation of baiting techniques for sampling wireworms (Coleoptera: Elateridae) infesting wheat in Washington, *J. Econ. Entomol.*, 76, 850, 1983.
130. Maskell, F. E., Wheat bulb fly—methods of control, internal report, MAFF, U.K.
131. McKinlay, R. G. and Franklin, M. F., A comparison of a corer and a shovel for sampling populations of wheat bulb fly (*Delia coarctata*) eggs, *Ann. Appl. Biol.*, 95, 279, 1980.
132. Oakley, J. N. and Uncles, J. J., Use of oviposition trays to estimate numbers of wheat bulb fly *Leptohylemia coarctata* eggs, *Ann. Appl. Biol.*, 85, 407, 1977.
133. Landis, B. J. and Onsager, J. A., Wireworms on irrigated lands in the west. How to control them, USDA Farmers Bull. 2220.
134. Day, A. and Reid, W. J., Jr., Response of adult southern potato wireworms to light traps, *J. Econ. Entomol.*, 62, 314, 1969.
135. Morrill, W. L., Emergence of click beetles (Coleoptera: Elateridae) from some Georgia grasslands, *Environ. Entomol.*, 7, 895, 1978.
136. Robinson, J. F., Smith, C. M., and Trahan, G. B., Rice water weevil: sampling larval populations, *Proc. 18th Rice Technical Working Group*, Davis, CA, 1980, 50.
137. Cave, G. L., Smith, C. M., and Robinson, J. F., Population dynamics, spatial distribution, and sampling of the rice water weevil on resistant and susceptible rice genotypes, *Environ. Entomol.*, 13, 822, 1984.
138. Allison, D. and Pike, K. S., An inexpensive suction trap and its use in an aphid monitoring network, *Agric. Entomol.*, 5, 103, 1988.
139. Taylor, L. R., The Rothamsted insect survey—an approach to the theory and practice of synoptic pest forecasting in agriculture, in *Movement of Highly Mobile Insects: Concepts and Methodology in Research*, Rabb, R. L. and Kennedy, G. G., Eds., North Carolina State University, Raleigh, 1979, 148.
140. Woiwood, I. P., Tatchell, G. M., and Barrett, A. M., A system for the rapid collection, analysis, and dissemination of aphid-monitoring data from suction traps, *Crop Prot.*, 3, 273, 1984.
141. Plumb, R. T., The infectivity index and barley yellow dwarf virus, *Proc. 10th Int. Congr. Plant Protection*, Brighton, England, 1983, 171.
142. Tatchell, G. M., Aphid-control advise to farmers and the use of aphid-monitoring data, *Crop Prot.*, 4, 39, 1985.
143. Plumb, R. T., Barley yellow dwarf virus in aphids caught in suction traps, 1969–1973, *Ann. Appl. Biol.*, 83, 53, 1976.
144. Plumb, R. T. and Carter, N., The use and validation of the infectivity index as a method of forecasting the need to control barley yellow dwarf virus in autumn-sown crops in the United Kingdom, *Acta Phytopathol. Entomol. Hung.*, 26, 59, 1991.
145. Kendall, D. A. and Smith, B. D., The significance of aphid monitoring in improving barley yellow dwarf virus control, *Proc. 1981 British Crop Protection Conference—Pests and Diseases*, Brighton, England, 1981, 399.
146. Kendall, D. A. and Chinn, N. E., A comparison of vector population indices for forecasting barley yellow dwarf virus in autumn sown cereal crops, *Ann. Appl. Biol.*, 116, 87, 1991.
147. Tatchell, G. M., Plumb, R. T., and Carter, N., Migration of alate morphos of the bird cherry aphid (*Rhopalosiphum padi*) and implications for the epidemiology of barley yellow dwarf virus, *Ann. Appl. Biol.*, 112, 1, 1988.
148. Kring, T. J., Gilstrap, F. E., and Michels, G. J., Jr., Role of indigenous coccinellids in regulating greenbugs (Homoptera: Aphididae) on Texas grain sorghum, *J. Econ. Entomol.*, 78, 269, 1985.
149. Rice, M. E. and Wilde, G. E., Experimental evaluation of predators and parasitoids in suppressing greenbugs (Homoptera: Aphididae) in sorghum and wheat, *Environ. Entomol.*, 17, 836, 1988.
150. Elliott, N. C. and Kieckhefer, R. W., The dynamics of aphidophagous coccinellid assemblages in small grains, *Environ. Entomol.*, 19, 1320, 1990.
151. Lapchin, L., Ferran, A., Iperti, G., Rabasse, J. M., and Lyon, J. P., Coccinellids (Coleoptera: Coccinellidae) and syrphids (Diptera: Syrphidae) as predators of aphids in cereal crops: a comparison of sampling methods, *Can. Entomol.*, 119, 815, 1987.
152. Seber, G. A. F. and Le Cren, E. D., Estimating population parameters from catches large relative to the population, *J. Anim. Ecol.*, 36, 631, 1967.
153. Taylor, L. R., Agregation, variance, and the mean, *Nature*, 189, 732, 1961.
154. Green, R. H., On fixed precision level sequential sampling, *Res. Popul. Ecol.*, 12, 249, 1970.

155. Iperti, G., Lapchin, L., Ferran, A., Rabasse, J. M., and Lyon, J. P., Sequential sampling of adult *Coccinella septempunctata* L. in wheat fields, *Can. Entomol.*, 120, 773, 1988.
156. Hein, G. L., Elliott, N. C., Michels, G. J., Jr., and Kieckhefer, R. W., A general method for estimating cereal aphid populations in small grain fields based on frequency of occurrence, *Can. Entomol.*, in press.
157. Jones, V. P., Developing sampling plans for spider mites (Acari: Tetranychidae): those who don't remember the past may have to repeat it, *J. Econ. Entomol.*, 83, 1656, 1990.
158. Teetes, G. L. and Sterling, W. L., A sequential sampling plan for a white grub in grain sorghum, *Southwest. Entomol.*, 1, 118, 1976.
159. Elliott, N. C. and Kieckhefer, R. W., Cereal aphid populations in winter wheat: spatial distributions and sampling with fixed levels of precision, *Environ. Entomol.*, 15, 954, 1986.
160. Elliott, N. C. and Kieckhefer, R. W., Spatial distributions of cereal aphids (Homoptera: Aphididae) in winter wheat and spring oats in South Dakota, *Environ. Entomol.*, 16, 896, 1987.
161. Ba-Angood, S. A. and Stewart, R. K., Occurrence, development, and distribution of cereal aphids on early and late cultivars of wheat, barley, and oats in southwestern Quebec, *Can. Entomol.*, 112, 615, 1980.
162. Ba-Angood, S. A. and Stewart, R. K., Sequential sampling for cereal aphids on barley in Southwestern Quebec, *J. Econ. Entomol.*, 73, 679, 1980.
163. Kring, T. J. and Gilstrap, F. E., Within-field distribution of greenbug (Homoptera: Aphididae) and its parasitoids in Texas winter wheat, *J. Econ. Entomol.*, 76, 57, 1983.
164. Ekbom, B. S., Spatial distribution of *Rhopalosiphum padi* (L.) (Homoptera: Aphididae) in spring cereals in Sweden and its importance for sampling, *Environ. Entomol.*, 14, 312, 1985.
165. Ward, S. A., Chambers, R. J., Sunderland, K., and Dixon, A. F. G., Cereal aphid populations and the relation between mean density and spatial variance, *Neth. J. Plant Pathol.*, 92, 127, 1986.
166. Ward, S. A., Rabbinge, R., and Mantel, W. P., The use of incidence counts for estimation of aphid populations. 2. Confidence intervals from fixed sample sizes, *Neth. J. Plant Pathol.*, 91, 100, 1985.
167. Ward, S. A., Sunderland, K. D., Chambers, R. J., and Dixon, A. F. G., The use of incidence counts for estimation of cereal aphid populations. 3. Population development and the incidence-density relationship, *Neth. J. Plant Pathol.*, 92, 175, 1986.
168. Feng, M. G. and Nowierski, R. M., Spatial distribution and sampling plans for four species of cereal aphids (Homoptera: Aphididae) infesting spring wheat in southwestern Idaho, *J. Econ. Entomol.*, 85, 830, 1992.
169. Sawyer, A. J. and Haynes, D. L., Allocating limited sampling resources for estimating regional populations of overwintering cereal leaf beetles, *Environ. Entomol.*, 7, 62, 1978.
170. Filipy, F. L., Burbutis, P. P., and Fuester, R. W., Sampling for the European wheat stem sawfly (Hymenoptera: Cephidae), *J. Econ. Entomol.*, 78, 493, 1985.
171. Golightly, W. H. and Woodville, H. C., Studies of recent outbreaks of saddle gall midge, *Ann. Appl. Biol.*, 77, 97, 1974.
172. Mukerji, M. K., Olfert, O. O., and Doane, J. F., Development of sampling designs for egg and larval populations of the wheat midge, *Sitodiplosis mosellana* (Gehin) (Diptera: Cecidomyliidae), in wheat, *Can. Entomol.*, 120, 497, 1988.
173. Bates, B. A., Weiss, M. J., Carlson, R. B., and McBride, D. K., Sequential sampling plan for *Limothrips denticornis* (Thysanoptera: Thripidae) on spring barley, *J. Econ. Entomol.*, 84, 1630, 1991.
174. Bates, B. A. and Weiss, M. J., The spatial distribution of *Limothrips denticornis* Haliday (Thysanoptera: Thripidae) eggs on spring barley, *Can. Entomol.*, 123, 205, 1991.
175. Legg, D. E., Hein, G. L., Nowierski, R. M., Feng, M. G., Peairs, F. B., Karn, M., and Cuperus, G. W., Binomial regression models for spring and summer infestations of the Russian wheat aphid (Homoptera: Aphididae) in the southern and western plains states and Rocky Mountain Region of the United States, *J. Econ. Entomol.*, 85, 1779, 1992.
176. Schotzko, D. J. and Smith, C. M., Effects of host plant on the between-plant spatial distribution of the Russian wheat aphid (Homoptera: Aphididae), *J. Econ. Entomol.*, 84, 1725, 1991.
177. Hein, G. L., Baxendale, F. P., Campbell, J. B., Hagen, A. F., and Kalisch, J. A., Russian wheat aphid, University of Nebraska Cooperative Extension Service Publ. G89-936, Lincoln, 1989.
178. Schaalje, G. B. and Butts, R. A., Binomial sampling for predicting density of Russian wheat aphid (Homoptera: Aphididae) on winter wheat in the fall using a measurement error model, *J. Econ. Entomol.*, 85, 1167, 1992.
179. Pesho, G. R., McGuire, J. U., Jr., and McWilliams, J. M., Sampling methods for surveys of damage caused by the wheat stem sawfly, *FAO Plant Prot. Bull.*, 19, 121, 1971.
180. Holmes, N. D., McKenzie, H., Peterson, L. K., and Grant, M. N., Accuracy of visual estimates of damage by the wheat stem sawfly, *J. Econ. Entomol.*, 61, 679, 1968.

181. Burton, R. L., Starks, K. J., and Peters, D. C., The army cutworm, Oklahoma State University Agricultural Experiment Station Publ. B-749, Stillwater, 1980.
182. Hagen, A., Campbell, J. B., and Keith, D. L., A guide to grasshopper control, University of Nebraska Cooperative Extension Service Publ. G86-791, Lincoln, 1986.
183. Singh, V. S. and Bhatia, S. K., Reduction in yield of some barley varieties due to brown wheat mite infestation, *Indian J. Entomol.*, 45, 190, 1983.
184. Singh, V. S. and Bhatia, S. K., Uneven distribution of the brown wheat mite on barley crop, *Indian J. Entomol.*, 45, 193, 1983.
185. Harvey, T. L. and Martin, T. J., Sticky-tape method to measure cultivar effect on wheat curl mite (Acari: Eriophyidae) populations in wheat spikes, *J. Econ. Entomol.*, 81, 731, 1988.
186. Staples, R. and Allington, W. B., The efficiency of sticky traps in sampling epidemic populations of the eriophyid mite *Aceria tulipae* (K.) vector of wheat streak mosaic virus, *Ann. Entomol. Soc. Am.*, 52, 159, 1959.
187. Zhong, S. J., Studies of the spatial pattern of striped stemborer *Chelo suppressalis* Walker and damaged rice stems in paddy fields, in 2nd Int. Conf. Plant Protection in the Tropics, Colorcom Grafix System, SDN BHD, Kuala Lumpur, Malaysia, 1986, 57.
188. Justo, H. D., Jr., Shepard, B. M., Perez, V. A., and Tsuboi, T., Comparison of sweep net sampling patterns for estimating population density of green leafhopper, *Int. Rice Res. Newsl.*, 13(6), 40, 1988.
189. Wada, T., Distribution pattern and sampling techniques of the rice leaf roller, *Cnaphalocrocis medinalis* in a paddy field, *Appl. Entomol. Zool.*, 20, 230, 1985.

Section V: Implementation of Sampling Programs

Chapter 22

TRAINING SPECIALISTS IN SAMPLING PROCEDURES

Gerrit W. Cuperus and Richard C. Berberet

TABLE OF CONTENTS

I. Introduction to Teaching Sampling Principles 670
   A. Role of Specialists 671

II. Basic Principles of Sampling: Role of Probability in Sampling 672
   A. The Concept of Risk 672
   B. Communicating the Probabilities (Risks) of Incorrect Decisions 673
      1. Type I Error 673
      2. Type II Error 674
   C. Communicating about Viability and Precision 674

III. Requirements for Adoption of Sampling Protocols 674
   A. Relative Advantage 675
   B. Compatibility 675
   C. Perceived Complexity 676
   D. Easily Tried 676
   E. Observable 676

IV. Improving Rate of Adoption 677

V. Examples of Sampling Systems 678
   A. Alfalfa 678
   B. Stored Grain 679

References 680

I. INTRODUCTION TO TEACHING SAMPLING PRINCIPLES

Although various definitions have been proposed for Integrated Pest Management (IPM), the common theme is for consolidation into "a unified program to manage pest populations so that economic damage is avoided and adverse side effects on the environment are minimized."[1] This chapter deals with the overall goal of maintaining profitable production through IPM and stresses implementation of statistically valid sampling protocols as an essential component of effective benefit/cost considerations. It is primarily the responsibility of researchers to describe the statistical relationships of pest densities vs. yield and/or quality reductions in agricultural commodities that are necessary for dynamic economic thresholds. However, researchers, extension specialists, and consultants must work together to develop efficient sampling protocols that will be accepted by agricultural producers. In the struggle to change from spray schedules and prophylactic applications to a system where pesticides are used only when there is a high probability that benefits will outweigh the costs, credibility of sampling and decision-making procedures must be proven. Although much progress is needed for widespread implementation of these procedures on a field-by-field basis, the ultimate goal is for use in a holistic system that goes beyond individual fields and momentary time frames (Figure 1). The rate at which this transition occurs can vary significantly depending upon the success of delivery programs.

Effective communication and cooperation among researchers, extension specialists, and consultants is essential to the task of information transfer and implementation (Figure 2). Joint workshops, field demonstrations, and other opportunities for interaction must be used to promote communication. Each group must have confidence in

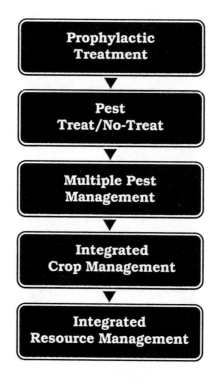

FIGURE 1. Traditional information dissemination system involving research, extension, consultants, and the producers.

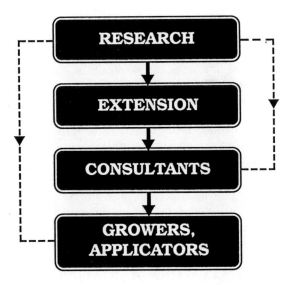

FIGURE 2. Evolution of producer decision-making process from individual pest to systems management.

the sampling and decision-making procedures being proposed for adoption. Challenges facing extension specialists and consultants in their attempts to implement improved procedures appear to have become greater as profit margins decline, treatment options become limited due to voluntary withdrawals and cancellations of pesticide registrations, and as concerns of the public about pesticide contamination increase. Ironically, the added challenges are coming at a time in the U.S. when state extension staffs are being reduced and the role of extension in the land grant system is being reviewed.

There is an extensive amount of information available on arthropod distributions, sampling, and decision making.[2,3,4] Although sampling and decision-making procedures have been refined for insect pests in cotton,[5] corn,[6] soybeans,[7] and alfalfa,[8] most have not been widely adopted.[9] This chapter addresses barriers to adoption, such as lack of education at the consultant and producer level and the perception that procedures are too time consuming and unreliable.[10,11]

Producers often make decisions affecting crop and livestock commodities worth thousands or millions of dollars without adequate information, seemingly because they are unaware of sampling protocols, or such protocols are perceived as being too complex and having an unacceptable risk of an erroneous decision. Although it is understandable that users consistently request simplistic sampling procedures that provide sound management information with limited time expenditure, it is questionable whether these expectations can be met in most instances. While researchers work with extension personnel and consultants to refine procedures that maintain the desired level of precision and acceptance by users, those involved with implementation must develop effective educational programs for those users.

A. ROLE OF SPECIALISTS

This chapter emphasizes imparting some basic principles of sampling when developing and delivering training programs. The discussion is targeted to roles of several groups including (1) extension personnel, (2) consultants, and (3) agribusiness personnel. Although each of these groups has unique views on management of the commodities with which they are involved, all require reliable sampling and decision-making procedures. The first educational challenge is to assure that all understand the critical importance of reliable sampling data in making economically and environmentally sound management decisions. The basic premise is that knowledge and acceptance of procedures by county and area extension personnel, consultants, and agribusiness personnel is essential to producer adoption. Primary responsibility for training currently rests with the researchers and state extension people. *However, as changes in the traditional land grant system occur, it is likely that consultants and agribusiness personnel will assume many training functions.*

Objectives of this chapter will be to (1) examine ways to improve efficiency and acceptance of sampling and decision-making protocols, (2) examine barriers to adoption of protocols, and (3) discuss how these barriers are being overcome using examples in alfalfa and stored grain systems.

II. BASIC PRINCIPLES OF SAMPLING: RISK AND PROBABILITY WHEN MAKING DECISIONS

A. THE CONCEPT OF RISK

One of the most important components of educational programs related to sampling and decision making is communication of the concepts of precision and probability. Decisions ranging from whether or not a particular pesticide application is needed to where and when to market a commodity all have probabilities for increasing profits when correct choices are made or, conversely, risks of losses with incorrect decisions. Although all stages of commodity production, processing, and distribution have elements of probability and risk, educational programs typically do not address these concepts in an effective manner. Our discussion will relate specifically to probability and risk at the consultant and producer level, where we will address pest management decisions.

It is not unexpected that producers desire protocols that minimize sampling time in the field, yet provide necessary information to minimize or eliminate the risk of an incorrect decision.[11] In some instances, it seems that producers believe that decisions made without sampling data have less risk of resulting in economic losses than those for which data are gathered. Explanations for this belief include the possibility that they fail to realize that *all* decisions have chances of increasing or decreasing profitability. Perhaps without realizing it, through decisions to apply pesticides without appropriate sampling, producers are assuming the unlikely event of 100% probability of losses due to pests exceeding the control costs. By contrast, the tacit assumption of those who never use pesticides is that losses due to pests never exceed the costs of control.

Through effective educational efforts, concepts of probability and risk can be communicated. When information on sampling and decision making is transferred to extension personnel, consultants, and finally to producers, it is essential that probabilities for correct decisions be explained. It should also be indicated whether decisions are based on static economic thresholds or dynamic thresholds that consider changing control costs and commodity values. Producer adoption is based on their being convinced that protocols will help increase profitability and maintain or reduce risk of losses in the agricultural enterprise. Considering the variety of pest problems in most commodities, time commitments for sampling may be extensive. Educational programs need to candidly address these time requirements. The users will ultimately decide how much time they will invest personally or the amount they will spend to hire someone to fulfill those time requirements.

An example of successful adoption of a protocol involves the tomato fruitworm, *Heliocoverpa zea* (Boddie), one species in a complex of pests that feed on tomatoes grown for the processing industry in California. Historically, most insecticide applications directed at this pest were based on visual estimates of damage that had limited reliability. However, a problem existed with rejection of produce lots because insect feeding injury was evident. New sampling techniques were developed based on seasonal life history and egg distribution of the pest. The improved sampling methods increased reliability in decision making to the extent that produce lots exceeding the threshold for cosmetic damage were eliminated. Once developed, this sampling protocol was delivered through the primary source of management information, the Pest Control Advisors (California licensed consultants). Greater reliability of fruitworm population estimates and an average reduction of insecticide costs of over $7/acre has resulted in the numbers of producers adopting the plan increasing by 10 × between 1981 and 1985.[12]

In contrast to successful adoption of sampling protocols for tomatoes in California, apple production in the U.K. is plagued by several major insect and disease pests that cause serious losses in yield and quality. Additionally, concerns for cosmetic appearance have encouraged prophylactic pesticide applications and prevented adoption of IPM systems. Sampling and decision-making procedures have not been widely adopted, and analyses have shown that the sampling procedures do not reduce risk of cosmetic damage to an acceptable level. The risk of losses has caused growers not to adopt these procedures.[13]

B. COMMUNICATING THE PROBABILITIES (RISKS) OF INCORRECT DECISIONS

Two types of errors may occur in sampling and decision making: (1) the probability of concluding that the pest population is above the economic threshold when the reverse is true (Type I error), and (2) the probability of concluding that the pest population is below the economic threshold when the reverse is true (Type II error).[14] To effectively deliver sampling information, it is important to communicate the implications of these two error types and allow the users of the protocols to design their approaches according to probabilities for these errors. The acceptable probability for each error type may be dependent on the person using the protocol.

1. Type I Error

Producer: If a Type I error is made, there will be an unnecessary pesticide application that reduces economic efficiency and potentially has some negative environmental impact. Many commodities have enough margin of profit to allow up to $20 or more in unneeded application costs without jeopardizing all potential profits. Thus, while a Type I error is not desirable, it may not destroy the enterprise. Those who follow spray schedules and use prophylactic treatments accept the likelihood of Type I error as routine. Improved sampling methods should provide enough information to minimize Type I error and make consultants and growers comfortable when the appropriate decision is not to apply pesticides, thereby improving profitability. An additional point is that those who attempt to minimize the risk of pest damage by using pesticides may ultimately increase the potential for losses by reducing beneficial insect populations or increasing resistance to chemical insecticides.

Consultant: If a Type I error is made, clients (producers) may express concern because of the wasted investment in pesticide. However, a consultant will likely not have his/her contract terminated as a result. However, the chance of reducing Type I error is often a reason to hire a consultant; thus, decreasing control costs becomes an important objective.

2. Type II Error

Producer: When a type II error is made, the potential economic impact of missing a needed pesticide application may be catastrophic. This mistake may reduce yield and threaten the economic integrity of the farming operation. Producers are understandably risk averse regarding pesticide applications. This is particularly true for high-value crops. Significant improvement can be made in determining the need for applications and selecting proper timing with reliable sampling protocols.

Consultant/field personnel: Minimizing the probability of Type II error is of utmost concern for consultants and agribusiness personnel. A Type II error may guarantee loss of the contract with the producer and loss of credibility for the consultant in that locality. Consultants are particularly interested in sampling information that can reduce probability of a Type II error. This provides a significant opportunity for transfer of information and implementation of protocols.

Future sampling and decision support systems need to respond to the end user's needs by having differing probabilities for Type I and Type II errors[15] or allowing the user to select the level of risk they wish to assume.[16] Research indicates that presenting the information in a fashion that allows the user to understand and balance the risk is most effective.[17] Researchers and extension specialists must be candid about what is known about arthropod distribution, sampling, and probabilities for errors in decision-making guidelines.

C. COMMUNICATING ABOUT VARIABILITY AND PRECISION

Define what is meant by variability and precision (repeatability) in sampling. An important factor to emphasize is that with more comprehensive sampling, the risk of erroneous decisions is reduced. Several factors that influence sampling results should be reviewed, including:

1. Variation between samplers[18]
2. Sampling methods[19]
3. Species sampled[20]
4. Number of samples taken
5. Climatic conditions[20]

Any variation in results make the sampling procedures appear less reliable and may reduce acceptance. Educational programs should stress means by which variability can be reduced. If not handled properly, discussion of variability can contribute to the perception that sampling protocols cannot provide the desired reliability.[10]

III. REQUIREMENTS FOR ADOPTION OF SAMPLING PROTOCOLS

Success in development and implementation of sampling and decision-making protocols requires sensitivity to time and economic constraints under which most consultants and producers operate.[21] For any hope of adoption, protocols must satisfy several attributes including relative advantage, compatibility with existing operations, lack of perceived complexity, ease of trial or testing, and readily observable results.[21,22] The importance of each of these attributes is described below.

A. RELATIVE ADVANTAGE

Extension specialists, consultants, and producers are unlikely to promote and adopt sampling procedures that have no apparent advantage over existing practices. When developing and attempting to implement a new protocol, advantage is the most important element to consider. A critical point is the relative advantage in limiting risks of erroneous decisions. An important principle to teach is the relationship between numbers of sample-units collected and precision of estimates for pest population density. Figure 3 shows the relationship between confidence intervals and the actual mean of a population. When sampling protocols are developed and delivered to clientele, the concept of sample numbers vs. accurate and precise answers must be an integral part of the program.

Example: Managers of Oklahoma elevators typically take just one sample for insects, moisture, and foreign material from each 600- to 1000-bushel truckload.[23] With one sample, there is less than a 50% probability of detecting an insect or mold infestation which could result in contamination of the facility (Table 1). In implementing a new sampling protocol, the advantage of multiple samples per load in reducing the risk of contamination was a major factor in decisions to adopt the new protocol.

B. COMPATIBILITY

Sampling for pests is a critical component in profitable and environmentally sound management systems for crop or livestock commodities.[24,25] Decisions regarding pest management must readily integrate with those on machinery management, harvest

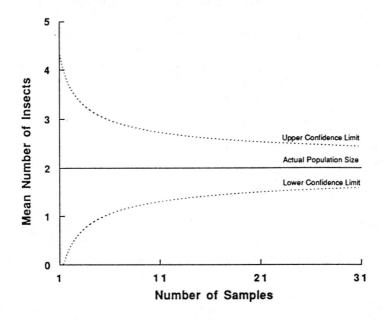

FIGURE 3. The relationship between sample size and the true mean of an insect population. (Adapted from Hagstrum, D., Flinn, P., and Fargo, W. S., in *Management of Grain, Bulk Commodities, and Bagged Products*, Krischik V., Cuperus, G. W., and Galliart, D., Eds., Oklahoma State University Cooperative Extension Service Circular E-912. With permission.)

TABLE 1
Probability for Detection of Insects in Stored Grain in Relation to Numbers of Samples and Insect Population Density

Number of samples	Mean number of insects / kg			
	0.02	0.2	2.0	6.0
1	0.02	0.19	0.76	0.95
2	0.04	0.34	0.94	1.00
5	0.10	0.68	0.99	1.00
10	0.19	0.87	1.00	1.00
25	0.42	0.99	1.00	1.00

management, tillage, irrigation, and other operations. Procedures most likely to be adopted are readily compatible with the overall operation, especially the time constraints of consultants, agribusiness personnel, and producers.

Example: Dramatic improvements have occurred in detection of insect infestations in stored grain through development and implementation of procedures that sample insects directly without need of taking samples of grain. These sampling systems have resulted from research on pheromone trapping systems.[26] Although the new traps were not effectively integrated into existing grain management systems initially, joint efforts of researchers, extension specialists, and industry personnel have allowed the sampling technologies to be implemented.[27] A key in this effort was making the procedures compatible with grain handling and storage operations.

C. PERCEIVED COMPLEXITY

A balance must be achieved between providing enough information about a new sampling procedure to encourage its adoption but not so much as to make it seem overly complex. Although we believe that communicating about probabilities and risks of erroneous decisions is important, the actual mechanics of the sampling process should be presented as simply as possible. It is inescapable that moving from a situation with no sampling to a sampling and decision-making protocol leads to increase complexity. However, careful preparation without great emphasis on theoretical aspects will reduce concerns about complexity of procedures.

D. EASILY TRIED

The actual sampling process should be made as simple as possible such as pulling stems, examining plant terminals, or using a sweep net. Field days or one-on-one meetings of researchers, extension specialists, consultants, and finally, producers should be scheduled to assure extensive "hands-on" experience at each step of the implementation process. Time should be allowed for questions and the opportunity for users to become comfortable with the procedures. Effective implementation of programs (see Chapter 22) allow users to experiment with sampling protocols in their own business or farming operations. The users need to understand the trade-offs between sample size, accuracy, and time requirements.

E. OBSERVABLE

Observability of sampling and decision-making protocols occurs at two levels. First, at the time of sampling, consultants and producers can see the pests and quantify pest abundance. Over the course of the season, demonstrations can be established to illustrate the consequences of correct and erroneous decisions. The educational program must be built around these demonstrable results. Comparison of benefits and costs with careful sampling for decisions on controls vs. spray schedules or complete lack of applications can provide observable results.

Educational opportunities often occur at times when there are economically damaging pest populations. Prior to such "crises", sound educational programs are needed to lay the framework with discussions on population dynamics, sampling protocols, and management strategies. In the outbreak situation, the previously laid foundation provides the opportunity to use the sampling methods to quantify pest abundance and work with criteria for decision making. These "teachable moments" often occur during relatively short time frames.

IV. IMPROVING RATE OF ADOPTION

One of the biggest impediments in implementing new protocols is the lack of effective delivery by extension specialists and consultants to producers. Diffusion research indicates that adoption of new concepts and technologies is an exponential occurrence, with a lag phase and then a period of increasing adoption. Without effective development and delivery, the lag phase can become quite lengthy and delay adoption significantly.[21,22] Agribusiness personnel and private producers often have greatly differing sources of information when dealing with similar problems (Table 2), and the same program may not be equally effective for these two groups. Information source varies significantly with the type of information (Table 3) and the traditional information systems that have existed.[24]

Expert systems and other computer technologies have been touted as the solution to delivery and the answer to improved management. Yet studies indicate that they are often not utilized. While up to 30% of producers may have access to microcomputers, they do not utilize them due to lack of experience, age, present time commitments, and they have not become convinced that computers are worth their time investment.[29] Many sampling and crop management programs fail because the information is presented at a technology level either above or below the targeted

TABLE 2
Sources of Stored Grain Management Information for Producers and Elevator Operators[23]

Source	Percent using sources listed	
	Private producers	Elevator operators
Chemical fieldmen	12	48
Fumigation manual	10	57
Newspaper	0	1
Commercial fumigator	0	31
CCC/USDA	0	2
FGIS personnel	0	3
Trade magazine	0	19
Extension personnel	54	24
Chemical supplier	25	47
Other farmers/ elevator operators	1	19

Note: From surveys of 89 and 112 producers and commercial managers, respectively. Sums of responses total more than 100% due to multiple sources of information.

TABLE 3
Sources of Information Used by Producers on Weed and Insect Management, Oklahoma, 1988[24]

Source	Weed	Insects
Ag sales rep	30.2	15.4
Extension newsletters	24.4	29.6
Extension agents	21.3	18.2
Neighbor	13.9	13.6
Farm magazines	13.0	12.7
Ag lenders	0.5	0.6
Other	3.4	9.8

Note: Numbers in table represent percentages of producers using sources.

audience or the educational effort is poorly planned. Expert systems, satellite videoconferences, and Fact Sheets all represent technologies that, if used alone, may be poor delivery techniques. However, use of these technologies may provide excellent delivery if they are tied together in an integrated educational system.

V. EXAMPLES OF SAMPLING SYSTEMS

The following are examples of sampling programs that have been developed, refined, and are being adopted throughout the U.S. We are using these examples to illustrate application of adoption requirements previously discussed. The examples contrast a sampling protocol for alfalfa weevil that has had a relatively high degree of acceptance with a protocol for stored grain insects that has gained more limited acceptance. The examples highlight successes and failures of delivering sampling protocols and evaluate reasons for adoption.

A. ALFALFA

The alfalfa weevil, *Hypera postica* (Gyllenhal), is a key pest of alfalfa throughout the U.S. This species completes one generation per year that typically begins with egg deposition in the fall by adults that have completed a summer estivation period. The extent of egg deposition and survival during fall and winter and the timing of subsequent larval infestations in the spring is dependent upon geographical location. In northern states, few eggs survive cold conditions during winter, and peak numbers of larvae occur near the time of first harvest.[30] In southern states, peak larval numbers may occur early in the growth of the first crop.[31] Because of the high degree of variability in seasonal occurrence of this pest from region to region, and often year to year, an effective scouting protocol is required for proper timing of insecticide applications and other management activities.

Visual estimates of damage that were used for many years in making decisions regarding insecticide applications for alfalfa weevil have been called into question.[32] However, when a more reliable alternative (now called the "shake-bucket" method) was first developed,[33] extension personnel and producers were slow to adopt this method. Among the concerns expressed by extension personnel when the protocol was first adapted for Oklahoma[34] were (1) the method was judged to be too technical in that it required a record of degree-day accumulation from January 1, estimation of stem heights, and counts of larvae in 30-stem samples, and (2) it was considered to be too time consuming in that it took 15 to 30 min/week for each field.

TABLE 4
Statewide Producer Ranking of Factors in Determining When to Spray for Alfalfa Weevil, Oklahoma, 1988[24]

Method	Percent
Visible damage	41.7
Degree-day based/scouting program	36.1
Spray by the date	8.9
Applicator recommendations	7.0
Other producers treating	5.0
Other	1.3

In the early 1980s, a concerted effort by researchers and extension specialists was begun to take this sampling protocol to producers. This included frequent demonstrations emphasizing degree-day calculations and the sampling procedure, involvement of scouts hired in counties with large acreages of alfalfa, and development of our expert system to illustrate the relationship of alfalfa weevil damage and yield reductions. The sampling and decision-making protocol has been accepted as the standard procedure by extension personnel, consultants, and many producers in Oklahoma. Nearly 40% of producers base their decisions for weevil control on the new protocol (Table 4). Acceptance has occurred largely due to improved education programs and recognition of the following advantages by producers:

- *Relative advantage* — This sampling protocol has greatly improved timing of pesticide applications. Previous visual methods consistently resulted in poorly timed applications, unneeded pesticide applications, or serious losses in yield. The relative advantage has been evident to producers through their own experiences.
- *Compatibility* — This system is directly compatible with sampling requirements for other pests such as aphids, and field visits can also be used to observe weed infestations, crop maturity, and other needed farm management operations.
- *Complexity* — After initial misunderstandings, this new sampling system has been accepted as a relatively simple procedure. Degree-day accumulations are being provided through newsletters, radio broadcasts, and electronic media to facilitate use of the protocol.
- *Easily tried and observable* — This sampling protocol has been effectively demonstrated by extension specialists and has a high degree of acceptance by consultants and producers. Throughout much of the U.S., and Oklahoma in particular, heavy infestations of weevil larvae have made sampling and the value of correct decisions highly observable. Extensive information on benefits vs. costs of alfalfa weevil control have resulted in increased adoption of the protocol.
- *Dissemination* — A key to dissemination in Oklahoma has been the unified voice of researchers and state and county extension personnel, all recommending the same sampling protocol as the key to effective decision making.

B. STORED GRAIN

The U.S. stores over 15 billion bushels of grain each year. Damage by insects, molds, heating, and sprouting is a common occurrence, with losses over $1 billion. Throughout the marketing system, sampling for arthropod pests has often been limited to counting insects in samples taken when grain is graded, and producers and elevator operators throughout the U.S. are often cited as not effectively managing this system. Producers do not own sampling equipment and often do not monitor insect populations.[23,35] Because appropriate sampling education has not been accomplished, effective pest management in grain distribution systems does not occur.[35]

There has been a significant research emphasis to improve sampling protocols for management decisions.[36] However, few producers, elevator operators, or mills presently utilize new trapping systems for insects in stored grain. There are several reasons why adoption has been limited:

- Those storing grain often do not realize the risk that they are taking without adequate sampling. Because there is no perceived risk, producers do not feel there is a comparative advantage over what they are presently doing. The educational effort must include information on the potential for pest damage, pesticide costs, market discounts, and long-term marketability. Without this information, producers will not accept these protocols.
- Limited educational efforts have been made to convey sampling principles. The annual summary of the USDA-ES IPM programs throughout the U.S. listed only five states working in the stored grain area. Part of the reason for this limited effort is that extension personnel do not have familiarity with problems in stored grain.
- *Industry reluctance* — In contrast to commodities such as vegetables and cotton which are intensively managed, stored grain management has not received a high priority from many local and regional elevator operators, or flour mills. Without their emphasis, the risk seems lower.
- *Compatability* — Once grain is stored, many producers do not routinely set up a monitoring program because it is not an area they would frequent.
- *Easily observed and tried* — This is not an ecosystem that can be easily tried/demonstrated from major highways or field days.
- Adoption has not occurred because of limited research in sampling programs. There have not been significant resources applied toward this research area.

REFERENCES

1. National Academy of Science, *Insect Pest Management and Control*, Publ. 1695, Washington, DC, 1989.
2. Southwood, T. R. E., *Ecological Methods*, Chapman and Hall, Eds., John Wiley & Sons, New York, 1978.
3. Strickland, A. H., Sampling crop pests and their hosts, *Annu. Rev. Entomol.*, 6, 201, 1961.
4. Kuno, E., Sampling and analysis of insect populations, *Annu. Rev. Entomol.*, 36, 285, 1991.
5. Sterling, W. L., Developing sampling technologies for IPM implementation in cotton, in *Proc. Integrated Pest Management on Major Agricultural Systems*, Frisbie, R. E. and Adkisson, P. L., Eds., Texas A & M University, College Station, 1985, 199.
6. Tollefson, J. J., Comparison of adult and egg sampling for predicting subsequent populations of western and northern corn rootworms (Coleoptera: Chrysomelidae), *J. Econ. Entomol.*, 83, 574, 1990.
7. Pedigo, L. P. and van Schaik, J. W., Time-sequential sampling: a new use of the sequential sampling probability ratio test for pest management decisions, *Bull. Entomol. Soc. Am.*, 30, 32, 1984.
8. Hutchison, W. D., Hogg, D. B., Ashraf Poswal, M. A., Berberet, R. C., and Cuperus, G. W., Implications of the stochastic nature of Kuno's and Green's fixed precision stop lines: sampling plans for the pea aphid (Homoptera: Aphididae) in alfalfa as an example, *J. Econ. Entomol.*, 81, 749, 1988.

9. Wilson, L. T., Estimating abundance and impact of arthropod natural enemies, in *Biological Control in Agricultural IPM Systems*, Hoy, M. A. and Herzog, D. C., Eds., Academic Press, Orlando, 1985, 303.
10. Reissig, H., Seem, R.C., and Nyrop, J., Integration of pesticide use into IPM systems, in *Proc. Int. Pest Management Symposium/Workshop*, Communications Services New York State Agricultural Experiment Station, Cornell University, Geneva, NY, 1989, 96.
11. Wearing, C. H., Evaluating the IPM implementation process, *Rev. Entomol.*, 33, 17, 1988.
12. Zalom, F. G., Weakley, C. V., Hoffmann, M. P., Wilson, L. T., Grieshop, J. I., and Miyao, G., Monitoring tomato fruitworm eggs in processing tomatoes, *Calif. Agric.*, 44, 12, 1990.
13. Fenemore, P. G. and Norton, G. A., Problems of implementing improvements in pest control: a case study of apples in the UK, *Crop Prot.*, 4, 51, 1985.
14. Barney, R. J. and Legg, D. E., Accuracy of a single 30-stem sample for detecting alfalfa weevil larvae (Coleoptera: Curculionidae) and making management decisions, *J. Econ. Entomol.*, 80, 512, 1987.
15. Young, J. H., Price, R. G., and Pinkston, K., Sequential sampling of the cotton fleahopper in Oklahoma, *Okla. State Univ. Coop. Ext. Serv.* Current Report CR-7171, 1979.
16. Stark, J. A., Williams, O. H., and Cuperus, G. W., Cotton decision aid, *Okla. State Univ. Coop. Ext. Serv.* Software Series (CSS-40), 1988.
17. Sandman, P. M., *Explaining Environmental Risk*, U.S. Environmental Protection Agency, Office of Toxic Substances, Washington, DC, 1986.
18. Shufran, K. A. and Raney, H. G., Influence of inter-observer variation on insect scouting observations and management decisions, *J. Econ. Entomol.*, 82, 180, 1989.
19. Fargo, W. S., Epperly, D., Cuperus, G. W., Clary, B. L., and Noyes, R. T., Effect of temperature and duration of trapping on four stored grain insect species, *J. Econ. Entomol.*, 82, 970, 1989.
20. Cuperus, G. W., Fargo, W. S., Flinn, P. W., and Hagstrum, D. W., Variables affecting capture of stored-grain insects in probe traps, *J. Kan. Entomol. Soc.*, 63, 486, 1990.
21. Lambur, M. T., Whalon, M. E., and Fear, F. A., Diffusion theory and integrated pest management: illustrations from the Michigan fruit IPM program, *Bull. Entomol. Soc. Am.*, 31, 40, 1985.
22. Rogers, E. M., Diffusion of Innovations, *The Free Press*, New York, 1983.
23. Cuperus, G. W., Noyes, R. T., Fargo, W. S., Clary, B. L., Arnold, D. C., and Anderson, K., Successful management of a high risk stored wheat system in Oklahoma, *Am. Entomol.*, 36, 129, 1990.
24. Stark, J. A., Cuperus, G. W., Ward, C., Huhnke, R., Rommann, L., Mulder, P., Stritzke, J., Johnson, G., Criswell, J. T., and Berberet, R. C., Integrated management programs: a case study of Oklahoma alfalfa management, *Okla. State Univ. Coop. Ext. Serv. Circ.* E-899, 1990.
25. Lipke, L. A., Ladewig, H. W., and Taylor-Powell, E., National assessment of extension efforts to increase farm profitability through integrated programs, *Tex. Agric., Ext. Serv.*, College Station, TX, 1987.
26. Burkholder, W., Stored product insect behavior and pheromone studies: keys to successful monitoring and trapping, in *Proc. 3rd Int. Working Conf. Stored Product Entomology*, Manhattan, KS, 1984, 20.
27. Hagstrum, D. W., Flinn, P. L., Subramanyan, B. H., Keever, D. W., and Cuperus, G. W., Interpretation of trap catch for detection and estimation of stored-product insect populations, *J. Kan. Entomol. Soc.*, 63, 550, 1990.
28. Flint, M. L. and Klonsky, K., IPM information delivery to pest control advisors, *Calif. Agric.*, 42, 18, 1989.
29. Iddings, R. K. and Apps, J. W., What influences farmers' computer use? *J. Ext.*, 36, 19, 1990.
30. Berberet, R. C., Senst, K. M., Nuss, K. E., and Gibson, W. P., Alfalfa weevil in Oklahoma: the first ten years, *Okla. Agric. Exp. Sta. Bull.*, B-751, 1980.
31. Fick, G. W., Alfalfa weevil effects on regrowth of alfalfa, *Agron. J.*, 68, 809, 1976.
32. Cothran, W. R. and Summers, C. G., Visual economic thresholds and potentials pesticide abuse: alfalfa weevils, and example, *Environ. Entomol.*, 3, 891, 1974.
33. Wedburg, J. L., Ruesink, W. G., Armbrust, E. J., and Bartell, D. P., Alfalfa weevil pest management program, *Ill. Coop. Ext. Serv. Circ.*, 1136, 1977.
34. Berberet, R. C. and Sholar, J. R., Sampling for alfalfa weevil in Oklahoma, *Okla. State Univ. Coop. Ext. Serv.*, 7177, 1980.
35. Harein, P., Gardener, R., and Cloud, H., 1984 Review of Minnesota stored grain management practices, *Univ. Minn. Agric. Exp. Sta. Bull.* AD-SB-2705, 1985.
36. Hagstrum, D., Flinn, P., and Fargo, W. S., How to sample grain for insects, in *Management of Grain, Bulk Commodities, and Bagged Products*, Krischik, V., Cuperus, G. W., and Galliart, D., Eds., Oklahoma State University Cooperative Extension Service Circular E-912.

Chapter 23

DESIGNING AND DELIVERING IN-THE-FIELD SCOUTING PROGRAMS

Edward J. Bechinski

TABLE OF CONTENTS

I. Introduction 684

II. Perspectives on Scouting and IPM Decision Making in the Real World 684

III. Social Theory Framework 687

IV. Public and Private Implementation of Scouting Programs 688
    A. The Role of Extension 688
    B. The Role of Other USDA Agencies 690
        1. USDA-ASCS ICM Program 690
        2. USDA-APHIS-PPQ CAPS Program 690
    C. The Role of Growers 691
    D. The Role of Grower Crop Management Associations 691
    E. The Role for Private IPM Consultants 692
    F. The Role of Industry Field Staff 693

V. Types of Scouting Programs 694

VI. Delivering Scouting Information 696
    A. Scouting Forms and Decision Aids 697
    B. Computer Software and Scouting 698

VII. Epilogue: A Scouting Call-to-Arms 701

References 702

I. INTRODUCTION

Field sampling or *scouting* is such an integral part of integrated pest management in the U.S. that it has become virtually synonymous with the very definition of IPM. Indeed, some critics argue that IPM is nothing more than count-and-spray pest control that pays but lip service to nonchemical controls.[1-2] Although this viewpoint ignores a large body of data documenting significant reductions pesticide use and other benefits through programs of field scouting and economic thresholds,[3] a rich irony is that considerable research effort has been given to the development of IPM sampling protocols that never see the light of day on a scout's clipboard in the field.

Consider that during an arbitrarily selected but, presumably, representative 15-year period beginning with 1977, some 240 papers were published about insect sampling and supporting topics in the *Journal of Economic Entomology* and *Environmental Entomology*. Of these papers, seven in ten were described by their authors either as an improvement to existing sampling practice or as labor-saving, cost-reducing, precision-maximizing tools ready for adoption. Yet, with few notable exceptions, a concurrent review of grower IPM guides and scouting manuals suggests that these advances largely have been ignored. Instead, instructions to farmers typically consist of the arbitrary, subjective "take five random samples in an X-shaped pattern" kind. This dichotomy begs the question: Why is scouting so unsophisticated on the farm when so much more detail is available in the literature?

This chapter attempts to provide some answers. The focus is on the pragmatic: development and delivery of scouting programs for IPM practitioners. Examples largely are drawn from field and row crops in the U.S., but the concepts and issues broadly apply to pest management sampling in general. At the risk of provincialism and bias, I have drawn on selected illustrative examples rather than exhaustively summarizing the literature. I especially use personal examples from my experiences as an Extension IPM Specialist, in part to provide fresh case histories, but also because design and delivery of scouting programs is more an art learned individually on the job rather than a science that follows well-established principles.

II. PERSPECTIVES ON SCOUTING AND IPM DECISION MAKING IN THE REAL WORLD

Many factors explain the relative simplicity of IPM scouting programs among U.S. farmers (see this chapter), including satisfaction with established methods, suspicion of unfamiliar technology, and inadequate communication between university researchers who develop scouting protocols and IPM advisors who implement them. All contribute to an inevitable time lag between the development and adoption of any new technology. Cynics can make a case that a publish-or-perish mindset among university faculty encourages irrelevant research and esoteric manuscripts for the sake of the career advancement rather than for advancement of IPM practice among producers, or that educational outreach programs within extension ineptly are transferring new scouting technology to potential users. I believe there is an elemental lesson to be gained about matching scouting programs to what farmers and pest advisors want: easy-to-use sampling tools and plans that address perceived pest problems and which satisfy their concerns about reliability and risk.

Surveys of growers and IPM advisors repeatedly have identified lack of simple sampling programs as serious constraints to IPM adoption.[4-5] Glass,[6] writing in 1975, recognized that "the need for reliable, uncomplicated and statistically valid sampling methods is great." Seventeen years later, Glass[7] still decried the "need to simplify

IPM methods, especially monitoring and sampling pests and beneficials." Frisbie[8] and Frisbie and McWhorter[9] state the issue succinctly: simple, quick, inexpensive scouting programs are more likely to gain immediate acceptance by farmers than methods or plans that are labor- or knowledge-intensive. To the extent that advances in IPM sampling technology either satisfy or ignore these prerequisites, they will be adopted or discarded.

"Easy-to-use" can be read both as the amount of time required to make a management decision and the level of specialized training or equipment needed to conduct the scouting program. Farmer adoption of IPM is voluntary and so especially depends on convincingly demonstrating cost efficiency incentives. Yet for many farmers, scouting and pest management decision making ranks low on their priority list of daily tasks. Their time, instead, can be spent more profitably attending to crop marketing or other financial management decisions. Private consultants and other IPM advisors may be responsible for scouting thousands of acres weekly and so, too, face pressing time constraints. The premium on scouting time especially is important when farmers alternatively can use pesticides preventively as part of normal farming operations without any pest sampling, such as "just-in-case" insurance treatments with granular, soil-applied insecticides at planting time to control soil-borne pests vs. field scouting and "wait-and-see" treatment later during the growing season. From a grower's perspective, it may not be worth the time or expense to regularly scout fields during the growing season only to later discover that another trip over the field is required for a pesticide application that, instead, could have been made more easily or in a more timely manner during planting.

Streamlining of scouting plans must receive the same attention as is now more widely given to maximizing sampling precision and accuracy. My own subjective rule of thumb is that scouting should require no more than 15 to 20 min/field if it is to be widely adopted. Satisfying this time constraint is no small task when the farmer must sample a patchily distributed insect across an expansive acreage. Sequential decision plans (particularly binomial plans that reduce sampling effort to determining pest presence or absence) long have been championed for their time efficiency.[10-11] However, the seeming complexity of sequential sampling plans diminishes their utility to some growers. Hence, while numerous sequential decision plans can be cited from the scientific literature, few, if any, have seen universal adoption among their intended users.

For some IPM practitioners, fixed-time sampling plans that limit scouting to an acceptable predetermined duration inherently are more intuitive and are perceived as less risky than sequential sampling. Simple-to-use plans can be derived by expressing sampling time as a function of pest density.[12] Bechinski and Hescock[13] adopted this approach and derived plans that determine from 30-min field surveys if insecticides are warranted for the alfalfa snout beetle, *Otiorhynchus ligustici* (L.) (Coleoptera: Curculionidae). And though variation in sampling efficiency or other inconsistencies among scouts sometimes call into question the accuracy and transportability of fixed-time plans[14] as well as other schemes, a pragmatic viewpoint is that many growers are willing to sacrifice some accuracy and precision in order to save scouting time.[15] A useful analogy is to envision a dual-pan, balance-beam scale. A scouting program that quickly delivers enough accuracy and precision to tip the scale in favor of the grower making the correct management decision is better than a highly precise and completely accurate scouting program which requires so much time that it never is adopted.

Many commonly used scouting techniques such as sweepnet and ground cloth require minimal technical training to employ. However, these methods nonselectively sample a crop fauna and so can quickly overwhelm the taxonomic expertise of a scout whose IPM training may be limited to an annual half-day farmers' meeting. "Automatic" scouting techniques such as pheromone traps or bait traps that more selectively monitor pest density or that identify fields requiring more intensive sampling can be important simplifying features. In Idaho, home-made traps constructed from materials locally available at hardware stores are the foundation of a scouting program for the sugarbeet root maggot, *Tetanops myopaeformis* (Roder) (Diptera: Otitidae).[16] Larvae are the damaging stage, but because they are soil borne, it is impractical for growers to sample them directly. Adult flies instead preferentially can be attracted to an orange-colored sticky trap; here fly captures are an index that predicts crop yield loss from subsequent larval feeding on the sugarbeet taproot. This simple system gives growers an alternative to automatic use of insecticides applied during planting for maggot control.

That scouting programs must address real pest problems if they are to be used is self-evident, but what often is much less obvious is that a farmer's notion of pest threat can be quite different than that of researcher or IPM advisor. Too frequently, scouting programs are developed without first conferring with the growers who are targeted as the eventual users. Rather than conducting post-mortem case studies after sampling protocols have been designed to explain why a new method was not adopted, grower perceptions and needs more productively can be assessed as the beginning step in scouting program design. Farmer surveys are an excellent tool for identifying and recognizing needs. Mail surveys of commercial potato producers in Idaho generated a ranked listing of insect pests that more often than not contradicted extension's own ranking of species importance. Here perception became reality; farmers reported they were treating fields for pests deemed unimportant by extension. These survey data provided the rationale for reprioritizing development and implementation of scouting programs that better matched the industry's stated needs. On-farm participatory research that directly involves growers during early program development likewise can go a long way in addressing perceived needs.

Attitudes about uncertainty, especially perceived financial risks of not using pesticides, play a critical role in deciding if scouting programs are adopted. Scouting and thresholds can be difficult to implement when pesticide costs are but a small fraction of crop value. Palti and Ausher[17] suggest that "enlightened growers will consider this (field scouting) an expense similar to crop insurance," yet the reality of how growers view the added cost of sampling often is quite different. For instance, a typical sugarbeet grower in Idaho annually can anticipate gross returns of $900 to $1300/acre. Compared with even the most expensive preventive application of insecticides (currently $25 to $35/acre), this expense literally is viewed by some as a cheap insurance policy that virtually guarantees an acceptable income. Mumford[18] observed that when choosing among pest management options, farmers often place greater emphasis on minimizing the probability of catastrophic losses rather than normal losses. Hence, some growers err to the conservative side and automatically protect their crop investment with insecticide rather than depend on scouting to identify a problem. A complicating issue is that automatic use of insecticides in lieu of sampling and decision rules sometimes simultaneously suppresses other important pests species. Some 15% of Idaho's sugarbeet industry uses root maggot traps, and, though additional growers, too, would like to us traps and thresholds, they do not because infestations of other soil-borne pests dictate application of insecticides which also control any root maggots present.

Uncertainty also can slow adoption when the rationale for a scouting plan cannot easily be scrutinized by the user. As alluded to earlier, this especially is true for mathematically complex (or seemingly complex) scouting programs where growers and IPM advisors have little possibility of personally reviewing the validity of any unstated assumptions or other theoretical considerations on which the program depends. As an example, after developing a binomial sequential sampling card to help grain producers in my state make aphid management decisions, I learned that farmers in one county modified the recommended sample-unit. Whereas instructions called for scouts to inspect individual plants and score them as either infested or not infested, these growers instead inspected five consecutive plants and recorded a positive score when any one of the five was infested. This modification reduced the economic threshold to one fifth the true value. Yet these same growers previously had accepted the true threshold without question when scouting involved checking 100 plants along a diagonal across the field. The difference was that the sequential decision card gave a spray/don't spray recommendation after as few as 20 plants had been inspected. Not only did the growers perceive there to be an unacceptably high risk of making an incorrect management decision after examining but 20 plants, more importantly, they could not personally check and verify the logic of how the decision card made its recommendations. Adoption required a leap of faith into a scouting program based on unknown statistical magic.

III. SOCIAL THEORY FRAMEWORK

It may surprise at first that there is any connection between the social sciences and insect scouting, yet all the foregoing issues perhaps best can be summarized from the larger perspective of behavioral sociology. A starting point is Bennett's[19] KASA model, which states that people adopt new practices (such as scouting programs) after progressing through a systematic hierarchy of changes in *k*nowledge, *a*ttitudes, *s*kills, and *a*spirations. The KASA model recognizes that mere development and availability of innovative scouting practices in no way guarantees they will be used, nor is technical training which adds to a farmer's knowledge and skills by itself sufficient to bring about adoption. Individual innovation and adoption depends on simultaneously recognizing and addressing motivations for change. Theories of behavioral motivation, models for adult learning, and principles of communication all are germane but beyond the intent of this chapter; Cole and Cole,[20] Griffith,[21] Praw1 et al.,[22] and Whale[23] are helpful introductions to these topics. An illustrative application of these theories is the work of Musser et al.,[24] who analyzed relationships among farmer beliefs, IPM adoption, and educational programs. The authors suggest a useful construct of IPM "belief measures" consisting of damage, expense, personnel, yield, environment, method, profit, and risk variables that explain and predict motivation for IPM adoption. An understanding of diffusion theory also is relevant, but again, the literature is too extensive to summarize here. Lamble[25] gives a good general introduction; Rogers[26] is the authoritative text. Two principles deserve comment: the attributes of innovation that affect adoption rates and types of adopters.

Rogers[26] identifies five attributes that most affect adoption of a new idea: *relative advantage*, or why the innovation is better than the current way of doing things in terms of profitability, prestige, convenience, and satisfaction; *compatibility*, how the innovation matches the experience, values, and needs of intended users; *complexity*, the level of difficulty required to understand and use the innovation; *trialability*, the degree to which an innovation can be tried on a limited or experimental basis; *observability*, or how visible the results of an innovation are to others. From the context of designing pest scouting programs, farmer adoption of new advances depends on maximizing relative advantage and compatability while minimizing complexity. The notion that people reinvent or selectively adopt new technology further suggests that rigid "all-or-nothing" scouting programs are less likely to be used than more flexible plans that allow users to modify and refine them to fit local conditions. In the context of delivering scouting programs, adoption can be enhanced by limited trials and demonstrations that can be easily observed and described.

Rogers[26] groups adopters into five general categories. From first users to last adopters, they are the "venturesome" innovators, the "respected" early adopters, the "deliberate" early majority, the "skeptical" late majority, and the "traditional" laggards. Together they account for the S-shaped curve that typically marks the diffusion of any new practice through a population. As their names suggest, each group differs distinctly with respect to behavioral and motivational profile, socioeconomic background, communication styles, and other attributes. The value for implementing pest scouting program is that each adopter group should be targeted with a unique educational and training strategy by the IPM advisor whose job it is to convert a research-based advance into practice on the farm. The work of Grieshop et al.[27] is an instructive model of how these concepts can be applied to scouting program delivery.

IV. PUBLIC AND PRIVATE IMPLEMENTATION OF SCOUTING PROGRAMS

Development of scouting programs and transfer to farmers in the U.S. depends on producers and private businesses working cooperatively with public and government agencies. Public leadership rests with the Cooperative Extension System (CES), a national network of federal employees and faculty at state land-grant universities charged with moving research-based knowledge into local communities.[28] A contributing role in the public sector is played by survey entomologists at state departments of agriculture and by staff at other state and federal regulatory agencies. Within the private sector, growers who have neither the technical skills nor management time to personally scout their fields look for assistance from crop management associations, for-hire private consultants, and agricultural industry fieldstaff. None of these groups provides scouting services to the mutual exclusion of others. Growers often simultaneously depend on extension personnel, industry field staff, consultants, and other scouts.

A. THE ROLE OF EXTENSION

Extension's role in IPM scouting has changed considerably since the first federally supported demonstration projects in North Carolina and Arizona during 1971. Since 1979 when extension IPM programs became national in scope,[29] each U.S. state and protectorate annually has received special funding from the Extension Service of the U.S. Department of Agriculture (ES-USDA) to conduct educational programs in IPM. When they began, an operational objective of extension IPM programs was to implement practical methods for monitoring pests in farmers' fields. The legislative intent was for federal support to serve as cost-share monies to encourage farmers to adopt scouting programs, but with the mandate that scouting eventually be privatized by requiring growers themselves to assume all costs after a trial period. Accordingly, most state IPM programs have included extension-sponsored pilot scouting services for growers that since have been turned over to the private sector.

Extension-sponsored scouting programs typically operate as follows.[30-32] Prospective participants learn about the pilot program by means of circular letters and farmer meetings. Extension then hires, trains, and supervises scouts who make weekly crop inspections for key pests. Scouts, who often are college students working during the summer growing season, work under a scout supervisor, who in turn is responsible for providing them with initial and continued seasonal scout training, checking their job performance, and serving as communications liaison to a senior extension program leader. Scouts prepare written reports both for the farmer and for the scout supervisor, but leave interpretation and management recommendations for the grower to work out, in consultation with experienced local extension personnel. As an incentive for adoption, extension-sponsored scouting services initially are offered to growers below their actual cost. During a 2- to 5-year demonstration period, growers assume incremental increases in scouting fees, and ultimately, pilot programs are turned over to private IPM advisors who provide scouting services for a fee. Rather than directly providing individualized services to select growers or competing with private IPM scouting businesses, extension assumes a training and support role in keeping with its formal educational mission of transmitting knowledge to the public. This model for privatization of IPM scouting exists yet in many states. A growing added dimension in other states is the conduct of adoptive research by extension faculty who directly develop and refine scouting plans.

The integrated pest management program for alfalfa seed in the Pacific Northwest chronicles this privatization process.[33] Formal extension IPM programs for the regional alfalfa seed industry began in Washington state during 1973 with a twofold focus: weekly field scouting of four key insect pests and their natural enemies, and management of the pollinating bees required for alfalfa seed production. Following Washington's lead, extension entomologists in Idaho introduced these methods in 1976 as a 4-year demonstration program in five counties. During that first year, 23 commercial seed growers in Idaho entered 945 acres into the pilot program for a $3.00/acre fee. Augmented with ES-USDA IPM funds, these fees paid for scout wages, travel expenses, supervisor salary, and related operating expenses. Acreage scouted by extension doubled the second year (1977) to 2500 acres and doubled again to 5300 acres in 1978. Simultaneously, fees assessed growers increased to $4.50/acre in 1978 and $6.50/acre in 1979, the last year of direct extension sponsorship. Privatization of scouting services in Idaho began during 1979 with one private IPM consulting business and one seed company scouting 1200 acres. By 1980, scouting services were provided by six private IPM consultants, five alfalfa seed companies, and five agrichemical supply and service companies. Extension withdrew its financial sponsorship in 1980 and thereafter provided educational programming by convening annual scout training schools and by producing IPM manuals and videos. These relationships continue today. Private consultants and seed companies still offer pest scouting while extension provides the educational programs.[34]

Nationwide, extension sponsored more than 900 IPM educational and demonstration projects during 1990 to 1991 on topics ranging from commercial agriculture to urban horticulture.[35-36] More than 11,000 scouts and 45,000 producers were trained in scouting methods and other IPM methods. State Extension IPM Specialists further estimated that private IPM consultants, agrichemical dealers, cooperatives and others trained by extension in turn reached an additional 370,000 producers. The private sector strongly has endorsed these delivery and education roles for extension and IPM businesses. As president of the National Alliance of Independent Crop Consultants (the professional association of independent crop management advisors), Bradshaw[37] wrote in 1990 that "NAICC members do not feel that tax money should be used to support unneeded government competition to our own private businesses." This concern echoes policy statements issued 12 years earlier by the National Agricultural Chemical Association,[38] which stated "scouts and pesticide consulting should be paid for by the user and not the government." The NACA statement went on to observe that "the greatest threat to establishing successful IPM programs may be the government itself."

B. THE ROLE OF OTHER USDA AGENCIES
1. USDA-ASCS ICM Program

The Agricultural Stabilization and Conservation Service of the U.S. Department of Agriculture (USDA-ASCS) began during 1989 a pilot program designated Special Practice 53, Integrated Crop Management (SP53 ICM), which mirrors and extends the goals of the extension-sponsored scouting programs. SP53 provides cost-sharing subsidies to farmers who adopt field scouting programs as well as other IPM methods and best management practices (BMPs) for pesticides, fertilizers, irrigation, and soil tillage. Cost-sharing reimburses participants up to 75% of their actual expenses to a maximum of $7.00/acre for field and row crops and $14.00/acre for vegetable, fruit, and specialty crops.

Originally targeting as a program goal a 20% reduction in pesticide or fertilizer use,[39] the goals of SP53 were revised in 1991 to emphasize "efficient use [of pesticides and fertilizers] and demonstration of ecological benefits."[40] Farmers follow a written management plan developed for their farm by the Soil Conservation Service (USDA-SCS), the state extension system, or by private consultants who have been certified as satisfying certain educational, technical, and business standards. As of 1992, program impact on farmer adoption of scouting programs has been uneven nationally; the strongest programs seem to have developed where private consultants are available to write ICM plans. SP53 remains in a pilot demonstration phase, which limits participation to 5 counties per state and 20 producers per county.

2. USDA-APHIS-PPQ CAPS Program

The Cooperative Agricultural Pest Survey (CAPS) program is a national pest monitoring and communication network coordinated and funded by the Plant Pest Quarantine unit of the Animal and Plant Health Inspection Service section of the U.S. Department of Agriculture (USDA-APHIS-PPQ). Originally christened the National Cooperative Plant Pest Survey and Detection Program (with a resultant mind-numbing acronym, USDA-APHIS-PPQ-NCPPSDP), but mercifully renamed CAPS during 1988, this project began in 1982 as a 16-state pilot effort to centrally collect and manage incidence and severity data for 3 types of pests: nonnative exotic species, endemic species, and pests that hinder international exports of U.S. agricultural products. The premise was that numerous surveys for pest and beneficial species were being conducted by federal, state, and local agencies, but there was no easy way to access the entire body of data being collected. The vision was to apply computer technology to all these data so as to nationally track current pest conditions, forecast pest status, issue early warnings, plan eradication programs, and delimit pest distribution or host range to aid export certification. To that purpose, APHIS developed a centralized computerized database system, NAPIS (*N*ational *A*gricultural *P*est *In*formation *S*ystem), as the national clearinghouse for U.S. pest information.[41]

Funding was provided to university extension and research faculty, state department of agriculture, and other state and federal plant regulatory agencies to begin entering pest and crop information to NAPIS from ongoing surveys. By the mid 1980s, local, regional and national scouting and survey data from all 50 states were being processed through NAPIS. Notwithstanding some successes in tracking pest movement (such as mapping the initial U.S. establishment and subsequent dispersal of the Russian wheat aphid, *Diuraphis noxia* [Mordvilko] [Homoptera: Aphididae]), by the late 1980s it was clear that no single system could satisfy the differing expectations and informational needs of regulatory personnel, researchers, individual growers, industry field staff, nd other IPM advisors.[42] Beginning in 1992, APHIS redirected the CAPS program to concentrate on pest surveys most closely aligned with the mission of PPQ, namely exotic species trapping and presence-absence distributional surveys in support of commodity exports. Endemic pests, which are the target of most IPM scouting programs, no longer are supported financially by the CAPS program.[43] The unwieldy NAPIS system simultaneously was reorganized as the Historical Distributional System, a simplified database of county-level presence-absence records for a limited number of PPQ program pests. As has always been the case, access to the database is restricted to a few authorized persons within each state. The impact of this redirection remains to be seen.

C. THE ROLE OF GROWERS

Many growers carry out pest scouting themselves, though it is difficult to know the extent to which their methods follow formal protocols or less formal field walks and "windshield surveys" (automobile drive-bys at 55 mi/h!). Yet growers and their families do seem to conduct the majority of scouting in the U.S. Case studies of IPM programs for nine agricultural and horticultural crops during 1985 indicated that 66% of all scouting was performed by growers.[3] Surveys in Idaho during 1992 showed 92% of potato producers and 89% of sugarbeet producers reporting that they or other family members "frequently" scouted their own fields during the growing season. Growers annually receive updates and training about scouting methods and other IPM technologies from extension at classroom-style commodity schools held before the growing season begins. Local in-field clinics and workshops give hands-on training and demonstrations during the growing season. Chapter 23 describes these training efforts in more detail.

D. THE ROLE OF GROWER CROP MANAGEMENT ASSOCIATIONS

Crop management associations typically take the form of grower-owned and -operated nonprofit corporations. Operating under Articles of Incorporation and Bylaws, members elect a Board of Directors that determine the scope of services offered. Members usually pay per-acre service fees, although some associations bill at an hourly rate and others assess "check-off" fees based on crop marketing receipts (such as member of the Idaho Potato Pest Management Association, who pay $\frac{1}{2}$¢ for every 100 pounds of potatoes marketed). These fees allow employment of scouts or managing IPM/crop consultants who provide members with scouting and crop management services. Because they are not affiliated with manufacturers of agrichemicals, grower crop management associations provide their members with independent pest management recommendations. This distinction compares with another type of grower-owned organization, the *IPM Cooperative*, which markets products (such as pesticides or, less frequently, biocontrol agents) as well as services. Guidelines for establishing grower IPM associations and perspectives on alternative operating structures are detailed by Vogelsang[44] and Good et al.[45]

National surveys during 1991 showed 162 pest and crop management associations were operating in 26 states.[46] These serviced 12,000 growers with 3.3 million acres for fees averaging $5.50/acre and ranging from $1.31 to $40/acre. Of 754 association employees, only 70 persons working in 16 states were full-year professionals; all others were seasonal employees, typically students. In addition to weekly pest scouting, many associations offered plant and soil nutrient sampling for fertility recommendations. Some additionally provided tillage and residue management, crop variety selection advice, water management, and record-keeping services. Associations were most common in Georgia and Texas, where 56 associations provided scouting to approximately 8900 growers.

The Texas Pest Management Association (TPMA) documents a successful case history of a producer-operated scouting service.[9,47-49] Organized IPM programs began in Texas during 1971 with growers subscribing to cotton scouting services sponsored by extension. As demand for scouting increased, it became necessary to separate the fiscal and personnel administration of the scouting program from extension's mission in education. This led to TPMA's chartering in 1977 as a statewide nonprofit corporation operated by growers and dedicated to delivering scouting and IPM services. Extension provides scout training, pest management consultation, and related educational programming through county-based IPM agents, who in turn are supported by regional specialists. The TPMA hires scouts and scout supervisors, sets and collects scouting fees, and ultimately has the sole authority for determining producer membership and participation. Scouting programs are operated either intensively or extensively. In the former, each scout collects field data weekly or twice weekly on 800 to 1500 acres; in the latter, scouting relies on pheromone traps and spot checks of fields, and each scout handles up to 7500 acres. Scouts prepare reports for the grower, the supervisor, and the county extension agent. Scout supervisors direct four scouts, recheck questionable fields, and serve as the primary scout for a small portion of acres. Scout supervisors especially serve as the communication link between the farmer and the local extension agent. Neither scouts nor supervisors make pest control recommendations. Organizationally, TPMA members operate county-based scouting programs through steering committees of local producers and agribusiness personnel. Regional commodity commissions provide operating assistance for corn, cotton, grain-sorghum and peanut growers; membership also includes producers of alfalfa, citrus, pecans, rice, soybeans, and wheat. Overall administration is provided by a board of directors consisting of a representative from each county-based program and each commodity group. Association membership in 1992 stood at 2500 growers who farmed 958,000 acres.

E. THE ROLE FOR PRIVATE IPM CONSULTANTS

Upwards of 2000 to 7000 professional consultants provide producers in 45 states with scouting and other IPM services.[35,50] Like grower crop management associations, most private consultants offer total crop production and management services rather than only IPM scouting. Focus on the total crop is a response both to customer demand and fiscal necessity; pest problems are only a subset of all the production and marketing questions that farmers must solve, and the fees most are willing to pay for pest scouting alone are too low to keep businesses solvent.[51] The range of services includes plant-tissue and soil-nutrient sampling for fertility recommendations, irrigation scheduling, and crop enterprise financial management. Some private consultants sell and apply pesticides or fertilizers while others have no business affiliation with agrichemical sales or marketing. The term *commercial consultant* usually is applied to the former, while *independent consultant* is preferred for the latter.[52]

Although Blair[53] earlier had characterized crop consultants as "a major force" in delivering pest management information to growers in the Midwest, consultants now seem most active in the south and southwest where they are employed by a high proportion of cotton producers and vegetable growers. Extensive concentration of high value crops are incentives that draw private consultants.[54] Consulting firms surveyed by Lambur et al.[50] served 18,077 clients, 77.5% of which produced field crops and 20% produced horticultural crops (primarily vegetables). The remainder served livestock producers and urban clients (apartments, golf courses, ornamentals, homeowners). Firms are small, with 90% employing one or fewer full-time employees. Weddle[55] relates from a consultant's perspective the day-to-day problems and issues of operating a private IPM business.

The rise of private IPM consulting is a recent development. Allen and Rajotte[56] reported a 1000% increase nationally in the number of private crop consulting firms between 1969 and 1985. This increase is an outgrowth of support and cultivation from extension as well as the active enterprise of consultants themselves. During the pilot demonstration phase of extension-sponsored scouting programs, special effort often is given to directly involving consultants in planning, implementation and evaluation.[57] In the Pacific Northwest, demand for scouting services created by the pilot alfalfa seed scouting program was satisfied when scouts originally hired by extension went into business for themselves. About one third of private consultants began their careers in extension.[3]

Consultants most often adopt scouting methods and educational materials developed by extension, though some firms do design in-house scouting programs. An analogy is that extension often operates as a wholesaler of IPM information which is refined and adopted to fit specific fields by consultants and other information retailers. One niche almost exclusively filled by private consulting businesses is remote sensing by aerial infrared photography. Rather than directly observing the insects themselves, remote sensing uses visual differences in false-color aerial maps to identify plant stress caused by insects. Because diseases, mineral deficiencies, and other agents also produce comparable color changes in photographs, remote sensing has not replaced conventional field scouting on the ground. Instead, it can save subsequent time and effort by rapidly pinpointing the location and size of infestations requiring more intensive sampling.[58] Riley[59] overviews the technology and application of remote sensing in entomology.

F. THE ROLE OF INDUSTRY FIELD STAFF

Agricultural chemical dealers and their sales and technical field staff often are among a grower's most common source of pest control advice. However, the "scouting" they provide probably better is described as ad hoc problem-solving inspection rather than regularly scheduled monitoring. Nonetheless, growers commonly cite pesticide sale persons as key providers of scouting assistance.[3] Although their service usually is free, it also is worth noting as do Eshel and Palti[60] that agrichemical industry personnel sometimes give pest management advice "far removed from the goal of integrated pest management." Because they are in the business of marketing pesticides, industry personnel especially promote those scouting plans that potentially contribute to sales of their particular product. There is nothing intrinsically wrong with this, but as illustrated in the following two cases, there is potential for conflicts of interest.

When traps for sugarbeet root maggots first were introduced in Idaho, some sales personnel recommended insecticide application as soon as flies were captured. This use violated the underlying concept that insecticide application was warranted only if captures exceeded economic thresholds. As virtually all sugarbeet fields in Idaho have detectable levels of root maggots, but only a fraction exceed economic levels, this advice resulted in unnecessary pesticide application, though it is difficult to know if recommendation to treat at first fly capture represented naive confusion about a new sampling technique or a deliberate effort to increase sales. In a different case, a sales representative objected to use of pheromone traps on the basis that current understanding of pest population dynamics was inadequate to accurately interpret trap captures and issue management recommendations. Although there was validity to that argument, what was left unsaid was the potential impact that use of traps could have on reducing insecticide sales. In particular, when traps indicated that control was needed, the only insecticide available was a competitor's product. More importantly, this product also suppressed a second pest species which comprised a significant market for the sale representative's own insecticide. Farmers who used pheromone traps potentially could eliminate their use of his product and replace it with a single application of the competitors'.

Farmer supply cooperatives also provide IPM scouting services that range from trouble-shooting field visits to preplanned systematic sampling, following formal protocols. By nature, supply cooperatives are organized to provide farmers with pesticides and other agrichemical products, so, like the preceding examples of company representatives with vested interest in pesticide sales, field staff from supply co-ops too are open to the perception of conflict of interest. Some co-ops avoid such potential conflicts by creating independent subsidiaries that separate product sales from the scouting and service portion of their business. Processed and packaged food industries, as well as specialty seed companies, also commonly employ field personnel who do IPM scouting and give pest management advice. Some have established formal IPM scouting and management programs, among them a tomato IPM program by Del Monte Corporation in California, a carrot program by Campbell Foods, Inc., in Texas; a peach IPM program by Gerber Foods, Inc., in Arkansas and South Carolina, and Ocean Spray Cranberries, Inc., in Massachusetts, New Jersey, Washington, and Wisconsin.[61] These field staff can add a significant multiplier effect to public IPM programs.

V. TYPES OF SCOUTING PROGRAMS

Southwood[62] opens his text on insect sampling with the seemingly obvious but often unheeded warning: "The object of a study will largely determine the methods used and thus this must be clearly defined at the outset." In the same manner, the purpose of IPM scouting must be clearly understood before practical tools and plans can be developed. Norton[63] categorized the information needs of farmers into "fundamental" biological and ecological information, "historical" pest incidence and abundance data, "real-time" or current pest status, and "forecast information" of future status. Scouting traditionally has satisfied the day-to-day need for current pest status information, but increasingly provides the data required by predictive models to forecast future pest status. In either instance, the ultimate purpose is the same: to provide the data which determines if pest control action is needed, whether it be preventive action based on forecasted pest density or corrective action based on present density.

A more practical distinction is scouting for initial detection of a new exotic pest vs. regular monitoring of an established pest. In the former instance, the objective usually is to determine if the pest is present rather than to estimate actual density. Accordingly, sampling purposively is biased to maximize the probability of pest detection.[64] Habitats where the pest most likely occurs are sampled to the exclusion of all others; hosts showing typical damage symptoms deliberately are sought out rather than sampling randomly; the easiest-to-detect life stage often is targeted regardless of whether it is the damaging stage; survey techniques such as traps that exploit pest behavior particularly are used. Specific sampling methods and protocols used are determined by "pathway studies" that identify where and how the species likely is introduced. In diametric contrast, regular scouting for established pests depends on minimizing bias; concepts of random and representative sampling (see Chapters 1 and 3) are essential to accurately determine abundance. Otherwise, if scouting deliberately selects hosts with signs and symptoms typical of pest infestation, pest status will be overestimated. Traps and devices that concentrate pests are used but as a means of assessing actual severity rather than only presence or absence.

Another practical distinction in scouting purpose comes from the perspective of geographic scale. Pest control decisions typically are made by individual farmers who take management action independently of their neighbors to protect single fields on their own farm. Hence, scouting largely occurs within individual fields on a farm-by-farm basis. But because many insects are highly mobile and so must be treated as "common property,"[65] scouting sometimes encompasses larger geographic scales. Regional scouting programs typically are conducted by government agencies, university extension services or other public institutions for the collective benefit of all producers within a production area. These areawide scouting programs often complement rather than replace field-level scouting programs by providing early warnings which signal when local scouting should begin. Some regional programs are used by themselves to coordinate pest management action within a production area. Two illustrative case histories follow: Idaho's BEACON program and the Pacific Northwest Aphid Suction Trap Network.

BEACON, the *BEA*n *C*utworm *O*utlook and *N*otification program, is a regional network of 20 electrical black-light traps strategically located within a 5-county area of south central Idaho. Since 1979, BEACON has provided bean and sweet corn growers with advance warning of damage expected from the western bean cutworm, *Loxagrotis albicosta* (Smith) [Lepidoptera: Noctuidae], a univoltine pest that emerges from overwintering sites in the soil as adults during July and subsequently damages bean pods and corn ears as larvae during August. Technicians working under the direction of an Extension Entomologist check light traps daily during the moth flight season from early July through mid-August. First moth detection and subsequent seasonal captures can be used to schedule field inspections for cutworm egg masses in sweet corn fields. Larvae are not sampled because control action must be taken soon after eggs hatch and before caterpillars bore into the corn ear. Egg mass scouting usually is performed by the field staff of the sweet corn processor as a free service to their contract growers.

In contrast to sweet corn fields, few if any bean fields directly are scouted for western bean cutworm. Most bean growers instead rely on regional damage predictions from a simple linear statistical model between cumulative moth captures in the BEACON light traps and subsequent larval damage.[66] Although pheromone traps were introduced to the industry during 1985 as an alternative scouting tool for individual growers, adoption has been minimal. As with the light trap network, a simple linear model relates moth captures in pheromone traps to damage expected from larval feeding on the bean pod.[67] Despite several incentives for using pheromone traps, such as substantially lower purchase and operating costs, local availability, portability (independence from an electrical supply), high species selectivity, and most importantly, the capability of predicting damage levels on a field-by-field basis rather than on a regional basis, historic satisfaction and familiarity with BEACON, as well as a perception among growers that pheromone traps are less reliable than light traps, explain why this superior technology largely has been ignored.

Bean cutworm status reports and management recommendations are mass disseminated to farmers and industry field staff by means of direct-mail newsletters, by alerts published in local newspapers, and at weekly extension-sponsored "fieldmen luncheon" meetings for industry field staff, consultants, and agrichemical industry personnel. The cost of operating the BEACON trap network is borne by the growers themselves through their statewide association, the Idaho Bean Commission, and by two vegetable processors, the Del Monte Corporation and Green Giant, Inc. Since

1979 when BEACON began, the acreage treated with insecticide for western bean cutworm has decreased from 45,000 to less than 6000 acres annually by identifying outbreak areas that actually require insecticide application.

Another regional scouting program monitors the flight activity of agronomically important aphids across the western U.S. Patterned after the Rothamsted network of aphid suction traps in Great Britain and mainland Europe,[68-70] Washington state began during 1983 to survey airborne aphids with a series of 8-m tall suction traps.[71-72] By 1991, 9 U.S. states and 1 Canadian province informally were operating on a 60-trap network. Idaho, Oregon, and Washington began in 1992 to pool resources for a 32-trap network within the tristate Pacific Northwest region. The network monitors the density and phenology of 12 economic aphids from the beginning of the flight season in mid-May through season's end in late October. Each trap is believed to be representative of flight activity within a 20- to 50-mi radius.

The most direct use of these regional data is for pest incidence: whether aphids are present or not. Farmers use incidence data to identify pest-free crop planting dates which minimize the probability of aphid colonization. For example, last seasonal capture of cereal-infesting aphids in the fall signals that winter-seeded wheat and barley safely can be sown without threat of subsequent infestation. Conversely, growers attempt to establish spring-seeded crops well before the first aphid is captured so that crops will have developed beyond the highly susceptible seeding stage before aphid colonization begins. First and last captures also indicate when local field scouting should begin and end. Rudimentary thresholds based on correlations between suction trap captures and aphid density in surrounding fields are gross predictors of local damage potential.[73]

Technicians service suction traps weekly and mail samples to a regional university specialist for processing and identification. Mass media is used to deliver data to farmers and their advisors. A printed newsletter, the *Pacific Northwest Aphid Flyer*,[74] is mailed weekly to 400 farmers, extension faculty, private consultants, pesticide dealers, and other pest management advisors. The newsletter summarizes trap captures by location and suggests management options. Owing to the logistics of trap operation and mailing delays, scouting data are 5 or more days old when newsletters reach the farmer. To speed dissemination, supplemental electronic delivery began in 1992 via a computer bulletin board system, IDEX, the *ID*aho *EX*tension Bulletin Board.[75-76] IDEX is a public access system that can be used 24 h/d, for the cost of a telephone call, by anyone equipped with a personal computer, modem, and communications software. Like the BEACON program, the Pacific Northwest Aphid Suction Trap Network is a joint private-public project funded by two state commodity groups, the Idaho Wheat Commission and the Idaho Barley Commission, as well as by public funds from the three cooperating universities, Oregon State University, the University of Idaho, and Washington State University. Operating costs (excluding professional salaries) approached $40,000 for the 1992 season.

VI. DELIVERING SCOUTING INFORMATION

As the foregoing examples suggest, scouting data (as well as training in scouting protocols) are communicated to farmers by means of individual and mass contact methods using a variety of media (see also Chapter 23).One-on-one consultation often is cited as most effective,[4] but time and financial constraints inevitably limit direct contact. Production meetings, videos, field clinics, demonstrations, and other group presentation methods supplement individual contact; so do printed materials such as pest status newsletters, IPM bulletins and circulars, scouting manuals and handbooks, and articles in farm magazines and trade journals. Content ranges from straightforward pest situation updates to in-depth instruction. Electronic mass media approaches move scouting information quickly among IPM advisors and growers; methods include recorded phone messages and hot lines, radio and television spots, and videotext services. These mass media methods also are useful in cultivating general public awareness of IPM programs. Computers are increasingly important both as tools for delivering scouting reports and as decision support aids.

Although extension produces much of the scouting and IPM training materials used in the U.S., none of these information delivery methods are the exclusive domain of public IPM advisors; all too are used by advisors in the private sector. Consultants write newsletters, sponsor grower educational meetings, and produce computer software that assists IPM scouting;[77] the agrichemical industry also publishes pest scouting instructions and other educational materials.[78] The most up-to-date listing of extension IPM materials currently available in the U.S. is compiled in PMMDb, the *Pest Management Materials Database*,[79] a database software package for DOS and Macintosh personal computers. PMMDb lists over 1550 print, audio-visual, software, and mass electronic IPM educational materials that can be custom searched by media type, crop or commodity, and other keywords.

A. SCOUTING FORMS AND DECISION AIDS

A distinction needs to be made between scouting forms and IPM decision aids. Both seek to deliver information as simply and concisely as possible. However, the former merely summarize the results of field scouting, while the latter additionally interpret the data and so assist with pest management decision making.[80] Scouting forms typically take the form of a tabular record keeping sheets. In addition to pest species incidence or abundance, information recorded on scouting forms usually includes scouting date and time, grower identification and field location, general crop growth condition and plant stage, and observations on weather conditions. Report forms often are appended with hand-drawn field maps that pinpoint the location of pest infestations. Data interpretation can be enhanced by designing scouting sheets as fill-in-the-blank graphs that more immediately depict pest seasonal phenology than tabular sheets. "Barn-charts" printed on heavy cardstock are useful for posting scouting results at remote field sites. Spencer and Jacobsen[81] describe and contrast scouting and related record-keeping forms currently used in the U.S. for integrated pest and crop management.

Seasonal scouting forms often are supplemented with general information about field history. Several states (e.g., Cooksey et al.[82]) have developed printed record-keeping systems in three-ring notebook format, complete with fill-in-the-blank field maps, scouting calendars, pesticide application sheets and similar recording forms. Others (e.g., Geiger[83]) have developed stand-alone database management software for personal computers that use menu-driven data entry screens to prompt the user for the appropriate data. Computerized systems offer the particular advantage of rapid retrieval and summarization of data through custom searches and reports. Some work also has been done on replacing printed scouting forms with portable personal computers.[84] Legg and Bennett[85] describe a "mobile workstation system" for Russian wheat aphid scouting that uses a battery-powered portable computer, printer, and custom software for data entry, report generation, and decision making. Compared with printed forms, the computerized system saved little field scouting time, but was more efficient in generating reports and creating a historical data base of scouting records.

Decision aids go beyond scouting forms by incorporating pictorial graphs, charts, or worksheets to organize and interpret data. Their purpose is to give pest management recommendations, based either upon formal algorithms and data from controlled experiments or by using informal rules of thumb based on the subjective experience of experts in the field. Sequential sampling plans (Chapter 8) in the format of "decision cards" are a familiar example of a decision aid. A promising decision aid that has seen some application among plant pathologists (e.g., see Cook and Veseth[86] pp. 126 to 127), but only limited use by entomologists (e.g., Webb et al.[87]), is the risk-assessment scheme. Formatted as score cards or tables, these schemes assign variable numeric points to environmental, agronomic, and economic factors such as cultivar susceptibility, current plant growth stage, crop yield potential, fertility management, and short-term weather forecast which together determine whether or not pest management action is needed. The sum of these points expresses as a single index the combined effect of each factor on pest status.

A crude example is a score card for management of green peach aphid, *Myzus persicae* (Sulzer) (Homoptera: Aphididae) developed by Bechinski.[88] The card combines scouting data with five modifying factors which together determine if potato fields should be sprayed with an insecticide to prevent aphids from transmitting leafroll virus. Tubers with the virus develop an internal rot termed "net necrosis" which renders them unmarketable. High scores are given to modifying factors that decrease the probability of green peach aphid infestations (e.g., previous application of a soil-applied, systematically transported insecticide or presence of natural enemies). High scores likewise are given to factors that either decrease the severity of disease expression in tubers (e.g., fields where farmers will kill potato vines preparatory to harvest within 21 d or older-than-normal plants at time of aphid infestation) or that reduce the probability of tubers exhibiting rot symptoms before they are marketed or processed (e.g., potatoes intended for the fresh market vs. long-term storage). The total numeric score, taken with field scouting data, determines whether an aphid infestation justifies management action. For instance, a total score of $+2$ or more calls for control if pest density prior to August 1 exceeds 40 aphids per 10 leaves, whereas a score of -2 warrants control if density exceeds 2.5 aphids per 10 leaves.

Unlike sequential sampling, there are no authoritative formulas or standard analytical methods for developing risk assessment schemes. In the foregoing example, the relative importance of each modifying factor and assignment of point ranges was determined subjectively based upon the 20-year experience of a veteran extension entomologist. Quantitative approaches using multivariate statistical methods and Bayesian decision theory[89] deserve attention; the approaches used by Watt[90] and Griffiths and Holt[91] to develop graphical pay-off matrices for grain aphid control are potential starting points. Bechinski and Hescock[13] describe an alternative quantitative approach derived from the *t*-statistic that may prove a useful approach for developing risk assessment tables. In addition, decision aid design requires more than quantitative skills or technical competence; attention too should be given to appropriate graphical design as an aid for communicating and interpreting scouting data. Tufte[92] is a useful introductory reference to aesthetic theory and design technique.

B. COMPUTER SOFTWARE AND SCOUTING

Application of computer technology to scouting data is evolving so rapidly that the nonspecialist quickly can be left behind in the nomenclatural dust of systems already passe to the gurus of computer science. What needs to be remembered is that this technology indeed is but a tool for repackaging and delivering information. "Repackaging" varies in complexity from simple data processing to complex decision making which integrates and interprets the information needed to solve pest management problems.[93] Functionally, computer applications for IPM scouting take the form of software that runs independently on personal computers or *on-line* systems which link the user to a centrally located computer. Although the distinction is not without exception, on-line systems act as electronic newsletters that deliver current pest status reports, pest forecasts and management advice. In contrast, stand-alone software commonly is commonly used either to manage scouting data or as computerized decision aids.

The on-line concept originated with Tummala et al.,[94] who coined the term for systems that use scouting data and local weather information to update computer models which predict pest status and immediately issue pest management advice. The Michigan PMEX system (*P*est *M*anagement *EX*ecutive system) was first to put this concept into practice.[95-96] PMEX provided apple growers and scouts with current information on regional pest and weather conditions as well as forecasts of pest phenology by combining a database of historical information on pest occurrence (PESTAPP, *PEST APP*lication) with a degree-day model (*PETE, P*est *E*xtension *T*iming *E*stimator). Forecasts included expected dates of pest emergence, occurrence of pest life stages, and predicted economic status, which growers and IPM advisors used both to schedule scouting programs and to plan management strategies.

Today, on-line generically is applied to any system that delivers information about current or real conditions. These include electronic mail networks such as Cornell University's CENET system (formerly called SCAMP), which links campus-based extension and research faculty with their field-based colleagues and is used to disseminate weekly reports about statewide pest conditions for fruit, vegetables, field crops, ornamentals and turf, and homes and grounds. Other systems, such as the University of California IMPACT program (*I*ntegrated *M*anagement of *P*roduction in *A*griculture using *C*omputer *T*echnology), extend the on-line concept to the public at large. Originally designed for extension in 1980, but opened to the public in 1986, IMPACT serves over 400 individuals and organizations outside the university. Scouting data from insect traps can be entered, analyzed, and graphed; degree-days can be calculated either from data stored in IMPACT's weather files or from data supplied by the user. A database of more than 70 phenological models (most of them describing insect development) also is included, as is a database of pest management guidelines.[97]

Electronic bulletin board systems (EBBS) organize information into "bulletins" (text files) that can be viewed and "downloaded" (electronically transferred) to the user's own personal computer. Depending on content, some bulletins remain on the system for a prescribed time and then are deleted; others regularly are updated. The attraction of bulletin boards is ease-of-access; anyone with a personal computer, modem, and communications software can connect by telephone line anytime, anywhere. Another attraction is ease of implementation. The software needed to build and operate a bulletin board can be purchased off the shelf from any number of commercial firms. Some bulletin boards specialize in pest management information, such as the University of Massachusetts' INFONET, which contains IPM information about tree and small fruits, turf and landscape, greenhouses, cranberries, biological control, and vertebrates. Other systems, such as the University of Missouri's AgEBB system (*A*gricultural *E*lectronic *B*ulletin *B*oard[98]), include a broader array of pest, market, and other current reports as well as downloadable software.

Although bulletin boards allow users to read and save information without restriction, data quality and integrity are safeguarded by only allowing certain authorized persons (such as staff from extension or state departments of agriculture) to add information to the system. Most systems do provide electronic mail services that allow any user to post messages or notes to other users. To prevent any one person from monopolizing access, each is allocated a maximum amount of daily connect time; 60 min is typical. Demand for some bulletin boards is so high to justify connect-time fees that both allocate time among users while providing operating dollars; RCEBBS (Rutgers Cooperative Extension Bulletin Board System[99]), which handles 10,000 to 15,000 calls annually, imposes a $15 annual subscription fee for 60 min of daily access.

Stand-alone software for managing scouting data takes several formats. One is custom templates for commercial programs, such as the Sorghum Pest Management Report Database System developed by Texas A & M University, which allows users to enter and store scouting results to databases created in DBase format. A more common alternative is software written in a programming language and compiled to run independently on personal computers. The University of Georgia FACET software series is written in BASIC and prompts users via menu-driven screens to enter data directly from field scouting forms. Built-in edit and retrieval functions generate reports for specific fields as needed. Complimentary graphing and mapping software (GRAPHIX and TRAPMAP) can be used to display data from a regional pheromone trap network; here regional pest data are maintained in a file updated weekly on a mainframe computer. Again, the distinction with on-line programs is not clear-cut; some programs for managing scouting data are available both on-line and stand-alone.

Like printed IPM decision aids, decision-aid software systems for personal computers interpret scouting information and issue management advice by using formal, quantitative algorithms or by using informal rules of thumb and experience. Two fundamentally different technological approaches are illustrated by TEXCIM,[100] a *decision support system* which combines databases with models to analyze and solve problems, and EASY-MACS,[101] an *expert or knowledge-based system* which applies technology from artificial intelligence to mimic the reasoning and problem-solving methods of a human expert.

TEXCIM is an interactive software package that uses scouting data to predict the economic benefits of pest control in cotton fields. Written in FORTRAN and compiled to run on IBM and compatible personal computers, TEXCIM links a computer model for cotton plant growth with models that forecast pest phenology and abundance with still other models that assess the impact of beneficial natural predators. Critical to the accuracy of model predictions is real-time data from field scouting. Current field counts of the cotton fleahopper (*Pseudatomoscelis seriatus* [Reuter] [Hemiptera: Mididae]), boll weevil (*Anthonomus grandis* Boheman [Coleoptera: Curculionidae]), bollworm-budworm complex (*Heliothis zea* [Boddie] and *H. virescens* [F.] [Lepidoptera: Noctuidae]), and ten predatory spiders, ants, hemipterans, lady beetles, and lacewings are among the input data that drive the insect models. Counts of adult weevils and bollworm-budworms from early-season surveys with pheromone traps are used to forecast the timing of regional population peaks and pest occurrence and density within individual cotton fields. Depending on the computer hardware configuration, a single run may require over an hour on an 8088 machine to less than 3 min on a 386 computer. In addition to using current data to schedule future scouting activities and to project crop yield losses, TEXCIM has the unique capability of estimating the current economic value of the predator complex. Another unique feature is the capability of forecasting sample size required for a specified level of precision. Here predicted pest density directly is used in lieu of preliminary sampling that otherwise would be required to compute optimal sample size. Finally, if real-time scouting data are not available, TEXCIM supplies databases of default values based on historical pest occurrence; and, though some users may be tempted to entirely forego field scouting and instead use default values, Hartstack et al.[100] issue this warning: "Do not think that by using a computer model, field scouting can be avoided. When using TEXCIM40, the need for reliable information increases." The same is true for other real-time information, particularly current weather data for degree-day forecasts. Here TEXCIM users either must use alternative on-line services (such as the Texas BUGNET computer bulletin board) or they personally must monitor and record the data needed.

EASY-MACS (*E*xpert *A*dvisory *S*ystem for *M*anaging *A*pple *C*ropping *S*ystems) too is a highly interactive, self-contained collection of computer programs ("modules") and databases that interprets scouting data and assists with pest management decisions. However, rather than using simulation models to forecast future pest densities, EASY-MACS instead depends on a set of rules representing the consensus knowledge and experience of IPM specialists about how best to interpret pest scouting data and prescribe management action. Exactly how expert systems make inferences and formulate advice from a knowledge-base of rules is unimportant here; consult Coulson and Sanders[93] for a general introduction and Plant and Stone[102] for a detailed discussion; Carrascaland Pau[103] comparatively survey agricultural expert systems available as of 1990. The salient distinction is that expert systems duplicate the wisdom and problem-solving abilities of experts in the field. Plant and Stone[102] further suggest that because knowledge-based systems attempt to emulate the process of consulting with a human expert, they inherently are easier to use than model-based systems, which ironically can be so powerful that they intimidate users with their complexity. A number of hybrid *expert support systems* combine the attributes of both decision support systems and expert systems. GOSSYM/COMAX is an exemplary expert support system used by commercial cotton producers in the U.S.[104]

EASY-MACS uses fill-in-the-blank on-screen queries and pop-up menus to coach user through the process of collecting, entering, editing and managing the appropriate scouting data as well as information about pesticide use, weather, and other supporting data. A recommendation system uses these data and the set of rules distilled from the human experts to identify a ranked list of insecticides that best maximizes effectiveness against the target pest while minimizing harm to beneficial natural enemies. An accompanying scouting manual[105] combines do-it-yourself sequential sampling forms with simplified instructions that allow even inexperienced growers to sample an orchard in 10 to 15 min.[106] For apple growers who wish to further reduce sampling effort, EASY-MACS can assess scouting needs within each orchard block (i.e., homogeneous tree groups or subdivisions within the orchard). Querying users to respond to a series of questions based on prior pesticide use and block history, a risk-assessment function identifies those blocks that are at low, intermediate, or high risk of pest infestation. Scouting recommendations are issued only for intermediate risk blocks. High risk blocks are presumed to be above threshold levels and control recommendations are given without scouting; low risk blocks require neither scouting nor treatment. A particularly useful feature of expert systems is the ability to explain the rationale for any advice given. For instance, EASY-MACS can tell users the reasoning the system used to assign pest risk categories.

VII. EPILOGUE: A SCOUTING CALL-TO-ARMS

The philosophy, if not the practice, of integrated pest management in the U.S. is shifting away from one of "prescriptive control" predominated by pesticide use as prescribed by scouting and thresholds to a "biologically intensive IPM"[107] that relegates pesticides to a tactic of last resort and instead relies on biological controls, host resistance, and cultural methods. In general, these biointensive alternatives either must be used preventively before infestations develop, or, if they can be used remedially after infestations occur, their control action tends to be slower or less complete than conventional chemical insecticides. An implication for future scouting programs is a greater-than-ever requirement for highly detailed data that will allow producers to better anticipate problems so that pest populations do not increase to the point where the only option is control with quick-acting conventional insecticides.[108] Another issue that will drive development of scouting programs is "precision farming", or adjusting production inputs to precisely match the landscape mosaic so as to minimize runoff and leaching of agrichemicals. Already "variable-input" fertilizer and herbicide application equipment is available (e.g., Lor-Al Soilection Systems, Box 265, Benson, MN 56215) which can apply varying rates and blends of chemicals on-the-go by means of on-board computers that integrate data from digitized infrared aerial photographs, soil survey maps and nutrient test result. Compared with the typical data now generated from IPM scouting programs, orders of magnitude increases in the spatial resolution of sampling data will be required to adopt these smart sprayer technologies to insect management.

Can scouting programs be developed that provide the requisite fine-grained data while still remaining uncomplicated enough to be adopted on the farm? That has been the central question here all along. In the final analysis, continued achievement of integrated pest management in no small part depends on taking up this scouting call-to-arms that challenges researchers and pest advisors to reduce the complicated to the simple.

REFERENCES

1. Anon., *The IPM Trap: Managing Pesticides Instead of Pests*, Greenpeace, Washington, DC, 1992, 3.
2. Hoy, M. A., Integrating biological control into agricultural IPM systems: reordering priorities, *Proc. Nat. Integrated Pest Management Symp./Workshop*, Las Vegas, April 25–28, 1989, 41.
3. Rajotte, E. G., Kazmierczak, R. F., Lambur, M. T., Norton, G. W., and Allen, W. A., The National Evaluation of Extension's IntegratedPest Management Programs, *Virginia Cooperative Extension Service Publication 491-010*, Virginia Polytechnic Institute and State University, Blacksburg, Virginia, 1987, 123.
4. Wearing, C. H., Evaluating the IPM implementation process, *Annu. Rev. Entomol.*, 33, 17, 1988.
5. Zalom, F. G. and Fry, W. E., Eds., *Food, Crop Pests and the Environment*, APS Press, St. Paul, 1992, 179.
6. Glass, E. H., Coordinator, *Integrated Pest Management: Rationale, Potential, Needs and Implementation*, Entomological Society of America Special Publication 75-2, 1975, 141.
7. Glass, E. H., Constraints to the implementation and adoption of IPM, in *Food, Crop Pests and the Environment*, Zalom, F. G. and Fry, W. E., Eds., APS Press, St. Paul, 1992, chap. 6.
8. Frisbie, R. E., Critical issues facing IPM technology transfer, *Proc. of the Nat. Integrated Pest Management Symp./Workshop*, Las Vegas, April 25–28, 1989, 157.
9. Frisbie, R. E. and McWhorter, G. M., Implementing a statewide pest management program for Texas, U.S.A., in *Advisory Work in Crop and Disease Management*, Palti, J. and Ausher, R., Eds., Springer-Verlag, Berlin, 1986, chap. 3.6.
10. Binns, M. R. and Nyrop, J. P., Sampling insect populations for the purpose of IPM decision-making, *Annu. Rev. Entomol.*, 37, 427, 1992.
11. Ives, P. M. and Moon, R. D., Sampling theory and protocol for insects, in *Crop Loss Assessment and Pest Management*, Teng, P. S., Ed., APS Press, St. Paul, 1987, chap. 6.
12. Nyrop, J. P., Reissig, W. H., Agnello, A. M., and Kovach, J., Development and evaluation of a control decision rule for first-generation spotted tentiform leafminer (Lepidoptera: Gracillariidae) in New York apple orchards, *Environ. Entomol.*, 19, 1624, 1990.
13. Bechinski, E. J. and Hescock, R., Fixed-time classification and detection plans for alfalfa snout beetle (Coleoptera: Curculionidae), *J. Econ. Entomol.*, 84, 1388, 1991.

14. Liebhold, A., Twardus, D., and Buonaccorsi, J., Evaluation of the timed-walk method of estimating gypsy moth (Lepidoptera: Lymantriidae) egg mass densities, *J. Econ. Entomol.*, 84, 1774, 1991.
15. Miller, G. R. and Coultas, J. S., Grower needs—IPM, *Proc. of the Nat. Extension Workshop on Organizing Grower IPM Organizations*, Kansas City, MO, November 11–13, 1980, 7.
16. Bechinski, E.J., Sugarbeet root maggot: life cycle, damage and options, *Proc. of the 1986 University of Idaho Cooperative Extension System Winter Commodity Schools*, 18, 232, 1986.
17. Palti, J. and Ausher, R., Crop value, economic damage thresholds, and treatment thresholds, in *Advisory Work in Crop and Disease Management*, Palti, J. and Ausher, R., Eds., Springer-Verlag, Berlin, 1986, chap. 2.3.
18. Mumford, J. D., Farmer's perceptions and crop protection decision making, in *Decision Making in the Practice of Crop Protection*, Austin, R. B., Ed., Monograph 25, British Crop Protection Council, Croydon, England, 13, 1982.
18. Bennett, C. F., Analyzing impacts of extension programs, *Publication ESC-575*, Extension Service, U.S. Department of Agriculture, Washington, DC, 1977, 21.
20. Cole, J. M. and Cole, M. F., *Advisory Councils: A Theoretical and Practical Guide for Program Planners*, Prentice-Hall, Englewood Cliffs, NJ, 1983, 207.
21. Griffith, W. S., Learning theory, in *Extension Handbook*, Blackburn D. J., Ed., University of Guelph, Ontario, 1984, chap. 2.
22. Prawl, W., Medlin, R., and Gross, J., Adult and Continuing Education Through the Cooperative Extension Service, *Publication UED 76*, Extension Division, University of Missouri, Columbia, 1984, 279.
23. Whale, B., Motivation, in *Extension Handbook*, Blackburn, D. J., Ed., University of Guelph, Ontario, 1984, chap. 3.
24. Musser, W.N., Wetzstein, M. E., Reece, S. Y., Varca, P. E., Edwards, D. M., and Douce, G. K., Beliefs of farmers and adoption of integrated pest management, *Agric. Econ. Res.*, 38, 34, 1986.
25. Lamble, W., Diffusion and adoption of innovations, in *Extension Handbook*, Blackburn, D. J., Ed., University of Guelph, Ontario, 1984, chap. 4.
26. Rogers, E. M., *Diffusion of Innovations*, The Free Press, New York, 1983, 453.
27. Grieshop, J. I., Zalom, F. G., and Miyao, G., Adoption and diffusion of integrated pest management innovations in agriculture, *Bull. Entomol. Soc. Am.*, 34, 72, 1988.
28. USDA-ES, The Cooperative Extension System, Publication PA 1412, Extension Service, U.S. Department of Agriculture, Washington, DC, 1989, 16.
29. Blair, B. D. and Edwards, C. R., Development and status of extension integrated pest management programs in the United States, *Bull. Entomol. Soc. Am.*, 26, 363, 1980.
30. Allen, W. A., *A Manual for Implementing an Insect Pest Management Program for Soybeans*, Virginia Cooperative Extension Service, Virginia Polytechnic Institute and State University, Blacksburg, VA, 1978, 11.
31. Bessin, R., Hershman, D., Hartman, J., Brown, J., Strang, J., Jones, T., and Johnson, D., *Kentucky Apple Management Program Scout Manual*, University of Kentucky IPM Program, Princeton, KY, 1978, 11.
32. Blair, B. D., Extension Pest Management Pilot Projects: an Evaluation, Publication ESC 579, Extension Service, U.S. Department of Agriculture, Washington, DC, 1976, 34.
33. Frolich, D. R., McNeal, C. D., Baird, C. R., and Bitner, R. M., Consultant's Guide to Integrated Pest Management in Alfalfa Seed, Miscellaneous Series No. 59, Agricultural Experiment Station, University of Idaho, Moscow, 1980, 11.
34. Baird, C. R., Berg, J., Bitner, R. M., Fisher, G., Johansen, C., Johnson, D. A., Kish, L. P., Krall, J., Lauderdale, R., Mayer, D. F., O'Bannon, J. H., Parker, R., Peaden, R., Rincker, C. M., Stephen, W. P., and Undurraga, J., Alfalfa Seed Production and Pest Management, Western Regional Extension Publication 12, Cooperative Extension, Washington State University, Pullman, 1991.
35. Anderson, J. and Fitzner, M., Co-chairs, Research and extension/education constraints to adoption of integrated pest management strategies, in *Discussion Papers of the Constraints Resolution Teams*, National Integrated Pest Management Forum, Washington, DC, June 1992, Part IV.
36. Fitzner, M. S., Can integrated pest management survive sustainable agriculture, in *18th Annu. Illinois Crop Protection Workshop Proc.*, Gray, M. E. and Ostroski, M. E., Eds., University of Illinois, Champaign, 1992, 23.
37. Bradshaw, D. E., Inclinations on IPM: Leaning Toward Private Enterprise, *Agrichem. Age*, November, 15, 1990.
38. Hart, A. T., *National Agricultural Chemicals Association Policy Statement on Pest Management with Supporting Papers*, NACA, Washington, DC, 1978, 53.
39. Shaw, D. L., Notice ACP-210, USDA-ASCS, Washington, DC, November 3, 1989.

40. Shaw, D. L., Notice ACP-231, USDA-ASCS, Washington, DC, August 20, 1990.
41. Anon., *National Agricultural Pest Information System*, USDA-APHIS-PPQ, Hyattsville, MD, 1987, 8.
42. McNeal, C. D., Why CAPS? Because of those of us who use the data!, *CAPS Comment.*, 7, 1, 1990.
43. Anon., Cooperative Agricultural Pest Survey, Presentation to CAPS 1992 National Meeting, USDA-APHIS-PPQ Evaluation Services, Hyattsville, MD, 1992, Section 5.
44. Volgelsang, D. L., Local Cooperatives in Integrated Pest Management, Report 37, Farmer Cooperative Service, U.S. Department of Agriculture, Washington, DC, 1977, 44.
45. Good, J. M., Hepp, R. E., Mohn, P. O., and Vogelsang, D. L., Establishing and Operating Grower-Owned Organizations for Integrated Pest Management, Publication PA 1180, Extension Service, U.S. Department of Agriculture, Washington, DC, 1977, 25.
46. Pruss, J., Musser, T., and Calvin, D., 1991 Nationwide Crop Management Association Telephone Survey [unpublished draft], Department of Agronomy, The Pennsylvania State University, University Park, 1991, 9.
47. Anderson, R., Texas pest management association producers implementing IPM in Texas, *Proc. of the Nat. Extension Workshop on Organizing Grower IPM Organizations*, Kansas City, MO, November 11–13, 1980, 79.
48. Frisbie, R. E., Crawford, J. L., Bonner, C. M., and Zalon, F. G., Implementing IPM in cotton, in *Integrated Pest Management Systems and Cotton Production*, Frisbie, R. E., El-Zik, K. M., and Wilson, L. T., Eds., Wiley-Interscience, New York, 1989, chap. 13.
49. Zalom, F. G., Ford, R. E., Frisbie, R. E., Edwards, C. R., and Tette, J. P., Integrated pest management: addressing the economic and environmental issues of contemporary agriculture, in *Food, Crop Pests and the Environment*, Zalom, F. G. and Fry, W. E., Eds., APS Press, St. Paul, 1992, chap. 1.
50. Lambur, M. T., Kazmierczak, R. F., Jr., and Rajotte, E. G., Analysis of private consulting firms in integrated pest management, *Bull. Entomol. Soc. Am.*, 35, 5, 1979.
51. Osgood, B. and Murtagh, T., Institutional Constraints: discussion paper, in *Discussion Papers of the Constraints Resolution Teams*, National Integrated Pest Management Forum, Washington, DC, June 1992, Part I.
52. Ferguson, J., Certification: progress, but some confusion, *Ag Consultant*, 48(7), 1, July 1992.
53. Blair, B. D., Dissemination of pest management information in the Midwest, U.S.A., in *Advisory Work in Crop nd Disease Management*, Palti, J. and Ausher, R., Eds., Springer-Verlag, Berlin, 1986, chap. 3.5.1.
54. Kimbrough, J. M., Services of professional agricultural consultants, *Proc. of the Nat. Extension Workshop on Organizing Grower IPM Organizations*, Kansas City, MO, November 11–13, 1980, 15.
55. Weddle, P. W., Privatization of IPM: problems and promise, *Proc. of the Nat. Integrated Pest Management Symp./Workshop*, Las Vegas, April 25–28, 1989, 166.
56. Allen, W. A. and Rajotte, E. G., The changing role of extension entomology in the IPM era, *Annu. Rev. Entomol.*, 35, 379, 1990.
57. DeWitt, J., Can you add an IPM service to your business?, Cooperative Extension Service Publication CE-1720r, Iowa State University, Ames, 1982, 2.
58. Blazquez, C. H., Hochberg, R., and Edelbaum, G., Remote sensing by aerial infrared colour photography as in aid in monitoring crops for pests and diseases, in *Advisory Work in Crop and Disease Management*, Palti, J. and Ausher, R., Eds., Springer-Verlag, Berlin, 1986, chap. 2.4.5.
59. Riley, J. R., Remote sensing in entomology, *Annu. Rev. Entomol.*, 34, 247, 1989.
60. Eshel, J. and Palti, J., Pest and disease control advice given by representative of pesticide manufacturers, in *Advisory Work in Crop and Disease Management*, Palti, J. and Ausher, R., Eds., Springer-Verlag, Berlin, 1986, chap. 2.6.2.
61. Wallace, M., The national coalition on integrated pest management, *Proc. of the Nat. Extension IPM Conference*, Washington DC, November 14–16, 1989, 78.
62. Southwood, T. R. E., *Ecological Methods with Particular Reference to the Study of Insect Populations*, Chapman and Hall, London, 1978, 524.
63. Norton, G. A., Crop protection decision making—an overview, in *Decision Making in the Practice of Crop Protection*, Austin, R. B., Ed., Monograph 25, British Crop Protection Council, Croydon, England, 3, 1982.
64. USDA-APHIS, *Exotic Pest Detection Survey Guidelines*, Animal and Plant Health Inspection Service, U.S. Department of Agriculture, Hyattsville, MD, 1989, 19.
65. Frisbie, R. E., Hardee, D. D., and Wilson, L. T., Biologically intensive pest management: future choices for cotton, in *Food, Crop Pests and the Environment*, Zalom, F. G. and Fry, W. E., Eds., APS Press, St. Paul, 1992, chap. 3.
66. Blickenstaff, C. C., History and Biology of the Western Bean Cutworm in Southern Idaho, 1942–1977, *Univ. Idaho Agric. Exp. Stn. Bull.*, 592, 1979, 23.

67. Mahrt, G. G., Stoltz, R. L., Blickenstaff, C. C., and Holtzer, T. O., Comparisons between blacklight and pheromone traps for monitoring the western bean cutworm (Lepidoptera: Noctuidae) in south central Idaho, *J. Econ. Entomol.*, 80, 242, 1987.
68. Robert, Y., Dedryver, C. A., and Pierre, J. S., Sampling techniques, in *Aphids: Their Biology, Natural Enemies and Control*, Vol. 2B, Minks, A. K. and Harrewign, P., Eds., Elsevier, Amsterdam, chap. 8.1.
69. Tatchell, G. M., Monitoring and forecasting aphid problems, in *Aphid-Plant Interactions: Populations to Molecules*, Peters, D. C., Webster, J. A., and Chlouber, C. S., Eds., Publication MP-32, Agricultural Experiment Station, Oklahoma State University, Stillwater, OK, 1990, 215.
70. Woiwod, I. P., Tatchell, G. M., and Barrett, A. M., A system for the rapid collection, analysis and dissemination of aphid-monitoring data from suction traps, *Crop Prot.*, 3, 273, 1984.
71. Allison, D. and Pike, K. S., An inexpensive suction trap and its use in an aphid monitoring network, *J. Agric. Entomol.*, 5, 103, 1988.
72. Halbert, S., Connelly, J., and Sandvol, L., Suction trapping of aphids in western North America (emphasis on Idaho), *Acta Phytopathol. Entomol. Hung.*, 25, 411, 1990.
73. Halbert, S., Elberson, L., and Johnson, J., Suction trapping of Russian wheat aphid: what do the numbers mean? *Proc. of the Fifth Russian Wheat Aphid Conf.*, Fort Worth, Texas, January 26–28, 1992, 282.
74. Halbert, S., Mowry, T., Sandvol, L., and Quisenberry, S., *Pacific Northwest Aphid Flyer*, 7(1), 1992, 2.
75. Bechinski, E. J., Trent, A., and Molnau, M. P., On-line IPM and agricultural information in Idaho, in *Computers in Agricultural Extension Programs*, Watson, D. G., Zazueta, F. S., and Bottcher, A. B., Eds., ASAE Publication 1-92, 1992, 601.
76. Niskanen, K. W., IDEX, Getting Started with the University of Idaho Extension Network, Publication MCUG-50, College of Agriculture, University of Idaho, Moscow, 1992, 15.
77. Easton, D. T. and Aschelman, R. E., Pesticide loss software, *Ag Consultant*, 47, 1991, 18.
78. Anon., *Mocap Potato Institute*, Rhome-Poulenc Ag Company, Research Triangle Park, NC, 1991, 16.
79. Geiger, C. R., *PMMDb Pest Management Materials Database User's Guide*, Purdue Pest Management Program, Purdue University, West Lafayette, IN, 1990, 53.
80. Seem, R. C. and Russo, J. M., Simple decision aids for practical control of pests, *Plant Dis.*, 68, 656, 1984.
81. Spencer, L. and Jacobsen, J. S., Record Keeping Systems for Integrated Crop Management: a State-by-State Survey, Extension Publication EB 107, Montana State University, Bozeman, 1992, 38.
82. Cooksey, D., Jacobsen, J., Baldridge, D., Baquet, A., Baringer, J., Bauder, J., Bowman, H., Inglis, D., Jensen, G., Junke, M., Nelson, J., and Riesselman, J., Integrated Crop and Pest Management Record keeping System, Extension Publication EB 12, EB 13, and EB 14, Montana State University, Bozeman, 1988, 120.
83. Geiger, C. R., *Field History Database 2.0 User's Manual*, Purdue Pest Management Program, Purdue University, West Lafayette, IN, 1988, 96.
84. Legg, D. E. and Barney, R. J., Use of portable computers to assess insect populations in advanced integrated pest management programs: alfalfa weevil as an example, *J. Econ. Entomol.*, 81, 995, 1988.
85. Legg, D. E. and Bennett, L. E., A mobile workstation for use in an integrated pest management program on the Russian wheat aphid, *Proc. of the Fifth Russian Wheat Aphid Conf.*, Fort Worth, Texas, January 26–28, 1992, 68.
86. Cook, R. J. and Veseth, R. J., *Wheat Health Management*, APS Press, St. Paul, 1991, 152.
87. Webb, R. E., Ridgway, R. L., Thorpe, K. W., Tatman, K. M., Wiber, A. M., Venables, L., Development of a specialized gypsy moth (Lepidoptera: Lymantriidae) management program for suburban parks, *J. Econ. Entomol.*, 84, 1320, 1991.
88. Bechinski, E. J., Integrated pest management: what it is and how to do it right, *Proc. of the 1992 University of Idaho Cooperative Extension System Winter Commodity Schools*, 24, 104, 1992.
89. Winkler, R. L., *An Introduction to Bayesian Inference and Decision*, Holt, Rinehart & Winston, New York, 1972, 563.
90. Watt, A. D., The influence of forecasting on cereal aphid control strategies, *Crop Prot.*, 2, 417, 1983.
91. Griffiths, E. and Holt, J., Economics of Sitbion avenae (Aphididae) forecasting and control in the U.K., *Crop Prot.*, 5, 238, 1986.
92. Tufte, E. R., *The Visual Display of Quantitative Information*, Graphics Press, Chesire, CT, 1986, 197.
93. Coulson, R. N. and Sanders, M. C., Computer-assisted decision-making as applied to entomology, *Annu. Rev. Entomol.*, 32, 415, 1987.
94. Tummala, R. L., Concept of on-line pest management, in *Modeling for Pest Management*, Tummala, R. L., Haynes, D. L., and Croft, B. A., Eds., Michigan State University, East Lansing, 1976, 28.
95. Battenfield, S. L., *Instruction Manual: Biological Monitoring in Apple Orchards*, Cooperative Extension Service, Michigan State University, East Lansing, 1983, 119.

96. Welch, S., Developments in computer-based IPM extension delivery systems, *Annu. Rev. Entomol.*, 29, 359, 1984.
97. Anon., IMPACT User's Manual, *UC IPM Publication 15*, Statewide IPM Project, University of California, Davis, CA, 1991.
98. Travlos, J., *AgEBB User's Guide*, University Extension, University of Missouri, Columbia, 1990, 34.
99. Anon., *Rutgers Cooperative Extension Bulletin Board Service*, The State University of New Jersey, 1992.
100. Hartstack, A. W., Sterling, W. L., and Dean, D. A., TEXCIM User's Guide Version 4.0, Publication MP-1646, Texas Agricultural Experiment Station, College Station, TX, 1990, 156.
101. Nyrop, J., Coordinator, *EASY-MACS Users Guide*, New York State Agricultural Experiment Station, Cornell University, Geneva, NY, 1991, 40.
102. Plant, R. E. and Stone, N. D., *Knowledge-Based Systems in Agriculture*, McGraw-Hill, New York, 1991, 364.
103. Carrascal, M. J. and Pau, L. F., A survey of expert systems in agriculture and food processing, *AI Appl.*, 6, 27, 1992.
104. McKinion, J. M., Baker, D. N., Whisler, F. D., and Lambert, J. R., Application of the GOSSYM/COMAX system to cotton crop management, *Agric. Syst.*, 31, 55, 1989.
105. Agnello, A., Kovach, J., Nyrop, J., Reissig, H., and Wilcox, W., Simplified Integrated Pest Management Program: A Guide for Apple Sampling Procedures in New York, Cornell Cooperative Extension, Publication IPM 201c, New York State Agricultural Experiment Station, Geneva, NY, 1991, 39.
106. Reissig, H., Seem, R. C., and Nyrop, J., Integration of pesticide use into IPM systems, *Proc. of the Nat. Integrated Pest Management Symp./Workshop*, Las Vegas, April 25–28, 1989, 96.
107. Frisbie, R. E. and Smith, J. W., Jr., Biologically intensive integrated pest management: the future, in *Progress and Perspectives for the 21st Century*, Menn, J. J. and Steinhauer, A. L., Eds., Entomological Society of America Centennial National Symposium, Washington, DC, September 26–27, 1989, 151.
108. Zalom, F. G. and Fry, W. E., Biologically intensive IPM for vegetable crops, in *Food, Crop Pests and the Environment*, Zalom, F. G. and Fry, W. E., Eds., APS Press, St. Paul, 1992, chap. 5.

INDEX

A

Abiotic components, agroecosystem, 18
Absolute estimates, 4–6, 100, 303
Acari: eriophyidae, 402
Acari: phytoseiidae, 403
Acari: tetranychidae, 398
Accuracy, population estimates and, 100–101
Aculus schlechtendali, 402
Adaptive frequency classification monitoring, 262–69
  figures/tables of, 264, 266–69
Adès distribution, dispersion evaluation with, 42
Adoption, improving rate of, 677–78
Adult sampling, 439–45
Aerial net technique, 82
Age structure, 2, 3
Aggregated dispersion, 106–7
Aggregation patterns, 48
Agricultural ecosystems, variability in, 16
Agroecosystem
  definition of, 18
  natural mortality diagram for, 2
  table of characteristics of natural ecosystem vs., 21
Alfalfa
  forage production, insect damage in, 358–61
  sampling and decision making for insect control in, 361–76
  sampling program example, 678–79
  seed production, 358–59
  table of forage production, 361
Amblyseius fallacis, 403
Aphids
  blue alfalfa, 372–73
  cereal
    biological control, 28
    sampling in wheat, 639–44
  green apple, 406–7
  patch associations and, 25
  pea, 369–72
  rosy apple, 405–6
  Russian wheat
    figure/tables of sampling efforts with, 57–59
    sampling in wheat, 644–45
    variance and, 56–59
  spiraea, 406
  spotted alfalfa, 373–74
Aphis pomi, 406
Aphis spiraecola, 406
Apple orchards
  sampling pest and beneficial arthropods of, 384–410
  sampling techniques for, 385–86

Armyworms
  sampling in rice, 649–50
  sampling in wheat and barley, 637
Arthropod acoustic signatures
  drawing of, 95
  indirect assessment and, 93, 95
Arthropod extraction, soil, 86
Arthropod-induced injury, 91, 93
Arthropod products
  indirect assessment and, 93
  photograph of frass for, 94
Autographa californica, 366
Average outgoing quality, 138
  figure of, 139

B

Baermann funnel, 91
Baits, 83
  drawing of, 84
Berlese extraction devices
  drawing of, 89
  heat lamp with, 89
Bias
  definition of, 56–59
  importance of, 59–60
  population estimates and, 100–101
  sample unit selection and, 103, 105
  sources of consistent error, 62–63
Binomial classification systems, 189–204
Binomial count tripartite plans, 260–61
  figure/table of, 261–62
Binomial distribution, 177–78
  time-sequential sampling plan and, 341–42
Binomial estimation procedures, 177–89
Binomial sampling
  enumerative vs., 505–6
  future studies in, 203
  population density, 34
  programs, 176–204
  variable intensity of, 200–203
Biological control, 27–28
  integrated pest management and, 2
  prey-natural enemy ratios in, 246–51
Biological error, 60, 101
Biologically relevant sample unit, 488
Biotic components, agroecosystem, 18
Black rice bugs, sampling in rice, 638
Blacklights
  photograph of visual trap with, 84
  visual trap with, 83, 105
Blister beetles, 367–68
Borers
  European corn

coexistence of, 451–53
economics of sampling and management for, 466–70
sampling to manage, 447–70
stem, sampling in rice, 646–47
Brushing technique, 79
BYDV vectors, migrant, monitoring, 653–55

C

Calendar date, initiation and, 121
California red scale, 422–24
parasitoids of, 427
Canopy light interception, 134
Carabidae, biological control of aphids with, 28
Carbon dioxide, chemical knockdown technique with, 80
Caterpillars
alfalfa, 366
looper, 8
Census, 4, 56, 100, 138
Centrifugation, 91
Chemical knockdown technique, 80–81
Choristoneura rosaceana, 388
Citrus pests
direct cosmetic, 420–21
indirect, 421–22
management of, 418–19
natural enemies, 426–27
representative programs, 422–27
sampling programs for, 419–20
secondary induced, 422
Cochlionmyia hominivorax, corridor movement by, 27
Codling moth, 386–88
Coleoptera: coccinellidae, 404
Coleoptera: curculionidae, 395
Collecting tools, 63
Commercial monitoring, sample size estimation and, 500–504
Computer programs, 150–72, 269–335, 349–52, 689–701
Confidence intervals, figure of, 178
Conotrachelus nenuphar, 395
Corn, field, sampling arthropod pests in, 434–70
Corn earthworm, sampling with, 16
Corn earworm, 366
Corn rootworm
drawing of flotation device for, 90
economics of sampling and management with, 445–47
figures of direct sample of, 76–77
Mexican, 435
northern, 435
patch associations and, 26
sampling to manage, 435–47
southern, 435
western, 435
diagram of fitting a common k for, 44
patch associations and, 25–26

Corridors
agricultural landscapes and, 19–20, 22
associated patches and, 27–28
movement and, 26–27
Costs
estimates, relative vs. absolute and related, 6
initiation and, 134
Cotton agroecosystem
arthropod species in, 478
cotton growth and development, 477–78
damage potential of herbivorous arthropods in, 478–80
tables of damage potential of herbivorous arthropods in, 479–80
Cotton production, future of pest management in, 507–8
Counting, corn rootworm and visual, 441–42
tables of, 442
Counting errors, 62–63
Counting grid, drawing of mite population estimation with, 80
Counts per trap, relative estimates by, 5
Counts per unit effort, relative estimates by, 5
Covering method, 82
Crop field, sampling with, 16
Crop loss assessment, 410
Crop phenology, European corn borer egg deposition and, 453–57
Crop rotation, temporal associations and, 25–26
Cropping systems, sampling programs for integrated pest management in, 3
Curatives, integrated pest management and, 2–3
Curvilinear models, 127–30
Cutworm
army, 365–66
variegated, 365
Cydia pomonella (L.), 386

D

Damaging stage of pest, time-sequential sampling plan for, 347–49
Decision estimation, sample size and, 500–504
Decisions, incorrect, probabilities of, 673–74
Defoliation, photograph of indirect assessment, 94
Defoliators, 362–68
Degree-day, 127–30, 561
Density, 2, 3
absolute, 4, 6, 100, 104
defintion of arthropod, 75
estimation, 2
precision in, 101–2
statistical foundation and assumptions for, 208–14
population, sample mean, 34
population measures, 4
relative, 4
relative estimates and, 5–6
sample units and estimates of mean, 102, 113
Detection sampling, 101
Diabrotica barberi, 435
patch associations and, 26

Diabrotica undecimpunctata howardi, 435
Diabrotica virgifera virgifera, 435
　diagram of fitting a common k for, 44
　patch associations and, 25–26
Diabrotica virgifera zeae, 435
Direct examination, livestock and host, 519–30
Direct observation techniques, 75–78
Direct techniques, sample units and, 77
Dispersion, 2, 3
　figure of examples of patterns of, 35
　mathematical distributions to describe, 36–42
　population, 34–36
　regression techniques for evaluation of, 47–49
　table of discrete distributions for insect, 42
Dispersion patterns, indices for classification of, 42–47
Distance measurements, spatial structure evaluation with, 49–50
Disturbance regimens, agricultural landscape, 19–20
Diuraphis noxia, variance and, 56–59
Dry extraction techniques, 88–89, 91
Dry sieving, 89
Dysaphis plantaginea, 405

E

Ecological studies, sampling framework for, 2
Economic injury levels, 2
　definition of, 74
　European corn borer and, 458–61
　netting techniques and, 81–82
　soybean sampling and, 580–81
　timing of sampling programs and, 120
Ecosystem, mechanized, diagram of natural agroecosystem vs., 22
Ecosystems, landscapes and, 18
Education, specialists in training procedures, 670–72
Efficiency, population estimates and, 100–101
Egg diapause, prolonged, 26
Egg sampling, 438–39
　table of, 440
EIL. *See* Economic injury levels
Electronic technology, arthropod acoustic signatures, 93, 95
Elements. *See* Ecosystems
Elutrification, 91
Emigration, 4
Empirical models, for sequential sampling, 212–15
Empoasca fabae, relative estimates of, 5–6
Enumeration plans, livestock and, 519–26
Enumerative sampling, binomial vs., 505–6
Environmental considerations, sampling and, 121, 134
Errors, type I and II, 673–74
Estimates
　absolute, 4–6, 100, 104
　converting relative to absolute, 6
　related, 4–6
　relative, 4–6, 100, 105
　　sampling units and, 8
　sweep-net, 6, 9
Euseius tularensis, 426–27
Euxoa auxiliaris, 365
Extension, scouting and, 688–90
Extensive programs, 9, 101–3

F

Fall armyworm, 366
Fidelity, population estimates and, 100–101
Field sampling. *See* Scouting
Field validation, 219
Flotation, 90–91
Fluid feeders, 368–74
Foliage- and fruit-feeding arthropods, 609–11
Foliage-feeding complex, 366
Forest carabid beetles, hedgerow network and, 26–27
Frass droppings, population indices and, 7
Fumigation technique, 80–81

G

Geostatistics, spatial structure evaluation with, 51–52
Grain, stored, sampling program example, 679–80
Grains, small
　pest management with, 628
　See also Rice, Wheat
Grasshoppers, absolute estimates of, 6
Green cloverworm eggs, poisson distribution with, 36
Green's index, dispersion classification with, 46
Grower crop management associations, scouting and, 691–92
Growers, scouting and role of, 691
Growing degree-days, 127
Growth, thermal development and, 124–30
Growth form, 3

H

Habitats, arthropods and, 4
Heat technique, 80, 89
Heat units, 127
Hedgerows, arthropod distribution and, 26–27
Helicoverpa zea, 366, 612–14
　sampling with, 16
Hemiptera: miridae, 396
Herbivorous insects, patch characteristics and, 23
Homoptera: aphididae, 405, 406
Homoptera: cicadellidae, 409
Homoptera: diaspididae, 394
Host/pest characteristics, sampling and, 121
Hosts, initiation and, 130–31, 368
Hypera brunbeipennis, 362
Hypera postica [Gyllenhal], figure of direct sample of, 76

I

$I\delta$ index
   dispersion classification with, 45, 47
   figure of quadrat size in, 45
Immigration, 4
Indirect assessment, arthropod, 74, 91–95
   photographs of, 92–93
Industry field staff, scouting and role of, 693–94
Initiation
   based on life history/environment, 121–30, 134
   based on ongoing monitoring, 130–33, 134
   based on pest impact, 133–34
   traps and, 131–32
Insect counts per units of time, 105
Insect egg stages, 4
Insect injury, population indices and, 8
Insecticide use, soybean sampling and, 570
Integrated pest management
   basic premises of, 74
   diagram of sampling role in, 2
   sampling importance in, 100
   sampling role in, 2
   scouting and role of private consultants, 692–93
Integrated systems program, diagram of, 2
Intensity, definition of arthropod, 5, 75
Intensive programs, 9, 101–2, 104
Interpersonal error, 62–63
IPM. *See* Integrated pest management
Irrigation, soybean sampling and, 570
Iwao's Confidence Interval Method, 143–72
   compared with Sequential Probability Ratio Test, 144–46
      figure/table of, 145
   computer programs for, 150–72
Iwao's patchiness, dispersion evaluation with, 47–49, 107

J

Jarring technique, 78–79

K

Keiferia lycopersicella, 614–15
Knockdown techniques
   chemical, 80–81
   drawing of, 82
   ground cloth, 78–79
   photographs of, 78–79
Kono-Sugino equations
   classification methods based on, 195–98
      figures of, 196–97
   figure of, 182, 186
   mean estimation with, 180, 182–86

L

Land surface
   absolute densities and, 4
   population estimates and, 5
   stage-specific survival and, 4
Landscapes
   agricultural, arthropods and characteristics of, 18
   arthropod sampling and characteristics of, 21
   definition of, 16–18
   ecosystems and, 18
   figure of corridor elements in agricultural, 19
   figure of Midwest agricultural, 17
Larval sampling, 436–38, 465–66
   table of, 437
Larval stages, 4
Leaffolders, rice, sampling in rice, 638
Leafhoppers
   patch associations and, 25
   potato, 368–69
   relative estimates of, 5–6
   white apple, 409
Leafminers, 615–17
   apple blotch, 407–9
   spotted tentiform, 407–9
   tentiform, 407–9
   western tentiform, 407–9
Leafrollers
   obliquebanded, 388–89
      figure of sequential sampling plan for, 388–89
   pandemis, 392
Lepidoptera: gracillariidae, 407
Lepidopteran caterpillars, 366
Lepidopteran foliage-feeding complex, 366–67
Lepidoptera: tortricidae, 386, 388, 389, 392
Life tables, absolute measure in, 4
Linear summation, 127–28
Liriomyza sativae, 615–17
Liriomyza trifolii, 615–17
Livestock
   arthropod pests and damage with, 517–18
   direct examination of host, 519–30
   off-host sampling, 530–34
Livestock production systems, 516–18
   sampling plans for, 518–34
Looper, alfalfa, 366
Lygus bugs, 374–76
Lygus lineolaris–Pratensis, 396

M

Maggot, apple, 392–94
Mean crowding, dispersion classification with, 45, 47
Mean crowding regression, dispersion evaluation with, 47–49, 107
Melanoplus differentialis, absolute estimates of, 6
Melanoplus femurrubrum, absolute estimates of, 6
Meloidae, 367
Metopolophium dirhodum, 639
Migration, 2
   corridor movement, 27
   figure of weather pattern during, 123
   initiation and, 122–24

Mites
  apple rust, 402–3
  citrus rust, 425–26
  figure of action thresholds for, 402
  phytoseiid predatory, 403–4
  spider, 618–20
  tetranychid, 398–402
  two-spotted spider, patch associations and, 26
  western predatory, 403–4
Monitoring, sample, 120–21, 130–33
Mortality, 3
Moth, tufted apple bud, 389–92
  figure of damage caused by, 391
Multistage sampling, 215–17

N

Nalepa, 402
Natality, 3
Natural ecosystem, diagram of mechanized agroecosystem vs., 22
Nearest-neighbor approach, figure of, 50
Negative binomial distribution, 38–40, 107, 110–11, 186–87
  time-sequential sampling plan and, 341–42
Negative binomial k, dispersion classification with, 43–44, 47, 107
Negative binomial probability model
  mean estimation with, 186–89
  methods based on, 198–200
  modifications of, 187
Nested analysis of variance tests, 108–9
  table of, 109
Netting techniques, 81–82
  photograph of, 83
Nondamaging stage of pest, time-sequential sampling plan for, 342–47
Nonrandom sampling, 60–63

O

Operating characteristic function, 138
  figure of, 139
Orchard block, sampling with, 16
Ostrinia nubilalis, sampling to manage, 447–70

P

Pandemis pyrusana kearfott, 392
Panonychus ulmi, 398
Papaipema nebris larvae, frequency distribution graph of, 36
Parameter estimation, sample size, 498–500
Patch characteristics, arthropod populations and, 22–24
Patches
  agricultural landscape, 18–21, 23
  diagram of host odor and patch shape, 24
Pattern vs. process, sampling and, 18
Pest impact, initiation based on, 133–34

Pest management, 3
Pest population sampling, 2
Pest populations, cultural practices to maintain, 2
Pesticides
  conventional, 2
  microbial, 2
Phenological model, codling moth and, 387–8
Phyllocoptruta oleivora, 425
Phyllonorycter blancardella, 407
Phyllonorycter crategella, 407
Phyllonorycter elmaella, 407
Pinworm, tomato, 614–15
Plant phenology, European corn borer feeding injury and, 457–61
Plant resistance, integrated pest management and, 2
Planthoppers, brown- and whitebacked, sampling in rice, 645
Plathypena scabra (F.) eggs, poisson distribution with, 36
Plathypena scabra, 366
Platynota idaeusalis, 389
Plum curculio, 395–96
Poisson distribution, 36–38, 107, 110–11
  table of sample calculations for fitting, 38
  time-sequential sampling plan and, 341–42
Population
  density
    binomial sampling, 34
    monitoring through time, 251–69
  dispersion, 106–7
  dynamics, 2–4
  ecology, 2
  estimates
    basic, 5, 100, 105
    types of, 4
  indices, 4, 6–8
  intensity estimates, 5, 100, 105
  movements, 4
  nature of, 3–4
  protective management, 3
  sampling
    major properties estimated through, 3
    sample size estimation in, 498–500
Power law, *See* Taylor's power law
Precision
  communicating about, 674
  definition of, 63–65
  figures of standard error/relative variation, 66, 68
  illustrations of, 64
  importance of, 65–68
  increasing, 65–69
  population estimates and, 100–101
  sample unit selection and, 103
  sampling, 105–6
Presence/absence counts, 56–58, 60–63, 67–68
Presence/absence treatment decision plans, livestock and, 526–30
Preventive tactics, integrated pest management and, 2

Prey-natural enemy ratios, predicting biological control via, 246–51
Probability, role in sampling, 672–74
Probability distributions, 209–12
Process vs. pattern, sampling and, 18
Program sampling, 222–39
Proportion
   classification of, 189–93
      figure of, 190
   sequential estimation of, 178–80
      figures/tables, 179–81
Prunus padus, patch associations and, 25
p_t-to-m relationships
   advantages/disadvantages of binomial sampling using, 176–77
   general classification using, 194–95
Pupal stage, 4
Pyrethrum, chemical knockdown technique with, 80

Q

Quadraspidiotus perniciosus, 394
Quality control, process of, 138

R

Random coordinate selection, 61
Random dispersion, 106–7
Random samples, definition of, 107–8
Random sampling, 60–62
Regression techniques, dispersion evaluation with, 47–49
Relative net precision, 104
Reliability. *See* Precision
Remote sensing, initiation and, 132–34
Reproduction, 2
Rhagoletis pomonella, 392
Rhopalosiphum padi, patch associations and, 25
Rice
   arthropod pests of, 630–33
   arthropod sampling methods, table of, 636
   arthropod sampling methods with, 633–53
Risk, sampling and, 672–73
Root-feeding arthropods, 605–7

S

Sample
   definition of, 8–9
   listing, 61
Sample mean, population density, 34
Sample size estimation, defining objectives of, 497–98
Sample units
   definition of, 8, 100
   direct techniques and, 77
   number of, 109–13
   selection of, 102–4
Sampling
   arthropod pests, overview of, 3
   basic principles of, 672–74
   binomial vs. enumerative, 505–6
   concepts, 8–10
   cost, optimizing reliability of, 504–5
   enumerative vs. binomial, 505–6
   error
      nonrandom sampling, 60–62
      random sampling, 60–62
   integrated pest management and use of, 100
   objectives, 208–9
   pattern/timing of, 107–8
   preliminary, 102–7
   prerequisites, 208
   program
      definition of, 9–10, 100
      objectives/types of, 101–2
      primary, 107–13
      treatment thresholds and design of, 420–22
   protocols, requirements for adoption of, 674–77
   quality control in, 138–40
   sequential. *See* Sequential sampling
   systems, examples of, 678
   table of computer-simulated assessments, 65
   technique, definition of, 9
   techniques, 74–95
      comparison of, 104–6
   universe
      definition of, 9, 102
      stratification of, 108–9
San Jose scale, 394–95
Scouting
   computer programs for, 698–701
   design and delivery of in-the-field programs, 684–702
   information in, 696–701
   integrated pest management decision making and, 684
   public/private implementation of, 688–94
   social theory framework for, 687–88
   types of programs, 694–96
Screwworm, corridor movement by, 27
Seasonal development pattern
   European corn borer and, 448–53
   figures/tables of European corn borer and, 448–53
Sedimentation, 91
Seed production pests, 374–76
Seed yield reduction potential, 457–48
Semivariogram
   figure of, 51
   geostatistics and, 51–52
Sequential classification, cascaded tripartite, 252–60
Sequential classification sampling plans, program for, 269–335
Sequential estimation of a proportion, 178–80
   figures/tables of, 179–81
Sequential Probability Ratio Test, 140–43
   compared with Iwao, 144–46
   computer programs for, 150–72

proportion classification using binomial, figures of, 190–91
table of formulas for slopes and intercepts of plans for, 142
tables of formulas for binomial decision lines, 192–93
three-decision, 146–48
table of, 148
Sequential ratio classification sampling plan, figures of, 247, 250
Sequential sampling
binomial, 176–204
for biological control decision making, 246–336
corn rootworm, 445
for one-time decision making, 139–51
performance/validation of, 217–22
in practice, 148–50
table of corn rootworm, 446
Sequential sampling plan, *See also* Time-sequential sampling plan
Sequential sampling plans, soybean, 581
Sex pheromones. *See* Traps, pheromone
Simulation, 219–22
Social theory, scouting and, 687–88
Soil arthropods, techniques for, 74, 86, 87–91
Soil-dwelling arthropod pests, wheat/rice, 652–53
Soil habitat collection, 88
Soybean
arthropods, 543–45
development/phenology, 542–43
production, 540–42
sampling
absolute, 573–75
density differences and geographic regions/fields, 564–71
density differences and time, 559–63
economics of, 545, 549, 554
fixed-precision, 577
future directions for, 585–88
improving efficiency of, 554–56
population dispersion pattern for, 580
population indices, 577–80
relative, 575–77
techniques/programs for, 571–85
where/when to sample, 556–71
tables of arthropods/natural enemies of, 546–53
Spatial aggregation
extrinsic factors affecting, 484, 486
as a function of density, 487–88
intrinsic factors affecting, 484
probability distribution functions in, 492–95
quantifying, 486–97
sample unit size and perceived, 488–90
sampler efficiency and perceived, 490–92
statistical hierarchy in, 495–97
Spatial associations, patches and, 25–26
Spatial patterns of distribution, 480–86
table of, 484–85
Species population, 3

Split core cutter
drawing of, 88
soil arthropod extraction with, 88
Spodoptea exigua, 612–14
SPRT. *See* Sequential Probability Ratio Test
Stage-specific survival, 4
Staining, selective, 78
Staphalinidae, biological control of aphids with, 28
Stem-feeding arthropods, 607–8
Stethorus punctum, 404–5
Stop limits, adaptive frequency classification monitoring and, 263–68
Strawberry sampling programs, economic, horticultural, and environmental information, 618–20
Sucking technigue, drawing of device for, 81
Sucking technique, 79–80
Sunn pests, sampling in wheat, 637–38
Suppressants, 2
Surface and above-ground arthropods, techniques for, 74–86
Surveys, initiation and, 132
Sweep net technique, 6, 9, 82, 102–5, 650–51

T

Tadpole shrimps, sampling in rice, 649
Target species, detection of, 101
Tarnished plant bug, 396–98
Taylor's power law
dispersion evaluation with, 48–49, 107, 110–11, 113
figures of sample unit numbers, 112
Teaching sampling principles, 670–80
Temperature
figure of rate summation effect at low, 127
initiation and, 124–27
Temporal associations, patches and, 25–26
Termites, drawing of acoustic frequency of mandibular movements of, 95
Tetranychus McDanieli, 398
Tetranychus urticae, 398, 618–20
Tetraynachus urticae, patch associations and, 26
Therapeutics, integrated pest management and, 2
Thermal development
estimates used in sampling, 129–30
estimation of, 127–29
figure of summation effect at low temperatures and rate of, 127
figure of thermal development curve, 125
initiation and, 124–30
limitations with, 129
Thermal units, 127
Thresholds
ARM, 403
codling moth and, 388
economic, 2, 101, 458, 580–83
ERM, 401
RAA, 405–6

tally, 187–88
  figures/tables of, 189, 194, 196, 198–200
  treatment, sampling program design and, 420–22
Thrips, citrus, 424–25
Time-sequential sampling
  concept of, 338
  development of, 338–42
  use of, 342
Time-sequential sampling plan
  advantages/disadvantages of, 349
  applications of, 342–49
  figures/tables of, 343–48
  program for, 349–52
Time-sequential sampling plans, soybean, 581
Time/space comparisons, relative estimates for, 6
Timing
  sample, 120–21, 134
  See also Initiation
Tomatoes
  arthropod pests of, 612–18
  economic, horticultural, and environmental information on, 611–12
  table of arthropod pests of, 613
Trapping techniques, 82–86
Traps
  aerial suction, 653–55
  bait, 131
  blacklight, figures of, 452
  boxed animal, 131
  carbon dioxide, 131
  cone, 131
  drawing of, 84
  emergence, 86, 444–45
  impaction, 651–52
  interception, 85
  kairomone, 131
  light, 8, 131
  malaise, 85, 131
  Manitoba, 131
  pattern/color, 131
  pheromone, 85, 105, 131–32
    photograph of, 85
  pitfall, 86, 105, 131
    drawing of, 87
  shelter, 131
  sticky, 131, 442–44
  suction, 86
    drawing of, 87
  tethered animal, 131
  visual, 82–83
  water, 131
  window, 85, 131
Trialcurodes vaporariorum, figure of power law in dispersion of, 49
Trichoplusia ni, 612–14
Tripartite sequential classification
  cascaded, 252–60
  figures/tables of, 253, 256–62

Typhlocyba pomaria, 409
Typhlodromus occidentalis, 403
Typhlodromus pyri, 403

U

Uniform dispersion pattern, 106–7
USDA agencies, scouting and, 690–91

V

Vacuum net technique, 82
Vacuum samplers, 650
Variable intensity sampling, 201–203
Variance
  definition of, 56–59
  importance of, 59–60
  sample unit selection and, 103, 105
  sources of inconsistent error, 60–62
Variance components, tables/figures of, 184–85, 196, 198
Variance-to-mean ratio, dispersion classification with, 43
Variation, relative, 65–68, 105–6
Vegetable production sampling
  key pests/major techniques in, 605–11
  representative programs in, 611–20
  tomatoes, 611–18
Viability, communicating about, 674
Voltinism, expression at geographic site, 448–51

W

Washing technique, 79
Weather pattern, figure of migration during, 123
Webworms, 366
Weevils
  alfalfa
    figure of binomial sampling with, 35
    figure of direct sample of, 76
    figure of mean crowding regression in dispersion of, 48
    tables of negative binomial distribution with, 41
  alfalfa weevil complex, 362–65
  pecan, photograph of emergence traps for, 86
  rice water, sampling, 647–48
Wet extraction techniques, 90–91
Wet sieving, 90
Wheat
  arthropod pests of, 630–32
    table of, 629–30
  arthropod sampling methods for, 633–53
    table of, 635
Wheat bulb fly, sampling, 650
Wheat/rice, natural enemies, 655–56
Whitefly, greenhouse, figure of power law in dispersion of, 49